Medical Geology of Africa:
A Research Primer

Medical Geology of Africa

A Research Primer

Theophilus Clavell Davies

Mangosuthu University of Technology, Umlazi/Durban,
KwaZulu Natal Province, Republic of South Africa

ELSEVIER

Elsevier
Radarweg 29, PO Box 211, 1000 AE Amsterdam, Netherlands
125 London Wall, London EC2Y 5AS, United Kingdom
50 Hampshire Street, 5th Floor, Cambridge, MA 02139, United States

ISBN: 978-0-12-818748-7

For Information on all Elsevier publications
visit our website at https://www.elsevier.com/books-and-journals

Publisher: Candice G. Janco
Acquisitions Editor: Peter Llewellyn
Editorial Project Manager: Ellie Barnett
Production Project Manager: Sajana Devasi P K
Cover Designer: Miles Hitchen

Typeset by MPS Limited, Chennai, India

Contents

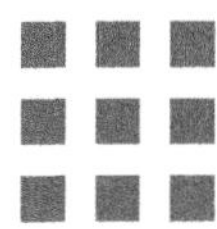

Part III   Conclusion

About the author

Theophilus Clavell Davies is a Research Professor (Medical Geology), a Chartered Geologist, and an Environmental Consultant, currently based at the Mangosuthu University of Technology in South Africa. He was formerly a Council Member and Trustee of the Geological Society of London (1996–2000), is presently a Regional Councillor for the Association of Applied Geochemists (AAG), Africa, a former Member of the Supervisory Board of the International Medical Geology Association (IMGA), a Member of the Scientific Committee of the UNESCO International Centre for Global-Scale Geochemistry (ICGG), and a former Member of the IUGS  Commission on Geosciences for Environmental Management (GEM). Prof. Davies' abiding interest centers around the role of geoenvironmental variables as causal cofactors in the etiology of some obscure diseases (of unknown etiology) in Africa. He has taught and carried out research, community engagement, and consultancy in Germany, Great Britain, Kenya, Nigeria, Sierra Leone, South Africa, and Zimbabwe. Prof. Davies' consultancy for the National Bank of Kenya in 1993 led to the discovery of huge gemstone deposits in the Tsavo National Park in southern Kenya and the commissioning of a mine. Prof. Davies has published over 200 papers (peer-reviewed manuscripts, edited books and book chapters, abstracts, conference papers, and reports) in course of these activities, some of which appear in the most reputable international geoscience journals. Prof. Davies is an Associate Editor, Guest Editor, or Reviewer for several high impact journals, including the *Journal of Geochemical Exploration*, *GEEA*, and *China Geology*. He has singly or jointly edited six "Special Issues" of these journals and is the author or coauthor of chapters in the inaugural books: "Essentials of Medical Geology—Impacts of the Natural Environment on Public Health" (2005) and "Medical Geology—A Regional Synthesis" (2010). Prof. Davies currently holds the status of "Experienced Researcher" according to the National Research Foundation (NRF) of South Africa "Rating System." Major awards include two from the prestigious Alexander von Humboldt Foundation of Germany (1986; 1988) and another from the NMGS/ Shell Petroleum Ltd. (2014) in recognition of "... persistent and consistent excellent research on the environmental geology of the African continent" [Sic].

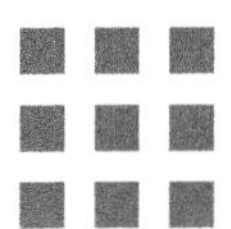

Foreword

The book "Medical Geology of Africa: A Research Primer" by Theophilus Clavell Davies reports updated information on some of the most pressing topics awaiting research on medical geology in Africa.

As the Deputy Vice-Chancellor: Research, Innovation and Engagements (DVC: RIE) at the Mangosuthu University of Technology (MUT), with a nursing background, I find this book would prove invaluable to graduate student researchers of medical geology, public health scholars, and others from allied fields, as well as anyone who is interested in the intersection of geology and medicine. The number of diseases of unknown etiology recognized in Africa to date is staggering; and the author has assembled an extensive literature documenting their existence and put forward starting points for researching their origin.

Professor Theophilus Davies has been the principal exponent of medical geology in Africa. He is one of the leading authors in this field. His abundant scientific contributions published in highly reputable international journals are all focused on elucidating the links between the geoenvironment and health in Africa.

Since the first African Workshop convened by Professor Davies in Nairobi in 1999 (Davies, 1999; Chapter 1, Part I), the flurry of research activities and the study of medical geology in Africa have been relentless and have led to exciting results on many of the topics covered in this primer. It is now considered opportune to reflect on the advances that have been made in research on geoenvironmental factors and disease in Africa and to chart the course of future research on the subject.

Through references to some of the most recent literature, the chapters attempt to summaries the present state of knowledge based on the work done and studies that are ongoing in medical geology. This, hopefully, will provide ready literature for researchers starting their investigations in any of the suggested areas of further research, and hence, shorten the time required to dig out the essentials of recent progresses in medical geology from the avalanche of publications in which they are emerging.

The original objective was to emphasize the importance of geology in health and disease in humans and animals. Professor Davies has immaculately fulfilled the objective of this compilation by combining new observations from recent research studies with existing data to give new insights into the main sources, entry, distribution, and metabolic interactions of the geomedically important chemical elements, and so, providing a wider latitude in the search for diagnostic and therapeutic certainties for geochemical diseases.

Leading causes of death in Africa in 2019, according to Statistica (2023), were neonatal conditions, lower respiratory infections, diarrheal diseases, HIV/AIDS, and ischemic heart disease, all of which conditions can be attributable to respective geoenvironmental variables as probable risk factors or cofactors (Part I). Outcomes of the geoscientific studies of health and medicine could therefore play a role in developing better public health policies by national and regional governments of Africa.

This book also stresses the necessity for greater interaction and communication between the geoscience and biomedical and public health communities to promote a better understanding of the role of minerals in diseases to be able to formulate cogent measures for protecting human health from the harmful effects of geomedical agents in the environment.

The reader is presented with a summary of available information including current hypotheses and an illustration of why improved understanding of these relationships, based on interdisciplinary studies, would lead to better diagnoses and therapy, which also underscores the need for the inclusion of medical geologists and allied specialists in medical teams studying diseases all across the continent.

Professor Davies also advocates for the inclusion of medical geology in higher education curricula of African tertiary institutions (Davies, 2019) so that students will be aware of the connection between geology and health and be motivated to pursue a career in medical geology.

I found this book to be organized into three major compartments. Part I deals with points to consider in selecting a project in medical geology, feasible and capable of bringing out potential breakthrough findings, inherent in a study still very much in its infancy. Three very pressing geomedical issues: "diseases of unknown etiology", "ionizing radiation exposure," and "geophagy" are espoused to expose the numerous gaps in knowledge of these issues and directions of future research to bridge them. The chapter narratives begin with the basic geoscience intersecting medicine but do not lose sight of the clinical implications in line with the underlying philosophy of the book that the material must be of practical value to medical researchers as well.

A second compartment of the book, Part II, presents an updated synthesis of the characteristics of eight of the more geomedically relevant chemical elements (As, F, I, Hg, Mn, Pb, Se, and Zn) to illustrate how their insufficiency or excess is related to the occurrence of myriad *geomedical diseases*. The need to address the types of data needed for drawing up meaningful correlations between patterns of trace element distribution and environmental diseases in humid tropical settings is strongly emphasized in Part II of this volume.

Part III provides cogent conclusions and takes a look at the future of research in medical geology of Africa with proposals on the direction along which such research should be predicated.

The rapidly emerging scientific discipline of medical geology holds promise for increasing our environmental health knowledge base, contributing to substantial and tangible improvements for the well-being of the African community. For this promise to be realized, however, we need to adopt a multi- and inter-disciplinary approach. Even in the present period of increased specialization, geoscientists must realize that it is by working closely with scientists in other fields that we can best contribute to this exciting yet demanding task.

One of the most challenging problems for the future will be the understanding and characterizing of the environmental and natural factors that contribute to emerging and reemerging infectious diseases.

The book is timely because there is now an unprecedented interest in deciphering unknown etiologies by researchers from across the geoscience and public health spectrum. How I wish these studies had come much earlier, in view of the continent's unique geoenvironmental condition that offers such an ideal setting for these studies.

Professor Davies must be congratulated for his excellent effort! Well done, Professor Davies.

Prof. M.N. Sibiya
Acting Vice Chancellor, Mangosuthu University of Technology, Republic of South Africa

References

Davies, T.C., 1999. The East and Southern Africa regional workshop on geomedicine. Episodes 22 (4), 303−304. Available from: file:///C:/Users/davies.theophilus/Downloads/IUGS022-04-09%20(4).pdf (accessed 04.09.2023).

Davies, T.C., 2019. A medical geology curriculum for African geoscience institutions. Scientific African 6, e00131. Available from: https://doi.org/10.1016/j.sciaf.2019.e00131 (accessed 04.09.2023).

Statistica 2023. Leading 10 causes of death in Africa in 2019 (in deaths per 100,000 population). https://www.statista.com/statistics/1029287/top-ten-causes-of-death-in-africa/ (accessed 03.09.2023).

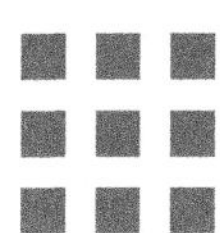

Preface

The book "Medical Geology of Africa: A Research Primer" by Professor Theophilus Davies presents an exquisitely customized guide for navigating the most recent bibliography of the more pressing issues pertinent to the medical geology of Africa, intended for researchers eager to make urgently needed contributions towards an understanding of how judicious exploitation of the interphase between geology and health can help unravel environmental disease causality, especially of diseases of unknown etiology that are so widespread on the African Continent.

The role of geoenvironmental factors in disease causation in man and other animals (medical geology) worldwide is still not fully understood. In Africa, this study is of special significance, by virtue of the uniqueness of the continent's surficial geochemical environment, its largely rural-based population (living close to the land), and having a significant proportion of diseases of hitherto unknown etiology.

Renewed awareness of the importance of medical geology studies in Africa coincided with the formalization of the subject in Uppsala, Sweden, in the year 2000 by the International Union of Geological Sciences' (IUGS) International Working Group on Geomedicine (IWGG) led by Olle Selinus, and the change of name from "Geomedicine" to "Medical Geology," a term that was considered more acceptable to the medical profession. The flurry of research activities and study of medical geology in Africa as well as globally has been relentless ever since. These endeavors culminated in the publication by IMGA in 2010 of a Springer volume (Selinus et al., 2010) summarizing medical geology knowledge in various regions of the world including the African Continent up to that time (2010).

Research activities in this field have continued since (2010 to present), reporting several successes in aiding the medical profession to improve diagnoses and therapy of some of the more obscure environmental diseases around the continent.

Ninety percent of the reference listings in this book post-date the year 2000, coinciding with the date of formalization of research in the field of medical geology, provide ready literature for graduate medical geology researchers and others from allied disciplines starting their investigations in any of the suggested areas of further research given at the end of the main chapters of this book. An attempt has been made to include the most pertinent and important references to reflect progress achieved mostly in the last 3—4 years in researching the link between geology and disease.

The reader is presented with a summary of available information including current hypotheses and an illustration of why an improved understanding of these relationships, based on interdisciplinary studies, would lead to better diagnoses and therapy.

The book is also intended to serve as a provisional route map for future scholars. Although the history of medicine has matured and expanded in recent decades, there remain critical issues to address particularly in relation to the history of disease, a history that has sometimes been diminished by territorial disputes between disciplines and unproductive arguments about the perils of retrospective diagnosis.

Just about three decades ago, when it became clearly evident that geoenvironmental variables play a significant role in the occurrence and geographical distribution of a number of diseases in Africa, the study of medical geology has continued to uncover "new" and important correlations between these variables, with corresponding invaluable contributions made by medical geologists in the diagnosis, therapy, and prognosis of many diseases. It is hoped that the advent of this primer will greatly enhance efforts at keeping this momentum going!

Reference

Selinus, O., Finkelman, R.B., Centeno, J. (Eds.), 2010. Medical Geology: A Regional Synthesis, 1st Edition. Springer, Dordrecht, The Netherlands. p. 392. ISBN: 978-90-481-3429-8.

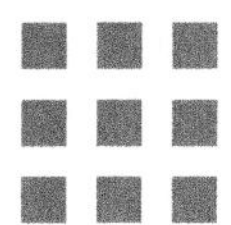

Acknowledgments

It is a pleasure to acknowledge the considerable help I have received from a number of colleagues and institutions who have directly or indirectly contributed to the preparation of this volume.

Firstly, my ever present and dependable host at the Technical University of Braunschweig (TUB) in Germany, Prof. Dr. Antje Schwalb, has, through her unselfish devotion and encouragement, contributed immensely to the completion of Part I of this book. This phase of the project was achieved through the award to me of a Research Fellowship by the Alexander von Humboldt Foundation of Germany during the Summer of 2020. The staff at the TUB Archive have been remarkable. They have provided all the needs that could possibly be expected from a library service. It is, however, only fair to add that I remain entirely responsible for any defects of fact, treatment, judgment, or style.

The numerous colleagues and students with whom I have interacted during the many International Medical Geology Association (IMGA) conferences have also made significant contributions in diverse ways. My interactions with Olle Selinus, Bob Finkelman and Jose Centeno, founders of contemporary Medical Geology, will forever remain cherished. They infused in me a lot of passion and enthusiasm for this study during those early days.

In 2020, I rejoined the Mangosuthu University of Technology's Faculty of Natural Sciences (FNS) in South Africa, where I continued my research and writing. My colleagues at FNS as well as those at the Research, Innovation and Engagement (RIE) Directorate have provided me with an extraordinary support base.

All financial resources covering data collection, analytical work, as well as writing and publication costs were provided by research grants from the South Africa "Highly Disadvantaged Institutions (HDI) Subvention" and the South Africa National Research Foundation (NRF) "Incentive Funding for Rated Researchers" (IFRR).

Any copyrighted material on these pages is used under "fair use" for the purpose of study and review. A thorough effort was made to clear any necessary reprint permissions. All illustrations and citations in the text are credited to their appropriate and respective sources, and a general expression of my indebtedness and thanks is given to the many institutions and authors whose images and photographs have been so freely called upon. Any required acknowledgment omitted is unintentional.

I thank the (anonymous) reviewers who read the manuscript for raising important queries, and for providing many insightful comments and suggestions. The careful attention paid towards addressing these concerns has resulted in a much improved manuscript quality.

Finally, I wish to express my thanks to the staff of my publishers, Elsevier, for their enthusiastic and helpful cooperation throughout the production of this book; and, in particular,

Ellie Barnett, Czarina Osuyos, Peter Llewellyn, Moises Carlo Catain, Subash Balakrishnan, Miles Hitchen, Sajana Devasi, Joshua Mearns, Charlotte Kent, Mohan Raj Rajendran, Hilary Carr, Amy Shapiro, Kavitha Balasundaran, Alexandra Ford, Ruby Gammell, and other editorial staff members, for reviving my hope when all was seemingly lost.

Some topical issues in medical geology of Africa: current knowledge and research gaps

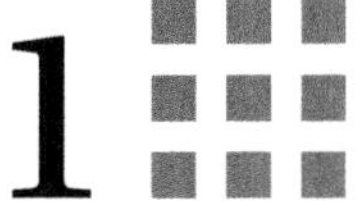

1

General introduction

"The power of Medical Geology lies in the ability to capture principles and concepts from several allied disciplines and unify them in a single place, thus providing a more holistic approach to disease diagnosis and therapy."

Key chapter features

1. Brief synopses on Medical Geology of Africa (MGA) topics for which significant research output has been generated in the last few years (up to 2023), and ongoing studies.
2. An updated list of references is presented to aid the search process of researchers wanting to explore *gaps in knowledge* of rapidly evolving areas of MGA.

Historical evolution of Medical Geology in Africa

The role of geoenvironmental variables in disease causation has long been recognized worldwide, but only relatively recently has the importance of this role been strongly demonstrated. The first systematic attempt at drawing a causal relationship between human diseases and environmental factors is credited to the Greek physician, Hippocrates, through the book: *Air, Waters, and Places* (Hippocrates of Cos II, 2009, Translation by Francis Adams) thought to have been written by him in the 5th or 4th century BCE. Since then, the significance of the role of environmental factors in disease causation has continued to grow, a feature reiterated in the position of "the environment" in Lalonde's *"Health Field Concept,"* which he propounded in 1974. The "Health Field" can be broken up into four broad elements: "human biology," "environment," "lifestyle," and "health care organization" (Lalonde, 1974; Tulchinsky, 2018).

Jul Låg of Sweden was one of the first contemporary scientists to bring to light the possibility that geological materials and processes could play a much more significant role as cofactors of certain diseases than the medical profession was hitherto aware. Låg's ideas were later documented in the book "Geomedicine" (Låg and Contributors, 1990). Coinciding approximately with the time that Låg was formulating his ideas, the International Union of Geological Sciences (IUGS) approved the inauguration of the International Geological Correlation Program (IGCP) on Geomedicine in 1987, with Olle Selinus of Sweden being the principal pioneer of the "Initiative." At a landmark conference of the IGCP Geomedicne Committee in Uppsala, Sweden in 2000, the name "Geomedicine" was replaced by the name "Medical Geology," a terminology that was deemed to be much more acceptable to the medical profession (Skinner and Berger, 2003; Selinus et al., 2008). During the period of just over

Medical Geology of Africa. DOI: https://doi.org/10.1016/B978-0-12-818748-7.00014-9

a score of years that has passed since the Uppsala meeting, much progress has been made worldwide, more so, in Africa, in understanding the links between the geographical distribution of diseases in man and animals and the geological environment.

Geology and health in Africa

In Africa, significant evidence exists on health effects resulting from our interactions with geological materials and processes, and we continue to encounter "new" and important correlations in this domain. These relationships have been observed for many years, but only relatively recently have any real attempts been made to formalize their study. Awareness of these links increased rapidly in Africa since the 1960s, coincident with the time that the principles of geochemical exploration were receiving widespread acclamation of effectiveness in mineral exploration programs across the Continent; pioneered by John Webb's Applied Geochemistry Research Group at the Imperial College, London. Organized efforts aimed at peering more deeply into the relationships between geology and health in Africa have triggered a number of exciting workshops and related training courses in the last three decades.

The first real attempt to coordinate research in this field (then referred to as "Geomedicine") in Africa took place in Nairobi, Kenya, in 1999, bringing together over 60 interdisciplinary scientists from the region (Davies, 1999; Davies and Schlüter, 2002). The keynote papers at this workshop revealed much of what had been learned then (1999) about the subject (Geomedicine) since the pioneering days of Hippocrates and Jul Låg, respectively, and concluded, inter alia, that considerable gaps existed in our knowledge of the identification and quantification of the mineral and chemical forms of certain nutritional and toxic trace elements in the rocks, soils, and sediments that constitute the natural sources of these elements entering water and the food chain (Davies, 1999; Davies and Schlüter, 2002).

The elements that were considered to need urgent research attention at that time, according to the Nairobi Workshop's findings were As, Cd, Hg, Se, Rn, Pb, Zn, F, and I (Davies and Schlüter, 2002). The different chemical forms or species have a marked influence on the solubility as well as the bioavailability and bioaccessibility of these elements. Today, just over two decades later, a mass of data has accumulated on the *Medical Geochemistry* of these elements as articulated in Part II of the present Volume.

Despite this enormous store of knowledge and wealth of medical data garnered in the last few decades, the world of medicine is still full of many unknowns. The past few years have seen an increasing realization of the importance of the possible role of geoenvironmental variables as causal cofactors of diseases of unknown *etiology*, as well as occasional enigmatic disease outbreaks in various geographical regions of Africa (e.g., Mao et al., 2020; Davies, 2022). The causes of diseases of unknown etiology (particularly the role of geochemical variables) are the subject of Chapter 3, Part I of this Primer. This relationship (geochemistry vs disease) has triggered a flurry of research initiatives on the Continent, directed mainly at an understanding of the geochemical fluxes of both nutritional and potentially toxic elements (PTEs) in the groundwater-soil-food crop continuum, that largely define the nutritional quality of the African diet, and hence the correct balance of micronutrients and vitamins in metabolic processes. The Medical Geology of some of the geomedically more important elements is the subject of Part II of this Primer.

A natural laboratory for Medical Geology research

In Africa, the relationships between geological materials and processes, and health are particularly marked. The continent presents a diverse geography. The range of altitudes is wide, and the hydrological network is quite distinctive. There are also large stretches of arid lands, such as the Sahara and the Mega Kalahari (Table 1−1). Rifting and the formation of the Great Rift Valley on the eastern flank of the continent has been accompanied by volcanic

Table 1–1   Some characteristics of Africa's surface environment.

Location	Africa, the World's second largest Continent is bounded on the west by the Atlantic Ocean, on the north by the Mediterranean Sea, on the east by the Red Sea and the Indian Ocean, and on the south by the Atlantic and Indian oceans.
Total land surface area	Africa's total land area is approximately 30,365,000 sq. km. and covers 6% of Earth's total surface area.
Climate	The climate ranges from tropical to subarctic on its highest peaks. The northern half is primarily desert, or arid. A hot climate prevails in most of the regions and for most of the year.
Natural resources	The array of natural resources is huge, and includes: diamonds, gold, copper, uranium, iron, cobalt, bauxite, silver, and petroleum products. We should also consider the possibility that many natural resources are yet undiscovered or barely harnessed.
Terrain	The surface area of Africa is very variable, marked by an alternation of tablelands or plateaus, rim of mountains, and shallow depressions or basins at varying elevation. The eight major geographical regions that influence Africa's landscape are: the Sahara Desert; the Savanna; the Sahel; the Ethiopian Highlands; the African Great Lakes; the Swahili Coast; the Rain Forest, and Southern Africa (encompassing the Mega Kalahari).
Soils	The soils are the product of a nexus of surficial processes, acting with such magnitude and intensity that when combined with the prevailing climatic conditions, produce soils of remarkably unique geochemical character. As such, African soil types vary very widely, and for each climatic type, there can be rich as well as poor soils.
Land use	Up to 70% of Africa's population, and approximately 80% of the Continent's poor live in rural areas. They traditionally practice small-scale cultivation or pastoralism, which is more common in the arid areas of northern, eastern, and southern Africa.
Irrigated land	As already indicated, the African people depend on agriculture and nonfarm rural enterprises for their livelihoods; yet, only approximately 6% of cultivated land is currently irrigated.
Hydrological network	The hydrological network is very poorly developed with limited coverage and short incomplete records, except for the Nile Basin and in a few countries in the northern and southern parts of the Continent.
Main countries with some geochemical survey data	South Africa, Nigeria, Botswana, Morocco, Zimbabwe, Uganda, Tanzania, Sierra Leone, Gabon, Ghana, Namibia, Liberia, and the Kingdom of Lesotho. It is postulated that a massive attempt by African geological surveys and universities to bring to completion the Africa Geochemical Database Program (AGD) by providing the necessary resources, will contribute greatly toward the advancement of Medical Geology research in Africa.

activity, with the release of various trace elements, mostly above background levels, into the environment. The marked separation of these chemical elements (both nutritional elements and PHEs) has resulted in more recent geological times, following extreme tropical conditions of weathering, leaching, and eluviation, secondary mineralization, *lateritization, posolization*, and *gleying* (See: "Glossary of Terms"). The sheer intensity of these processes has left large stretches of the continent containing element deficiencies or toxicities, which are closely related to the local geology and/or geographical location. As rural communities are still largely dependent on the land, from where water and food sources are locally derived, and as stretches of cultivated land in Africa are arid, semiarid, or lack essential trace elements for healthy plant growth (Table 1−1), the probability of detecting relationships between the geological environment and certain diseases is greatly increased. Such a setting thus provides a *natural laboratory* for studying the influence of geochemical factors on the distribution of diseases in man and animals, a key objective of research in Medical Geology/ Geochemistry.

Medical Geology and the leading causes of mortality in Africa

A cursory look at the top 10 causes of death in Africa (Fig. 1−1) justifies additional research on the Medical Geology of Africa; as well as the investment of more resources to conduct the research. Many of the leading causes of death in Africa, as in other developing regions of the world, could be linked to factors of the geoenvironment. As can be seen in Fig. 1−1, geogenic dust (e.g., from mining, ore processing, and other human activities) causes respiratory infections including asthma, the second leading cause of death (LCD) in Africa; water pollution causes diarrheal diseases, the third LCD in Africa, as well as other water-related illnesses. The position of Se in the incidence of HIV/AIDS (the fourth LCD in Africa) is still considered important as a possible cofactor (Pourmoradian et al., 2023; Sun et al., 2023), as is the role of Ca and Mg in the incidence of ischemic heart disease and stroke, respectively (items 5 and 6 in the ranking in Fig. 1−1).

Researchable topics in Medical Geology of Africa

The Book is written primarily for postgraduate researchers and other scholars of Medical Geology, providing opportunities for spotting novel research topics and current pressing thematic issues in Medical Geology for their thesis writing, journal publications, or other forms of result dissemination. Government policy makers on environmental and public health issues would find much useful information upon which tangible recommendations on disease control and prevention could be predicated. The process of project selection for scientific research in Medical Geology is the subject of Chapter 2 of Part I of this Volume.

Many of the diseases studied within the compass of Medical Geology could be described as *geochemical diseases,* for the involvement of *geochemical variables* is proven or suspected

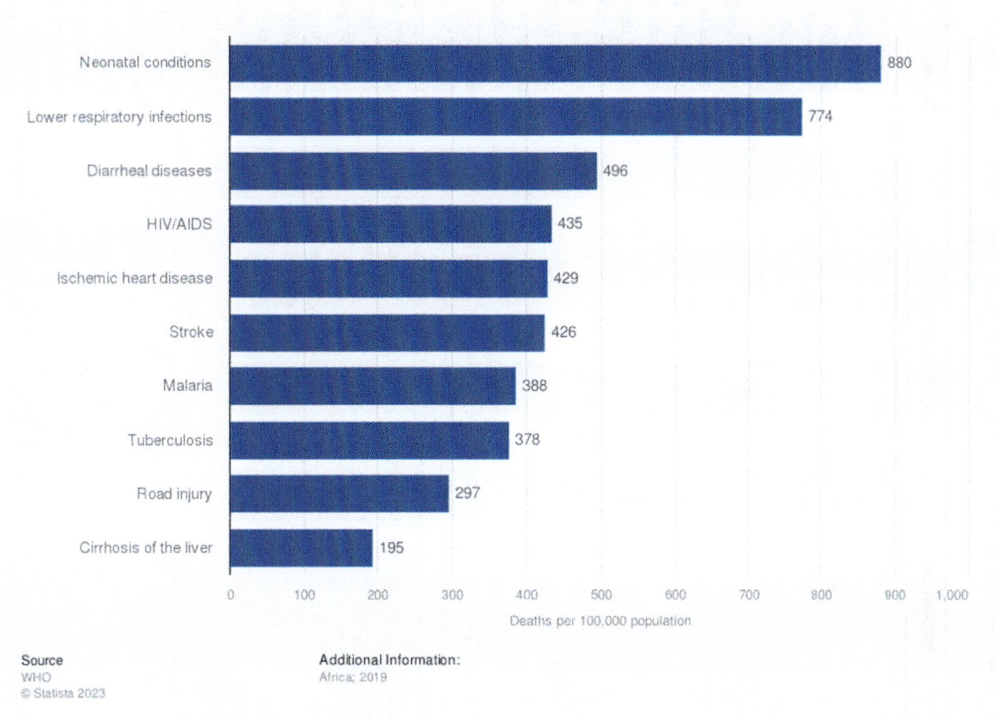

FIGURE 1–1 The 10 leading causes of death in Africa in 2019 (in deaths per 100,000 population). Credit: *Statistica. https://www.statista.com/ (Accessed 09 September 2023).*

as a cofactor(s) in their etiology (Table 1−2). Through short summaries characterizing the range of researchable health conditions linked with some geoenvironmental factor(s), the rest of this Chapter illustrates the enormity of Medical Geology issues and emphasizes aspects where tremendous opportunities exist for pursuit by the next generation of researchers in this emerging and important field.

The most pressing environmental topics on Medical Geology in Africa over the last few years are those relating to "diseases of unknown etiology" (Chapter 3, Part I), "Pollution of the Soil, Water, and Air Environments" (Chapter 4, Part I), "Ionizing Radiation Exposure" (Chapter 5, Part I) and "Geophagy" (Chapter 6, Part I), among others; and research results from these topics have generated the most voluminous literature to date. As this trend is envisaged to continue for some time, these are therefore the topics that are dealt with in Part I to illustrate the scope of the task facing the Medical Geologist in terms of the huge gaps that still remain in our understanding of the links between geology and disease in Africa.

The exposure of pregnant women and children to ionizing radiation from mining in Sub-Saharan Africa has long been a source of concern, and many questions related to this

Table 1–2 Some examples of proven or suspected correlations between geochemical factor and environmental disease/public health condition in Africa Davies (2010).

Geochemical factor	Possible health effects	Distribution	Mitigation	References	Remarks
Fluoride deficiency (<0.5 mg/L)	Dental caries and dental cavities may occur; weak bones and teeth	A continent-wide problem	Increased fluoride intake can be achieved by controlled direct additions to drinking water or to specific foods	Kut et al. (2016), WHO (2017), Kerdoun et al. (2022), Podgorski and Berg (2022), Sawangjang and Takizawa (2023)	As at present (2023), the World Health Organization (WHO) guideline value for fluoride concentration in drinking water is 1.5 mg/L (WHO, 2017). Adding fluorides to soils that produce food crops is ineffective because fluorides in the soil are inactivated and do not usually find their way into the food chain (See: Chapter 3, Part II for further details on "Medical Geology of Fluorine," including gaps in knowledge of F- nutrition)
Fluoride toxicity (>1.5 mg/L)	Mottling of teeth and dental fluorosis may occur; association with skeletal fluorosis at higher concentrations	Populations living within the African rift zones and associated volcanic centers (Ethiopia, Kenya, Tanzania, etc.); areas of crystalline bedrock (e.g., Ghana, parts of Tanzania and Malawi), and some sedimentary basins with high-fluoride groundwaters (e.g., Senegal and parts of North Africa)	Defluoridation, for which there are a number of simple and readily available techniques, for example, the Nalgonda technique	Wanke et al. (2017), Onipe et al. (2020), Wu et al. (2020), Maghanga et al. (2022), Chaudhary et al. (2023), Lubojanski et al. (2023), Mohandas et al. (2023), Njau et al. (2023), Zhang et al. (2023)	Davies (2013) considers that a revised scheme of therapy for *prolapsed intervertebral disk* (PID) (See: Wu et al., 2020), *Paget's disease of bone (PDB)* (See: Banaganapalli et al., 2023) and *spinal stenosis* (See: Kwon et al., 2022; Walter and O'Toole, 2022) would benefit from considerations involving the maintenance of the correct F^- (and probably also Mg) balance within the body; thus strengthening the call for further research to determine reliable biomarkers for the burden of F^- in the body (See: Chapter 3, Part II for further details on "Medical Geology of Fluorine")

Iodine deficiency	Iodine deficiency disorders (IDD), namely, goiter and its sequelae	These disorders are found in almost every country right across the Continent	Mainly fortification of table salt by addition of potassium iodate and use of oil are the only recommended strategies for adequate iodine exposure	Saha et al. (2019), Gorstein et al. (2020), Lisco et al. (2023)	There is now a marked reduction of IDD in African countries, thanks to the effective implementation of the universal salt iodization program of the WHO. However, iodine insufficiency has recently been observed to be reemerging in industrialized countries such as the United States, United Kingdom, and Australia (See, e.g., Hatch-McChesney and Lieberman, 2022)
Selenium deficiency in soils and food crops	IDD and high diffusion rate of HIV-1 in southern Africa	IDD incidence is continent-wide; HIV-1 incidence is highest in southern Africa	Selenium supplementation; research on-going on the most effective means of supplementation	Muzembo et al. (2019), John-Olabode et al. (2023), Pourmoradian et al. (2023), Sun et al. (2023)	Research on the presumed link between Se deficiency and HIV infection, and predisposition to tuberculosis, continues. However, evidence supporting the beneficial effects of Se supplementation in patients with HIV remains inconclusive (See: Chapter 8, Part II)
Mg and Ca concentrations in drinking water—Coronary heart disease (CHD), stroke and cerebrovascular diseases (CV) rank among the top 10 causes of death in Africa (Fig. 1−1). Major risk factors do not entirely explain variability of morbidity and mortality due to these conditions. It is therefore reasonable to submit that a further look at the role of geoenvironmental (co)factors in their etiology might contribute significantly toward the design of improved diagnostic and therapeutic schemes	Discussions still rife on the possible inverse correlations between water hardness (concentration of Mg and Ca ions) and cardiovascular diseases (CVD)	Cardiovascular diseases are widespread in Africa, particularly in Sub-Saharan Africa (SSA) which is today experiencing an epidemic of CVD on an unimaginable scale	Therapeutic lifestyle changes underlie the nondrug management of CVD, including cessation from smoking, reduced intake of sodium diet, body weight management, limiting alcohol consumption and regular physical exercises	Yuyun et al. (2020), Minja et al. (2022), Bykowska-Derda et al. (2023), Noubiap et al. (2023), Vogiatzi et al. (2023)	Minja et al. (2022) point out important knowledge gaps in cardiovascular medicine in Africa and priority areas of research for improving the current rate of progress toward achievement of a 33% reduction in CVD mortality

(Continued)

Table 1–2 (Continued)

Geochemical factor	Possible health effects	Distribution	Mitigation	References	Remarks
Other micronutrient deficiencies, for example, Fe, Zn, Se, Ca, P, K, and Mn	Possible examples are *endemic osteoarthritis [Mseleni joint disease (MJD)]*, *rickets* and *geophagy*	The only documented cases of MJD are found in Maputaland in South Africa; "Geophagy" is practiced continent-wide; and rickets have been reported in northern Nigeria	MJD is treated with nonsteroid containing drugs and antiinflammatory medication; sometimes a total hip replacement is needed. Dietary Ca supplementation is proposed for rickets	Ceruti et al. (2003), Thacher et al. (2012), Dinkele et al. (2020), Kihara et al. (2020), Davies (2023), Haffner et al. (2022), Ogunmwonyi et al. (2022), Mays and Brickley (2023), Vlok et al. (2023)	A fuller account on *geophagy*, a Continent-wide practice and one of importance in the debate on food security, is given in Chapter 6, Part I. Further examples of diseases created by element deficiencies (and excesses) (*geochemical diseases*) are presented in Part II
Nutrient element (e.g., Mg) metabolism; climatic factor (e.g., cold weather)	*Psoriasis*, a chronic autoimmune malady in which skin cells build up and form inflamed, red, raised areas that often develop into silvery scales and itchy, dry patches rapidly; most often on the scalp, elbows, or knees, though other parts of the body can be affected as well	Probably widespread in Africa as it is in the rest of the world; but is thought to be widely under-reported	In Africa, there is a wide variety of available treatment options, ranging from: topical treatment (e.g., application of steroid creams or ointments such as topical corticosteroids, commonly used to *treat* mild to moderate *psoriasis* in most areas of the body; *phototherapy*, for moderate disease), to systemic therapy or biological agents for more advanced form of the disease Bathing in spas containing mineral salts such as $MgCl_2$ has been considered an important treatment modality. Repeated bathing in the Dead Sea (a *hypertonic salt solution*) (See: "Glossary of Terms") for a certain length of time has a therapeutic effect on a number of skin diseases including psoriasis	Davies (2013), Emmanuel et al. (2020), Jacobs and Kgokolo (2020), Zheng et al. (2021), Dai et al. (2023), Wang et al. (2023a)	The importance of psoriasis as a worldwide dermatological problem is underlined by the devotion of an entire international journal (by Dove Press) to this malady: https://www.dovepress.com/psoriasis-targets-and-therapy-journal# (Accessed 08 September 2023)

| Mineral dust | (1) Asbestosis (mesothelioma) (2) Silicosis; can predispose to tuberculosis (3) Coal worker's pneumoconiosis (CWP) | Populations in and around asbestos mining and processing plants, mainly in southern Africa, for example, Swaziland and Zimbabwe (chrysotile); South Africa (amphibole fiber)
Found among miners and mining communities in countries where mining and processing of ore deposits associated with free crystalline silica, take place, for example, gold and gemstones (Ghana and South Africa); diatomite (Algeria, Ethiopia, Kenya and South Africa)
Coal-producing countries of Africa, of which South Africa is by far the largest | There are no treatment regimens for asbestosis; but there are certain precautions that can help retard the progression of the disease and relieve symptoms, such as avoidance of further exposure to asbestos and abstinence from cigarette smoking
There is no *cure* for *silicosis* since the lung damage that has been done cannot be reversed. Just as in the case of asbestosis, treatment centers on retardation of disease progression and relieving symptoms. Avoidance of further exposure to silica and other irritants such as cigarette smoke, is crucial
There is no cure for CWP. The damage done by coal dust cannot be reversed; but certain steps can be taken to slow down progression of the disease relieve symptoms and improve quality of life | Vorster et al. (2022), Mutetwa et al. (2022), Peña-Castro et al. (2023), Rees and Murray (2020), Ehrlich et al. (2021), Ehrlich et al. (2022), Jamshidi et al. (2023), Maboso et al. (2023), Mbuya et al. (2023), Naidoo et al. (2005), du Toit (2010), Mbedzi et al. (2020), Kamanzi et al. (2023) | Most asbestos mines worldwide have now closed, but many African countries still use asbestos in one form or another. Besides, the latency period for asbestosis can be up to 40 years; and asbestos is deemed to be the cause of the recent global increase in the incidence of "idiopathic" pulmonary fibrosis (IPF) (See: Reynolds et al., 2023). Health effects of asbestos in Africa are dealt with in Chapter 4, Part I
The occupational exposure limit (OEL) for silica dust of 0.1 mg/m^3 seems not to be protective against silicosis, as a large burden of silicosis among older black South African gold miners is likely to worsen
Given the sheer importance of coal as a primary energy source in South Africa, more research effort on the environmental health impacts of coal mining appears to be desirable |
| High environmental levels of cerium | Endomyocardial Fibrosis (EMF) | Since identification in Uganda in 1947, reports of this restrictive cardiomyopathy have come from other parts of tropical Africa | Prognosis is generally poor. Medical *management* can alleviate *symptoms* for a while, but has no curative effect | Grimaldi et al. (2016), Mbanze et al. (2020), Aliku et al. (2022), Cuenca et al. (2022) | Immunosuppressive therapies have a role as a control strategy, but medicare in general remains challenging, as many patients with advanced disease die within 2 years |

(*Continued*)

Table 1–2 (Continued)

Geochemical factor	Possible health effects	Distribution	Mitigation	References	Remarks
Fine reddish-brown volcanic soils high in elements such as Be and Zr	*Podoconiosis* (nonfilarial elephantiasis)	Endemic in Africa; most commonly in Ethiopia, but has been reported in many other African countries especially among populations living in areas of heavy exposure to volcanic clay minerals, such as around the African Rift Valley	Key *podoconiosis control strategies center around the prevention of contact with irritant soil through footwear use*, simple but effective limb hygiene measures, including deep cleaning of limbs with diluted antiseptic solution, soap, and water, bandaging, application of emollient on the skin *and covering floors inside traditional huts*. The main preventive measure is wearing protective shoes from childhood	Deribe et al. (2020), Getachew and Churko (2022), Wanji et al. (2021), Cooper and Nick (2023), Debele et al. (2023), Deribe et al. (2023)	Endemic Kaposi's sarcoma (KS), a neoplasm common in eastern and central Africa is believed to arise in the lymphatic endothelium and is associated with chronic lymphoedema. As such, KS bears a superficial resemblance to podoconiosis in both its presentation and its geographical epidemiology, suggesting a common etiology for both conditions (viz., exposure to red-clay soil derived from alkali volcanic rock) (Price, 1976; Ziegler, 1993; Ziegler et al., 2003)
High environmental levels of Fe taken in as inorganic Fe in the diet is referred to as African Fe overload or *siderosis*	*Siderosis* involves the deposition of iron, occurring mainly in the liver and *reticuloendothelial system*	Described in Sub-Saharan African populations; most notable example found among the South African Bantu living in areas of Fe-rich soils in Johannesburg	*Venesection or phlebotomy is a* therapy for disorders associated with excess Fe in the blood; and this is the common management regimen usually recommended for African Fe overload	Gordeuk (2002), Ginzburg and Finberg (2021), Salomao (2021), Hsu et al. (2022)	The iron-rich soils are thought to have been formed by the metamorphism of ultrabasic rocks by the Johannesburg Dome, a granitic intrusion lying between Johannesburg and Pretoria
Cancer and the geochemical environment	Examples include: (1) Lung cancer due to exposure of U miners and associated communities to radon gas and its decay products (2) High arsenic concentrations in domestic water supply; skin disorders and malignancy, internal cancers	Uranium deposits occur in several countries all around the Continent but South Africa, Niger, Namibia and Gabon have the largest resources Populations living in the vicinity of metal sulfide mines (e.g., South Africa, Zimbabwe) and those consuming arsenic-rich groundwaters from mineralized Proterozoic basement rocks (e.g., Burkina Faso)	Adherence to the code of practice and safety guide for radiation protection and radioactive waste management in mining and mineral processing, including (1) hermetic sealing of galleries; (2) elimination of infiltrating water; (3) powerful ventilation and (4) normal dust prevention techniques. Methods for removal of As from contaminated water by use of activated alumina, iron oxide adsorption, membrane, and ion exchange; chelation therapy	Moshupya et al. (2019), Moshupya et al. (2022), Zupunski et al. (2023), Ahoulé et al. (2015), Irunde et al. (2022), Khosravi-Darani et al. (2022), WHO (2022), Dilpazeer et al. (2023), Dintsi et al. (2023), Kapwata et al. (2023), Patel et al. (2023), Pezeshki et al. (2023)	Combatting the effects of exposure to ionizing radiation around U and Au mines in Sub-Saharan Africa is the subject of Chapter 5, Part I of this discourse. The new recommended limit of As in drinking-water has been set at 10 µg/L, although this guideline value is deemed provisional on account of the practical difficulties encountered in removing As from drinking-water (WHO, 2022) (See Chapter 2, Part II)

Beneficial effects of geological materials and processes	The term *balneotherapy* is used to describe a decades-old physical therapy that involves bathing in hot mineral water (hot springs, thermal waters), *peloids* or gases for health promotion and prevention or treatment of certain diseases. Medicinal mineral resources with considerable potential for development in Africa are: *spas*, *muds*, and *peat deposits* *Peloids* are natural substances (mud or clays) that are *admixtures of inorganic and organic substances in different* proportions; and used therapeutically, as part of balneotherapy, or therapeutic bathing. The most popular peloids are *peat pulps*, various *medicinal clays*, mined in various locations in Africa and around the world, and a variety of plant substances	Balneotherapy is practiced in several African countries, particularly those countries with ancient volcanic history, for example, South Africa, Kenya, Nigeria, Ethiopia, and Egypt	Balneotherapy has shown significant effects for some diseases in Africa, such as cardiovascular conditions and skin lesions. Clinical evidence based on epidemiological studies emphasizes the benefits of pelotherapy on degenerative and inflammatory rheumatism taking advantage of the peloid's analgesic, antiinflammatory, antioxidant and antimicrobial properties. There is also evidence of positive effects of peloids on dermatological infections, especially *psoriasis*, and on skin care functions (cleansing, degreasing, exfoliation, hydrating, tonifying, and reaffirming)	Davies (2013), Pedrazzini et al. (2016), Sekome and Maddocks (2019), Williams (2019), Mangwane and Ntanjana (2020), Ncube et al. (2020), Oluwaseun et al. (2020), Gebretsadik et al. (2021), Haji et al. (2021), Kgabi and Ambushe (2023), Wang et al. (2023b)	Despite the abundance of natural hot springs in Africa, their therapeutic value has not been thoroughly studied (See, e.g., Gebretsadik, 2023). Curative properties of geothermal waters are highly valued in several African countries, especially those countries located around the African Rift Valley, such as Kenya and Ethiopia
Medicinal clays and cosmetic products. Zeolites, which have properties similar to clays, also have a range of applications in health and medicine	The use of medicinal clays or healing clays, such as bentonite and kaolin, is also notable, for these clays have large specific area, high adsorptive capacity, favorable rheological characteristics and relative chemical inertness, making them amenable to use as active ingredients or excipients in therapeutical and pharmacological formulations as well as in beauty therapy (See, e.g., Kgabi and Ambushe, 2023)	The medicinal and cosmetic uses of various clays have been reported in various countries across the Continent	—	Mpuchane et al. (2010), Masango et al. (2017), Gomes and de Sousa (2018), Williams (2019), Incledion et al. (2021), Morekhure-Mphahlele et al. (2017), Buthelezi-Dube et al. (2022)	Further research is needed to distinguish between "healing clays" and those identified as antibacterial clays. According to Williams and Haydel (2010), "… the highly adsorptive properties of many clays may contribute to healing a variety of ailments, although they are not antibacterial." See also: (Moosavi, 2017; Nomicisio et al., 2023). According to Cao et al. (2022) "Functionalization of fibrous clays and their organic modified derivatives for high value-added dermopharmaceutical and cosmetic applications is

(Continued)

Table 1–2 (Continued)

Geochemical factor	Possible health effects	Distribution	Mitigation	References	Remarks
					currently an active area of research."
Metallic implants and *prostheses*—Medical implants are devices or tissues that are placed inside or on the surface of the body. Many implants are *prosthetics*, intended to replace missing body parts	Metals and alloys such as Al, Ti, stainless steel and Co-Cr are commonly used in the construction of implants and *prostheses* due to their biocompatibility and high mechanical strength	Worldwide	—	Kulkarni et al. (2023), Meng et al. (2023), Saha and Roy (2023), Wu et al. (2023)	Metallic implant materials have gained immense clinical importance in the medical field for a long time. The use of surgical implants and prosthetic devices to replace the original function of different components of the human biological system is a well established tradition in the history of medicine. However, these new technologies need to be further researched to gauze the full extent to which they can enable human beings not just to survive, but thrive in the face of any conceivable physical difficulty
The soil as a reservoir of disease pathogens	An example is the transmission of *prion disease*, an invariably fatal condition	Sporadic Creutzfeldt-Jakob disease (sCJD), the most common prion disease occurs worldwide, and sporadically in Africa	There are currently no treatments for prion diseases, but an experimental genetic treatment has dramatically extended the lifespans of infected mice (Mead et al., 2022)	Adam and Akuku (2005), Babelhadj et al. (2018), UCL (2022)	Further details on "prion diseases" are given in Chapter 3, Part I

Nutrient imbalances in soil that are of consequence to crop, livestock and wildlife development	Examples are: Cu deficiency in wheat and Cu and Co deficiencies in cattle. Cases of *"molybdenosis"* (molybdenum toxicity), a disease in ruminants (stiffness of legs, loss of hair) feeding on forage containing more than 10–20 mg/kg Mo have been reported from Kenya (McDowell, 2003). Low concentrations of Cu and Co have also been reported in soils developed on volcanic ash and lake sediments in the Kenya Rift Valley (e.g., Maskall and Thornton, 1991). Ruminants require Co in order to synthesize vitamin B12. Cobalt deficiency arises on soils of diverse geological origin and climatic conditions, including coarse volcanic soils around the town of Nakuru in the Kenya Rift Valley, where the disease is referred to by the local name of *"nakuruitis."* The most appropriate scientific designation for the condition is *enzootic marasmus* (McDowell, 2003)	At Lake Nakuru National Park in Kenya, the elevated pH of soils has been linked with the high Mo content of several plant species (Maskall and Thornton, 1991). Mo-induced Cu deficiency has been reported in other wild animals including Grant's gazelle (*Gazelle granti*) from the Kenya Rift Valley; in this case the plant Mo content ranged from 0.5 to 5.6 µg/g (Hedger et al., 1964)	Because diseased animals (Nakuruitis) could be cured by a change to healthy ground, it was rightly assumed that this was a nutritional deficiency disease. Cobalt deficiency in ruminants can be cured or prevented by treatment of the soils or pastures with Co-containing fertilizers or by direct oral administration of Co to the animal (Underwood, 1981)	Maskall and Thornton (1991), McDowell, (2003), Wichard et al. (2009), González-Montaña et al. (2020), Ranaweera et al. (2023) —

Source: Modified and updated after Davies, T.C., 2010. Chapter 8: Medical geology in Africa. In: Selinus, O., Finkelman, R.B., Centeno, J. (Eds.), Medical Geology - A Regional Synthesis, first ed. Springer Verlag, Amsterdam, The Netherlands, pp. 199–219, ISBN-10: 9048134293.

problem still remain unanswered. We know that there is a high percentage of neonatal deaths due to fetal abnormalities or congenital malformations from ionizing radiation exposure in major U and Au mining countries where good morbidity estimates exist, for example, South Africa. There is therefore an urgent need to study the incidence of neonatal, maternal, and child deaths in areas of the subcontinent with high exposure to ionizing radiation exposure due to mining. This subject is dealt with in Chapter 5, Part I.

Geophagy (or *geophagia*), the habit of eating earth material is a common practice in many African countries. There are many theories about why people practice geophagy. Aside from the benefits that eaten soils can impart to the consumer, very serious health problems may result. Practical behavioral adjustments that would obviate the undesirable consequences of the practice, while at the same time promoting optimum derivation of its benefits, and gaps in knowledge of various other aspects of the phenomenon constitute the subject matter of Chapter 6 in Part I of this Volume.

Of the topics covered in Part I, "Pollution of the Soil, Water, and Air Environments" (Chapter 4) accounts for the largest volume of research literature on Medical Geology of Africa over the last 3−4 years (2020−23), significantly contributing to our understanding of the geochemical circulation of nutritional and PTEs in these environments. This is consistent with the growing realization of the important role of nutritional elements in shaping the diet, the criticality of their "optimal range" of intake in metabolic processes and hence, in warding off disease.

In the same way, knowledge of the fluxes of PTEs in the water, soil, and air environments is essential for their (PTE) exclusion from the food chain. Projects involving pollution of the water, soil, and air environments have been conducted with respect to the source of pollution-mining activities, in particular, leachates from dumpsites, agricultural activities, industrial activities, and so on. These studies have been widely supported since the beginning of the last decade, thanks to a favorable shift in international funding policies. For example, results from the UNESCO/SIDA funded IGCP 594 and 606 projects on: "Environmental Health Impacts of Mining (on the soil, water, and air environments) in Africa," which were completed in 2017, have been compiled in two special issues of the Journal of Geochemical Exploration (JGE): (Kříbek et al., 2014, JGE, Vol. 144, Part C; and Toteu et al., 2019, JGE Vol. 209), both of which have, since publication, been experiencing an extraordinary download rate.

The information and discussion presented in this book are based largely on the most recently published scientific and technical literature on some of the more current aspects of Medical Geology and contain new conclusions or interpretations made up to the year 2023. However, older seminal works and those that have laid the groundwork for this science are inevitably included. Knowledge gaps are identified and a look is taken at the current challenges and future perspectives (Part III) of Medical Geology of Africa. This approach would offer graduate students and other researchers major directions for future research in Medical Geology.

The book is written in a novel style using simple, nontechnical language, easily understood by a wide readership that reflects the multidisciplinary nature of the subject. A "Glossary of Terms" is nevertheless provided at the end of each chapter, to aid those users with a geoscience background who may not be familiar with some of the medical terminologies or even the less frequently used geoscience terminologies. The subject matter is presented with a minimum of jargon, with frequent reference to observational evidence. Different perspectives are examined, pinpointing trends and gaps in the literature, in order offer graduate students and allied researchers major directions for future research on Medical Geology.

The timing of this volume is auspicious, appearing just as realization of the link between geological materials and processes, and health of man and animals is reaching sky-high proportions in Africa, as well as worldwide; and diseases of unknown etiology continue to emerge everywhere.

The book has taken a long time to complete. However, its completion now, rather than a few years before now can be considered timely, in view of the increasing number of "proven" correlations between the geoenvironment and health in Africa in the last few years. Despite using the word "proven," an attempt has been made to present a balanced view of any "new" correlations that have been observed.

Finally, it is hoped that this primer will go some way toward addressing the current dearth of dedicated and customized volumes that are adaptable to guiding Medical Geology research across the geoscience, environmental, and public health spectrum in Africa.

Glossary of Terms

- *Gleying*: A type of hydric (saturated-, flooded-, or ponded-) soil, sticky, greenish to bluish gray in color, and low in oxygen content.
- *Hypertonic solution* is any external solution that has a higher concentration of solute and low water concentration than another solution (body fluid), meaning that water will flow into it.
- *Lateritization*: A weathering process that produces red residual soil by the leaching of silica and the enrichment with aluminum and iron oxides, especially in humid tropical climates.
- *Paget's disease of bone* is a chronic disorder that causes bones to break down (resorption) and then regrow, becoming weaker than normal. It most commonly occurs in the pelvis, skull, spine, and legs.
- *Phototherapy* is a major effective therapeutic treatment modality in dermatology, whereby UV light is used to slow skin cell growth and inflammation.
- *Podsolization* is an extreme form of leaching, whereby eluviation of iron, lime, and alumina occurs and become concentrated in the lower horizons.
- *Prolapsed intervertebral disk* is a term used to describe the condition that occurs when the outer fibers of the intervertebral disk bulge out or rupture. The damaged disk can then put pressure on the spinal cord or on a single nerve fiber, causing pain, numbness, weakness, and other symptoms.
- *Reticuloendothelial System (RES)* consists of a heterogeneous population of phagocytic cells in systemically fixed tissues, which are able to perform phagocytosis of foreign materials and particles. Ninety percentage of the RES are located in the liver.
- *Spinal stenosis* occurs when the spinal canal narrows, putting pressure on the spinal cord and nerves. It may cause back pain, weakness, and other nerve-related problems.
- *Venesection (Phlebotomy)* is a minor procedure of drawing out or removing blood from the vein for the purpose of blood donation or treatment of a variety of blood disorders.

References

Adam, A.M., Akuku, O., 2005. Creutzfeldt-Jakob disease in Kenya. Tropical Medicine and International Health 10, 710−712. Available from: https://doi.org/10.1111/j.1365-3156.2005.01435.x.

Ahoulé, D., Lalanne, F., Mendret, J., Brosillon, S., Maïga, A., 2015. Arsenic in African waters: a review. Water, Air, and Soil Pollution 226 (9), 1−13. Available from: https://doi.org/10.1007/s11270-015-2558-4.

Aliku, T.O., Rwebembera, J., Lubega, S., Zhang, W., Lugero, C., Namuyonga, J., et al., 2022. Trends in annual incidence rates of newly diagnosed endomyocardial fibrosis cases at the Uganda Heart Institute: a 14-year review. Frontiers in Cardiovascular Medicine 9, 841346. Available from: https://doi.org/10.3389/fcvm.2022.841346.

Babelhadj, B., Di Bari, M., Pirisinu, L., Chiappini, B., Gaouar, S., Riccardi, G., et al., 2018. Prion disease in dromedary camels, Algeria. Emerging Infectious Diseases 24 (6), 1029−1036. Available from: https://doi.org/10.3201/eid2406.172007.

Banaganapalli, B., Fallatah, I., Alsubhi, F., Shetty, P.J., Awan, Z., Elango, R., et al., 2023. Paget's disease: a review of the epidemiology, etiology, genetics, and treatment. Frontiers in Genetics 14, 1131182. Available from: https://doi.org/10.3389/fgene.2023.1131182.

Buthelezi-Dube, N.N., Muchaonyerwa, P., Hughes, J., Modi, A., Caister, K., 2022. Properties and indigenous knowledge of soil materials used for consumption, healing and cosmetics in KwaZulu-Natal, South Africa. Soil Science Annual 73 (4), 157408. Available from: https://doi.org/10.37501/soilsa/157408.

Bykowska-Derda, A., Spychala, M., Czlapka-Matyasik, M., Sojka, M., Bykowski, J., Ptak, M., 2023. The relationship between mortality from cardiovascular diseases and total drinking water hardness: systematic review with meta-analysis. Foods 12 (17), 3255. Available from: https://doi.org/10.3390/foods12173255.

Cao, L., Xie, W., Cui, H., Xiong, Z., Tang, Y., Zhang, X., et al., 2022. Fibrous clays in dermopharmaceutical and cosmetic applications: traditional and emerging perspectives. International Journal of Pharmaceutics 625, 122097. Available from: https://doi.org/10.1016/j.ijpharm.2022.122097.

Ceruti, P.O., Fey, M., Pooley, J., 2003. Chapter 24: Soil nutrient deficiencies (Mseleni Joint Disease) and dwarfism in Maputaland, South Africa. In: Skinner, H.C.W., Berger, A.R. (Eds.), Geology and Health: Closing the Gap. Oxford University Press, New York, pp. 151−154.

Chaudhary, M., Rawat, S., Maiti, A., 2023. Defluoridation by bare nanoadsorbents, nanocomposites and nanoadsorbent loaded mixed matrix membranes. Separation & Purification Reviews 52 (2), 135−153. Available from: https://doi.org/10.1080/15422119.2022.2045610.

Cooper, J.N., Nick, K.E., 2023. A geochemical and mineralogical characterization of soils associated with podoconiosis. Environmental Geochemistry and Health 45 (11), 7791−7812. Available from: https://doi.org/10.1007/s10653-023-01625-5.

Cuenca, S., Bitchou, M.P., Morales-Jiménez, G., 2022. Tropical endomyocardial fibrosis, a neglected disease. Case Series in Cameroon. Revista Espanola de Cardiologia (English Ed.) 75 (8), 690−692. Available from: https://doi.org/10.1016/j.rec.2022.02.011.

Dai, D., Ma, X., Yan, X., Bao, X., 2023. The biological role of Dead Sea Water in skin health: A review. Cosmetics 10 (1), 21. Available from: https://doi.org/10.3390/cosmetics100100.

Davies, T.C., 1999. The East and Southern Africa regional workshop on geomedicine. Episodes 22 (4), 303−304.

Davies, T.C., 2010. Chapter 8: Medical geology in Africa. In: Selinus, O., Finkelman, R.B., Centeno, J. (Eds.), Medical Geology - A Regional Synthesis, first edition. Springer Verlag, Amsterdam, The Netherlands, pp. 199−219.

Davies, T.C., 2013. Geochemical variables as plausible aetiological cofactors in the incidence of some *common environmental diseases* in Africa. Journal of African Earth Sciences 79, 24−49.

Davies, T.C., 2022. The position of geochemical variables as causal co-factors of diseases of unknown aetiology. Springer Nature Applied Sciences 4, 236. Available from: https://doi.org/10.1007/s42452-022-05113-w.

Davies, T.C., 2023. Current status of research and gaps in knowledge of geophagic practices in Africa. Frontiers in Nutrition 9, 1084589. Available from: https://www.frontiersin.org/articles/10.3389/fnut.2022.1084589/full.

Davies, T., Schlüter, T., 2002. Current status of research in geomedicine in East and Southern Africa. Environmental Geochemistry and Health 24, 99−102. Available from: https://doi.org/10.1023/A:1014208817450.

Debele, G.R., Shifera, E., Dessie, Y.L., Jaleta, D.D., Borena, M.U., Kanfe, S.G., et al., 2023. From neglected to public health burden: factors associated with podoconiosis in resource limited setting in case of southwest Ethiopia: a community based cross sectional study. Research and Reports in Tropical Medicine 14, 49−60. Available from: https://doi.org/10.2147/RRTM.S412624.

Deribe, K., Mackenzie, C.D., Newport, M.J., Argaw, D., Molyneux, D.H., Davey, G., 2020. Podoconiosis: key priorities for research and implementation. Transactions of the Royal Society of Tropical Medicine and Hygiene 114 (12), 889−895. Available from: https://doi.org/10.1093/trstmh/traa094.

Deribe, K., Sultani, H.M., Okoyo, C., Omondi, W.P., Ngere, I., Newport, M.J., et al., 2023. Geostatistical modelling of the distribution, risk and burden of podoconiosis in Kenya. Transactions of the Royal Society of Tropical Medicine and Hygiene 117 (2), 72−82. Available from: https://doi.org/10.1093/trstmh/trac092.

Dilpazeer, F., Munir, M., Baloch, M.Y.J., Shafiq, I., Iqbal, J., Saeed, M., et al., 2023. A comprehensive review of the latest advancements in controlling arsenic contaminants in groundwater. Water 15, 478. Available from: https://doi.org/10.3390/w15030478.

Dinkele, E.S., Ballo, R., Fredlund, V., Ramesar, R., Gibbon, V., 2020. Mseleni joint disease: an endemic arthritis of unknown cause. The Lancet Rheumatology 2 (1), e8−e9. Available from: https://doi.org/10.1016/S2665-9913(19)30104-3.

Dintsi, B.S., Letsoalo, M.R., Ambushe, A.A., 2023. Bioaccumulation and human health risk assessment of arsenic and chromium species in water−soil−vegetables system in Lephalale, Limpopo Province, South Africa. Minerals 13, 930. Available from: https://doi.org/10.3390/min13070930.

du Toit, A.Z., 2010. Description of Coal Associated Diseases and Coal Dust Concentrations in Mpumalanga Coal Mines (M.Sc. dissertation). Faculty of Health Sciences, School of Public Health, University of the Witwatersrand, Johannesburg, South Africa. https://wiredspace.wits.ac.za/server/api/core/bitstreams/9940057c-a7ec-42bf-b691-a45e8b71cf72/content. (Accessed 12 September 2023).

Emmanuel, T., Lybæk, D., Johansen, C., Iversen, L., 2020. Effect of Dead Sea climatotherapy on psoriasis; a prospective cohort study. Frontiers in Medicine 7, 83. Available from: https://doi.org/10.3389/fmed.2020.000.

Ehrlich, R., Akugizibwe, P., Siegfried, N., Rees, D., 2021. The association between silica exposure, silicosis and tuberculosis: A systematic review and meta-analysis. BMC Public Health 21 (1), 953. Available from: https://doi.org/10.1186/s12889-021-10711-1.

Ehrlich, R., Barker, S., te Water Naude, J., Rees, D., Kistnasamy, B., Naidoo, J., et al., 2022. Accuracy of computer-aided detection of occupational lung disease: silicosis and pulmonary tuberculosis in ex-miners from the South African gold mines. International Journal of Environmental Reearch and Public Health 19, 12402. Available from: https://doi.org/10.3390/ijerph191912402.

Getachew, T., Churko, C., 2022. Prevalence of podoconiosis and its associated factors in Gamo zone, Southern Ethiopia, 2021. Journal of Foot Ankle Research 15, 13. Available from: https://doi.org/10.1186/s13047-022-00517-8.

Gebretsadik, A., 2023. Effect of balneotherapy on skin lesion at hot springs in southern Ethiopia: a single-arm prospective cohort study. Clinical Cosmetic Investigations in Dermatology 2023 (16), 1259−1268. Available from: https://doi.org/10.2147/CCID.S413926.

Gebretsadik, A., Taddesse, F., Melaku, N., Haji, Y., 2021. Balneotherapy for musculoskeletal pain management of hot spring water in southern Ethiopia: perceived improvements. Inquiry: A Journal of Medical Care Organization, Provision and Financing 58, 469580211049063. Available from: https://doi.org/10.1177/00469580211049063.

Ginzburg, Y., Finberg, K., 2021. Iron Metabolism and Related Disorders, pp. 445–499. ISSN: 9780128125359. 10.1016/B978-0-12-812535-9.00012-1.

Gomes, C., de Sousa, F., 2018. Healing and edible clays: a review of basic concepts, benefits and risks. Environmental Geochemistry and Health 40, 1739–1765. Available from: https://doi.org/10.1007/s10653-016-9903-4.

González-Montaña, J.-R., Escalera-Valente, F., Alonso, A.J., Lomillos, J.M., Robles, R., Alonso, M.E., 2020. Relationship between Vitamin B12 and cobalt metabolism in domestic ruminant: an update. Animals 10, 1855. Available from: https://doi.org/10.3390/ani10101855.

Gordeuk, V.R., 2002. African iron overload. Seminers in Hematology 39 (4), 263–269. Available from: https://doi.org/10.1053/shem.2002.35636.

Gorstein, J.L., Bagrianski, J., Pearce, E.N., Kupka, R., Zimmermann, M.B., 2020. Estimating the health and economic benefits of universal salt iodization programs to correct iodine deficiency disorders. Thyroid: Official Journal of the American Thyroid Association 30 (12), 1802–1809.

Grimaldi, A., Mocumbi, A.O., Freers, J., Lachaud, M., Mirabel, M., Ferreira, B., et al., 2016. Tropical endomyocardial fibrosis: natural history, challenges, and perspectives. Circulation 133 (24), 2503–2515. Available from: https://doi.org/10.1161/CIRCULATIONAHA.115.021178.

Haffner, D., Leifheit-Nestler, M., Grund, A., Schnabel, D., 2022. Rickets guidance: part II-management. Pediatric Nephrology (Berlin, Germany) 37 (10), 2289–2302. Available from: https://doi.org/10.1007/s00467-022-05505-5.

Haji, Y., Taddesse, F., Serka, S., Gebretsadik, A., 2021. Effect of balneotherapy on chronic low back pain at hot springs in southern Ethiopia: perceived improvements from pain. Journal of Pain Research 14, 2491–2500. Available from: https://doi.org/10.2147/JPR.S322603.

Hatch-McChesney, A., Lieberman, H.R., 2022. Iodine and iodine deficiency: a comprehensive review of a re-emerging issue. Nutrients 14 (17), 3474. Available from: https://doi.org/10.3390/nu14173474.

Hedger, R.S., Howard, D.A., Burdin, M.L., 1964. The occurrence in goats and sheep in Kenya of a disease closely similar to sway-back. Veterinary Record 76, 493–497.

Hippocrates of Cos II, 2009. In: Adams, F. (Ed.), On Airs, Waters, and Places. Dodo Press.

Hsu, C.C., Senussi, N.H., Fertrin, K.Y., Kowdley, K.V., 2022. Iron overload disorders. Hepatology Communications 2022 (6), 1842–1854. Available from: https://doi.org/10.1002/hep4.2012.

Incledion, A., Boseley, M., Moses, R.L., Moseley, R., Hill, K.E., Thomas, D.W., et al., 2021. A new look at the purported health benefits of commercial and natural clays. Biomolecules 11, 58. Available from: https://doi.org/10.3390/biom11010058.

Irunde, R., Ijumulana, J., Ligate, F., Maity, J.P., Ahmad, A., et al., 2022. Arsenic in Africa: potential sources, spatial variability, and the state of the art for arsenic removal using locally available materials. Groundwater for Sustainable Development 18, 100746. Available from: https://doi.org/10.1016/j.gsd.2022.100746.

Jacobs, T., Kgokolo, C.M., 2020. Treatment of psoriasis. South African Pharmaceutical Journal Incorporating Pharmacy Management 87 (6), 23–26.

Jamshidi, P., Danaei, B., Arbabi, M., Mohammadzadeh, B., Khelghati, F., Akbari Aghababa, A., et al., 2023. Silicosis and tuberculosis: a systematic review and meta-analysis. Pulmonology S2531-0437(23)00092-2. Available from: https://doi.org/10.1016/j.pulmoe.2023.05.001.

John-Olabode, S.O., Akintan, P., Okunade, K.S., Ajie, I., 2023. Comparative assessment of serum selenium status in HIV-infected and non-infected children: a pilot study in a Tertiary Hospital in Nigeria. Cureus 15 (5), e39626. Available from: https://doi.org/10.7759/cureus.39626.

Kgabi, D.P., Ambushe, A.A., 2023. Characterization of South African bentonite and kaolin clays. Sustainability 15, 12679. Available from: https://doi.org/10.3390/su151712679.

Kříbek, B., Davies, T.C., De Vivo, B.(Eds.), 2014. Impacts of mining and mineral processing on the environment and human health in Africa. Journal of Geochemical Exploration 44 (C), 387−580. Available from: http://www.sciencedirect.com/science/journal/03756742/144/part/PC.

Kulkarni, P.G., Paudel, N., Magar, S., Santilli, M.F., Kashyap, S., Baranwal, A.K., et al., 2023. Overcoming challenges and innovations in orthopedic prosthesis design: an interdisciplinary perspective. Biomedical Materials & Devices 1−12. Available from: https://doi.org/10.1007/s44174-023-00087-8.

Kamanzi, C., Becker, M., Jacobs, M., Konečný, P., Von Holdt, J., Broadhurst, J., 2023. The impact of coal mine dust characteristics on pathways to respiratory harm: investigating the pneumoconiotic potency of coals. Environmental Geochemistry and Health 45 (10), 7363−7388. Available from: https://doi.org/10.1007/s10653-023-01583-y.

Kapwata, T., Wright, C.Y., Reddy, T., Street, R., Kunene, Z., Mathee, A., 2023. Relations between personal exposure to elevated concentrations of arsenic in water and soil and blood arsenic levels amongst people living in rural areas in Limpopo, South Africa. Environmental Science and Pollution Research International 30 (24), 65204−65216. Available from: https://doi.org/10.1007/s11356-023-26813-9.

Kerdoun, M.A., Mekhloufi, S., Adjaine, O.E.K., Bechki, Z., Gana, M., Belkhalfa, H., 2022. Fluoride concentrations in drinking water and health risk assessment in the south of Algeria. Regulatory Toxicology and Pharmacology 128, 105086. Available from: https://doi.org/10.1016/j.yrtph.2021.105086.

Khosravi-Darani, K., Rehman, Y., Katsoyiannis, I.A., Kokkinos, E., Zouboulis, A.I., 2022. Arsenic exposure via contaminated water and food sources. Water 14, 1884. Available from: https://doi.org/10.3390/w14121884.

Kihara, J., Bolo, P., Kinyua, M., Rurinda, J., Piiki, K., 2020. Micronutrient deficiencies in African soils and the human nutritional nexus: opportunities with staple crops. Environmental Geochemistry and Health 42, 3015−3033. Available from: https://doi.org/10.1007/s10653-019-00499-w.

Kut, K.M.K., Sarswat, A., Srivastava, A., Pittman Jr, C.U., Mohan, D., 2016. A review of fluoride in African groundwater and local remediation methods. Groundwater for Sustainable Development 2-3, 190−212. Available from: https://doi.org/10.1016/j.gsd.2016.09.001.

Kwon, J.W., Moon, S.H., Park, S.Y., Park, S.J., Park, S.R., Suk, K.S., et al., 2022. Lumbar spinal stenosis: review update 2022. Asian Spine Journal 16 (5), 789−798. Available from: https://doi.org/10.31616/asj.2022.0366.

Låg, J.Contributors, 1990. Geomedicine. CRC Press, p. 288. Available from: https://books.google.co.za/books/about/Geomedicine_1990.html?id = rHX4swEACAAJ&redir_esc = y (Accessed 28 August 2023).

Lalonde, M., 1974. A new perspective on the health of Canadians. https://nccdh.ca/resources/entry/new-perspective-on-the-health-of-canadians. (Accessed 11 September 2023).

Lisco, G., De Tullio, A., Triggiani, D., Zupo, R., Giagulli, V.A., De Pergola, G., et al., 2023. Iodine deficiency and iodine prophylaxis: an overview and update. Nutrient 15 (4), 1004. Available from: https://doi.org/10.3390/nu15041004.

Lubojanski, A., Piesiak-Panczyszyn, D., Zakrzewski, W., Dobrzynski, W., Szymonowicz, M., Rybak, Z., et al., 2023. The safety of fluoride compounds and their effect on the human body - a narrative review. Materials (Basel, Switzerland) 16 (3), 1242. Available from: https://doi.org/10.3390/ma16031242.

Maboso, B., te Water Naude, J., Rees, D., Goodman, H., Ehrlich, R., 2023. Difficulties in distinguishing silicosis and pulmonary tuberculosis in silica-exposed gold miners: a report of four cases. American Journal of Industrial Medicine 66, 339−348. Available from: https://doi.org/10.1002/ajim.23460.

Maghanga, J.K., Okello, V.A., Michira, J.A., Ojwang, L.M.B., Segor, F.K., 2022. Fluoride in water, health implications and plant-based remediation strategies. Physical Sciences Reviews . Available from: https://doi.org/10.1515/psr-2022-0123.

Mangwane, J., Ntanjana, A., 2020. Wellness tourism in south africa: development opportunities. In: Rocha, Á., Abreu, A., de Carvalho, J., Liberato, D., González, E., Liberato, P. (Eds.), Advances in Tourism,

Technology and Smart Systems. Smart Innovation, Systems and Technologies, Vol. 171. Springer, Singapore. Available from: https://doi.org/10.1007/978-981-15-2024-2_50 (Accessed 08 September 2023).

Mao, R.J., Moa, A., Chughtai, A., 2020. The epidemiology of unknown disease outbreak reports globally. Global Biosecurity 2 (1).

Masango, C.A., Ekosse, G.I., Netshandama, V., 2017. Indigenous knowledge use of clay within an African context: possible documentation of entire clay properties? Southern African Journal for Folklore Studies 27 (1), 92−104. Available from: https://doi.org/10.25159/1016-8427/2497.

Maskall, J., Thornton, I., 1991. Trace element geochemistry of soils and plants in Kenyan conservation areas and implications for wildlife nutrition. Environmental Geochemistry and Health 13, 93−107. Available from: https://doi.org/10.1007/BF01734300.

Mays, S., Brickley, M.B., 2023. Dietary calcium versus vitamin D in rickets: a response to Vlok et al. 2023. American Journal of Humam Biology 35, e23872. Available from: https://doi.org/10.1002/ajhb.23872.

Mbanze, J., Cumbane, B., Jive, R., Mocumbi, A., 2020. Challenges in addressing the knowledge gap on endomyocardial fibrosis through community-based studies. Cardiovascular Diagnosis and Therapy 10 (2), 279−288. Available from: https://doi.org/10.21037/cdt.2019.08.07.

Mbedzi, M.D., van der Poll, H.M., van der Poll, J.A., 2020. Enhancing a decision-making framework to address environmental impacts of the South African Coal Mining Industry. Energies 13, 4897. Available from: https://doi.org/10.3390/en13184897.

Mbuya, A.W., Mboya, I.B., Semvua, H.H., Mamuya, S.H., Msuya, S.E., 2023. Prevalence and factors associated with tuberculosis among the mining communities in Mererani, Tanzania. PLoS One 18 (3), e0280396. Available from: https://doi.org/10.1371/journal.pone.0280396.

McDowell, L.R., 2003. Copper and molybdenum. In: Cunha, T.J. (Ed.), Minerals in Animal and Human Nutrition, second ed. Elsevier, pp. 235−276. Available from: https://doi.org/10.1016/B978-0-444-51367-0.50011-8.

Mead, S., Khalili-Shirazi, A., Potter, C., Mok, T., Nihat, A., Hyare, H., et al., 2022. Prion protein monoclonal antibody (PRN100) therapy for Creutzfeldt-Jakob disease: evaluation of a first-in-human treatment programme. The Lancet Neurology 21 (4), 342−354. Available from: https://doi.org/10.1016/S1474-4422(22)00082-5.

Meng, M., Wang, J., Huang, H., Liu, X., Zhang, J., Li, Z., 2023. 3D printing metal implants in orthopedic surgery: methods, applications and future prospects. Journal of Orthopaedic Translation 42, 94−112. Available from: https://doi.org/10.1016/j.jot.2023.08.004.

Minja, N.W., Nakagaayi, D., Aliku, T., Zhang, W., Ssinabulya, I., Nabaale, J., et al., 2022. Cardiovascular diseases in Africa in the twenty-first century: gaps and priorities going forward. Frontiers in Cardiovascular Medicine 9, 1008335. Available from: https://doi.org/10.3389/fcvm.2022.1008335.

Mohandas, S.A., Janardhanan, S., Rasheed, P.A., Gangadharan, P., 2023. Improved defluoridation and energy production using dimethyl sulfoxide modified carbon cloth as bioanode in microbial desalination cell. Heliyon 9 (6), e16614. Available from: https://doi.org/10.1016/j.heliyon.2023.e16614.

Moosavi, M., 2017. Bentonite clay as a natural remedy: a brief review. Iranian Journal of Public Health 46 (9), 1176−1183.

Morekhure-Mphahlele, R., Focke, W.W., Grote, W., 2017. Characterisation of vumba and ubumba clays used for cosmetic purposes. South African Journal of Science 113 (3-4), 1−5. Available from: https://doi.org/10.17159/sajs.2017/20160105.

Moshupya, P., Abiye, T., Mouri, H., Levin, M., Strauss, M., Strydom, R., 2019. Assessment of radon concentration and impact on human health in a region dominated by abandoned gold mine tailings dams: a case from the West Rand Region, South Africa. Geosciences 9, 466. Available from: https://doi.org/10.3390/geosciences9110466.

Moshupya, P.M., Mohuba, S.C., Abiye, T.A., Korir, I., Nhleko, S., Mkhosi, M., 2022. In situ determination of radioactivity levels and radiological doses in and around the gold mine tailing dams, Gauteng Province, South Africa. Minerals 12, 1295. Available from: https://doi.org/10.3390/min12101295.

Mpuchane, S.F., Ekosse, G.-I.E., Gashe, B.A., Morobe, I., Coetzee, S.H., 2010. Microbiological characterisation of southern African medicinal and cosmetic clays. International Journal of Environmental Health Research 20 (1), 27−41. Available from: https://doi.org/10.1080/09603120903254025.

Mutetwa, B., Moyo, D., Brouwer, D., 2022. Prediction of asbestos-related diseases (ARDs) and chrysotile asbestos exposure concentrations in asbestos-cement (AC) manufacturing factories in Zimbabwe. International Journal of Environmental Research and Public Health 20 (1), 58. Available from: https://doi.org/10.3390/ijerph20010058.

Muzembo, B.A., Ngatu, N.R., Januka, K., Huang, H.L., Nattadech, C., Suzuki, T., et al., 2019. Selenium supplementation in HIV-infected individuals: a systematic review of randomized controlled trials. Clinical Nutrition ESPEN 34, 1−7. Available from: https://doi.org/10.1016/j.clnesp.2019.09.005.

Naidoo, R.N., Robins, T.G., Murray, J., 2005. Respiratory outcomes among South African coal miners at autopsy. American Journal of Industrial Medicine 48 (3), 217−224. Available from: https://doi.org/10.1002/ajim.20207.

Ncube, S., Mlunguza, N.,Y., Dube, S., Ramganesh, S., Ogola, H.J.O., Nindi, M.M., et al., 2020. Physicochemical characterization of the pelotherapeutic and balneotherapeutic clayey soils and natural spring water at Isinuka traditional healing spa in the Eastern Cape Province of South Africa. The Science of the Total Environment 717, 137284. Available from: https://doi.org/10.1016/j.scitotenv.2020.137284.

Njau, O.E., Otter, P., Machunda, R., Rugaika, A., Wydra, K., Njau, K.N., 2023. Removal of fluoride and pathogens from water using the combined electrocoagulation-inline-electrolytic disinfection process. Water Supply 23 (7), 2745−2757. Available from: https://doi.org/10.2166/ws.2023.146.

Nomicisio, C., Ruggeri, M., Bianchi, E., Vigani, B., Valentino, C., Aguzzi, C., et al., 2023. Natural and synthetic clay minerals in the pharmaceutical and biomedical fields. Pharmaceutics 15 (5), 1368. Available from: https://doi.org/10.3390/pharmaceutics15051368.

Noubiap, J.J., Millenaar, D., Ojji, D., Wafford, Q.E., Ukena, C., Böhm, M., et al., 2023. Fifty years of global cardiovascular research in Africa: a scientometric analysis, 1971 to 2021. Journal of the American Heart Association 12 (3), e027670. Available from: https://doi.org/10.1161/JAHA.122.027670.

Ogunmwonyi, I., Adebajo, A., Wilkinson, J.M., 2022. The genetic and epigenetic contributions to the development of nutritional rickets. Frontiers in Endocrinology 13, 1059034. Available from: https://doi.org/10.3389/fendo.2022.1059034.

Oluwaseun, O.T., Makama, A.Y., Ibrahim, S.Z., Rahman, S.A., Oladimeji, O., 2020. Balneotherapy/spa therapy: potential of Nigerian thermal hot spring waters for musculoskeletal disorders and chronic health conditions. International Journal of Advanced Research and Publications 4 (3), 95−100.

Onipe, T., Edokpayi, J.N., Odiyo, J.O., 2020. A review on the potential sources and health implications of fluoride in groundwater of Sub-Saharan Africa. Journal of Environmental Science and Health. Part A, Toxic/Hazardous Substances and Environmental Engineering 55 (9), 1078−1093. Available from: https://doi.org/10.1080/10934529.2020.1770516.

Patel, K.S., Pandey, P.K., Martín-Ramos, P., Corns, W.T., Varol, S., Bhattacharya, P., et al., 2023. A review on arsenic in the environment: contamination, mobility, sources, and exposure. RSC Advances 13 (13), 8803−8821. Available from: https://doi.org/10.1039/d3ra00789h.

Pedrazzini, A., Delsignore, R., Martelli, A., Tocco, S., Vaienti, E., Ceccarelli, F., et al., 2016. Thermal balneotherapy in Antsirabe-Madagascar: water analysis and its applications in an African context. Acta Biomedica: Atenei Parmensis 87 (Suppl. 1), 25−33.

Peña-Castro, M., Montero-Acosta, M., Saba, M., 2023. A critical review of asbestos concentrations in water and air, according to exposure sources. Heliyon 9 (5), e15730. Available from: https://doi.org/10.1016/j.heliyon.2023.e15730.

Pezeshki, H., Hashemi, M., Rajabi, S., 2023. Removal of arsenic as a potentially toxic element from drinking water by filtration: a mini review of nanofiltration and reverse osmosis techniques. Heliyon 9 (3), e14246. Available from: https://doi.org/10.1016/j.heliyon.2023.e14246.

Podgorski, J., Berg, M., 2022. Global analysis and prediction of fluoride in groundwater. Nature Communications 13, 4232. Available from: https://doi.org/10.1038/s41467-022-31940-x.

Pourmoradian, S., Rezazadeh, L., Tutunchi, H., Ostadrahimi, A., 2023. Selenium and zinc supplementation in HIV-infected patients. International Journal for Vitamin and Nutrition Research. Internationale Zeitschrift fur Vitamin- und Ernahrungsforschung. Journal International de Vitaminologie et de Nutrition Advance online publication. Available from: https://doi.org/10.1024/0300-9831/a000778.

Price, E.W., 1976. The association of endemic elephantiasis of the lower legs in East Africa with soil derived from volcanic rocks. Transactions of the Royal Society of Hygiene and Tropical Medicine 70, 288−295.

Ranaweera, S., Hewawardhana, S., Manatunga, D., 2023. Chapter 15: Cobalt and copper deficiency and molybdenosis. In: Prasad, M.N.V., Vithanagepage, M. (Eds.), Medical Geology: En Route to One Health, pp. 235−252. 10.1002/9781119867371.ch15.

Rees, D., Murray, J., 2020. Silica, silicosis and tuberculosis. Occupational Health Southern Africa 26 (5). Available from: https://hdl.handle.net/10520/ejc-ohsa-v26-n5-a17.

Reynolds, C.J., Sisodia, R., Barber, C., Moffatt, M., Minelli, C., De Matteis, S., et al., 2023. What role for asbestos in idiopathic pulmonary fibrosis? Findings from the IPF job exposures case-control study. Occupational and Environmental Medicine 80 (2), 97−103. Available from: https://doi.org/10.1136/oemed-2022-108404.

Saha, S., Abu, B.A.Z., Jamshidi-Naeini, Y., Mukherjee, U., Miller, M., Peng, L.L., et al., 2019. Is iodine deficiency still a problem in sub-Saharan Africa?: a review. The Proceedings of the Nutrition Society 78 (4), 554−566. Available from: https://doi.org/10.1017/S0029665118002859.

Saha, S., Roy, S., 2023. Metallic dental implants wear mechanisms, materials, and manufacturing processes: a literature review. Materials 16, 161. Available from: https://doi.org/10.3390/ma16010161.

Salomao, M.A., 2021. Pathology of hepatic iron overload. Clinical Liver Disease 17, 232−237. Available from: https://doi.org/10.1002/cld.1051.

Sawangjang, B., Takizawa, S., 2023. Re-evaluating fluoride intake from food and drinking water: effect of boiling and fluoride adsorption on food. Journal of Hazardous Materials 443, Part A, 130162. Available from: https://doi.org/10.1016/j.jhazmat.2022.130162.

Sekome, K., Maddocks, S., 2019. The short-term effects of hydrotherapy on pain and self-perceived functional status in individuals living with osteoarthritis of the knee joint. South African Journal of Physiotherapy 75 (1), a476. Available from: https://doi.org/10.4102/sajp.v75i1.476.

Selinus, O., Finkelman, R.B., Centeno, J.A., Cave, M.R., 2008. Medical geology - the European perspective. Central European Geology 51 (2), 133−151. Available from: https://doi.org/10.1556/CEuGeol.51.2008.2.3.

Skinner, H.C.W., Berger, A.R. (Eds.), 2003. Geology and Health: Closing the Gap. Oxford University Press, New YorkOnline Edition, Oxford Academic, November 12, 2020. Available from: https://doi.org/10.1093/oso/9780195162042.001.0001. (Accessed 03 September 2023).

Sun, Y., Wang, Z., Gong, P., Yao, W., Ba, Q., Wang, H., 2023. Review on the health-promoting effect of adequate selenium status. Frontiers in Nutrition 10, 1136458. Available from: https://doi.org/10.3389/fnut.2023.1136458.

Thacher, T.D., Fischer, P.R., Isichei, C.O., Zoakah, A.I., Pettifor, J.M., 2012. Prevention of nutritional rickets in Nigerian children with dietary calcium supplementation. Bone 50 (5), 1074−1080. Available from: https://doi.org/10.1016/j.bone.2012.02.010.

Toteu, S.F., Mahe, G., Moritz, R., Sracek, O., Davies, T.C., Ramasamy, J., 2019. Environmental, health and social legacies of mining activities in sub-saharan Africa. Journal of Geochemical Exploration 209, 106441. Available from: https://doi.org/10.1016/j.gexplo.2019.10644.

Tulchinsky, T.H., 2018. Marc Lalonde, the health field concept and health promotion. Case Studies in Public Health 523−541. Available from: https://doi.org/10.1016/B978-0-12-804571-8.00028-7.

UCL (University College London), 2022. Creutzfeldt-Jakob disease treatment shows promising early results. *ScienceDaily*. http://www.sciencedaily.com/releases/2022/03/220317111917.htm (Accessed 10 September 2023).

Underwood, E.J., 1981. The incidence of trace element deficiency diseases. Philosophical Transactions of the Royal Society of London. Series B, Biological Sciences 294 (1071), 3−8. Available from: https://doi.org/10.1098/rstb.1981.0085.

Vlok, M., Snoddy, A.M.E., Ramesh, N., Wheeler, B.J., Standen, V.G., Arriaza, B.T., 2023. The role of dietary calcium in the etiology of childhood rickets in the past and the present. American Journal of Human Biology: The official Journal of the Human Biology Council 35 (2), e23819. Available from: https://doi.org/10.1002/ajhb.23819.

Vogiatzi, G., Lazaros, G., Oikonomou, E., Kostakis, M., Kypritidou, Z., Christoforatou, E., et al., 2023. Impact of drinking water hardness on carotid atherosclerosis and arterial stiffness: insights from the "Corinthia" study. Hellenic Journal of Cardiology: HJC = Hellenike Kardiologike Epitheorese 74, 32−38. Available from: https://doi.org/10.1016/j.hjc.2023.04.006.

Vorster, T., Mthombeni, J., teWaterNaude, J., Phillips, J.I., 2022. The association between the histological subtypes of mesothelioma and asbestos exposure characteristics. International Journal of Environmental Research and Public Health 19 (21), 14520. Available from: https://doi.org/10.3390/ijerph192114520.

Walter, K.L., O'Toole, J.E., 2022. Lumbar spinal stenosis. JAMA: The Journal of the American Medical Association 328 (3), 310. Available from: https://doi.org/10.1001/jama.2022.6137.

Wang, Y., Li, S., Bai, J., Cai, X., Tang, S., Lin, P., et al., 2023a. Bimekizumab for the treatment of moderate-to-severe plaque psoriasis: a meta-analysis of randomized clinical trials. Therapeutic Advances in Chronic Disease 14. Available from: https://doi.org/10.1177/20406223231163110.

Wang, P.C., Song, Q.C., Chen, C.Y., Su, T.C., 2023b. Cardiovascular physiological effects of balneotherapy: focused on seasonal differences. Hypertension Research 46, 1650−1661. Available from: https://doi.org/10.1038/s41440-023-01248-4.

Wanji, S., Deribe, K., Minich, J., Debrah, A.Y., Kalinga, A., Kroidl, I., et al., 2021. Podoconiosis - from known to unknown: obstacles to tackle. Acta Tropica 219, 105918. Available from: https://doi.org/10.1016/j.actatropica.2021.105918.

Wanke, H., Ueland, J.S., Hipondoka, M.H.T., 2017. Spatial analysis of fluoride concentrations in drinking water and population at risk in Namibia. Water SA 43 (3), 413−422. Available from: https://doi.org/10.4314/wsa.v43i3.06.

WHO (World Health Organisation), 2017. Guidelines for drinking-water quality, fourth ed., Incorporating the First Addendum. https://www.who.int/publications/i/item/9789241549950. (Accessed 17 September 2023).

WHO (World Health Organisation), 2022. Arsenic. https://www.who.int/news-room/fact-sheets/detail/arsenic. (Accessed 19 September 2023).

Wichard, T., Mishra, B., Myneni, S.C., Bellenger, J.-P., Kraepiel, A.M.L., 2009. Storage and bioavailability of molybdenum in soils increased by organic matter complexation. Nature Geoscience 2, 625−629. Available from: https://doi.org/10.1038/ngeo589.

Williams, L.B., 2019. Natural antibacterial clays: historical uses and modern advances. Clays Clay Minerals 67, 7−24. Available from: https://doi.org/10.1007/s42860-018-0002-8.

Williams, L.B.Haydel, S.E., 2010. Evaluation of the medicinal use of clay minerals as antibacterial agents. International Geology Reviews 52 (7/8), 745−770. Available from: https://doi.org/10.1080/00206811003679737.

Wu, P.H., Kim, H.S., Jang, I.-T., 2020. Intervertebral disc diseases, part 2: a review of the current diagnostic and treatment strategies for intervertebral disc disease. International Journal of Molecular Science 21, 2135. Available from: https://doi.org/10.3390/ijms21062135.

Wu, Y., Liu, J., Kang, L., Tian, J., Zhang, X., Hu, J., et al., 2023. An overview of 3D printed metal implants in orthopedic applications: present and future perspectives. Heliyon 9 (7), e17718. Available from: https://doi.org/10.1016/j.heliyon.2023.e17718.

Yuyun, M.F., Sliwa, K., Kengne, A.P., Mocumbi, A.O., Bukhman, G., 2020. Cardiovascular diseases in Sub-Saharan Africa compared to high-income countries: an epidemiological perspective. Global Heart 15 (1), 15. Available from: https://doi.org/10.5334/gh.403.

Zhang, K., Lu, Z., Guo, X., 2023. Advances in epidemiological status and pathogenesis of dental fluorosis. Frontiers in Cell and Developmental Biology 11, 1168215. Available from: https://doi.org/10.3389/fcell.2023.1168215.

Zheng, X., Wang, Q., Luo, Y., Lu, W., Jin, L., Chen, M., et al., 2021. Seasonal variation of psoriasis and its impact in the therapeutic management: a retrospective study on chinese patients. Clinical, Cosmetic and Investigational Dermatology 14, 459−465. Available from: https://doi.org/10.2147/CCID.S312556.

Ziegler, J.L., 1993. Endemic Kaposi's sarcoma in Africa and local volcanic soils. Lancet 342 (8883), 1348−1351. Available from: https://doi.org/10.1016/0140-6736(93)92252-o.

Ziegler, J., Newton, R., Bourboulia, D., Casabonne, D., Beral, V., Mbidde, E., et al., 2003. Risk factors for Kaposi's sarcoma: a case-control study of HIV-seropositive people in Uganda. International Journal of Cancer 103 (2), 233−240.

Zupunski, L., Street, R., Ostroumova, E., Winde, F., Sachs, S., Geipel, G., et al., 2023. Environmental exposure to uranium in a population living in close proximity to gold mine tailings in South Africa. Journal of Trace Elements in Medicine and Biology: Organ of the Society for Minerals and Trace Elements (GMS) 77, 127141. Available from: https://doi.org/10.1016/j.jtemb.2023.127141.

2

Project selection, proposal writing, and project execution in medical geology

"Every project has challenges, and every project has its rewards."

Stephen Schwartz

Key chapter features

1. Criteria for research project selection in Medical Geology.
2. Proposal/thesis/dissertation/report writing process from idea to publication.
3. How to identify and address ethical issues through the formal approval process.
4. Submission and (postexamination) review of proposal/thesis/dissertation/report.

BOX 2−1   Learning outcomes

- Students acquire an objective tool for prioritizing projects, and selecting the one that is most in line with their individual circumstances, but also taking into account the importance of the project to their community and how achievable the project is.
- Students can list and describe the chapters and subsections of a proposal/thesis/dissertation and research report in their proper order.
- Students get a firm grasp of the styles appropriate for (1) a research proposal, (2) a thesis or dissertation, (3) a research report, and (4) a journal article.

Introduction

In graduate Medical Geology education, whether full- or part-time, students are required to complete projects of one kind or another in the subject area. In many instances, institutional requirements dictate that the project forms a relatively major part of the course or is virtually the entire basis on which the degree award is made. Davies (2019) for instance, has recommended that a weighting of 40% of the entire program be allocated to the field project component of the MSc Medical Geology degree at African geoscience institutions.

Medical Geology of Africa. DOI: https://doi.org/10.1016/B978-0-12-818748-7.00005-8

The primary purpose of doing the research project as part of their degree qualification in Medical Geology is to foster the personal development of the student, but also, perhaps more importantly, to ensure that a significant contribution to Medical Geology knowledge is made, particularly in the case of a PhD study.

With a research project, at whatever level—Master's or PhD, the agenda is set by the student to a larger extent than is possible in the ordinary taught course. Similarly, the student takes responsibility for the quality of learning acquired from the research project and for the eventual written outcome.

The purpose of this Chapter is to help graduate students in Medical Geology go through the thesis and dissertation process from beginning to end more successfully. An attempt is made to provide the student with a clear, comprehensive, and useful guide, briefly describing the process of finding, choosing, executing, and writing a high-quality proposal/thesis/dissertation/project report (PTDR), taking into consideration a whole set of circumstances under which the project would be carried out, not least, the potential relevance of the study toward explaining or resolving some geoenvironmental or public health issue(s) of specified communities, and to improve their chances of a successful outcome in their job search prospects.

The chapter also reflects the changes that have affected student research in graduate studies in African tertiary institutions in recent years. The ability to conduct a research project successfully and write their PTDR from beginning to end with little impediment has now come to be seen as a key transferable skill required of all higher education students (Chadha, 2006), and the management of a student research project speaks directly to the skill element of this.

Detailed guidelines are provided to those learning how to write the major elements (sections) of a PTDR and specific information is provided about how to arrange the different sections commonly found in the document such as how to write research questions or hypotheses, how to select a sample for the study, how to write descriptions of instruments, how to write results of data analyses, how to interpret the results, and so forth, as well as definitions and explanations of the rationale for the common elements of each section of the PTDR.

Furthermore, the research process at all levels is now far more systematized than in the past, with the single most important change (as from the 1990s) being the impact of the World Wide Web on student research. This Chapter is written and developed in response to this evolution.

The model used is the traditional multiple chapter PTDR. It is realized that there are many variations to this model. Students and faculty who are working on *proposals* should feel free to modify the approach to reflect advising style, unique area of Medical Geology, and institutional requirements. Students from other disciplines (beside Medical Geology) will also find the Chapter useful as a guide in carrying out their PTDR preparation. Variations in approach depending on whether the write-up is a proposal or preparation of a thesis, dissertation, or project report, are explained in the Section: "The finished product (thesis, dissertation, or research project report)," in this Chapter.

The guidelines also allow students to work on a research project throughout its duration, and determine how the final submitted PTDR document is assessed. A structured approach to basic research training in Medical Geology allows one to explore an area of research with the potential of making a significant contribution to solving geology and health problems in one's community, but also provides the opportunity to develop links with cognate research projects at the local, national, regional, or international level.

Thesis or dissertation?

To obtain a graduate degree (in Medical Geology) a candidate is generally required to complete either a *thesis* or a *dissertation*. The two words are often used *interchangeably*, implying that they have the same meaning. This brings about some confusion in academia regarding what each individual word actually means. However, there are a number of differences between the two usages, as interpreted from the American versus the European understanding.

In the United States, your research work would lead to the writing of a *thesis* if you are doing a Master's degree; while you write a *dissertation*, if you would have enrolled in a PhD program. Thus in the American context, you would normally be awarded with a Master's degree after successfully completing and submitting your thesis, whereas after successfully completing and submitting your dissertation, you would be awarded with a doctoral level degree   [https://www.quora.com/What-is-the-difference-between-a-thesis-and-dissertation (accessed 09 March 2019)].

From the American understanding, the focus in a Master's thesis, according to Enago Academy (2019), is on the student's ideas and his ability to arrange and express these ideas clearly. It is a compilation of research that proves that the student is knowledgeable about the information learned throughout her/his graduate program.

If a student wishes to advance further in academia in the United States, she/he could pursue a dissertation (for a PhD), which is a particular kind of academic task where the candidate will usually be asked to generate a topic for herself/himself, to plan and execute a project investigating that topic, and write-up what her/his findings were (University of Leicester, n.d.). A *dissertation* is your opportunity during a doctorate program to contribute new knowledge, theories, or practices to your field (Medical Geology). The point is to come up with an entirely new concept (or interpret an existing one differently), develop it, defend its worth, and show how it has made a significant contribution to knowledge in your field.

In Europe, on the other hand, the definition of a *thesis* is almost the opposite of that in the United States. The original distinction between *thesis* and *dissertation* is largely retained, so that the definitions (MSc vs PhD) become divergent. You have to write a *thesis* if you are doing a PhD, while you have to write a *dissertation* if you are enrolled in a Master's program (*cf.*, the American understanding). According to Enago Academy (2019), a *dissertation* is part of a broader postgraduate research project, whereas a doctoral *thesis* is a focused piece of original research, which is performed in order to obtain a PhD. This original research requires plenty of background literature survey, engendering extensive citations and references to earlier work in the project write-up, although the focus remains on the originality of the work and its significance in terms of its contribution to knowledge in one's field (Medical Geology).

A *diploma* study in either context is generally viewed as a subset of a Master's or doctoral program, with course content aligned to the intended career trajectory.

In many African institutions, the structure of academic programs including terminology systems generally follows that used by institutions of their historical colonial administrators, who, for many African countries, were from Europe or the United States. African institutions today have tended to retain these structures, including curriculum design, development, and even terminology usage.

What is project selection?

Project selection is a process by which you scrutinize each project idea that appeals to you during your lectures and tutorials (or from other sources, such as from textbooks, professional journals, thesis, dissertations, reports, theories, current employment, or existing database), and choosing the project with the highest priority, based on several other criteria, which vary greatly according to the circumstances of the project.

The *project selection process* (PSP) may not at all be so straightforward and is often not the most fun task; but it is extremely important to choose a current and relevant topic within the *niche area* of your department/faculty/institution. A "niche area" in this context refers to the Medical Geology area of focus or expertize of your department/faculty/institution.

A well-chosen research topic ensures that you can more easily write a qualitatively good thesis, dissertation, or project report (see Section: "The finished product (thesis, dissertation, or research project report)," in this Chapter). This also makes the writing process more satisfying for you.

Why do project selection?

During your PSP, you may come across a number of project ideas, but not enough resources or time to undertake one or the other of these projects. You may find that you have a mix of projects that look straightforward and others looking difficult, and do not know where to begin, or which of these projects to choose from. You should realize that projects are still only suggestions at this stage, so the choice is often made based on only brief descriptions of the project. As some projects will only be ideas, you may need to write a brief description of each project before conducting the PSP.

The PSP should be followed step-by-step and in an objective way in prioritizing projects and eventually selecting the one you would eventually settle for. The process involved in the PSP can be used to explain to reviewers or potential employers the reasoning behind why you selected a particular project over others.

The PSP will also assist you by providing a basis for comparing the importance of the projects and how achievable they are (*feasibility*), as a predicate for selecting the most suitable project to undertake. Another important benefit of completing the PSP stage is the obtainment of a transparent and documented record of why a particular project was selected.

Finding the right research project

Project selection criteria

Research project selection (decision-making process) is often seen by students to be quite stressful and time-consuming (Poock and Love, 2001). Undertaking the project itself can be challenging, but also exciting. It is challenging because a tremendous amount of self-discipline, time, and effort needs to be put into it.

Despite seeming quite daunting challenges at first, there are ways by which students can overcome the initial challenges posed by the PSP. For instance, many departments/faculties/institutions maintain a project database that contains a number of project suggestions made by members of the department/faculty/institution, which students can consult. Note that these are just suggestions for suitable topics and there is no requirement that you select one of them. In fact, students are always encouraged to formulate their own ideas for a project. The suggestions are provided only as a guide and are meant to help you understand the scale of the project you would be attempting.

Taking into account the developmental agenda of most African countries, graduate project selection in the environmental sciences (including of course, Medical Geology) should principally be directed toward those projects that augur with the betterment of the lives and livelihoods of the community (local/national/regional/international level) including their environmental health benefits.

Also, your thesis, dissertation, or research project will take many months to complete. Therefore it is extremely important to choose a topic that you find interesting. Maybe you will find a topic that is focused on your career goal. Or maybe you will be inspired by a subject in another module on your course? Whatever your motivation for your thesis, dissertation, or research project, it will be much easier to maintain if you have passion for the selected research topic.

It is perhaps obvious that, in the context of Medical Geology, the most important criterion for project selection would be the project's relevance in terms of its potential in tackling some geoenvironmental or public health issue in your community (local/national/regional international). However, there are a number of other important criteria, some of them, on logistics, and questions of feasibility, that need to be considered—your individual circumstances, including those provisions your department/faculty/institution can make toward fulfillment of the fieldwork, laboratory work, or other components of the research agenda, and your career aims, are some of these.

A nonexhaustive list of the more important questions to consider for your Medical Geology project selection follows:

- What are your [(department's/faculty's/institution's) major interests in Medical geology (*cf., niche areas*)]? Which of these interests is likely to make a significant contribution in terms of elucidating or resolving some geoenvironmental health issues in your community? Inherent in these questions is what is often described as the *benefits* of the project—a measure of the positive outcomes of the project. These outcomes are often described as "the reasons why you are undertaking the project" (e.g., environmental health, economic, social and cultural, and fulfilling commitments made as part of local, national, regional, or international plans and agreements).

- Is this idea stimulating and important enough to me and my community so that I would spend considerable time thinking and reading about it?
- What personal experiences have you had that were particularly significant or meaningful for you that are relevant to your discipline? What coursework did you take that you found the most exciting?
- Has a great deal of research already been conducted in this topic area? Could my studies fill a gap or lead to greater understanding?
- What theories and concepts are most interesting to you? Are there some theories you want to avoid?
- To what extent does this project augur well with your intended career goals after completing your degree?
- What bodies of literature have you encountered that intrigued you?
- Do you have ideas for specific data or texts you would like to study? With what kind of data do you enjoy working?
- What are the important research questions in this subject area?
- What kinds of methods do you like to use when you research?
- How feasible would it be to execute this project (logistics, availability of analytical facilities and other resources, and so on)?
- Which funding agencies would be interested in funding this study?

Suitable answers to these questions can translate into successful strategies for finding an appropriate topic for your research. To find answers to these questions, you require a partner. Your supervisor is the preferred partner, as she/he will guide you through the dissertation/thesis/report execution and the PTDR writing process.

Once you and your supervisor have begun cycling through certain ideas over and over again in your conversation, repeating the same ideas with fewer and fewer additions, then you have reached the state of readiness to formalize the key pieces of what you want to be a part of your PTDR. These elements will be used to formulate the proposal for your thesis/dissertation/report.

How *feasible* would it be to undertake the selected project?

Feasibility is a measure of the likelihood of the project being successfully completed (i.e., achieving its objectives). Because Medical Geology projects vary greatly in complexity and risk, considering feasibility when selecting projects is absolutely vital. In this connection, the easiest projects with the greatest benefits should be given priority.

A detailed review of a project's feasibility is conducted in the *feasibility study stage* (FSS) and entails a realistic thought about the practical implications of your project choice, in terms of:

- The time requirement. The research project component of an MSc thesis/dissertation, for instance, may require a minimum of 6 months; a PhD project, is much longer. Note that certain projects, by their very nature, require considerably more time to accomplish the desired results (e.g., observing a disease condition over an extended period of time to put forward a prognosis). Further discussion on time management strategies in project execution is presented in Section on: "Plan of Work and Time Schedule," in this Chapter.
- Necessary traveling; estimated distance to the study area;
- Accessibility of the study area. How easily can one get there? (*cf.*, poor road conditions, tremendous distance from base, and so on);
- Access to the population of interest;
- Access to equipment or room space;
- Possible costs.
- Safety considerations (e.g., high incidence of banditry, civil war, wild animals, rugged topography, hazardous laboratory procedures for processing and analysis of samples);
- Ready availability of resources such as financial tools and equipment for sampling and analytical work.

Main features of a good medical geology research project?

A summary statement encapsulating the key features of a *well-selected Medical Geology research project* could run as follows: It is *original, focused,* and *relevant* [potentially offering a significant contribution to knowledge of geology and health issues in your community (local/regional/national/international)], and takes cognizance of career aspirations, while also demonstrating a high probability that the project would be feasibility to undertake. A good research project will show a high probability that it would be feasible to execute.

Originality

It is important to choose a unique topic for your research project. However, it has to be noted that research is never totally original (e.g., Hedge and Salvatore, 2020), and finding a unique area of research may be close to impossible. According to Hayton (2012), "research is built on the edge of what is already known; moving forward but still connected to and dependent on that which has been done before."

The judicious application of ideas that are unoriginal and skillful application of well-established techniques give you a reliable foundation to work from. You could, for instance, consider approaching an already-researched area from a different angle? Or maybe you could develop a unique idea from a smaller topic that has not already been exhaustively researched.

The Student Learning Development Division of the University of Leicester (n.d.) has noted that a research study can:

- Replicate a previous study in a different setting;
- Explore an underresearched area;
- Extend a previous study;
- Review the knowledge thus far in a specific area;
- Develop or test out a methodology or method;
- Address a research question in isolation, or within a wider program of work; or
- Apply a theoretical idea to a real-world problem in Medical Geology.

This list is not exhaustive, and it is the student's responsibility to check whether her/his department/faculty/institution has a preference for particular kinds of research study (*cf.*, your department/faculty/institution's "niche areas").

Relevance

A thesis or dissertation has to be useful; and one would be much more comfortable about doing a thesis or dissertation, knowing that somebody will use it. The relevance of your thesis or dissertation depends on the nature of your audience, as some projects will be fascinating to some but completely uninteresting to others.

As a point of emphasis, perhaps the most important criterion for project selection in Medical Geology is its pertinence to your community (local/regional/national/regional/ international) in explaining or resolving some geology and health issues. The subject of your research needs to be of interest to other academics in the field, as well.

Finally, it is also important to select research projects whose results will stand the test of time. It should not be a snapshot of information that would soon become dated.

Career considerations

A strategic project selection could influence your career trajectory in many ways. Choosing a subject that might also benefit your future career is something that tutors would definitely recommend; it will give you a greater understanding of the in-depth area of Medical Geology you seek specialty in, while also giving you additional strength when writing a future job application statement.

Employers take a very close look at the dissertation topic, and so you should be considerate enough before choosing your project. You must try your best to make sure that the thesis/dissertation can help you get a great job or a promotion.

Select a topic that creates a trajectory. Have a look at what technical skills are being asked for by the employers in your domain. By examining the range of possible topics carefully with respect to your professional direction you have a much better chance of selecting a topic that is not only of interest to you, but also advances your career plan. The thesis/dissertation/or research project report tells the employer what you can bring to the table. More often than not, you will be asked to describe your thesis/dissertation/research project during your job interview.

There are many career opportunities open to graduates in Medical Geology, such as in the Department of Environmental Affairs, Water Affairs and Agriculture and Forestry; but also in industry and consulting companies (Davies, 2019). In the Water Affairs Ministry, for example, a Medical Geologist would be expected to plan, coordinate, and implement the enforcement of state and local laws and regulations regarding the compliance, monitoring, and prevention of contamination and pollution of the public water supply. Professional medical geologists can also work in public health consultancies, and health spar resorts. Others become professors or research staff at universities and colleges, engaged in curricula, such as public health, toxicology, environmental pollution, and forensic geology.

When developing your research topic, it is therefore a good idea to consider these career possibilities and identify the pathway you would like to follow in the future after graduation. Yes, grades are important, but sometimes, employers do not care much about the grades; especially when you have a Master's degree. If your research is good enough and your grades are not too low, then you still have a chance of being deemed appointable.

At which stage of your project planning do you do "project selection"?

Ideas for project selection will continue to crop up as you go through your lectures and tutorials, which is why the modules on project selection are recommended for early delivery in your curricular program (e.g., Davies, 2019). Ideas for project selection would also keep coming up as you do your literature survey and contemplate your intended career pathway.

Project selection should normally be undertaken when you have more ideas than the number of projects you conceive of as feasible and need to select the project that should be given priority. If your interest revolves only around one project, it may still be useful to score the project against a set of criteria to identify the strengths and weaknesses of the project. The results may be useful later in the FSS.

How does the project selection process differ for an MSc versus PhD thesis/dissertation?

The general principles and characteristics of the *PSP* outlined in the Section: "What is Project Selection?" in this Chapter, apply to both MSc and PhD. Fine distinctions between a MSc and a PhD research project requirements for a thesis/dissertation engender stipulations made by individual institutions. However, some general characteristics are considered to be inherent in most institutional specifications.

The Master's degree program

An MSc project is considered a substantial and extensive investigation of a challenging topic in the subject area (Medical Geology). A Master's program is not just a continuation of an undergraduate program, but one where candidates are expected to specialize in a particular subject field within the Medical Geology compass. Hence it is a process of acculturation. So, it can certainly increase your job prospects in a niche domain of Medical Geology. Within the Master's program, the dissertation or thesis component is extremely important. It is what you have to show for your time in a Master's degree program, especially, if you do not have any prior work experience. The course develops student skills in critical thinking, research planning, academic writing and experimental design appropriate to their chosen project within the Medical Geology compass. It is intended to give an MSc student a major opportunity to exercise her/his new understanding and advanced skills acquired on the program by applying them to a significant and advanced practical problem.

Most institutional regulations for the submission of an MSc thesis/dissertation would normally embody a statement that reads approximately as this example from the University of Warwick (2012): "A dissertation submitted in part fulfillment of the requirements for the award of a Master's degree shall constitute an ordered, critical, and reasoned exposition of knowledge in an approved field and shall afford evidence of knowledge of the relevant literature, and be submitted in accordance with the appropriate course regulations."

The PhD degree program

The attributes of a good PhD project are originality, relevance ("significant contribution to knowledge"), clarity, and manageability (feasibility), which are the criteria that most universities have at the core of their PhD assessment. These criteria are similar to that used by many refereed journals and conferences for assessing articles for publication. Such publications represent the way the research community communicates and continues to build knowledge. Therefore you can provide evidence of "originality" and "significance," in advance of submission of your PhD thesis/dissertation by publishing your work in refereed journals or conferences.

Note that research in Medical Geology is sometimes focused on resolving issues related to the sudden occurrence of strictly locality-related diseases or health conditions in your community that requires intervention by rapid response teams that include Medical Geologists. Such investigations are not normally among the list of possible project undertakings in the *portfolio selection* of graduate research in Medical Geology. The project selection guidelines giving in this Chapter, therefore, do not include such interventions.

Selecting your supervision committee

Select and prepare your supervision committee carefully. Although *content expertise* is important, do not select committee members based on this criterion alone. Select faculty for your committee who are supportive of you and are willing to assist you in completing your research. You want a committee that you can turn to for help and you know that help will be forthcoming. Remember that you can always access content experts, but you rely on your committee members for guidance and encouragement. If you do your enquiries well you would be able to select a supervision committee that would be truly helpful.

Very early in your proposal preparation stage, set up a formal meeting with your full committee to discuss your research ideas. Provide the committee members with a well-written proposal in advance of the meeting. Check with them to see how much time they are willing to spend reading your proposal.

Confirm that your supervisor and committee members are fully supportive of your chosen project before you begin. This initial proposal meeting presents a wonderful opportunity for you and your supervision committee to reach an agreement on the fundamental goals and procedures for your research.

The project supervision process

Although a thesis or dissertation affords the student an opportunity to work independently, generations of students have benefitted immensely by working closely with their supervisor, who monitors each project and would normally assess the project and thesis or dissertation separately from the modules. Supervisors are there to help you shape your ideas and give you advice on how to conduct the research and the process of writing your thesis or dissertation.

Some institutions require the appointment of a second supervisor, called a *technical supervisor*, or, for industry-related or externally-based projects, an *industrial supervisor*. Your technical supervisor will advise on any technical aspects of completing your chosen project. The industrial supervisor would be a suitably qualified, senior individual within the company (where you are undertaking your research) who has knowledge of the circumstances surrounding the project and who can judge the relevance of methods used and conclusions drawn in relation to normal company practices and current and future company policies.

Depending on institutional stipulations, the general research topic is selected by you, in consultation with your research supervisor, although students are typically allowed to have considerable latitude in selecting their own research topic or program of work. In the latter case, students approach a prospective supervisor and explain their requirements.

If you are expected to select a topic more or less independently, you will undoubtedly engage in a period of thinking about and considering various topics. Tips for judicious project selection have already been elaborated on in the Section: "Finding the Right Research Project," in this Chapter. You should obtain an agreement on the content of the project and the supervisor's agreement to fulfill the role.

Another crucial role that supervisors play is to read and comment on draft versions of the reports you intend to submit.

The way in which supervisions are organized will vary depending on your supervisor. Nowadays, meetings generally are organized by email, which is accepted as the primary means for supervisors to contact supervisees; so, it is important that you check your email regularly.

Because academics are known to be very busy people, you will need to be organized and take responsibility for the relationship with your supervisor, to get the most out of her/him. It is not your supervisor's job to chase you into completing your thesis/dissertation, or to tell you how to manage the different stages of the project.

Assuming you have done a good job of "thinking about" your research project, you are ready to write your research proposal. Here we go!

Writing your research proposal

Depending on the institutional requirements for entering a Master's or PhD program, you may have to write a *research proposal*. Some institutions require that you write the proposal before being accepted as a student, others require this at a much later stage. It is up to you to find out how it works in your particular institution.

The purpose of this section is to assist graduate students in Medical Geology developing and pursuing high-quality proposals *after* selecting a suitable research program for their thesis, dissertation, or project report for submission to the elected supervisor; or, for faculty researchers who are developing competitive proposals for submission to a grant-awarding agency(-ies). It provides the necessary basic guidelines to produce proposals that meet proposal reviewers'/ supervisor's expectations of "good quality proposals," and on how to undertake research on geology and health topics that are worthy of in-depth investigation, in a commendable manner. These guidelines do not constitute a text in research methodology and statistics; nor are they a "recipe" or a "how-to-do" companion. They are for reference purposes only. The section is also not exhaustive in its treatment of the information provided; but only seeks to sensitize graduate students and faculty researchers to some of the basics of proposal writing and research in Medical Geology.

For more detailed instructions on *proposal writing*, you should consult your institution's policy document roughly titled: "Regulations Governing Proposal Writing and Thesis/ Dissertation Preparation and Submission" that would normally be provided by your department/faculty/institution. Interested students and faculty wishing to learn much more about *proposal writing* are referred to recent works in the subject, such as that of Erickson (2015) and Hash and Hash (2013).

Note that some of the statements on proposal writing in the following sections also apply to thesis and dissertation writing, with the use of the mnemonic, PTDR, wherever this is the case.

What is a research proposal?

A *research proposal* is a plan of investigation into a researchable problem, a more detailed description of the project you have selected to undertake. It is a communication instrument, that outlines the researcher's specific and clear intentions to the proposal reviewer(s) or supervisor. A research proposal prepared for a grant awarding body can also be regarded as a promotional or selling document intended to impress.

Attributes of a well-written proposal are clarity, simplicity, originality, and parsimony. Clarity is the key. It should be immediately obvious exactly what you are trying to do, and this is only possible to communicate if you first have clarity in your own mind. Do not attempt to write down everything you could possibly imagine doing, nor everything you know about the subject.

In addition, you must convince your reviewers of your ability to carry out the plan and the scientific worth of the proposal. It should be immediately obvious exactly what you are trying to do, and this is only possible to communicate if you first have clarity in your own mind. Avoid verbosity at all costs. Do not attempt to write down everything you could possibly imagine doing, nor everything you know about the subject. Reviewers often get irritated going through voluminous, discursive proposals. A proposal is not the platform to brainstorm, to think aloud about one's research interest. This intellectual activity (discussed in the Section: "Finding the Right Research Project," in this Chapter) must be completed before writing the proposal, which embodies the research plan.

Putting together the *research plan* in the body of the proposal demands *patience, humility, flexibility (but maintaining focus), originality*, and *passion*. Patience is required to understand that your research question can, and probably will, change during the initial stages of your research. It is not written in stone (unless of course the research question has been set by your funding body), and the beauty of research is its ability to change things, including your very own research process.

Humility is the "buzzword" for keeping your chosen topic to a manageable level, even if that means that, at times, you are not being faithful to your initial ambitions. Experience shows that it is more than likely that you are not going to be. And that is not necessarily a bad thing. Humility also matters in realizing that although you already know a lot about your topic before you start, you may actually know very little. That is what will allow you to grow and be challenged intellectually. There is after all, no point doing a thesis or dissertation, if you already know the answer?

Focus your research very specifically, but try to be flexible. Students sometimes feel that a narrow focus will distort what they want to do. However, a broadly defined project can be unmanageable as a research project. When you complete your research, it is important that you have something specific and definitive to say. Otherwise, you may be left with broad, vague conclusions that provide little guidance to scholars who follow you.

Flexibility can be a very important attribute during a research project, and even more so during a long project, such as a PhD. It is normal to come across obstacles during your thesis/dissertation work. Many of these will be irresolvable and you will need to be flexible with your topic. Perhaps you planned to use a particular conceptual framework but you realize that it will not work with your particular case (or cases). A decision has to be made to either change the case or change the framework. Being flexible in such instances will not be a reflection of weakness, but rather one of intellectual strength.

Originality
The guidelines on "originality" in project selection (Section on: "Main Features of a Good Medical Geology Research Project?" in this Chapter) also apply in formulating your proposal document.

Passion

If you are not passionate about the topic, you might struggle to finish it off, and even fail in the *viva voce*. So, pick something that you find really exciting. After all, you will need to sell it when you go for your job interview. When you demonstrate excitement about your thesis or dissertation, you basically demonstrate your engagement and passion about the field of study.

The required manuscript format

There are conceivably many ways of arranging the various topics in a proposal, but the classic format is similar. The following approximates the basic format commonly adopted by most departments/faculties/institutions/donors, but the nature of an individual problem as well as the *"Guide to Authors"* given by some donors or institutions will dictate the details of each proposal format. If, on the contrary, there are no institutional stipulations, it is proposed that the document be kept short, typed double-spaced, using 12-point, and written as one continuous document, with sections (see Section: "Stylization and Structure in Document Preparation," in this Chapter). It is recommended that the Harvard system of referencing be adopted in your PTDR document, although a number of institutions do prefer use of the APA or IEEE systems (see Section: "References/Bibliography," in this Chapter). Above all, adhere strictly to the guidelines (formatting, language, page limits, type size and font, margins, reference style, etc.), if given by the institution or donor agency.

Cover page and "table of contents"

The cover page

The *cover page* should include the title of the research project, name of candidate, student number (where applicable) your degree course and department, name(s) of project supervisor and the calendar year of submission. Space should be provided for supervisor's signature. A statement worded approximately as: "A Master's/PhD/Honors proposal submitted to the Department of..., Faculty of..., University of..." should be included. The date of submission should be indicated at the bottom of the page. The title of the research project should preferably be highlighted, while the rest of the page is written using the stipulated font size. The *table of contents* should also be included and neatly presented.

The table of contents

The *table of contents* is very helpful to the reader, and valuable to the writer. The reader can use it to instantly locate any section of the proposal she/he wishes to reference. This "table" gives a detailed breakdown of the contents of the proposal document with main headings, section headings, and maybe subsection headings, each with page numbers. The "table of contents" can also help you to improve your manuscript. For instance, it can be used to see if you have left something out, if you are presenting your sections in the most logical order, or if you need to clarify your wording.

The development of computer technology now makes it possible to easily copy and paste each heading from the document into the *table of contents*. It is then possible to see if the "table of contents" makes logical sense to the reader. It is amazing how easy it is to see areas that need more attention. Do not wait until the end to do your "table of contents." Do it early enough so that you can benefit from the information it will provide.

A typical *contents page* would comprise the following topics, listed in a sequence that is standard for most institutions' thesis/dissertation or report submissions (again the actual listing followed would be that stipulated by your department/faculty/institution):

- Cover Page
- Signature Page
- Table of Contents (with page numbers), neatly presented.
- Declaration: A signed statement of originality together with an overview of any intellectual property rights agreements that you have made. See typical wording of statement of originality (declaration) in the Section: "Researcher Declaration," in this Chapter.
- Abstract
- Chapters
- List of Tables, Figures, Plates, Other Illustrations
- Acknowledgments
- References
- Appendices

Every page in the thesis/dissertation must be numbered with the exception of the title page. The title page is not numbered but is counted in the pagination as "i," which is considered a silent number.

Certain headings may be difficult to fill in the content page right at the start of your project. However, you can use the gaps to help identify where to begin work. If, for example, you are unsure about the limitations of your methodology, you should talk to your supervisor and read a bit more about that methodology before you start.

Title selection

Project *topic* selection versus *title* selection

There appears to be a subtle difference between selecting a *suitable research project topic* and selecting a *fitting and appropriate title* of your PTDR. Explanation of the PSP, followed by customized guidelines for carrying out this intellectual activity, has already been presented in Section: "What is Project Selection?" in this Chapter.

The research project *title*, on the other hand, according to Bruwer (2016), expands on the research project topic and falls within the compass of your "niche area."

Research project *title* selection

Graduate students are often in a dilemma when making the important decision of which Master's or doctoral *research project title* they will select, and present the best fit in describing their research project. A good PTDR has a good "title," and it is the first thing to help the reader begin to understand the nature of your work. Work on your "title" early in the process and revisit it many times, if necessary.

There are many different ways the research project title can be framed to reflect the scientific content of the work. These versions, however, may not necessarily embody other aspects of a good "research project title." The "title" should be as short as possible, cutting-edge and captivating to the reader, while at the same time accurately conveying to the reader the scientific content of the proposed study, and the work's potential to make a meaningful contribution to your subject (Medical Geology). It should give a clear indication of your proposed research approach or key question.

A good "title" has the most important words toward the beginning. It avoids ambiguous or confusing words. It is broken into a title and subtitle when you have too many words and includes *keywords* or phrases to give a clear and concise description of the scope and nature of the report (e.g., Van Dalen, 1979). Keywords will also help other researchers to identify your work and allow bibliographers to index the study in proper categories.

The title of the paper by Davies (2013): "Geochemical variables as plausible etiological cofactors in the etiology of common diseases in Africa," with keywords: "Geochemical variables; Optimal range; Beneficial uses; Diseases in Africa," can be regarded as a suitable title of a Medical Geology research project, as it relates to:

- Disease causal factors (geochemical etiology).
- Potential for mitigation, since this (mitigation) begins with an identification of etiology.

The implication of a *geochemical etiology* for diseases apparently renders the title captivating, in the sense that the usage of the term "geochemical etiology" is still of quite uncommon usage.

Despite incorporating all the above attributes, the selected title is initially described as a *working title*, because of the high likelihood that it may change as the project evolves. The main application of a working title is its use in choosing examiners for your project. Master's and doctoral students are generally advised to try to retain the project title throughout their research, since changing the title during the course of the project may involve complex and protracted institutional protocols, especially at advanced stages of project execution. So, think carefully about the suitability of a new proposed title before requesting a change from your institution.

Abstract

An *abstract* should always be included in your PTDR document. The *proposal abstract* is a much abridged version of the entire proposal document.

It is sometimes considered the equivalent of a *concept paper*, though some consider a "concept paper" to be one that only reflects your current thinking, and tends to be a bit more detailed. A concept paper helps you narrow down your topic and forces you to describe your idea systematically. It can also be used to obtain feedback from colleagues and potential funding agencies.

The abstract essentially summarizes the PTDR, highlighting major points and describing the scope and conclusions of the research. In *reports*, this section is sometimes often referred to as an *executive summary*.

A good *abstract* gives the background and purpose of the research, it answers questions such as: (1) What is the study all about (background, scope)? (2) How was the study conceived (eliciting conceptual framework, hypotheses, and feasibility)? (3) Why is it worthwhile (importance/justification) to carry out such a study? (4) How would the study be carried out (methods, approaches, procedures, strategies to obtain the objectives); and what are the expected results? (5) Of what significance would be the results of this study; what contribution would it make to your field; and can the results be practically applied for the benefit of the identified end users [community(-ies)]?

Introduction

The *introduction* gives the motivation for the project, the theoretical foundation, and a brief background to the research problem.

The aims of the project should at least be stated in the first paragraph, but preferably in the first sentence. You could also explain in the first paragraph, the structure of the proposal document, and give a broad overview of the topic and method. This Section is preferentially written after remainder of the proposal document is completed.

Background to the study

The "introduction section" is sometimes divided into subsections, with titles such as *background to the study, theoretical framework, study setting, the study area*, and so on. However, the purpose of this Section is to provide the general framework for the conduct of the study geared toward addressing the research problems. Provide the reviewer with necessary background and setting to put the problem in proper context. Provide a logical lead-in to a clear and concise statement of the problem. A well-grounded theoretical basis (theoretical framework) for your study or project is needed. Include definition of key terms [probably a "Glossary" section at the beginning or at the end (see Section: "Definitions of Terms used in your Proposal/Thesis/Dissertation/Report (PTDR)," in this Chapter)].

Differentiate between those statements that are your own and those excerpted from the literature. Where appropriate, provide support with reference to literature. Specify the objectives of the project and the needs of your intended users [community (-ies)] that are achievable in terms of time available and your experience. It should introduce both the problem area (remember your reader may not know anything about the particular problem you have chosen) and give an overview of the rest of the proposal. Normally, you should start with a short introduction to the problem you are addressing and your aims and objectives, give a short review of the context, and describe what follows in the main body of the proposal. In a funding proposal, you need to address the capabilities and capacity of individuals and agency/institution in this section.

Provide a logical lead-in to a clear and concise statement of the problem and justify and convince the reviewer that the study is needed. Let the reviewer see the basis for the study. Be factual. Statements, opinions and points of view should be documented. Place the work in context.

The introduction section can be divided into subheadings, such as *theoretical background, background to the research problem, literature review*, and *study setting*.

Under "theoretical background," place the work in context. You need to convince the reviewer that you are planning to test hypotheses, not simply collect data to confirm your favorite hypotheses, and that you are open-minded enough to reject your hypotheses, if the experimental results do not support your hypothesis.

A well-grounded theoretical basis (theoretical framework) for your study or project with strong theoretical underpinnings tends to entice reviewers to look favorably upon your project; and in the specific case of Medical Geology projects, reviewers would be particularly attracted to projects that are potentially impactful on the environmental health of your community (local, national, regional, or international).

Statement of the problem, subproblems, and research questions

A *statement of the research problem* is usually a statement that makes a declaration about what is wrong, iffy, or unsettling. It is important that you establish a research problem at, or close to, the start of your project. It is one of the key tools you have to ensure that your project keeps going in the right direction. Every task you undertake should begin with you checking your research problem and asking "will this help me address this problem?"

Background to the research problem

Questions that could be addressed under *background to the research problem* may include: What issues would your research address? What are the key concepts that will guide the study? How were these conceived (conceptual framework)? Provide reader with the necessary background and setting to put the problem in proper context. Let the reader see the basis for the study. Justify and convince the reader that the study is needed. Where appropriate, provide support for your statements with reference to literature. In one sense, usually the problem is to expand the body of knowledge examined in the literature review (see Section: "The Literature Review," in this Chapter).

Once your topic has been accepted by your department/faculty/institution, you need to begin the process of refining the topic and turning it into something that is focused enough to guide your project. Try describing it as a "research problem" that sets out:

- The issue that you are going to be investigating;
- Your argument or thesis (what you want to prove, disprove, or explore); and
- The limits of your research (i.e., what you are not going to be investigating).

A popular Study Guide (University of Leicester, n.d.) notes that students should be willing to revise their research problem as they find out more about their topic. They may, for example, discover that the data they were hoping to analyze are no longer available, or they may encounter a new piece of information or a new concept while undertaking a literature search, that engenders a rethink of the basis of the research problem. You should always discuss with your supervisor before you make any substantial revision to your plans, and explain why you think the change is necessary.

Describe clearly the characteristics of properly stated problems that you intend to research, e.g., (1) "Selenium deficient soils in southern Africa contribute to the relatively high rates of diffusion of HIV/AIDS in the region"; (2) "The widespread occurrence of goitrogens in the African diet is the most important factor militating against the attainment of a complete eradication of goiter and its sequelae among the Continent's population"; or (3) "Improved diagnoses of certain *diseases of unknown etiology* can be better achieved by inclusion of Medical Geologists in professional teams investigating such conditions." But in these forms, these statements are too broad to inform a research plan. For purposes of clarity and specificity, a *research problem* has to be resolved into *research question statements*.

Obviously, one "research problem" would engender several "research questions," some of which can be addressed together in a single research plan, while others demand different research plans or proposals. State the research problem broadly, in a question form. Give subquestions. Explain carefully.

Developing a research question

It is important that your research proposal be organized around a guiding set of questions. When selecting those guiding questions, put them down in such a way that they frame your research and put it into perspective with the literature. Those questions serve to link your research with the research that preceded yours.

You should formulate the "research questions" clearly, giving an explanation as to what problems and issues you wish to explore and why they are worth exploring. Your questions should clearly show the relationship of your work to your field of study (Medical Geology).

For example, "the selenium deficient soils problem" may be translated into its many constituent researchable questions. (1) What is the role of the element selenium in the functioning of the immune system? (2) What are the factors controlling the distribution and behavior of selenium in the surface environment of areas with high- and low-HIV/AIDS diffusion rates, respectively? (3) What lines of evidence exist to suggest that the diffusion rate of HIV-1 could be retarded by increasing dietary levels of selenium? (4) Why has the country of Senegal in West Africa maintained such a low HIV/AIDS prevalence level (relative to countries in southern Africa), despite having such a high level of promiscuity? (5) What is the efficacy of the various methods of selenium interventions that have been applied?

You should not attempt to force the research problem into a single-research plan. Try to ask proper research questions and concentrate on the specific ones you can handle comfortably. If possible, choose those that you can comfortably address in one research plan. You should practice matching research questions, activities, and methods.

A well-thought-out and focused *research question* leads directly into *research hypotheses*.

The research hypotheses: basic assumptions

Hypotheses are more specific predictions or rather, basic assumptions, or proposed explanations regarding the nature and direction of the relationship between two or more variables in such a way that the relationships are testable empirically. Hypotheses are the very subject of the research problem. So, do not make assumptions about them. Assumptions are not hypotheses or objectives.

Ideally, a hypothesis should (1) give insight into a research question; (2) Be testable and measurable by the proposed research methodology; (3) Spring logically from the experience of the researchers.

Make sure that you: (1) Provide a *rationale* for your hypotheses, explaining how they were derived and why they are strong? (2) Provide alternative possibilities for the hypotheses that could be tested, and explain why you choose the ones you did over others?

Basic assumptions are accepted without thought of immediate proof, but they should not be made about procedures; these are propositions for which no information can (or will) be made available within the scope of the study. They are propositions that *the reasonable person* would be ready to adopt but which cannot be proven. Assumptions that are most commonly stated explicitly are limited in their nature, and serve to hold the size or scope of an investigation within prescribed boundaries (i.e., they put parameters around the study). According to the University of Minnesota (2000), "These statements are usually made when the argument rests on *a priori* reasoning, but can be made on basis of present knowledge on research which is as yet incomplete (specific qualifications must be made in the conclusions in the proposal document in which assumptions are made)."

Aims, objectives, and purposes

Aim(s) of your project

An *aim* is a broad statement of desired outcomes, and describes the *purpose* and intention(s) of your study. It includes a description of your motivations for undertaking a particular project.

Be sure to understand the meaning of *objective* with reference to the scope [i.e., the aspect(s) or subtheme(s) you want to focus on]. "Objective" and "purpose" are often used interchangeably. In the first part of your proposal, clearly state what the "purpose" of the study is, and explain the study's significance. A study may involve more than one purpose.

"Objectives" are the steps you are going to take to test your hypotheses or to answer your research question(s). The *research objective statement* specifies the intentions of the researcher regarding exactly what deliverables are to be provided. The statement relates to the expected outcome(s) of the project and should be broken down into *primary objectives*, which would lead to the achievement of your aims, and *secondary objectives*, which make a direct contribution to meeting the primary objectives. *Extensions* are objectives that will only be implemented if time and resource availability allow. According to the University of Sussex (2020), the "primary objectives" should be clearly specified, but the "extensions" may be vaguer. Do not hesitate to specify more extensions than you will be able to implement.

The "research objectives" section must convince the reviewers that the objective is attainable and that the *research question* is researchable. In other words, the "research objectives" should match the *project plan* or *workplan* (see Section: "Plan of Work and Time Schedule," in this Chapter).

The "aims and objectives section" is best located after the "statement of the problem." Make sure that the objectives are listed in proper sequence.

How does the "conceptual framework" relate to the "objectives"?

Make sure that each hypothesis is matched with a specific objective, and provide a summary of the *short-* and *long-term* goals of the project. Indicate clearly the problems the project will help to resolve. These will serve to determine the *scope* of the project.

Your "objectives" should be measurable, highly focused and feasible, given the time and resources (including the budget) you are requesting in the grant application. Be realistic about what you can accomplish in the duration of the grant and within the budget requested. You should indicate clearly the direct connection between "specific objectives" and *related literature* and *theory*.

For example, some "broad objectives" in designing an MSc Medical Geology curriculum for geoscience students, according to Davies (2019), could be paraphrased as:

- Development of cutting-edge researches that will help us understand better the role of Earth materials, Earth systems, and processes in the occurrence of environmental diseases in Africa;
- Provide education and training for the next generation of researchers that would lead in this effort (Medical Geology research) in Africa in general;
- Application of the research findings in a way that would allow a broadening of the diagnostic spectrum as well as therapy for specific diseases of geoenvironmental etiology; and
- Contribution toward the substantial improvement of the environmental health, lives, and livelihoods of the African people by reducing their environmental disease burden.

Make sure the *research objectives* convince the reviewers that they are attainable or that the research question is researchable. In other words, the "research objectives" should match the *project plan* or *work plan* (see Section: "Plan of Work and Time Schedule," in this Chapter)—no ambiguity is allowed.

- Indicate how the "conceptual framework" relates to the "objectives"?
- Indicate the gaps in knowledge that would be addressed.
- Make clear the direct connection between specific objectives and related literature and theory.
- Be sure to list objectives in proper sequence. Follow that sequence throughout the remainder of the write-up.

Significance and justification (rationale) of the study

Statements on the *significance* and *justification* of the work can be made in the *Background and Setting* section, and does not need to be a special section. However, some grant awarding bodies require this section to be written separately, because it is probably the most important attribute by which the worth of a proposal may be judged.

Significance of the problem

The *significance* is addressed by discussing how the study adds to the theoretical body of knowledge in Medical Geology and the study's practical significance for communicating to professionals in environmental and public health.

The PhD students must explain how their research makes an original contribution to the body of knowledge in Medical Geology. It is especially critical that this section be well developed. Without a clearly defined purpose and strong theoretical grounding, the proposal is fundamentally flawed from the outset.

The statements in this section convey answers to questions such as:

- What is the justification for the main focus of the study?
- Why is this project so important that you have decided to undertake it?
- Why is this research relevant and necessary for the target population (individuals or groups)?
- How can the target population use the new knowledge or information yielded by the research to change or improve the present situation?

You should provide information on broad and long-term impact of the project within the context of environmental health, social, economical, and technological benefits. Indicate the direct beneficiaries of the project.

Justification

What is the justification for the main focus of the study? Why is this project so important that you have decided to undertake a study on it? Why is this research relevant and necessary for the target population? Use statistics and prevalence rates to emphasize the need for this study.

The justification should revolve around the intrinsic value of the research method chosen in terms of yielding the data that will enable the student to answer the research questions but could also address issues like limited time, the fact that it is a preliminary study, financial constraints, etc.

Use statistics and prevalence rates to emphasize the need for the research. Cite sufficient references to support key statements in your *rationale*. Indicate how the results can be generalized beyond the bounds of specific work. You can use the arguments of others (expert opinion) who call for an investigation of the problem (properly documented, of course). You can use conflict in findings of related research as justification for the study. Be sure it is documented in the review of literature.

The literature review

Literature review or *literature survey* is a review of material that has already been published, either in hard copy or electronically, that may be relevant for your research project. The literature review will normally include a critical review of relevant literature, describing and analyzing previous research on the topic, so the reader understands what you are building on. The *literature review* could also include a summary of key debates and developments in the field, and identification of gaps in knowledge relating to your project.

A thorough literature review enables you to demonstrate the *rationale* for your research, and to describe how it fits within the wider research context in your research area, and thus providing tentative solutions to the problem or tentative answers to the questions.

The literature review should not be used merely to string together what other researchers have found. Rather, you should discuss and analyze the body of knowledge with the ultimate goal of determining what is known and gas in knowledge about the topic. This determination leads to your research questions and/or hypotheses. In some cases, of course, you may determine that replicating previous research is needed (see Section: "Replication," in this Chapter). The Review of Literature can be utilized to verify the concepts/theory under study and the scope of the methods to assess the concepts. A good literature review could be publishable as a review paper.

A critical and thorough review of the relevant literature (journal articles, books, periodicals, bulletins, brochures, magazines, newspapers, unpublished reports, etc.), including highlights of ongoing research, exposes gaps in knowledge of the subject, and hence points out areas of enquiry from where new results are likely to emerge quickly. Citations older than about 10 years should be used only when it is absolutely necessary (e.g., seminal works; in making a case for the proposed study).

It is a good idea to make an appointment to see the librarian specializing in your subject. An information librarian should be able to give you advice on your literature search, and on how to manage the information that you generate and write up your literature search without returning to the books you have read. Avoid plagiarism by all means. If you copy more than half a line directly from a source without quoting and citing it, then it is considered plagiarism. If something is so good you want to cite it literally, then, quotation marks have to be used. Checks for plagiarism use techniques like looking for any seven successive words that are the same in the examined text and also occur in another text.

The research design

The goal of the *research design* and methods section is to minimize the number of assumptions reviewers must make about your project. Here, the researcher must provide in detail the general methodology and the specific methods that will be used to collect data; also, describe the methods to be used for analysis of the data and display of the data. Show that you are using scientifically sound approaches.

Indicate clearly, which *design* is adopted to address each objective, if you have different purposes and designs. Schematic (graphic) diagrams, or models, often aid in understanding the design. Define the symbols you use.

The researcher must give details of (1) The design elements of the study (e.g., descriptive, comparative, longitudinal, case-control, quasiexperimental, randomized, and purposive) and explain why that design was chosen; (2) What criteria were used in selecting the study sites? (3) How was the sample size determined. Indicate whether this is appropriate or adequate in the light of the expected difference within the control and test groups. What power does this sample size give you for addressing the objectives of the study? How long would it take to obtain this sample size? (4) What data collection procedures are considered appropriate for this kind of study?

In describing the *research design*, make sure you also highlight information pertaining to:

- Internal validity;
- External validity;
- Description of population and description of and justification for, type of sample used or method for selecting units of observation;
- Development of instrument or method for making observations (e.g., question guide, categories for content analysis);
- Pretest;
- Reliability and validity of analytical instrumentation or method (quality assurance testing);
- Administration of tools or method for making observations (e.g., interviews, observation, content analysis);
- Coding of data;
- Description of data analysis;
- Statistical analysis and tests performed;
- Identification of themes/categories (qualitative or historical research).

Research method and methodology

What is the difference between *research method* and *research methodology*?

Research methods is not quite the same as *research methodology* (e.g., Sileyew, 2019).

According to Paltridge and Starfield (2016), *research methods* refer to the actual research instruments and materials used, whereas, *research methodology* describes the broad philosophical underpinning to your chosen research methods, including whether you are using qualitative or quantitative methods, or a mixture of both, and why. A key part of your proposal is the *research methodology*, since the chosen methodology will inform the choice of methods and what count as data.

Selection of research methods

The selection of research methods in Medical Geology research or whatever scientific investigation is an important and crucial issue, as it revolves around the researcher's ability to collect the necessary data to deal with the research problem. The overall success of a research project is in large part a function of a judicious choice of a *research method* or strategy. In contrast, according to Hariri (2016), this selection will affect the result and outcomes of the project in a negative way, if not done judiciously.

Begin this Section by describing *the method* you chose and why this method was the most appropriate. In doing so, you should cite references about the method, and justify your choice using the literature.

The researcher must make sure that the study described corresponds with the specific objectives listed earlier in the proposal and that the underlying science and methods behind the research plan are sound, feasible, and as complete as possible.

Details must be given of (1) The design of the study (e.g., descriptive, comparative, longitudinal, case-control, quasiexperimental, randomized) and explain why that design was chosen; (2) What criteria were used in selecting the study sites? (3) The sample size. Indicate whether this is (was) sufficient in the light of the expected difference within the control and test groups. What power does this sample size give you for addressing the objectives of the study? How long would it take to obtain this sample size? (4) Data collection procedures—Describe and justify the use of method(s) for data collection; Who will collect the data? What procedures will be used? For example, if a classification system will be used, how would you identify class boundaries? If questionnaires are to be used, include the questionnaires in the "Appendix." If a form will be used to record measurements from field observations, include a copy of this, too; (5) The procedures for training of researchers or interviewers; (6) Access to specialized facilities or equipment where applicable; (7) Procedures for handling of participants (*cf.*, participants with disability and those with special needs) and confidentiality issues; (8) Procedures and approval for working with animals where applicable (see Section: "Dealing with human subjects or animals," in this Chapter); (9) Possible hazards to research personnel and study participants and procedures to prevent dangerous situations; (10) Describe the methods of data analysis to be employed—Describe how you plan to analyze the data and arrive at conclusions. Explain the specific treatment of data and variables for each subproblem, research question, or objective. Specify methods used to assess validity and reliability; (11) Data display— How would the data be portrayed?—Maps, diagrams, figures, tables, pie charts, histograms, regression curves, correlation matrices, etc. (12) Methodological assumptions. Discuss limitations they impose (see Section: "Delimitations of the research," in this Chapter).

Kerlinger (1986) describes maximizing the differences between the levels of the independent variable. A typical shortcoming is comparing a "new" method with a "traditional" or "conventional" method of doing something. When this happens, the researcher will often describe at length the "new" method but not the "traditional" method. How, really, are they different?

If attribute variables are used in the design, identify them, and the number of levels of each and briefly describe the rationale for the selection of the attribute variables.

The research methodology

Provide sufficiently detailed description of the *methodology* of your proposed (or completed) research. Describe the approaches and methods you will use (was used) to achieve the desired outputs of the project, including the procedure(s) used in analyzing your data. The *methodology* should be linked back to the literature to explain why you are using certain methods, and the academic basis of your choice.

Choose your methodology wisely

Methodological considerations are a core issue as you develop and refine your research topic. Consider questions such as the following: What are the most common research methods used in your discipline? Which methods are most strongly supported within your program and by your supervisor and supervisory committee members? What are the leading methodological debates within your discipline, particularly in relation to your research topic or problem? What methodological issues have been raised in recent research literature in your area? You need to be thoroughly acquainted with effective principles and practices of choosing research methods for your thesis or dissertation. Be sure to discuss methodological questions and issues with your supervisor and committee in the early stages of your proposal development.

According to Sharp et al. (2012), there are four main objectives in the judicious choice of *methodology* for executing the research:

- "To identify problems, which, with hindsight, could have been avoided and to suggest anticipatory action that should be adopted;
- To suggest ways of coping with unavoidable or unexpected problems, which may arise (see Section: "Delimitations of the research," in this Chapter).
- To make positive suggestions, which will facilitate research progress, and
- To discuss in depth the role and responsibilities of the research supervisor and how she or he can be used effectively in executing the research."

What to include in your methodology

The *methodology section* should describe and justify the data gathering method used. If you are submitting your PTDR in sections with *the methodology* submitted before you actually undertake the research, you should use this section to set out exactly what you plan to do. If you are submitting as a single PTDR, then the methodology should explain what you did, with any refinements that you made as your work progressed. Again, it should have a clear academic justification of all the choices that you made and be linked back to the literature.

Provide an outline of:

- The theoretical resources to be drawn on;
- The research approach (*cf.,* theoretical framework in Section: "Background to the study," in this Chapter);
- The research methods appropriate for the proposed research;

- A discussion of advantages as well as limits of particular approaches and methods (see Section: "Delimitations of the research," in this Chapter).
- A description of the size of the sample, how it was determined and the rationale for the size.
- An identification of sampling units.

The researcher must provide in detail the general *methodology* and the specific methods that will be (was) used to collect data; also, describe the methods to be (was) used (for analyses and portrayal of the data).

Data gathering mechanisms and procedures

For most research projects the *data collection phase* is perhaps the most important part. However, students should not jump straight into this phase without having adequately defined the research problem, and the extent and limitations of the research. Otherwise, you risk collecting data that are unusable. To avoid this kind of situation, it is advisable to conduct what is known as a *pilot study*, which might involve making *presite visits*.

Pilot studies and presite visits

A *pilot study* is a preliminary data collection exercise, using your planned methods, but with a very small sample. The aim is to test out your approach, and identify any details that need to be re-examined before the main data collection goes ahead. According to In (2017), a pilot study is important for the improvement of the quality and efficiency of the main study.

Spend time to consider the implications that your pilot study might have for your research project, and adjust your plan accordingly. Even if it is not possible to run a formal pilot study, because of time or other limitations, you should try and revisit your methods after you have started generating some data.

After completing your pilot study you should exercise much caution about reading too much into the results that you have generated (although these may sometimes be interesting). The real value of your pilot study is what it tells you about your method. According to the University of Leicester (2019), the following questions underline the information you may gain from your pilot study:

- "Was it easier or harder than you thought it was going to be?
- Did it take longer than you thought it was going to?
- Did participants, chemicals, processes behave in the way you expected?
- What impact did it have on you as a researcher?"

Presite visits

The objectives of presite visits are to review the nature of the site in respect of possible imponderables, such as poor accessibility, lack of amenability to planned research, and/or methodological design, access permit requirements, faulty instrumentation, and so on. Such visits enable adjustments to be made to the research plan or methodological design or acquisition of additional materials or documents that were not included in the original research plan. A presite visit also enables the testing of efficacy and justification of planned method(s) for data collection and other field procedures.

Instrumentation and validation

In developing your instrumentation, it is important that the *validity* and *reliability* be established. All items of equipment should be pilot- and/or field-tested (see Section: "Pilot Studies and Presite Visits," in this Chapter). Describe how this was done. A field test can be used to identify potential suitability problems. Append copies of the instrument specifications to the proposal.

A paragraph or two on quality assurance assessment would be in place here. Information on choice of instrument operating parameters, instrument sensitivity (minimum detection limits), accuracy, precision, reproducibility, interlaboratory comparisons, use of certified standards, and so on are worthy of mention.

Treatments

In the *treatments section*, you need to explain what would be (was) done to the subjects? Describe all levels so that they are replicable. Were any methods employed and abandoned because they were not worthwhile?

Dealing with human subjects or animals

Many national laws and university regulations require researchers who conduct research involving human and/or animal subjects to obtain institutional clearance before beginning such work. If you plan, use humans or animals as subjects to test your hypothesis, as is often the case in Medical Geology, you must prepare a protocol that clearly shows how you will protect the subjects from unacceptable risk or harm, gain their consent (in the case of human subjects), and ensure confidentiality. You must get the protocol approved by an institutional review committee. You must read the appropriate material in these institutional guidelines and obtain your approvals to use human subjects or animals prior to collecting data for your PTDR.

Handling tissue samples

Certain projects in Medical Geology may require the preparation and analysis of human tissues (e.g., serum, urine, hair, toenails), which must be obtained under consent. Since many disease processes have their origin within tissues, these represent an important source of material necessary to study pathogenesis. This is a very specialized area of activity, and the researcher would have to obtain necessary ethical clearance (see Section: "Ethical considerations," in this Chapter), seek the collaboration of a relevant specialist(s) in this area. The logistics involved in these activities need to be highlighted in your PTDR write-up.

Replication

Replication (reproducibility), one of the key principles of the scientific method, must be kept in mind when writing this section. Multiple experimentation is more typical of science than once-and-for-all definitive experiment, and according to Yu (2018), "Experiments really need replication and cross-validation at various times and conditions before the results can be theoretically interpreted with confidence."

You must provide accurate, detailed descriptions of how the project would be (was) done so it could be replicated (repeated) by others. You should provide explanations that will enable the reader to reproduce the exact conditions of the original experiments including *validity* and *reliability* of the instrumentation (to be) used. A rather extensive explanation should be provided so that readers are *au fait* with why and how you are going to do (did) the research (in a final report). Your procedures should answer questions or objectives as efficiently, economically, and validly as possible.

A huge replication crisis is brewing in the field of medical science (e.g., Bossuyt, 2019; Perera, 2019). Medical Geologists should be aware of this, and consider building internal replications into their research methodology.

Data management and analysis

According to Sharp et al. (2012), many types of analysis (and, therefore, of research projects) are difficult or impossible to carry out because suitable data gathering techniques are not in place. The type of analysis employed and its purpose determine to a large extent, the nature of the data needed.

The invention of new methods of collecting data or the improvement of existing ones can therefore have a substantial impact on the research done in a (Medical Geology) research assignment, as witness the impact of the World Wide Web on the ease with which various agency reports can now be obtained. You should therefore provide an overview of how your data will be collected and managed.

The higher levels of research have seen the development of modern analytical instrumentation that have quickened the route to success; so also is the introduction of new approaches to data gathering and major refinements to analytical techniques already in use. There are two categories of data: *primary data*, that the researcher collects from first hand sources, *viz.*, through observation, experiment, and so on; and *secondary data* that have been collected by someone other than the researcher.

Data storage and retrieval

There are many tools that support effective data collection, data storage, and data retrieval. The University of Leicester (n.d.) lists a wide range of data collection tools, from card indexes and cross-referenced exercise books, through electronic tools like spreadsheets, databases, and bibliographic software, to discipline-specific tools. You should talk about how you plan to store your data with your supervisor or an information librarian. The use of index cards, a dedicated notebook, or an electronic file is recommended for recording the ideas that come up as you try to do your project selection. You can then refer back to this list of project alternatives when you start writing. They may be useful as ideas in themselves, and may be useful as a record of how your thinking developed through the research process.

In considering, how you are going to store and retrieve your data, you may decide to set up a system that allows you to:

- Record data accurately as you collect them;
- Retrieve data quickly and efficiently;
- Analyze and compare the data you collect; and
- Create appropriate outputs for your thesis/dissertation, for example, tables and graphs, if appropriate.

Analyzing the data

Many students are tempted to embark on literature searches and massive programs of data evaluative gathering before thinking about how the results are to be analyzed. All too often this leads to considerable wasted effort, since, according to Sharp et al. (2012), essential items of data may not be collected, or the first attempts at analysis would be too trivial. So, a complete rethink is necessary.

First, describe the methods of data analysis to be employed and how you plan to analyze the data and arrive at conclusions. Next, detail every step of the data gathering mechanism and analytical protocol(s) used. Note that this section varies depending on method and analytical technique chosen. Explain how different variables will be analyzed. Until a feasible outline of the type of analysis to be undertaken has been determined, the research plan must be considered incomplete.

Subject Selection (Population) is very important in Medical Geology. The population to be studied is first identified and an explanation given of why this population is appropriate for this study. Data describing the characteristics of the subjects should be provided; and, if available, data from the population to enable the reader to judge the representativeness of the sample.

Explain the specific treatment of data and variables for each subproblem, research question, or objective.

Presentation of the research findings

This Section is concerned with the presentation of the results from your data analysis only, and does not include a discussion of other research literature or the implications of your findings.

For qualitative research, this Section is usually organized by the themes or categories identified in your research. If you have conducted focus groups or interviews, it is often appropriate to provide a brief descriptive (e.g., demographic) profile of the participants first.

Direct quotation and paraphrasing of data from focus groups and interviews, and use these to support the generalizations made. In some cases, this analysis also includes information from field notes or other interpretative data.

Numerical data portrayal

How should numerical data be portrayed? Usually you begin by outlining any descriptive, exploratory, or confirmatory analyses (e.g., reliability tests, factor analysis) that were conducted. You next address the results of the tests of hypotheses, and go on to discuss any attendant analysis.

Illustrate and summarize all numerical information as tables, maps, diagrams, histograms, frequency curves, regression curves, correlation matrices, pie-charts, and so on. Researchers that do not have a sound working knowledge of statistics are advised to consult with a statistician to ensure that the procedures for sampling, data collection, data analysis, and data display are scientifically valid.

Expected outcomes

Expected outcomes refer to realization of originally targeted focal points of delivery of the project.

Refer to the original aims and objectives of the project, and indicate how measurement of the effects of the project would show (shows) that these would be (were) achieved. Highlight the techniques or instruments you will use (used) to measure the success level of the Project. For example, if you plan to use interviewers or observers, give the success level you expect, particularly with reference to their *interrater* and *intrarater reliabilities*.

Consider the following questions that relate to the *deliverables* or *outcomes* expected:

- Why are you doing this research?
- What will happen with the research findings?
- What would be the ultimate application or the long-term implications?
- Who will benefit from these findings (and who might be deprived or harmed) as a result of the study?

Give practical examples of how the expected results will improve the lives, livelihoods, and environmental
condition of communities and other stakeholders—a results-based management approach.

- Would there be capacity building and training in use of proven geomedical research techniques and technology? What categories of personnel would have received some sort of training as a direct result of this research project?
- Any issues to be dealt with regarding sustainability of effective mitigation measures that would be developed?
- What are the knowledge transfer and dissemination mechanisms you expect to develop?

Discussion

The purpose of this section is not just to recapitulate your findings but rather to discuss the meaning of those findings in relation to the theoretical body of knowledge on the topic and the Medical Geology profession as a whole. Typically, students understate this section even though it may be the most important one because it answers the "So what?" question.

Start by discussing your findings in relation to the theoretical framework put forward in the introduction and literature review sections. In some cases, you may need to introduce new literature (particularly with qualitative research). When discussing implications, deal with both the theoretical and the practical implications. Present only interpretations of the findings, not opinions. Structure the Section as follows:

- Brief overview.
- Discussion of results of application of method.
- Discussion of descriptive analysis.
- Discussion of tests of hypothesis.
- Post hoc analysis.

This Section also should address what your findings mean for communicating to professionals in the field being examined (Medical Geology).

Delimitations of the research

Once you have started the data generation phase, you may find that the research project is not developing as you had hoped. Briefly summarize the limitations brought about by the procedures of the proposed study, and put forward alternative methodologies for carrying out the proposed research plan, if these limitations impact negatively on your ability to conduct the study as planned.

Are there factors that would militate against the extrapolation of the results from this study area to a wider region? What other challenges or limitations are expected in the course of the research? Describe what intervening conditions might affect the outcomes of the experimental results.

A particularly disturbing delimitation is the making of *assumptions*, especially propositions for which no information can (or will) be made available within the scope of the study.

Analyze the validity of the overall project. How do you know that you are going to be doing (have done) what you say you are? Also, discuss threats to validity. Describe how your study will measure or control these threats.

Dealing with problems

Do not be upset that you have encountered a problem. According to Cziko (1989), research is, by its very nature, unpredictable. You would have to analyze the situation critically. Think about what the problem is and how it arose. Try going back a few steps and see whether that may resolve it? Or, if the problem is more fundamental, estimate its significance in answering your research question, and try to calculate what it will take to resolve the situation. Changing the title is not normally the answer, although modification of some kind may be useful.

If a problem seems intractable, you should arrange to meet your supervisor as soon as possible. Give her or him a detailed analysis of the problem, and always value their suggestions for resolving it. Never try to ignore a problem, or hope that it will go away. Also do not think that by seeking help you are failing as a researcher. It is all the nature of the research experience.

Finally, it is worth remembering that every problem you encounter, and successfully solve, is potentially useful information in writing up your research. So, do not be tempted to skim over any problems you encountered when you come to do your write-up. Rather, bring out these problems and show your supervisors/reviewers/examiners, how you overcame them.

Project management

Successful *project management* hinges simultaneously on the four basic elements of a project: resources, time, money, and most importantly, scope. In practice, these elements are interrelated, but each must be managed effectively.

As stated by Phillips (2003), "The primary challenge of project management is to achieve all of the project goals and objectives within the given constraints and time." The practice of initiating, planning, executing, controlling, and closing the activities and meet specific success criteria at the specified timelines should be diligently laid out.

It is hardly possible to complete your project successfully just by embarking on a period of intense activity just prior to the final deadline, as you may be tempted (ill-advisedly) to do for other modules. You should work consistently and effectively throughout the duration of the project. It is often worthwhile writing drafts of the final report while carrying out the work.

You will find that you are time tabled to attend *project management sessions* (*cf.*, Section on: "Project monitoring and evaluation," this Chapter), which are usually delivered by your supervisor committee, who would offer advice and support on various important, but non-technical topics, such as constructing a good project proposal, project planning, and time management, writing your progress reports. Plan these proposal meetings well. A well-planned meeting will help your committee understand that you are prepared to move forward with well-planned research.

If graphic presentations are necessary to help the committee, make sure they are clear and attractive. Rehearse your presentation well. Do not assume that all your committee members read the proposal carefully; so be sure to cover all the important facts and issues.

Plan of work and time schedule

The project management plan

Having selected a suitable topic for research, the next questions to ask are the following: What would you do to ensure the thesis or dissertation is completed timeously? The sequence of events to achieve this is referred to as a *project management plan* (Fig. 2–1).

FIGURE 2–1 The project management plan (PMP). From *https://organizationalexcellence.virginia.edu/project-management. (Accessed 19 January 2020).*

Completing a PTDR document in an extended project undertaking that requires judicious time management and performance of a variety of tasks. Some graduate programs schedule the thesis or dissertation at the end, while in other programs, it runs concurrently with other modules. Whatever the structure of the course schedule, you would have to create a *project workplan* that helps you allocate enough time to each task you have to complete.

The project workplan

A *project workplan* is a breakdown structure of the work to be done (tasks and other activities) in phases, with estimates of time to complete each phase. It will identify interdependencies of tasks, which are critical work elements to be completed in a logical sequence. You should indicate the work completed so far and what you intend to do in the next phase. One of these tasks should be writing of *progress reports (draft-/interim-/final-).*

Your research plan should also include information about what equipment you will need to complete your project, and any travel costs or other expenses that you are likely to incur through the pursuit of your research. You should also think about whether you are dependent on any one else to complete your project, and think about an alternative course of action if, for some reason, they are unable to help you.

Give a timeline for tasks to be completed during different phases of the project period. Identify the major activities of your project at the defined timelines, and the output expected at the end of each activity. All outputs are expected to be completed at well-defined timelines and contribute to the achievement of the overall objectives of the project.

The timelines must accurately reflect what was planned for in the study, and should be consistent with the requested budget. Present clearly the various tasks or activities using a 7-column matrix format (Table 2–1), or a Gantt chart template (Fig. 2–2), or other format recommended by your department/faculty/institution.

You should include an outline of the various stages and corresponding time lines for developing and implementing the research, including writing up your thesis. Clearly state the *indicator(s) of achievement* of each major activity of the project.

Also, provide an interim log, indicating meetings with your supervisor and reflecting the phases of the research plan completed so far. The complete log may be given as an appendix to your final report.

Table 2–1   The project workplan, incorporating schedule of major project phases/activities with approximate timelines.

S/N	Activity agenda	Actors	Objectives	Timeframe (tentative)	Expected outcomes/deliverable(s)/ indicators of achievement	Remarks

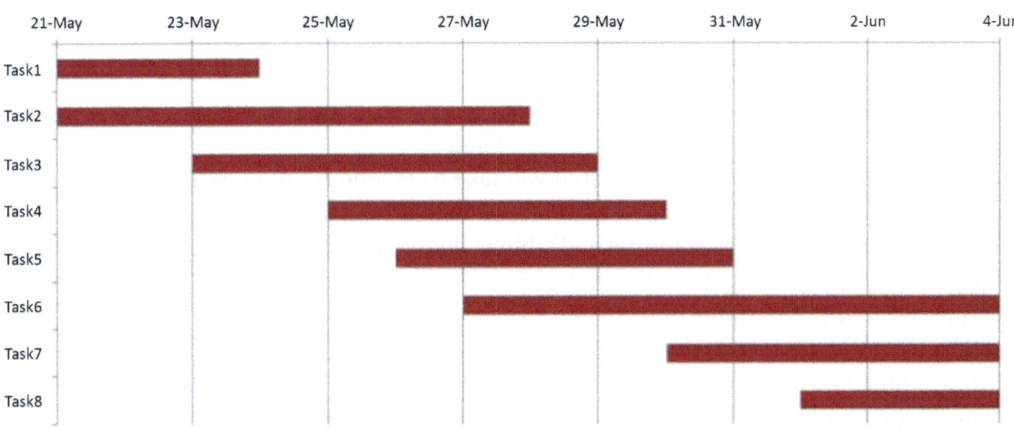

FIGURE 2–2 An example of a Gantt chart. *From https://appfluence.com/productivity/gantt-chart-in-excell. (Accessed 30 July 2019).*

Procrastination/poor time management

Procrastination in this context refers to poor *time management* in executing your research.

Some people do find that they procrastinate more than they would like. This is a common problem, so it is probably best to be well-prepared to identify it and deal with it if it does start to happen. People procrastinate for various reasons for example:

- Daunted by the scale of the task;
- Loss of motivation;
- Perfectionism;
- Fear of success;
- Fear of failure;
- Lack of interest;

- Difficulty concentrating;
- Need to feel under pressure;
- Personal problems;
- Poor organizational skills, and many more.

It is necessary to determine in a realistic way how long each task is likely to take. Some focused thought at the beginning, then at the planning stage of each phase, could save hours later on. Write down the resources needed for each stage. It could be time in the library; the resource of your working hours; or the use of equipment or room space that needs to be booked in advance.

Curbing procrastination

To improve the prospect of completing on time, and avoiding procrastination, Korrapati (2016) suggests the following steps:

1. "Be realistic about when you can/will start;
2. Devote time to planning and revising your plan;
3. Try to work out if any of your research will take a set amount of time to complete;
4. Allocate appropriate time for any traveling you need to do for your research;
5. Include other (nondissertation related) things that you have to do between now and then;
6. Have clear and achievable objectives for each week;
7. Focus on one thing at a time;
8. Leave time for editing and correcting;
9. Reward yourself when you complete objectives that you have timetabled; and
10. If you fall behind make sure you spend time reworking your plan."

Above all, be organized and methodical while conducting and writing up your research.

Early detection of the signs of procrastination will give you the best chance of obviating any negative effects. Once you suspect that you are procrastinating, it can be helpful to review what you are expecting of yourself, and check that those expectations are realistic. This is where planning is vital.

Estimated budget

For proposals requesting funding, candidates should provide justification for all categories of funds requested. Include all other sources of funding for the proposed study.

Reviewers can recommend budget cuts when they think that expenses are overly high or unwarranted. The budget must accurately reflect the plan for data collection, data analysis, and data write-up. Include in the "Appendix," quotations, invoices, instrument specifications, and other documents to show you have ascertained the exact identification, descriptions, and cost of all items you need for carrying out your research.

Many grant awarding bodies require some local budgetary input from the applicant's institution. This can take the form of an in-kind contribution, such as donated equipment or furniture or provision of office space, staff time, or services.

If you are just beginning as an independent investigator, do not ask for a very large grant in your proposal. Demonstrate that you can complete a good small project for a relatively smaller amount of money and establish a good track record before applying for larger research grants.

A 5-column matrix format (Table 2–2) can be used.

However, the actual format for presenting budget estimate should be that stipulated by your institution or donor agency.

Project monitoring and evaluation

As a check on performance quality, a *project monitoring and evaluation team* should be set up, typically comprising the candidate (and other project participants where applicable), her/his supervisor, a few colleagues/classmates of the candidate, a couple of national/international experts in your project area, and any other stakeholders, such as representative(s) from the funding agency.

Project monitoring comprises a check on all project-related activities, such as operational efficiency of project participants and time management, identifying potential problems and taking remedial actions necessary to ensure that the project is within scope, on budget and meets the specified deadlines. If there is need to modify the original project implementation plan and *"modus operandi"* of the project implementation, the review panel will advise.

Project evaluation refers more to the research purpose than to a specific method. You should select evaluation tools that answer the question(s) involved with the study and its objectives. Specify what means of evaluation will be used for each objective.

Table 2–2 Budget estimate.

S/N	Item	Deployment	Specification/quotation/invoice	Estimated
1				
2				
3				
4				

Note: Grand total (Less 5% contingency cost) to give actual total = **** (in appropriate currency).

Stylization and structure in document preparation

In your research plan, you need to state clearly when you are going to stop researching and begin the writing process; a stage which many students consider as the most daunting part of project undertaking. However, you should aim to stick to this plan, barring a compelling reason why you need to continue your research longer.

Thesis/dissertation writing should be clear, concise, and put in an as objective fashion as possible. Indeed, these are some of the hallmarks of a good scientist. Your personal impressions and feelings should rarely come into it. Avoid using alternate words to refer to the same thing. This could be frustrating to the reader.

While the passive voice is normal for scientific writing, it is not used universally. However, where you are genuinely voicing an individual opinion, you may use the first person.

Review a few high quality, well-organized theses/dissertations produced by former graduate students in your department/faculty/institution. Examine their use of headings, style, typeface, and organization. They should assist you to begin to write with a clear idea of what the final product should look like.

Above all, to write well, you would have to create a customized writing technique. You could start by preparing a list of keywords that highlight your research, and then use that set of keywords throughout.

Other students start writing at Chapter One and write straight through. This is seldom a practicable way to go about your writing. An approach that has benefitted many students is one where you begin writing those parts that are most comfortable or convenient. Then complete the various sections as you think of them.

Note that the guidelines given in this Section may vary somewhat in accordance with institutional specifications for thesis/dissertation writing, or *instructions to contributors* in writing journal articles.

Segmenting, scheduling, and rewarding your PTDR writing

The three techniques: *segmenting, scheduling,* and *rewarding* are known to have benefitted generations of students (Hon et al., 2007−2008). In this approach, the whole thesis is divided up into small sections. Tackle just one section at a time. Avoid concentrating on doing the entire thesis all at once. Instead, focus clearly on just one small piece at a time.

Whatever technique works for you should ensure that you have enough time not only to write up your research, but also to review it critically, then spend time editing and improving it.

Acknowledge your progress to yourself. At the end of each day's writing, you may even decide to treat yourself!

Document preparation tools

Several high-quality typesetting systems are now available to support document preparation, from LaTeX to tools built into Microsoft Word. Find out about them and use them. It is possible to obviate all spelling errors by using readily available spelling checkers. The same applies to errors in referencing and poorly laid out graphics for which you could use a simple tools to insert them for you. The many simple tools to help you manage and organize citations include CiteULike, EndNote, Mendeley, and Zotero.

Tables and illustrated material

Photographs should be mounted on good bond paper. Copy paper is unsuitable for this purpose. Photographs, maps, graphs, and other statistical tables should be located where they are first mentioned in the text. Great care should be taken in folding maps, diagrams, or tables larger than paper size.

The lists of tables and illustrations shall follow the "table of contents" and should list all tables, photographs, diagrams, etc., in the order in which they occur in the text (see Section: "The Table of Contents," in this Chapter). Page references in respect of diagrams included in the text, shall be to the page on which the diagram occurs or faces. In the "table of figures," the location, number, and legend of all figures in the document should be provided.

Length of proposal/dissertation/thesis/report

While your supervisor will encourage you to be succinct, your thesis/dissertation still needs to meet the required word count. However, there is no universal regulation on length for a PTDR document, since such stipulations are the prerogative of respective institutions. Whatever the length stipulated, however, all institutions would normally judge a document's merit on quality and not on its length.

The finished product (thesis, dissertation or research project report)

In the field of written scientific communication, a *research project proposal, thesis,* or *dissertation,* and *research project report* are all similar versions of scientific writing. All of these documents can be classified as *scientific report,* which is a natural sequence to a successfully executed research plan. A well-written proposal should be transformed into a PTDR document with only minor changes, such as going from "will be done" in the proposal to what "was done" in the PTDR document, the main difference between the two kinds of document being the fact that the *report* must include the discussion of actual results of the study.

So, if you prepared a comprehensive proposal you will now, (on completion of your thesis, dissertation or project report write-up) be rewarded! Pull out the proposal and check your proposed plan. According to Leshan (n.d.) you should, in converting your proposal to your thesis, dissertation or project report, "... change the tense from future tense to past tense and then make additions or changes so that the *methodology* section truly reflects what you did." You have now been able to change sections from the proposal to sections for the thesis, dissertation or research project report, move on to the *statement of the problem* and the *literature review* in the same manner.

Further details on style in technical writing is discussed in more detail in Strunk (2005).

Institutional approval

Ensure that your proposal has the necessary ethics and institutional approval before submitting to your department/faculty/institution/funding agency. Note that some funding institutions stipulate that approval should be given by the head of the host department/faculty/institution, to:

- Confirm that the researcher is a staff of the institution and collaborating institution(s) where appropriate.
- The institution will provide space for the successful conduct of the research.
- The institution will guarantee the proper usage of the grant for project execution in line with the approved budget.
- The head of institution signs and stamps the grant application form.

Ethical considerations

Be aware of ethical considerations and procedures, since ethical questions are of very high importance in Medical Geology. Standards governing the conduct of research with human and animal subjects or handling of tissue samples are normally given in institutional policy documents with title such as "Code of Conduct of Research" and all students are expected to familiarize themselves with these.

A project consisting of basic Medical Geology research may be little affected by ethical issues, but even in such cases some of the points in the relevant policy document are likely to apply. Topics that deal with human or animal subjects or handling of tissue samples have much wider ethical implications, and for these you should give particular awareness to technological procedures and standards as well as knowledge, understanding, and compliance with legislation. You should include in this section of your proposal, and subsequent reports a description of ethical issues that you will need to take (took) account of while you are (were) working on your project. In your final write-up you should discuss how you addressed these issues.

Your conclusions

Often, students use the *conclusions/implications section* to merely reiterate the research findings. This is not right! The "conclusions section" is a key section of the PTDR document and so, should be given considerable attention. It is sometimes best done after you have had a few days to step away from your research and put it into perspective.

This Section embodies a summary of the entire PTDR in a few paragraphs/pages, and should include an assessment of the success of the finished product. Were you able to achieve your objectives? If not, why not?

You should state the implications of the work, and may include your opinion (clearly stated as such). Label speculations clearly. You may refer to the literature review, but more importantly, state how the results of the study may be applied, speculating about broadest possible consequences, both theoretical and practical. Recapitulate on the limitations (theory, method), and say how these were handled. Muse on suggestions for future research and extensions, or alternative methodologies that, with hindsight, might have led to better outcomes.

Acknowledgments

The writing style for thesis, dissertation, research report, or a journal article, *acknowledgments* tend to be more or less informal, as this is not part of the actual academic or scientific work itself. It is your chance to write something more personal. For that reason, you may, depending on institutional stipulations, use first person pronouns in this section.

The *acknowledgments* section (optional) usually includes any individuals or institutions you wish to thank. These may be your supervisor(s), other students, if part of a related project and any other person(s) or establishment(s) that have helped and supported you in any way in the conduct of the research and documentation process. Clear acknowledgment should be given to any work upon which you have built, both published and unpublished.

The acknowledgment section in a dissertation or thesis normally appears directly after the title page and before the abstract, and is usually no longer than one page (Gahan, 2019). It can also come after the abstract page and before the table of contents, or put at the end, between the last appendix and the bibliography. In a scientific paper for a journal, the usual position is after the conclusions section, just before the list of cited references. However the exact positioning of the acknowledgment section, whether in a dissertation, thesis, research project report, or scientific paper is again determined by institutional requirements or *author guidelines* in the cast of a journal article.

It is important not to overlook anyone who has made an input or rendered support of some kind, particularly those in the professional sphere.

Further information on writing good "acknowledgments" for your PTDR document is provided in several authentic Web resources, such as Gahan (2019).

References/bibliography

According to the University of Birmingham (2020), "A *reference list* is the detailed list of references that are cited in your work," whereas a *bibliography* (generally considered as optional), "... is a detailed list of references cited in your work, plus the background or recommended readings, or other material that you may have read, but not actually cited." *Bibliographic essays* usually include sources searched, such as indexes, abstracts, and bibliographies.

Despite the differences outlined for these terminologies, the terms "reference list" and "bibliography" are sometimes used interchangeably; so it is important to check your "institutional guidelines" or "journal guide to authors" (in the case of a scientific paper for a journal) to make sure you know what is required for your project write-up.

Why is referencing/bibliography important?

In academic work, it is quite acceptable and appropriate to quote the work of others, provided that you give accurate acknowledgment to the sources (*viz.*, book chapters, journal articles, magazines, bulletins, newspaper reports, standards, patents, conference proceedings, web pages, photographs, emails, and so on). Good referencing is an academic convention that shows that you can select, collate, and synthesize information from a variety of sources. It enables your supervisors/reviewers and other interested readers to check the accuracy and validity of the evidence presented, and to trace the sources you cite. Another important point about good referencing is that it obviates the accusation of plagiarism and acknowledges the work of other researchers. Examiners/reviewers pay great attention to accurate referencing, which can also enhance the quality of your thesis or dissertation, and contribute to the credits you gain in the assessment of your work.

Style of references

References in your PTDR document should include works (books, journal articles, periodicals, brochures, reports, papers, or web sites) that you have used, plus other relevant documents. Special care must be taken in citing certain documents, such as newspapers, some web pages, and unpublished reports, even though they may have provided useful information. This is so, because such documents are, in general, not peer reviewed, or vetted for accuracy, so that some of the information they contain may be misleading.

References must be given correctly. Full references with page numbers are normally required. If you consult or cite a similar PTDR document done in previous years it must be referenced [using the Candidate (Author) Name, Student Number, Title, Year, and Institution]. The references should be cited in the body of your PTDR document where appropriate. In the case of web pages the title/topic of the article and the last time it was accessed, must be given, not just the URL.

The referencing systems commonly recommended for adoption for most PTDR documents include the Harvard system, the APA system, and the EEEP system.

The Harvard (author-date) style

In the Harvard (author-date) style, [(University of Birmingham, 2020)] the author's name and the date of publication are used in the body of the text when citing sources, for example, Davies (2019). Variations are possible, for example, we can say that Davies (2019) has developed a new curriculum for graduate studies in Medical Geology. The bibliography is given alphabetically by author. Journal and book names are italicized, for example, Pokras, M. (2005), *Book Review: Essentials of Medical Geology: Impacts of the Natural Environment on Public Health. Env. Health Pers.* 113 (11). Notice that there is a lot of information about the article cited (Year of Publication of Journal, Name of Journal, Volume Number, Pages), not only the title and author. Thus the reader can easily locate the article in question. Find out what is expected for different types of article (e.g., books, conference papers, dissertation, thesis) and give as complete information as possible.

Further details on use of the Harvard System are given in Mendeley Elsevier (2019a). This is a complete guide to Harvard in-text and reference list citations.

The APA style

The American Psychological Association (APA) Citation Guide (APA, 2010) provides the general format for in-text citations and the reference page. This is an "author-date" style, whereby the citation in the text consists of the author(s) name(s) and year of publication given wholly or partly in round brackets. It uses only the surname of the author(s) followed by a comma and the year of publication. Further details on use of the APA system are given in Mendeley's easy-to-use, comprehensive guide (Mendeley Elsevier, 2019b).

IEEE style

The IEEE style is numeric. References are given a number, the *citation number*, which corresponds to the order in which references appear in the text. This number is used when citing the document in the body of the text. A full corresponding "list of references" is given at the end of the PTDR document, next to the respective citation number. Further details of the use of the IEEE system of referencing are given in Victoria University (2019).

Which of the referencing style should I use?

You should select which style to use (as recommended in the institutional guidelines for PTDR writing) and use it consistently. Look up how to reference different kinds of sources, taking particular care with electronic sources. Give as much information about these as possible (title, author, date, if possible) and consider just using footnotes for nonauthoritative electronic sources, or possible inauthentic sources, such as some newspapers and web pages.

If you are writing a thesis or dissertation and want to use another style apart from Harvard, APA, or IEEE you should discuss it with your supervisor.

Whatever style you use the *references section* should come between the main text and the appendices. Normally references should start on a fresh page, and should not have a chapter or section number, just the heading "References." Some word processing tools, such as *EndNote's Cite-While-You-Write (CWYW)* tools provide help with referencing—consider using these. However, the main thing is to give proper thought to how and what you cite.

Suggestions for the list of appendices

In the list of *Appendices*, you should include your project logs and any additional relevant system or test data. You may also include any technical material, which you estimate as too detailed for the main body of the report. The University of Sussex (2020) reminds students to always make sure that any technical material is appropriately annotated and consistently presented. The source code of your project (including documentation) must be submitted electronically and should not be included in the Appendix. Also, any material that does not fit in the report's Appendix that you nevertheless deem important can be included in the electronic submission. Note, however, that it will be left to the markers'/examiners' discretion whether they wish to look at such extra material in the electronic submission.

Other important materials to include in your Appendix are:

- Questionnaires.
- Research permits, ethical clearance documents; other permits.
- Letters of consent.
- Your curriculum vitae and curriculum vitae of team members (faculty and staff expertise), if appropriate, with some emphasis on the individual's experiences relevant to the work she/he/they will do on the project, such as research skills and experiences, management/supervision experience, publications and/or paper presentations.
- Raw data or tabular material from pilot studies.
- Maps, flyers and clips that are too large or too bulky to be placed within the body of the PTDR document.
- Diagrams of research or statistical models.
- Equipment specifications; inventories; quotations, invoices, etc.
- Schematics for constructed equipment.
- Supplementary bibliographies.
- Project proposal documents (in the case of thesis/dissertation).
- Facilities; library and computer resources, other special facilities that could contribute to a successful study.

Definitions of terms used in your proposal/thesis/dissertation/report

This section gives some suggestions on your nomenclature system. These suggestions are for guidance only, since the actual institutional stipulations may vary somewhat.

- Define terms in the context where they will be used. Provide operational definitions as well as constitutive definitions.
- Include a list of definitions for terms and concepts that have significant meaning for the project.
- Construct your list like a dictionary, not in prose form.
- Do not define generally understood concepts, principles, and concerns.
- Much of the specific information about the terms will be presented in other appropriate sections of the PTDR document.

Abbreviations

Use of abbreviations is discretionary, but be judicious in the way they are used. Where abbreviations are used, a key should be provided. Abbreviate consistently throughout the thesis/dissertation, but do not use them excessively, as too many abbreviations make your document more difficult to read. Do not abbreviate terms only used a few times in the thesis/dissertation. Provide a table of abbreviations used throughout the PTDR so that the reader can quickly interpret an abbreviation they have forgotten. Do not include common abbreviations in this table. For an abbreviation not in common use, give the term in full at the first instance followed by the abbreviation in brackets.

Submission of project proposal/thesis/dissertation/project report

Again, this Section gives only suggestions, which are meant to guide you through the submission process. Your project will be assessed primarily from the submitted PTDR documents, so it is essential that it gives a full account of your work and is clearly presented. The detailed structure will be dependent on the type of project, and you should obtain advice from your supervisor. Your supervisor can also be expected to comment and correct your outlines and drafts. Be aware that individual institutional stipulations may vary somewhat, and that you must always refer to specifications given in the relevant institutional document, such as "Guidelines for Preparation and Submission of Project Proposal/Thesis/Dissertation/ Project Report."

Project presentation

In addition to writing your PTDR, you will be assessed on your ability to present your project. The project presentation can take different forms, depending on institutional requirements. It may take the form of: (1) A poster presentation at an institutional event such as "faculty research day"; or (2) a short talk [(oral defense or *viva voce*)] to an assessment or examinations panel of your project.

In the case of a poster presentation, details of the required poster format, submission procedure, and assessment criteria are normally given on the project presentation page of your institutional website.

In the case of an oral defense or *viva voce*, the assessment session usually starts with a short talk lasting around 15–20 minutes. In the presentation, you are normally required to give an overview of your project and describe what you achieved. If you have written software you should demonstrate it. Your task is to convince the assessment panel that your project work has the potential of making (has made) a significant contribution in Medical Geology and is of a high quality. It is an excellent idea to prepare your talk in advance and practice it on a friend or small group of fellow students.

The poster presentation or the short talk is an important component of the project assessment, and many universities have regulations regarding their delivery. The University of Sussex (2020), for instance, notifies that students would lose credit points if they do not submit a poster by the deadline or fail to attend the poster event in their assigned time slot. If you do not turn up to your short talk at the scheduled time you will be marked as absent and forfeit the associated marks. Therefore managing your time so as to keep to deadlines is vital (see Section: "Procrastination/Poor Time Management," in this Chapter). You must anticipate technical problems, such as printer breakdowns and corrupted disks. Make sure you back your work up regularly. Personal computer problems, theft, or data loss are never accepted as grounds for an exceptional circumstances claim.

You will be expected to submit a specified number of copies of spiral bound hard copies of your PTDR document in addition to any required for the student's own purposes. Some institutions (e.g., Heriot-Watt University, 2013) require one electronic copy burned onto a CD together with code listings, software demonstration, etc., and a specified number of the PTDR documents and appendices. The CD should contain an electronic copy in PDF format of the whole of your PTDR and all its appendices as a single document. Students are required to submit this PDF file electronically via the web. This PDF will be checked for *plagiarism* using an appropriate software, such as *TurnItIn*. Your document should include a signed and dated declaration that the work is your own (see Section: "Researcher Declaration," in this Chapter).

Remember to properly acknowledge all sources from which you have quoted as well as cited, and the source of any excerpted material (see Section: "Acknowledgments," in this Chapter). Some institutions provide a standard page for inclusion, which contains this text, and is available from the appropriate administrative office; and, depending upon which institution provided along with the covers for your document. Only this text is exempt from the requirement of quoting and citation to avoid committing *plagiarism*.

Proposal/ thesis/dissertation defense

The steps to be taken to ensure a successful *proposal defense* and final *oral defense* (*viva voce*) of the thesis or dissertation include:

- Preparation of a well-written document;
- Knowing the format of the defense;
- Preparing the presentation;
- Practicing the presentation;
- Anticipating questions.

The assessment or examinations panel would normally provide explanation of how decisions are made by thesis/dissertation committees at the defense, what students should do after a decision has been reached in a *postexamination review*.

Dissertation/thesis assessment

Assessment forms a vital part of the learning process, and presents students both an opportunity to learn and an opportunity for learning to be evaluated. Different institutions would have their own guidelines for assessment design, delivery and feedback; but there are certain requirements that could be considered as generic in awarding examination grades.

Marking guidelines

There are many factors to consider when designing assessment tasks in order to ensure that a wide range of knowledge, skills, and abilities are developed and tested in a variety of settings. In addition, institutions normally require that assessment reflects the goals of respective (Medical Geology) programs (Davies, 2019).

To ensure that assessment contributes to learning, it is also essential that useful and constructive feedback is provided. This feedback needs to be anchored onto the assessment criteria, demonstrating to the student the ways in which they have satisfied these, as well as where they could have improved their work. Feedback should also be structured in a manner that helps students to identify ways to improve their performance for future tasks.

All institutions have their own criteria for grading different components of the examination package for graduate studies. Generic criteria used for grading the *oral examination component*, for instance, may comprise of the assignment of a *pass/fail* grade, considering the structure of the presentation and demonstration of grasp of problem during the question/answer session. Using the University of Warwick (2012) grading system for the "oral examination" for instance, to be awarded a "pass level": "a candidate should demonstrate an understanding of the work presented and be able to answer questions on both the work presented and on the subject area in general, and defend the work undertaken and its suitability for the degree in question." A "fail grade" is assigned, "when the candidate shows an incomplete understanding of the area of work and general difficulty in handling questions without help; and not able to convince the assessors that the work presented was her/his own."

Post-examination review: revision of project/thesis/dissertation report and dissemination

After the oral examination results are out, the next stage is the implementation of the examiners's comments, binding and dissemination of the PTDR. Be prepared for revisions after the defense. You can expedite clearance by the graduate school by letting the staff examine a draft of the thesis or dissertation before you defend.

It is a general practice to provide your chair and committee members with a bound copy of the final version of the thesis or dissertation.

Binding of project proposal/thesis/dissertation/report

Many institutions prefer that the PTDR document be submitted in a permanent form of binding, for example, thermal or spiral binding, not ring binders. Most institutions do not stipulate that PTDR document be hard bound, but they should be neat and presentable. The individual university house style of binding is normally acceptable.

Dissemination strategies

Mode of dissemination

Indicate the steps you will take to ensure the project outcomes are brought to the attention of key stakeholders. This can be through seminars, journal articles, or other publications, workshops, conferences, and so on.

Institutional and subject repositories

Institutional and subject repositories are becoming increasingly important in providing academic publication and information resources around the world; and according to Copeland and Penman (2004) and Rasuli et al. (2019), *Electronic Theses and Dissertations* programs are increasingly being recognized as one of the most effective outlets through which theses and dissertations can be made to reach academic communities and beyond. They are being developed as a consequence of the availability of scholarly resources in digital formats, and in response to Open Access policies and mandates (See: Pampel, 2023).

Many African countries now boast of several Digital Research Repositories [De Mutis and Kitchen, 2016; Recent compilations by the International African Institute (IAI International African Institute, 2020)] constitute a service to African studies research and scholarship in line with the remit of promoting the dissemination of knowledge from, and on, Africa. They are still not yet comprehensive, but are being regularly reviewed and updated (De Mutis and Kitchen, 2016).

According to IAI International African Institute (2020), these existing data bases form part of a larger future project to enable and support more sophisticated search of existing repositories, as well as the further digitization and online publication of Masters' and doctoral theses and dissertations from the African continent. This initiative truly represents a crucial resource for research and knowledge exchange in Africa and beyond!

Researcher declaration

All institutions require a *declaration* be made, *viz.*, that the PTDR is your own work. This embodies a signed statement of originality together with an overview of any intellectual property rights agreements that you have made. This usually comes somewhere after the cover page of the thesis or dissertation, or immediately following the "acknowledgment," under a separate heading depending on institutional requirements.

The exact wording of the statement of originality varies for different institutions, but essentially embodies a signed and dated statement that the work produced is the author's own work, and that the information provided is to the best of the author's knowledge, complete and correct. If the thesis or dissertation is based on joint research the nature and extent of the author's individual contribution should be indicated.

The accuracy of the text of a *declaration* is often a serious matter and examiners may refer any thesis or dissertation with suspected plagiarism to the institution's *Disciplinary Committee*.

Chapter conclusion

Students and faculty who have gone through this Chapter are probably at a stage where they are developing their project proposal document or considering starting their MSc or PhD project work (the most important component of your graduate degree program), but are still unsure about what topic to choose. This Chapter has highlighted the factors that inform an independent investigation of a specific problem (in Medical Geology).

Regardless of how different institutions use the terms *thesis* and *dissertation*, the purpose remains broadly similar for both projects. A thesis or dissertation should be a formal, extensive, and well-focused piece of research that addresses a specific, well-defined problem or question. A Medical Geology proposal, thesis or dissertation, should, in addition, show that some significant contribution to knowledge in the field would be (has been) made, and frequently, that a new kind of skill (would be) has been acquired.

The decision-making process for selecting an ideal thesis or dissertation topic; however, complex it might seem at first, is a process that can be accomplished with perseverance, patience, and the right attitude. Therefore while it is important that the topic you choose has been selected freely and out of your own interest, rather than your supervisor's, there are many issues involving critical factors and requirements that both you and your supervisor should consider strongly, and that will impact on your ability to complete your thesis or dissertation timeously and successfully.

A research proposal should be a clear, well-written document that follows a simple and logical train of thought. You have to think very carefully about your topic and make sure that it is well-focused. Write a detailed research proposal to help you anticipate the geoenvironmental health issues or problems that you are going to deal with. Identify a cutting-edge and captivating title, which at the same time reflects accurately, the scientific content of the work, its significance, and application.

A cutting-edge title in Medical Geology is also one that shows a high level of originality and underlines the study's potential of making a significant contribution to knowledge in this field, *viz.*, contribution toward the explanation or resolution of some geoenvironmental health problem, and by extension, is impactful on the lives and livelihoods of your community (local, national, regional, or international).

You should include a thorough review of the literature in your proposal, conducted before proposal preparation, to identify the problems. It should define the extent to which solution or answer is already known, and thus, help to discern the existing gaps in knowledge. A good literature review can also be used to verify the concepts and theories you have advanced, and the scope of the methods to assess the concepts.

According to Hon et al. (2007−2008), "Most research begins with a question." So, think about which topical questions and theories in Medical Geology that interest you, and what you would like to know about them. Think about the topics and theories you have studied in your program. Is there some question that you feel the body of knowledge in your field does not answer adequately? Think about environmental diseases in your community or region, especially those for which the etiology is still in conjecture. Once you have a question in mind, begin looking for information relevant to the topic and its theoretical framework.

You must make sure that the study described aligns with the specific objectives listed earlier in the research proposal. The success of a research project is in large part predicated on a judicious choice of a research method or strategy that allows the researcher to collect the necessary data to deal with the research problem. Therefore make sure that the underlying science and methods behind the research plan are sound, feasible, and as complete as possible. Be flexible and be prepared to modify your research questions, if necessary.

Though your ability to execute the research will depend on your specific research skills (existing and developing), access to other resources such as equipment, funding, technical support, and time would also characterize a potentially successful project. Since these factors vary greatly, the viability of a project should be established in a very circumspective manner.

Your supervisor is your mentor and guide throughout the process of preparing your proposal, executing the research and writing your thesis or dissertation. She/he is there to help you with any question you have, no matter how big or small. When you have an idea for your thesis, dissertation or research project, and you have carried out some preliminary research yourself, schedule some time to talk to your supervisor to ask for her/his advice.

Choose a topic that will help you sail through the career pathway you envisage. Employers will sometimes ask about your thesis or dissertation or even want to see it—especially if you go into the field of *environmental disease etiology*. Your choice of thesis or dissertation can pave the way to a job or let you hold one.

Many institutions and individuals are now taking advantage of recent developments in the field of information technology to improve access to scholarly publications. Many universities now encourage the submission of theses and dissertations in electronic format. The *African Digital Research Repositories* project should encourage all universities in Africa to consider depositing full text theses and dissertations in a repository as an integral part of their research process.

Executing a well-chosen Medical Geology project can be exciting because a successful project rewards with great satisfaction [*inter alia*, that you have contributed in some way toward explaining/resolving some geoenvironmental health issue(s) in your community] and have achieved experiential learning. However, such a project undertaking can also be challenging, because a tremendous amount of self-discipline, time, and effort needs to be put into it.

Finally, if successfully completed, the project gives you something you can talk about knowledgeably and enthusiastically to the larger academic community, prospective employers and many other stakeholders, not least, your government policy makers. A well-executed project is an excellent indicator of your overall ability to carry out a serious piece of work, and therefore, all are impressed by such a project. It has presented you with a great opportunity to demonstrate your creative abilities and independence!

Review questions and suggested areas for further research

Potentially, the most preposterous part of the thesis or dissertation is the *Suggestions for Further Research* section. This section is usually written at the very end when little vigor is left to make it meaningful. The biggest dilemma is that the suggestions are often ones that could have been made prior to conducting the work. This section should be read and reread until you are certain that you have made suggestions that emerge from your experiences and findings. According to the University of British Columbia (n.d.): "students should make sure that "suggestions for further research" link their project with future projects and provide a further opportunity for the reader to understand the significance of what they have done."

At the time of writing, it was noted that observations from the "1999 East and Southern Africa Association of Medical Geology (ESAAMEG) Workshop" (Davies and Schlüter, 2000) still hold true today, *viz.*, "... the great need for further collaborative and interdisciplinary studies to unravel the hitherto unclear relationships between the geoenvironment and cardiovascular disease, cancer, geophagy, and certain endocrinal disorders." Progress on research in these areas since 2000 are given in relevant chapters of the book.

Other examples of needed research from which African Medical Geology project selection can be drawn include:

- The role of confounding geoenvironmental factors in influencing studies in disease attribution?
- The impact of contaminant mobility from extreme weather events, such as flooding.
- Lead and other heavy metal exposure resulting from dust and other particulates.
- Asbestos exposure such as amphibole asbestos dusts in some Sub-Saharan African countries, for example, South Africa.
- Fungal infection resulting from airborne dust, such as Rift Valley Fever or coccidioidomycosis.
- Development of methodologies to determine bioaccessibility and/or bioavailability of potentially harmful elements PHEs.
- Applications of geoengineering to Medical Geology issues. Recently, the new concept of geomedical engineering has been introduced in Medical Geology through a paper titled: "*Geomedical Engineering: A new and captivating prospect*" (Ur-Rehman, 2009). It provides the fundamentals of engineering applications to Medical Geology issues.
- Combating neonatal, maternal and child deaths due to fetal abnormalities or congenital malformations from ionizing radiation exposure in major uranium and gold mining countries of Africa, for example, Niger, Namibia, and South Africa.

Glossary of terms

- *Content expertise (base profession)*: The formally recognized knowledge, skills, and abilities a faculty member possesses in a chosen field (base profession) by virtue of advanced training, education, and/or experience.
- *Deliverables*: Tangible "things" that the project produces.
- *Interrater reliability*: (Also referred to as *interrater agreement, interrater concordance, interobserver reliability,* and so on), *interrater reliability* (IRR) is the level of agreement among raters (or observers, coders, examiners). It is a score of how much homogeneity or consensus exists in the ratings given by various raters and addresses the issue of consistency

of the implementation of a rating system. If everyone agrees, IRR is 1 (or 100%) and if everyone disagrees, IRR is 0 (0%). IRR can be evaluated by using a number of different statistics, such as percentage agreement, kappa, product-moment correlation, and intraclass correlation coefficient. High IRR values refer to a high degree of agreement between two examiners. Low IRR values refer to a low degree of agreement between two examiners.

- *Intrarater reliability*, in contrast, is a score of the consistency in ratings given by the same person across multiple instances, or put more simply, a measure of how consistent an individual is at measuring a constant phenomenon.

Interrater and intrarater reliability are aspects of test validity.

- *Issues*: Risks that have happened.
- *Milestones*: Dates by which major project activities are performed.
- *Medical Geology niche area*: The Medical Geology area of focus or expertise of your department/faculty/institution.
- *Replication crisis*: Also called *the replicability crisis* or *reproducibility crisis* is, as of 2020, an ongoing methodological crisis in which it has been found that many scientific studies are difficult or impossible to replicate or reproduce. The replication crisis affects the social sciences and medicine most severely (Bossuyt, 2019).
- *Risks*: Potential problems that may arise, with the possibility of loss or injury (Merriam-Webster Online, 2020); *Project risk* is an uncertain event or condition that, if it occurs, has an effect on at least one project objective (PMI, 2008).
- *Scope*: The aspect(s) or subtheme(s) on which you would put the focus. The different subthemes, theory and knowledge that pertain to the subject under investigation.
- *Stakeholder*: Any person or group of people who may be affected by your project.
- *Tasks*: Also called *Actions*. Activities undertaken during the project.

References

APA (American Psychological Association), 2010. Publication Manual of the American Psychological Association, sixth ed. American Psychological Association. https://www.apa.org/pubs/books/4200066. (Accessed 26 December 2019).

Bossuyt, P.M., 2019. Laboratory measurement's contribution to the replication and application crisis in clinical research. Clinical Chemistry 65 (12), 1479–1480. Available from: https://doi.org/10.1373/clinchem.2019.311605.

Bruwer, J.-P., 2016. The Difference Between a Research Topic, Title and Research Problem (Revised). https://www.academia.edu/26977481/The_difference_between_a_research_topic_title_and_research_problem_revised. (Accessed 17 January 2020).

Chadha, D., 2006. A curriculum model for transferable skills development. Engineering Education: A Journal of the Higher Education Academy 1 (1).

Copeland, S., Penman, A., 2004. The development and promotion of electronic theses and dissertations (ETDs) within the UK. New Review of Information Networking 10 (1), 19–32.

Cziko, G.A., 1989. Unpredictability and indeterminism in human behavior: arguments and implications for educational research. Educational Researcher 18 (3), 17–25.

Davies, T.C., 2013. Geochemical variables as plausible aetiological cofactors in the incidence of some common environmental diseases in Africa. Journal of African Earth Sciences 79, 24−49. Available from: https://doi.org/10.1016/j.Elsevier.

Davies, T.C., 2019. A medical geology curriculum for African geoscience institutions. Scientific African 6 (3), e00131. Available from: https://doi.org/10.1016/j.sciaf.2019.e00131 (Accessed 08.02.24).

Davies, T.C., Schlüter, T., 2000. Current status of research in geomedicine in East and Southern Africa. Journal of Environmental Geochemistry and Health 24 (2), 99−102.

De Mutis, A., Kitchen, S., 2016. African Digital Research Repositories. Survey Report. Africa Bibliography, 2015, VII−XXV. Available from: https://doi.org/10.1017/S0266673116000027, https://www.internationala-fricaninstitute.org/repositories.phtml. (Accessed 14 January 2020).

Enago Academy, 2019. Thesis vs. Dissertation. https://www.enago.com/academy/thesis-vs-dissertation/. (Accessed 31 December 2019).

Erickson, D., 2015. Effective Proposal Writing: Making Your Words Impact Reviewers. Evergent Technologies. samples.leanpub.com > effectiveproposalwriting-sample (Accessed 10 January 2020).

Gahan C., 2019. Dissertation Acknowledgements. https://www.scribbr.com/dissertation/acknowledgements/. (Accessed 14 January 2020).

Hariri, M.M., 2016. Importance of methods of selection in geosciences studies and exploration. International Journal of Earth and Environmental Sciences 1 (114), 1:IJEES-114. Available from: https://doi.org/10.15344/2456-351X/2016/114.

Hash, W.R., Hash, F., 2013. Award Winning Grant Proposal Writing. Lulu.com; Amazon.com. https://books.google.com.ng/books?id = CFeCDwAAQBAJ&pg = PA2&lpg = PA2&dq = Hash + and + Hash. (Accessed 10 January 2020).

Hayton, J., 2012. How To Choose a Thesis Topic. https://jameshaytonphd.com/quick-tips/how-to-choose-a-thesis-topic. (Accessed 16 January 2020).

Hedge, M.N., Salvatore, A.P., 2020. Clinical Research in Communication Disorders, fourth ed. Plural Publishing, San Diego, CA.

Heriot-Watt University, 2013. Post Graduate Programme Handbook. Academic Year 2013−2014. School of Mathematical and Computer Sciences (Dubai Campus). https://studylib.net/doc/9717936/mathematical-and-computer-sciences---heriot (Accessed 08 February 2024).

Hon, L.C., Kent, K., Edwardson, M., 2007−2008. Guidelines for Writing a Thesis or Dissertation. https://www.jou.ufl.edu/grad/forms/Guidelines-for-writing-thesis-or-dissertation.pdf. (Accessed 14 January 2020).

IAI (International African Institute), 2020. African Digital Research Repositories. https://www.internationala-fricaninstitute.org/repositories.phtml. (Accessed 14 January 2020).

In, J., 2017. Introduction of a pilot study. Korean Journal of Anesthesiology 70 (6), 601−605. Available from: https://doi.org/10.4097/kjae.2017.70.6.601, https://www.ncbi.nlm.nih.gov/pmc/articles/PMC5716817/ (Accessed 13.01.2020).

Kerlinger, F.N., 1986. Foundations of Behavioural Research, third ed. Holt, Reinhart and Winston, New York.

Korrapati, R., 2016. Five Chapter Model for Research Thesis Writing: 108 Practical Lessons. Diamond Pocket Books Pvt. Ltd.

Leshan, D., (n.d.). Strategic Communication. https://books.google.com.ng/books?id = MT4syoIKn-. (Accessed 14 January 2020).

Mendeley (Elsevier), 2019a. Harvard Format Citation Guide. https://www.mendeley.com/guides/harvard-cita-tion-guide. (Accessed 09 January 2020).

Mendeley (Elsevier), 2019b. APA Format Citation Guide. https://www.mendeley.com/guides/apa-citation-guide. (Accessed 09 January 2020).

Merriam-Webster Online, 2020. Risk. http://www.merriam-webster.com/dictionary/Risk. (Accessed 17 January 2020).

Paltridge, B., Starfield, S., 2016. English for Specific Purposes. Ebook. Routledge and CRC Press. Available from: https://doi.org/10.4324/9781315716893.ch5.

Pampel, H., 2023. Promoting Open Access in Research-Performing Organizations: Spheres of Activity, Challenges, and Future Action Areas. Publications 11 (3), 44. Available from: https://doi.org/10.3390/publications11030044 (Accessed 08.02.24).

Perera, B.J.C., 2019. Replication in medical research: a modern day enigma. Sri Lanka Journal of Child Health 48 (2), 97−99. Available from: https://doi.org/10.4038/sljch.v48i2.8699.

Phillips, J., 2003. PMP Project Management Professional Study Guide. McGraw-Hill Professional.

PMI (Project Management Institute Inc.), 2008. A Guide to the Project Management Body of Knowledge (PMBOK Guide), fourth ed. Project Management Institute, Inc., Newtown Square, PA.

Poock, M.C., Love, P.G., 2001. Factors influencing the program choice of doctoral students in higher education administration. NASPA Journal 38 (2), 203−223.

Rasuli, B., Solaimani, S., Alipour-Hafezi, M., 2019. Electronic theses and dissertations programs: a review of the critical success factors. College and Research Libraries 80 (1), 60−75.

Sharp, J.A., Peters, J., Howard, K., 2012. The Management of a Student Research Project, third ed. Gower Ebook Publishing, p. 278.

Sileyew, K.J., 2019. Research Design and Methodology [Online First]. IntechOpen. Available from: https://doi.org/10.5772/intechopen.85731, https://www.intechopen.com/online-first/research-design-and-methodology. (Accessed 06 January 2020).

Strunk, W., 2005. The Elements of Style: A Style Guide for Writers. http://academic.csuohio.edu/simond/courses/elos3.pdf.

University of Birmingham, 2020. Harvard: Reference List and Bibliography. https://intranet.birmingham.ac.uk/as/libraryservices/library/referencing/icite/harvard/referencelist.aspx. (Accessed 14 January 2020).

University of British Columbia, n.d. Getting Started With Your Thesis or Dissertation. https://www.grad.ubc.ca/current-students/professional-development/getting-started-your-thesis-or-dissertation. (Accessed 15 January 2020).

University of Leicester, n.d. Student Learning Development Division: Planning and conducting a dissertation research project. https://www2.le.ac.uk/offices/ld/resources/writing/writing-resources/planning-dissertation. (Accessed 01 January 2020).

University of Leicester, 2019. Writing Skills: Pilot Studies. https://www.le.ac.uk/oerresources/ssds/writingskills/page_87.htm. (Accessed 13 January 2020).

University of Minnesota, 2000. The Thesis Proposal: Designing Curriculum. http://www.umn.edu > ~kgilbert. (Accessed 16 January 2020).

University of Sussex, 2020. Final Year Project Information for Students. http://www.sussex.ac.uk/ei/internal/for-students/informatics/undergraduate/finalyearprojects/informationforstudents. (Accessed 12 January 2020).

University of Warwick, 2012. Guidelines on Industrial Projects. https://warwick.ac.uk > sci > wmg > tas > mscproject > pt_project_guide_-_tas. (Accessed 14 January 2020).

Ur-Rehman, A., 2009. Geomedical engineering: a new and captivating prospect. Medical Geology Newsletter 14, 22−23. http://www.medicalgeology.org/. (Accessed 14 January 2020).

Victoria University, 2019. IEEE Referencing: Getting Started with IEEE Referencing. http://libraryguides.vu.edu.au/ieeereferencing/gettingstarted. (Accessed 09 January 2020).

Yu, C., 2018. Threats to Validity of Research Design. https://www.creative-wisdom.com/teaching/WBI/threat.shtml. (Accessed 13 Januray 2020).

Geoenvironmental variables as causal cofactors of diseases of unknown etiology

"The power of Medical Geology lies in its potential to synthesize knowledge from multiple scientific disciplines (epidemiology, geochemistry, immunology, public health, toxicology, virology, and so on) and unify them in a single place, to provide a more circumspective approach to unraveling environmental disease etiology and therapy."

Theo Davies (2020)nd

Key chapter features

1. Geoenvironmental risk factors associated with diseases of unknown etiology.
2. Familiarity with tools for investigating the causes of diseases of unknown etiology and other enigmatic disease outbreaks.
3. Interrelationship of trace elements/metals/metalloids with the immune system, and how these interactions may produce toxicity and the risk of autoimmune disease.
4. Climate change, the immune system, and infectious disease.
5. Perinatal vulnerability and developmental immunotoxicity.
6. Disease clusters and environmental risk maps.
7. Further research—new challenges.

(Box 3−1)

BOX 3−1   Learning objectives.

1. Acquire the skills necessary to bring greater awareness to the medical profession, relevant government officials, and other stakeholders, the huge potential contribution of Medical Geologists in teams investigating the causes of diseases of unknown etiology and sudden disease outbreaks.
2. Understand the relationship of certain trace elements/metals/metalloids with the immune system.
3. Ability to present examples of current evidence of associations between environmental risk factors, such as climate change, and immune system impairment.
4. Familiarity with the most recent research literature supporting the suggestion that consideration of geoenvironmental factors would enhance efforts in unraveling diseases of unknown etiology.

Medical Geology of Africa. DOI: https://doi.org/10.1016/B978-0-12-818748-7.00013-7

Introduction

Despite the enormous advances made by modern medicine, there are many diseases whose causes are still unknown. Mao et al. (2020) estimate that 76% of unknown disease outbreaks remain undiagnosed. In Africa, the number of diseases of unknown etiology reported in the literature is not staggering, but this may well be a result, either of under-reporting or of misdiagnosis (See, e.g., Badoe et al., 2011; Morar and Feldman, 2015a; Morar and Feldman, 2015b).

According to Rappaport (2012), "Although chronic diseases are primarily environmental (i.e., not genetic) in origin, the particular environmental causes of these diseases are poorly understood." A recent WHO study of worldwide cancer mortality identified ten diverse environmental risk factors, including some with links to the geological environment, such as air pollution and ionizing and ultraviolet radiation (UVR) (WHO (World Health Organisation), 2018a). However, the joint effect of these ten factors accounted for only a small proportion of cancer mortality, indicating that the majority of cancer types are of unknown etiology. Rappaport (2012) provides some reasons why the precise unravelment of the environmental causes of many categories of cancer remains problematic.

The contribution of the geoenvironmental factor in a multifactor explanation of diseases of unknown etiology is emphasized in this Chapter, as well as developmental frameworks and developmental pathways in disease progression. The expectation is that a firm understanding of the possible role of geoenvironmental cofactors will help greatly in unraveling the etiology of diseases of unknown etiology when this knowledge is applied in a circumspective way.

Such knowledge is within the compass of the Medical Geologist who can provide "… a more holistic understanding of the occurrence, mobility, bioavailability, bioaccessibility, exposure, and transfer mechanisms of *geochemicals* to the food-chain and humans, and the related ecotoxicological impacts and health effects." (Bundschuh et al., 2017).

An understanding of the source of the antigen/pathogen, including trace elements/metals/metalloids and their migration pathways into the body, is required in order for credible diagnosis to be made, as well as to guide appropriate management. These considerations underline the need for the Medical Geologist's inclusion in a multidisciplinary team in such investigations, for optimum research and management strategies.

Such an approach is integrative, and recognizes that most diseases of unknown etiology's and other enigmatic diseases have multifactorial causes, engendering a complex interplay between genetic factors (polygenic), immunological mediators (trace elements/metals/metalloids) and various (geo)environmental factors, none of which factors would cause the disease on its own.

This approach is buttressed by Panelli's (2017) statement "The focus on multisystem diseases of unknown cause highlights not only the immediate need in medicine to research the origin of an unknown disease with multiple symptoms, but also the necessity to find novel ways to diagnose and discriminate diseases that have no clear etiology but represent a dramatic reality affecting large populations of patients."

A note on causality

In the field of medicine, *cause*, also sometimes referred to as *etiology* is the reason or origination of a disease, or of a *pathology* (essential nature of disease) (Ross, 2018). The word "etiology" is derived from the Greek αἰτιολογία, *aitiologia*, "giving a reason for" (αἰτία, *aitia*, "cause"; and -λογία, - *logia*) (Simpson and Weiner, 2002).

The search for the etiologies of human diseases goes back to antiquity. Hippocrates, a Greek physician of the fourth and fifth centuries BCE, is credited with being the first to adopt the concept that disease is not a visitation of the gods but rather is caused by earthly influences (Britannica, 2020). Medieval European doctors were generally of the view that disease was related to the *air* and adopted a miasmatic approach to disease etiology (Berry, 2007). Scientists have since continually searched for the causes of disease and, indeed, have discovered the causes of many. Where no etiology can be ascertained, the disorder is said to be *idiopathic* (CMD (Concise Medical Dictionary), 2010).

Traditional accounts of the causes of disease may point to the *evil eye*, a phenomenon elucidated by Abu-Rabia in Abu-Rabia (2005), in describing the rituals of diagnosis, treatment, and prevention among the Bedouin tribes of the Negev in the Middle East.

Etiological discovery in medicine also has a history in Robert Koch's demonstration that the tubercle bacillus (*Mycobacterium tuberculosis* complex) causes the disease tuberculosis (TB), *Bacillus anthracis* causes anthrax, and *Vibrio cholerae* causes cholera (Cambau and Drancourt, 2014). This line of thinking and evidence is summarized in Koch's postulates. In the epidemiology of infectious diseases, proof of causation is limited to individual cases where experimental evidence of etiology can be advanced. To infer causation, we require several lines of evidence, taken together.

The *damage-response framework* (See "Glossary of Terms," in this Chapter) of microbial *pathogenesis* is intriguing. Several questions regarding certain individuals' susceptibility to epidemics are difficult to answer—questions, such as 'Why do some individuals become sick'? 'Why do some even die, whereas others recover from illness and still others show no signs or symptoms of disease'? (See Casadevall and Pirofski, 2018). These authors have examined the attributes of susceptibility, and identified 11 attributes that, although not independent, are sufficiently distinct to be considered separately. These are microbiome, inoculum, sex, temperature, environment, age, chance, history, immunity, nutrition, and genetics. *Combinatorial diversity* (See "Glossary of Terms," in this Chapter) among the 11 attributes makes identifying "signatures" of susceptibility possible. In these discussions, Casadevall and Pirofski (2018) conclude that "However, with their inevitable uncertainties and propensity to change, there may still be a low likelihood for prediction with regard to individual host-microbe interactions, although probabilistic prediction may be possible."

Chain of causation and correlation

We need to distinguish between *causation* and *association* or *statistical correlation*. Events may occur simultaneously simply due to *chance, bias*, or *confounding* (See: "Glossary of

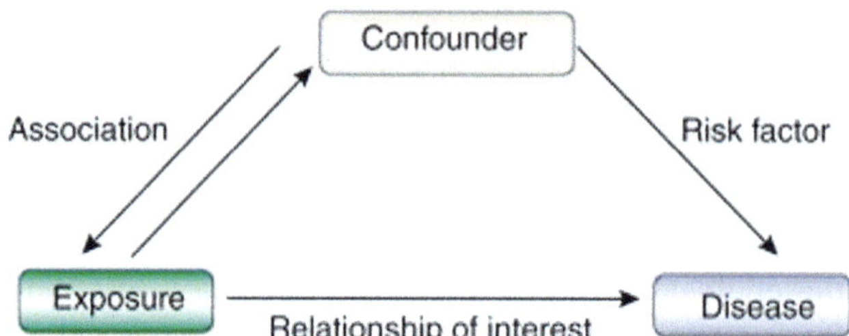

FIGURE 3–1 The structure of confounding (Jager et al., 2008). *From Jager, K.J., Zoccali, C., Macleod, A., Dekker, F. W., 2008. Confounding: what it is and how to deal with it. Kidney International 73 (3), 256–260. https://doi.org/ 10.1038/sj.ki.5002650.*

Terms," in this Chapter, for definitions), instead of one event being precipitated by the other. It is also necessary to decipher which event is the cause. Confounding is said to occur when exposure to a probable disease causative agent or cofactor and an outcome have an apparent but false correlation (Fig. 3–1). It is important to control for the confounder, otherwise, it would appear that there is a link between the exposure and the outcome, when in fact both are due to the confounding effect and bear no relationship at all (or as strongly) (Schuster et al., 2023). Careful sampling and analyses should be the *sine qua non*, rather than complex statistical analysis to establish causation. Evidence from experimental studies involving interventions (providing or removing the supposed cause) provides the most convincing evidence of etiology.

It is also necessary to state that there are times when several symptoms appear together, sometimes more than could be expected, though it is known that one cannot cause the other. These situations are referred to as *syndromes* (See "Glossary of Terms," in this Chapter). The assumption is that an underlying condition exists that explains all the symptoms. Quite often, however, a single cause for a disease cannot be found, but rather, we find a *chain of causation* from an initial trigger to the development of the clinical disease. An etiological agent of disease may require an independent cofactor and be subject to a *promoter* (See "Glossary of Terms," in this Chapter) to cause disease.

Besides the mere exposure to a causative agent, there are a number of other host factors that determine whether the agent will induce disease or not. In the pathogenesis of disease, therefore, factors, such as the resistance, immunity, age, and nutritional state of the person exposed, in addition to the virulence or toxicity of the agent and the level of exposure, all play a role in determining whether disease develops. Many chronic diseases of unknown etiology may be studied within this framework to explain multiple epidemiological associations or risk factors, which may or may not be causally related; and to seek the actual etiology.

This corroborates well with Stein et al.'s (2008) observation of 2008, *viz.*, that: "To be more fully understood, causal webs of disease must ultimately be considered not only as a collection of individual strands, but also in their integrated complexity." *Complex humanitarian emergencies (CHEs)* often involve the unravelment of the causes of communicable disease outbreaks. Descriptions of *risk factors* for such outbreaks are often nonspecific and not easily generalizable to similar situations (Hammer et al., 2018).

Understanding health risks

The terms *health risk* and *health hazard* are commonly used to describe aspects of the potential harm to health; but are often used incorrectly. According to HSA (2017), the term *risk* is the "likelihood that a person may be harmed or suffers adverse health effects if exposed to a *hazard*." The level of risk is often classified according to the potential harm or adverse health effect that the hazard may cause, how many times a person is exposed, and the number of persons exposed (HSA, 2017).

Risk factors

The term *risk factor*, coined by Kannel et al. (1961) is a variable associated with increased risk of disease or infection, or any attribute, characteristic or exposure of an individual that increases the likelihood of developing a disease or injury (Parritz and Troy, 2013; Sabel et al., 2007; Yu et al., 2023; WHO (World Health Organisation), 2017). Risk factors can encompass an attribute, such as family history, age, sex, or ethnicity, an exposure to an environmental variable, or the presence of a prior health condition. In the framework of *HRA*, (See Section: "Health risk assessment," in this Chapter), health risk factors, and their main parameters in *built environments* (See Section: "Health hazards in the built environment," in this Chapter) were recently identified and classified by Dovjak and Kukec in 2019.

Understanding our risk factors for different diseases constitutes part of the knowledge we need for taking charge of our health. There are certain risk factors that we have little or no control over. Examples of risk factors we may be born with or exposed to, through no fault of our own, include: family history of a disease, sex/gender (male or female), and ancestry or ethnicity.

A list of risk factors you can control includes:

- What you eat
- How much physical activity you get
- Whether you are a smoker
- Quantity of alcohol you consume
- Whether you use hard drugs
- Whether seat belt is worn when driving.

Causal attribution of risk factors

A key for preventing disease and injury engenders a reliable and comparable analysis of risks to health (Ezzati et al., 2002) and perhaps, also, for narrowing down on accurate diagnosis. Indeed, substantial proportions of global disease burden are attributable to the major risk factors, even to an extent greater than previously estimated (Ezzati et al., 2002).

Quite often, use is made of statistical methods to assess the strength of an association and to provide causal evidence (for example, in assessing the link between smoking and lung cancer). Statistical analysis used in conjunction with the biological sciences can establish that risk factors can indeed be causal. Analyzing the 'Second Comparative Risk Assessment for South Africa

(SACRA2)', for instance, Bradshaw et al. (2022) noted the value of quantifying the contribution of modifiable risk factors for identification and prioritization of areas of concern for population health and opportunities for health promotion and disease prevention interventions.

Because correlation does not prove causation, risk factors, or determinants, being *correlational*, cannot necessarily be *causal*. Many diseases are caused by multiple risk factors, and individual risk factors may interact in their impact on the *overall risk of disease*. As a result, attributable fractions of deaths and burden for individual risk factors usually overlap and often add up to more than 100% (WHO (World Health Organisation), 2009).

In almost all listings of major risk factors responsible for the global and regional burden of disease, unsafe water and micronutrient deficiency (e.g., Ezzati et al., 2002; Lopez et al., 2006) as well as the place factor (geographical location) (e.g., Cluver, 2019; Elliott, 2014; Niohuru, 2023; Oh et al., 2018; Toms et al., 2019) and environmental toxins, all of which can be considered to have links to geoenvironmental variables, are always mentioned.

In their work on disease causation, Ezzati et al. (2002) estimated the contributions of selected major risk factors to global and regional burden of disease in a unified framework and noted that African countries and other developing nations suffer most from the disease burden due to pervasion of many of the leading risks. According to Ezzati et al. (2002), "Strategies that target these known risks can provide substantial and underestimated public-health gains."

Determinants of health

The term *determinant* is sometimes used as a synonym for "risk factor," due to a lack of harmonization across disciplines. In the context of community health policy, Durch et al. (1997) define a "determinant" as:

"... a health risk that is general, abstract or pertains to inequalities and is difficult for an individual to control ..."

An interplay between multiple associated variables is what usually determines the probability of an outcome. In epidemiological studies to evaluate one or more determinants for a specific outcome, the other determinants may act as *confounding factors* (See Section: "Chain of causation and correlation," in this Chapter), and need to be controlled for, e.g., by stratification. Among the determinants that are most commonly encountered in epidemiological associations are age, sex or gender (male or female), and ethnicity (race); these are also the determinants that are most commonly controlled for confounders (See: Blumenthal et al., 2001; Filardo et al., 2011; Bovbjerg, 2019). Other determinants (less commonly adjusted for possible confounders) include genetic predisposition, social status/income, geographic location, occupation, level of chronic stress, diet, level of physical exercise, alcohol consumption, and tobacco smoking.

Of the determinants that are less commonly adjusted for possible confounders, we shall look briefly at "genetic predisposition."

Role of genetics as causal cofactor

A *gene* is the basic physical unit of heredity. Genes are made up of deoxyribonucleic acid (*DNA*) and act as instructions to synthesize molecules called *proteins*. Many proteins are actually *enzymes*, and are responsible for carrying out all cellular functions. Salzberg (2018) estimated the number of genes in the *human genome* (genetic complement) to be 20,000−25,000. Genes are passed on from parents to offspring and contain the information needed to specify traits.

Certain human diseases result from mutations in the genetic complement contained in the DNA of chromosomes. More than 5000 distinct diseases have been ascribed to *mutations* that result in deficiencies of some very important enzymes (Griffiths, 2020). Genetic diseases that are the result of a mutation in one gene are inherited in either a *dominant* or *recessive* fashion. In the former case (dominant), only one mutant *allele* (See: "Glossary of Terms," in this Chapter), which codes for a defective protein or does not produce a protein at all, is required for the disorder to occur. In *recessively inherited disorders*, two copies of a mutant gene must be present for the disorder to appear; if only one copy is inherited, the progeny is unaffected, but the malady may continue to be passed on to future offspring. In addition to dominant or recessive transmission, genetic disorders may be inherited in an *autosomal* or *X-linked manner* (See: "Glossary of Terms," in this Chapter).

Although mutations occurring in the DNA of somatic (body) cells cannot be inherited, they can cause *congenital malformations* (existing at birth) and cancers. However, mutations that occur in germ cells—i.e., the *gametes*, ova and sperm—are passed on to offspring causing inherited diseases.

Most genetic disorders can be discerned at birth, since the child is born with characteristic defects. Thus these abnormalities are congenital genetic disorders. A few genetic defects, however, do not appear until later in life. It could thus be said that most, but not all genetic diseases, are congenital. Conversely, not all congenital diseases are genetic in origin. For instance, a congenital disease may arise from some direct injury to the developing fetus.

Another clarification needs to be made regarding use of the terms *genetic* and *familial*. A *familial disease* is an inherited disease, passed on from one generation to the next. It is the result of a genetic mutation that is passed on by mother or father (or both) through the gametes to their offspring. Not all genetic diseases are familial, however, because the mutation may arise for the first time during the formation of the gametes or during the early development of the fetus. Such an infant will have some genetic malformation, though the parents themselves do not. *Down syndrome* is an example of a genetic disease that is not familial (Hernandez and Fisher, 1996), with the role of an environmental agent considered realistic by some (See, e.g., Sperling et al., 2023). In another example, the case of *Kawasaki Disease* (KD), ethnic-specific rates have revealed very striking differences, which Rowley and Shulman (2018) consider to be indicative of a very strong genetic basis of susceptibility.

Studies on how environmental exposures modify the expression of genes without directly changing the genetic code stored in DNA belong to the field of environmental epigenetics, that is being researched by US NIEHS (2019), and was appraised by Rappaport (2016) and more recently by Perera et al. (2019), Skinner (2023) and King et al. (2024).

Health hazards in the built environment

The *built environment* refers to the spaces in which we live, work and play; and it affects us through associated land use strategies, natural resource consumption and patterns of waste disposal. Health hazards frequently occur in all built environments.

WHO (World Health Organization's) (2017) identification of health risk factors from built environments includes genetics, income, education level, and relationships with friends and family, with all of these factors considered to have marked impacts on health. Other more commonly considered factors, such as access and use of health care services are often considered to have less of an impact on health. WHO's 2017 tabular categorization of determinants of health in built environments (Table 3–1) does not specifically include *geoenvironmental variables*, but it could be inferred that this is covered in the "physical environment" category of Table 3–1.

Geoenvironmental hazards in the built environment may be in the form of chemical hazards (e.g., volcanic gas escape) or physical hazards (e.g., exposure to ionizing radiation). Identification of health hazards and risk factors in built environments is a key activity for effective control and prevention of illness, and is the first step in a process called *HRA* (See Section: "Health risk assessment," in this Chapter). Indeed, as noted by Dovjak and Kukec in 2019, proper identification and classification of health risk factors and their parameters constitute the basis for the identification of single and multigroup interactions among them. The proper identification and characterization of risk factors are also important because strategies that target these risk factors can provide substantial and underestimated public-health gains (See, e.g., Ezzati et al., 2002; Briggs, 2003) including improvements in diagnosis and therapy of many enigmatic diseases, such as diseases of unknown etiologies (Table 3–2).

Table 3–1   Determinants of health in built environments.

Determinants of health	Links to health
Physical environment	Safe water and clean air, healthy workplaces, safe houses, communities, and roads all contribute to good health
Income and social status	Higher income and social status are linked to better health. The greater the gap between the richest and poorest people, the greater the differences in health
Education	Low education levels are linked to poor health, more stress, and lower self-confidence
Employment and working conditions	People with employment are healthier, particularly those who have more control over their working conditions
Social support networks	Greater support from families, friends and communities is linked to better health
Personal behavior and coping skills	Eating habits, level of physical activity, smoking, drinking, and how we deal with life's stresses and challenges all affect health
Culture	Customs, traditions, and the beliefs of the family and community all affect health
Genetics	Inheritance plays a part in determining lifespan, healthiness, and the likelihood of developing certain illnesses
Gender	Men and women suffer from different types of diseases at different ages
Health services	Access to, and use of services that prevent and treat disease influence health

Source: WHO (World Health Organisation), 2017. Health Impact Assessment (HIA). The Determinants of Health. http://www.who.int/hia/evidence/doh/en/. (Accessed 04 December 2020).

Table 3–2 Probable geoenvironmental co-factor(s) of diseases of unknown etiology.

Diseases of unknown etiology	Presentation	Geographical/ demographical patterns/ distribution/incidence	Geochemical (including trace element-mediated immune response)	Climatic or seasonal variations	Remarks
Acrocyanosis	*Acrocyanosis* is a peripheral vascular disorder which presents as a persistent bluish or cyanotic discoloration of the extremities, most commonly occurring in the hands	Geographical locale (latitude; urban versus rural setting) uncertain—Kurklinsky et al. (2011)	Chronic As toxicity—Tseng (1977); Mak (1988)	Cold climate—Crocq (1896); Carpentier (1998); Cold environments—Das and Maiti (2013); Kent and Carr (2020)	"Diagnosis remains mostly clinical, and pathological mechanisms vary, suggesting that acrocyanosis may not be a single entity"—Kurklinsky et al. (2011)
Acute febrile illness	*Acute febrile illness* (AFI) is characterized by malaise, myalgia (pain in muscle or group of muscles) and a raised temperature that is a nonspecific manifestation of infectious diseases in the tropics (Bhaskaran et al., 2019)	Sub-Saharan Africa; tropics and Subtropics	Viral respiratory tract infections—Crump et al. (2013); D'Acremont et al. (2014); Fe, Zn, Mg, and Ca not significantly related to febrile convulsion—Amouian et al. (2013)	Clear seasonal trend observed—Kaboré et al. (2020)	–
Acute severe asthma	*Asthma* is characterized by chronic airway inflammation, resulting in periodic wheeze, cough, and breathlessness (See Carlsson and Bayes, 2020)	Worldwide	Respiratory tract infections—Johnston et al. (1995); Jackson and Johnston (2010); Air pollution—Guarnieri and Balmes (2014); Shmool et al. (2014)	Cold weather—Hyrkäs et al. (2016); Weather changes—D'Amato et al. (2018)	The mechanisms by which these environmental stimuli and viruses initiate asthma or cause worsening of the disease are still unknown—Kostakou et al. (2019)

(Continued)

Table 3–2 (Continued)

Diseases of unknown etiology	Presentation	Geographical/ demographical patterns/ distribution/incidence	Geochemical (including trace element-mediated immune response)	Climatic or seasonal variations	Remarks
Alzheimer's disease (AD)	*Alzheimer's disease* (AD) is the most common neurodegenerative disorder and the leading cause of dementia (i.e., the particular group of symptoms shown). It becomes worse with time (degenerative). The symptoms expressed are as a result of the damage or destruction of nerve cells (neurons) in parts of the brain involved in thinking, learning and memory (cognitive function)	By 2009, the global prevalence of dementia was estimated at 3.9% in people aged 60 + years, with the regional prevalence being 1.6% in Africa, 4.0% in China, and Western Pacific regions, 4.6% in Latin America, 5.4% in Western Europe, and 6.4% in North America (Qiu et al., 2009)	Significantly different ($P < .05$) mean concentrations of Br, Cl, Ce, Hg, N, Na, P, and Rb were observed in AD bulk brain samples compared to controls—Ehmann et al. (1986); Varying trace element relationships with AD severity, with aluminum deposits greater in severely affected AD brain—Jagannatha Rao et al. (1999); Loef and Walach (2012); Bagheri et al. (2018); Liu et al. (2019); Mocanu et al. (2020) Anomalous concentration levels of metals in metal-binding proteins have growth inhibition functions on neurons—Constantinidis (1991); Richarz and Brätter (2002); Trace metals and abnormal metal metabolism influence protein aggregation, synaptic signaling pathways, mitochondrial function, oxidative stress levels, and inflammation, ultimately resulting in synapse dysfunction and neuronal loss in the AD brain—De Benedictis et al. (2019); Wang et al. (2020)	Long-term exposure to O_3 and $PM_{2.5}$ above the current US EPA standards are associated with increased the risk of AD —Jung et al. (2015); Association between Alzheimer Dementia mortality rate and altitude —Thielke et al. (2015); Hu et al. (2016); Lall et al. (2019); Koester-Hegmann et al. (2019); Associations with seasonal temperature—Wie et al. (2019); Global warming and neurodegenerative disorders—Habibi et al. (2014)	In 2019 Alzheimer's Disease International (ADI) estimates that there are over 50 million people living with dementia globally, a figure set to increase to 152 million by 2050—ADI (Alzheimer's Disease International) (2019). In both developed and developing nations, Alzheimer's disease has had tremendous impact on the affected individuals, caregivers, and society. Because developing countries are projected to see the largest increase in absolute numbers of older persons, their share of the worldwide aging population will increase from 59% to 71%—Qiu et al. (2009). "Because occurrence of AD is strongly associated with increasing age, it is anticipated that this dementing disorder will pose huge challenges to public health and elderly care systems in all countries across the world"—Qiu et al. (2009)

| Chronic fatigue syndrome/*myalgic encephalomyelitis* (CFS/ME) | A disabling, debilitating and complex disease characterized by profound fatigue, sleep abnormalities, pain and other symptoms that are worsened by exertion | Worldwide prevalence (Panelli, 2017; Lim et al., 2020), with 17–24 million people having the disease (EME CFS, 2020) | Metal hypersensitivity—Manousek et al. (2014); Some nutrient deficiencies (vitamin C, vitamin B complex, Na, Mg, Zn, folic acid, L-carnitine, L-tryptophan, essential fatty acids, and coenzyme Q10) appear to be important in the severity and exacerbation of CFS Bjørklund et al. (2019a); Ca associated with some of the neurological findings described in ME/CFS—Pacini et al. (2012); Insufficient Ca inflow into cells that perform intracellular functions—Nguyen et al. (2017) | — | "Researchers have not yet found what causes ME/CFS, and there are no specific laboratory tests to diagnose ME/CFS directly"—US CDC (2018a). It cannot be fully explained by an underlying medical condition. "There is a need for more epidemiologic studies on the prevalence and sociodemographic characteristics of CFS in developing countries."—Njoku et al. (2007). Sierpina and Carter (2002) suggested that 200 mcg of *chromium picolinate* (taken with meals) may have the potential to reduce any reactive hypoglycemia that may aggravate symptoms |
| Chronic kidney disease of unknown etiology (CKDu) | The predominant feature of CKD is tubular atrophy and interstitial fibrosis (thickening and scarring of the tiny air sacs and interstitial tissues in the lungs) | Reported in many parts of the world, especially among rural farming communities. High incidence in low- and middle-income countries over last two decades—Mills et al. (2015); Bikbov et al. (2020); "In 2017 the global prevalence of chronic kidney disease | Synergistic reaction between Cd and diabetic-related hyperglycemia—Edwards and Prozialeck (2009); Consumption of (polluted) well water suggested; need for investigating role of Cd—Wanigasuriya et al. (2011); Too high Ca intake?—Schiffrin et al. (2007); "Geographical mapping | Altitude—Almaguer et al. (2014); Heat stress nephropathy due to global warming—Glaser et al. (2016); Sorensen and Garcia-Trabanino (2019); Climatic patterns—Wimalawansa et al. (2020); A quintessential climate-sensitive disease—Salas et al. (2019) | "There have been several global epidemics of CKDu. Some, such as Itai-Itai disease in Japan and Balkan endemic nephropathy, have been explained, whereas the etiology of others remains unclear"—Gifford et al. (2017). "Those affected do not |

(Continued)

Table 3–2 (Continued)

Diseases of unknown etiology	Presentation	Geographical/ demographical patterns/ distribution/incidence	Geochemical (including trace element-mediated immune response)	Climatic or seasonal variations	Remarks
		(CKD) was 9.1% (95% uncertainty interval [UI] 8.5–9.8), which is roughly 700 million cases"—Cockwell and Fisher (2020)	showed that villages with a high prevalence of CKDu are often related to irrigation water sources and/or located below the level of the water table" (Jayasekara et al., 2013); Toxins/heavy metals—Jha et al. (2013); Groundwater geochemistry (high levels of F^-, Cd, arsenic)—Wijetunge et al. (2015); Exposure to low levels of Cd—Redmon et al. (2014); High ionicity of drinking water due to fertilizer runoff—Dharma-Wardana et al. (2015); Rajapakse et al. (2016); Toxic metal exposure; water pollution—Lunyera et al. (2016); Synergistic reaction between fluoride and water hardness—Sengupta (2013); Wickramarathna et al. (2017); Balasooriya et al. (2019); Chemical species, such as Ca, phosphate, oxalate, and F^- form intrarenal nanomineral particles initiating the CKD of multifactorial origin (CKDmfo)—Wimalawansa and Dissanayake (2019); Total dissolved solids and As in drinking water have a positive correlation with CKDu—Gobalarajah et al. (2020)		have common risk factors or underlying conditions that lead to CKD, such as diabetes, immune-mediated glomerulonephritis, or structural renal disease"—Caplin et al. (2019)

| Geographic tongue | Geographic tongue (also known as benign migratory glossitis) is an inflammatory disorder that usually appears in a map-like (geographic) pattern on the dorsum and margins of the tongue. Typically, affected tongues have a bald, red area of varying sizes that is surrounded, at least in part, by an irregular white border | A common condition, affecting 2%–3% of the adult general population, worldwide (Ship et al. 2003; Picciani et al., 2020) | Fe and Zn deficiency; Vitamin B12—Ogueta et al. (2019); Nandini et al. (2016); Picciani et al. (2016); Picciani et al. (2017); Low levels of salivary Zn compared to control groups—Khayamzadeh et al. (2019) | — | — |
| Ill-thrift or unthriftiness as it is called in South Africa | Ill-thrift or unthriftiness is an ill-defined condition of young sheep. Affected animals show mild to severe depression of growth rate and anemia | In South Africa, the disease occurs in the coastal areas of the Eastern Cape Province. The condition has been reported from a number of other countries, including Australia, New Zealand, France, and Norway (Stewart and Burroughs, n.d.) | Cu, Co, Se and I, being essential component of the diet of beef cattle for maintaining their health and productivity, their deficiency in these elements can cause ill-thrift and infertility, among other causes—Crawshaw and Caldow (2005); A state of suboptimal growth (ill-thrift) in buffalo-calves was largely attributed to trace element deficiency, in particular Cu, Co, Fe, Se, and Zn deficiency that may cause reduction in the total antioxidant capacity, with a lower ability to reduce oxidative compounds—Ali et al. (2015); Ismael et al. (2015); MLA (2020) | Mainly reported from coastal areas of high rainfall—[Australia (MLA, 2020); South Africa (Stewart and Burroughs, n.d.)] | — |

(Continued)

Table 3–2 (Continued)

Diseases of unknown etiology	Presentation	Geographical/ demographical patterns/ distribution/incidence	Geochemical (including trace element-mediated immune response)	Climatic or seasonal variations	Remarks
Kawasaki disease (KD)	*Kawasaki syndrome* is an acute, self-limited *vasculitis* (inflamed blood vessels) of infants and children with unknown etiology. Signs of KD include prolonged fever associated with rash, red eyes, mouth, lips and tongue, and swollen hands and feet with peeling skin. The disease causes damage to the coronary arteries in a quarter of untreated children and may lead to serious heart problems in early adulthood	KD occurs worldwide; most prominently in Japan, Korea, and Taiwan, reflecting increased genetic susceptibility among Asian populations (Rowley and Shulman, 2018). The epidemiology of KD in Africa is very ill-defined, which inevitably leads to misdiagnosis and the reporting of very few cases. This gives the impression that the condition is rare in Africa —Badoe et al. (2011); Animasahun et al. (2017); Noorani and Lakhani (2018). The presentation of KD is similar to that of measles (which is very prevalent in Africa), so the exact prevalence (of KD in Africa) is difficult to ascertain See, e.g., Badoe et al. (2011); Davaalkham et al. (2011)	Environmental exposure to Hg —Orlowski and Mercer (1980); Mutter and Yeter (2008); Yeter et al. (2016); Portman et al. (2018); Airborne pathogens or toxins —Rodo et al. (2014); Rodo et al. (2016)	Seasonality of KD, with winter peaks and winter-spring predominance in Japan and the United States, respectively, and in many other temperate areas—Rowley and Shulman (2018); Decades of research have been unable to pinpoint the cause of the disease, but its distinct seasonality can hardly be in doubt— UCSDH (2013); Lin and Wu (2017); Rypdal et al. (2018); Kim (2019); Elakabawi et al. (2020)	i. "… Thus these twin studies suggest that environmental factors contribute more to the development of KD than genetic factors among individuals with the same ethnicity."—Hara et al. (2016). ii. The temporal association between the COVID-19 pandemic and the results of RT-PCR and antibody testing suggest a causal link between Kawasaki disease and COVID-19—Toubiana et al. (2020). (i) As Rowley and Schulman remarked in 2018, "The occurrence of epidemics and geographic wave-like spread of KD during epidemics supports a presently unknown single agent or closely related group of agents as the etiology."

| Systemic *lupus erythematosus* (SLE) | *Systemic lupus erythematosus* is a chronic inflammatory autoimmune disease of multifactorial origin (Pedro et al., 2019) | Worldwide; highest prevalence rate in North America (Rees et al., 2017). Once thought to be of low prevalence rate in Sub-Saharan Africa (due to under-reporting?) SLE prevalence rate is now found to 1.7% (Essouma et al., 2020), as a result of availability of improved diagnostic capacity | Low serum levels of albumin, Zn, Se and Zn/Cu ratio; negative correlation between serum Cu levels and lupus disease activity—Sahebari et al. (2014); SLE patients have different profiles of trace elements and toxic metals compared to healthy controls—Pedro et al. (2019) | "Active SLE has the characteristics of seasonal distribution and is associated with temperature. The mechanism remains to be further studied"—Hua-Li et al. (2011) | The etiology of SLE is complex and incompletely understood (See Justiz Vaillant et al., 2020). "More epidemiological studies in Africa are warranted."—Rees et al. (2017) |
| Multiple sclerosis (Ms) | *Multiple sclerosis* is a demyelinating disease (a nervous system disease in which the insulating covers of nerve cells in the brain and spinal cord are damaged). This damage disrupts the ability of parts of the nervous system to transmit signals, resulting in a range of signs and symptoms, including physical, mental, and sometimes psychiatric problems (Compston and Coles, 2002; Compston and Coles, 2008; Murray et al., 2012). Kister et al. (2013) list 11 specific symptom domains commonly affected in multiple sclerosis: mobility, hand function, vision, fatigue, cognition, bowel/bladder function, sensory, spasticity, pain, depression, and tremor/coordination | Distribution is worldwide. The age-standardized Ms prevalence estimate per 100,000 population for eastern Sub-Saharan Africa is put at 3.3 by WGBD (2019); but such an estimate would always be influenced by misdiagnosis and under-reporting (e.g., Bhigjee et al., 2007) | Metabolic imbalance of trace elements/metals—Rieder et al. (1983); Smith et al. (1989); Melø et al. (2003); Tamburo et al. (2015); Bredholt and Frederiksen (2016); Janghorbani et al. (2017); Sarmadi et al. (2020); the therapeutic potential of antioxidant [reactive oxygen species (ROS)] protection in the pathogenesis of Ms—van Horssen et al. (2011) | People living in higher geographical *latitudes* may receive lower levels of sunlight, and therefore have lower vitamin D levels, which probably explains the reason for a higher incidence of Ms in countries with higher *latitudes*—Ms International Federation (2016); "There is a striking latitudinal gradient in multiple sclerosis (Ms) prevalence …" (Simpson et al., 2011) | — |

(Continued)

Table 3–2 (Continued)

Diseases of unknown etiology	Presentation	Geographical/ demographical patterns/ distribution/incidence	Geochemical (including trace element-mediated immune response)	Climatic or seasonal variations	Remarks
Nodding disease	*Nodding disease* is characterized by an occasional nodding of the head, as in epilepsy, with seizures, stunted growth, and mental retardation sometimes occurring	This is an emerging disease in South Sudan, southern Tanzania, and northern Uganda. The exact prevalence and geographic distribution of the disease in the affected countries is unknown—Korevaar and Visser (2013)	Deficiency of vitamin B6 (pyridoxine) and other micronutrients such as vitamin A, Se, and Zn—GU/WHO (2012)	Climate change—Donnelly (2012). Cold weather—Kaiser et al. (2000); Nyungura et al. (2011); Living in the vicinity of fast-flowing streams, the breeding habitat of the black fly—Donnelly (2012)	"Nutritional toxicity remains a possible cause of nodding syndromes."—Dowell et al. (2013)
Noma	*Noma* (*cancrum oris* or *gangrenous stomatitis*), is a severe and progressive gangrenous infection (body tissues die as a result of infection or inadequate blood supply) that affects the mouth and face	Mainly observed in tropical countries, particularly Sub-Saharan Africa. True global incidence unknown; but estimated incidence of 30,000—40,000 has been suggested by Srour et al. (2017)	Deficiencies of trace elements and amino acids influencing the efficacy of the immune system: Fe, Zn, cysteine, methionine, serine, and glycine—Baratti-Mayer et al. (2003); Malnutrition—Srour et al. (2017); Srour and Baratti-Mayer (2020)	–	Patients generally live in extremely poor conditions, frequently located in remote rural areas
Parkinson's disease (PD)	*Parkinson's disease* (PD) is a progressive heterogeneous, multisystem and neurodegenerative nervous system disorder that affects movement. The cardinal features of Parkinson's disease are (1) tremor, mainly at rest; (2) muscular rigidity, which leads to difficulties in walking, writing, speaking and masking of	Worldwide occurrence. According to the 2016 Dorsey and GBD Collaborators Study published in Dorsey, the 2016 Global Burden of Disease Collaborators (2018), 6.1 million [95% uncertainty interval (UI) 5.0—7.3] individuals had Parkinson's disease globally, compared with 2.5 million (2.0—3.0) in 1990	Elevated trace metals (namely, Cu, Zn, Mg and Fe) found compared to controls ($P < .001$) In Nigerian Parkinson's Disease patients (Ogunrin et al., 2013); Association with metal and trace element concentration in urine, serum, whole blood and cerebrospinal fluid—Bocca et al. (2004); Forte et al. (2004); Gellein et al. (2008); Zhao et al. (2013); Very high or very low levels	Existence of seasonality (related to temperature) in Parkinson's disease symptoms—Rowell et al. (2017); Improvement of PD symptoms at high altitude—Capcha et al. (2018); Thomas (2019)	Regional maps depicting correlations between the distribution of PD and soil geochemistry which would be helpful in this etiological debate, are very rare in the published literature (See an example in: Sun (2018) for PD distribution in the United States)

	facial expression; (3) *bradykinesia*, a slowness in initiating and executing movements; and (4) stooped posture and instability (Sian et al., 1999). Parkinson's disease occurs when nerve cells, or dopamine-rich neurons an area of the brain that controls movement called the *substantia nigra*. Become impaired and/or die. But the complete series of steps leading to this cell death is still vague, and the underlying causes remain one of medicine's greatest mysteries	of Se—Ellwanger et al. (2016); Combination of Mo deficiency and purine ingestion—Bourke (2018); Significant association between the PD mortality rates and soil concentrations of Se, Sr, and Mg—Sun (2018); Adani et al. (2020); Lemelle et al. (2020)			
Sarcoidosis	*Sarcoidosis* is a multisystem, granulomatous, inflammatory disease that affects one or more organs, but most commonly affects the lungs and lymph glands. The inflammation may change the normal structure and possibly the function of the affected organ(s). The causes of sarcoidosis are still unknown and	*Sarcoidosis* is observed throughout the *world* and affects all races and ages, with an average incidence of 16.5 per 100,000 in men and 19 per 100,000 in women (Patil et al., 2015). Sarcoidosis in not a rare condition in Africa (Awotedu et al., 1987; Kaloga et al., 2015), the apparent infrequency of reporting being probably	Exposure to heavy metals and rare earth elements (REEs) in the environment—Newman (1998); Beijer et al. (2020); Denisova et al. (2020); Metal dusts—Judson (2020); Crystalline silica—Arkema and Cozier (2018); Industrial exposure to Be—Culver et al. (2007)	Race and geographical location are considered as factors in the incidence and prevalence of sarcoidosis, which has consistently been observed to be highest in Nordic countries and in African Americans (Arkema and Cozier, 2018). Geographic clustering of disease in many parts of the world has long been noted	"Although the etiology of this condition is unclear, environmental and genetic factors may be substantial in its pathogenesis." *Sic.*—Ganeshan et al. (2018). According to Ahmadzai et al. (2013), in sarcoidosis, conventional sampling techniques or cultures of noncaseating granulomas cannot detect tissue micro-

(Continued)

Table 3–2 (Continued)

Diseases of unknown etiology	Presentation	Geographical/ demographical patterns/ distribution/incidence	Geochemical (including trace element-mediated immune response)	Climatic or seasonal variations	Remarks
	epidemiological data are often discordant (Pietra et al., 1988; Beghè et al., 2017)	a result of misdiagnosis (as, e.g., tuberculosis) (Morar and Feldman, 2015a, b). Further research is therefore necessary in Africa to unravel the various clinical aspects of this mysterious and complex disease		[e.g., in the US (Sartwell and Edwards, 1974)], and has ignited further speculations concerning weather, soil, and foliage in the pathogenesis of sarcoidosis. More recently (2019), Ramos-Casals et al. (2019) have asserted that local weather is a key environmental factor influencing the incidence of sarcoidosis in a specific geographical area, with the peak of diagnosed cases following a specific seasonal distribution pattern	organisms; although as Newman earlier (1998) proposed, clinicians should use a systematic approach to investigating the occupational and environmental history and immunologic responses of patients with sarcoidosis, for discriminating metal-induced granulomatosis from sarcoidosis
Spastic paraparesis	*Hereditary spastic paraplegia (HSP)* is a group of inherited disorders whose main feature is a progressive gait disorder. The disease presents with progressive stiffness (spasticity) and contraction in the lower limbs. HSP is also known as familial spastic paraplegia (US NINDS, 2014); however, *"paraparesis"* indicates weakness in both legs, and is of lesser severity than *"paraplegia"*	Cases of sarcoidosis have been reported worldwide. It affects individuals from diverse ethnic groups. The prevalence of all hereditary spastic paraplegias combined is estimated to be 2–6 in 100,000 people (US NIH, 2019); northern Mozambique (Taibo et al., 2017)	Nutritional disorders, including Cu deficiency, vitamin B_{12} and E deficiencies—Mitchell et al. (1986); Hedera (2016); Hedera (2000)	High intake of HCN^- engendered by drought conditions Cliff et al. (1984)	See Taibo et al. (2017) for report of a real-life outbreak of spastic paraparepsis investigation undertaken in Mozambique in 1981

Sudden infant death syndrome (SIDS)	*Sudden infant death syndrome (SIDS)* is the sudden and unexplained death of a baby younger than 1 year old, after thorough case investigation (Duncan and Byard, 2018)	Global distribution (Müller-Nordhorn et al., 2020); a significant cause of mortality in Africa (See: Ogbu, 2003; Ndu, 2016; Dempers et al., 2018)	Increased tissue Pb levels in SIDS infants—Erickson et al. (1983); Potassium levels significantly lower in less than 6-month old SIDS infants than in non-SIDS infants—Steele et al. (1984); Soft water with low Mg and Ca and with high concentration of Na, linked to higher SIDS rates—Caddell (1992); Recharge of groundwater, which increases its nitrate content —George et al. (2001)	Cold wet weather—Deacon and Williams (1982); Overheating or disordered thermoregulation—Sawczenko and Fleming (1996); Exposure to increased ambient *temperature* associated with an increased risk of *SIDS*—Schluter et al. (1998); Jhun et al. (2017)	Scarce attention is given to SIDS research in Africa, though the syndrome may be widespread since many of the risk factors do exist (See, e.g., Ogbu, 2003; Ndu, 2016)

A listing of the steps for the prevention and control of health risk factors in built environments, originally detailed by HSA in HSA (2017), is summarized here as:

- "Identification of hazards and risk factors that have the potential to cause harm (hazard identification) to humans and/or ecological systems regarding health.
- Analysis and evaluation of the risk in association with that hazard (risk analysis and risk evaluation).
- Determining appropriate ways to eliminate the hazard or control the risk when the hazard cannot be eliminated (risk control)" (HAS, 2017).

Health risk assessment

In 2009 the Centers for Disease Control and Prevention (US CDC) defined HRA as "... a systematic approach to collecting information from individuals that identifies risk factors, provides individualized feedback, and links the person with at least one intervention to promote health, sustain function, and/or prevent disease." (US CDC, 2009). Much more recently (2016), the Environmental Protection Agency (US EPA) defined HRA as "... the process of estimating the nature and probability of adverse health effects in humans who may be exposed to chemicals in contaminated environmental media, now or in the future." (US EPA, 2016), a definition that lent a more circumspective approach to the subject. The definition of the US EPA (2016) uses HRAs to characterize the nature and magnitude of health risks to humans (e.g., residents, workers, recreational visitors) and ecological receptors (e.g., birds, fish, wildlife) from chemical contaminants and other stressors that may be present in the environment.

The issue surrounding both of these definitions is how to assess the comparative importance of risks to health and their outcomes in different demographic groups of the population. What we really need is a framework definition for integrating, validating, analyzing, and disseminating the fragmentary, and at times contradictory information that is available on a population's health, coupled with knowledge of how that population's health is changing, so that the information is more relevant for health policy and planning purposes.

The geo-environment as an agent of illness

"Struggles over environmentally induced diseases are struggles over the very nature of what exists and how we know the nature of the phenomenon"—Kroll-Smith et al. (2000).

In this Section, we examine suggestions that the geoenvironmental mileu (geographical patterns, seasonal variations, and geochemical variables) can have a significant influence on the occurrence and development of disease, with particular reference to diseases of unknown etiology.

This is a relationship that has captivated scholarly attention across a number of disciplinary and policy domains. Mehri (2020), for instance, discuss how geoenvironmental conditions provide infectious agents that activate *innate* and *adaptive* immune system (See "Glossary of Terms," in this Chapter, for definitions) and provoke diseases of unknown etiology in genetically susceptible patients.

Writing on one of the more well-known diseases of unknown etiology (chronic kidney disease, CKD), Hara et al. (2016) remarked on the significance of the contribution of environmental factors compared to genetic factors in the development of CKD among individuals with the same ethnicity. In 2017 Senanayake and King, reviewing recent research done on emerging health-environment relationships, categorized the studies into three themes, *viz*: complexity, uncertainty, and bodies. Although there have been robust contributions to these thematic areas from geography and the social sciences, Senanayake and King (2017) argue that integrating them into an analytical framework can extend geographic perspectives on scale, knowledge production, and human-environment relations, while also incorporating valuable insights from cognate fields.

The cardinal thesis here is that geoenvironmental cofactors can significantly contribute to causation of diseases of unknown etiology, probably to an extent greater than what has hitherto been conceived. Some examples of geoenvironmental and related cofactors to be considered are:

- Water, soil, and air pollution (geogenic contaminants—GCs) emanating from diverse sources (mining, agriculture, industry, and so on).
- Entry of potentially toxic elements (PTEs) (e.g., As, F, Hg, Pb) from geogenic sources, such as volcanic emissions and contaminated groundwater into the food chain (through consumption of food crops, drinking water, and through other dietary intake pathways); and the consequent modulating influence of the balance (insufficiency/excess) of both micronutrient elements and PTEs on immune system functioning.
- Geogenic dust particles (from mining, ore processing, or other geogenic activities, vehicular transportation on untarred roads, volcanic dust emanations, and so on).
- Over-exposure to ionizing radiation in mining, ore processing, and tailings handling of uranium and gold ores.
- Other radionuclides present in the water, soil, and air environments
- Initially undetected release of a chemical from the Earth's subsurface, such as CO_2 release in the Lake Nyos (Cameroon) disaster of the 1980s (See, e.g., Rouwet et al., 2016).
- Geographical patterns, seasonal variations, and climate change effects.
- Factors of *geopathic stress (GS)* and *heat stress*.

Geogenic contaminants

Geochemicals, such as metal(loid)s, radioactive metals and isotopes as well as transuraniums, referred to as GCs by Bundschuh et al. (2017), occur naturally in geogenic sources (minerals, rocks, ground and surface waters, and volcanic emissions).

The accelerated release of GCs globally has been attributed to rapid population growth and economic growth, and the associated increase in demand for water, energy, food, and mineral resources. When GCs occur in near-surface environments, their release can be triggered into the environment by natural biogeochemical processes and/or by anthropogenic activities, such as mining and oil exploitation, fossil fuel combustion, as well as exploitation of geothermal resources. The result is the contamination of soil, water, air, and biota, and their subsequent entry into the food chain with often deleterious health consequences that are mostly underestimated and poorly recognized; not to say that other categories of cofactors of environmental pollution are to be discounted, when studying environmental toxicology and pathogenesis. Indeed, Garchitorena et al. (2017) demonstrate clearly that "Control of environmentally transmitted diseases can be more effective when human treatment is complemented with interventions targeting the *environmental reservoir* of the pathogen."

Environmental soil factors

Soil can be a reservoir for a variety of disease-causing organisms for both plants and animals. Pathogens may be transmitted from the soil via insect vectors or by soil organisms, such as nematodes.

A number of recent studies have supported the observation that soil conditions can influence the occurrence and progression of several diseases through the survival and activities of micro-organisms (bacteria and viruses) in the soil. Factors that influence bacterial growth in the soil include: physical requirements such as oxygen availability, pH, osmosis; sources of energy, carbon, nitrogen; and nutritional requirements—water, micronutrients/trace elements; and growth factors, such as amino acids, purines, pyrimidines, and vitamins that a cell must have for growth but cannot synthesize on its own (See, e.g., Demoling et al., 2007; Rousk and Bååth, 2007; Rousk and Bååth, 2011). In the case of virus survival in the soil, temperature, and virus adsorption to soil appear to be the most important factors (See, e.g., Hurst et al., 1980).

The following are the examples of soil-borne diseases and the soil environmental mileu that favor their prevalence; *a relationship that we still do not fully understand.*

1. The mysterious prion diseases

Transmissible spongiform encephalopathies (TSEs) or *prion* diseases are a family of rare progressive and fatal neurodegenerative brain disorders that affect both humans and other mammals. These diseases include bovine spongiform encephalopathy in cattle, scrapie in sheep and goats, chronic wasting disease (CWD) in North American members of the deer family (cervids), transmissible mink encephalopathy in farmed mink, and Creutzfeldt-Jakob disease, and kuru in humans (See Greenlee and Greenlee, 2015; US CDC, 2018b).

Less than two decades ago, the etiological agents for TSEs had not been fully characterized (Prusiner, 1998), but available evidence today (2020), suggests that prions [(proteinaceous infectious particles lacking nucleic acid (Prusiner, 1998)] are the etiological agents of TSEs (See e.g., Ma, 2012; Kuznetsova et al., 2020), and that a misfolded form of the host-encoded prion protein (PrP) constitutes the primary, if not the only component of the prion (See Lathe and Darlix, 2020).

Attachment of prions to soil particles

Several studies, including the use of laboratory experiments and epidemiological modeling have shown that prion diseases, including CWD and scrapie, can be transmitted *via* indirect environmental routes (See, e.g., Johnson et al., 2006; Schramm et al., 2006; Saunders et al., 2011; Smith et al., 2011; Zabel and Ortega, 2017; Saunders et al., 2012). In 2011 Smith et al. wrote "Soil represents a plausible environmental reservoir of scrapie and CWD agents, which can persist in the environment for years. Attachment to soil particles likely influences both the persistence and infectivity of prions in the environment…." We now know, for instance, that the local soil type may be a key factor that determines the incidence of prion (See, e.g., Kuznetsova et al., 2020), and thus, serves as a reservoir for the natural prion transmission, as well as a potential prion exposure route for humans (See, e.g., Johnson et al., 2006; Saunders et al., 2012).

Animals habitually ingest soil. Humans inadvertently, but also sometimes deliberately, ingest soil (See Chapter 6, "Geophagic practices in Africa", this Volume). Prions have the ability to bind to a wide range of soils and soil minerals, whereby they retain the ability to replicate, and remain infectious (See, e.g., Johnson et al., 2006). Epidemiological modeling by Saunders et al. (2012) has led to the inference that soil parameters, such as soil texture may influence the incidence of prion disease. Results from epidemiological modeling are buttressed by those from experimental work that show variance in prion interactions with soil, including variance in prion soil adsorption and soil-bound prion replication with respect to soil type.

According to Schramm et al. (2006), for soil to serve as environmental reservoir, the following conditions must hold "(1) prions must enter the environment; (2) prions must persist in soil; (3) prions in soil must retain infectivity; (4) prions must remain near the soil surface where they can be accessed by animals; (5) naïve animals must be exposed; and (6) the dose must be sufficient to cause infection." For soil to represent a plausible reservoir of TSE agents in the environment, we should add the requirement that: soil particle-bound prions must remain *bioavailable* and infectious to susceptible animals.

Prion research in Africa

Among the few reported researches to date (2020) on prion disease occurrence in Africa, or accounts by researchers in Africa, are those by Obi and Nwanebu (2008), who presented a general review of the subject; Teferedegn et al. (2019), whose work on the extent of the importance of prion-related studies in Ethiopia also considered livestock management, food quality, food security, and food safety; Babelhadj et al. (2018), who first reported the detection of a new prion disease (*Camelus dromedarius*) in dromedary camels in Algeria; and WOAH World Organisation for Animal Health (2019), who commented that "The identification of a new prion disease in dromedary camels in Algeria and Tunisia, called *camel prion disease* (*CPD*), extends the spectrum of animal species naturally susceptible to prion diseases and *opens up new research areas for investigation*. CPD was identified in 2018 in adult camels showing clinical signs at the ante mortem inspection at slaughterhouses in the region of Ouargla (Algeria), and in 2019 in the region of Tataouine (Tunisia) …"

The work of Babelhadj et al. (2018) also made mention that: "Our identification of this prion disease in a geographically widespread livestock species requires urgent enforcement of surveillance and assessment of the potential risks to human and animal health." As at 2008, there was still no treatment that can cure or control TSEs in livestock or humans in Africa (Obi and Nwanebu, 2008).

Rift Valley Fever (RVF) virus ecology

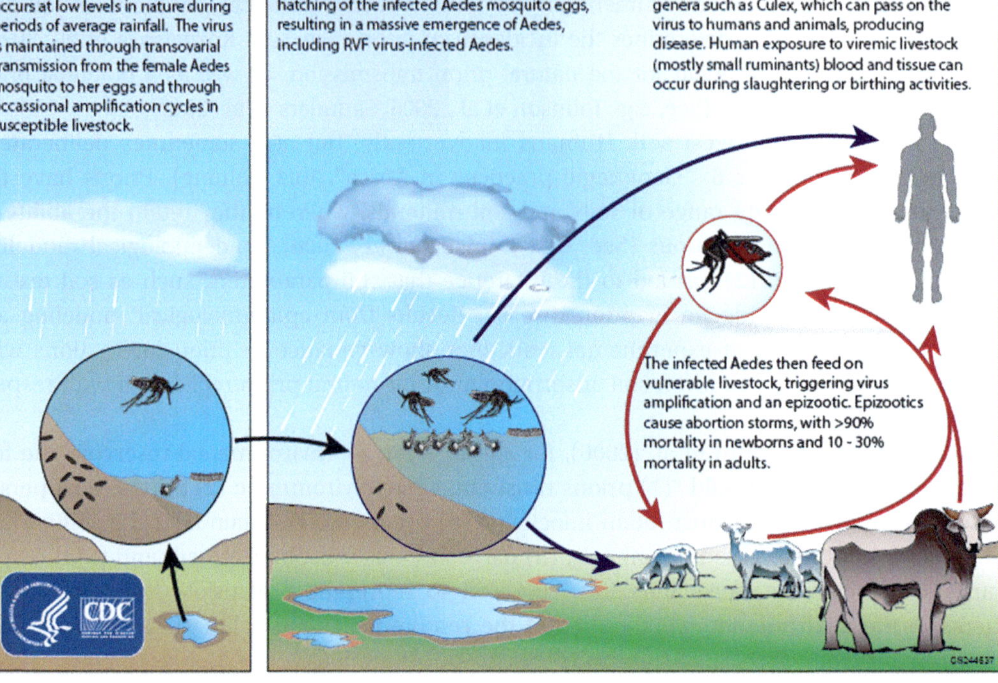

FIGURE 3–2 The ecological cycle of the Rift Valley Fever virus (Greenhalgh, 2015). *Credit: US Centers for Disease Control. From Greenhalgh, E., 2015. El Niño, East Africa, and Rift Valley Fever. NOAA Climate.gov. https://www. climate.gov/news-features/understanding-climate/el-ni%C3%B1o-east-africa-and-rift-valley-fever. (Accessed 22 January 2021).*

2. Rift Valley fever

Rift Valley fever (RVF) is a viral disease first identified in 1931 during an investigation into an epidemic among sheep on a farm in the Rift Valley of Kenya (Daubney et al., 1931). It is most prevalent in domesticated animals, such as cattle, buffalo, sheep, goats, and camels in a number of countries in Sub-Saharan Africa. People can contract RVF through contact with the blood, body fluids, or tissues of infected animals, or through bites from infected mosquitoes (Fig. 3–2). This disease can have detrimental impacts on ruminants, humans, as well as on regional and national economies (Verster et al., 2020).

In 2016 Williams et al. carried out a retrospective analysis of the 2008–11 RVF epidemics in South Africa, which study revealed a pattern of continuous and widespread seasonal rainfall causing substantial soil saturation followed by explicit rainfall events that flooded *dambos* (seasonally flooded depressions) that triggered outbreaks of the disease. Incorporation of rainfall and soil saturation data into a prediction model for major RVF outbreaks, resulted in the correct identification of risk in nearly 90% of instances at least

1 month before outbreaks occurred. All indications from the Williams et al. (2016) study were that irrigation was of major importance in the remaining 10% of outbreaks.

Numerous other recent studies exist on the outbreak of RVF in Africa and its impact on human and animal health (See, e.g., Mosomtai et al., 2016; Bashir and Hassan, 2019; Javelle et al., 2020; Petrova et al., 2020; WHO World Health Organisation, 2020a), but despite the magnitude of research effort exerted on the subject, *relatively little is known about the role of environmental factors, especially the nature of the soil (soil conditions, soil characteristics, and soil parameters), on the estivation (state of dormancy) of the virus.* The research effort of Verster et al. (2020), for instance, was only an initial attempt to predict sites prone to RVF livestock mortality from soil properties, *which only served as a basis for broader research on the interaction between soil, mosquitoes, and RVF virus.* These authors (Verster et al., 2020) appropriately recommended that *"Future research should include other environmental components, such as vegetation, climate, and water properties as well as correlating soil properties with floodwater Aedes spp. abundance and RVF virus prevalence."*

3. Valley fever

Valley fever (VF) is a pulmonary disease caused by inhalation of wind-borne fungal spores of *Coccidioidomycosis* (*cf.* RVF). Although most infected individuals are asymptomatic, the disease can cause flu-like symptoms, pneumonia, and, in some cases, death. The fungus that causes VF—*Coccidioides immitis* or *Coccidioides posadasii*—live in the soil in parts of the United States, where roughly 10,000 cases are reported yearly, mostly from the states of Arizona and California (US CDC, 2020a). The extent of VF prevalence in Africa is not reported in the literature.

A significant part of its life cycle of *Coccidioidomycosis* is spent in the soil. Frank et al. (2010) studied the habitat of this fungus as part of an investigation of the interrelationship between its environment and human health. These researchers (Frank et al., 2010), developed a model for the spread and survival of the fungus based on analyses of the historic climate data in the Tucson area (USA). The work revealed that factors affecting the establishment and growth of new *Coccidioidomycosis* colonies included wind direction and intensity; soil geology, texture, and moisture content; surface and soil temperatures; and a sufficient time interval of favorable conditions for colony survival. *These results, together with supporting studies, show that both geologic and environmental soil factors control the location of infectious sites.* Frank et al.'s (2010) model explains the factors controlling the distribution of actual sites and provides clues useful in mapping potentially infected sites and in mitigation work to control the disease.

4. Ticks

In 2011 Naicker noted that soil moisture is a factor in the developmental stages of *ticks* (small blood-sucking arachnids), with mortality related to dry conditions and soil evaporation (See also: Valcárcel et al., 2020). *Hyalomma ticks,* the vector for *Crimean-Congo hemorrhagic fever,* are particularly well-adapted to surviving in dry conditions than other ticks. Duygu et al. (2018) and Nili et al. (2020) have reviewed the relationship between Crimean-Congo Hemorrhagic fever and climate.

5. Anthrax

The distribution of the spores of *B. anthracis* (the bacterium that causes anthrax) has also been shown to correlate with the composition of the soil (Hugh-Jones and Blackburn (2009), who noted that optimal soil conditions, including humus rich-, high calcium- and alkaline (pH > 6.1) conditions are necessary for spore survival. According to Naicker (2011), susceptible vertebrate hosts and certain human factors are also necessary requirements for disease to occur.

Geographical patterns, seasonal variations, and climate change

As recently as 2008, Dummer recapitulated the intrinsic link between geography and health, and went on to further elucidate that where we were born, where we live, where we work, the food we eat, the viruses we are exposed to, among other circumstances, all directly influence our health experiences (Dummer, 2008). The spatial location (the geographic context of places and the connectedness between places) plays a major role in shaping environmental risks as well as many other health effects; therein lies the key focus of any research on the geography of disease, *viz.*, determination of the extent to which health is influenced by geographical factors. Buttressing these observations was a noteworthy study by NCSV (2010) that factored climate and geographical variables, population data, disease-control data, pathogen data, and human history data into statistical models that attempted to show which factors had stronger correlations to disease.

As Ruiz (2017) further noted, carefully designed geographical investigations of the spatial variation of diseases can provide useful clues on what contributes to disease outcomes and options for disease prevention. *The occurrence of diseases in some places but not in others or the variation of disease rates from place to place, should lead the Medical Geologists and others in disease investigation teams to a closer look at the characteristics that differentiate those places in providing clues to etiology (See Section: "Investigation of disease clusters," this Chapter).* The importance of the "place factor" in unraveling diseases of unknown etiology is well-illustrated in the study of spastic paraparesis in Mozambique by Cliff et al. (1984) (Fig. 3–3).

In this connection, it is necessary to state that the geography of disease in Africa is quite different from that in the Western world—highly industrialized, wealthy, with a higher literacy rate, and relatively secure type of community. Most of the population of Africa is still rural and many parts of the Continent are not yet at an advanced stage of development. Malnutrition, illiteracy, poverty and access to clean water and air are still realistic challenges.

Again too, apart from accounting for the diseases themselves, teams investigating the geographical drivers of diseases must take cognizance of the characteristics and dynamics of pathogens; vectors and host populations; the presence of xenobiotic chemicals; and the movement of vectors, hosts, or toxic agents (See Ruiz, 2017). This would lend to a more holistic approach in the etiological characterization of enigmatic diseases.

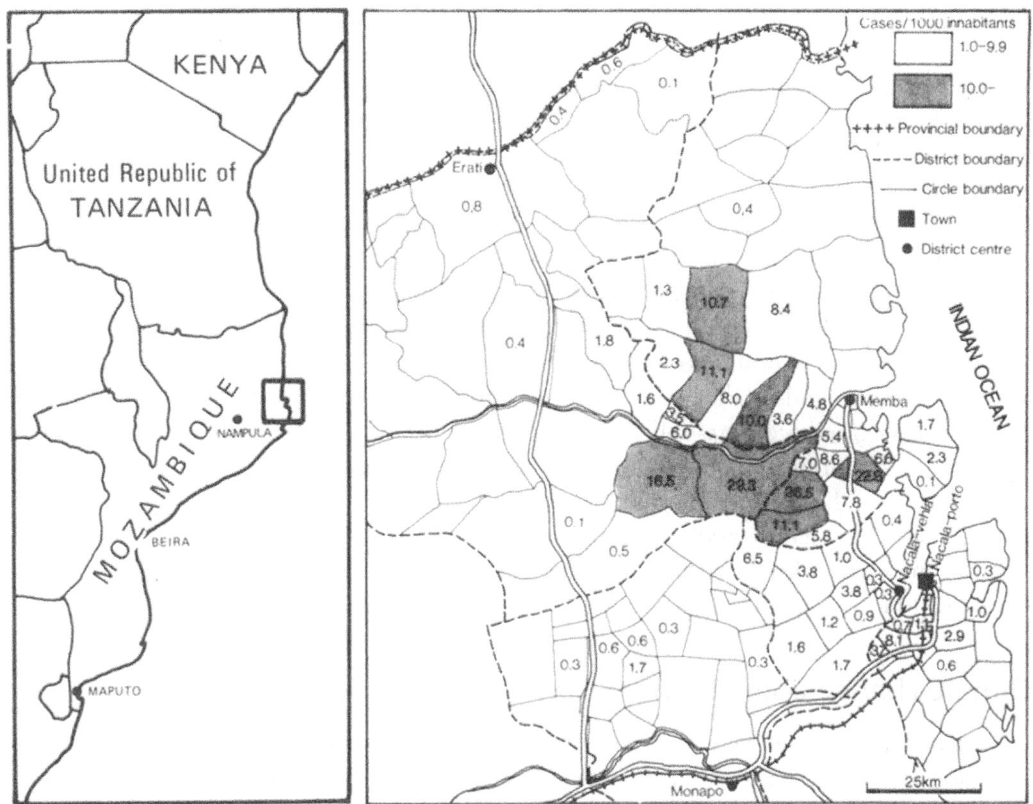

FIGURE 3–3 Map of Mozambique, showing the affected area and incidence rates of "mantakassa" (spastic paraparesis) per 1000 inhabitants.
Please see Excerpt from: https://www.who.int/about/policies/publishing/copyright. (Accessed 07 November 2121)
Reproduced from Cliff, J., Martelli, A., Molin, A., Rosling, H., Ministry of Health, 1984. Mantakassa: an epidemic of spastic paraparesis associated with chronic cyanide intoxication in a cassava staple area of Mozambique. 1. Epidemiology and clinical and laboratory findings in patients. Bulletin of the World Health Organization 62 (3), 477–484.

Unraveling the seasonality/spaciotemporal associations of disease

Seasonal change in the incidence of infectious diseases is a common phenomenon in Africa, as is also the case in other continents having temperate or tropical climates. *However, the mechanisms responsible for seasonal disease incidence, and the epidemiological consequences of seasonality are, with only a few exceptions, poorly understood* (Grassly and Fraser, 2006).

The several mechanistic hypotheses that have been advanced to explain seasonality of various directly transmissible diseases include human activity, pathogen infectivity, seasonal variability in human immune system function, seasonal variations in vitamin D levels, and seasonality of melatonin. The influence of these factors in the seasonal patterns of infectious diseases, and whether or not seasonal immune modulation takes place in humans, have been discussed by Grassly and Fraser (2006), Fisman (2012), Fares (2013), and Paynter et al. (2015).

According to Ruiz (2017), geographical factors of disease are related to the weather and geo-hydrology of a region, the built and the natural environment, and population differences in age and cultural and social habits. Spatial patterns are often linked to temporal factors. Seasonal differences in rainfall and temperature, as well as agricultural activities can affect disease patterns and transmission rates (US NRC, 2001; Ruiz, 2017). Under these circumstances, Ruiz (2017) observed that "When the number of cases of a disease changes across time periods, spatiotemporal analysis can reveal the direction and speed that the disease has spread, and declining disease rates may indicate the effect of preventative actions or development of immunity." Though an individual may remain geographically constrained, their environmental exposures may vary dramatically depending on seasonality (MacGillivray and Kollmann, 2014). However, changing temperature, rainfall, exposure to PTEs, food availability, diet, and exposure to infectious agents may also be compounded or alleviated by local cultural and lifestyle practices.

Sometimes disease occurs only after long exposure and can even be absent in ensuing generations; and prior immunity, nutritional health, and genetic differences can have significant impacts on the spatial patterns of diseases. Given this variety, interdisciplinary effort is vital in research aimed at tackling complex disease problems. Ruiz (2017) attempts a presentation of methodological and theoretical threads across a variety of disease systems that are all related to the geography of diseases.

Climate change and health

The relationship between climate change and health is becoming increasingly clear (Fig. 3—4) and current knowledge is well-documented. Developments in this area of research can be followed up in a multitude of recent publications. Many scientific journals are devoted exclusively to this subject or have sections addressing it [e.g., The Journal of Climate Change and Health (Elsevier); Climate Change and Public Health (Elsevier); Journal of Climate (American Meteorological Society); International Journal of Climatology (Royal Meteorological Society); Nature Climate Change (Springer Nature); American Journal of Climate Change (Scientific Research Publishing); Climate (MDPI, Basel, Switzerland); Environmental Health (Springer Nature)]. With diseases of unknown etiology, specifically, however, the relationship with climate is much less clear and relatively few publications exist on it.

According to WHO (2003; 2018b), climate change affects many of the social and environmental determinants of health—clean air, safe drinking water, sufficient food, and secure shelter, and so on (Fig. 3—4). The US Centers for Disease Control and Prevention have warned that "Climate change threatens human health and well-being in many ways, including impacts from increased extreme weather events, wildfire, decreased air quality, and illnesses transmitted by food, water, and diseases carriers, such as mosquitoes and ticks." (US CDC, 2020b).

In recent years, attempts in grappling with the challenges of global climate change have revealed unexpected findings on immune system mediation by toxic trace elements (See Section: "Climate change and the immune system," in this Chapter), infectious disease (re)emergence, and the expanding field of epigenetics. These findings have helped us recast the environment as an agent of illness. Reflecting this shift, leading international bodies assessing the

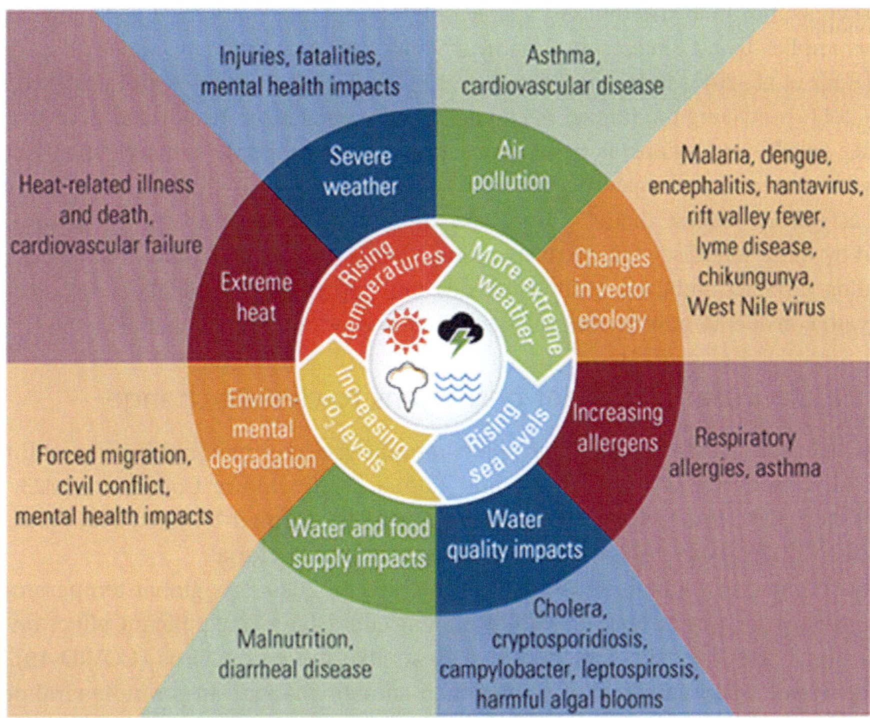

FIGURE 3–4 Impact of climate change on human health. *Credit: George Luber, CDC. From Ebi et al. (2017).*

science related to climate change, such as the WHO and the Intergovernmental Panel on Climate Change (IPCC) [through its Fifth Assessment Report], respectively, have begun to focus attention on contingent, nonlinear, and cross-scalar cause and effect relationships between the environment and human health (WHO (World Health Organisation), 2011a; Smith et al., 2014).

The IPCC Report (Smith et al., 2014) along with several other subsequent studies on climate change (e.g., Denton et al., 2014; Copeland et al., 2017; Morris et al., 2017; NRC, 2019; Aleuy and Kutz, 2020) has revealed that the possible effects of climate change on health and well-being would be realized via multiple, overlapping, and often unpredictable pathways. As Portier et al. (2010) noted earlier, climate related health issues would include increased risks of injury, disease, and death from extreme weather events. Added to these effects are those that are wrought by biophysical systems such as the distribution of disease vectors, water-borne diseases, and air pollution (e.g., Smith et al., 2014).

By virtue of its unique environmental setting and other variables in its life support system, there are indications that Africa would be one of the most vulnerable continents to health effects of climate change (e.g., Kasotia, n.d.; WHO (World Health Organisation), 2012; Rohr and Palmer, 2013; Ilevbare, 2019). The impacts are predicted to be multiple, but also spatially heterogeneous and historically contingent. According to the IPCC (Smith et al., 2014), they (the impacts) can potentially engender both parasitic extinctions and unprecedented rates of disease diffusion as vectors, such as mosquitoes and ticks, capable of traversing previously fixed geographical boundaries. This implies that a single causal pathway could produce different outcomes in dissimilar settings (Rohr et al., 2011; Rohr and Palmer, 2013). Thus in accordance with Rohr et al.'s observation in 2011, predicting the impact of climate change impacts on health and disease in Africa, as in other developing regions, is beset by confounding variables, is context dependent, and is fraught with uncertainties regarding the drivers and outcomes of change. In this connection, Rohr et al. (2011) suggest that: "… forecasts of climate-change impacts on disease can be improved by more interdisciplinary collaborations, better linking of data and models, addressing confounding variables and context dependencies, and applying metabolic theory to host-parasite systems with consideration of community-level interactions and functional traits, …."

Response of viruses and other pathogens to climate change

The links between climate change and infectious disease are manifold but are still not well understood (Casadevall, 2020; El-Sayed and Kamel, 2020; Nature Collection, 2023; Ogden, 2018; US NRC, 2001; Wu et al., 2016; Boukerche and Mohammed-Roberts, 2020) and according to Hofer (2019), these links are both complex and multifactorial.

As Boukerche and Mohammed-Roberts remarked in 2020, "As global temperatures rise, long-term changes in climate and wildlife habitat could have a significant effect on human health and increase the risk of infectious diseases like the coronavirus (COVID-19)." These authors went on to stipulate four ways by which climate change can promote viral or pathogen transmission to produce disease and thus, the reasons why the combat against infectious diseases and pandemics is also one against climate change, *viz.*:

1. The increasing risk of infectious diseases everywhere due to changing weather patterns.
2. The role of air pollution in causing viruses to become airborne and more deadly.
3. The emergence of ancient diseases due to melting of ice and permafrost, and
4. Promotion by global warming of viral mutations, thus resisting our defenses for fighting diseases.

In this regard, a novel area of knowledge, and one potentially of high etiological significance, is developing, *viz.*, how rising temperatures are making our natural immune systems less effective. As pathogens get attuned to gradually warmer temperatures in the natural world, they become better able to withstand the high temperature inside our body.

Climate change, our immune system, and infectious diseases

According to Casadevall (2020), climate change will bring about major changes to the human defense that works against microbial disease. The action of an advanced immune system engendering *innate* and *adaptive* arms as well as *endothermy* (See "Glossary of Terms," in this Chapter), creates a thermal restriction zone for many microbes. However, given that microbes can adapt to increased temperatures, it is possible to proffer that global warming will favor the growth of microbes with higher heat tolerance that can overwhelm our endothermy defenses and bring new infectious disease (Casadevall, 2020).

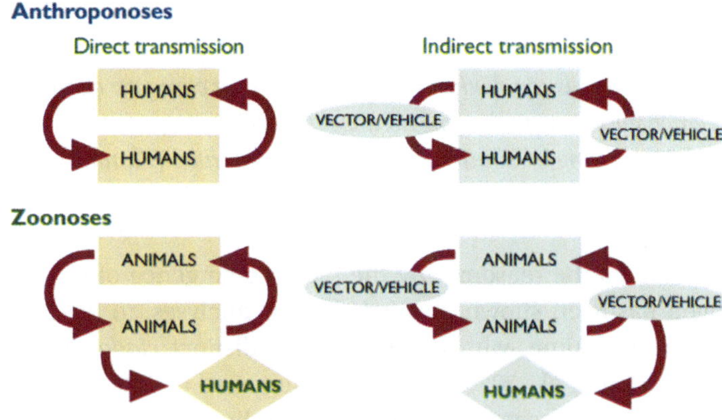

FIGURE 3–5 The four main types of transmission cycles for infectious diseases (Wilson, 2001). *Reproduced from Wilson, M.L., 2001. Ecology and infectious disease. In: Aron, J.L., Patz, J.A. (Eds.), Ecosystem Change and Public Health: A Global Perspective. John Hopkins University Press, Baltimore, USA, pp. 283–324.*

New microbial threats in a changing climate

Nearly three decades ago, scientists began to create a stern awareness of the possibility that climate change could bring about changes in the epidemiology of infectious diseases (See, e.g., Anwar et al., 2019; Nature Collection, 2023; Patz et al., 1996; Boukerche and Mohammed-Roberts, 2020; US CDC, 1992). There has since been a flurry of research on the subject, with a focus on how climate will influence the locations of pathogenic microbes and vectors of infectious diseases (Fig. 3–5). Fig. 3–5 provides a framework on infectious disease transmission types, including human to human, animal to animal, and animal to human. Climate change is increasing the global emergence, resurgence, and redistribution of infectious diseases risks across all of these. See Section: "Climate change and health" in this Chapter for examples of peer reviewed journals covering this topic.

Based on recent research, we can today say with some confidence that: the survival and reproduction of infectious microorganisms, their vectors, and their animal reservoirs are all directly influenced by climate variables (See, e.g., Coates and Norton, 2020). In addition, climate change has caused an expansion in the geographical range of certain pathogenic microbes as a result of sustained warmer temperatures at higher latitudes (See, e.g., El-Sayed and Kamel, 2020).

Garcia-Solache and Casadevall (2010) hypothesize that global warming will increase the prevalence of fungal diseases in mammals by two mechanisms, "(i) increasing the geographic range of currently pathogenic species and (ii) selecting for adaptive thermotolerance for species with the significant pathogenic potential but currently not pathogenic by virtue of being restricted by mammalian temperatures."

A modern example of an infectious disease—*Candida auris*

Candida auris is a new fungal species that has made a mysterious and simultaneous appearance on five continents (Vila et al., 2020), including Africa, where its presence has been reported in countries, such as Kenya (Adam et al., 2019; Heath et al., 2019) and South Africa (Magobo et al., 2014; Govender et al., 2018; Szekely et al., 2019). *The origins of C. auris are currently unknown, and it is a matter of conjecture as to whether environmental and climactic changes were contributory factors in its recent emergence as a pathogen. Because its precise diagnosis remains illusory, it is possible to suggest that the spreading of C. auris is likely underestimated (Zamith-Miranda et al., 2019).*

Origin

In 2020 Du et al. pondered the reasons why the recent emergence of *C. auris* is so puzzling. This fungal pathogen has been shown, through genetic analyses, to have emerged in different continents at the same time, with five genetically distinct *clades* isolated from distinct geographical locations (Du et al., 2020). Since its emergence in 2009 (Hofer, 2019; Du et al., 2020), the scientific and health care communities have been witnessing an exponential rise of infection rates and outbreaks in hospital settings throughout the world. But *C. auris* is difficult to identify and is multidrug resistant. These features, together with evolution of virulence factors, associated high mortality rates in patients, and long-term survival on surfaces in the environment make *C. auris* particularly difficult to study in clinical settings.

Adaptation to environmental stresses

Strategies for survival and adaptation to different ecological niches are necessary requirements for successful pathogen colonization. In the case of *C. auris,* persistence in stressful environmental conditions is a key feature that distinguishes it from most of the other human fungal pathogens (Du et al., 2020). For example, while most fungi are incapable of surviving at human physiological temperatures (36.5°C–37.5°C and up to 40°C during a fever), and therefore do not have the ability to colonize humans and cause infections (Garcia-Solache and Casadevall, 2010; Du et al., 2020), at 37°C and 40°C, respectively, *C. auris* grows comparatively with *C. albicans* (Kean et al., 2020), temperatures at which the more genetically similar *C. haemulonii* will wax and wane.

In a recent study comparing the temperature tolerance of *C. auris* to other *Candida* species, Casadevall et al. (2019) hypothesized that climate change, specifically global warming, may have contributed to the evolution of *C. auris* as a human pathogen and to its ability to grow at high temperatures (*thermotolerance*). Another characteristic feature of *C. auris* is its ability to tolerate high salt concentrations (> 10% NaCl, wt./vol.) (*osmotolerance*) compared to other *Candida* species (Welsh et al., 2017; Wang et al., 2018).

Thermotolerance and osmotolerance are characteristics that may contribute to the persistence and survival of *C. auris* on biotic and abiotic surfaces for long periods of time (See: Biswal et al., 2017; Kean et al., 2018; Sekyere, 2018). Finally, Du et al. (2020) held that *C. auris* can thrive on human skin and environmental surfaces for up to several weeks and can even withstand some commonly used disinfectants.

Spread of viruses and other pathogens by air streams and wind

Some studies have suggested that there exists a link between different weather conditions (e.g., associated with the physics or the chemical characteristics of air masses) and the spread of pathogens from one region to another (See, e.g., Rodo et al., 2016; Boukerche and Mohammed-Roberts, 2020; WHO (World Health Organisation), 2020b).

Complex biometeorological weather analyses have shown the plausibility of distinct weather conditions having a significant influence on the occurrence of ruptured abdominal aortic aneurysms (e.g., Schuld et al., 2013) and type A acute aortic dissection (AAD) (e.g., Luo et al., 2020). However, a number of previous studies on this theme have not been successful in attempting to relate single weather parameters, such as temperature and pressure to disease outcomes, such as AAD (See Mehta et al., 2002; Benouaich et al., 2010; Verberkmoes et al., 2012).

Fine particulate pollution, such as black carbon (a potent climate-warming component of particulate matter), sulfates, and nitrates penetrate deep into the bloodstream and lungs, rendering the immune system less effective (Genc et al., 2012). Several studies (e.g., Giani et al., 2020; Karan et al., 2020; Nurshad and Farjana, 2020; Mostafa et al., 2021) suggest that air pollution increases the risk of COVID-19 spreading faster and becoming deadlier. Pozzer et al. (2020) have shown that an increase in fine particulate pollution of just $1 \, \mu g/cm^3$ results in a 15% increase in COVID-19 deaths. Therefore massive efforts aimed at air quality improvement by reducing emissions, especially in cities, could yield substantial gains in the fight against both viral and climate risks. This holds particularly true in the case of African cities, where improved air quality control mechanisms are most desirable (See, e.g., Davies, 2015).

The aquatic environment as reservoir host of pathogens

The WHO (World Health Organisation) (2020b) lists "water supply" as one of the most important of environmental factors that can influence the spread of communicable diseases that are prone to cause epidemics; other important environmental factors are known to be: sanitation facilities, food and climate.

Heavy precipitation favors a rapid rise in abundance of vector mosquitoes. Writing on what can be described as a surveillance paradox of communicable diseases, such as the *arboviral* (arthropod-borne viruses) diseases, Chikungunya, and West Nile virus in Africa, Chanda (2020) noted correlation of their prevalence with heavy rainfall events wrought by global warming. Indeed, recent changes in climatic conditions, particularly fluctuations in rainfall amounts and increased ambient temperature are thought to be major contributing factors to the endemization process of these diseases in various locations in Africa (See, e.g., Prinsloo, 2006; Paz, 2015; Faustine et al., 2017; Braack et al., 2018; Marchi et al., 2018; Mensah and El Zowalaty, 2018; Ndiaye et al., 2018; Oluwayelu et al., 2018; Fouque et al. (2019); Nyaruaba et al., 2019; Kading et al., 2020). The concentration of reservoir hosts in stagnant water locations, such as wells, roof gutters, cisterns, domestic wells, and water tanks also facilitates the ease of disease transmission. Droughts too, can decrease the mosquito predators, and thus increasing vector abundance post-drought (Mayer et al., 2017; Braack et al., 2018; Bellone and Failloux, 2020).

The influence of melting of ice and permafrost on the (re)emergence of ancient diseases

The rapid rise of temperatures in the Arctic Circle, reckoned to be about three times faster than in the rest of the world (IPCC (Intergovernmental Panel on Climate Change), 2019), will not only accelerate climate change further, but will also trigger the (re)emergence of infectious agents as ice and permafrost melt.

Because permafrost in cold, dark, and devoid of oxygen, it has the capacity for excellent preservation of microbes and viruses. Indeed, numerous studies (e.g., Jansson and Tas, 2014; Margesin and Collins, 2019; Vigneron et al., 2019) have shown the survival of certain pathogens, such as bacteria, viruses and fungi for hundreds, thousands, even millions of years after being frozen in permafrost.

Fragments of RNA from the 1918 Spanish flu virus, for instance, have been found in corpses buried in mass graves in Alaska's tundra (Watanabe and Kawaoka, 2011; Boukerche and Mohammed-Roberts, 2020). There are concerns that the melting of ice wrought by climate change could lead to the release of pathogens, the repulsion of which, our immune systems are quite unprepared.

Other extreme weather events

El Niño Southern Oscillation and disease

El Niño Southern Oscillation (ENSO) is a fluctuation of the global ocean-atmosphere system that originates in the tropical Pacific. It consists of hot and warm phases that contribute to increased extreme weather events (WHO (World Health Organization), 2015). The warm phase is commonly referred to as El Niño and the cold phase, La Niña.

The effects of ENSO are felt in many parts of the world, but its most intense impacts are realized in tropical regions (e.g., in Africa) that are particularly prone to natural hazards. By altering climate conditions, ENSO can have severe effects on major determinants of health (WHO (World Health Organization), 2015).

While the precise impact of global warming on the ENSO is as yet uncertain (WMO (World Meteorological Organisation), 2020), we do now know that ENSO's characteristic heavy rainfalls and the accompanying widespread disease outbreaks, such as RVF in East Africa, altered transmission patterns of vector-borne, rodent-borne, and waterborne diseases, are induced by rising temperatures (See, e.g., Anyamba et al., 2001; Anyamba et al., 2019).

Other common regional influences of El Niño in Africa include: Drier conditions in southern part of the Continent and some areas in its Sahel region; and wetter conditions in equatorial East Africa during the short rainy season (October–December) (WHO (World Health Organization), 2015; Fig. 3–6).

Vectors, such as mosquitoes, that are responsible for the transmission of vector-borne diseases in Africa (e.g., malaria and dengue fever), with variable local impacts, are sensitive to changes in temperature, rainfall, and humidity, which determine the suitability of ecosystems for vector reproduction, development, and activity (Caminade et al., 2019).

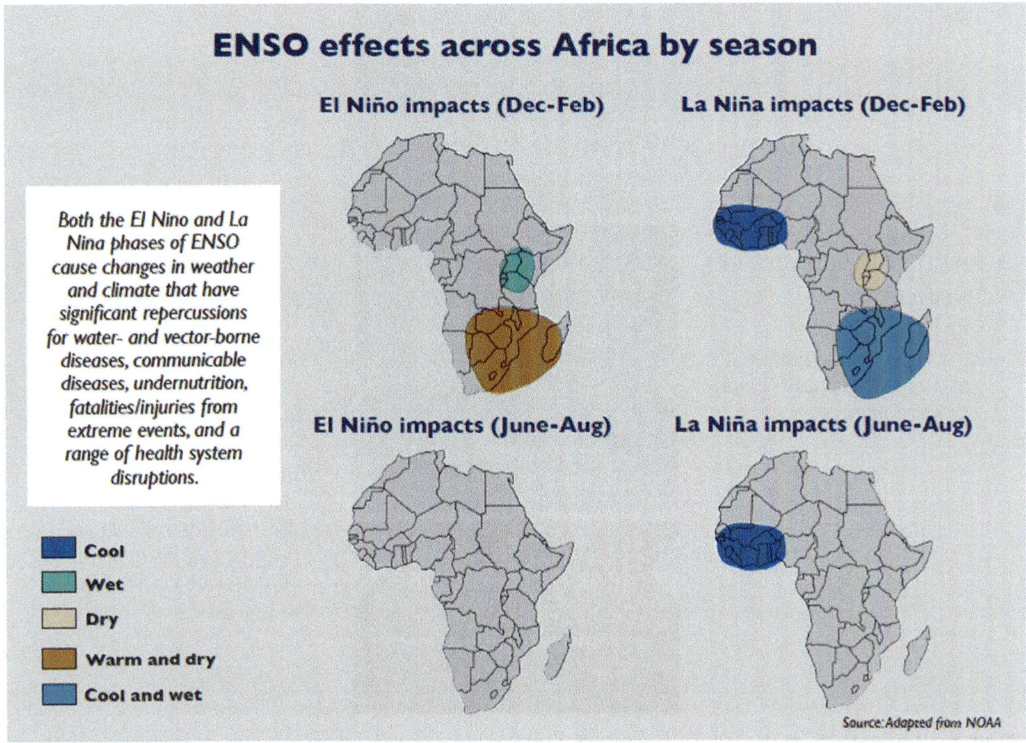

FIGURE 3–6 Health impacts of ENSO across Africa. *From Zermoglio et al.'s 2019 adaptation from NOAA.*

As McGregor and Ebi (2018) conclude in their recent paper: "… there is still some way to go before a thorough understanding of the association between ENSO and health is achieved, with a need to move beyond analyses undertaken through a purely statistical lens, with due acknowledgment that ENSO is a complex noncanonical phenomenon, and that simple ENSO health associations should not be expected."

The spectrum of illness during heat waves and heat stress
The increasing frequency, duration, and magnitude of heat waves have been noted as one category of extreme temperature events (WHO (World Health Organisation), 2021). *Heat wave* or *heatwave* refers to a period of sustained, abnormally high, surface temperatures relative to those normally expected (Robinson, 2001). This phenomenon is observed worldwide and is associated with climate change (Smoyer-Tomic et al., 2003; WHO (World Health Organisation), 2021). Human exposure to heat waves can lead to a wide range of physiological impacts, such as heat exhaustion, dehydration, and heatstroke (Kilbourne, 1999; Ye et al., 2012; Fig. 3–7). Heat waves may be characterized by low humidity, which may precipitate drought, or high humidity, which may amplify the health effects of *heat-related stress*. Existing health conditions may also be exacerbated, causing premature death and disability.

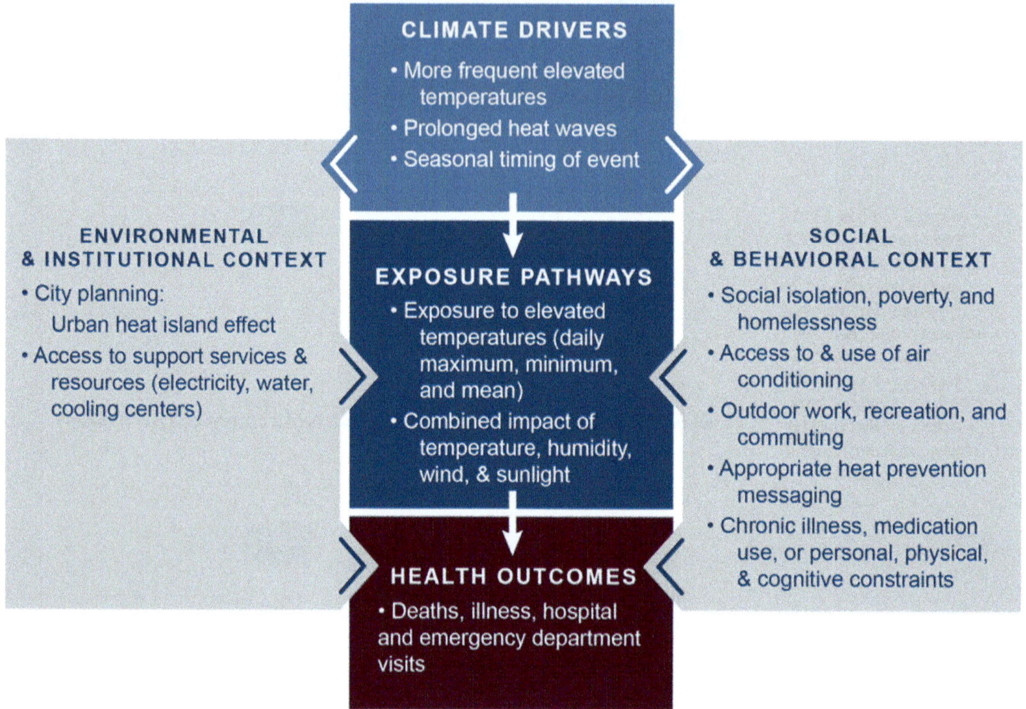

FIGURE 3–7 Schematic illustrating the key pathways by which climate change influences human health during an extreme heat event, and potential resulting health outcomes (centre boxes). *From US GCRG (US Global Change Research Group), 2016.*

Heat stress occurs when the body cannot cool down properly through sweating, to maintain a healthy temperature (37°C). The increase in global temperatures *wrought by climate change* will increase the prevalence of *heat stress* (WHO, 2018b; Li et al., 2020). Heat stress can take the form of relatively mild conditions, such as heat rash and heat cramps to more serious conditions, such as heat exhaustion (Li et al., 2020).

Glaser et al. (2016) have considered the effect of heat stress (resulting from climate change) on the emergent epidemic of CKD in rural communities. These authors propose that "... heat stress nephropathy may be a major cause of CKD, representing an overlooked disease in neglected populations in hot communities."

Effect of climate change on the geoenvironmental determinants of health in Africa

In Africa, climate change affects the geo*environmental* determinants of *health* in diverse ways. The effect of climate perturbation on availability of clean air, safe drinking water, and sufficient amount of essential nutrient elements in the diet, among others, remain major challenges for the health of populations in many areas of the Continent. Maladies, such as malaria, diarrhea and heat stress, all of which can be directly or indirectly linked to climate-modulating cofactors, also still abound.

The scale and magnitude of the climate and health problems in Africa has prompted a flurry of research leading to a voluminous set of publications. A selection of recent (post-2000) peer-reviewed references on the subject includes noteworthy and comprehensive accounts by WHO (2003), Byass (2009), UNECA (2011), WHO (World Health Organisation), 2012, ACPC (African Climate Policy Centre), 2013, Zaitchik (2017), WHO (2018b), UNCC (2020), Atwoli et al. (2023) and Moyo et al. (2023). Among the several existing peer reviewed regional- or country-based studies, particularly for South Africa, those of Ramin and McMichael (2009) and USAID (2017) (for Sub-Saharan Africa); Myers et al. (2011) and Serdeczny et al. (2017) (for southern Africa); Chersich et al. (2018) and Chersich and Wright (2019) (for South Africa), and Adebisi (2020) (for West Africa) are quite informative. Further pertinent references are given in the Section: "Further reading," in this Chapter.

Geopathic stress

The original definition of GS is: "... the study of *Earth* energies and their effect on human health." (Dovjak and Kukec, 2019). The word "geopathic" is derived from two Greek words: *geo*, meaning "of the Earth" and *pathos*, meaning "suffering" or "disease." Interest in the phenomenon was first generated in Germany back in the 1920s (Freshwater, 1997). In the early days of study of GS, it was noted that very few people were aware of its association with specific localities, and how a careful selection of locations for civilizations and sites for construction of structures and dwellings were important in avoiding GS (Dharmadhikari et al., 2011). The rationale for judicious site selection for construction purposes was that, certain localities overlay *geopathic stress zones*, defined as: "places on the surface of the Earth having a natural frequency of Earth's magnetic field (Fig. 3−8), and known to cause health problems." The term also encompasses other weak electromagnetic fields produced by innate geological features like underground water streams or deposits of coal, oil, or gas. Man-made structures, such as roads, bridges, railways, electricity, water mains, and sewers can also cause disruption in the Earth's electromagnetic field, resulting in GS.

It is now fairly well-established that *stress* is a factor in ill health, as it can affect the body's ability to perform certain functions (Freshwater, 1997). These functions include those that rely on the body's subtle energy system (*the etheric body, chakras*, and *meridians*) (See "Glossary of Terms," in this Chapter for definitions) and the body's electrical system (brain, heart, and muscles), thus delaying healing and recovery.

The spectrum of causes of GS, however, is still not yet well-defined. Freshwater's (1997) account challenges all professionals in health-care who are dedicated to the development of holistic care to "... *further their understanding and awareness of this phenomenon.*"

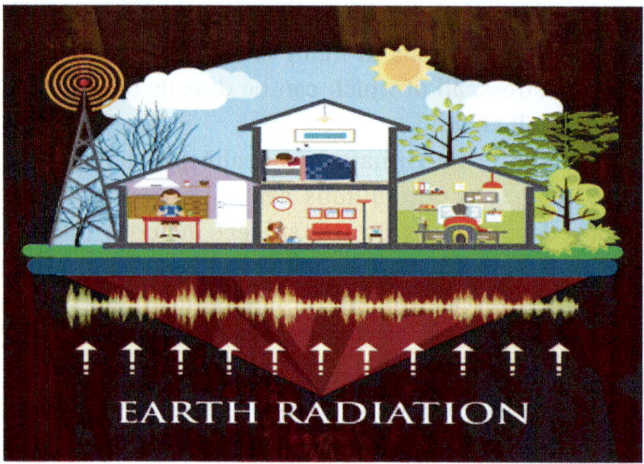

FIGURE 3–8 Geopathic stress generation. *From DVNHC (Da Vinci Natural Health Centre). n.d., Geopathic Stress and Its Effects on Health. https://www.naturaltherapycenter.com/geopathic-stress-and-its-effects-on-health/. (Accessed 22 January 2021).*

Geoenvironmental variables and the human immune system

The immune system is a complex network of cells and proteins that finds and attacks infectious agents, such as bacteria, viruses and fungi. The white blood cells are a key component, with each cell type having specialized functions. For instance, *neutrophils* are important to fight bacteria and fungi, while *lymphocytes* generally fight viruses.

Numerous geoenvironmental factors can modulate human immunity (Touil et al., 2023). An understanding of the geoenvironmental interactions involved in the development of the immune system is necessary in order to be able to address specific aspects of diseases, such as unraveling the etiology of diseases of unknown etiology; also, to identify the strategies needed to change trajectories toward long-term, life-long protection from disease. The purpose of this section is to try and amalgamate existing data into a cohesive vision that illustrates how exposure to geoenvironmental variables can leave a lasting impression on the human immune system, and how this impression can either have beneficial or potentially deleterious effects.

Influence of geographical patterns, seasonal variations and climate events on the immune system

A lot of information, based on population studies, is currently being garnered on the relationship between geography or climate and autoimmune disease (AD) prevalence; however, *the mechanisms underlying these associations remain poorly understood and the link between seasonality and immune status, still conjectural* (See, e.g., Nelson and Demas, 1996; Nelson et al., 2002; Fares, 2013; MacGillivray and Kollmann, 2014).

Warren and Warren (2001) and Staples et al. (2003) have noted a gradient of increasing prevalence of diseases, such as multiple sclerosis and insulin-dependent diabetes with increasing latitude. A number of related studies (e.g., Chaplin, 2010; MacGillivray and Kollmann, 2014; Boudiaf and Achir, 2016; Watad et al., 2017; Manlhiot et al., 2018; Mehri, 2020; Nwaogu et al., 2020) have shown how geographical patterns, seasonal variations, and extreme weather and climate events provide infectious agents that activate the innate and adaptive immune system and provoke diseases of unknown etiology in genetically susceptible patients.

One of the more recent lines of evidence and probably the most compelling so far, to show seasonal variation of our immune systems was that provided by the University of Cambridge study of 2015 (Dopico et al., 2015; UoC, 2015). The study shows that the activity of almost a quarter of our genes (5136 out of 22,822 genes tested) differs according to the time of year, with some more active in winter and others more active in summer. The cells of our immune system as well as the composition of our blood and adipose tissue (fat) are all affected by seasonality (Dopico et al., 2015). Findings such as these would go a great way in helping to explain why certain conditions, such as heart disease and rheumatoid arthritis are worsened in winter while people tend to be healthier in the summer.

Martineau et al. (2011) provided an assessment of the influence of seasonality, vitamin D, and infectious disease with regards to TB in South Africa; and revealed a reciprocal seasonal variation in serum 25(OH)D concentration and TB notification. This study also emphasized the need for consideration of "seasonality," now known to be a multifactorial parameter, in studies of human immunity (See Section "Suggestions for future research," this Chapter. According to Behrman et al. (2018), a proper understanding of the rapid, cyclic response to seasonal environmental pressure is urgent, since it will provide a broader perception of the complex ecological and genetic interactions that determine the evolution of immune defense in natural populations.

Autoimmune diseases

ADs are a heterogeneous group of chronic conditions that affect specific target organs or multiple organ systems. *Their exact etiopathogenesis is still ill-defined (Getts et al., 2020).*

Descotes (2004) recapitulated on the importance of autoimmunity as an important area of immunotoxicology, especially because ADs are frequent diseases in the general population, and some experimental data suggesting a possible link between chemical exposures and autoimmunity.

Many questions remain as how pathogenic challenge, for instance, may interfere with immune system regulation and give rise to autoimmunity; and it is likely that other apparently novel immune modulatory mechanisms (e.g., trace element/metal interaction) also contribute to clinical AD (Fig. 3−9).

Among the different environmental factors that are known to influence the development of AD are: infections, low vitamin D levels, UV radiation, and melatonin (Smith and Germolec, 1999; Watad et al., 2017), which factors are also known to exhibit seasonal variation patterns that could influence disease development, severity, and progression. Autoimmune disorders may cause destruction of body tissues, abnormal organ development, or changes in organ function.

Dopico et al. (2015) give, as an example of evolutionary adaption of humans to different environments, the loss of skin pigmentation as humans migrated out of Africa to more

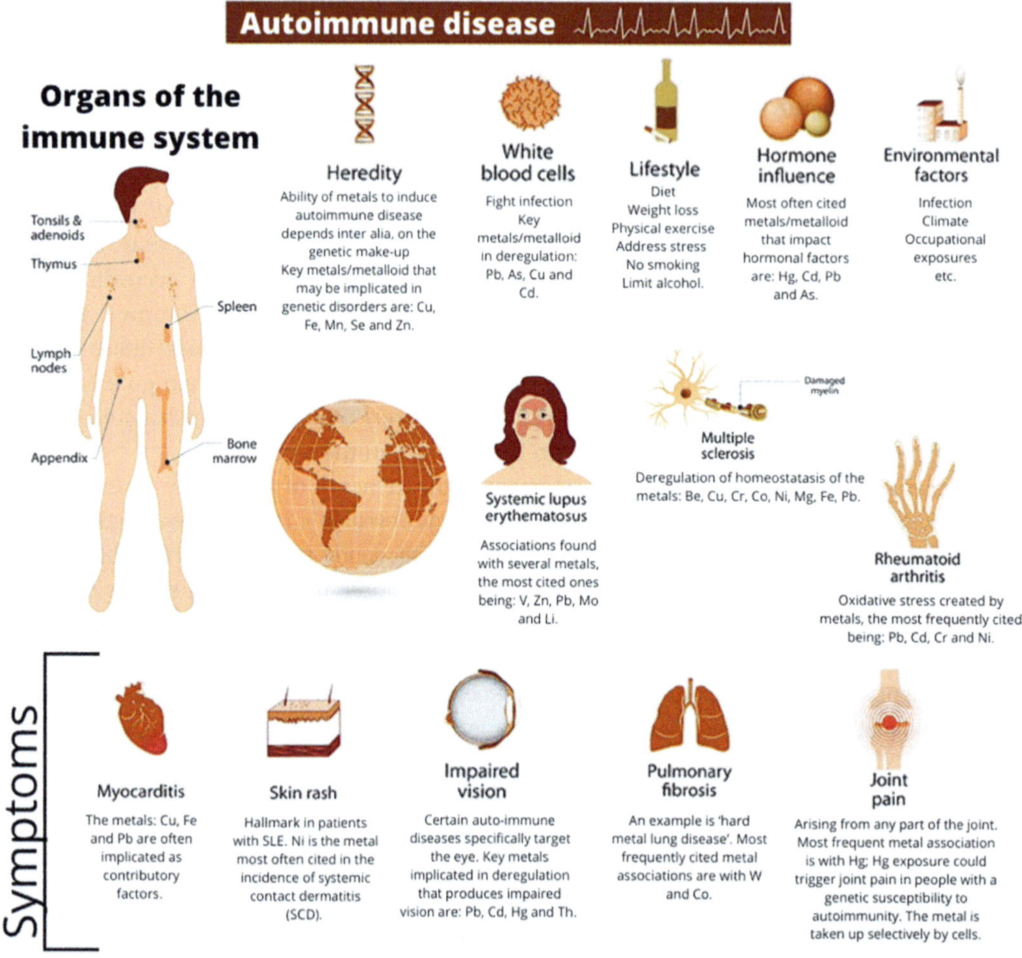

FIGURE 3–9 Schematic illustration of the link between the trace element (particularly metal) deregulation in organs of the immune system and development of autoimmune diseases. The main trace elements/metals involved in deregulation are given for each corresponding autoimmune disease portrayed. *From U.S. National Institute of Environmental Health Sciences.*

temperate and colder zones to increase sunlight-driven vitamin D production. Despite such observations, however, the mechanisms by which seasons might more broadly influence the underlying molecular details of human physiology remains unknown (See, e.g., Nelson, 2004). For further discussion on the effects of seasonality on the onset, relapses, and activity of various ADs (See Watad et al., 2017).

Effect of sunlight on the incidence of autoimmune diseases

Sunlight, the radiation and heat from the sun (UVR), is an essential source of light and warmth for all living organisms. It can, however, have a number of undesirable effects on the health of man "… such as promoting the malignant transformation of skin cells and suppressing the ability of the human immune system to efficiently detect and attack malignant cells." (González-Maglio et al., 2016).

However, consideration of seasonal influences on both disease incidence and clinical course as well as recent analytical studies at the individual level have lend support for a possible protective role for UVR in certain diseases (See, e.g., Moozhipurath et al., 2020, in relation to the novel corona virus). But Ponsonby et al. (2002) note that the available data upon which such notion is made are largely inconclusive.

González-Maglio et al. (2016) have put forward some ideas in perspective and posed some questions within the field of photoimmunology based on established and new information. These authors presume that their ideas may lead to new experimental approaches and hence, to a better understanding of the effects of sunlight on the human immune system. See also Neale et al. (2023) evaluation of the effects of ultraviolet (UV) radiation on human health within the context of the Montreal Protocol (the landmark multilateral environmental agreement that regulates the production and use of the ozone depleting substances), and its Amendments.

Developmental immunotoxicology

Despite a bulk of published data that exist on the subject today (2020), we are only just beginning to understand how some exposures may affect pediatric populations and human development (See, e.g., McElroy, 2020; WHO (World Health Organisation), 2020c). We do know, however, that in early life, there are numerous environmental factors that act in concert to modulate the immune system and to influence its competence; although as MacGillivray and Kollmann reiterated in 2014, evidence for long-term immune programming remains uncertain. These factors range from abiotic chemical exposures and nutritional status to biotic insults from infectious diseases and microbial or parasitic colonization; and interact in a network of components that include maternal cells, commensal microbes, and pathogens (Scanlon, 2020). The factorial interactions present windows of both vulnerability and opportunities that impact the developing immune system (MacGillivray and Kollmann, 2014; Kollmann et al., 2017; Maggini et al., 2018), with both short-term and long-term implications for lifelong health.

As MacGillivray and Kollmann (2014) noted, the outcomes of immune factorial interactions also depend on the magnitude and duration of prenatal and perinatal environmental exposures, including potential inherited programming (genetic and epigenetic), improper development due to altered metabolism (nutrition and toxicology), and inapposite immune system decision-making (for both adaptive and innate arms). With increasing age, the risk and severity of infections would vary in line with immune competence, which also depends on how the immune system develops, how it matures, and how it declines (Maggini et al., 2018).

Immune programming in early life: African examples
Many changes occur in the immune system over a life span—maturing gradually during childhood, potentially attaining optimal function in early adulthood, and declining slowly as

old age bekons (Goenka and Kollmann, 2015; Simon et al., 2015). Each life stage is characterized by distinct immune features and specific factors differentially affect immune function. This results in differences in the type, prevalence, and severity of infections with age (Nicholson, 2016; Rodriguez, 2017).

In Africa, perturbations in immune system function in early life have been observed in many instances to have important short-term impacts. For example, both birth season and nutritional status are commonly observed to decrease humoral response to *pertussis* (whooping cough) vaccination (See Chen and He, 2017; Akinola et al., 2018; Zimmermann and Curtis, 2019). Observations by Paynter et al. in 2015 also indicate that "... in the tropical setting of West Africa, both cell-mediated and humoral immune responses appear to be reduced in children during the rainy season." Lisse et al. (1997) and Moore et al. (2008) have also shown that in West Africa, birth during the wet season increases the proportion of T-cells in the CD8 + compartment, decreasing the CD4 + :CD8 + ratio. In rural Gambia, season of birth has also been associated with infection-related adult mortality and postulated to be due to perinatal nutritional status or pathogen exposure during early immune programming (Moore et al., 2008). However, no association between seasonality and immune status have been detected so far in subsequent studies of youth (See Moore et al., 2001) or young adults See Ghattas et al. (2007). As Ghattas et al. (2007) observed in a rural Gambian setting, "...independent of nutrition or pathogen exposure, seasonality can also change diurnal exposure to sunlight, and levels of circulating vitamin D."

An adequate supply of micronutrients is an absolute essential throughout life, since it plays key roles in supporting immune function; and mineral deficiency-induced abnormalities in the immune system are particularly profound when they occur during early development (Keen et al., 2004). Sadly, though, micronutrient deficiencies are still legion throughout the world, not least in Africa, with the likelihood increasing with age.

A "Special Issue" of the Journal, SCIENCE on: "Early Life Immunology" [edited by Tsevis (2020)] surveys recent advances in the field of early life immunology. "Review articles highlight distinctive features of human fetal immune system development elucidated by unbiased multiomics analysis, the impact of commensal metabolites and *xenobiotics* (in this context: toxic trace elements, including metals) on immunity before and after birth, how maternal and fetal immune components work together to combat viral infections during pregnancy (and what happens when these mechanisms fail), and the potential of vaccination approaches to boost fetal−maternal immunity and protect neonates against the pathogens that most frequently cause them harm." (Scanlon, 2020).

Why are children more susceptible to certain diseases than are adults?
The influence of nutritional status (micronutrient deficiencies) on immunity and the effects on the risk and severity of infection and xenobiotic/pathogen exposure during immune programming are most evident in children. This is so, because, taking body mass into consideration, children are unique in their susceptibility since their organ systems are in an early stage of development, with differences in the rate of micronutrient metabolism (See, e.g., McCarty et al., 2011; Maggini et al., 2018). "The innate immune system is relatively susceptible to pathogens, while the adaptive immune system is less able to quickly respond to T-cell-dependent antigens"

(Maggini et al., 2018). In contrast, however, see Carsetti et al. (2020) for an explanation of why the pediatric immune system is apparently prepared and fit to react to novel infectious agents, a function that might be diminished in adults and ineffective in the elderly. For further recent accounts on the effectiveness of childrens' immune response against diseases, see WHO (2011a, 2011b); Tesher (2019) and relevant references in the "Suggested List of References for Further Reading," in this Chapter.

Children are also at a higher risk of exposure to xenobiotics hence, disease, because they put objects in their mouths, eat dirt (*geophagy*: See Chapter 6), and spend much of their time outdoors. Exposure of children to Cd, for example, demonstrates dose-dependent suppression of circulating IgG (an antibody; See expanded definition in: "Glossary of Terms," this Chapter), which is associated with adult diabetes and impaired fasting glucose (Ritz et al., 1998; Schwartz et al., 2003). Miles et al. (2008) also noted that "African infants born in the wet season have stronger CD154 vaccine responses to BCG than those of infants born in the dry season."

According to Maggini et al. (2018): "... tailored supplementation based on the specific needs of each age group may help to provide an adequate basis for optimal immune function. *However, much more research is required into the effects of micronutrient supplementation on immune functions and on clinical outcomes.*"

Geochemical dynamics of immune system response during pathogenesis

The effects of trace elements/metals/metalloids involvement in many human metabolisms have been somewhat underestimated (See, e.g., Chen et al., 2016; Davies, 2022).

The human immune system is an advanced defense mechanism that protects its host from infectious diseases. This mechanism includes innate and adaptive arms and physical defenses used to detect and eliminate pathogenic microbes that are themselves constantly evolving (Chaplin, 2010), as well as eliminate toxic or allergic substances. The immune system has the ability to distinguish self from nonself. A disturbed immune function aggravates tissue injury (Chaplin, 2010) despite existing uncertainties on whether currently used clinical parameters accurately reflect immune responses (See: Yang et al., 2023).

Trace element/metal modulation of the immune system

Several trace elements have been shown to be essential for proper functioning of the immune system. These elements include Cu, Zn, and Se. Both deficiencies and excess amounts of trace elements can influence various parameters of the immune system, such as antibody responses, cell-mediated immunity, and natural killer (NK) cell activity (Beck, 1999; See also, Section: "Criticality of the optimal range of intake and the occurrence of nutrient toxicities," in this Chapter).

We now know that, depending on concentration, trace elements/metals, such as Cd, Hg, Pb can be either *immunopotentiating* or *immunosuppressive* (See, e.g., Cabassi, 2007; and Section: "Criticality of the "optimal range of intake" and the occurrence of nutrient toxicities," in this Chapter).

Trace elements (including trace metals and metalloids) are capable of crossing the placenta, where they produce effects that depend on the dose, duration, and timing of exposure (Galask et al., 2008; Gundacker and Hengstschläger, 2012; Xenofon et al., 2014). Other xeno-toxic chemicals can have similar effects. In utero exposure to Hg (in its organometallic form, methylmercury), for instance, can decrease T-cell functionality and alter circulating immuno-globulin levels (See, e.g., Bjørklund et al., 2019b). Such interactions have been implicated in the low birth weight and preterm birth so often experienced during pediatric consultations (Belles-Isles et al., 2002; Bose-O'Reilly et al., 2010; Chen et al., 2014). Neonatal immunotoxicity can also occur well below adult safety thresholds (Miller et al., 1998). Another example is As exposure, which may lead to direct effects on immunity through epigenetic programing involving leukocytes and metabolic physiology (Intarasunanont et al., 2012; Kile et al., 2012; Ahmed et al., 2012; Dangleben et al., 2013; Koestler et al., 2013; Lu et al., 2014).

Immunotoxicity due to metals

The role of the "metallome" is of critical importance in discussing the etiology of diseases of unknown etiology.

As Wang et al., stated recently (2020): "The homeostasis of metal ions is critical for the physiological functions of the brain. In AD *(Alzeimer's Disease)* patients or the animal models, the imbalanced metal ions and their transporters have been widely observed. The deposition of metal ions in different brain regions impairs mitochondrial functions and thus causes oxidative stress, which can result in cascade pathological reactions."

Also, according to Nriagu and Skaar (2015), "Many parts of the world in which common infectious diseases are endemic also have the highest prevalence of trace metal deficiencies or rising rates of trace metal pollution. Infectious diseases can increase human susceptibility to adverse effects of metal exposure (at suboptimal or toxic levels), and metal excess or deficiency can increase the incidence or severity of infectious diseases."

Metals and metalloids influence the function of immunocompetent cells by a variety of mechanisms (Fig. 3−10). Several of these are known to be *immunotoxic*, including: Al, As, Be, Cd, Co, Cr, Cu, Fe, Hg, Mg, Mn, Ni, Pb, Se, Sn, Va, and Zn. Depending on the particular metal, its speciation, concentration, and bioavailability, and a number of other factors, a continuous metal/metalloid exposure will result in an immunosuppression or *immunoenhancement effect* (Kakuschke and Prange, 2007). Immunotoxicity occurs "... either direct action of the free metal on the cell membrane or other organs of immunocytic components or by catalysis or inhibition of numerous enzyme reactions that are essential to cellular metabolism" (Cabassi, 2007). These interactions interfere with expression of the immune response. In this connection, Cabassi (2007) notes that: "... certain metals at a low exposure concentration manifest immunopotentiating effects, while at high concentrations they cause immunosuppression." This feature had earlier been observed by Theron et al. (2012) who pointed out that it held true particularly for PTEs, such as Cd, Hg, and Pb due to their cytotoxic effects, which induce apoptosis and/or necrosis of immune cells with consequential weakening of the immune defenses to infection.

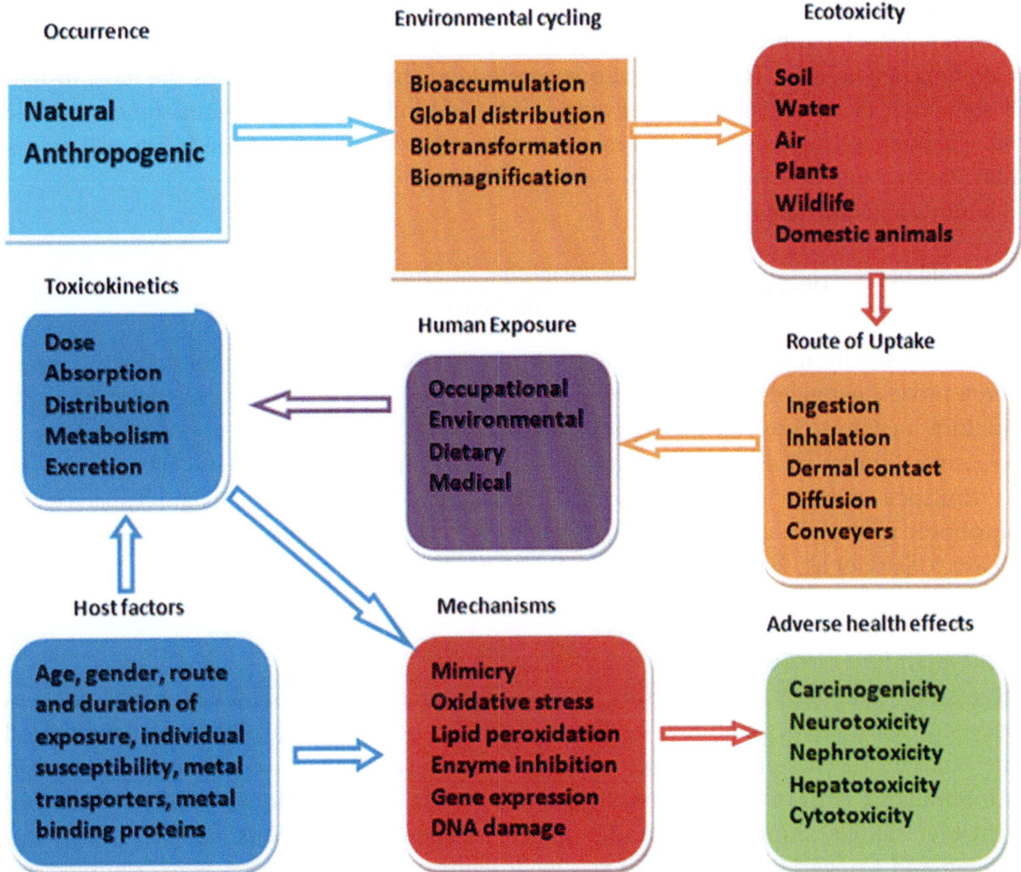

FIGURE 3–10 Schematic of routes of exposure, from source to outcomes for metal fluxes in humans (Anyanwu et al., 2018). *Excerpted from Anyanwu, B.O., Ezejiofor, A.N., Igweze, Z.N., Orisakwe, O.E., 2018. Heavy metal mixture exposure and effects in developing nations: an update. Toxics 2018 (6), 65. https://doi.org/10.3390/ toxics6040065*

Cabassi (2007) went on to describe some of the immunosuppression effects earlier identified by Descotes (1999) that xenobiotics (including trace elements/metals/metalloids) can produce. Effects described include: "... changes in leucocyte cellularity, lymphocyte subpopulation, reduced resistance of the organism to immune specific alterations, immunosuppression with increased susceptibility to infection, and tumor development, immunostimulation with hypersensibility, and development of ADs." (See Fig. 3—9).

Principles of infection and immunity

According to Galask et al. (2008), virtually any organism may behave as a pathogen under the right set of conditions; and therefore, it is more instructive to place organisms along a continuum from lesser to greater virulence, rather than classifying them as either pathogens or nonpathogens. Galask et al. (2008) also contended that "… among individual human hosts, there is a continuum in the intrinsic ability of each host to resist infection."

As long ago as 1934, Theobald Smith suggested, in what is now, perhaps the most insightful statement of the relationship between microbial virulence and host resistance to infection. This relationship is considered to accurately reflect the nature of the infectious process today, despite modern changes in the ecology of infections.

Smith's equation states:

$$\text{Disease} = \frac{\text{Number of organisms} \times \text{Virulence of organisms}}{\text{Host's resistance to infection}} \tag{3.1}$$

One can see from Eq. (3.1) that the result of a host's encounter with an infectious agent, even a proven pathogen, will not necessarily be an infectious disease. However, if the host's immunity becomes lowered for some reason or if the host becomes overwhelmed by an increasing number of organisms, disease may appear, even with an organism of relatively low virulence. Another noteworthy point about Eq. (3.1), is its practical significance, which contributes to the clinician's knowledge about the role of the individual host in infectious disease (Galask et al., 2008).

Immune system mechanisms

There are numerous mechanisms by which trace elements/metals/metalloids are absorbed, distributed, modified, and stored in the body, and subsequently eliminated. This section presents only a brief look at immune system mechanisms with reference to its interactions with trace elements, including metals and metalloids. Readers interested in further details should consult the many excellent publications on the topic (e.g., Sullivan and Krieger, 2001; Coico et al., 2003: Failla, 2003; Keen et al., 2004; Plumlee and Ziegler, 2006; Plumlee et al., 2006; Galask et al., 2008; Chaplin, 2010; Winans et al., 2011; Marshall et al., 2018; Paludan et al., 2020); and others in the Section: "Selected List of References for Further Reading," in this Chapter).

Toll-like receptors (TLRs), which are located either on the cell surfaces or within *endosomes* (See: "Glossary of Terms," in this Chapter), are type I integral transmembrane receptors involved in the recognition and conveyance of pathogens (including trace elements/metals/metalloids) to the immune system (El-Zayat et al., 2019). Some micronutrients (vitamins and trace elements) may be considered as important TLR regulators, as they have immunomodulatory functions. Vitamins D, B12, and A, Zn, Cu, and Fe, for instance, have important role on innate immune responses (El-Zayat et al., 2019).

Thurnham's (2004) review summarizes work on, inter alia, "… interactions between nutrients and genes, the influence of gene polymorphisms on micronutrients, the impact of immune responses on micronutrients and specific interactions of antioxidant micronutrients in disease processes to minimize potential pro-oxidant damage." Mineral deficiency-induced abnormalities in the immune system are particularly profound when they occur during early development (Keen et al., 2004).

In addition to the effect of trace elements on immune function, several studies have shown that, at certain levels, some of these elements, such as Se can influence the genetics of a viral pathogen. Thus trace element nutrition influences not only the host response to a pathogen but also the pathogen itself (See, e.g., Beck, 1999; Lingamaneni et al., 2015).

Factors that influence the toxicities of substances that encounter the body in *bioaccessible* form (those that are readily released from Earth materials into the body fluids) include: the exposure route, the dose, the chemical form of the substance at exposure, and the processes that chemically transform the substance during absorption, transport, and metabolism. (Plumlee et al., 2006; Finkelman et al., 2018).

Particle clearance mechanisms

The body's immune system is complex. This system is what offers us protection against foreign substances, pathogens, parasites, and mutated cells (Plumlee et al., 2006). One way by which the immune system achieves this is through *particle clearance mechanisms*, whose complexity is manifested in the way Earth materials and their contained toxicants are handled.

Particle clearance mechanisms for Earth materials and their contained toxicants include:

- "Protective barriers. The skin is one example, but the mucous membranes can also be considered as a physical barrier that helps prevent access of foreign substances and pathogens into the body.
- Chemical barriers. These include, for example, the acidic gastric fluids and the digestive enzymes contained in saliva, tears, and gastric fluids. The body's increased production of fluids in the nasal passages and upper respiratory tract in response to the influx of foreign substances is also a chemical defense mechanism that helps dilute chemical impacts of the substance.
- Microbiological barriers. The intestines contain a number of different bacteria that compete with pathogens for sustenance, thereby helping to diminish multiplication of pathogens into sufficiently large numbers to trigger illness.
- Phagocytic activity. The respiratory macrophages discussed previously one of several types of autonomous phagocytic cells that patrol the body to identify and engulf foreign particles or microbes. The macrophages are in part physical barriers, as they can help isolate foreign substances or pathogens and transport them away from physiologically sensitive areas. They also are chemical barriers in that they can help digest substances with their acidic lysosomal fluids rich in cytotoxic acid proteases and reactive oxygen species." (Plumlee et al., 2006). *Sic.*

The myriad ways in which toxicants released from Earth materials can cause chronic and/or acute toxicity can include, for example, generation of *reactive oxygen species (ROS)*; interference with the body's antioxidant processes and mechanisms; impairment of cellular function; replacing essential elements in key physiological chemicals (such as Cd replacing Ca in bone); and many others (Plumlee et al., 2006). Detailed explanation of how the mechanism of ROS interactions in metabolic processes can contribute to an understanding of the etiology of certain diseases (Diseases of unknown etiologies), are given in several recent publications, e.g., Checa and Aran (2020), Tavassolifar et al. (2020) and Yang and Lian (2020).

The quintessential role of nutritional elements in immune system function: Cu, Zn, Mg, Mn, Fe and Se

The relationship between nutritional status and immunological competence is a complex and highly debatable subject (See, e.g., MacGillivray and Kollmann, 2014; Kollmann et al., 2017; Calder, 2020; Calder et al., 2020; Mao et al., 2020). One thing, though, is clear: there are known nutritional deficiencies that can influence immune activity early in life.

Among the elements essential for life, Cu, Zn, Mg, Mn, Fe and Se stand out, both for the cruciality of their immunomodulatory functions and in the general nutritional maze. They can cause disease in case of deficiency, imbalance, or toxicity (See, e.g., Ciftci et al., 2003; Hoffmann and Berry, 2008). Therefore an evaluation of the pathological value of determining the fluxes of these elements (Cu, Zn, Mg, Mn, Fe, and Se) as well as the Cu/Zn ratio in metabolic reactions is instructive. So also, is the monitoring of their contents in whole blood, serum, hair, and/or other accessible fluids and tissues in subjects presenting with enigmatic environmental diseases.

Dietary deficiencies of these trace elements have been associated with multiple anomalies in immune function in both humans and experimental animals. Ciftci et al. (2003) outline clinical criteria for the recognition of their deficiency states, which include decreases in metal content of whole blood, serum, hair, and/or other accessible fluids and tissues, changes in the activities of metalloenzymes, and other characteristic signs and symptoms.

Criticality of the "optimal range of intake" and the occurrence of nutrient toxicities

Writing on the criticality of the optimal range for the micronutrient elements, Mao et al. (2020), remarked: "The dietary requirement for an essential trace element is an intake level which meets a specified criterion for adequacy and thereby minimizes risk of nutrient deficiency or excess. Disturbances in trace element homeostasis may result in the development of pathologic states and diseases."

A desirable dietary intake of essential elements (micronutrients) generally falls within a fairly narrow range (Table 3−3); and the level of intake has to be absolutely right for them to perform optimally in metabolic processes. However, many nutrients have an antagonistic relationship to one another, which can mean that when one is too high, it causes the other to become too low; and this could increase one's susceptibility to infectious disease, which may be acute or chronic. No pair of elements better exemplifies this relationship than Cu and Zn, which is a result of their complex interactions in metabolic processes. A high intake of Cu may adversely affect the absorption or utilization of Zn, and *vice versa*. In other words, when your Cu−Zn ratio becomes out of balance, many health problems can occur.

In 1998 the U.S. Institute of Medicine (USIM) published a reference table of values called the *Tolerable Upper Intake Level* (*UL*) for selected nutrients. Many countries, including some in Africa still use this Table as a model. The UL is the maximum acceptable level of daily nutrient intake (*safe intake*) that is considered to pose no risk of adverse health effects for almost all individuals in the general population. USIM (1998) states that: "... is not meant to apply to

Table 3–3 Dietary reference intakes (DRI's)/tolerable upper intake level (UL) for selected nutrients in the human body and rich dietary sources (WHO, 1996; Kulkarni et al., 2014; Park, 2011).

Trace element	Recommended daily intake (RDI)	Recommended dietary allowance (RDA)	Tolerable upper intake level (UL)	Dietary sources
Cu	2000 μg	Children 1−3 years old: 340 μg/day; 4−8 years old: 440 μg/day; 9−13 years old: 700 μg/day; 14−18 years old: 890 μg/day; Men and women aged 19 years and older: 900 μg/day; Pregnancy: 1000 μg/day; Lactation: 1300 μg/day	Children 1−3 years old: 1 mg/day; 4−8 years old: 3 mg/day; 9−13 years old: 5 mg/day; 14−18 years old: 8 mg/day; Adults 19 years old and above (including lactation): 10 mg/day; Pregnancy: 8 mg/day	Oysters, other shellfish, whole grains, beans, nuts, potatoes, organ meats (kidney, liver), dark leafy greens, dried fruits, and yeast
Fe	18 mg	Children 1−3 years old: 7 mg/day; 4−8 years old: 10 mg/day; 9−13 years old: 8 mg/day; Boys 14−18 years old: 11 mg/day; Girls 14−18 years old: 15 mg/day; Adults: 8 mg/day for men aged 19 years and older and women aged 51 years and older; Women 19−50 years old: 18 mg/day Pregnant women: 27 mg/day Lactating mothers: 10 mg/day	Infants and children from birth to the age of 13 years: 40 mg/day; Children aged 14 years and adults (including pregnant and lactating mothers): 45 mg/day	Hem iron: liver, meat, poultry, and fish; Nonhaem iron: cereals, green leafy vegetables, legumes, nuts, oilseeds, *jaggery*, and dried fruits
Zn	15 mg	Infants and children 7 months old to 3 years old: 3 mg/day; 4−8 years old: 5 mg/day; 9−13 years old: 8 mg/day; Girls 14−18 years old: 9 mg/day; Boys and men aged 14 years and older: 11 mg/day; Women 19 years old and above: 8 mg/day; Pregnant women: 11 mg/day; Lactating women: 12 mg/day	Infants: 4−5 mg/day; Children 1−3 years old: 7 mg/day; 4−8 years old: 12 mg/day; 9−13 years old: 23 mg/day; 14−18 years old: 34 mg/day; Adults 19 years old and above (including pregnancy and lactation): 40 mg/day	Animal food: meat, milk, and fish; Bioavailability of Zn in vegetable food is low
Co	6 μg	Infants: 0.5 mcg; Children 1−3 years old: 0.9 μg; 4−8 years old: 1.2 μg; 9−13 years old: 1.8 μg; Older children and adults: 2.4 μg; Pregnant women: 2.6 μg; Lactating mothers: 2.8 μg	Not known	Fish, nuts, green leafy vegetables (broccoli, spinach), cereals, and oats

(*Continued*)

Table 3–3 (Continued)

Trace element	Recommended daily intake (RDI)	Recommended dietary allowance (RDA)	Tolerable upper intake level (UL)	Dietary sources
Cr	120 μg	Children 1–3 years old: 11 μg; 4–8 years old: 15 μg; Boys 9–13 years old: 25 μg; Men 14–50 years old: 35 μg; Men 51 years old and above: 30 μg; Girls 9–13 years old: 21 μg; 14–18 years old: 24 μg; Women 19–50 years old: 25 μg; 51 years old and above: 20 μg; Pregnant women: 30 μg Lactating women: 45 μg	Doses larger than 200 μg are toxic	Best sources: processed meats, whole grains, and spices
Mo	75 μg	Children 1–3 years old: 17 μg/day; 4–8 years old: 22 μg/day; 9–13 years old: 34 μg/day; 14–18 years old: 43 μg/day; Men and women aged 19 years and above: 45 μg/day; Pregnancy and lactation: 50 μg/day	Children: 300–600 μg/day; Adults (including pregnancy and lactation): 1100–2000 μg/day	Animal food: liver; vegetables: lentils, dried peas, kidney beans, soybeans, oats, and barley
Se	70 μg	Children 1–3 years old: 20 μg/day; Children 4–8 years old: 30 μg/day; Children 9–13 years old: 40 μg/day; Adults and children 14 years old and above: 55 μg/day; Pregnant women: 60 μg/day; Breastfeeding women: 70 μg/day	The safe upper limit for Se is 400 μg/day in adults	Liver, kidney, seafood, muscle meat, cereal, cereal products, dairy products, fruits, and vegetables
I	150 μg	Children 1–8 years old: 90 μg/day; 9–13 years old: 120 μg/day; Children aged 14 years and adults: 150 μg/day; Pregnant women:209 μg/day; Lactating mothers: 290 μg/day	Children 1–3 years old: 200 μg/day; 4–8 years old: 300 μg/day; 9–13 years old: 600 μg/day; 14–18 years old: 900 μg/day; Adults above the age of 19 years including pregnant and breastfeeding women: 1100 μg/day	Best sources: seafoods (sea fish and sea salt) and cod liver oil Small amounts: milk, vegetables, and cereals

(Continued)

Table 3–3 (Continued)

Trace element	Recommended daily intake (RDI)	Recommended dietary allowance (RDA)	Tolerable upper intake level (UL)	Dietary sources
F	In drinking water: 0.5–0.8 mg	Children 1–3 years old: 0.7 mg; 4–8 years old: 1 mg; 9–13 years old: 2 mg; 14–18 years old: 3 mg; Men 19 years old and above: 4 mg; Women 14 years old and above (including pregnant or breastfeeding women): 3 mg	0.7–9 mg for infants; 1.3 mg for children 1–3 years of age; 2.2 mg for children 4–8 years of age; 10 mg for children above 8 years old, adults, and pregnant and breastfeeding women	Drinking water, foods (sea fish and cheese), and tea

Source: Bhattacharya, P.T., Misra, S.R., Hussain, M., 2016. Nutritional aspects of essential trace elements in oral health and disease: an extensive review. Scientifica (Cairo) 2016, 5464373. https://doi.org/10.1155/2016/5464373. (Compiled using data from: WHO (World Health Organization), 1996. Trace Elements in Human Nutrition and Health. World Health Organization, Geneva, Switzerland. ISBN: 92 4 156173 4. https://www.who.int/nutrition/publications/micronutrients/9241561734/en/. (Accessed 02 February 2021); Kulkarni, N., Kalele, K., Kulkarni, M., Kathariya, R., 2014. Trace elements in oral health and disease: an updated review. Journal of Dental Research and Review 1 (2), 100–104; Park, K., 2011. Park's Textbook of Preventive and Social Medicine, twenty-first ed. Banarsidas Bhanot, Jabalpur, India.

people under medical supervision." "Safe intakes" apply only to adults, with most ULs for infants, children, and adolescents being considerably lower. Table 3–3 is a more recent compilation of "safe limits" (including for children and adolescents) by Bhattacharya et al. (2016) from data obtained from various sources, for selected nutritional elements and PTEs.

Excesses or deficiencies of trace elements/metals/metalloids and infectious diseases often co-occur and are the result of complex metabolic interactions. Most of our essential nutrient intake is from our diet, though, thankfully, this portion alone is unlikely to bear excessive element intake levels. However, the consumption of fortified foods or supplements can also raise the level of increase the level of trace elements/metals/metalloids and hence increase the chance of toxicity.

Environmental or occupational exposure to potentially toxic levels of elements/metals/metalloids can induce concentrations that are bioavailable to immune cells, high enough to affect their function. Such an imbalance of the immune system caused by pollutants may play a significant role in the incidence of infectious diseases (See, e.g., Erickson et al., 2000; Chaturvedi et al., 2004; Hara et al., 2016). In any case, our bodies have an elaborate system for managing and regulating the amount of key trace elements and limiting or eliminating the PTEs circulating in blood and stored in cells (Osredkar and Sustar, 2011). It is when this system fails to function correctly, that we have abnormal levels and ratios of trace elements/metals/metalloids developing and paving the way for occurrence of infectious disease.

This has led to the concept of *nutritional immunity* in the context of host defense against pathogens (Djoko et al., 2015). This concept perceives a role for mechanisms by which a host organism sequesters trace elements/metals/metalloids to limit invading pathogens during infection. Calprotectin, for example, can restrict the acquisition of Zn or Mn (Kehl-Fie et al., 2011). The question remains, however, as to whether the host is able to exploit the toxic properties of transition metal ions and use them as bactericides? (See Djoko et al., 2015).

The antagonistic effects of the important micronutrients Cu and Zn, and why the Cu/Zn ratio is more relevant, has already been described at the beginning of the Section on: "Criticality of the "optimal range" and the occurrence of nutrient toxicities." It is this ratio in the body that makes it possible for enzymes to function properly. In Box 3−2, a brief account is given of the involvement of the Cu and Zn (Cu/Zn ratio) (at the optimum intake level) as an example of the process of intracellular communication involving the essential elements in immune cells.

Immune system research in Africa

It is now well-established that the immune system can be highly active in some people but much less so in others; and that there are important determinants, according to evolutionary theory, that account for differences in host immune responsiveness between human populations. Such observations have major implications for the protection of the organism against pathogens

---

BOX 3−2   Role of Cu and Zn (Cu/Zn ratio): Immunomodulatory Interactions.

The essentiality of Cu and Zn in humans' cellular metabolism has never been in doubt and it is clear that dietary fluxes of these nutrient elements do influence the susceptibility of the host to bacterial and other infections (See: Osredkar and Sustar, 2011; Djoko et al., 2015). Copper deficiency inhibits *interleukin (IL)-2* production by affecting gene transcription, IL-2 is a *cytokine* involved in normal T-cell metabolism (Keen et al., 2004). Copper has antimicrobial properties and supports *neutrophil, monocyte,* and *macrophage* function and NK cell activity (Calder, 2020; Gombart et al., 2020). Djoko et al., (2015) present a general review and assessment of the potential role for Cu and Zn intoxication in innate immune function and their direct bactericidal function.

The immunomodulatory functions of Zn, and in particular how this element helps in preserving physical tissue barriers, have been described by Name et al. (2020), with perspectives for the novel COVID-19 pandemic.

Few recent African studies have reported on the role of Cu in immunity and host susceptibility to infection (See, e.g., Montgomery and Shaw, 2015; Calder, 2020; Gombart et al., 2020). Of these studies, hardly any have referred to Cu imbalances in the immune system due to malnutrition in Africa. But the importance of Zn in immunity is well-demonstrated by the high prevalence of infectious diseases in the Continent due to Zn-poor diets (See, e.g., Black, 2003; Endalifer et al., 2020).

The work of Joy et al. (2014), based on dietary supply estimates, indicates that the risk of deficiencies of Cu and Zn as well as other vital dietary micronutrients (including Ca, Fe, I, Mg, and Se) in Africa, is high. Other recent literature on the relationship between the Cu/Zn ratio and the incidence of infectious diseases in Africa and elsewhere include the works of Malavolta et al. (2015), Bahi et al., 2017, Escobedo-Monge et al. (2020), and Laine et al. (2020).

and offer clues that can help explain the reason why some populations are particularly susceptible to diseases, such as COVID-19, which is more common in the West than in Africa.

A large-scale study performed by scientists from the Institute Pasteur and the CNRS and reported by Quach in 2016, attempted to uncover evolutionarily important determinants of differences in host immune of populations from different geographical locations. This study analyzed the immune responses of 200 African and European individuals, and showed that "… there is indeed a difference in the way these populations respond to infection; that this response is largely controlled by genetics, and that natural selection has played an important role in shaping such immune profiles." The study also offered proof that the genetic legacy passed on by Neanderthal man to Europeans has significantly influenced their ability to respond to viral challenges.

In Africa, only a handful of recent studies have looked at the influence of undesirable uptake levels of trace elements/metals/metalloids, or of seasonal variations on immune system function; nor of the implication of these factors on the susceptibility to pathogenic infection.

Ayanshina et al. (2020) looked at how changes in weather will affect host immune response to the novel COVID-19 infection and how conditions, such as rainfall, humidity, and sunlight influence the host susceptibility to infectious diseases in Nigeria and other developing countries. These researchers (Ayanshina et al., 2020) have also expressed the importance of understanding the impact of weather changes in transmission dynamics and in boosting of immune response to the novel COVID-19 virus through vitamin supplementation in infected patients during the rainy seasons in Nigeria. *(Nigeria has two rainy seasons. The first rainy season begins around March and last to the end of July; the second rainy season is shorter, starting around early September; and lasting to mid-October).*

Glennie et al. (2012) have attributed the high levels of morbidity and mortality associated with infectious diseases in Sub-Saharan Africa to "… the constant immune pressure from a range of environmental factors, such as undernutrition and multiple concurrent infections from birth through to adulthood." These authors (Glennie et al., 2012) also observed that "… the greatest risk of invasive disease is in the young, the malnourished and HIV-infected individuals." In order to acquire natural immunity for protection against infectious diseases during adulthood, occasional microbial exposure during childhood is required. But, as Glennie et al. (2012) observed, given the poor health care settings in many African countries, "… the heavy burden of malarial, diarrheal, and respiratory infections in childhood may subvert or suppress immune responses rather than protect, resulting in suboptimal immunity."

Disease clusters

The term *cluster* (as used in epidemiology) has been defined in various ways (See, e.g., Last, 1988; US CDC, 1990; Antó and Cullinan, 2001; Porta, 2008; Dolk et al., 2015; PHE, 2019). The common thread through all these definitions is an integration of the following features: (1) An unusually large aggregation, real, or perceived, of events or diseases, over and above the number that could be expected by chance; (2) The events or diseases are grouped together in time and space (i.e., a particular geographical location, the place of residence, or the location used for a common activity, e.g., workplace). The definition of Porta (2008) encapsulates all of these features, i.e., "Aggregations of relatively uncommon events or diseases in space and/or time in amounts that are believed or perceived to be greater than could be expected by chance. Putative disease clusters are often perceived to exist on the basis of anecdotal evidence, and much effort may be expended by epidemiologists and biostatisticians in assessing whether a true cluster of disease exists."

Communicable diseases often occur in clusters, whereas *noncommunicable disease* clusters are observed to be very rare *(See: "Glossary of Terms," in this Chapter, for definitions)*. Recognition of a cluster depends on its size; being reckoned to be greater than would be expected by the play of chance. According to Kumar et al. (2004), "clustering" may provide clues about the underlying etiology of a disease (*cf.*, diseases of unknown etiology), though this may have initially been based on anecdotal evidence. It is therefore incumbent on epidemiologists and biostatisticians to assess whether or not the suspected cluster corresponds to an actual increase of disease in the area. Usually, when clusters are recognized, they are reported to public health authorities in the local area, who may then re-evaluated them as outbreaks, if they (the clusters) are of sufficient size and importance.

John Snow's pioneering investigation of the 1854 cholera outbreak in Soho, London (Tulchinsky, 2018), is considered as a classic example of the study of "disease clusters."

Categories of disease clusters

The Ministry of Health of New Zealand (MoH) (2015) reported the identification of the following three categories of clusters:

- *time clusters*—when an unusual number of cases of a disease(s) occurs within a defined period of time;
- *space clusters*—when an unusual number of cases of a disease(s) occurs within a defined area; and
- *time-space clusters*—when an unusual number of cases of a disease(s) occurs within a defined time period and area." (Read and Borman, 2015).

Another classification of "clusters" could be according to the nature of the "outbreak" that leads to ill health, such a when exposures to contaminants in the environment occur from the atmosphere, water, soil, land, or consumer products. Such environmental exposures can lead to the occurrence of "clusters" that may be identified as physical, chemical or radiological in nature. Methods for linking such exposures to potential health effects and how such "clusters" may be investigated, have been outlined by PHE (2019).

Investigating disease clusters

Whenever disease clusters are identified and reported to local or public health authorities by concerned citizens or health care professionals, a response team is usually set up and sent immediately to investigate the cluster. Such an investigation follows a well-established process based on national, regional, or international guidance.

What is the value of investigating disease clusters?

There is hardly any doubt regarding the desirability for disease cluster investigations (including those that are potentially related to environmental or occupational exposures) to be conducted, and for the outcomes of these investigations to be transmitted to relevant public health authorities. Amid heightened public concerns and attention. There are many gains in performing a good cluster investigation, but there are also challenges as well.

The investigation often begins with determining whether a suspected cluster is a true cluster or not. Allied health professionals, industrial managers, and other stakeholders are able to address community concerns as well as undertake the appropriate level of scientific study. Epidemiologists can identify the appropriate scientific approach and organize a data collection plan that synthesizes the available health data. In this way, the identification of clusters and small outbreaks may eventually lead to the isolation of relevant risk factors (See later Section on: "Composition of cluster investigative team," this Chapter) and provide clues about the underlying etiology (*cf.*, diseases of unknown etiology).

Despite the views of Rothman (1990) that there is little scientific value in cluster investigations, there are several researchers who contend that "clusters" should be looked upon solely from their scientific merit.

Methodology for performing disease cluster investigation

Most reports of cluster investigations are of noninfectious illnesses, particularly cancer (CAWG, 2014). However, all the methodological components involved in the investigation of cancer clusters are also applicable to the investigation of other noninfectious diseases and conditions (CAWG, 2014).

The original "guidelines for investigating clusters of health events" given by the US Centers for Disease Control and Prevention (US CDC, 1990) remain valid today. A four-stage process is advocated in these guidelines, *viz.*, *initial response, assessment, major feasibility study*, and *etiologic investigation*, with each step providing opportunities for collecting data and making decisions. According to US CDC (1990), although it may not always be possible to follow this approach sequentially, it does provide a systematic basis for making decision as to whether to terminate or continue the investigation; and also that, although a systematic approach is of utmost importance, health agencies should be judicious in their method of analysis and tests of statistical significance depending on the particular attributes or characteristics of the potential cluster. An update on cancer cluster activities at the US CDC was presented by Kingsley et al. (2007).

Similar versions of steps of a cluster investigation are given by: Birnbaum et al. (1996), GoA (Government of Alberta) (2011), CAWG (2014), Read and Borman (2015), and PHE (2019).

In 2014 the Cancer/Cluster Analysis Work Group of the Cancer Data Registry of Idaho (CDRI) put forward procedures for investigating noninfectious disease clusters as a whole, with an emphasis on cancer, breaking down the "Investigative Procedure" into the following several distinct phases, with decision points at each phase:

- "Receipt of Report and Initial Evaluation (Phase I);
- Case Verification and Rate Comparison (Phase II);
- Determining Feasibility of Conducting an Epidemiological Study (Phase III); and
- Targeted Surveillance or Epidemiologic Study (Phase IV)" CAWG (2014).

Public Health England (2019) outlines the difficulties that are normally encountered in investigating potential clusters, how time-consuming the whole process could be, and the need for adopting a systematic, integrated approach in responding to such clusters. The guidance also describes some resources that can be utilized to aid cluster investigations, such as computer software packages, geographic information systems (GIS), and mapping. The guidance also stresses the importance of regular communication and reporting of results throughout the investigation and provides some examples of enquiries to illustrate the type of situations where this guidance can be used.

Finally, Chang et al./Exponent (Updated) pose the following questions, which are believed to be worthy of consideration by any cluster investigation team:

- "Are the rates of disease greater than expected in person, place, and time?
- If the observed rates of disease are greater than expected, is there documented exposure that could explain this cluster?
- Does the proposed exposure reflect a biologically plausible relationship between exposure and disease?
- Are there alternative explanations for the perceived cluster of disease?
- Are there specific subgroups that may or may not be at an elevated risk of disease?
- Are improvements needed in the workplace or environmental health surveillance systems to accurately characterize potential clusters?"

Combining cluster methodologies with GIS

Today, the use of GIS tools to delineate and display potential clusters is made routinely, particularly in cancer research. More complicated applications of GIS have included the active surveillance of cancer data to detect cancer clustering, a procedure that requires systematic implementation of GIS mapping and statistics-based spatiotemporal cluster analyses (See: Rushton, 2003; Sheehan et al., 2000). A good example is the NCI Cancer Atlas (for the United States) (US NCI, 2020), which contains images that are visually compelling and can be downloaded for research purposes.

However, there are a number of pitfalls associated with the use of GIS in cancer cluster research. For instance, there are reports of analytical complexities not readily addressed by current conventional GIS techniques and software packages (Jacquez, 2009; Jarup, 2004). Again too, there are the inherent problems with data accuracy and quality, metadata standards, inconsistent resolution of data points, aggregated data usage, and ethical problems regarding privacy and confidentiality (Elliott and Wartenberg, 2004; Jacquez et al., 2005; Nuckols et al., 2004; Waller, 2015).

Other opponents of the way that GIS is used in cluster analysis studies contend that this technology was originally intended for use in generating static "snapshots" of locational information in business settings (See, e.g., Thrall, 1999). They proffer that a more useful approach in applying the technique would be in investigating the etiology of chronic diseases if it were re-engineered to evaluate geographic locations in a dynamic time series (Jacquez, 2009; Jacquez et al., 2005; Sabel et al., 2000, 2007). Recently, Foster et al. (2018) have reviewed GIS applications as they pertain to 10 steps of an epidemiological field investigation.

The case of co-occurrences

Co-occurrences of cases often come up in cluster research, and are frequently reported as potential clusters. Various environmental exposures, such as can occur in the community or in the workplace, are frequently reported as sources of potential clusters. However, not all co-occurrences of disease cases are clusters per se. Preliminary analytical work, if performed systematically, can often resolve the question of whether a true cluster exists, or not. This may involve comprehensive epidemiological and statistical measurements and thorough assessment of the exposure characteristics to confirm whether a true cluster exists.

Clusters of diseases of unknown etiology

Rodo et al. (2016) reviewed the relevance of environmental factors to health outcomes of ailments whose causes are still poorly understood; and listed several examples of emerging diseases belonging to this category, and surprisingly sharing some common epidemiological features, "... such as their appearance in "clusters" (grouped geographically; and temporary progress in nonrandom sequences that repeat every year in a similar way). They also show concurrent trend changes within regions in countries and among different world regions.)" Rodo et al.'s (2016) list included: rheumatic diseases, such as vasculitides, some inflammatory diseases, or even severe childhood acquired heart diseases, KD, Henoch-Schönlein purpura, Takayasu's aortitis, and ANCA-associated vasculitis.

Clusters of diseases of unknown etiology do occur occasionally in various geographical localities in Africa. However, there hardly exists the requisite framework in African countries for undertaking disease cluster investigations that might contribute toward unraveling the underlying causes of diseases of unknown etiology.

Composition of investigation teams

Investigating a disease cluster involves the assembly of a multidisciplinary *cluster investigation team* and the construction of a collaborating network comprised of national and regional health agencies, law enforcement agencies and others. Certain investigations can proceed comfortably with only few resources needed in terms of time and money; others may be much more protracted and costly.

The composition of investigative teams of disease clusters and the role of respective team members are given in a number of publications (See, e.g., US CDC, 1990; Wartenberg, 2001; Carr et al., 2010; GoA (Government of Alberta), 2011; US CDC, 2012; CAWG, 2014; Read and Borman, 2015). However, none of the earlier publications mentions the contribution of a Medical Geologist, though this role is probably partly captured in the compass of the epidemiologist and environmental health scientist (e.g., GWA, 2017). It is emphasized that the most important role of the Medical Geologist in such a team would be, first, to identify and characterize any suspected geoenvironmental agent that may be involved and offer information on the source and migration pathways of its (geoenvironmental agent or toxin) into the human body, and the subsequent fate of the bioavailable fraction in metabolic processes (See: Box 3−3). Such information is vital in demonstration that an association does or does not exist between exposure and disease.

The PHE (2019) guidance suggests the membership composition of disease cluster investigation teams, and the various individual and collective roles and responsibilities of the members. Since reports of potential clusters require a public health response and often further investigation (PHE, 2019), an investigation team must also include, in addition to an epidemiologist and statistician, a social scientist, who understands the social dimensions of a cluster, *viz.*, the community's perception of risk, potential legal ramifications and the role or influence of the media (US CDC, 1990). It is also important to nurture credibility and understanding of the investigation by the public. To ensure this, it is important to recruit into the team, a relevant professional who would address communication activities at each stage of the cluster investigation, thus developing and maintaining community relationships and trust. In addition to these professionals, it is now preferred that the inclusion of Medical Geologists in cluster investigation teams will considerably enhance their capacity for success (See Box 3−4).

Multimorbidity disease clusters

The term *disease clusters* are often used in another sense, differing from that described in the foregoing sections. *Multimorbidity* refers to the occurrence in a patient of two or more medical conditions simultaneously, although some health care professionals prefer the more intuitive term *"multiple health conditions."* (Aiden, 2018). Morbidities that cluster together are referred to as *concordant multimorbidities*, since they can share a common etiology. Conditions, such as: depression, cardiometabolic disorders, cerebrovascular disease, and musculoskeletal disorders are most commonly present within multimorbidity clusters (Willadsen et al., 2016).

According to UKRI/NIHR (2020), multimorbidity has received limited scientific attention to date (2020) having previously been regarded as a random assortment of diseases; and hence, difficult to address. However, as recent research begins to unveil the impact of various factors [including biological, social, behavioral, (geo)environmental and others], we now understand multimorbidity to be a series of nonrandom *clusters* of disease. Some clusters, such as obesity with cardiovascular disease-osteoarthritis and several cancers are obvious. Other multimorbidity clusters, however, need further identification (particularly physical and mental health comorbidities) and follow-on mechanistic exploration.

> **BOX 3–3   The role of the medical geologist in the cluster investigation team.**
>
> Whenever there is a rare and strictly locality related disease outbreak whose etiology cannot be readily and clearly established, a suspected geoenvironmental factor(s) [cofactor(s)] is often implicated. Wherever such clusters occur, the Medical Geologist's first step in tracing etiology is the obtention of local knowledge of the nature of the soil (composition, chemistry, texture, and so on) and related geoenvironmental variables, such as changes in water composition and air pollution levels. The transdisciplinary nature of the questions begged can be answered through collaboration between the Medical Geologist and collaborators in the team of ready responders that include public health specialists, toxicologists, and data scientists and modelers and across the disciplines of epidemiology, ecology, climate science, and so on [See Taibo et al.'s (2017) for multidisciplinarity in investigation teams working on such outbreaks].
>
> Such a mix of professionals enhances greatly the task of the collaborating network of first responders, which is to investigate the problem so as to identify causes and risk factors, implement prevention, and control measures, and communicate with all stakeholders. The specific roles of a Medical Geologist in such an investigative team would be: (1) to provide pertinent geoscientific information to the public health specialists to help ascertain whether or not a geoenvironmental factor (cofactor) is indicated (2) to determine the nature of the geological materials and/or processes that may be involved; and (3) take appropriate action to prevent further contact/exposure to the implicated geological materials or processes.
>
> It follows that inclusion of the Medical Geologist in the investigation team would no doubt increase the likelihood of a cluster investigation detecting the source or origin (source apportionment) of a causal agent, as well as its nature (composition, form, speciation, etc.) as well as its migration characteristics. The Medical Geologist has the potential of making an invaluable contribution regarding etiology and therapy by providing better understanding of underlying mechanisms that produce disease clusters and by taking a broad approach that can encompass the biological, and wider determinants as appropriate. Medical Geologists will make clinical, public health, and scientific progress more achievable.
>
> In a similar vein, the inclusion of Medical Geochemists and Toxicological Geochemists (See: "Glossary of Terms," this Chapter for definitions) of these scientists in teams investigating *disease clusters* could ensure the provision to both scientists from health disciplines and allied researchers in a cluster investigating team with an overview of the potential geochemical mechanisms (from source apportionment to outcomes from metabolic interactions) by which Earth materials can influence human health (See: Plumlee et al., 2006; Hasan, 2020). It is evident that increased research collaborations between Medical Geologists/Medical Geochemists/Toxicological Geochemists and their counterparts in the cluster investigation team will herald massive advancement in the pursuit of cluster detection, its assessment, and implementation of follow-up measures that will lead to improved diagnoses and therapy.

The call for more research on "multimorbity" is justified by the fact that the proportion of patients who have two or more medical conditions simultaneously is rising steadily in all age groups in Africa and worldwide (Osakunor et al., 2018; Whitty et al., 2020). Such research would need the application of new multidisciplinary approaches to mapping and tackling

BOX 3–4 Parkinson's disease: an example to illustrate disease clustering.

"Clusters of *Parkinson's* cases occur from time to time, when, for example, a number of people in a neighborhood or small town develop the *disease.* But they often go unnoticed or are ignored because scientists lack the time and money to look into them."—Excerpt from Duenwald (2002).

Parkinson's disease occurs when cells in the *substantia nigra,* a darkly pigmented part of the midbrain, begin to die off. These cells produce *dopamine,* a chemical messenger that is essential for normal muscle movement. The cell death occurs gradually, which is why Parkinson's disease can go unnoticed for a long time.

According to Duenwald (2002) "Parkinson's does not seem to be primarily a genetic disorder since it runs in the families of only about 10%−15% of patients ..." So, in recent years, the search for pathogenic mechanisms for PD has been geared to environmental factors.

The association between environmental agents, such as trace elements and PD has been made several times in the literature (e.g., Bocca et al., 2004; Forte et al., 2004; Gellein et al., 2008; Ogunrin et al., 2013; Zhao et al., 2013; Bourke, 2018; Sun, 2018; Adani et al., 2020; Lemelle et al., 2020). *It is an association that probably gives rise to clustering,* which may provide clues about the underlying etiology of this enigmatic disease. However, at the time of writing (2021), only a few reports on clustering of PD had been published (e.g., Herishanu et al., 1989; Goldsmith et al., 1990; Kumar et al., 2004). Some of the more recent of these accounts, such as the study by Kumar et al. (2004) document three PD clusters in Canada and reviews evidence for PD causation in the light of their findings, which indicated the existence of a possible role for environmental factors, such as viral infection and toxins.

Working in the Negev (southern region) of Israel, back in 1990, Goldsmith et al. identified clusters of PD cases that strongly suggest that common environmental factors (drinking water and agricultural chemicals) were the likely causative factors.

Another noteworthy, relatively recent review of cluster involving PD, among other *neurodegenerative diseases,* is that of Miller and Sanzolone (2003) [whose account is based on earlier publications by Kurland and Mulder (1954) and Herishanu et al. (1989)], which noted the occurrence of high incidences of these diseases, mainly dementia, parkinsonism, and amyotrophic lateral sclerosis on the island of Guam (in the Micronesia subregion of the western Pacific Ocean). Parkinson's disease was found to be more prevalent along the southern coast, underlain mostly by Tertiary volcanic rocks, whereas the northern part of Guam, with lower incidences of the disease, consists of carbonate rocks. Since epidemiological studies have so far failed to conclusively establish the cause to be genetic (Steece-Collier et al., 2002; Shulman et al., 2011; Lill et al., 2012; Zhang et al., 2018; Zhao et al., 2020; Bandres-Ciga et al., 2020), the environmental causative paradigm presently reigns supreme.

Studies by Yase (1972) had also suggested that the neurodegenerative diseases in Guam may be related to accumulation of trace elements, such as Mn and Al, both of which may cause neurodegeneration. Later studies (Yanagihara et al., 1984; Yasui et al., 1990) have suggested that a deficiency in Ca and Mg in the soil and water along with readily available Al could be linked with occurrence of the diseases. Other studies have investigated metal exposure, particularly Al and Mn, and again, deficiencies in Ca and Mg (e.g., Yasui et al., 1992; Yasui et al., 1995).

One of the more convincing recent studies that illustrates the association of PD with trace elements is that of Sun (2018), who analyzed 41 soil elements from thousands of sites across the

(Continued)

BOX 3–4   (Continued)

contiguous United States. This study (Sun, 2018) concluded that: "Given that soil elemental concentration in a region is broad indicator of the trace element intake from food, water, and air by people, implications of the results are that high soil Se and Mg concentrations helped reduce the PD mortality rates and benefited the PD patients in the 48 states."

Finally, Bourke's 2018 study in the United States contends that "The combination of Mo deficiency and purine ingestion could explain the movement disorder, the distribution, the latent period and the dementia association.... Adenosine ingestion in molybdenum-deficient humans will lead to adenosine loading and potentially a disturbance to the A2a adenosine receptors in the nigro-striatum. This could result in Parkinson's disease."

Bourke (2018) concluded that "Consistent with the role proposed for Mo deficiency in PD is the observation that affected individuals have elevated sulfur amino acid levels, depressed sulfate levels, and depressed uric acid levels. Likewise, the geographical distribution of Parkinson's dementia complex on Guam corresponds with the distribution of molybdenum-deficient soils hence molybdenum-deficient food gardens on that island." Since hardly any conclusive evidence has so far not been adduced to show that it is infectious or associated with a widespread environmental toxin, Bourke (2018) concludes that nutrition could likely be a credible potential cause, and that "The present perspective presents the opinion that the salvage of dietary sources of the purines adenosine and guanosine in Mo-deficient humans could initiate the process that leads to PD with dementia." (Bourke, 2018).

---

multimorbidity, at scale (UKRI/NIHR, 2020). This kind of research was proposed and sponsored by UKRI/NIHR (2020) starting in 2019. Seghieri et al. (2024) have put forward a new methodology based on the joint implementation of data reduction and clustering algorithms to determine patterns of chronic diseases that are likely to co-occur in multichronic patients.

Finally, writing on the *modus operadi* of effectively tackling "multimorbidity," Whitty et al. (2020) contend that "Clustering of diseases, and how we might better tackle management of coexisting physical and mental health problems, should be embedded into medical training and continuous professional development, including for specialists ..."

Assessment of disease clusters

In recent years, there has been considerable progress in the use of modern methods for the assessment of clusters. The magnitude of such development has resulted in the recognition of *epidemiological and statistical analysis of clusters* as a specialized field that has found a place in several journals.

Following, is a selection from the many templates and formats for assessment of disease clusters that have been proposed by various authors and institutions, with some notable exemplars.

- Mansfield et al. (n.d.) have outlined standardized stepwise process for receiving and evaluating cluster reports.

- Queensland Health (n.d.) has attempted to address the various frequently asked questions regarding cancer clusters.
- Waller et al. (1995) looked at methods for detecting and assessing of clusters of disease and illustrated their application to nuclear power plant facilities and childhood leukemia in Sweden.
- Antó and Cullinan (2001) have reported on two conceptual and methodological issues concerning interstitial lung disease (ILD) about which there is a lot of misunderstanding and contradiction. These authors have illustrated how epidemics and clusters of ILD can be effectively investigated and their classification systematically carried out.
- Wartenberg (2001) reviews different methods available for investigating disease clusters.
- McElvenny et al. (2003) discussed points on investigating and analyzing workplace clusters of diseases.
- Song and Kulldorff (2003) have discussed many different statistics tests that have been proposed to *test* for spatial *clustering*.
- Aamodt et al. (2006) have done a simulation study of three methods for detecting disease clusters.
- Porta (2008) has commented on effort expended by epidemiologists and biostatisticians in assessing whether a true cluster of disease exists.
- Guajardo and Oyana (2009) present a critical assessment of geographic clusters of breast and lung cancer incidences among residents living near the Tittabawassee and Saginaw Rivers, Michigan, USA.
- Coory and Jordan (2013) have advocated for the removal of assessment of chance from protocols for investigating cancer clusters.
- Gómez-Rubio et al. (2005) looks at detecting clusters of disease with R (See: "Glossary of Terms," this Chapter, for definition).
- Forouzanfar et al. (2016) have done a comparative risk assessment of 79 behavioral, environmental and occupational, and metabolic risks or clusters of risks that date from 1990 to 2015, within the ambit of the "Global Burden of Disease Study of 2015." A similar assessment was performed later (2017) and reported by Stanaway et al. (2018), this time, involving a look at 84 behavioral, environmental, and occupational, and metabolic risks or clusters of risks for 195 countries and territories from 1990 to 2017.
- Kronenfeld and Wong (2017) have described the visualization of statistical significance of disease clusters using *cartograms* (See: "Glossary of terms," in this Chapter).
- The largest and most comprehensive study of disease clusters known so far was undertaken by the Natural Resources Defense Council (NRDC) and the National Disease Clusters Alliance (NDCA) of the United States using information from federal, state, and local officials (Navarro et al., 2011). Forty-two disease clusters were found in 13 US states. These institutions subsequently urged federal coordination to better investigate these clusters, determine the causes, and work to protect residents in the affected areas. An interactive map showing the distribution of disease clusters in this study is presented in Fig. 3−11.

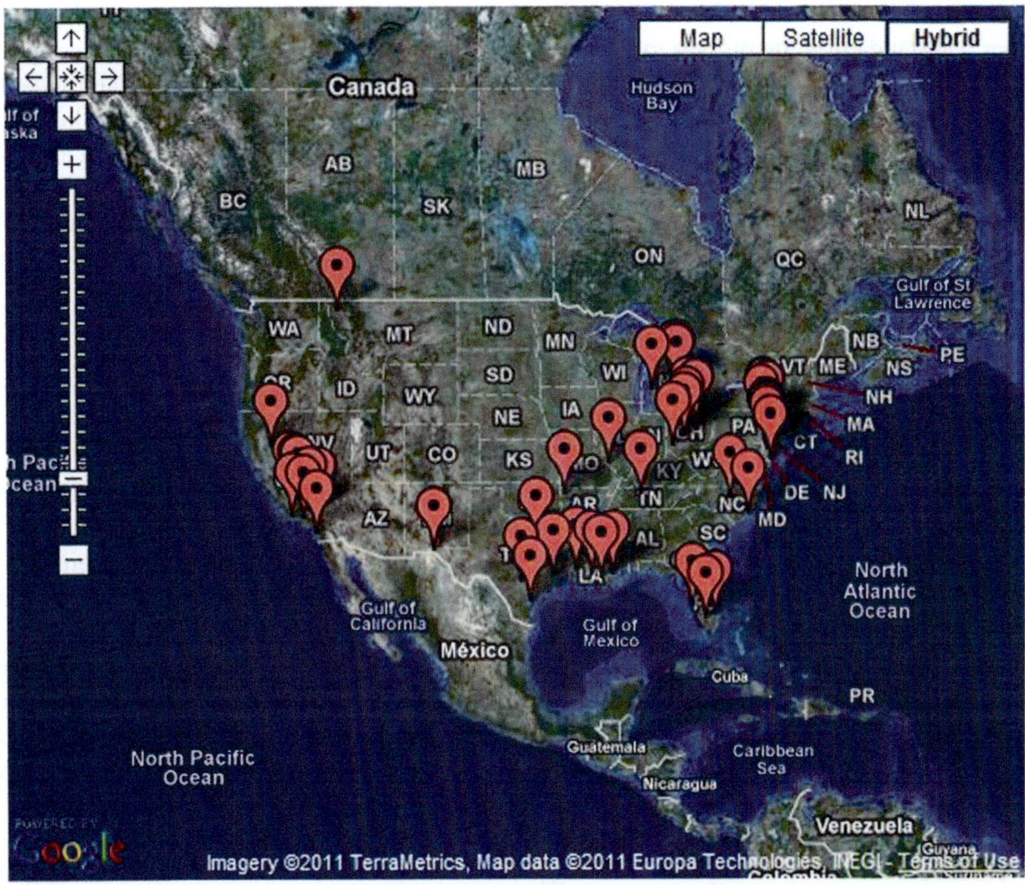

FIGURE 3–11 Disease clusters in some selected states of the US (The NRDC/NDCA Study, Navarro et al., 2011).
*This is an interactive map that can be navigated at: https://medicalxpress.com/news/2011-03-disease-clusters-states.
html. (Accessed 23 January 2021). *Credit: TerraMetrics/Europa Technologies, 2011.*

Disease clusters research: Africa

Very few disease cluster investigations in African countries are reported in the literature. These
have been performed on a case by-case basis, each one typically following a process similar to
that outlined by the United States Centers for Disease Control and Prevention (US CDC) in
1990 (See, e.g., Darikwa and Manda, 2020; Mudau et al., 2014). The US CDC (1990) protocol is
a useful guide; however, it does not account for data access and data quality available in
African countries, nor does it address specific communication needs in these countries. As
Davis noted in Davis (2001), "Frequently, vertically oriented disease surveillance programs at
the national level and above in Africa often result in too much paperwork, too many different
instructions, different terminologies, too many administrators, and conflicting priorities..." The
absence of a standardized country-specific document has occasionally created confusion

making stakeholder communication and education very difficult. Established national guidelines will standardize the process to promote efficient and informed communication among all stakeholders including provincial and local public health professionals, provincial, and federal health agencies, as well as the general public. Additionally, the guidelines will help instruct, support, and educate individuals completing surveillance analytic work required in such investigations. These suggestions are encapsulated in the works of Mudau et al. (2014) on "... identification of space-time clusters of multidrug resistant TB (MDR-TB) in South Africa, 2006–12" and Darikwa and Manda (2020) on "Spatial Co-Clustering of Cardiovascular Diseases and Select Risk Factors among Adults in South Africa."

The value of environmental risk maps

What are "environmental risk maps"?

According to Lahr and Kooistra (2010), "Risk maps help risk analysts and scientists to explore the spatial nature of the effects of environmental stressors such as pollutants." Environmental risk maps are used as a means for conveying the results of complex environmental risk assessments to public health authorities, policy makers, urban planners, and other stakeholders, such as the general public.

Pinto et al. (2014) wrote "Geochemical mapping is the base knowledge to identify the regions of the planet with critical contents of PTEs from either natural or anthropogenic sources. Sediments, soil, and waters are the vehicles, which link the inorganic environment to life through the supply of essential macro and micro nutrients. The chemical composition of surface geological materials may cause metabolic changes, which may favor the occurrence of endemic diseases in humans." *Sic.*

In the above context, it is possible to suggest that, for us to create a better understanding of the relationship between surficial geochemistry and public health it is necessary, first, to construct complete geochemical maps at appropriate scales across national boundaries, depicting the surficial distribution of all nongaseous chemical elements (See Darnley et al., 1995). Such maps have already been drawn for China [(See: Wang et al., 2007; Xie et al., 2008; Cheng et al., 2014), England and Wales (See Rawlins et al., 2012), Australia (See Reimann and de Caritat, 2017) the United States (See Smith et al., 2019), and a few other countries]. An overlay of epidemiological maps (of disease distribution) on these geochemical maps would make possible the depiction of areas where disease clusters overlie anomalous element distribution (in water or soil), and so permit an evidence-based statistical assessment of the magnitude of any geochemical component in the disease causative web.

Maps encompassing geochemical databases have been proposed for Africa (AGD), intended to, among other environmental applications, not least, in the field of Medical Geology, allow for overlays with epidemiological maps to be made, for identifying risk areas of geochemical disease. Several "attempts" have been made to complete such maps

(See, e.g., GEO, 2016; Davies, 2019), but a number of challenges have, to a large extent, stymied them (these "attempts"). A detailed account of the history of the AGD Project and other aspects of its evolution to date (2023) are presented in Chapters 3 and 17, respectively.

In response to the call for sustainable use of groundwater for the benefit of the poor across Sub-Saharan Africa, and in response to the UNICEF/WHO Report on "Safely Managed Drinking Water" (UNICEF/WHO, 2017), much effort (through modeling studies and intensive literature survey) has recently been put into the construction of environmental risk maps of water pollution, using various water quality parameters, such as anionic pollution levels (e.g., nitrate) (Fig. 3–12).

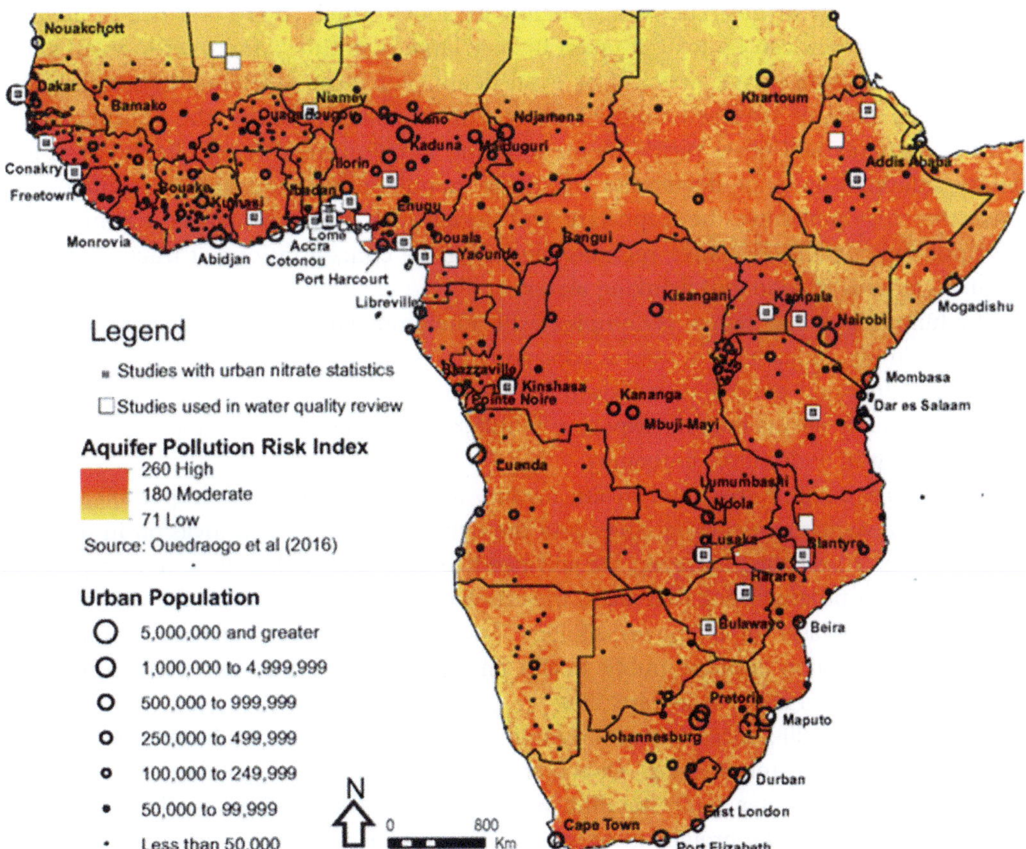

FIGURE 3–12 Relationship between urban centers in sub-Saharan Africa (SSA) and estimated aquifer pollution risk using an intrinsic aquifer modeling approach (Ouedraogo et al., 2016).
Note: The location of studies in the major cities in Sub-Saharan Africa are shown and are from the ESRI cities dataset 2006 (ESRI, 2007). *From Ouedraogo, I., Defourny, P., Vanclooster, M., 2016. Mapping the groundwater vulnerability for pollution at the pan African scale. Science of the Total Environment 544, 939–953.*

Conclusions

This Chapter has advanced reasons for a keener look to be taken at why causality cofactors linked to the geoenvironmental mileu deserves better attention.

A functional immune system able to prevent or limit infections of the host, is particularly important for many rural populations in Africa, where exposure to novel infectious occur frequently. From the abundance of evidence adduced in the Chapter, it is becoming increasingly clear that the amount of trace elements/metals/metalloids taken up largely through the African diet, and its outcome in metabolic processes (leading either to accumulation or to deficiency in human tissues), has a significant control on whether the exerted effects are toxic or beneficial. There is, thus, the need for each micronutrient to fall within a safe or desirable range, below or above which, there is a risk of adverse effects. Any nutrient can be toxic if taken in very large quantities, and overdoses of some can cause poisoning (acute toxicity) or even death. For most nutrients, habitual excess intake poses a risk of adverse health effects (chronic toxicity). Clearly, the optimum level of trace elements and other nutrients for immune function needs to be included in any dietary intervention plan developed for these populations. Food safety regulators in Africa therefore now have the important and urgent task of reconsidering, harmonizing, and updating current legislative regimes regarding the concentrations of trace elements/metal/metalloids in food and drinking water.

The human immune system is complex, with numerous environmental factors modulating it early in life. As such the system is constantly in a state of flux, trying to adapt to various local constraints and conditions imposed by selective pressures of our environment. This inherent plasticity means that our exposure to different geochemicals and pathogenic organisms can result in undesirable outcomes.

To promote immune-mediated health for life, we must consider the importance of our exposure to geoenvironmental variables and the dynamics of pathogen invasion in immune programming. To do this, however, we still need to seek knowledge on several aspects of immune system programming that starts in early life, and its influence on the risk of developing various diseases of unknown etiology. Such research would generate information needed for articulation of future public health initiatives and for drawing renewed attention to the vulnerability of children in early life.

The reasons for differences in host immune responsiveness between human populations are now becoming clearer (See, e.g., Quach et al., 2016; Barreiro and Quintana-Murci, 2020), as a result of our greater understanding of evolutionary theory. Such explanations have major implications for the protection of the organism against pathogens and offer clues that can help explain the reason why some populations are particularly susceptible to diseases, such as the novel COVID-19, which is more common among the Western nations than in Africa (See also: Onyilagha and Uzonna, 2019 and Alfituri et al., 2020).

On the other hand, Bentwich et al.'s (1995) explanation of the difference in HIV diffusion rates between Africans and western populations, back in 1995, lies in the "... overactivation of the immune system in the African population, owing to the extremely high prevalence of infections, particularly helminthic, in Africa. Such activation shifts the cytokine balance toward a T helper 0/2 (Th0/2)-type response, which makes the host more susceptible to infection with human immunodeficiency virus (HIV) and less able to cope with it."

Despite the unprecedented rate of global climate change, links between weather and infectious disease have not received much research attention even in high income countries. Improved knowledge on the linkages between environment and infection risk in Africa would help sensitize national authorities, policy makers, and other stakeholders on the need for investment in robust public health surveillance systems that are able to detect changing disease burdens. An understanding of the divers effects of climate change on the immune system, coupled with recent results from the emerging science of epigenetics would help unravel the etiology of diseases of unknown etiology, which can never be explained or resolved by isolating independent cause and effect relationships.

After decades of research on the complexity and developmental trajectory of the fetal-neonatal immune system (See, e.g., Amarasekera et al., 2020; Jain, 2020; Scanlon, 2020), we are only just beginning to acquire knowledge and insights on the participation of trace elements/metals/metalloids in the selection, maturation, and early activation events of the immune cells. Judicious use of modern analytical tools in cell biology and molecular genetics research, and array technology, will no doubt hasten our understanding of the outcomes in these metabolic processes.

Determinants of disease emergence and persistence interact in a highly complex way within the environment-ecosystem network. Consequently, model-based projections, as useful as they may be, are fraught with great uncertainty (See: "Suggestions Areas of Future Research," in this Chapter).

Cluster investigations are an important public health strategy for responding to public concern about possible associations between disease and environmental exposures. Roomaney et al. (2023) highlighted disease clusters in South Africa that are not well-researched; and advocated the need to assess disease associations, and garner information that could be used to advocate for better screening and management of patients. A proper investigation of a noncommunicable disease cluster may help generate hypotheses that can be tested in another study population and time period. It is submitted that the efficiency of cluster investigation teams would be enhanced by inclusion of Medical Geologists, from whom significant geoenvironmental exposure information can be obtained, as well as for an increased potential for unravelment of environment and disease relationships.

Increase in public awareness and concern about environmental exposures will continue to spur the need to follow up on reported clusters of disease. This is an important task that we need to carry out within a customized and systematic framework in the light of glowing publicity and under pressing and stressful circumstances. The overarching need for development of techniques for recognizing the grouping of cases of a particular disorder in space and time (disease clusters), is that this may provide clues about the underlying etiology (*cf.*, diseases of unknown etiology).

There is little doubt that GIS technology will continue to find widespread application in cluster investigations for noninfectious diseases registries, and by cluster responders. However, to achieve optimal results, new developments such as including more systematic approaches to cluster investigations, improved accuracy in data acquisition, and consideration of geoenvironmental cofactors in disease incidence may be necessary.

Finally, a lot of research is being devoted to the resolution of apparently tractable questions in genetics, gene therapy, and immunology to trace the cause of diseases (diseases of unknown etiology) that are still poorly diagnosed. For those (diseases) occurring under geographically localized conditions, more attention should be given to the geoenvironmental hypotheses and an examination made of their scientific strength (See Davies, 2013). It is clear that improving knowledge in the area of pathogen-metal interactions using state-of-the-art investigating techniques will be instrumental to the development of novel therapeutic measures (diagnostic tools) against infectious diseases.

Suggested areas of further research

- Not much research has been done to investigate the form of the Earth material with which a particular pathogen is associated, and how this may influence the pathogen's ability to persist and thrive *in vivo* (See Plumlee et al., 2006).
- CHEs, such as communicable disease outbreaks are a common threat to public health in Africa. Descriptions of risk factors for such outbreaks are often nonspecific and often cannot generally be applied in similar situations. Hammer et al. (2018) have attempted: "… to capture relevant evidence and explore whether it is possible to better generalize the role of risk factors and risk factor cascades these factors may form." *More rigorous research on the risk of disease outbreaks in CHEs is needed, and the composition of investigating teams reviewed, to include Medical Geologists.*
- Even though the recognition of seasonal patterns in infectious disease occurrence was made at least as far back as the Hippocratic era (Fares, 2013), *we still do not understand fully, the mechanisms underlying the development of these patterns and the association between seasonality and immune status still remains conjectural* (See, e.g., Nelson and Demas, 1996; Nelson et al., 2002; MacGillivray and Kollmann, 2014).
- Diseases, such as West Nile fever, Chikungunya fever and Lyme disease are all examples of how climate change, other geoenvironmental variables and anthropogenic factors have combined to result in the emergence of *zoonoses* (See: "Glossary of Terms," in this Chapter; See, e.g., Naicker, 2011). However, their epidemiology, including the *modus operandi* of their emergence and spread in Africa has not, as yet, been well-documented.
- Much research has been done in unraveling seasonal fluctuations in infectious diseases, but few have documented long-term trends in climate-disease associations.
- According to Pfavayi et al. (2020), "The prevalence of allergic diseases in the African continent has received limited attention with the allergic diseases due to fungal allergens

being among the least studied." This gives the impression that the prevalence of allergic disease in Africa is low, contrary to recent reports from different African countries. *Thus there is need to understand both the etiology and pathogeneses of these diseases, particularly the neglected fungal allergic diseases.*

- As at the time of writing (2020), the origins of *C. auris* were still unclear; and it has been difficult to determine whether environmental and climactic changes were contributing factors in its recent emergence as a pathogen (See Vila et al., 2020). *Increased research effort is warranted to unravel the origin of this pathogen, the exponential rise of infections and whether environmental and climactic changes were contributory factors in its recent emergence as a pathogen.*

- Also, Forsberg et al. (2019) noted that methods for rapid and reliable identification of *C. auris* are still not yet in place, for allowing quick screening of patients and identification of new cases. Vila et al. (2020) have called for "… *a robust response involving rapid diagnostics, prompt interventions and implementation of precautions, are paramount in curtailing the spread of infections by this fungal species.*" And, according to Du et al. (2020), "*Future efforts should be directed toward understanding the mechanistic details of its biology, epidemiology, antifungal resistance, and pathogenesis with a goal of developing novel tools and methods for the prevention, diagnosis, and treatment of C. auris infections.*" Despite this call, by 2023, the environmental phase of *C. auris'* life cycle and the transmission pathways associated with it were still not fully understood (Akinbobola et al., 2023).

- Patz et al. (2003) have highlighted the evidence linking climatic factors, such as temperature, precipitation, and sea level rise, to the lifecycles of infectious diseases, including both direct and indirect associations via ecological processes. Also, according to Liang and Gong (2017), "The life cycles and transmission of most infectious agents are inextricably linked with climate. In spite of a growing level of interest and progress in determining climate change effects on infectious disease, the debate on the potential health outcomes remains polarizing, which is partly attributable to the varying effects of climate change, different types of pathogen-host systems, and spatiotemporal scales." *Sic.*

- The seasonality of certain diseases may be as a result of enhanced wintertime survival of associated pathogens, but also the result of increased host susceptibility to relative "wintertime immune suppression." Geoenvironmental influences (e.g., hydrological variables) on vector or reservoir abundance, and vector biting rates, are probably very important. *However, numerous areas of uncertainty exist, making this an exciting topic for future research.*

- The etiopathogenesis of ADs remain obscure. We are only just beginning to gain insight into the role of different environmental factors in their development. Research into the geological links of probable environmental factors of autoimmunity deserves some attention. Consideration of seasonal variation patterns of ADs can possibly provide clues to diseases pathogenesis and lead to development of new approaches in treatment and preventive care.

- Paynter et al. (2015) have noted the variability of human immunity by season, and that: "In the tropical setting of West Africa, both cell mediated and humoral immune responses appear to be reduced in children during the rainy season." *Further research to assess the extent of seasonal immune modulation is required.*

- For performing further research to assess the extent of seasonal immune modulation, Paynter et al. (2015) have outlined a number of recommendations to minimize bias in this domain.
- "Although it is widely recognized that essential trace elements are required for the differentiation, activation and performance of numerous functions of immune cells, the specific roles of these inorganic micronutrients in these processes remain largely undefined." (Failla, 2003).
- According to Thurnham (2004), "More research is needed on interactions between nutrients and genes, the influence of gene polymorphisms on micronutrients, the impact of immune responses on micronutrients and specific interactions of antioxidant micronutrients in disease processes to minimize potential pro-oxidant damage."
- The crucial roles of several elements including the metallome in viral infections, in virus survival, in the activation of the host's immune system, and in maintenance of its (the immune system's) overall integrity, have now been firmly established. These roles largely depend on the presence of micronutrients. The role of trace elements (both micronutrients and PTEs) in the immune system and the state of the art in metallomics' involvement in diseases of unknown etiology etiology are topics that are urgently in need of further research. This subject has gained even more attention in the light of the novel Corona virus infection that continues to plague the world as we advance into the year 2021.
- According to Maggini et al. (2018): "The available clinical data suggest that micronutrient supplementation can reduce the risk and severity of infection and support a faster recovery. However, much more research is required into the effects of micronutrient supplementation on immune functions and on clinical outcomes."
- Transmission patterns of infectious disease in Africa are constantly changing, with the likely major cause being climate change. We need to keenly research the underlying complex causal relationships and apply the findings to the prediction of future climate change impacts, using more complete, better validated, tightly integrated models.
- Existing models are no doubt helpful in predicting likely health outcomes in relation to projected climatic conditions; and indeed, several different kinds (of model) have been designed to simulate climate changes and future outbreaks of disease. However, many models have failed to demonstrate strong relationships between observables and projections. Uncertainties in some of these climate models and in predictive epidemiology, influence their accuracy. Experimental models that demonstrate dose dependent effects between climate change and disease are largely absent. There are a number of other existing knowledge gaps that future initiatives must address in certain aspects of modeling climate change and disease outbreaks (See: Patz et al., 2005; Flato et al., 2013; Zelinka et al., 2020; Zhu et al., 2020). The use of *multispecies* (arising or occurring between species) approaches under highly standardized experimental conditions are needed to be able to predict responses of immunological traits to environmental changes (Meister et al., 2017).

- According to Flato et al. (2013), "A climate model's credibility is increased if the model is able to simulate past variations in climate, such as trends over the 20th century and paleoclimatic changes." Many of the existing models do not control successfully for important sociodemographic and environmental influences. The lack of long-term, high-quality data sets as well as the large influence of socio-economic factors and changes in immunity and drug resistance precludes credible attribution of expansion or resurgence of diseases to climate change (Patz et al., 2003; Patz et al., 2005).
- Rohr et al. (2011) reiterated: "We suggest that forecasts of climate-change impacts on disease can be improved by more interdisciplinary collaborations, better linking of data and models, addressing confounding variables and context dependencies, and applying metabolic theory to host-parasite systems with consideration of community-level interactions and functional traits. Finally, although we emphasize host-parasite interactions, we also highlight the applicability of these points to climate-change effects on species interactions in general."
- Studying the consequences of extreme heat events in Africa, Manyuchi et al. (2022) emphasized the urgent need for the development of heat-health plans and implementation of interventions for combatting the high ambient temperatures and human morbidity and mortality rates on the Continent.
- The African population is highly dependent on animals and animal products as food. *The likelihood of prion entry into the food chain and the severity of prion disease, underscore the urgent need for epidemiological, genetic and clinical studies on prions (See e.g., Teferedegn et al., 2019).*
- There are many reports suggesting that prions can be trapped in soil for several years. However, the overall effect of such entrapment, and prion persistence in the environment are still not well understood (See Smith et al., 2011). Nor has there been much progress in unraveling the mechanism by which PrPTSE attaches to soil particles. Further studies are therefore needed to determine the implications of prion's entrapment in the soil, and on the environmental transmission of scrapie and CWD (Smith et al., 2011).
- The newly (2018) discovered CPD extends the list of prion diseases in Africa. Further research of its dynamics within the dromedary population is required to assess the risks it poses to animal and public health.
- The coclustering of diseases of unknown etiology with trace element/metal deficiency or toxicity is observed (but not fully investigated) in many instances in Africa. The web of interactions producing such clusters and their geospacial associations are poorly understood and poorly researched at present. Despite being complex, expensive and often inconclusive, investigations of disease clusters should be encouraged, especially those (clusters) occurring in rural African populations, as they so often bring out the importance of the "place factor" in disease causation; and potentially strong contributors to better diagnosis and therapy for diseases of unknown etiology. A paradigm shift in cluster response is therefore advocated, and for which, we need to develop new scientific tools for investigating cause-and-effect relationships in small populations.

Glossary of Terms

- *Acute disease/illness* is any disease or illness that develops quickly, is intense or severe and lasts a relatively short period of time, or, any condition, e.g., infection, trauma, fracture with a short (often <1 month) clinical course.
- An *allele*, also called *allelomorph*, is any one of two or more genes that may occur alternatively at a given site (locus) on a chromosome.
- *Antigens* are anything that causes an immune response. Antigens can be entire pathogens, like bacteria, viruses, fungi, and parasites, or smaller proteins that pathogens express.
- *Autosomal* means that the gene in question is located on one of the numbered, or nonsex, chromosomes. An autosome is any chromosome that is not a sex chromosome (an allosome).
- *Bias: In the field of statistics, bias* refers to the tendency of a *statistic* to overestimate or underestimate a parameter.
- A *cartogram* is a thematic map of a set of features, in which their geographic size is altered to be directly proportional to a selected ratio-level variable, such as travel time, population, or GNP [Wikipedia, 2021. Cartogram. https://www.google.com/search?... (Accessed 20 January 2021)].
- *Chakra* (pl. *chakras*), a concept is found in the early traditions of Hinduism, refers to various focal points used in a variety of ancient meditation practices, collectively denominated as Tantra, or the esoteric or inner traditions of Hinduism. [Wikipedia, 2021. https://en.wikipedia.org/wiki/Chakra. (Accessed 20 January 2021)].
- A *chronic condition* is a human health condition or disease that is persistent or otherwise long-lasting in its effects or a disease that comes with time. The term chronic is often applied when the course of the disease lasts for more than 3 months. Common chronic diseases include arthritis, asthma, cancer, chronic obstructive pulmonary disease, diabetes, Lyme disease, and some viral diseases, such as hepatitis C and acquired immunodeficiency syndrome.
- A *clade* consists of an organism and all its descendants (living and extinct).
- *Combinatorial diversity* is additional diversity resulting from the random recombination of separate V, D, and J gene segments to form a complete V-region exon (Janeway et al., 2001).
- *Communicable diseases* are those that can be spread from person to person via an infectious agent, such as bacteria, viruses, fungi, or parasites.
- A *confounding factor* also called a *confounding variable*, or *confounder* is a third variable in a study examining a potential cause-and-effect relationship. A *confounding variable* is related to both the supposed cause and the supposed effect of the study.
- *Correlational research* is a type of nonexperimental research method in which a researcher measures two variables, understands and assesses the statistical relationship between them with no influence from any extraneous variable. "*Correlation* is not *causation*" means that just because two things *correlate does not* necessarily *mean* that one causes the other.

- *Cytokines* are a large group of proteins, peptides, or glycoproteins that are secreted by specific cells of immune system.
- The *damage-response framework* (according to Pirofski and Casadevall, 2018) is: "... an integrated theory of microbial pathogenesis that puts forth the view that microbial pathogenesis reflects the outcome of an interaction between a host and a microbe, with each entity contributing to the nature of the outcome, which in turn depends on the amount of host damage that results from the host-microbe interaction."
- *Dominant* refers to the relationship between two versions of a gene.
- *Down's syndrome* is a genetic disorder caused when abnormal cell division results in extra genetic material from chromosome 21.
- *Endosomes* are membrane-bound vesicles, formed via a complex family of processes collectively known as endocytosis, and found in the cytoplasm of virtually every animal cell.
- *Endothermy*, briefly put, is "the state of being warm-blooded." Maintenance of the body at a metabolically favorable temperature, without shivering, through heat generation by internal bodily functions.
- *The etheric body, ether-body* or *æther body*, is a name given by neo-Theosophy to a vital *body* or subtle *body* coined by esoteric philosophers to describe the first or lowest layer in the "human energy field" or *aura*. It is thought to be in immediate contact with the physical *body*, to sustain it and connect it with "higher" *bodies*.
- *Gametes* (or sex cells) are an organism's reproductive cells.
- *Geoclimate*: A climate that has been affected by a geological event (typically a volcano).
- *Health hazard* is defined as something that is dangerous to health. The most common definition of a health hazard is: "a potential source of harm or adverse health effect on a person(s)." For example, a place is structurally unsafe (HSA, 2017).
- *Health risk* is defined as the chance or likelihood that a given exposure or series of exposures could cause harm or otherwise affect one's health.
- *Heredity* is the transmission of physical characteristics from parents to children through their genes. It influences all aspects of physical appearance, such as height, weight, body structure, the color of the eye, the texture of the hair, and even intelligence and aptitudes.
- *IL-2* is one of a group of related proteins made by *leukocytes* (white blood cells) and other cells in the body. It affects the development of the immune system.
- *Medical Geochemistry* is considered by some to represent a branch of Medical Geology. Studies in this field "... have focused on how chemical elements in rocks, soils, and sediments are transmitted via water or vegetation into the food chain, and how regional geochemical variations can result in "disease clusters" either through dietary deficiency of essential elements or dietary excess of toxic elements." (See Plumlee et al., 2006).
- *Meridian* (as used in acupuncture and Chinese medicine) refers to each of a set of pathways in the body along which vital energy is said to flow.
- *Metallome:* In biochemistry, the *metallome* is the distribution of metal ions in a cellular compartment.
- *Monocytes* are the largest type of leukocyte (white blood cells). As a part of the vertebrate innate immune system monocytes also influence the process of adaptive immunity.

- *Neurodegenerative disorders* are illnesses that involve the death of certain parts of the brain.
- *Noncommunicable diseases*: Noncommunicable diseases are the conditions or diseases, which are not caused by transmission of infections like that in communicable diseases.
- *IgG* is the main type of antibody found in blood and extracellular fluid, allowing it to control infection of body tissues. By binding many kinds of pathogens such as viruses, bacteria, and fungi, IgG protects the body from infection. [Wikipedia, 2020. Immunoglobulin G. https://en.wikipedia.org/wiki/Immunoglobulin_G#:~:text = IgG%20is%20the%20main%20type,protects%20the%20body%20from%20infection. (Accessed 26 December 2020)].
- *Infectious diseases* are disorders caused by organisms, such as bacteria, viruses, fungi, or parasites. Many organisms live in and on our bodies. They are normally harmless or even helpful. But under certain conditions, some organisms may cause disease.
- *Jaggery* is an unrefined sugar (molasses) product that is extracted from the juice of sugarcane or sap of palm or sap of coconut. It is sometimes referred to as a "noncentrifugal sugar," because it is not spun during processing to remove the molasses which is thought to be nutritious.
- The *immune system* is a series of complex defense mechanisms found in humans and other vertebrates, that helps to combat and destroy pathogenic organisms such as bacteria, fungi, viruses, and parasites. The immune system consists of two types of response mechanisms: (1) An *antigen-specific adaptive immune response mechanism*, also referred as the *acquired immune system*, which is composed of specialized, systemic cells, and processes that eliminate pathogens by preventing their growth; and (2) The *innate immune system* is a collection of cells and proteins that are functionally diverse and that defend against invasion by foreign organisms. An *innate immune response mechanism*, also called *natural*, is the set of processes that operate to protect the host from the surrounding environment in which a variety of toxins and infectious agents are found.
- *Immunopotentiation* represents the shifting to increased immune response.
- *Immunosuppression* refers a state of decreased immunity.
- *Lymphocytes* are white blood cells that are also one of the body's main types of immune cells.
- *Macrophages* are large specialized cells that detect, engulf, and destroy bacteria and other harmful organisms.
- *Mutations* are permanent changes in the DNA sequence, and they are a main cause of diversity among organisms.
- *Neutrophils* are a type of white blood cell. Most of the white blood cells that lead the immune system's response are neutrophils.
- *Pathogenesis* refers to the way (biological mechanism) in which a disease develops.
- *Pathogenicity* is the ability of an agent to cause disease (i.e., to harm the host).
- In genetics, a *promoter* is a sequence of deoxyribonucleic acid (DNA) to which proteins bind that initiate transcription of a single ribonucleic acid (RNA) from the DNA downstream of it. [Wikipedia, 2021. Promoter (genetics). https://en.wikipedia.org/wiki/Promoter_(genetics). (Accessed 26 January 2021].

- *R*: A statistical and graphical modeling technique (software) that can be used to illustrate disease progress over time for different systems.
- ROS: An unstable molecule that contains oxygen and that easily reacts with other molecules in a cell. ROS are the contributors of oxidative stress which leads to various diseases and disorders.
- *Recessive:* relating to or denoting heritable characteristics controlled by genes which are expressed in offspring only when inherited from both parents.
- A *syndrome* is a set of medical signs and symptoms, which are correlated with each other and often associated with a particular disease or disorder. [Wikipedia, 2020. Syndrome. https://en.wikipedia.org/wiki/Syndrome. (Accessed 10 January 2021)].
- *Toxicological Geochemistry*, regarded by Plumlee et al. (2006) as a somewhat narrower area of Medical Geochemistry, studies the chemical interactions between Earth materials and body fluids, and how these interactions might influence toxicity.
- "A *xenobiotic* is a chemical substance found within an organism that is not naturally produced or expected to be present within the organism. It can also cover substances that are present in much higher concentrations than are usual." [Wikipedia, 2020. https://en. wikipedia.org/wiki/Xenobiotic. (Accessed 26 January 2021)].
- *X-linked* refers to *genes* that are on the *X chromosome*, one of the two sex chromosomes that we have in humans.
- *Zoonosis* (or *zoonotic diseases*) refers to diseases that can be passed from animals to humans.

References

Aamodt, G., Samuelsen, S.O., Skrondal, A., 2006. A simulation study of three methods for detecting disease clusters. International Journal of Health Geographics 5, 15. Available from: https://doi.org/10.1186/1476-072X-5-15.

Abu-Rabia, A., 2005. *The evil eye* and *cultural beliefs* among the Bedouin tribes of the Negev, Middle East. Folklore 116 (3), 241−254. Available from: https://doi.org/10.1080/00155870500282677.

ACPC (African Climate Policy Centre), 2013. Climate Change and Health in Africa: Issues and Options. https://www.uneca.org/sites/default/files/PublicationFiles/policy_brief_12_climate_change_and_health_i-n_africa_issues_and_options.pdf. (Accessed 29 November 2020).

Adam, R.D., Revathi, G., Okinda, N., Fontaine, M., Shah, J., Kagotho, E., et al., 2019. Analysis of *Candida auris fungemia* at a single facility in Kenya. International Journal of Infectious Diseases 85, 182−187.

Adani, G., Filippini, T., Michalke, B., Vinceti, M., 2020. Selenium and other trace elements in the etiology of Parkinson's Disease: a systematic review and meta-analysis of case-control studies. Neuroepidemiology 54, 1−23. Available from: https://doi.org/10.1159/000502357.

Adebisi, Y., 2020. Understanding the health impact of climate change in West Africa: a review of current knowledge and research gaps. Global Health Focus, Africa. In: Conference Proceedings, Planetary Health Conference 2020, Banjul, The Gambia/MRC; The Gambia at LSTMH. Available from: https://doi.org/10.13140/RG0.2.2.22970.75202.

ADI (Alzheimer's Disease International), 2019. World Alzheimer Report 2019: Attitudes to dementia. Alzheimer's Disease International, London. https://www.alzint.org/u/WorldAlzheimerReport2019-Summary.pdf. (Accessed 29 January 2021).

Ahmadzai, H., Thomas, P.S., Wakefield, D., 2013. Laboratory investigations and immunological testing in sarcoidosis. Available from: https://doi.org/10.5772/55294. https://www.intechopen.com/books/sarcoidosis/laboratory-investigations-and-immunological-testing-in-sarcoidosis. (Accessed 31 October 2020).

Ahmed, S., Ahsan, K.B., Kippler, M., Mily, A., Wagatsuma, Y., Hoque, A.M.W., et al., 2012. *In utero* arsenic exposure is associated with impaired thymic function in newborns possibly via oxidative stress and apoptosis. Toxicological Science 129 (2), 305–314. Available from: https://doi.org/10.1093/toxsci/kfs202.

Aiden, H., 2018. Multimorbidity: Understanding the Challenge. A report for the Richmond Group of Charities. https://richmondgroupofcharities.org.uk/sites/default/files/multimorbidity_-_understanding_the_challenge.pdf. (Accessed 05 January 2021).

Akinbobola, A.B., Kean, R., Hanifi, S.M.A., Quilliam, R.S., 2023. Environmental reservoirs of the drug-resistant pathogenic yeast *Candida auris*. PLoS Pathogens 19 (4), e1011268. Available from: https://doi.org/10.1371/journal.ppat.1011268 (accessed 14.02.24).

Akinola, F., Muloiwa, R., Hussey, G.D., Dirix, V., Kagina, B., Amponsah-Dacosta, E., 2018. Assessment of humoral and cell-mediated immune responses to pertussis vaccination: a systematic review protocol. BMJ Open 9 (6), e028109. Available from: https://doi.org/10.1136/bmjopen-2018-028109.

Aleuy, O.A., Kutz, S., 2020. Adaptations, life-history traits and ecological mechanisms of parasites to survive extremes and environmental unpredictability in the face of climate change. International Journal for Parasitology: Parasites and Wildlife 12, 308–317. Available from: https://doi.org/10.1016/j.ijppaw.2020.07.006.

Alfituri, O., Quintana, J., MacLeod, A., Garside, P., Benson, R., Brewer, J., et al., 2020. To the skin and beyond: the immune response to African trypanosomes as they enter and exit the vertebrate host. Frontiers in Immunology 11, 1250. Available from: https://doi.org/10.3389/fimmu.2020.01250.

Ali, M.A., El-Khodery, S.A., El-Said, W.E., 2015. Potential risk factors associated with ill-thrift in buffalo calves (Bubalus bubalis) raised at smallholder farms in Egypt. Journal of Advanced Research 6 (4), 601–607. Available from: https://doi.org/10.1016/j.jare.2014.02.005.

Almaguer, M., Herrera, R., Orantes, C.M., 2014. Chronic kidney disease of unknown etiology in agricultural communities. MEDICC Review 16 (2), 9–15.

Amarasekera, M., Prescott, S.L., Palmer, D.J., 2020. Nutrition in early life, immune-programming and allergies: the role of epigenetics. Asian Pacific Journal of Allergy and Immunology 31 (3), 175–182.

Amouian, S., Mohammadian, S., Behnampour, N., Tizrou, M., 2013. Trace elements in febrile seizure compared to febrile children admitted to an Academic Hospital in Iran, 2011. Journal of Clinical and Diagnostic Research (JCDR) 7 (10), 2231–2233. Available from: https://doi.org/10.7860/JCDR/2013/5548.3478.

Animasahun, A.B., Adekunle, M.O., Yejide, K., Fadipe, C., 2017. The diagnosis of Kawasaki disease among Nigerian children: A nightmare for the caregivers and the doctors. Journal of Publlic Health and Emergency 1 (7), 69.

Antó, J., Cullinan, P., 2001. Clusters, classification and epidemiology of interstitial lung diseases: Concepts, methods and critical reflections. The European Respiratory Journal 18, . Available from: https://erj.ersjournals.com/content/18/32_suppl/101s accessed 17.11.2020.

Anwar, A., Anwar, S., Ayub, M., Nawaz, F., Hyder, S., Khan, N., et al., 2019. Climate change and infectious diseases: Evidence from highly vulnerable countries. Iranian Journal of Public Health 48 (12), 2187–2195.

Anyamba, A., Chretien, J.P., Britch, S.C., Soebiyanto, R.P., Small, J.L., Jepson, R., et al., 2019. Global disease outbreaks associated with the 2015–2016 El Niño Event. Science Reports 9, 1930. Available from: https://doi.org/10.1038/s41598-018-38034-z.

Anyamba, A., Linthicum, K.J., Tucker, C.J., 2001. Climate-disease connections: Rift Valley fever in Kenya. Cadernos Saude Publica 17 (Suppl), 133–140. Available from: https://doi.org/10.1590/s0102-311x2001000700022.

Anyanwu, B.O., Ezejiofor, A.N., Igweze, Z.N., Orisakwe, O.E., 2018. Heavy metal mixture exposure and effects in developing nations: An update. Toxics 2018 (6), 65. Available from: https://doi.org/10.3390/toxics6040065.

Arkema, E.V., Cozier, Y.C., 2018. Epidemiology of sarcoidosis: current findings and future directions. Therapeutic Advances in Chronic Disease 9 (11), 227−240. Available from: https://doi.org/10.1177/2040622318790197.

Atwoli, L., Erhabor, G.E., Gbakima, A.A., Haileamlak, A., Kayembe Ntumba, J.M., Kigera, J., et al., 2023. COP27 climate change conference: urgent action needed for Africa and the world. Biomolecules and Biomedicine 23 (3), 352−353. Available from: https://doi.org/10.17305/bjbms.2022.8137 (accessed 13.02.24).

Awotedu, A.A., George, A.O., Oluboyo, P.O., Alabi, G.O., Onadeko, B.O., Ogunseyinde, O., et al., 1987. Sarcoidosis in Africans: 12 cases with histological confirmation from Nigeria. Transactions of the Royal Society of Tropical Medicine and Hygiene 81 (6), 1027−1029. Available from: https://doi.org/10.1016/0035-9203(87)90387-7.

Ayanshina, O.A., Adeshakin, A.O., Afolabi, L.O., Adeshakin, F.O., Alli-Balogun, G.O., Yan, D., et al., 2020. Seasonal variations in Nigeria: understanding COVID-19 transmission dynamics and immune responses. Journal of Global Health Reports 4, e2020084. Available from: https://doi.org/10.29392/001c.14600.

Babelhadj, B., Di Bari, M., Pirisinu, L., Chiappini, B., Gaouar, S., Riccardi, G., et al., 2018. Prion disease in Dromedary Camels, Algeria. Emerging Infectious Diseases 24 (6), 1029−1036. Available from: https://doi.org/10.3201/eid2406.172007.

Badoe, E.V., Neequaye, J., Oliver-Commey, J.O., Amoah, J., Osafo, A., Aryee, I., et al., 2011. Kawasaki disease in Ghana: case reports from Korle Bu Teaching Hospital. Ghana Medical Journal 45 (1), 38−42. Available from: https://doi.org/10.4314/gmj.v45i1.68922.

Bagheri, S., Squitti, R., Haertlé, T., Siotto, M., Saboury, A.A., 2018. Role of copper in the onset of Alzheimer's disease compared to other metals. Frontiers in Aging Neuroscience 9, 446. Available from: https://doi.org/10.3389/fnagi.2017.00446.

Bahi, G.A., Boyvin, L., Méité, S., M'Boh, G.M., Yeo, K., N'Guessan, K.R., et al., 2017. Assessments of serum copper and zinc concentration, and the Cu/Zn ratio determination in patients with multidrug resistant pulmonary tuberculosis (MDR-TB) in Côte d'Ivoire. BMC Infectious Diseases 17, 257. Available from: https://doi.org/10.1186/s12879-017-2343-7.

Balasooriya, S., Munasinghe, H., Herath, A.T., Diyabalanage, S., Ileperuma, O.A., et al., 2019. Possible links between groundwater geochemistry and chronic kidney disease of unknown etiology (CKDu): an investigation from the Ginnoruwa region in Sri Lanka. Exposure and Health 12, 3. Available from: https://doi.org/10.1007/s12403-019-00340-w (accessed 20.01.2021)..

Bandres-Ciga, S., Diez-Fairen, M., Kim, J.J., Singleton, A.B., 2020. Genetics of Parkinson's disease: an introspection of its journey towards precision medicine. Neurobiology of Disease 137, 104782. Available from: https://doi.org/10.1016/j.nbd.2020.104782.

Baratti-Mayer, D., Pittet, B., Montandon, D., Bolivar, I., Bornand, J.-E., Hugonnet, S., et al., 2003. Noma: an "infectious" disease of unknown aetiology. The Lancet Infectious Diseases 3 (7), 419−431.

Barreiro, L.B., Quintana-Murci, L., 2020. Evolutionary and population (epi)genetics of immunity to infection. Human Genetics 139, 723−732. Available from: https://doi.org/10.1007/s00439-020-02167-x.

Bashir, R.S.E., Hassan, O.A., 2019. A One Health perspective to identify environmental factors that affect Rift Valley fever transmission in Gezira state, Central Sudan. Tropical Medicine and Health 47, 54. Available from: https://doi.org/10.1186/s41182-019-0178-1.

Beck, M.A., 1999. Trace minerals, immune function, and viral evolution. In: Beck, M. (Ed.), Military Strategies for Sustainment of Nutrition and Immune Function in the Field, Chapter 16. Committee on Military Nutrition Research. Institute of Medicine, National Academies Press. Available from: http://www.nap.edu/catalog/6450.html.

Beghè, D., Garavelli, C., Pastorelli, A.A., Muscarella, M., Saccani, G., Aiello, M., et al., 2017. Sarcoidosis in an Italian province. Prevalence and environmental risk factors. PLoS One 12 (5), e0176859. Available from: https://doi.org/10.1371/journal.pone.0176859.

Behrman, E.L., Howick, V.M., Kapun, M., Staubach, F., Bergland, A.O., Petrov, D.A., et al., 2018. Rapid seasonal evolution in innate immunity of wild *Drosophila melanogaster*. Proceedings of the Royal Society B: Biological Sciences 285 (1870), 20172599.

Beijer, E., Meek, B., Bossuyt, X., Peters, S., Vermuulen, R.C.H., Kromhout, H., et al., 2020. Immunoreactivity to metal and silica associates with sarcoidosis in Dutch patients. Respiratory Research 21 (2020), 141. Available from: https://doi.org/10.1186/s12931-020-01409-w.

Belles-Isles, M., Ayotte, P., Dewailly, E., Weber, J.-P., Roy, R., 2002. Cord blood lymphocyte functions in newborns from a remote maritime population exposed to organochlorines and methylmercury. Journal of Toxicology and Environmental Health. Part A 65 (2), 165−182. Available from: https://doi.org/10.1080/152873902753396794.

Bellone, R., Failloux, A.-B., 2020. The role of temperature in shaping mosquito-borne viruses transmission. Frontiers in Microbiology 11, 2388. Available from: https://doi.org/10.3389/fmicb.2020.584846.

Benouaich, V., Soler, P., Gourraud, P.A., Lopez, S., Rousseau, H., Marcheix, B., 2010. Impact of meteorological conditions on the occurrence of acute type A aortic dissections. Interactive Cardiovascular and Thoracic Surgery 10 (3), 403−406.

Bentwich, Z., Kalinkovich, A., Weisman, Z., 1995. Immune activation is a dominant factor in the pathogenesis of African AIDS. Immunology Today 16 (4), 187−191. Available from: https://doi.org/10.1016/0167-5699 (95)80119-7.

Berry, R.J. (Ed.), 2007. Environmental Dilemmas: Ethics and decisions. Springer Science & Business Media, p. 271.

Bhaskaran, D., Chadha, S.S., Sarin, S., Sen, R., Arafah, S., Arafah, S., 2019. Diagnostic tools used in the evaluation of acute febrile illness in South India: a scoping review. BMC Infectious Diseases 19 (1). Available from: https://doi.org/10.1186/s12879-019-4589-8 (accessed 11.02.24).

Bhattacharya, P.T., Misra, S.R., Hussain, M., 2016. Nutritional aspects of essential trace elements in oral health and disease: an extensive review. Scientifica (Cairo) 2016, 5464373. Available from: https://doi.org/10.1155/2016/5464373.

Bhigjee, A., Moodley, K., Ramkissoon, K., 2007. Multiple sclerosis in KwaZulu Natal, South Africa: an epidemiological and clinical study. Multiple Sclerosis 13 (9), 1095−1099. Available from: https://doi.org/10.1177/1352458507079274.

Bikbov, B., Purcell, C.A., Levey, A.S., Smith, M., 2020. Global, regional, and national burden of chronic kidney disease, 1990 - 2017: a systematic analysis for the Global Burden of Disease Study 2017. The Lancet 395 (10225), 709−733.

Birnbaum, D., Jacquez, G.M., Waller, L.A., Grimson, R., Wartemberg, D., 1996. The analysis of disease clusters, Part I: state of the art. Infection Control and Hospital Epidemiology 17 (5), 319−327.

Biswal, M., Rudramurthy, S.M., Jain, N., Shamanth, A.S., Sharma, D., Jain, K., et al., 2017. Controlling a possible outbreak of *Candida auris* infection: lessons learnt from multiple interventions. Journal of Hospital Infections 97 (4), 363−370.

Bjørklund, G., Dadar, M., Pen, J.J., Chirumbolo, S., Aaseth, J., 2019a. Chronic fatigue syndrome (CFS): suggestions for a nutritional treatment in the therapeutic approach. Biomedicine and Pharmacotherapy 109, 1000−1007. Available from: https://doi.org/10.1016/j.biopha.2018.10.076.

Bjørklund, G., Chirumbolo, S., Dadar, M., Pivina, L., Lindh, U., Butnariu, M., et al., 2019b. Mercury exposure and its effects on fertility and pregnancy outcome. Basic and Clinical Pharmacology and Toxicology 125, 317−327. Available from: https://doi.org/10.1111/bcpt.13264.

Black, R.E., 2003. Zinc deficiency, infectious disease and mortality in the Developing World. The Journal of Nutrition 133 (5), 1485S−1489S. Available from: https://doi.org/10.1093/jn/133.5.1485S.

Blumenthal, U., Fleisher, J., Esrey, S.A., Peasey, A., 2001. Chapter 7: Epidemiology: a tool for the assessment of risk. In: Fewtrell, L., Bartram, J. (Eds.), World Health Organization (WHO) - Water Quality: Guidelines, Standards and Health. IWA Publishing, London, UK.

Bocca, B., Alimonti, A., Petrucci, F., Violante, N., Sancesario, G., Forte, G., et al., 2004. Quantification of trace elements by sector field inductively coupled plasma mass spectrometry in urine, serum, blood and cerebrospinal fluid of patients with Parkinson's disease. Spectrochimica Acta. Part B: Atomic Spectroscopy 59 (4), 559−566. Available from: https://doi.org/10.1016/j.sab.2004.02.007.

Bose-O'Reilly, S., McCarty, K.M., Steckling, N., Lettmeier, B., 2010. Mercury exposure and children's health. Current Problems in Pediatric and Adolescent Health Care 40 (8), 186−215. Available from: https://doi.org/10.1016/j.cppeds.2010.07.002.

Boudiaf, H., Achir, M., 2016. Clinical profile of Kawasaki disease in Algerian children: a single institution experience. Journal of Tropical Paediatrics 62 (2), 139−143.

Boukerche, S., Mohammed-Roberts, R., 2020. Fighting infectious diseases: the connection to climate change. The World Bank. https://blogs.worldbank.org/climatechange/fighting-infectious-diseases-connection-climate-change. (Accessed 26 November 2020).

Bourke, C.A., 2018. Astrocyte dysfunction following molybdenum-associated purine loading could initiate Parkinson's disease with dementia. npj Parkinson's Disease 4, 7. Available from: https://doi.org/10.1038/s41531-018-0045-5.

Bovbjerg, M., 2019. Foundations of Epidemiology, Chapter 1: What is epidemiology? https://open.oregonstate.education/epidemiology/chapter/what-is-epidemiology/. (Accessed 04 December 2020).

Braack, L., Gouveia de Almeida, A.P., Cornel, A.J., et al., 2018. Mosquito-borne arboviruses of African origin: review of key viruses and vectors. Parasites and Vectors 11, 29.

Bradshaw, D., Pillay van-Wyk, V., Neethling, I., Roomaney, R.A., Cois, A., Joubert, J.D., et al., 2022. Overview: Second Comparative Risk Assessment for South Africa (SACRA2) highlights need for health promotion and strengthened surveillance. South African Medical Journal 112 (8B), 556−570. Available from: https://samajournals.co.za/index.php/samj/article/view/208 (accessed 13.02.24).

Bredholt, M., Frederiksen, J.L., 2016. Zinc in multiple sclerosis: a systematic review and meta-analysis. ASN Neuro 8 (3). Available from: https://doi.org/10.1177/1759091416651511.

Briggs, D., 2003. Environmental pollution and the global burden of disease. British Medical Bulletin 68, 1−24. Available from: https://doi.org/10.1093/bmb/ldg019.

Britannica, 2020. The Causes of Disease. https://www.britannica.com/science/human-disease/The-causes-of-disease. (Accessed 03 December 2020).

Bundschuh, J., Maity, J.P., Mushtaq, S., Vithanage, M., Seneweera, S., et al., 2017. Medical geology in the framework of the sustainable development goals. Science of the Total Environment 581−582, 87−104. Available from: https://doi.org/10.1016/j.scitotenv.2016.11.208.

Byass, P., 2009. Climate change and population health in Africa: where are the scientists? Global Health Action 2. Available from: https://doi.org/10.3402/gha.v2i0.2065.

Cabassi, E., 2007. The immune system and exposure to xenobiotics in animals. Veterinary Research Communications 31 (Supplement 1), 115−120. Available from: https://doi.org/10.1007/s11259-007-0074-8.

Caddell, J.L., 1992. Hypothesis: new concepts concerning the pathophysiology of the sudden infant death syndrome due to magnesium deficiency shock. Magnesium Research 5 (3), 165−172.

Calder, P.C., 2020. Nutrition, immunity and COVID-19. BMJ Nutrition, Prevention and Health 3, 74−92. Available from: https://doi.org/10.1136/bmjnph-2020-000085.

Calder, P.C., Carr, A.C., Gombart, A.F., Eggersdorfer, M., 2020. Optimal nutritional status for a well-functioning Immune system is an important factor to protect against viral infections. Nutrients 12 (4), 1181. Available from: https://doi.org/10.3390/nu12041181.

Cambau, E., Drancourt, M., 2014. Steps towards the discovery of *Mycobacterium tuberculosis* by Robert Koch, 1882. Clinical Microbiology and Infection 20 (3), 196−201. Available from: https://doi.org/10.1111/1469-0691.12555.

Caminade, C., McIntyre, K.M., Jones, A.E., 2019. Impact of recent and future climate change on vector-borne diseases. Annals of the New York Academy of Sciences 1436 (1), 157−173. Available from: https://doi.org/10.1111/nyas.13950.

Capcha, K.M., Pezo, A.P., Cosentino, C., Ramirez, L.E.T., 2018. Presentation of Parkinson's disease in patients originating of different geographical altitudes. Neurology 90 (15), 2.080. Available from: https://n.neurology.org/content/90/15_Supplement/P2.080.

Caplin, B., Yang, C.-W., Anand, S., Levin, A., Madero, M., Saran, R., et al., 2019. The International Society of Nephrology's International Consortium of Collaborators on Chronic Kidney Disease of Unknown Etiology: report of the Working Group on approaches to population-level detection strategies and recommendations for a minimum dataset. Kidney International 95 (1), 4−10. Available from: https://doi.org/10.1016/j.kint.2018.08.019 [Published correction appears in Kidney International 2019, 95 (4), 997−998].

Carlsson, J.A., Bayes, H.K., 2020. Acute severe asthma in adults. Medicine 48 (5), 297−302. Available from: https://doi.org/10.1016/j.mpmed.2020.02.008.

Carpentier, P.H., 1998. Définition et épidémiologie des acrosyndromes vasculaires [Definition and epidemiology of vascular acrosyndromes]. Revue du Praticien 48 (15), 1641−1646.

Carr, R., Warren, R., Towers, L., Bartholomew, A., Duggal, H.V., Rehman, Y., et al., 2010. Investigating a cluster of Legionnaires' cases: Public health implications. Public Health 124 (6), 326−331. Available from: https://doi.org/10.1016/j.puhe.2010.03.001.

Carsetti, R., Quintarelli, C., Quinti, I., Piano Mortari, E., Zumla, A., Ippolito, G., et al., 2020. The immune system of children: the key to understanding SARS-CoV-2 susceptibility? The Lancet Child and Adolescent Health 4 (6), 414−416. Available from: https://doi.org/10.1016/S2352-4642(20)30135-8.

Casadevall, A., 2020. Climate change brings the specter of new infectious diseases. Journal of Clinical Investigation 130 (2), 553−555.

Casadevall, A., Pirofski, L.A., 2018. What is a host? attributes of individual susceptibility. Infection and Immunity 86 (2), e00636-17. Available from: https://doi.org/10.1128/IAI.00636-17.

Casadevall, A., Kontoyiannis, D.P., Robert, V., 2019. On the emergence of *Candida auris*: climate change, azoles, swamps, and birds. mBio: American Society for Microbiology 10 (4), e01397-19. Available from: https://doi.org/10.1128/mBio.01397-19.

CAWG [Cancer/Cluster Analysis Work Group, Cancer Data Registry of Idaho (CDRI)/Idaho Hospital Association], 2014. Investigation of Non-infectious Disease Clusters. https://www.idcancer.org/Documents/Resources/CAWG%20Investigation%20of%20Non-infectious%20Disease%20Clusters%202014.pdf. (Accessed 12 November 2020).

Chanda, E., 2020. Disregarding reservoirs of disease vectors: a surveillance paradox in Africa. EClinicalMedicine 29-30, 100629. Available from: https://doi.org/10.1016/j.eclinm.2020.100629.

Chaplin, D.D., 2010. Overview of the immune response. Journal of Allergy and Clinical Immunology 125 (2 Supplement 2), S3−S23. Available from: https://doi.org/10.1016/j.jaci.2009.12.980.

Chaturvedi, U.C., Shrivastava, R.,, Upreti, R.K., 2004. Viral infections and trace elements: a complex interaction. Current Science 87 (11), 1536−1554.

Checa, J., Aran, J.M., 2020. Reactive oxygen species: drivers of physiological and pathological processes. Journal of Inflammation Research 13, 1057−1073. Available from: https://doi.org/10.2147/JIR.S275595.

Chen, Z., He, Q., 2017. Immune persistence after pertussis vaccination. Human Vaccines and Immunotherapeutics 13 (4), 744−756. Available from: https://doi.org/10.1080/21645515.2016.1259780.

Chen, P., Miah, M.R., Aschner, M., 2016. Metals and neurodegeneration. F1000Research 5, 366. Available from: https://doi.org/10.12688/f1000research.7431.1 (accessed 12.02.24).

Chen, Z., Myers, R., Wei, T., Bind, E., Kassim, P., Wang, G., et al., 2014. Placental transfer and concentrations of cadmium, mercury, lead, and selenium in mothers, newborns, and young children. Journal of

Exposure Science and Environmental Epidemiology 24 (5), 537−544. Available from: https://doi.org/10.1038/jes.2014.26.

Cheng, Z., Xie, X., Yao, W., Feng, J., Zhang, Q., Fang, J., 2014. Multi-element geochemical mapping in Southern China. Journal of Geochemical Exploration 139 (100), 183−192. Available from: https://doi.org/10.1016/j.gexplo.2013.06.003.

Chersich, M.F., Wright, C.Y., 2019. Climate change adaptation in South Africa: a case study on the role of the health sector. Globalization and Health 15 (2019), 22. Available from: https://doi.org/10.1186/s12992-019-0466-x.

Chersich, M.F., Wright, C.Y., Venter, F., Rees, H., Scorgie, F., Erasmus, B., 2018. Impacts of climate change on health and wellbeing in South Africa. International Journal of Environmental Research and Public Health 15 (9), 1884. Available from: https://doi.org/10.3390/ijerph15091884.

Ciftci, T.U., Ciftci, B., Yis, O., Guney, Y., Bilgihan, A., Ogretensoy, M., 2003. Changes in serum selenium, copper, zinc levels and Cu/Zn ratio in patients with pulmonary tuberculosis during therapy. Biology of Trace Element Research 95, 65−71.

Cliff, J., Martelli, A., Molin, A., Rosling, H.Ministry of Health, 1984. Mantakassa: an epidemic of spastic paraparesis associated with chronic cyanide intoxication in a cassava staple area of Mozambique. 1. Epidemiology and clinical and laboratory findings in patients. Bulletin of the World Health Organization 62 (3), 477−484.

Cluver, E.H., 2019. The influence of geographical factors on human health. South African Geographical Journal 3 (1), 49−55. Available from: https://doi.org/10.1080/03736245.1919.11882209.

CMD (Concise Medical Dictionary), 2010. Idiopathic. Concise Medical Dictionary, Eighth ed. Oxford University Press. Available from: 10.1093/acref/9780199557141.001.0001.

Coates, S.J., Norton, S.A., 2020. The effects of climate change on infectious diseases with cutaneous manifestations. International Journal of Women's Dermatology 7 (1), 8−16.

Cockwell, P., Fisher, L.-N., 2020. The global burden of chronic kidney disease. The Lancet 395 (10225), 662−664.

Coico, R., Sunshine, G., Benjamini, E., 2003. Immunology: A Short Course, 5th ed. City University of New York.

Compston, A., Coles, A., 2002. Multiple sclerosis. Lancet 359 (9313), 1221−1231. Available from: https://doi.org/10.1016/S0140-6736(02)08220-X.

Compston, A., Coles, A., 2008. Multiple sclerosis. Lancet 372 (9648), 1502−1517. Available from: https://doi.org/10.1016/S0140-6736(08)61620-7.

Constantinidis, J., 1991. The hypothesis of zinc deficiency in the pathogenesis of neurofibrillary tangles. Medical Hypotheses 35 (4), 319−323.

Coory, M.D., Jordan, S., 2013. Assessment of chance should be removed from protocols for investigating cancer clusters. International Journal of Epidemiology 42 (2), 440−447. Available from: https://doi.org/10.1093/ije/dys205.

Copeland, S.M., Bradford, J.B., Duniway, M.C., Schuster, R.M., 2017. Potential human impacts of overlapping land-use and climate in a sensitive dryland: a case study of the Colorado Plateau, USA. Ecosphere 8 (5), 1−25. Available from: https://doi.org/10.1002/ecs2.1823.

Crawshaw, M., Caldow, G., 2005. Trace Elements in Beef Cattle. Technical Note, No.: TN 572. ISBN: 0142 7695. file:///D:/SAC-TN572-Trace-element-disorders-in-beef-cattle.pdf. (Accessed 30 October 2020).

Crocq, M., 1896. De l' "acrocyanose.". Semaine Medicale 16, 298.

Crump, J., Morrissey, A.B., Nicholson, W.L., Massung, R.F., Stoddard, R.A., Galloway, R.L., et al., 2013. Etiology of severe non-malaria febrile illness in Northern Tanzania: a prospective cohort study. PLoS Neglected Tropical Diseases 7 (7), e2324.

Culver, D.A., Newman, L.S., Kavuru, M.S., 2007. Gene environment interactions in sarcoidosis: challenge and opportunity. Clinical Dermatology 25, 267–275. Available from: https://doi.org/10.1016/j.clindermatol.2007.03.005.

D'Acremont, V., Kilowoko, M., Kyungu, E., Philipina, S., Sangu, W., Kahama-Maro, J., et al., 2014. Beyond malaria - causes of fever in outpatient Tanzanian children. New England Journal of Medicine 370, 809–817.

D'Amato, M., Molino, A., Calabrese, G., Cecchi, L., Annesi-Maesano, I., D'Amato, G., 2018. The impact of cold on the respiratory tract and its consequences to respiratory health. Clinical and Translational Allergy 8, 20. Available from: https://doi.org/10.1186/s13601-018-0208-9.

Dangleben, N.L., Skibola, C.F., Smith, M.T., 2013. Arsenic immunotoxicity: a review. Environmental Health 12 (1), 73. Available from: https://doi.org/10.1186/1476-069X-12-73.

Darikwa, T.B., Manda, S.O., 2020. Spatial co-clustering of cardiovascular diseases and select risk factors among adults in South Africa. International Journal of Environmental Research and Public Health 17 (10), 3583. Available from: https://doi.org/10.3390/ijerph17103583.

Darnley, A.G., Björklund, A., Bølviken, B., Gustavsson, N., Koval, P.V., Plant, J.A., et al., 1995. A global geochemical database for environmental and resource management: final report of IGCP Project 259. Earth Sciences 19, 122.

Das, S., Maiti, A., 2013. Acrocyanosis: an overview. Indian Journal of Dermatology 58 (6), 417–420. Available from: https://doi.org/10.4103/0019-5154.119946.

Daubney, R., Hudson, J.R., Garnham, P.C., 1931. Enzootic hepatitis or rift valley fever. An undescribed virus disease of sheep cattle and man from east Africa. Journal of Pathology and Bacteriology 34, 545–579.

Davaalkham, D., Nakamura, Y., Baigalmaa, D., Chimedsuren, O., Sumberzul, N., et al., 2011. Kawasaki disease in Mongolia: Results from 2 nationwide retrospective surveys, 1996-2008. Journal of Epidemiology 21, 293.

Davis, J., 2001. Chapter 3: Emerging infections in Africa. In: Davis, J.R. and Lederberg, J. (Editors), Emerging Infectious Diseases from the Global to the Local Perspective: A Summary of a Workshop of the Forum on Emerging Infections; Washington (DC): National Academies Press (US), Institute of Medicine (US) Forum on Emerging Infections. https://www.ncbi.nlm.nih.gov/books/NBK99567/. (Accessed 24 January 2021).

Davies, T.C., 2013. Geochemical variables as plausible aetiological cofactor in the incidence of some common environmental diseases in Africa. Journal of African Earth Sciences 79, 24–49.

Davies, T.C., 2015. Urban geology of African megacities. Journal of African Earth Sciences 110, 188–226. Available from: https://doi.org/10.1016/j.jafrearsci.2015.06.01.

Davies, T.C., 2019. Regional Report, Africa. In: Smith, D. et al. (Eds.), 2018 Annual Report of the International Union of Geological Sciences Commission on Global Geochemical Baselines, pp.. 64–69. https://www.global-geochemicalbaselines.eu/datafiles/file/IUGS-CGGB_2018_Annual_Report_final.pdf. (Accessed 07 January 2021).

Davies, T.C., 2022. The position of geochemical variables as causal co-factors of diseases of unknown aetiology. Springer Nature Applied Sciences 4, 236. Available from: https://doi.org/10.1007/s42452-022-05113-w (accessed 27.07.22).

De Benedictis, C.A., Vilella, A., Grabrucker, A.M., 2019. Chapter 6: The role of trace metals in Alzheimer's Disease. In: Wisniewski, T. (Ed.), Alzheimer's Disease,. Codon Publications, Brisbane (AU), https://www.ncbi.nlm.nih.gov/books/NBK552144/. (Accessed 13 September 2020).

Deacon, E.L., Williams, A.L., 1982. The incidence of the sudden infant death syndrome in relation to climate. International Journal of Biometeorology 26, 207–218. Available from: https://doi.org/10.1007/BF02184936.

Demoling, F., Figueroa, D., Bååth, E., 2007. Comparison of factors limiting bacterial growth in different soils. Soil Biology and Biochemistry 39 (10), 2485–2495.

Dempers, J.J., Burger, E.H., Du Toit-Prinsloo, L., Verster, J., 2018. Chapter 17: A South African perspective. In: Duncan, J.R., Byard, R.W. (Eds.), SIDS - Sudden Infant and Early Childhood Death: The Past, the Present and the Future. University of Adelaide Press, https://www.ncbi.nlm.nih.gov/books/NBK513389/. (Accessed 31 October 2020).

Denisova, O., Chernogoryuk, G., Baranovskaya, N., Rikhvanov, L., Shefer, N., Chernjavskaya, G., et al., 2020. Trace elements in the lung tissue affected by Sarcoidosis. Biological Trace Element Research 196 (1), 66–73. Available from: https://doi.org/10.1007/s12011-019-01915-z.

Denton, F., Wilbanks, T.J., Abeysinghe, A.C., Burton, I., Gao, Q., Lemos, M.C., et al., 2014. Climate-resilient pathways: adaptation, mitigation, and sustainable development. In: Field, C.B., Barros, V.R. Dokken, D.J. Mach, K.J. Mastrandrea, M.D. Bilir, T.E. Chatterjee, M., Ebi, K.L. Estrada, Y.O., Genova, R., Girma, C., Kissel, B., Levy, E.S., MacCracken, A.N., Mastrandrea, S.P.R. and White, L.L. (Eds.), Climate Change 2014: Impacts, Adaptation, and Vulnerability. Part A: Global and Sectoral Aspects. Contribution of Working Group II to the Fifth Assessment Report of the Intergovernmental Panel on Climate Change. Cambridge University Press, Cambridge, United Kingdom and New York, NY, USA, pp. 1101–1131.

Descotes, J., 1999. An Introduction to Immunotoxicology. Taylor and Francis, London.

Descotes, 2004. Chapter 1: Definition, history, and scope of immunotoxicology. In: Descotes, J. (Ed.), Immunotoxicology of Drugs and Chemicals: An Experimental and Clinical Approach, vol. 1. Elsevier, pp. 1–18.

Dharmadhikari, N.P., Meshram, D.C., Kulkarni, S.D., Kharat, A.G., Pimplikar, S.S., 2011. Effect of geopathic stress zone on human body voltage and skin resistance. Journal of Engineering and Technology Research 3 (8), 255–263.

Dharma-Wardana, M.W., Amarasiri, S.L., Dharmawardene, N., Panabokke, C.R., 2015. Chronic kidney disease of unknown aetiology and groundwater ionicity: study based on Sri Lanka. Environmedntal Geochemistry and Health 37 (2), 221–231. Available from: https://doi.org/10.1007/s10653-014-9641-4.

Djoko, K.Y., Ong, C.L., Walker, M.J., McEwan, A.G., 2015. The role of copper and zinc toxicity in innate immune defense against bacterial pathogens. The Journal of Biological Chemistry 290 (31), 18954–18961. Available from: https://doi.org/10.1074/jbc.R115.647099.

Dolk, H., Loane, M., Teljeur, C., Densem, J., Greenlees, R., McCullough, N., et al., 2015. Detection and investigation of temporal clusters of congenital anomaly in Europe: seven years of experience of the EUROCAT surveillance system. European Journal of Epidemiology 30 (11), 1153–1164. Available from: https://doi.org/10.1007/s10654-015-0012-y.

Donnelly, J., 2012. CDC planning trial for mysterious nodding syndrome. The Lancet 379 (9813) 287–384, e20–e26. Available from: https://doi.org/10.1016/S0140-6736(12)60126-3.

Dopico, X.C., Evangelou, M., Ferreira, R.C., Guo, H., Pekalski, M.L., Smyth, D.J., et al., 2015. Widespread seasonal gene expression reveals annual differences in human immunity and physiology. Nature Communications 6, 7000. Available from: https://doi.org/10.1038/ncomms8000.

Dorsey, E.R.The 2016 Global Burden of Disease Collaborators, 2018. Global, regional and national burden of parkinson's disease, 1990-2016. The Lancet Neurology 17 (11), 939–953.

Dovjak, M. and Kukec, A. 2019. Chapter 3: Identification of health risk factors and their parameters. In: Dovjak, M. and Kukec, A. (Eds.), Creating Healthy and Sustainable Buildings: An Assessment of Health Risk Factors. Springer, Cham (CH). ISBN-13: 978-3-030-19411-6. https://www.ncbi.nlm.nih.gov/books/NBK553923/. (Accessed 05 December 2020).

Dowell, S.F., Sejvar, J.J., Riek, L., Vandemaele, K.A., Lamunu, M., Kuesel, A.C., et al., 2013. Nodding syndrome. Emerging Infectious Diseases 19 (9), 1374–1384. Available from: https://doi.org/10.3201/eid1909.130401.

Du, H., Bing, J., Hu, T., Ennis, C.L., Nobile, C.J., Huang, G., 2020. Candida auris: epidemiology, biology, antifungal resistance, and virulence. PLoS Pathogens 16 (10), e1008921. Available from: https://doi.org/10.1371/journal.ppat.1008921.

Duenwald, M., 2002. Parkinson's 'Clusters' Getting a Closer Look. The New York Times, Section F, p. 6. https://www.nytimes.com/2002/05/14/health/parkinson-s-clusters-getting-a-closer-look.html. (Accessed 02 September 2020).

Dummer, T.J., 2008. Health geography: supporting public health policy and planning. Canadian Medical Association Journal = Journal de l'Association Medicale Canadienne 178 (9), 1177−1180. Available from: https://doi.org/10.1503/cmaj.071783.

Duncan, J.R., Byard, R.W. (Eds.), 2018. SIDS - Sudden Infant and Early Childhood Death: The Past, the Present and the Future. University of Adelaide Press, p. 846.

Durch, J.S., Bailey, L.A., Stoto, M.A. (Eds.), 1997 1. Improving Health in the Community: A Role for Performance Monitoring. 2. Understanding Health and Its Determinants. Institute of Medicine (US) Committee on Using Performance Monitoring to Improve Community Health; National Academies Press (US). https://www.ncbi.nlm.nih.gov/books/NBK233009/. (Accessed 19 November 2020).

Duygu, F., Sari, T., Kaya, T., Tavsan, O., Naci, M., 2018. The relationship between Crimean-Congo hemorrhagic fever and climate: does climate affect the number of patients? Acta Clinica Croatica 57 (3), 443−448. Available from: https://doi.org/10.20471/acc.2018.57.03.06.

Edwards, J.R., Prozialeck, W.C., 2009. Cadmium, diabetes and chronic kidney disease. Toxicology and Applied Pharmacology 238, 289−293. Available from: https://doi.org/10.1016/j.taap.2009.03.007.

Ehmann, W.D., Markesbery, W.R., Alauddin, M., Hossain, T.I., Brubaker, E.H., 1986. Brain trace elements in Alzheimer's disease. Neurotoxicology 7 (1), 195−206.

Elakabawi, K., Lin, J., Jiao, F., Guo, N., Yuan, Z., 2020. Kawasaki disease: global burden and genetic background. Cardiology Research 11 (1), 9−14.

Elliott, S., 2014. Health geography. In: Michalos, A.C. (Ed.), Encyclopedia of Quality of Life and Well-Being Research. Springer, Dordrecht, https://doi.org/10.1007/978-94-007-0753-5_1248. (Accessed 02 February 2021).

Elliott, P., Wartenberg, D., 2004. Spatial epidemiology: current approaches and future challenges. Environmental Health Perspectives 112 (9), 998−1006. Available from: https://doi.org/10.1289/ehp.6735.

Ellwanger, J.H., Franke, S.I., Bordin, D.L., Prá, D., Henriques, J.A., 2016. Biological functions of selenium and its potential influence on Parkinson's disease. Annals of the Brazilian Academy of Sciences 88 (3 Suppl.), 1655−1674. Available from: https://doi.org/10.1590/0001-3765201620150595.

El-Sayed, A., Kamel, M., 2020. Climatic changes and their role in emergence and re-emergence of diseases. Environmental Science and Pollution Research 27 (18), 22336−22352. Available from: https://doi.org/10.1007/s11356-020-08896-w.

El-Zayat, S.R., Sibaii, H., Mannaa, F.A., 2019. Micronutrients and many important factors that affect the physiological functions of toll-like receptors. Bulletin of the National Research Centre 43, 123. Available from: https://doi.org/10.1186/s42269-019-0165-z.

EME CFS (Encyclopaedia of Myalgic Encephalomyelitis), 2020. Epidemiology of myalgic encephalomyelitis and chronic fatigue syndrome. https://mepedia.org/wiki/Epidemiology_of_myalgic_encephalomyelitis_and_chronic_fatigue_syndrome#:∼:text = Statistics%20on%20the%20prevalence%20of,that%20estimate%20at%2030%20million. (Accessed 29 October 2020).

Endalifer, M.L., Azeze, G.G., Diress, G., Bizuneh, A.D., Demelash, H., 2020. A systematic review protocol on the epidemiology of zinc deficiency and associated factors during pregnancy in Africa. Journal of Global Health Reports 4, e2020021. Available from: https://doi.org/10.29392/001c.12501.

Erickson, M.M., Poklis, A., Gantner, G.E., Dickinson, A.W., Hillman, L.S., 1983. Tissue mineral levels in victims of sudden infant death syndrome I. Toxic metals - lead and cadmium. Pediatric Research 17 (10), 779−784. Available from: https://doi.org/10.1203/00006450-198310000-00002.

Erickson, K.L., Medina, E.A., Hubbard, N.E., 2000. Micronutrients and innate immunity. Journal of Infectious Diseases 182 (Supplement 1), 5−10. Available from: https://doi.org/10.1086/315922.

Escobedo-Monge, M.F., Barrado, E., Alonso Vicente, C., Escobedo-Monge, M.A., Torres-Hinojal, M.C., Marugán-Miguelsanz, J.M., et al., 2020. Copper and copper/zinc ratio in a series of cystic fibrosis patients. Nutrients 12, 3344.

ESRI, 2007. ESRI Data and Maps 2006. An ESRI ® White Paper, May 2007. https://www.esri.com/arcgis-blog/products/product/mapping/esri-data-maps/. (Accessed 22 January 2021).

Essouma, M., Nkeck, J.R., Endomba, F.T., Bigna, J.J., Singwe-Ngandeu, M., Hachulla, E., 2020. Systemic *Lupus erythematosus* in Native sub-Saharan Africans: a systematic review and meta-analysis. Journal of Autoimmunity 106, 102348. Available from: https://doi.org/10.1016/j.jaut.2019.102348.

Ezzati, M., Lopez, A.D., Rodgers, A., Hoorn, S.V., Murray, C.J.L., 2002. The Comparative Risk Assessment Collaborating Group, 2002. Selected major risk factors and global and regional burden of disease. The Lancet 360 (9343), 1347−1360. Available from: https://doi.org/10.1016/S0140-6736(02)11403-6.

Failla, M.L., 2003. Trace elements and host defense: recent advances and continuing challenges. The Journal of Nutrition 133 (5), 1443S−1447S. Available from: https://doi.org/10.1093/jn/133.5.1443S.

Fares, A., 2013. Factors influencing the seasonal patterns of infectious diseases. International Journal of Preventive Medicine 4 (2), 128−132.

Faustine, N., Sabuni, E., Ndaro, A., Paul, E., et al., 2017. Chikungunya, dengue and West Nile virus infections in Northern Tanzania. Journal of Advances in Medicine and Medical Research 24 (4), 1−7. Available from: https://doi.org/10.9734/JAMMR/2017/37234.

Filardo, G., Adams, J., Ng, H.K.T., 2011. Statistical methods in epidemiology. In: Lovric., M. (Ed.), International Encyclopaedia of Statistical Science. Springer, Berlin, Heidelberg, https://doi.org/10.1007/978-3-642-04898-2_547. (Accessed 04 December 2020).

Finkelman, R.B., Orem, W.H., Plumlee, G.S., Selinus, O., 2018. Chapter 17: Applications of geochemistry to medical geology. In: De Vivo, B., Belkin, H.E., Lima, A. (Eds.), Environmental Geochemistry: Site Characterization, Data Analysis and Case Histories, second ed. Elsevier, pp. 435−465.

Fisman, D., 2012. Seasonality of viral infections: mechanisms and unknowns. Clinical Microbiology and Infection 18 (10), 946−954. Available from: https://doi.org/10.1111/j.1469-0691.2012.03968.x.

Flato, G., Marotzke, J., Abiodun, B., Braconnot, P., Chou, S.C., Collins, W., et al., 2013. Evaluation of Climate Models. In: Stocker, T.F., Qin, D. Plattner, G.-K. Tignor, M. Allen, S.K. Bosch, J., et al. (Eds.), Climate Change 2013: "The Physical Science Basis." Contribution of Working Group I to the Fifth Assessment Report of the Intergovernmental Panel on Climate Change. Cambridge University Press, Cambridge, United Kingdom and New York, NY, USA. https://pure.mpg.de/rest/items/item_1977534/component/file_3040450/content. (Accessed 14 January 2021).

Forouzanfar, M.H., Afshin, A., Alexander, L.T., Anderson, Bhutta, Z.A., et al., 2016. Global, regional, and national comparative risk assessment of 79 behavioural, environmental and occupational, and metabolic risks or clusters of risks, 1990 - 2015: a systematic analysis for the Global Burden of Disease Study 2015. The Lancet 388 (10053), 1659−1724.

Forsberg, K., Woodworth, K., Walters, M., Berkow, E.L., Jackson, B., Chiller, T., et al., 2019. *Candida auris*: the recent emergence of a multidrug-resistant fungal pathogen. Medical Mycology 57, 1−12. Available from: https://academic.oup.com/mmy/article/57/1/1/5062854.

Forte, G., Bocca, B., Senofonte, O., Petrucci, F., Brusa, L., Stanzione, P., et al., 2004. Trace and major elements in whole blood, serum, cerebrospinal fluid and urine of patients with Parkinson's disease. Journal of Neural Transmission (Vienna) 111 (8), 1031−1040. Available from: https://doi.org/10.1007/s00702-004-0124-0.

Foster, S., Adams, E., Dunn, I., Dent, A., 2018. The CDC Field Epidemiology Manual: Geographic Information System Data. Centers for Disease Control and Prevention, Epidemic Intelligence Service. https://www.cdc.gov/eis/field-epi-manual/chapters/GIS-data.html. (Accessed 19 November 2020).

Fouque, F., Reeder, J.C., 2019. Impact of past and on-going changes on climate and weather on vector-borne diseases transmission: a look at the evidence. Infectious Diseases of Poverty 8, 51. Available from: https://doi.org/10.1186/s40249-019-0565-1.

Frank, D.G., Wallace, A.R., Schneider, J.L., 2010. Western Mineral and Environmental Resources Science Center - Providing Comprehensive Earth Science for Complex Societal Issues. Circular 1363. US Department of the Interior/US Geological survey, 32 p.

Freshwater, D., 1997. Geopathic stress. Complementary Therapies in Nursing & Midwifery 3 (6), 160−162. Available from: https://doi.org/10.1016/s1353-6117(05)81003-0.

Galask, R., Larsen, B., Ohm, M.J., 2008. Infection in maternal-fetal medicine: an overview. Welfare of Women, Global Health Programme. Global Library of Women's Medicine. doi: 10.3843/GLOWM.10173. ISSN: 1756−2228.

Ganeshan, D., Menias, C.O., Luber, M.G., Pickhardt, P.J., Sandrasegaran, K., Bhalla, S., 2018. Sarcoidosis from head to toe: what the radiologist needs to know. Radiographics: A Review Publication of the Radiological Society of North America, Inc. 38 (4), 1180−1200. Available from: https://doi.org/10.1148/rg.2018170157.

Garchitorena, A., Sokolow, S.H., Roche, B., Ngonghala, C.N., Jocque, M., Lund, A., et al., 2017. Disease ecology, health and the environment: a framework to account for ecological and socio-economic drivers in the control of neglected tropical diseases. Philosophical Transactions of the Royal Society, Section B 372, 20160128. Available from: http://doi.org/10.1098/rstb.2016.0128.

Garcia-Solache, M.A., Casadevall, A., 2010. Global warming will bring new fungal diseases for mammals. mBio (American Society for Microbiology) 1 (1), e00061-10. Available from: https://doi.org/10.1128/mBio.00061-10.

Gellein, K., Syversen, T., Steinnes, E., Nilsen, T.I.L., Dahl, O.P., Mitrovic, S., et al., 2008. Trace elements in serum from patients with Parkinson's disease - a prospective case-control study: The Nord-Trøndelag Health Study (HUNT). Brain Research 1219, 111−115. Available from: https://doi.org/10.1016/j.brainres.2008.05.002.

Genc, S., Zadeoglulari, Z., Fuss, S.H., Genc, K., 2012. The adverse effects of air pollution on the nervous system. Journal of Toxicology 2012, 782462. Available from: https://doi.org/10.1155/2012/782462.

GEO (Group on Earth Observations), 2016. "2017−2019 Work Programme: African Geochemical Baselines. GEO-XIII-5.3, pp. 12−13. https://www.earthobservations.org/documents/geo_xiii/GEO-XIII-5-3_2017-19_Work%20Programme.pdf. (Accessed 07 January 2021).

George, M., Wiklund, L., Aastrup, M., Pousette, J., Thunholm, B., Saldeen, T., et al., 2001. Incidence and geographical distribution of sudden infant death syndrome in relation to content of nitrate in drinking water and groundwater levels. European Journal of Clinical Investigations 31 (12), 1083−1094. Available from: https://doi.org/10.1046/j.1365-2362.2001.00921.x.

Getts, D.R., Spiteri, A., King, N.J.C., Miller, S.D., 2020. Chapter 21: Microbial infection as a trigger of t-cell autoimmunity. In: Rose, N., Mackay, I. (Eds.), The Autoimmune Diseases, sixth ed. Academic Press, pp. 363−374. http://www.sciencedirect.com/science/article/pii/B9780128121023000021X. (Accessed 23 December 2020).

Ghattas, H., Wallace, D.L., Solon, J.A., Henson, S.M., Zhang, Y., Ngom, P.T., et al., 2007. Longterm effects of perinatal nutrition on T lymphocyte kinetics in young Gambian men. American Journal of Clinical Nutrition 85 (2), 480−487.

Giani, P., Castruccio, S., Anav, A., Howard, D., Hu, W., Crippa, P., 2020. Short-term and long-term health impacts of air pollution reductions from COVID-19 lockdowns in China and Europe: a modelling study. Lancet Planet Health 4 (10), e474−e482. Available from: https://doi.org/10.1016/S2542-5196(20)30224-2.

Gifford, F.J., Gifford, R.M., Eddleston, M., Dhaun, N., 2017. Endemic nephropathy around the World. Kidney International Reports 2 (2), 282−292. Available from: https://doi.org/10.1016/j.ekir.2016.11.003.

Glaser, J., Lemery, J., Rajagopalan, B., Diaz, H.F., García-Trabanino, R., Taduri, G., et al., 2016. Climate change and the emergent epidemic of CKD from heat stress in rural communities: the case for heat stress nephropathy. Clinical Journal of the American Society of Nephrology 11 (8), 1472−1483. Available from: https://doi.org/10.2215/CJN.13841215.

Glennie, S.J., Nyirenda, M., Williams, N.A., Heyderman, R.S., 2012. Do multiple concurrent infections in African children cause irreversible immunological damage? Immunology 135 (2), 125−132. Available from: https://doi.org/10.1111/j.1365-2567.2011.03523.x.

GoA (Government of Alberta), 2011. Guidelines for the Investigation of Clusters of Non-Communicable Health Events. Alberta Health Services. https://open.alberta.ca/dataset/2f113033-825d-4a24-a045-6bed269fe841/resource/797abfa8-cc8c-4bbc-9aa3-b52693576551/download/investigation-clusters-guidelines-2011.pdf. (Accessed 11 November 2020).

Gobalarajah, K., Prabagar, S., Jayawardena, U., Rasiah, G., Rajendra, S.,, Prabagar, J., 2020. Impact of water quality on Chronic Kidney Disease of unknown etiology (CKDu) in Thunukkai Division in Mullaitivu District, Sri Lanka. BMC Nephrology 21 (1), 507. Available from: https://doi.org/10.21203/rs.3.rs-19873/v2.

Goenka, A., Kollmann, T., 2015. Development of immunity in early life. The Journal of Infection 71 (Supplement 1), 112−120.

Goldsmith, J.R., Herishanu, Y., Abarbanel, J.M., Weinbaum, Z., 1990. Clustering of Parkinson's disease points to environmental etiology. Archives of Environmental Health 45, 88−94.

Gombart, A.F., Pierre, A., Maggini, S., 2020. A review of micronutrients and the immune System - working in harmony to reduce the risk of infection. Nutrients 12, E236. Available from: https://doi.org/10.3390/nu12010236.

Gómez-Rubio, V., Ferrándiz-Ferragud, J., López-Quílez, A., 2005. Detecting clusters of disease with R. Journal of Geographical Systems 7, 189−206. Available from: https://doi.org/10.1007/s10109-005-0156-5.

González-Maglio, D.H., Paz, M.L., Leoni, J., 2016. Sunlight effects on immune system: Is there something else in addition to UV-induced immunosuppression? Biomedical Research International 2016, 1934518. Available from: https://doi.org/10.1155/2016/1934518.

Govender, N.P., Magobo, R.E., Mpembe, R., Mhlanga, M., Matlapeng, P., Corcoran, C., et al., 2018. Candida auris in South Africa, 2012-2016. Emerging Infectious Diseases 24 (11), 2036−2040.

Grassly, N.C., Fraser, C., 2006. Seasonal infectious disease epidemiology. Proceedings. Biological sciences 273 (1600), 2541−2550. Available from: https://doi.org/10.1098/rspb.2006.3604.

Greenhalgh, E., 2015. El Niño, East Africa, and Rift Valley Fever. NOAA Climate.gov. https://www.climate.gov/news-features/understanding-climate/el-ni%C3%B1o-east-africa-and-rift-valley-fever. (Accessed 22 January 2021).

Greenlee, J.J., Greenlee, M.H., 2015. The transmissible spongiform encephalopathies of livestock. ILAR Journal 56 (1), 7−25. Available from: https://doi.org/10.1093/ilar/ilv008.

Griffiths, A.J.F., 2020. Mutation. Encyclopaedia Britannica. https://www.britannica.com/science/mutation-genetics. (Accessed 23 November 2020).

GU/WHO [(The Government of Uganda (GU)/World Health Organization (WHO)], 2012. Uganda adopts a multisectoral response to nodding syndrome. Press Release Kampala, March 2, 2012. http://reliefweb.int/sites/reliefweb.int/files/resources/kampala-nodding-press-release-02032012.pdf. (Accessed 25 August 2020).

Guajardo, O.A., Oyana, T.J., 2009. A critical assessment of geographic clusters of breast and lung cancer incidences among residents living near the Tittabawassee and Saginaw Rivers, Michigan, USA. Journal of Environmental and Public Health 2009, 316249. Available from: https://doi.org/10.1155/2009/316249.

Guarnieri, M., Balmes, J.R., 2014. Outdoor air pollution and asthma. Lancet 3 383 (9928), 1581.

Gundacker, C., Hengstschläger, M., 2012. The role of the placenta in fetal exposure to heavy metals. Wiener Medizinische Wochenschrift 162 (9-10), 201−206. Available from: https://doi.org/10.1007/s10354-012-0074-3.

GWA (Government of Western Australia, Department of Health), 2017. Guidelines for the Investigation of Cancer Clusters in Western Australia. https://ww2.health.wa.gov.au/-/media/Files/Corporate/general-documents/Epidemiology/Cancer-Cluster-Guidelines-PDF.pdf. (Accessed 14 November 2020).

Habibi, L., Perry, G., Mahmoudi, M., 2014. Global warming and neurodegenerative disorders: speculations on their linkage. Bioimpacts 4 (4), 167−170. Available from: https://doi.org/10.15171/bi.2014.013.

Hammer, C.C., Brainard, J., Hunter, P.R., 2018. Risk factors and risk factor cascades for communicable disease outbreaks in complex humanitarian emergencies: a qualitative systematic review. BMJ Global Health. 3 (4), e000647. Available from: https://doi.org/10.1136/bmjgh-2017-000647.

Hara, T., Nakashima, Y., Sakai, Y., Nishio, H., Motomura, Y., Yamasaki, S., 2016. Kawasaki disease: a matter of innate immunity. Clinical and Experimental Immunology 186 (2), 134−143. Available from: https://doi.org/10.1111/cei.12832.

Hasan S.E., 2020. Medical Geology. Reference Module in Earth Systems and Environmental Sciences, B978-0-12−409548-9.12523-0. https://doi.org/10.1016/B978-0-12-409548-9.12523-0. (Accessed 30 January 2021).

Heath, C.H., Dyer, J.R., Pang, S., Coombs, G.W., Gardam, D.J., 2019. *Candida auris* sternal osteomyelitis in a man from Kenya visiting Australia, 2015. Emerging Infectious Diseases 25 (1), 192−194. Available from: https://doi.org/10.3201/eid2501.181321.

Hedera, P., 2000. Hereditary spastic paraplegia overview. In: Adam, M.P., Ardinger, H.H., Wallace, S.E., Pagon, R.A., Bean, L.J.H., Mirzaa, G., Amemiya, A. (Eds.), GeneReviews®. University of Washington, Seattle (W.A.), pp. 1993−2021. Available from: https://www.ncbi.nlm.nih.gov/books/NBK1509/ (accessed 30.10.20).

Hedera, P., 2016. Hereditary and metabolic myelopathies. Handbook of Clinical Neurolology 136, 769−785.

Herishanu, Y.O., Goldsmith, J.R., Abarbanel, J.M., Weinbaum, Z., 1989. Clustering of Parkinson's disease in southern Israel. Canadian Journal of Neurological Sciences 16, 402−405.

Hernandez, D., Fisher, E.M.C., 1996. Down syndrome genetics: unravelling a multifactorial disorder. Human Molecular Genetics 5 (1), 1411−1416. Available from: https://doi.org/10.1093/hmg/5.supplement_1.141.

Hofer, U., 2019. *Candida auris'* potential link to climate change. Nature Reviews Microbiology 17, 588. Available from: https://doi.org/10.1038/s41579-019-0254-x.

Hoffmann, P.R., Berry, M.J., 2008. The influence of selenium on immune responses. Molecular Nutrition and Food Research 52 (11), 1273−1280.

HSA (Health and Safety Authority), 2017. Healthy, Safe and Productive Lives and Enterprises: Hazard and Risk. https://www.hsa.ie/eng/Topics/Hazards/. (Accessed 05 December 2020).

Hu, S.L., Xiong, W., Dai, Z.Q., Zhao, H.L., Feng, H., 2016. Cognitive changes during prolonged stay at high altitude and its correlation with C-reactive protein. PLoS One 11 (1), e0146290. Available from: https://doi.org/10.1371/journal.pone.0146290.

Hua-Li, Z., Shi-Chao, X., De-Shen, T., Dong, L., Hua-Feng, L., 2011. Seasonal distribution of active systemic lupus erythematosus and its correlation with meteorological factors. Clinics (Sao Paulo, Brazil) 66 (6), 1009−1013. Available from: https://doi.org/10.1590/s1807-59322011000600015.

Hugh-Jones, M., Blackburn, J., 2009. The ecology of *Bacillus anthracis*. Molecular Aspects of Medicine 30, 356−367.

Hurst, C.J., Gerba, C.P., Cech, I., 1980. Effects of environmental variables and soil characteristics on virus survival in soil. Applied Environmental Microbiology 40 (6), 1067−1079. Available from: https://doi.org/10.1128/AEM.40.6.1067-1079.1980.

Hyrkäs, H., Ikäheimo, T.M., Jaakkola, J.J., Jaakkola, M.S., 2016. Asthma control and cold weather-related respiratory symptoms. Respiratory Medicine 113, 1−7. Available from: https://doi.org/10.1016/j.rmed.2016.02.005.

Ilevbare, F.M., 2019. Investigating effects of climate change on health risks in Nigeria. Environmental Factors Affecting Human Health. IntechOpen. Available from: 10.5772/intechopen.86912.

Intarasunanont, P., Navasumrit, P., Woraprasit, S., Chaisatra, K., Suk, W.A., Mahidol, C., et al., 2012. Effects of arsenic exposure on DNA methylation in cord blood samples from newborn babies and in a human lymphoblast cell line. Environmental Health 11 (1), 31. Available from: https://doi.org/10.1186/1476-069X-11-31.

IPCC (Intergovernmental Panel on Climate Change), 2019. Special Report on the Ocean and Cryosphere in a Changing Climate. https://www.ipcc.ch/srocc/. (Accessed 10 January 2021).

Ismael, M., El-Sayed, M.S., Metwally, A.M., Abdullaziz, I.A., 2015. Trace elements status and antioxidants profile in ill-thrift buffalo calves. Alexander Journal of Veterinary Sciences 44 (1), 130−135. Available from: https://doi.org/10.5455/ajvs.170301, https://www.alexjvs.com/?mno = 170301.

Jackson, D.J., Johnston, S.L., 2010. The role of viruses in acute exacerbations of asthma. Journal of Allergy and Clinical Immunology 125, 1178–1187. Available from: https://doi.org/10.1016/j.jaci.2010.04.021.

Jacquez, G.M., 2009. Cluster morphology analysis. Spat Spatiotemporal Epidemiology 1 (1), 19–29. Available from: https://doi.org/10.1016/j.sste.2009.08.002.

Jacquez, G.M., Kaufmann, A., Meliker, J., Avruskin, G.A., Kaufmann, A., Goovaerts, P., et al., 2005. Global, local and focused geographic clustering for case-control data with residential histories. Environmental Health 4, 4.

Jagannatha Rao, K.S., Ranganath Rao, V., Shanmugavelu, P., Menon, R.B., 1999. Trace elements in Alzheimer's disease brain: a new hypothesis. In Alzheimer's disease brain: a new hypothesis. Alzheimer's Reports 2 (4), 211–216.

Jager, K.J., Zoccali, C., Macleod, A., Dekker, F.W., 2008. Confounding: what it is and how to deal with it. Kidney International 73 (3), 256–260. Available from: https://doi.org/10.1038/sj.ki.5002650.

Jain, N., 2020. The early life education of the immune system: moms, microbes and (missed) opportunities. Gut Microbes 12 (1), 1824564. Available from: https://doi.org/10.1080/19490976.2020.1824564.

Janeway, C.A.Jr, Travers, P., Walport, M., Shlomchik, M. (Eds.), 2001. Immunobiology: The Immune System in Health and Disease. fifth ed. Garland Science, New York.

Janghorbani, M., Shaygannejad, V., Hakimdavood, M., Salari, M., 2017. Trace elements in serum samples of patients with multiple sclerosis. Athens Journal of Health 4 (2), 145–154.

Jansson, J.K., Tas, N., 2014. The microbial ecology of permafrost. Nature Reviews Microbiology 12 (6), 414–425. Available from: https://doi.org/10.1038/nrmicro3262.

Jarup, L., 2004. Health and environment information systems for exposure and disease mapping, and risk assessment. Environmental Health Perspectives 112 (9), 995–997. Available from: https://doi.org/10.1289/ehp.6736.

Javelle, E., Lesueur, A., Pommier de Santi, V.P., de Laval, F., Lefebvre, T., Holweck, G., et al., 2020. The challenging management of Rift Valley Fever in humans: literature review of the clinical disease and algorithm proposal. Annals of Clinical Microbiology and Antimicrobiology 19, 4. Available from: https://doi.org/10.1186/s12941-020-0346-5.

Jayasekara, J.M., Dissanayake, D.M., Adhikari, S.B., Bandara, P., 2013. Geographical distribution of chronic kidney disease of unknown origin in North Central Region of Sri Lanka. Ceylon Medical Journal 58, 6–10.

Jha, V., Garcia-Garcia, G., Iseki, K., et al., 2013. Chronic kidney disease: global dimension and perspectives. Lancet 382, 260–272.

Jhun, I., Mata, D.A., Nordio, F., Lee, M., Schwartz, J., Zanobetti, A., 2017. Ambient temperature and sudden infant death syndrome in the United States. Epidemiology (Cambridge, MA) 28 (5), 728–734. Available from: https://doi.org/10.1097/EDE.0000000000000703.

Johnson, C.J., Phillips, K.E., Schramm, P.T., McKenzie, D., Aiken, J.M., Pedersen, J.A., 2006. Prions adhere to soil minerals and remain infectious. PLoS Pathology 2 (4), e32. Available from: https://doi.org/10.1371/journal.ppat.0020032.

Johnston, S.L., Pattemore, P.K., Sanderson, G., Smith, S., Lampe, F., Josephs, L., et al., 1995. Community study of role of viral infections in exacerbations of asthma in 9 - 11 year old children. British Medical Journal 310, 1225–1229. Available from: https://doi.org/10.1136/bmj.310.6989.1225.

Joy, E.J., Ander, E.L., Young, S.D., Black, C.R., Watts, M.J., Chilimba, A.D., et al., 2014. Dietary mineral supplies in Africa. Physiologia Plantarum 151 (3), 208–229. Available from: https://doi.org/10.1111/ppl.12144.

Judson, M.A., 2020. Environmental risk factors for sarcoidosis. Frontiers in Immunology 11, 1340. Available from: https://doi.org/10.3389/fimmu.2020.01340.

Jung, C.R., Lin, Y.T., Hwang, B.F., 2015. Ozone, particulate matter, and newly diagnosed Alzheimer's disease: a population-based cohort study in Taiwan. Journal of Alzheimer's Disease 44 (2), 573–584. Available from: https://doi.org/10.3233/JAD-140855.

Justiz Vaillant, A.A., Goyal, A., Bansal, P., et al., 2020. Systemic lupus erythematosus. StatPearls [Internet]. StatPearls Publishing, Treasure Island (FL). Available from: https://www.ncbi.nlm.nih.gov/books/NBK535405/.

Kaboré, B., Post, A., Lompo, P., Bognini, J.D., Diallo, S., Kam, B.T.D., et al., 2020. Aetiology of acute febrile illness in children in a high malaria transmission area in West Africa. Clinical Microbiology and Infection: The Official Publication of the European Society of Clinical Microbiology and Infectious Diseases 27, 590−596. Available from: https://doi.org/10.1016/j.cmi.2020.05.029.

Kading, R.C., Brault, A.C., Beckham, J.D., 2020. Editorial: Global perspectives on arbovirus outbreaks: a 2020 snapshot. Tropical Medicine and Infectious Disease 5, 142. Available from: https://doi.org/10.3390/tropicalmed5030142.

Kaiser, C., Benninger, C., Asaba, G., Mugisa, C., Kabagambe, G., Kipp, W., et al., 2000. Clinical and electro-clinical classification of epileptic seizure in west Uganda. Bulletin de la Société de Pathologie Exotique 93, 255−259.

Kakuschke, A., Prange, A., 2007. The influence of metal pollution on the immune system: a potential stressor for marine mammals in the North Sea. International Journal of Comparative Psychology 20, 179−193.

Kaloga, M., Gbéry, I.P., Bamba, V., Kouassi, Y.I., Ecra, E.J., Diabate, A., et al., 2015. Epidemiological, clinical, and paraclinic aspect of cutaneous sarcoidosis in Black Africans. Dermatology Research and Practice 2015, 802824. Available from: https://doi.org/10.1155/2015/802824.

Kannel, W.B., Dawber, T.R., Kagan, A., Revotskie, N., Stokes the 3rd., J., 1961. Factors of risk in the development of coronary heart disease: six-year follow-up experience. The Framingham Study. Annals of Internal Medicine 55, 33−50. Available from: https://doi.org/10.7326/0003-4819-55-1-33.

Karan, A., Ali, K., Teelucksingh, S., Sakhamuri, S., 2020. The impact of air pollution on the incidence and mortality of COVID-19. Global Health Research Policy 5, 39. Available from: https://doi.org/10.1186/s41256-020-00167-y.

Kasotia, P., n.d. The Health Effects of Global Warming: Developing Countries Are the Most Vulnerable. United Nations Chronicle. https://www.un.org/en/chronicle/article/health-effects-global-warming-developing-countries-are-most-vulnerable. (Accessed 25 November 2020).

Kean, R., Sherry, L., Townsend, E., McKloud, E., Short, B., Akinbobola, A., et al., 2018. Surface disinfection challenges for *Candida auris*: an *in-vitro* study. Journal of Hospital Infections 98 (4), 433−436.

Kean, R., Brown, J., Gulmez, D., Ware, A., Ramage, G., 2020. *Candida auris*: a decade of understanding of an enigmatic pathogenic yeast. Journal of Fungi 6 (1), 30.

Keen, C.L., Uriu-Adams, J.Y., Ensunsa, J.L., Gershwin, M.E., 2004. Trace elements/minerals and immunity. In: Gershwin, M.E., Nestel, P., Keen, C.L. (Eds.), Handbook of Nutrition and Immunity. Humana Press, Totowa, NJ. Available from: https://doi.org/10.1007/978-1-59259-790-1_6 (Accessed 26 Decmeber 2020).

Kehl-Fie, T.E., Chitayat, S., Hood, M.I., Damo, S., Restrepo, N., Garcia, C., et al., 2011. Nutrient metal sequestration by calprotectin inhibits bacterial superoxide defense, enhancing neutrophil killing of Staphylococcus aureus. Cell Host and Microbe 10 (2), 158−164.

Kent, J.T., Carr, D., 2020. A visually striking case of primary acrocyanosis: a rare cause of the blue digit. The American Journal of Emergency Medicine 40, 227.e3−227.e4. Available from: https://doi.org/10.1016/j.ajem.2020.07.064.

Khayamzadeh, M., Najafi, S., Sadrolodabaei, P., Vakili, F., Kharrazi Fard, M.J., 2019. Determining salivary and serum levels of iron, zinc and vitamin B_{12} in patients with geographic tongue. Journal of Dental Research, Dental Clinics, Dental Prospects 13 (3), 221−226. Available from: https://doi.org/10.15171/joddd.2019.034.

Kilbourne, E.M., 1999. The spectrum of illness during heat waves. American Journal of Preventive Medicine 16 (4), 359−360. Available from: https://doi.org/10.1016/s0749-3797(99)00016-1.

Kile, M.L., Baccarelli, A., Hoffman, E., Tarantini, L., Quamruzzaman, Q., Rahman, M., et al., 2012. Prenatal arsenic exposure and DNA methylation in maternal and umbilical cord blood leukocytes. Environmental Health Perspectives 120 (7), 1061−1066. Available from: https://doi.org/10.1289/ehp.1104173.

Kim, G.B., 2019. Reality of Kawasaki disease epidemiology. Korean Journal Pediatric 62 (8), 292−296. Available from: https://doi.org/10.3345/kjp.2019.00157.

King, D.E., Lalwani, P.D., Mercado, G.P., Dolan, E.L., Frierson, J.M., Meyer, J.N., Murphy, S.K., 2024. The use of race terms in epigenetics research: considerations moving forward. Frontiers in Genetics 1348855. Available from: https://doi.org/10.3389/fgene.2024.1348855.

Kingsley, B.S., Schmeichel, K.L., Rubin, C.H., 2007. An update on cancer cluster activities at the Centers for Disease Control and Prevention. Research mini-monograph. Environmental Health Perspectives 115 (1), 165−171.

Kister, I., Bacon, T.E., Chamot, E., Salter, A.R., Cutter, G.R., Kalina, J.T., et al., 2013. Natural history of multiple sclerosis symptoms. International Journal of MS Care 15 (3), 146−158. Available from: https://doi.org/10.7224/1537-2073.2012-053.

Koester-Hegmann, C., Bengoetxea, H., Kosenkov, D., Thiersch, M., Haider, T., Gassmann, M., et al., 2019. High-altitude cognitive impairment is prevented by enriched environment including exercise via VEGF signalling. Frontiers in Cellular Neuroscience 12, 532. Available from: https://doi.org/10.3389/fncel.2018.00532.

Koestler, D.C., Avissar-Whiting, M., Houseman, E.A., Karagas, M.R., Marsit, C.J., 2013. Differential DNA methylation in umbilical cord blood of infants exposed to low levels of arsenic *in utero*. Environmental Health Perspectives 121 (8), 971−977. Available from: https://doi.org/10.1289/ehp.1205925.

Kollmann, T.B., Kampmann, B., Mazmanian, S.K., Marchant, A., Levy, O., 2017. Protecting the newborn and young infant from infectious diseases: lessons from immune ontogeny. Immunity 46 (3), 350−363.

Korevaar, D.A., Visser, B.J., 2013. Reviewing the evidence on nodding syndrome, a mysterious tropical disorder. International Journal of Infectious Diseases 17 (3), e149−e152. Available from: https://doi.org/10.1016/j.ijid.2012.09.015.

Kostakou, E., Kaniaris, E., Filiou, E., Vasileiadis, I., Katsaounou, P., Tzortzaki, E., et al., 2019. Acute severe asthma in adolescent and adult patients: current perspectives on assessment and management. Journal of Clinical Medicine 8 (9), 1283. Available from: https://doi.org/10.3390/JCM8091283.

Kroll-Smith, S., Brown, P.M., Gunter, V., 2000. Illness and the Environment: A Reader in Contested Medicine. ISBN: 9780814747292, 464 pp. https://www.thebookstall.com/book/9780814747292. (Accessed 30 January 2021).

Kronenfeld, B.J., Wong, D.W.S., 2017. Visualizing statistical significance of disease clusters using cartograms. International Journal of Health Geographics 16, 19. Available from: https://doi.org/10.1186/s12942-017-0093-9.

Kulkarni, N., Kalele, K., Kulkarni, M., Kathariya, R., 2014. Trace elements in oral health and disease: an updated review. Journal of Dental Research and Review 1 (2), 100−104.

Kumar, A., Calne, S.M., Schulzer, M., et al., 2004. Clustering of Parkinson disease: shared cause or coincidence? Archives of Neurology 61 (7), 1057−1060. Available from: https://doi.org/10.1001/archneur.61.7.1057.

Kurklinsky, A.K., Miller, V.M., Rooke, T.W., 2011. Acrocyanosis: the flying dutchman. Vascular Medicine (London, England) 16 (4), 288−301. Available from: https://doi.org/10.1177/1358863X11398519.

Kurland, L.T., Mulder, D.W., 1954. Epidemiologic investigations of amyotrophic lateral sclerosis. I. Preliminary report on geographic distribution, with special reference to the Mariana Islands, including clinical and pathologic observations. Neurology 4 (5), 355−378. Available from: https://doi.org/10.1212/wnl.4.5.355.

Kuznetsova, A., McKenzie, D., Cullingham, C., Aiken, J.M., 2020. Long-term incubation PrP^{CWD} with soils affects prion recovery but not infectivity. Pathogens 9 (4), 311. Available from: https://doi.org/10.3390/pathogens9040311.

Lahr, J., Kooistra, L., 2010. Environmental risk mapping of pollutants: state of the art and communication aspects. Science of the Total Environment 408 (18), 3899−3907. Available from: https://doi.org/10.1016/j.scitotenv.2009.10.045.

Laine, J.T., Tuomainen, T.P., Salonen, J.T., et al., 2020. Serum copper-to-zinc-ratio and risk of incident infection in men: The Kuopio Ischaemic Heart Disease Risk Factor Study. European Journal of Epidemiology 35, 1149–1156. Available from: https://doi.org/10.1007/s10654-020-00644-1.

Lall, R., Mohammed, R., Ojha, U., 2019. What are the links between hypoxia and Alzheimer's Disease? Neuropsychiatric Disease and Treatment 15, 1343–1354.

Last, J.M., 1988. A Dictionary of Epidemiology, second ed. Oxford University Press, New York.

Lathe, R., Darlix, J.L., 2020. Prion protein PrP nucleic acid binding and mobilization implicates retroelements as the replicative component of transmissible spongiform encephalopathy. Archives of Virology 165, 535–556. Available from: https://doi.org/10.1007/s00705-020-04529-2.

Lemelle, L., Simionovici, A., Colin, P., Knott, G., Bohic, S., Cloetens, P., et al., 2020. Nano-imaging trace elements at organelle levels in substantia nigra overexpressing α-synuclein to model Parkinson's disease. Communications Biology 3, 364. Available from: https://doi.org/10.1038/s42003-020-1084-0.

Li, D., Yuan, J., Kopp, R., 2020. Escalating global exposure to compound heat-humidity extremes with warming. Environmental Research Letters 15, 064003. Available from: https://doi.org/10.1088/1748-9326/ab7d04.

Liang, L., Gong, P., 2017. Climate change and human infectious diseases: a synthesis of research findings from global and spatio-temporal perspectives. Environment International 103, 99–108.

Lill, C.M., Roehr, J.T., McQueen, M.B., Kavvoura, F.K., Bagade, S., Schjeide, B.-M.M., et al., 2012. Comprehensive research synopsis and systematic meta-analyses in Parkinson's Disease genetics: the PD gene database. PLoS Genetics 8 (3), e1002548. Available from: https://doi.org/10.1371/journal.pgen.1002548.

Lim, E., Ahn, Y.-C., Jang, E.-S., Lee, S.W., Lee, S.-H., Son, C.-G., 2020. Systematic review and meta-analysis of the prevalence of chronic fatigue syndrome/myalgic encephalomyelitis (CFS/ME). Journal of Translational Medicine 18, 100.

Lin, M.T., Wu, M.H., 2017. The global epidemiology of Kawasaki disease: review and future perspectives. Global Cardiology Science and Practice 3, e201720. Available from: https://doi.org/10.21542/gcsp.2017.20.

Lingamaneni, P., Kumar, K.K., Teja, C.R., Reddy, B.V.R., Krishna, P.L., 2015. A review on role of essential trace elements in health and disease. Journal of Dr. NTR University of Health Sciences 4 (2), 75–85.

Lisse, I.M., Aaby, P., Whittle, H., Jensen, H., Engelmann, M., Christensen, L.B., 1997. Tlymphocyte subsets in West African children: impact of age, sex, and season. The Journal Pediatrics 130 (1), 77–85. Available from: https://doi.org/10.1016/S0022-3476(97)70313-5171.

Liu, Y., Nguyen, M., Robert, A., Meunier, B., 2019. Metal ions in Alzheimer's Disease: a key role or not? Accounts of Chemical Research 52 (7), 2026–2035. Available from: https://doi.org/10.1021/acs.accounts.9b00248.

Loef, M., Walach, H., 2012. Copper and iron in Alzheimer's disease: a systematic review and its dietary implications. British Journal of Nutrition 107 (1), 7–19. Available from: https://doi.org/10.1017/S000711451100376X.

Lopez, A.D., Mathers, C.D., Ezzati, M., Jamison, D.T., Murray, C.J.L., 2006. Global Burden of Disease and Risk Factors. A copublication of The World Bank and Oxford University Press. https://openknowledge.worldbank.org/bitstream/handle/10986/7039/364010PAPER0Gl101OFFICIAL0USE0ONLY1.pdf?sequence = 1. Barcel. (Accessed 19 November 2020).

Lu, K., Abo, R.P., Schlieper, K.A., Graffam, M.E., Levine, S., Wishnok, J.S., et al., 2014. Arsenic exposure perturbs the gut microbiome and its metabolic profile in mice: an integrated metagenomics and metabolomics analysis. Environmental Health Perspectives 122 (3), 284–291. Available from: https://doi.org/10.1289/ehp.1307429.

Lunyera, J., Mohottige, D., Von Isenburg, M., Jeuland, M., Patel, U.D., Stanifer, J.W., 2016. CKD of uncertain etiology: a systematic review. Clinical Journal of the American Society of Nephrology (CJASN) 11 (3), 379–385. Available from: https://doi.org/10.2215/cjn.07500715.

Luo, Z.R., Yu, L.-L., Huang, S.-T., Chen, L.-W., Chen, Q., 2020. Impact of meteorological factors on the occurrence of acute aortic dissection in Fujian Province, China: a single-center seven-year retrospective study. Journal of Cardiothoracic Surgery 15, 178. Available from: https://doi.org/10.1186/s13019-020-01227-7.

Ma, J., 2012. The role of cofactors in prion propagation and infectivity. PLoS Pathogens 8 (4), e1002589. Available from: https://doi.org/10.1371/journal.ppat.1002589.

MacGillivray, D.M., Kollmann, T.R., 2014. The role of environmental factors in modulating immune responses in early life. Frontiers in Immunology 5, 434. Available from: https://doi.org/10.3389/fimmu.2014.00434.

Maggini, S., Pierre, A., Calder, P.C., 2018. Immune function and micronutrient requirements change over the life course. Nutrients 10 (10), 1531. Available from: https://doi.org/10.3390/nu10101531.

Magobo, R.E., Corcoran, C., Seetharam, S., Govender, N.P., 2014. *Candida auris*-associated candidemia, South Africa. Emerging Infectious Diseases 20 (7), 1250−1251. Available from: https://doi.org/10.3201/eid2007.131765.

Mak, O.T., 1988. Prostacyclin production in vascular endothelium of patients with Blackfoot disease. Advances in Experimental Medical Biology 242, 119−125.

Malavolta, M., Piacenza, F., Basso, A., Giacconi, R., Costarelli, L., Mocchegiani, E., 2015. Serum copper to zinc ratio: Relationship with aging and health status. Mechanisms of Ageing and Development 151, 93−100. Available from: https://doi.org/10.1016/j.mad.2015.01.004.

Manlhiot, C., Mueller, B., O'Shea, S., Majeed, H., Bernknopf, B., Labelle, M., et al., 2018. Environmental epidemiology of Kawasaki disease: linking disease etiology, pathogenesis and global distribution. PLoS One 13 (2), e0191087. Available from: https://doi.org/10.1371/journal.pone.0191087.

Manousek, J., Privarova, L., Pavkova, G.M., 2014. Metal hypersensitivity as the cause of chronic fatigue syndrome: case report. In: Hudson, C. (Ed.), Chronic Fatigue Syndrome - Risk Factors, Management and Impacts on Daily Life. Neuroscience Research Progress. Nova Science Publishers, New York, p. 154.

Mansfield, A.J., Stehr-Green, J.K., Stehr-Green, P.A. (n.d.). Cluster Investigations of Non-Infectious Health Events. Focus on Field Epidemiology 5, 4. North Carolina Center for Public Health Preparedness-The North Carolina Institute for Public Health. https://nciph.sph.unc.edu/focus/vol5/issue4/5-4ClusterInvestigations_issue.pdf. (Accessed 11 November 2020).

Manyuchi, A.E., Chersich, M., Vogel, C., Wright, C.Y., Matsika, R., Erasmus, B., 2022. Extreme heat events, high ambient temperatures and human morbidity and mortality in Africa: a systematic review. South African Journal of Science 118 (11/12). Available from: https://doi.org/10.17159/sajs.2022/12047 (accessed 13.02.24).

Mao, R.J., Moa, A., Chughtai, A., 2020. The epidemiology of unknown disease outbreak reports globally. Global Biosecurity 1 (4). Available from: https://doi.org/10.31646/gbio.62.

Marchi, S., Trombetta, C.M., Montomoli, 2018. Emerging and re-emerging arboviral diseases as a global health problem. In: Majumder, A.A., Kabir, R., Rahman, S. (Eds.), Emerging and Re-emerging Issues. IntechOpen. Available from: https://doi.org/10.5772/intechopen.77382. https://www.intechopen.com/books/public-health-emerging-and-re-emerging-issues/emerging-and-re-emerging-arboviral-diseases-as-a-global-health-problem. (Accessed 01 February 2021).

Margesin, R., Collins, T., 2019. Microbial ecology of the cryosphere (glacial and permafrost habitats): current knowledge. Applied Microbiology and Biotechnology 103, 2537−2549. Available from: https://doi.org/10.1007/s00253-019-09631-3.

Marshall, J.S., Warrington, R., Watson, W., Kim, H.L., 2018. An introduction to immunology and immunopathology. Allergy Asthma and Clinical Immunology 14, 49. Available from: https://doi.org/10.1186/s13223-018-0278-1.

Martineau, A.R., Nhamoyebonde, S., Oni, T., Rangaka, M.X., Marais, S., Bangani, N., et al., 2011. Reciprocal seasonal variation in vitamin D status and tuberculosis notifications in Cape Town, South Africa. Proceedings of the National Academy of Science, USA 108 (47), 19013−19017. Available from: https://doi.org/10.1073/pnas.1111825108.

Mayer, S.V., Tesh, R.B., Vasilakis, N., 2017. The emergence of arthropod-borne viral diseases: a global prospective on dengue, chikungunya and zika fevers. Acta Tropica 166, 155−163. Available from: https://doi.org/10.1016/j.actatropica.2016.11.020.

McCarty, K.M., Hanh, H.T., Kim, K.W., 2011. Arsenic geochemistry and human health in South East Asia. Reviews on Environmental Health 26 (1), 71−78. Available from: https://doi.org/10.1515/reveh.2011.010.

McElroy, J., 2020. Environmental exposures and child health: what we might learn in the 21st Century from the National Children's Survey. Environmental Health Insights 2 (1), 105−109.

McElvenny, D.M., Mounstephen, A.H., Hodgson, J.T., Osman, J., Elliott, R.C., Williams, N.R., 2003. Investigating and analysing workplace clusters of diseases: a health and safety executive perspective. Occupational Medicine 53 (3), 201−208. Available from: https://doi.org/10.1093/occmed/kqg039.

McGregor, G., Ebi, K.L., 2018. El Niño Southern Oscillation (ENSO) and health: an overview for climate and health researchers. Atmosphere 9 (7), 282. Available from: https://doi.org/10.3390/atmos9070282.

Mehri, A., 2020. Trace elements in human nutrition (II) - an update. International Journal of Preventive Medicine 11, 2. Available from: https://doi.org/10.4103/ijpvm.IJPVM_48_19.

Mehta, R.H., Suzuki, T., Hagan, P.G., Bossone, E., Gilon, D., Llovet, A., et al., 2002. Predicting death in patients with acute type A aortic dissection. Circulation 105 (2), 200−206. Available from: https://doi.org/10.1161/hc0202.

Meister, H., Tammaru, T., Sandre, S.-L., Freitak, D., 2017. Sources of variance in immunological traits: evidence of congruent latitudinal trends across species. Journal of Experimental Biology 220 (14), 2606−2615. Available from: https://doi.org/10.1242/jeb.154310.

Melø, T.M., Larsen, C., White, L.R., et al., 2003. Manganese, copper, and zinc in cerebrospinal fluid from patients with multiple sclerosis. Biological Trace Element Research 93, 1−8. Available from: https://doi.org/10.1385/BTER:93:1-3:1.

Mensah, E., El Zowalaty, M., 2018. Arboviruses in South Africa, known and unknown. Future Virology 13 (8). Available from: https://doi.org/10.2217/fvl-2018-0090.

Miles, D.J., van der Sande, M., Crozier, S., Ojuola, O., Palmero, M.S., Sanneh, M., et al., 2008. Effects of antenatal and postnatal environments on CD4 T-cell responses to Mycobacterium bovis BCG in healthy infants in the Gambia. Clinical Vaccine and Immunology 15, 995−1002. Available from: https://doi.org/10.1128/CVI.00037-08.

Miller, W.R., Sanzolone, R.F., 2003. Investigation of the possible connection of rock and soil geochemistry to the occurrence of high rates of neurodegenerative diseases on guam and a hypothesis for the cause of the diseases. United States Geological Survey (USGS) Open-File Report 03-126. U.S. Department of the Interior USGS, Denver. https://pubs.usgs.gov/of/2003/ofr-03-126/OFR-03-126-508.pdf. (Accessed 06 September 2020).

Miller, T.E., Golemboski, K.A., Ha, R.S., Bunn, T., Sanders, F.S., Dietert, R.R., 1998. Developmental exposure to lead causes persistent immunotoxicity in Fischer 344 rats. Toxicological Sciences 42 (2), 129−135. Available from: https://doi.org/10.1093/toxsci/42.2.129.

Mills, K., Xu, Y., Zhang, W., Bundy, J.D., Chen, C., Kelly, T., et al., 2015. A systematic analysis of world-wide population-based data on the global burden of chronic kidney disease in 2010. Kidney International 88, 950−957.

Mitchell, J.D., East, B.W., Harris, I.A., Prescott, R.J., Pentland, B., 1986. Trace elements in the spinal cord and other tissues in motor neuron disease. Journal of neurology, neurosurgery, and psychiatry 49 (2), 211−215. Available from: https://doi.org/10.1136/jnnp.49.2.211.

MLA (Meat and Livestock Australia), 2020. Mineral deficiencies. https://www.mla.com.au/research-and-development/animal-health-welfare-and-biosecurity/diseases/nutritional/mineral-deficiencies. (Accessed 31 October 2020).

Mocanu, C.S., Jureschi, M., Drochioiu, G., 2020. Aluminium binding to modified amyloid-β peptides: implications for Alzheimer's Disease. Molecules (Basel, Switzerland) 25 (19), 4536. Available from: https://doi.org/10.3390/molecules25194536.

Montgomery, R.R., Shaw, A.C., 2015. Paradoxical changes in innate immunity in aging: recent progress and new directions. Journal of Leukocyte Biology 98, 937–943. Available from: https://doi.org/10.1189/jlb.5MR0315-104R.

Moore, S.E., Collinson, A.C., Prentice, A.M., 2001. Immune function in rural Gambian children is not related to season of birth, birth size, or maternal supplementation status. American Journal of Clinical Nutrition 74 (6), 840–847.

Moore, S.E., Morgan, G., Prentice, A.M., 2008. Birth season and environmental influences on blood leucocyte and lymphocyte subpopulations in rural Gambian infants. BMC Immunology 9 (1), 18. Available from: https://doi.org/10.1186/1471-2172-9-18.

Moozhipurath, R.K., Kraft, L., Skiera, B., 2020. Evidence of protective role of ultraviolet-B (UVB) radiation in reducing COVID-19 deaths. Science Reports 10, 17705. Available from: https://doi.org/10.1038/s41598-020-74825-z.

Morar, R., Feldman, C., 2015a. Comorbid illnesses in South African patients with sarcoidosis. European Respiratory Journal 46 (Supplement 59). Available from: https://doi.org/10.1183/13993003.co9ngress-2015.PA839PA839.

Morar, R., Feldman, C., 2015b. Sarcoidosis in Johannesburg, South Africa: a retrospective study. European Respiratory Journal 46 (Supplement 59), . Available from: https://doi.org/10.1183/13993003.congress-2015PA841.

Morris, G.P., Reis, S., Beck, S.A., et al., 2017. Scoping the proximal and distal dimensions of climate change on health and wellbeing. Environmental Health 16, 116. Available from: https://doi.org/10.1186/s12940-017-0329-y.

Mosomtai, G., Evander, M., Sandström, P., Ahlm, C., Sang, R., Hassan, O.A., et al., 2016. Association of ecological factors with Rift Valley fever occurrence and mapping of risk zones in Kenya. International Journal of Infectious Diseases 46, 49–55. Available from: https://doi.org/10.1016/j.ijid.2016.03.013.

Mostafa, M.K., Gamal, G., Wafiq, A., 2021. The impact of COVID 19 on air pollution levels and other environmental indicators - a case study of Egypt. Journal of Environmental Management 277, 111496. Available from: https://doi.org/10.1016/j.jenvman.2020.111496.

MS International Federation, 2016. Geographical latitude and the onset of MS https://www.msif.org/news/2016/12/05/geographical-latitude-and-the-onset-of-ms/. (Accessed 31 August 2020).

Moyo, E., Nhari, L.G., Moyo, P., Murewanhema, G., Dzinamarira, T., 2023. Health effects of climate change in Africa: a call for an improved implementation of prevention measures. Eco-Environment and Health (Online) 2 (2), 74–78. Available from: https://doi.org/10.1016/j.eehl.2023.04.004 (accessed 13.02.24).

Mudau, M., Machingaidze, S., Ismail, N., Nanoo, A., Ihekweazu, C., 2014. Geospatial analysis and identification of space-time clusters of MDR-TB in South Africa, 2006–2012. International Journal of Infectious Diseases 21 (Supplement 1), 59.

Müller-Nordhorn, J., Schneider, A., Grittner, U., Neumann, K., Keil, T., Willich, S.N., et al., 2020. International time trends in sudden unexpected infant death, 1969 - 2012. BMC Pediatrics 20, 377. Available from: https://doi.org/10.1186/s12887-020-02271-x.

Murray, E.D., Buttner, E.A., Price, B.H., 2012. Depression and psychosis in neurological practice. In: Daroff, R., Fenichel, G., Jankovic, J., Mazziotta, J. (Eds.), Bradley's Neurology in Clinical Practice, sixth ed. Elsevier/Saunders, Philadelphia, PA.

Mutter, J., Yeter, D., 2008. Kawasaki's disease, acrodynia, and mercury. Current Medical Chemistry (Weinheim an der Bergstrasse, Germany) 15 (28), 3000–3010. Available from: https://doi.org/10.2174/092986708786848712.

Myers, J., Young, T., Galloway, M., Manyike, P., Tucker, T., 2011. A public health approach to the impact of climate change on health in southern Africa: identifying priority modifiable risks. South African Medical Journal 101, 817–820.

Naicker, P.R., 2011. The impact of climate change and other factors on zoonotic diseases. Archives of Clinical Microbiology 2 (2), 4.

Name, J.J., Remondi, S.A.C., Rodrigues, V.A., Sacramento, P.P., Martins, P.C.P., 2020. Zinc, vitamin D and vitamin C: Perspectives for COVID-19 with a focus on physical tissue barrier integrity. Frontiers in Nutrition (Nutritional Immunology) 7, 295. Available from: https://doi.org/10.3389/fnut.2020.606398.

Nandini, D.B., Bhavana, S.B., Deepak, B.S., Ashwini, R., 2016. Paediatric geographic tongue: a case report, review and recent updates. Journal of Clinical and Diagnostic Research 10 (2), ZE05−ZE09. Available from: https://doi.org/10.7860/JCDR/2016/16452.7191.

Navarro, K., Janssen, S., Nordbrock, T., Gina Solomon, G., 2011. Disease Clusters Spotlight the Need to Protect People from Toxic Chemicals. https://www.nrdc.org/sites/default/files/diseaseclusters_issuepaper. pdf. (Accessed 23 January 2021).

NCSV (North Carolina State University), 2010. Geography of human disease: Environment has much to do with surrounding pathogens. Science Daily, 15 April 2010. http://www.sciencedaily.com/releases/2010/ 04/100415080854.htm. (Accessed 19 November 2020).

Nature Collection, 2023. Microbes and climate change. Nature Microbiology 8. Available from: https://www. nature.com/collections/hdggabhfch#:∼:text = Microbes%20are%20integral%20members%20of,or% 20applied%20in%20waste%20recycling (accessed 13.02.24).

Ndiaye, E.H., Diallo, D., Fall, G., Ba, Y., Faye, O., Dia, I., et al., 2018. Arboviruses isolated from the Barkedji mosquito-based surveillance system, 2012-2013. BMC Infectious Disease 18 (1), 642.

Ndu, I.K., 2016. Sudden infant death syndrome: an unrecognized killer in developing countries. Pediatric Health, Medicine and Therapeutics 7, 1−4. Available from: https://doi.org/10.2147/PHMT.S99685.

Neale, R.E., Lucas, R.M., Byrne, S.N., Hollestein, L., Rhodes, L.E., Yazar, S., et al., 2023. The effects of exposure to solar radiation on human health. Photochemical and Photobiological Sciences 22, 1011−1047. Available from: https://doi.org/10.1007/s43630-023-00375-8 (accessed 13.02.24).

Nelson, R.J., 2004. Seasonal immune function and sickness responses. Trends in Immunology 25 (4), 187−192.

Nelson, R.J., Demas, G.E., 1996. Seasonal changes in immune function. Quarterly Reviews in Biology 71 (4), 511−548. Available from: https://doi.org/10.1086/419555.

Nelson, R.J., Demas, G.E., Klein, S.L., Kriegsfeld, L.J., Bronson, F., 2002. Seasonal changes in immune function. In: Demas, R.J., Demas, G.E. (Eds.), Seasonal Patterns of Stress, Immune Function and Disease. Cambridge University Press, pp. 89−114.

Newman, L.S., 1998. Metals that cause sarcoidosis. Seminars in Respiratory Infections 13 (3), 212−220.

Nguyen, T., Johnston, S., Clarke, L., Smith, P., Staines, D., Marshall-Gradisnik, S., 2017. Impaired calcium mobilization in natural killer cells from chronic fatigue syndrome/myalgic encephalomyelitis patients is associated with transient receptor potential melastatin 3 ion channels. Clinical and Experimental Immunology 187 (2), 284−293. Available from: https://doi.org/10.1111/cei.12882.

Nicholson, L.B., 2016. The immune system. Essays in Biochemistry 60 (3), 275−301. Available from: https:// doi.org/10.1042/EBC20160017.

Nili, S., Khanjani, N., Jahani, Y., Bakhtiari, B., 2020. The effect of climate variables on the incidence of Crimean Congo Hemorrhagic Fever (CCHF) in Zahedan, Iran. BMC Public Health 20, 1893. Available from: https://doi.org/10.1186/s12889-020-09989-4.

Niohuru, I., 2023. Disease Burden and Mortality.. Healthcare and Disease Burden in Africa. Springer Briefs in Economics. Springer, Cham., pp. 35−85. Available from: https://doi.org/10.1007/978-3-031-19719-2_3 (Accessed 13 February 2024).

Njoku, M.G.C., Jason, L.A., Torres-Harding, S.R., 2007. The prevalence of chronic fatigue syndrome in Nigeria. Journal of Health Psychology 12 (3), 461−474.

Noorani, M., Lakhani, N., 2018. Kawasaki disease: two case reports from the Aga Khan Hospital, Dar es Salaam-Tanzania. BMC Pediatrics 18, 334. Available from: https://doi.org/10.1186/s12887-018-1306-5.

NRC (Norwegian Red Cross), 2019. Overlapping Vulnerabilities: The Impacts of Climate Change on Humanitarian Needs. Norwegian Red Cross, Oslo. Available from: https://www.rodekors.no/globalassets/globalt/rapporter-program-avtaler/humanitar-analyse-rapporter/norwegian-redcross_report_overlapping-vulnerabilities.pdf, Accessed 19 Decmber 2020.

Nriagu, J.O., Skaar, E.P., (Eds.), 2015. Trace Metals and Infectious Diseases. ISBN: 9780262029193, 504 pp.

Nuckols, J.R., Ward, M.H., Jarup, L., 2004. Using geographic information systems for exposure assessment in environmental epidemiology studies. Environmental Health Perspectives 112 (9), 1007−1015. Available from: https://doi.org/10.1289/ehp.6738.

Nurshad, A., Farjana, I., 2020. The effects of air pollution on COVID-19 infection and mortality - a review on recent evidence. Frontiers in Public Health 8, 779. Available from: https://doi.org/10.3389/fpubh.2020.580057.

Nwaogu, C.J., Cresswell, W., Tieleman, B.I., 2020. Geographic variation in baseline innate immune function does not follow variation in aridity along a tropical environmental gradient. Scientific Reports 10, 5909. Available from: https://doi.org/10.1038/s41598-020-62806-1.

Nyaruaba, R., Mwaliko, C., Mwau, M., Mousa, S., Wei, H., 2019. Arboviruses in the East African Community partner states: a review of medically important mosquito-borne Arboviruses. Pathogens and Global Health 113 (5), 209−228. Available from: https://doi.org/10.1080/20477724.2019.1678939.

Nyungura, J.L., Akim, T., Lako, A., Gordon, A., Lejeng, L., William, G., 2011. Investigation into the nodding syndrome in Witto Payam, Western Equatoria State, 2010. Southern Sudan Medical Journal 4, 3−6.

Obi, R.K., Nwanebu, F.C., 2008. Prions and prion diseases. African Journal of Clinical and Experimental Microbiology 9 (1), 38−52. Available from: https://doi.org/10.4314/ajcem.v9i1.7481.

Ogbu, C.N., 2003. Sudden infant death syndrome (SIDS) or cot death: a review. West African Journal of Medicine 22 (1), 88−91. Available from: https://doi.org/10.4314/wajm.v22i1.27988.

Ogden, L.E., 2018. Climate change, pathogens, and people: the challenges of monitoring a moving target. Bioscience 68 (10), 733−739. Available from: https://doi.org/10.1093/biosci/biy101.

Ogueta, C.I., Ramírez, P.M., Jiménez, O.C., Cifuentes, M.M., 2019. Geographic tongue: what a dermatologist should know. Actas Dermo-Sifiliográficas (English Edition) 110 (5), 341−346. Available from: https://doi.org/10.1016/j.ad.2018.10.022.

Ogunrin, O., Sanya, E., Komolafe, M., Osubor, C.C., 2013. Trace metals in patients with Parkinson's Disease: a multi-center case-control study in Nigerian patients. Parkinsonism & Related Disorders 18. Available from: https://doi.org/10.12974/2309-6179.2013.01.01.4.

Oh, W.S., Yoon, S., Noh, J., Sohn, J., Kim, C., Heo, J., 2018. Geographical variations and influential factors in prevalence of cardiometabolic diseases in South Korea. PLoS One 13 (10), e0205005. Available from: https://doi.org/10.1371/journal.pone.0205005.

Oluwayelu, D., Adebiyi, A., Tomori, O., 2018. Endemic and emerging arboviral diseases of livestock in Nigeria: a review. Parasites and Vectors 11, 337. Available from: https://doi.org/10.1186/s13071-018-2911-8.

Onyilagha, C., Uzonna, J.E., 2019. Host immune responses and immune evasion strategies in African trypanosomiasis. Frontiers in Immunology 10, 2738.

Orlowski, J.P., Mercer, R.D., 1980. Urine mercury levels in Kawasaki disease. Pediatrics 66 (4), 633−636.

Osakunor, D.N.M., Sengeh, D.M., Mutapi, F., 2018. Coinfections and comorbidities in African health systems: at the interface of infectious and noninfectious diseases. PLoS Neglected Tropical Diseases 12 (9), e0006711. Available from: https://doi.org/10.1371/journal.pntd.0006711.

Osredkar, J., Sustar, N., 2011. Copper and zinc, biological role and significance of copper/zinc imbalance. Journal of Clinical Toxicology S3 (001). Available from: https://doi.org/10.4172/2161-0495.S3-001.

Ouedraogo, I., Defourny, P., Vanclooster, M., 2016. Mapping the groundwater vulnerability for pollution at the pan African scale. Science of the Total Environment 544, 939−953.

Pacini, S., Fiore, M.G., Magherini, S., Morucci, G., Branca, J.J., Gulisano, M., et al., 2012. Could cadmium be responsible for some of the neurological signs and symptoms of myalgic encephalomyelitis/chronic fatigue syndrome. Medical Hypotheses 79 (3), 403−407. Available from: https://doi.org/10.1016/j.mehy.2012.06.007.

Paludan, S.R., Pradeu, T., Masters, S.L., et al., 2020. Constitutive immune mechanisms: mediators of host defence and immune regulation. Nature Reviews Immunology 21, 137−150.

Panelli, M.C., 2017. JTM advances in uncharted territories: diseases and disorders of unknown etiology. Journal of Translational Medicine 15 (1), 192. Available from: https://doi.org/10.1186/s12967-017-1293-6.

Park, K., 2011. Park's Textbook of Preventive and Social Medicine, 21st ed. Banarsidas Bhanot, Jabalpur, India.

Parritz, R.H., Troy, M.F., 2013. Disorders of Childhood: Development and Psychopathology. Courage Learning, p. 480.

Patil, P., Jain, H., Mishra, V., Sharma, A., 2015. Sarcoidosis: an update for the oral health care provider. Journal of Cranio-Maxillary Diseases 4 (1), 69−75. Available from: https://doi.org/10.4103/2278-9588.151908.

Patz, J.A., Epstein, P.R., Burke, T.A., Balbus, J.M., 1996. Global climate change and emerging infectious diseases. JAMA: The Journal of the American Medical Association 275 (3), 217−223.

Patz, J.A. Githeko A.K., McCarty, J.P. Hussein, S. Confalonieri, U. and de Wet, N., 2003. Chapter 6: Climate change and infectious diseases. In: Climate Change and Human Health. WHO, Geneva, Switzerland. https://www.who.int/globalchange/publications/climatechangechap6.pdf. (Accessed 14 January 2021).

Patz, J.A., Campbell-Lendrum, D., Holloway, T., Foley, J.A., 2005. Impact of regional climate change on human health. Nature 17 438 (7066), 310−317. Available from: https://doi.org/10.1038/nature04188.

Paynter, S., Ware, R.S., Sly, P.D., Williams, G., Weinstein, P., 2015. Seasonal immune modulation in humans: observed patterns and potential environmental drivers. Journal of Infection 70 (1), 1−10. Available from: https://doi.org/10.1016/j.jinf.2014.09.006.

Paz, S., 2015. Climate change impacts on West Nile virus transmission in a global context. Philosophical Transactions of the Royal Society of London. Series B, Biological Sciences 370 (1665), 20130561. Available from: https://doi.org/10.1098/rstb.2013.0561.

Pedro, E.M., da Rosa Franchi Santos, L.F., Scavuzzi, B.M., Iriyoda, T.M.V., et al., 2019. Trace elements associated with systemic *Lupus erythematosus* and insulin resistance. Biology of Trace Element Research 191, 34−44. Available from: https://doi.org/10.1007/s12011-018-1592-7.

Perera, B.P.U., Svoboda, L.K., Dolinoy, D.C., 2019. Genomic tools for environmental epigenetics and implications for public health. Current Opinion in Toxicology 18, 27−33. Available from: https://doi.org/10.1016/j.cotox.2019.02.008.

Petrova, V., Kristiansen, P., Norheim, G., Yimer, S.A., 2020. Rift valley fever: diagnostic challenges and investment needs for vaccine development. BMJ Global Health 5, e002694.

Pfavayi, L.T., Sibanda, E.N., Mutapi, F., 2020. The pathogenesis of fungal-related diseases and allergies in the African Population: the state of the evidence and knowledge gaps. International Archives of Allergy and Immunology 181, 257−269. Available from: https://doi.org/10.1159/000506009.

PHE (Public Health England), 2019. Guidance for Investigating Non-infectious Disease Clusters from Potential Environmental Causes. https://assets.publishing.service.gov.uk/government/uploads/system/uploads/attachment_data/file/781573/INIDC_guidance_v1.0.pdf. (Accessed 12 November 2020).

Picciani, B.L., Domingos, T.A., Teixeira-Souza, T., Santos Vde, C., Gonzaga, H.F., Cardoso-Oliveira, J., et al., 2016. Geographic tongue and psoriasis: clinical, histopathological, immunohistochemical and genetic correlation - A literature review. Brazilian Annals of Dermatology [Anais Brasileiros de Dermatologia (ABD)] 91 (4), 410−421.

Picciani, B., Santos, V.C., Teixeira-Souza, T., Izahias, L.M., Curty, Á., Avelleira, J.C., et al., 2017. Investigation of the clinical features of geographic tongue: unveiling its relationship with oral psoriasis. International Journal of Dermatology (Basel, Switzerland) 56 (4), 421−427.

Picciani, B.L.S., Santos, L.R., Teixeira-Souza, T., Dick, T.N.A., Carneiro, S., Pinto, J.M.N., et al., 2020. Geographic tongue severity index: a new and clinical scoring system. Oral Surgery, Oral Medicine, Oral Pathology and Oral Radiology 129 (4), 330−338. Available from: https://doi.org/10.1016/j.oooo.2019.12.007.

Pietra, R., Edel, J., Sabbioni, E., Rizzato, G.P., 1988. Sarcoidosis and trace metals as investigated by neutron activation analysis. International Nuclear Information System (INIS) 19 (23), Excerpta Medica, The Netherlands.

Pinto, M.M.S.C., da Silva, E.A.F., Silva, M.M.V.G., Melo-Gonçalves, P., Candeias, C., 2014. Environmental risk assessment based on high-resolution spatial maps of potentially toxic elements sampled on stream sediments of Santiago, Cape Verde. Geosciences 4, 297−315. Available from: https://doi.org/10.3390/geosciences4040297.

Pirofski, L.A., Casadevall, A., 2018. The damage-response framework as a tool for the physician-scientist to understand the pathogenesis of infectious diseases. Journal of Infectious Diseases 218 (Supplement 1), S7−S11. Available from: https://doi.org/10.1093/infdis/jiy083.

Plumlee, G.S., Ziegler, T.L., 2006. Chap. 9.07: The medical geochemistry of dusts, soils, and other Earth materials. In: Sahai, N., Schoonen, M.A.A. (Eds.), Reviews in Mineralogy and Geochemistry, 64. US Geological Survey, Denver.

Plumlee, G., Mormon, S.S., Ziegler, T.L., 2006. The toxicological geochemistry of Earth materials: an overview of processes and the interdisciplinary methods used to understand them. In: Sahai, N., Schoonen, M.A.A. (Eds.), Reviews in Mineralogy and Geochemistry, 64. US Geological Survey, Denver, p. 7.

Ponsonby, A.L., McMichael, A., van der Mei, I., 2002. Ultraviolet radiation and autoimmune disease: insights from epidemiological research. Toxicology 181-182, 71−78. Available from: https://doi.org/10.1016/s0300-483x(02)00257-3.

Porta, M. (Ed.), 2008. A Dictionary of Epidemiology. Oxford University Press, USA, pp. 42−43.

Portier, C., Thigpen-Tart, K., Carter, S., Dilsworth, C., Grambsch, Gohlke, J.M., et al., 2010. A Human Health Perspective on Climate Change: A Report Outlining the Research Needs on the Human Health Effects of Climate Change. Environmental Health Perspectives. National Institute of Environmental Health Sciences. Available from: http://doi.org/10.1289/ehp.1002272.

Portman, M.A., Yeter, D., Kuo, H.-C., 2018. Ethnic variations in mercury exposure from seafood consumption and the risk of Kawasaki disease in young children. The FASEB Journal (Pathology) 31 (1), 982.5.

Pozzer, A., Dominici, F., Haines, A., Witt, C., Münzel, T., Lelieveld, J., 2020. Regional and global contributions of air pollution to risk of death from COVID-19. Cardiovascular Research 116 (14), 2247−2253. Available from: https://doi.org/10.1093/cvr/cvaa288.

Prinsloo, B., 2006. Arboviral diseases in Southern Africa. South African Family Practice 48 (8), 25−28. Available from: https://doi.org/10.1080/20786204.2006.10873441.

Prusiner, S.B., 1998. Prions. Proceedings of the National Academy of Sciences USA 95, 13363−13383.

Qiu, C., Kivipelto, M., von Strauss, E., 2009. Epidemiology of Alzheimer's disease: occurrence, determinants, and strategies toward intervention. Dialogues in Clinical Neuroscience 11 (2), 111−128. Available from: https://doi.org/10.31887/DCNS.2009.11.2/cqiu (accessed 11.02.24).

Quach, H., Rotival, M., Pothlichet, J., Loh, Y.E., Dannemann, M., Zidane, N., et al., 2016. Genetic adaptation and Neandertal admixture shaped the immune system of human populations. Cell 167 (3), 643−656. Available from: https://doi.org/10.1016/j.cell.2016.09.024. e17.

Queensland Health, (n.d.). Cancer cluster Frequently asked questions. 010 - Queensland. Healthhttps://www.health.qld.gov.au/__data/assets/pdf_file/0025/443275/cancer-cluster-factsheet.pdf. (Accessed 17 November 2020).

Rajapakse, S., Shivanthan, M.C., Selvarajah, M., 2016. Chronic kidney disease of unknown etiology in Sri Lanka. International Journal of Occupational and Environmental Health 22 (3), 259−264. Available from: https://doi.org/10.1080/10773525.2016.1203097.

Ramin, B.M., McMichael, A.J., 2009. Climate change and health in sub-Saharan Africa: a case-based perspective. Ecohealth 6 (1), 52−57. Available from: https://doi.org/10.1007/s10393-009-0222-4.

Ramos-Casals, M., Kostov, B., Brito-Zerón, P., Sisó-Almirall, A., Baughman, R.P., 2019. How the frequency and phenotype of sarcoidosis is driven by environmental determinants. Lung 197 (4), 427−436. Available from: https://doi.org/10.1007/s00408-019-00243-2.

Rappaport, S.M., 2012. Discovering environmental causes of disease. Journal of Epidemiology and Community Health 66 (2), 99−102. Available from: https://doi.org/10.1136/jech-2011-200726.

Rappaport, S.M., 2016. Genetic factors are not the major causes of chronic diseases. PLoS One 11 (4), e0154387. Available from: https://doi.org/10.1371/journal.pone.0154387.

Rawlins, B.G., McGrath, S.P., Scheib, A.J., Breward, N., Cave, M., Lister, T.R., et al., 2012. The Advanced Soil Geochemical Atlas of England and Wales. British Geological Survey, Keyworth, Nottingham, http://ukso.org/static-maps/advanced-soil-geochemical-atlas-of-england-and-wales.html#: ~ :text = The%20Advanced%20Soil%20Geochemical%20Atlas,National%20Soil%20Resources%20Institute%2C%20Cranfield. (Accessed 27 January 2021).

Read, D., Borman, B.MoH (Ministry of Health, New Zealand), 2015. Investigating Clusters of Non-communicable Disease Guidelines for Public Health Units. https://www.health.govt.nz/system/files/documents/publications/investigating-clusters-non-communicable-disease-guidelines-may15-v3.pdf. (Accessed 12 November 2020).

Redmon, J.H., Elledge, M.F., Womack, D.S., Wickremashinghe, R., Wanigasuriya, K.P., Peiris-John, R.J., et al., 2014. Additional perspectives on chronic kidney disease of unknown aetiology (CKDu) in Sri Lanka - lessons learned from the WHO CKDu population prevalence study. BMC Nephrology 15, 125. Available from: https://doi.org/10.1186/1471-2369-15-125.

Rees, F., Doherty, M., Grainge, M.J., Lanyon, P., Zhang, W., 2017. The worldwide incidence and prevalence of systemic Lupus erythematosus: a systematic review of epidemiological studies. Rheumatology (Oxford, England) 56 (11), 1945−1961. Available from: https://doi.org/10.1093/rheumatology/kex260.

Reimann, C., de Caritat, P., 2017. Establishing geochemical background variation and threshold values for 59 elements in Australian surface soil. Science of The Total Environment 578, 633−648.

Richarz, A., Brätter, P., 2002. Speciation analysis of trace elements in the brains of individuals with Alzheimer's disease with special emphasis on metallothioneins. Analytical and Bioanalytical Chemistry 372, 412−417.

Rieder, H.P., Schoettli, G., Seiler, H., 1983. Trace elements in whole blood of multiple sclerosis. European Neurology 22 (2), 85−92. Available from: https://doi.org/10.1159/000115542.

Ritz, B., Heinrich, J., Wjst, M., Wichmann, E., Krause, C., 1998. Effect of cadmium body burden on immune response of school children. Archives of Environmental Health 53 (4), 272−280. Available from: https://doi.org/10.1080/00039899809605708.

Robinson, P.J., 2001. On the definition of a heat wave. Journal of Applied Meteorology 40 (4), 762−775.

Rodo, X., Curcoll, R., Robinson, M., Ballester, J., Burns, J.C., Cayan, D.R., et al., 2014. Tropospheric winds from northeastern China carry the etiologic agent of Kawasaki disease from its source to Japan. Proceedings of the National Academy of Science of the U.S.A. 111 (22), 7952−7957.

Rodo, X., Ballester, J., Curcoll, R., Morguí, J., 2016. Revisiting the role of environmental and climate factors on the epidemiology of Kawasaki disease. Annals of the New York Academy of Sciences 1382 (1), 84−98. Available from: https://doi.org/10.1111/nyas.13201.

Rodriguez, M.J., 2017. Changes in the immune system at different stages of life. Journal of Cell Biology and Immunology 1 (1), 102.

Rohr, J.R., Palmer, B.D., 2013. Climate change, multiple stressors, and the decline of ectotherms. Conservation Biology 27 (4), 741−751. Available from: https://doi.org/10.1111/cobi.12086.

Rohr, J.R., Dobson, A.P., Johnson, P.T., Kilpatrick, A.M., Paull, S.H., Raffel, T.R., et al., 2011. Frontiers in climate change-disease research. Trends in Ecology and Evolution 26 (6), 270−277. Available from: https://doi.org/10.1016/j.tree.2011.03.002.

Ross, L.N., 2018. The Doctrine of Specific Etiology. PhilSci Archive 15079. http://philsci-archive.pitt.edu/15079/1/DSE.pdf. (Accessed 03 December 2020).

Roomaney, R.A., van Wyk, B., Cois, A., Pillay van-Wyk, V., 2023. Multimorbidity patterns in South Africa: a latent class analysis. Frontiers in Public Health 10, 1082587. Available from: https://doi.org/10.3389/fpubh.2022.1082587 (accessed 14.02.24).

Rothman, K.J., 1990. A sobering start for the cluster busters' conference. American Journal of Epidemiology 132 (1 Suppl.), S6−S13. Available from: https://doi.org/10.1093/oxfordjournals.aje.a115790.

Rousk, J., Bååth, E., 2007. Fungal and bacterial growth in soil with plant materials of different C/N ratios. Federation of European Microbiological Societies (FEMS) Microbiology Ecology 62 (3), 258−267. Available from: https://doi.org/10.1111/j.1574-6941.2007.00398.x.

Rousk, J., Bååth, E., 2011. Growth of saprotrophic fungi and bacteria in soil. Johannes Rousk1,2 Mini Review. Federation of European Microbiological Societies (FEMS) Microbiology Ecology 78, 17−30.

Rouwet, D., Tanyileke, G., Costa, A., 2016. Cameroon's Lake Nyos gas burst: 30 years later. EOS (Rome, Italy) 97. Available from: https://doi.org/10.1029/2016eo055627.

Rowell, D., Nghiem, S., Ramagopalan, S., Meier, U.C., 2017. Seasonal temperature is associated with Parkinson's disease prescriptions: an ecological study. International Journal of Biometeorology 61 (12), 2205−2211. Available from: https://doi.org/10.1007/s00484-017-1427-9.

Rowley, A.H., Shulman, S.T., 2018. The epidemiology and pathogenesis of Kawasaki Disease. Frontiers in Pediatrics 6, 374. Available from: https://doi.org/10.3389/fped.2018.00374.

Ruiz, M.O., 2017. Geography of Disease. Oxford Bilbliographies. doi: 10.1093/OBO/9780199874002-0153. https://www.oxfordbibliographies.com/. (Accessed 19 November 2020).

Rushton, G., 2003. Public health, GIS, and spatial analytic tools. Annual Reviews of Public Health 24, 43−56. Available from: https://doi.org/10.1146/annurev.publhealth.24.012902.140843.

Rypdal, M., Rypdal, V., Burney, J.A., Cayan, D., Bainto, E., Skochko, S., et al., 2018. Clustering and climate associations of Kawasaki Disease in San Diego County suggest environmental triggers. Scientific Reports 8, 16140. Available from: https://doi.org/10.1038/s41598-018-33124-4.

Sabel, C.E., Gatrell, A.C., Löytönen, M., Maasilta, P., Jokelainen, M., 2000. Modelling exposure opportunities: estimating relative risk for motor neurone disease in Finland. Social Science Medicine 50 (7-8), 1121−1137. Available from: https://doi.org/10.1016/s0277-9536(99)00360-3.

Sabel, C.E., Wilson, J.G., Kingham, S., Tisch, C., Epton, M., 2007. Spatial implications of covariate adjustment on patterns of risk: Respiratory hospital admissions in Christchurch, New Zealand. Social Science and Medicine 65 (1), 43−59. Available from: https://journals.lww.com/epidem/fulltext/2008/11001/Spatial_Implications_of_Covariate_Adjustment_on.326.aspx (accessed 11.02.24).

Sahebari, M., Abrishami-Moghaddam, M., Moezzi, A., Ghayour-Mobarhan, M., Mirfeizi, Z., Esmaily, H., et al., 2014. Association between serum trace element concentrations and the disease activity of systemic *Lupus erythematosus*. Lupus 23 (8), 793−801. Available from: https://doi.org/10.1177/0961203314530792.

Salas, R.N., Malina, D., Solomon, C.G., 2019. Prioritizing health in a changing climate. The New England Journal of Medicine 381, 773−774. Available from: https://doi.org/10.1056/NEJMe1909957.

Salzberg, S.L., 2018. Open questions: how many genes do we have? BMC Biology 16, 94. Available from: https://doi.org/10.1186/s12915-018-0564-x.

Sarmadi, M., Bidel, Z., Najafi, F., Ramakrishnan, R., Teymoori, F., AzhdariZarmehri, H., et al., 2020. Copper concentration in multiple sclerosis: a systematic review and meta-analysis. Multiple Sclerosis and Related Disorders 45, 102426. Available from: https://doi.org/10.1016/j.msard.2020.102426.

Sartwell, P.E., Edwards, L.B., 1974. Epidemiology of sarcoidosis in the U.S. Navy. American Journal of Epidemiology 99, 250−257.

Saunders, S.E., Shikiya, R.A., Langenfeld, K., Bartelt-Hunt, S.L., Bartz, J.C., 2011. Replication efficiency of soil-bound prions varies with soil type. Journal of Virology 85 (11), 5476−5482. Available from: https://doi.org/10.1128/JVI.00282-11.

Saunders, S.E., Bartz, J.C., Bartelt-Hunt, S.L., 2012. Soil-mediated prion transmission: Is local soil-type a key determinant of prion disease incidence? Chemosphere 87 (7), 661−667. Available from: https://doi.org/10.1016/j.chemosphere.2011.12.076.

Sawczenko, A., Fleming, P.J., 1996. Thermal stress, sleeping position, and the sudden infant death syndrome. Sleep 19 (10), S267−S270.

Scanlon, S.T., 2020. The immune system's first Steps. Science 368 (6491), 598−599. Available from: https://doi.org/10.1126/science.abc3140.

Schiffrin, E.L., Lipman, M.L., Mann, J.F.E., 2007. Chronic kidney disease: effects on the cardiovascular system. Circulation 116, 85−97. Available from: https://doi.org/10.1161/CIRCULATIONAHA.106.678342.

Schluter, P.J., Ford, R.P., Brown, J., Ryan, A.P., 1998. Weather temperatures and sudden infant death syndrome: a regional study over 22 years in New Zealand. Journal of Epidemiology and Community Health 52 (1), 27−33. Available from: https://doi.org/10.1136/jech.52.1.27.

Schramm, P.T., Johnson, C.J., Mathews, N.E., McKenzie, D., Aiken, J.M., Pederson, J.A., 2006. Potential role of soil in the transmission of prion disease. In: Sahai, N., Schoonen, M.A.A. (Eds.), Medical Mineralogy and Geochemistry - Reviews in Mineralogy and Geochemistry, 64. Geochemical Society of America, Washington, DC, pp. 71−77. , 3.

Schuld, J., Kollmar, O., Schuld, S., Schommer, K., Richter, S., 2013. Impact of meteorological conditions on abdominal aortic aneurysm rupture: evaluation of an 18-year period and review of the literature. Vascular and Endovascular Surgery 47 (7), 524−531. Available from: https://doi.org/10.1177/1538574413497109.

Schuster, N.A., Rijnhart, J.J.M., Bosman, L.C., Twisk, J.W.R., Klausch, T., Heymsns, M.W., 2023. Misspecification of confounder-exposure and confounder-outcome associations leads to bias in effect estimates. BMC Medical Research Methodology 23 (11). Available from: https://doi.org/10.1186/s12874-022-01817-0 (accessed 13.02.24).

Schwartz, G.G., Il'yasova, D., Ivanova, A., 2003. Urinary cadmium, impaired fasting glucose, and diabetes in the NHANES III. Diabetes Care 26 (2), 468−470. Available from: https://doi.org/10.2337/diacare.26.2.468.

Seghieri, C., Tortù, C., Tricò, D., Leonetti, S., 2024. Learning prevalent patterns of co-morbidities in multi-chronic patients using population-based healthcare data. Scientific Reports 14, 2186. Available from: https://doi.org/10.1038/s41598-024-51249-7 (accessed 12.02.24).

Sekyere, J.O., 2018. *Candida auris*: a systematic review and meta-analysis of current updates on an emerging multidrug-resistant pathogen. Microbiology Open 7 (4), e00578.

Senanayake, N., King, B., 2017. Health-environment futures: complexity, uncertainty, and bodies. Progress in Human Geography 43 (4), 711−728. Available from: https://doi.org/10.1177/0309132517743322.

Sengupta, P., 2013. Potential health impacts of hard water. International Journal of Preventive Medicine 4 (8), 866−875.

Serdeczny, O., Adams, S., Baarsch, F., Coumou, D., 2017. Climate change impacts in Sub-Saharan Africa: from physical changes to their social repercussions. Regional Environmental Change 17, 1585−1600.

Sheehan, T.J., Gershman, S.T., MacDougall, L.A., Danley, R.A., Mroszczyk, M., Sorensen, A.M., et al., 2000. Geographic assessment of breast cancer screening by towns, zip codes, and census tracts. Journal of Public Health Management and Practice 6, 48−57. Available from: https://doi.org/10.1097/00124784-200006060-00008.

Ship, J.A., Phelan, J., Kerr, A.R., 2003. Chapter 112: Biology and pathology of the oral mucosa. In: Freedberg, et al., (Eds.), Fitzpatrick's Dermatology in General Medicine, sixth ed.) McGraw-Hill., p. 1208.

Shmool, J.L., Kubzansky, L.D., Newman, O.D., Spengler, J., Shepard, P., Clougherty, J.E., 2014. Social stressors and air pollution across New York City communities: a spatial approach for assessing correlations among multiple exposures. Environmental Health 13, 91. Available from: https://doi.org/10.1186/1476-069X-13-91.

Shulman, J.M., De Jager, P.L., Feany, M.B., 2011. Parkinson's disease: genetics and pathogenesis. Annual Review of Pathology: Mechanisms of Disease 6 (1), 193−222.

Sian, J., Youdim, M.B.H., Riederer, P., Gerlach, M., 1999. Chapter 45: Parkinson's disease. In: Siegel, G.J., Agranoff, B.W., Albers, R.W., et al.,Basic Neurochemistry: Molecular, Cellular and Medical Aspects, 6th edition Lippincott-Raven, Philadelphia.

Sierpina, V.S., Carter, R., 2002. Alternative and integrative treatment of fibromyalgia and chronic fatigue syndrome. Clinics in Family Practice 4 (4), 853−872.

Simon, A.K., Hollander, G.A., McMichael, A., 2015. Evolution of the immune system in humans from infancy to old age. Proceedings of the Royal Society B: Biological Sciences 282 (1821), 20143085. Available from: https://doi.org/10.1098/rspb.2014.3085.

Simpson, J., Weiner, E. (Eds.), 2002. The Oxford English Dictionary. Second Edition Oxford University Press.

Simpson Jnr., S., Blizzard, L., Otahal, P., Van der Mei, I., Taylor, B., 2011. Latitude is significantly associated with the prevalence of multiple sclerosis: a meta-analysis. Journal of Neurology and Neurosurgical Psychiatry 82 (10), 1132−1141. Available from: https://doi.org/10.1136/jnnp.2011.240432.

Smith, D.A., Germolec, D.R., 1999. Introduction to immunology and autoimmunity. Environmental Health Perspectives 107 (Suppl 5), 661−665. Available from: https://doi.org/10.1289/ehp.99107s5661.

Smith, D.K., Feldman, E.B., Feldman, D.S., 1989. Trace element status in multiple sclerosis. American Journal of Clinical Nutrition 50 (1), 136−140. Available from: https://doi.org/10.1093/ajcn/50.1.136.

Skinner, M.K., 2023. Environmental epigenetics and climate change. Environmental Epigenetics 9 (1), dvac028. Available from: https://doi.org/10.1093/eep/dvac028 (accessed 13.02.24).

Smith, C.B., Booth, C.J., Pedersen, J.A., 2011. Fate of prions in soil: a review. Journal of Environmental Quality 40 (2), 449−461. Available from: https://doi.org/10.2134/jeq2010.0412.

Smith, K.R., Woodward, A., Campbell-Lendrum, D., Chadee, D.D., Honda, Y., Liu, Q., et al., (Eds.), 2014. Climate Change 2014: Impacts, Adaptation, and Vulnerability. Part A: Global and Sectoral Aspects. Contribution of Working Group II to the Fifth Assessment Report of the Intergovernmental Panel on Climate Change. Cambridge University Press, Cambridge, United Kingdom and New York, NY, USA, pp. 709−754.

Smith, D.B., Solano, F., Woodruff, L.G., Cannon, W.F., Ellefsen, K.J., 2019. Geochemical and Mineralogical Maps, With Interpretation, for Soils of the Conterminous United States. US Geological Survey Scientific Investigations Report 2017-5118. https://pubs.er.usgs.gov/publication/sir20175118. (Accessed 03 Decmber 2020).

Smoyer-Tomic, K.E., Kuhn, R., Hudson, A., 2003. Heat wave hazards: an overview of heat wave impacts in Canada. Natural Hazards 28, 465−486. Available from: https://doi.org/10.1023/A:1022946528157.

Song, C., Kulldorff, M., 2003. Power evaluation of disease clustering tests. International Journal of Health Geographics 2 (1), 9. Available from: https://doi.org/10.1186/1476-072X-2-9.

Sorensen, C., Garcia-Trabanino, R., 2019. A new era of climate medicine - addressing heat-triggered renal disease. New England Journal of Medicine 381, 693−696. Available from: https://doi.org/10.1056/NEJMp1907859.

Sperling, K., Scherb, H., Neitzel, H., 2023. Population monitoring of trisomy 21: problems and approaches. Molecular Cytogenetics 16 (6). Available from: https://doi.org/10.1186/s13039-023-00637-1 (accessed 13.02.24).

Srour, M.L., Baratti-Mayer, D., 2020. Why is noma a neglected-neglected tropical disease? PLoS Neglected Tropical Diseases 14 (8), e0008435. Available from: https://doi.org/10.1371/journal.pntd.0008435 (accessed 12.02.24).

Srour, M.L., Marck, K., Baratti-Mayer, D., 2017. Noma: overview of a neglected disease and human rights violation. The American Journal of Tropical Medicine and Hygiene 96 (2), 268−274. Available from: https://doi.org/10.4269/ajtmh.16-0718.

Stanaway, J.D., Afshin, A., Gakidou, E., Lim, S.S., Abate, D., Abate, K.H., et al., 2018. Global, regional, and national comparative risk assessment of 84 behavioural, environmental and occupational, and metabolic risks or clusters of risks for 195 countries and territories, 1990 - 2017: a systematic analysis for the Global Burden of Disease Study 2017. The Lancet: Global Health Metrics 392 (10159), 1923−1994.

Staples, J.A., Ponsonby, A.L., Lim, L.L., McMichael, A.J., 2003. Ecologic analysis of some immune-related disorders, including type 1 diabetes, in Australia: latitude, regional ultraviolet radiation, and disease prevalence. Environmental Health Perspectives 111 (4), 518−523. Available from: https://doi.org/10.1289/ehp.5941.

Steece-Collier, K., Maries, E., Kordower, J.H., 2002. Etiology of Parkinson's disease: genetics and environment revisited. Proceedings of the National Academy of Sciences of the United States of America 99 (22), 13972−13974. Available from: https://doi.org/10.1073/pnas.242594999.

Steele, R.J., Fogerty, A.C., Willcox, M.E., Clancy, S.L., 1984. Metal content of the liver in sudden infant death syndrome. Journal of Paediatrics and Child Health 20 (2), 141−142.

Stein, J., Schettler, T., Rohrer, B., Valenti, M., 2008. Chapter 2: The Changing Environment and Disease Patterns. In: Myers, N. (Ed.), Environmental Threats to Healthy Aging - With a Closer Look at Alzheimer's and Parkinson's Diseases, 202 pp. http://www.agehealthy.org/pdf/GBPSRSEHN_HealthyAging_sec1.pdf. (Accessed 02 February 20219).

Stewart, C.G., Burroughs, G.W., (n.d.). Infectious Diseases of Livestock, Part 3: Disease Complexes - Unknown Aetology: Ill-thrift. Anipedia. https://www.anipedia.org/resources/disease-complexes-/-unknown-aetiology-ill-thrift/982. (Accessed 31 October 2020).

Sullivan, J.B., Krieger, G.R., 2001. Clinical Environmental Health and Toxic Exposures, second ed. Lippincott Williams and Wilkins. Available from: https://pdfhow.com/downloads/clinical_environmental_health_toxic_exposures_sullivan.pdf, Accessed 28 January 2021.

Sun, H., 2018. Association of soil selenium, strontium, and magnesium concentrations with Parkinson's disease mortality rates in the USA. Environmental Geochemistry and Health 40, 349−357. Available from: https://doi.org/10.1007/s10653-017-9915-8.

Szekely, A., Borman, A.M., Johnson, E.M., 2019. *Candida auris* isolates of the southern Asian and South African lineages exhibit different phenotypic and antifungal susceptibility profiles *in vitro*. Journal of Clinical Microbiology 57 (5), e02055-18. Available from: https://doi.org/10.1128/JCM.02055-18.

Taibo, C.L.A., Cliff, J., Rosling, H., Hall, C.D., Park, M.M., Frimpong, J.A., 2017. An epidemic of spastic paraparesis of unknown aetiology in Northern Mozambique. Pan African Medical Journal 27 (Supplement 1), 6. Available from: https://doi.org/10.11604/pamj.supp.2017.27.1.12623.

Tamburo, E., Varrica, D., Dongarrà, G., Grimaldi, L.M.E., 2015. Trace elements in scalp hair samples from patients with relapsing-remitting multiple sclerosis. PLoS One 10 (4), e0122142. Available from: https://doi.org/10.1371/journal.pone.0122142.

Tavassolifar, M.J., Vodjgani, M., Salehi, Z., Izad, M., 2020. The influence of reactive oxygen species in the immune system and pathogenesis of multiple sclerosis. Autoimmune Diseases 2020, 5793817. Available from: https://doi.org/10.1155/2020/5793817.

Teferedegn, E.Y., Tesfaye, D., Ün, C., 2019. Valuing the investigation of Prion diseases in Ethiopia. International Journal of Agricultural Science and Food Technology 5 (1), 001−005. Available from: https://doi.org/10.17352/2455-815x.000034.

Tesher, M.S., 2019. Pediatric autoimmune diseases. Pediatric Annals 48 (10), e385−e386. Available from: https://doi.org/10.3928/19382359-20190924-02.

Theron, A., Tintinger, G.R., Anderson, R., 2012. Harmful interactions of non-essential heavy metals with cells of the innate immune system. Journal of Clinical Toxicology (Supplement 3), . Available from: https://doi.org/10.4172/2161-0495.S3-005.

Thielke, S., Slatore, C.G., Banks, W.A., 2015. Association between Alzheimer Dementia mortality rate and altitude in California Counties. JAMA Psychiatry 72 (12), 1253–1254. Available from: https://doi.org/10.1001/jamapsychiatry.2015.1852.

Thomas, J., 2019. Increasing the altitude to decrease the symptoms of Parkinson's Disease. High Altitude Health. https://highaltitudehealth.com/2019/04/15/increasing-the-altitude-to-decrease-the-symptoms-of-parkinsons-disease/. (Accessed 13 September 2020).

Thrall, G.I., 1999. The future of GIS in public health management and practice. Journal of Public Health Management and Practice 5 (4), 75–82.

Thurnham, D.I., 2004. An overview of the interactions between micronutrients and of micronutrients with drugs, genes and immune mechanisms. Nutrition Research Reviews 17 (2), 211–240.

Toms, R., Bonney, A., Mayne, D.J., Feng, X., Walsan, R., 2019. Geographic and area-level socioeconomic variation in cardiometabolic risk factor distribution: a systematic review of the literature. International Journal of Health Geographics 18, 1. Available from: https://doi.org/10.1186/s12942-018-0165-5.

Toubiana, J., Poirault, C., Corsia, A., Bajolle, F., Fourgeaud, J., Angoulvant, F., et al., 2020. Kawasaki-like multisystem inflammatory syndrome in children during the covid-19 pandemic in Paris, France: prospective observational study. BMJ (Clinical Research ed.) 369, m2094.

Touil, H., Mounts, K., De Jager, P.L., 2023. Differential impact of environmental factors on systemic and localized autoimmunity. Frontiers in Immunology 14, 1147447. Available from: https://doi.org/10.3389/fimmu.2023.1147447 (accessed 14.02.24).

Tseng, W.P., 1977. Effects and dose-response relationships of skin cancer and blackfoot disease with arsenic. Environmental Health Perspectives 19, 109–119.

Tsevis, C. (Ed.), 2020. Early Life Immunology. Special Issue, Science 368 (6491). Published by the American Association for the Advancement of Science.

Tulchinsky, T.H., 2018. John Snow, cholera, the Broad Street Pump: Waterborne diseases then and now. Case Studies in Public Health 2018, 77–99.

UCSDH (UC San Diego Health), 2013. Data from Across Globe Defines Distinct Kawasaki Disease Season. https://health.ucsd.edu/news/releases/Pages/2013-09-23-data-defines-kawasaki-disease-seasonal.aspx. (Accessed 28 August 2020).

UKRI/NIHR (UK Research and Innovation/National Institute for Health Research) Project, 2020. Tackling multimorbidity at scale: Understanding disease clusters, determinants and biological pathways. https://mrc.ukri.org/funding/browse/tackling-multimorbidity/tackling-multimorbidity-at-scale/. (Accessed 03 January 2021).

UNCC (United Nations Climate Change), 2020. Climate Change Is an Increasing Threat to Africa. https://unfccc.int/news/climate-change-is-an-increasing-threat-to-africa. (Accessed 29 November 2020).

UNECA (United Nations Economic Commission for Africa), 2011. Climate Change and Health Across Africa: Issues and Options. African Climate Policy Centre Working Paper 20. https://www.uncclearn.org/wp-content/uploads/library/uneca15.pdf. (Accessed 28 November 2020).

UNICEF/WHO (United Nations Children's Fund/World Health Organisation), 2017. Report on 'Safely Managed Drinking Water'. https://data.unicef.org/resources/safely-managed-drinking-water/#. (Accessed 22 January 2021).

UoC (University of Cambridge), 2015. Seasonal immunity: Activity of thousands of genes differs from winter to summer. *ScienceDaily*. https://www.sciencedaily.com/releases/2015/05/150512112356.htm. (Accessed 15 November 2020).

US CDC (United States Centres for Disease Control and Prevention), 1990. Guidelines for Investigating Clusters of Health Events. Morbidity and Mortality Weekly Reports 39(RR-11); 1–16.

US CDC (United States Centers for Disease Control and Prevention), 1992. Emerging Infections: Microbial Threats to the United States. Institute of Medicine, Washington, DC, USA.

US CDC (US Centers for Disease Control and Prevention), 2009. Health risk appraisals. https://www.cdc.gov/workplacehealthpromotion/tools-resources/workplace-health/assessment-tools.html. (Accessed 29 January 2021).

US CDC (United States Centres for Disease Control and Prevention), 2012. In: Introduction to Investigating an Outbreak - Uncovering Outbreaks. Lesson 6: Investigating an Outbreak, Section 1. https://www.cdc.gov/csels/dsepd/ss1978/lesson6/section1.html. (Accessed 14 November 2020).

US CDC (United States Centres for Disease Control and Prevention), 2018a. Myalgic Encephalomyelitis/ Chronic Fatigue Syndrome: Diagnosis of ME/CFS. https://www.cdc.gov/me-cfs/symptoms-diagnosis/ diagnosis.html. (Accessed 29 October 2020).

US CDC (United States Centers for Disease Control and Prevention), 2018b. Identified prion diseases. https:// www.cdc.gov/prions/index.html. (Accessed 14 December 2020).

US CDC (US Centers for Disease Control and Prevention), 2020a. Valley Fever Awareness. https://www.cdc. gov/fungal/features/valley-fever.html. (Accessed 12 December 2020).

US CDC (US Centers for Disease Control and Prevention), 2020b. Climate Effects on Health; and cited bibliography within. https://www.cdc.gov/climateandhealth/effects/default.htm. (Accessed 25 November 2020).

US EPA (United States Environmental Protection Agency), 2016. Human Health Risk Assessment. https:// www.epa.gov/risk/human-health-risk-assessment#:~:text = EPA%20begins%20the%20process%20of, assessment%20with%20planning%20and%. (Accessed 02 February 2021).

US NCI (United States National Cancer Institute), 2020. NCI Cancer Atlas. https://gis.cancer.gov/canceratlas/. (Accessed 03 January 2021).

US NIEHS (National Institute of Environmental Health Sciences), 2019. Environmental Epigenetics. https:// www.niehs.nih.gov/research/supported/health/envepi/index.cfm. (Accessed 12 December 2020).

US NIH (United States National Institute of Health), 2019. Hereditary Spastic Paraplegia Information Page. https://www.ninds.nih.gov/Disorders/All-Disorders/Hereditary-Spastic-Paraplegia-Information-Page. (Accessed 29 October 2020).

US NINDS (United States National Institute of Neurological Disorders and Stroke), 2014. Hereditary Spastic Paraplegia Information Page. https://web.archive.org/web/20140221102852/http://www.ninds.nih.gov/ disorders/hereditary_spastic_paraplegia/hereditary_spastic_paraplegia.htm. (Accessed 29 October 2020).

US NRC (US National Research Council Committee on Climate, Ecosystems, Infectious Diseases, and Human Health), 2001. Under the Weather: Climate, Ecosystems, and Infectious Disease. National Academies Press (US), Washington (DC). Available from: https://www.ncbi.nlm.nih.gov/books/NBK222258/.

USAID (United States Agency for International Development), 2017. Risk Expands, But Opportunity Awaits: Emerging Evidence on Climate Change and Health in Africa. https://www.climatelinks.org/sites/default/ files/asset/document/2017.11.30_USAID%20ATLAS_Emerging%20Evidence%20on%20Climate%20Change %20and%20Health%20in%20Africa_ENG.pdf. (Accessed 29 November 2020).

USIM (US Institute of Medicine Food and Nutrition Board), 1998. Dietary Reference Intakes: A Risk Assessment Model for Establishing Upper Intake Levels for Nutrients. National Academies Press (US), Washington (DC). Available from: 10.17226/6432 (Accessed 28 January 2021).

Valcárcel, F., González, J., González, M.G., Sánchez, M., Tercero, J.M., Elhachimi, L., et al., 2020. Comparative ecology of *Hyalomma lusitanicum* and *Hyalomma marginatum* Koch, 1844 (Acarina: Ixodidae). Insects 11 (5), 303. Available from: https://doi.org/10.3390/insects11050303.

van Horssen, J., Witte, M.E., Schreibelt, G., de Vries, H.E., 2011. Radical changes in multiple sclerosis pathogenesis. Biochimica et Biophysica Acta (BBA) - Molecular Basis of Disease 1812 (2), 141–150. Available from: https://doi.org/10.1016/j.bbadis.2010.06.011.

Verberkmoes, N.J., Hamad, M.S., ter Woorst, J.F., Tan, M.E.S.H., Peels, C.H., van Straten, A.H.M., 2012. Impact of temperature and atmospheric pressure on the incidence of major acute cardiovascular events. Netherlands Heart Journal 20, 193–196. Available from: https://doi.org/10.1007/s12471-012-0258-x.

Verster, A.M., Liang, J.E., Rostal, M.K., Kemp, A., Brand, R.F., Anyamba, A., et al., 2020. Selected wetland soil properties correlate to Rift Valley fever livestock mortalities reported in 2009-2010 in central South Africa. PLoS One 15 (5), e0232481. Available from: https://doi.org/10.1371/journal.pone.0232481.

Vigneron, A., Lovejoy, C., Cruaud, P., Kalenitchenko, D., Culley, A., Vincent, W.F., 2019. Contrasting winter versus summer microbial communities and metabolic functions in a permafrost thaw lake. Frontiers in Microbiology 10, 1656. Available from: https://doi.org/10.3389/fmicb.2019.01656.

Vila, T., Sultan, A.S., Montelongo-Jauregui, D., Jabra-Rizk, M.A., 2020. Candida auris: A fungus with identity crisis. Pathogens and Disease 78 (4), ftaa034. Available from: https://doi.org/10.1093/femspd/ftaa034.

Waller, L.A., 2015. Discussion: statistical cluster detection, epidemiologic interpretation, and public health policy. Statistics and Public Policy 2 (1), 1−8. Available from: https://doi.org/10.1080/2330443X.2015.1026621.

Waller, L.A., Turnbull, B.W., Gustafsson, G., Hjalmars, U., Andersson, B., 1995. Detection and assessment of clusters of disease: an application to nuclear power plant facilities and childhood leukaemia in Sweden. Statistics in Medicine 14 (1), 3−16. Available from: https://doi.org/10.1002/sim.4780140103.

Wang, X., Zhang, Q., Zhou, G., 2007. National-scale geochemical mapping projects in China. Geostandards and Geoanalytical Research 31 (4), 311−320. Available from: https://doi.org/10.1111/j.1751-908X.2007.00128.x.

Wang, X., Bing, J., Zheng, Q., Zhang, F., Liu, J., Yue, H., et al., 2018. The first isolate of *Candida auris* in China: clinical and biological aspects. Emerging Microbes and Infection 7 (1), 93. Available from: https://doi.org/10.1038/s41426-018-0095-0.

Wang, L., Yin, Y.L., Liu, X.Z., Peng, S., Zheng, Y.-G., Lan, X.-R., et al., 2020. Current understanding of metal ions in the pathogenesis of Alzheimer's disease. Translational Neurodegeneration 9 (10). Available from: https://doi.org/10.1186/s40035-020-00189-z.

Wanigasuriya, K.P., Peiris-John, R.J., Wickremasinghe, R., 2011. Chronic kidney disease of unknown aetiology in Sri Lanka: is cadmium a likely cause? BMC Nephrology 12, 32. Available from: https://doi.org/10.1186/1471-2369-12-32.

Warren, S.,, Warren, K.G., 2001. Multiple Sclerosis. World Health Organisation, Geneva. Available from: https://apps.who.int/iris/bitstream/handle/10665/42394/924156203X_en.pdf (Accessed 17 January 2021).

Wartenberg, D., 2001. Investigating disease clusters: why, when and how? Journal of the Royal Statistical Society. Series A (Statistics in Society) 164 (1), 13−22.

Watad, A., Azrielant, S., Bragazzi, N.L., Sharif, K., David, P., Katz, I., et al., 2017. Seasonality and autoimmune diseases: the contribution of the four seasons to the mosaic of autoimmunity. Journal of Autoimmunity 82, 13−30. Available from: https://doi.org/10.1016/j.jaut.2017.06.001.

Watanabe, T., Kawaoka, Y., 2011. Pathogenesis of the 1918 pandemic influenza virus. PLoS Pathogens 7 (1), e1001218.

Welsh, R.M., Bentz, M.L., Shams, A., Houston, H., Lyons, A., Rose, L.J., et al., 2017. Survival, persistence, and isolation of the emerging multidrug-resistant pathogenic yeast *Candida auris* on a plastic health care surface. Journal of Clinical Microbiology 55 (10), 2996−3005. Available from: https://doi.org/10.1128/JCM.00921-17.

WGBD (World Global Burden of Disease) 2016 Multiple Sclerosis CollaboratorsWallin, M.T., Culpepper, W.J., Nichols, E., Bhutta, Z.A., Gebrehiwot, T.T., Hay, S.S., et al., 2019. Global, regional, and national burden of multiple sclerosis 1990-2016: a systematic analysis for the Global Burden of Disease Study 2016. The Lancet Neurology 18 (3), 269−285. Available from: https://doi.org/10.1016/S1474-4422(18)30443-5.

Whitty, C.J.M., MacEwen, C., Goddard, A., Alderson, D., Marshall, M., et al., 2020. Rising to the challenge of multimorbidity: we need to combine generalist and specialist skills. BMJ 368, 16964. Available from: https://doi.org/10.1136/bmj.l6964.

WHO (World Health Organization), 1996. Trace Elements in Human Nutrition and Health. World Health Organization, Geneva, Switzerland, ISBN: 92 4 156173 4. https://www.who.int/nutrition/publications/micronutrients/9241561734/en/. (Accessed 02 February 2021).

WHO (World Health Organisation), 2003. Climate Change and Human Health - Risks and Responses. Summary. WHO/WMO/UNEP. [E], ISBN: 92 4 159081 5, 37 pp. https://www.who.int/globalchange/environment/en/ccSCREEN.pdf?ua = 1. (Accessed 25 November 2020).

WHO (World Health Organisation), 2009. Global health risks. Mortality and burden of disease attributable to selected major risks. http://www.who.int/healthinfo/global_burden_disease/GlobalHealthRisks_report_full.pdf?ua = 1&ua = 1. (Accessed 12 December 2020).

WHO (World Health Organisation), 2011a. Public Health and Environment. Global Strategy Overview. https://www.who.int/phe/publications/PHE_2011_global_strategy_overview_2011.pdf. (Accessed 24 January 2021).

WHO (World Health Organisation), 2011b. Immune Diseases and Children. Training for the Health Sector. https://www.who.int/ceh/capacity/immune_diseases.pdf. (Accessed 26 December 2020).

WHO (World Health Organisation), 2012. Adaptation to Climate Change in Africa: Plan of Action for the Health Sector, 2012-2016. WHO, Regional Office for Africa, Brazzaville. file:///D:/Adaptation-to-Climate-Change-in-Africa.pdf. (Accessed 29 November 2020).

WHO (World Health Organization), 2015. Fact Sheet: El Niño Southern Oscillation (ENSO) and Health. https://www.who.int/globalchange/publications/factsheets/el-nino-and-health/en/#: ~ :text. (Accessed 10 January 2021).

WHO (World Health Organisation), 2017. Health Impact Assessment (HIA). The Determinants of Health. http://www.who.int/hia/evidence/doh/en/. (Accessed 04 December 2020).

WHO (World Health Organisation), 2018a. Cancer. https://www.who.int/news-room/fact-sheets/detail/cancer. (Accessed 11 December 2020).

WHO (World Health Organisation), 2018b. Climate Change and Health. https://www.who.int/news-room/fact-sheets/detail/climate-change-and-health#: ~ :text = Climate%20change%20affects%20the%20social, malaria%2C%20diarrhoea%20and%20heat%20stress. (Accessed 25November 2020).

WHO (World Health Organisation), 2020a. Rift Valley Fever. https://www.who.int/health-topics/rift-valley-fever#tab = tab_1. (Accessed 12 December 2020).

WHO (World Health Organisation), 2020b. Environment, Climate Change and Health. https://www.who.int/teams/environment-climate-change-and-health/emergencies/disease-outbreaks/. (Accessed 26 November 2020).

WHO (World Health Organisation), 2020c. The Environment and Health for Children and Their Mothers. https://www.who.int/ceh/publications/factsheets/fs284/en/. (Accessed 23 December 2020).

WHO (World Health Organisation), 2021. Climate Change and Human Health - Information and Public Health Advice: Heat and Health. https://www.who.int/globalchange/publications/heat-and-health/en/. (Accessed 11 January 2021).

Wickramarathna, S., Balasooriya, S., Diyabalanage, S., Chandrajith, R., 2017. Tracing environmental aetiological factors of chronic kidney diseases in the dry zone of Sri Lanka - a hydrogeochemical and isotope approach. Journal of Trace Elements in Medicine and Biology: Organ of the Society for Minerals and Trace Elements (GMS) 44, 298–306. Available from: https://doi.org/10.1016/j.jtemb.2017.08.013.

Wie, Y., Wang, Y., Lin, C.K., et al., 2019. Associations between seasonal temperature and dementia-associated hospitalizations in New England. Environment International 126, 228–233. Available from: https://doi.org/10.1016/j.envint.2018.12.054.

Wijetunge, S., Ratnatunga, N.V., Abeysekera, T.D., Wazil, A.W., Selvarajah, M., 2015. Endemic chronic kidney disease of unknown etiology in Sri Lanka: correlation of pathology with clinical stages. Indian Journal of Nephrology 25 (5), 274–280. Available from: https://doi.org/10.4103/0971-4065.145095.

Willadsen, T.G., Bebe, A., Køster-Rasmussen, R., Jarbøl, D.E., Guassora, A.D., Waldorff, F.B., et al., 2016. The role of diseases, risk factors and symptoms in the definition of multimorbidity - a systematic review. Scandinavian Journal of Primary Health Care 34 (2), 112–121. Available from: https://doi.org/10.3109/02813432.2016.1153242.

Williams, R., Malherbe, J., Weepener, H., Majiwa, P., Swanepoel, R., 2016. Anomalous high rainfall and soil saturation as combined risk indicator of Rift Valley fever outbreaks, South Africa, 2008-2011. Emerging Infectious Diseases 22 (12), 2054−2062. Available from: https://doi.org/10.3201/eid2212.151352.

Wilson, M.L., 2001. Ecology and infectious disease. In: Aron, J.L., Patz, J.A. (Eds.), Ecosystem Change and Public Health: A Global Perspective. John Hopkins University Press, Baltimore, USA, pp. 283−324.

Wimalawansa, S.J., Dissanayake, C.B., 2020. Factors affecting the environmentally induced, chronic kidney disease of unknown aetiology in dry zonal regions in tropical countries - novel findings. Environments 7 (1), 2. Available from: https://doi.org/10.3390/environments7010002.

Winans, B., Humble, M.C., Lawrence, B.P., 2011. Environmental toxicants and the developing immune system: a missing link in the global battle against infectious disease? Reproductive Toxicology (Elmsford, N. Y.) 31 (3), 327−336. Available from: https://doi.org/10.1016/j.reprotox.2010.09.004.

WMO (World Meteorological Organisation), 2020. ENSO and Global Seasonal Climate Updates. https://public.wmo.int/en/our-mandate/climate/el-ni%C3%B1ola-ni%C3%B1a-update. (Accessed 10 January 2021).

WOAH (World Organisation for Animal Health), 2019. Camel prion disease: a possible emerging disease in dromedary camel populations? OIE Bulletin. file:///D:/OIE-News-December-2019-Camel-prion-disease.pdf. (Accessed 06 December 2020).

Wu, X., Lu, Y., Zhou, S., Chen, L., Xu, B., 2016. Impact of climate change on human infectious diseases: Empirical evidence and human adaptation. Environment International 86, 14−23.

Xenofon, M., Chrisostomos, S., Antonios, K., Prabha, S., 2014. Toxicological impact of heavy metals on the placenta: a literature review. HJOG 13 (4), 115−119.

Xie, X., Wang, X., Zhang, Q., Zhou, G., Cheng, H., Liu, D., et al., 2008. Multi-scale geochemical mapping in China. Geochemistry: Exploration, Environment, Analysis 8, 333−341.

Yanagihara, R., Garruto, R.M., Gajdusek, D.C., Tomita, A., Uchikawa, T., Konagaya, Y., et al., 1984. Calcium and vitamin D metabolism in Guamanian Chamorros with amyotrophic lateral sclerosis and parkinsonism-dementia. Annals of Neurology 15 (1), 42−48.

Yang, S., Lian, G., 2020. ROS and diseases: role in metabolism and energy supply. Molecular and Cellular Biochemistry 467 (1-2), 1−12. Available from: https://doi.org/10.1007/s11010-019-03667-9. accessed 18.01.2021.

Yang, Y.W., Wu, C.H., Tsai, H.T., Chen, Y.R., Chang, Y.P., Han, Y.Y., et al., 2023. Dynamics of immune responses are inconsistent when trauma patients are grouped by injury severity score and clinical outcomes. Scientific Reports 13 (1), 1391. Available from: https://doi.org/10.1038/s41598-023-27969-7 (accessed 14.02.24).

Yase, Y., 1972. The pathogenesis of amyotrophic lateral sclerosis. Lancet 2 (7772), 292−296. Available from: https://doi.org/10.1016/s0140-6736(72)92903-0.

Yasui, M., Yano, I., Yase, Y., Ota, K., 1990. Distribution of magnesium in central nervous system tissue, trabecular and cortical bone in rats fed with unbalanced diets of minerals. Journal of the Neurological Sciences 99 (2-3), 177−183.

Yasui, M., Kihira, T., Ota, K., 1992. Calcium, magnesium and aluminum concentrations in Parkinson's disease. Neurotoxicology 13 (3), 593−600.

Yasui, M., Ota, K., Garruto, R.M., 1995. Effects of calcium-deficient diets on manganese deposition in the central nervous system and bones of rats. Neurotoxicology 16 (3), 511−517.

Ye, X., Wolff, R., Yu, W., Vaneckova, P., Pan, X., Tong, S., 2012. Ambient temperature and morbidity: a review of epidemiological evidence. Environmental Health Perspectives 120 (1), 19−28.

Yeter, D., Portman, M.A., Aschner, M., Farina, M., Chan, W.C., Hsieh, K.S., et al., 2016. Ethnic Kawasaki disease risk associated with blood mercury and cadmium in U.S. children. International Journal of Environmental Research and Public Health 13 (1), 101.

Yu, W.W., Pooni, R., Ardern, C.I., Kuk, J.L., 2023. Is anyone truly healthy? Trends in health risk factors prevalence and changes in their associations with all-cause mortality. PLoS One 18 (6), e0286691. Available from: https://doi.org/10.1371/journal.pone.0286691.

Zabel, M., Ortega, A., 2017. The ecology of prions. Microbiology and Molecular Biology Reviews 81 (3), e00001−17. Available from: https://doi.org/10.1128/MMBR.00001-17.

Zaitchik, B.F., 2017. Climate and Health Across Africa. Climate Science. Oxford Research. Available from: https://doi.org/10.1093/acrefore/9780190228620.013.555 (Accessed 29 November 2020).

Zamith-Miranda, D., Heyman, H.M., Cleare, L.G., Couvillion, S.P., Clair, G.C., Bredeweg, E.L., et al., 2019. Multi-omics signature of *Candida auris*, an emerging and multidrug-resistant pathogen. mSystems: American Society for Microbiology 4 (4), e00257−19. Available from: https://doi.org/10.1128/mSystems.00257-19.

Zelinka, M.D., Myers, T., MacCoy, D.T., Po-Chedley, S., Caldwell, P.M., Ceppi, P., et al., 2020. Causes of higher climate sensitivity in CMIP6 models. Geophysical Research Letters 47 (1). Available from: https://doi.org/10.1029/2019GL085782.

Zhang, P.L., Chen, Y., Zhang, C.H., Wang, Y.X., Fernandez-Funez, P., 2018. Genetics of Parkinson's disease and related disorders. Journal of Medical Genetics 55 (2), 73−80. Available from: https://doi.org/10.1136/jmedgenet-2017-105047.

Zhao, H.W., Lin, J., Wang, X.B., Cheng, X., Wang, J.Y., Hu, B.L., et al., 2013. Assessing plasma levels of selenium, copper, iron and zinc in patients of Parkinson's disease. PLoS One 8 (12), e83060. Available from: https://doi.org/10.1371/journal.pone.0083060.

Zhao, Y., Qin, L., Pan, H., Liu, Z., Jiang, L., Guo, J., et al., 2020. The role of genetics in Parkinson's disease: a large cohort study in Chinese mainland population. Brain 143 (7), 2220−2234. Available from: https://doi.org/10.1093/brain/awaa167.

Zhu, J., Poulsen, C.J., Otto-Bliesner, B.L., 2020. High climate sensitivity in CMIP6 model not supported by paleoclimate. Nature Climate Change 10, 378−379. Available from: https://doi.org/10.1038/s41558-020-0764-6.

Zimmermann, P., Curtis, N., 2019. Factors that influence the immune response to vaccination. Clinical Microbiology Reviews 32 (2), e00084-18. Available from: https://doi.org/10.1128/CMR.00084-18.

Further Reading

Alonso, P., Engels, D., Reeder, J., 2017. Renewed push to strengthen vector control globally. Lancet 389, 2270−2271.

Anonymous (1989), National Conference on Clustering of Health Events. Atlanta, Georgia, February 16-17, 1989. *American Journal of Epidemiology* 132, s1-s202.

Antó, J.M., Sunyer, J., 1998. Urban epidemic asthma. In: Banks, D.E., Parker, J.E. (Eds.), Occupational Lung Disease: An International Perspective. Chapman and Hall Medical, London, pp. 401−407.

Beach, R.S., Gershwin, M.E., Hurley, L.S., 1981. Zinc, copper, and manganese in immune function and experimental oncogenesis. Nutrition and Cancer 3 (3), 172−191. Available from: https://doi.org/10.1080/01635588109513719.

Bertinato, J., Wu Xiao, C., Ratnayake, W.M., et al., 2015. Lower serum magnesium concentration is associated with diabetes, insulin resistance, and obesity in South Asian and white Canadian women but not men. Food and Nutrition Research 59 (1), 25974.

Bhaskaran, D., Chadha, S.S., Sarin, S., Sen, R., Arafah, S., Dittrich, S., 2019. Diagnostic tools used in the evaluation of acute febrile illness in South India: a scoping review. BMC Infectious Diseases 19, 970. Available from: https://doi.org/10.1186/s12879-019-4589-8.

Bjørklund, 2013. The role of zinc and copper in autism spectrum disorders. Acta Neurobiologiae Experimentalis 73 (2), 225−236.

Bjørklund, G., Chartrand, M.S., Aaseth, J., 2017. Manganese exposure and neurotoxic effects in children. Environmental Research 155, 380−384. Available from: https://doi.org/10.1016/j.envres.2017.03.003.

Böckerman, P., Bryson, A., Viinikainen, J., Viikari, J., Lehtimäki, T., Vuori, E., et al., 2016. The serum copper/zinc ratio in childhood and educational attainment: a population-based study. Journal of Public Health 38 (4), 696−703. Available from: https://doi.org/10.1093/pubmed/fdv187.

Campana, F., Vigarios, E., Fricain, J.C., Sibaud, V., 2019. Geographic stomatitis with palate involvement. Anais Brasileiros de Dermatologia 94 (4), 449−451.

Cannas, D., Loi, E., Serra, M., Firinu, D., Valera, P., Zavattari, P., 2020. Relevance of essential trace elements in nutrition and drinking water for human health and autoimmune disease risk. Nutrients 12, 2074.

Chang, E., Mezei, G., Odo, N.U., n.d. Cluster Investigation Analysis. https://www.exponent.com/services/practices/environmental-sciences/health. (Accessed 02 January 2021).

Chen, T.K., Knicely, D.H., Grams, M.E., 2019. Chronic kidney disease diagnosis and management: a review. JAMA: The Journal of the American Medical Association 322, 1294−1304. Available from: https://doi.org/10.1001/jama.2019.14745.

Correa-Rotter, R., Wesseling, C., Johnson, R.J., 2014. CKD of unknown origin in Central America: the case for a Mesoamerican nephropathy. American Journal of Kidney Diseases 63, 506−520.

Couser, W.G., Remuzzi, G., Mendis, S., Tonelli, M., 2011. The contribution of chronic kidney disease to the global burden of major noncommunicable diseases. Kidney International 80, 1258−1270.

de Caritat, P., Main, P.T., Grunsky, E.C., Mann, A.W., 2017. Recognition of geochemical footprints of mineral systems in the regolith at regional to continental scales. Australian Journal of Earth Sciences 64 (8), 1033−1043. Available from: https://doi.org/10.1080/08120099.2017.1259184.

DiNicolantonio, J.J., O'Keefe, J.H., Wilson, W., 2018. Subclinical magnesium deficiency: a principal driver of cardiovascular disease and a public health crisis. Open Heart 5 (1), e000668. Available from: https://doi.org/10.1136/openhrt-2017-000668.

Duyff, R.L., 2011. American Dietetic Association Complete Food and Nutrition Guide, third ed. John Wiley and Sons Inc., New Jersey, p. 688. Available from: http://154.68.126.6/library/Food%20Science%20books/batch2/Complete%20Food%20and%20Nutrition%20Guide/Complete%20Food%20and%20Nutrition%20Guide.pdf.

Duyff, R.L., 2012. American Dietetic Association Complete Food and Nutrition Guide, fourth ed. John Wiley and Sons Inc., New Jersey, p. 720.

DVNHC (Da Vinci Natural Health Centre). n.d., Geopathic Stress and Its Effects on Health. https://www.naturaltherapycenter.com/geopathic-stress-and-its-effects-on-health/. (Accessed 22 January 2021).

Ekrikpo, U.E., Kengne, A.P., Bello, A.K., Effa, E.E., Noubiap, J.J., Salako, B.L., et al., 2018. Chronic kidney disease in the global adult HIV-infected population: a systematic review and meta-analysis. PLoS One 13, e0195443.

Fries, B.C., Mayer, J., 2009. Climate change and infectious disease. Interdisciplinary Perspectives on Infectious Diseases 2009, 976403. Available from: https://doi.org/10.1155/2009/976403.

Gaayeb, L., Pinçon, C., Cames, C., Sarr, J.B., Seck, M., Schacht, A.M., et al., 2014. Immune response to *Bordetella pertussis* is associated with season and undernutrition in Senegalese children. Vaccine 32 (27), 3431−3437. Available from: https://doi.org/10.1016/j.vaccine.2014.03.086.

GBD 2017 Causes of Death Collaborators 2018, 2017. Global, regional, and national age-sex-specific mortality for 282 causes of death in 195 countries and territories, 1980 - 2017: A systematic analysis for the Global Burden of Disease Study 2017. Lancet 392, 1736−1788.

GEBN (Genetic Engineering and Biotechnology News), 2017. Impact of climate change on human health. https://www.genengnews.com/topics/translational-medicine/spread-of-infectious-disease-due-to-climate-change-may-be-greater-than-previously-thought/. (Accessed 22 January 2021).

Grober, J.S., Kisters, K., 2015. Magnesium in prevention and therapy. Nutrients 7 (9), 8199−8226.

Hedera, P., 2000. Hereditary spastic paraplegia overview. In: Adam, M.P., Ardinger, H.H., Pagon, R.A., Wallace, S.E., Bean, L.J.H., Mirzaa, G., et al.,GeneReviews® [Internet]. University of Washington, Seattle (WA). Available from: https://www.ncbi.nlm.nih.gov/books/NBK1509/.

Huang, P.-M., Wang, M.-K., Chiu, C.-Y., 2005. Soil mineral-organic matter-microbe interactions: Impacts on biogeochemical processes and biodiversity in soils. Pedobiologia 49 (6), 609–635.

Huang, Y.S., Higgs, S., Vanlandingham, D.L., 2019. Arbovirus-mosquito vector-host interactions and the impact on transmission and disease pathogenesis of arboviruses. Frontiers of Microbiology 10, 22. Available from: https://doi.org/10.3389/fmicb.2019.00022.

Ismail, A.A.A., Ismail, Y., Ismail, A.A., 2018. Chronic magnesium deficiency and human disease; time for reappraisal? QJM: An International Journal of Medicine 111 (11), 759–763. Available from: https://doi.org/10.1093/qjmed/hcx186.

Jahnen-Dechent, W., Ketteler, M., 2012. Magnesium basics. Clinical Kidney Journal 5 (Supplement 1), i3–i14. Available from: https://doi.org/10.1093/ndtplus/sfr163.

Jomova, K., Valko, M., 2011. Advances in metal-induced oxidative stress and human disease. Toxicology 283 (2-3), 65–87. Available from: https://doi.org/10.1016/j.tox.2011.03.001.

Joubert, B.R., Mantooth, S.N., McAllister, K.A., 2020. Environmental health research in Africa: important progress and promising opportunities. Frontiers in Genetics 10, 1166. Available from: https://doi.org/10.3389/fgene.2019.01166 (accessed 14.02.24).

Khalifa, I.I., Hassan, M.F., AL-Deri, S.M., Gorial, F.I., 2016. Determination of some essential and non-essential metals in patients with Fibromyalgia Syndrome (FMS). International Journal of Pharmaceutical Sciences and Research 8 (5), 306–311.

Koch-Henriksen, Sørensen, P.S., 2010. The changing demographic pattern of multiple sclerosis epidemiology. Lancet Neurology 9, 520–532.

Langer-Gould, A., Brara, S.M., Beaber, B.E., Zhang, J.L., 2013. Incidence of multiple sclerosis in multiple racial and ethnic groups. Neurology 80, 1734–1739.

Lapworth, D., Knights, K., Key, R., Johnson, C.C., Ayoade, E., Adekanmi, M., et al., 2012. Geochemical mapping using stream sediments in west-central Nigeria: implications for environmental studies and mineral exploration in West Africa. Applied Geochemistry 27, 1035–1052. Available from: https://doi.org/10.1016/j.apgeochem.2012.02.023.

Lee, H., Aronson, J.K., Nunan, D., 2021. Catalogue of Bias - association or Causation? How Do We Ever Know? https://catalogofbias.org/2019/03/05/association-or-causation-how-do-we-ever-know/. (Accessed 21 January 2021).

Levine, K.E., Redmon, J.H., Elledge, M.F., Wanigasuriya, K.P., Smith, K., Munoz, B., et al., 2016. Quest to identify geochemical risk factors associated with chronic kidney disease of unknown etiology (CKDu) in an endemic region of Sri Lanka-a multimedia laboratory analysis of biological, food, and environmental samples. Environmental Monitoring and Assessment 188 (10), 548. Available from: https://doi.org/10.1007/s10661-016-5524-8.

Lombard, M., De Bruin, D., Elsenbroek, J.H., 1999. High density geochemical mapping of soils and stream sediments in South Africa. Journal of Geochemical Exploration 66 (1-2), 145–149. Available from: https://doi.org/10.1016/S0375-6742(99)00032-1.

Martin, L.B., Weil, Z.M., Nelson, R.J., 2008. Seasonal changes in vertebrate immune activity: mediation by physiological trade-offs. Philosophical Transactions of the Royal Society of London. Series B, Biological Sciences 363 (1490), 321–339. Available from: https://doi.org/10.1098/rstb.2007.2142.

Mash, E.J., Wolfe, D., 2019. Abnormal Child Psychology. Courage Learning. Wolfe, D.A. (David Allen), 1951, seventh ed. Boston, MA, 624 p.

McKenzie, R.C., Rafferty, T.S., Beckett, G.J., 1998. Selenium: an essential element for immune function. Trends in Immunology 19 (8), 342–345. Available from: https://doi.org/10.1016/S0167-5699(98)01294-8.

Mitha, M., 2017. A review of pulmonary sarcoidosis. African Journal of Thoracic and Critical Care Medicine 23 (4), 100–105. Available from: https://doi.org/10.7196/SARJ.2017.v23i4.168.

Morris, G., Berk, M., Galecki, P., Maes, M., 2014. The emerging role of autoimmunity in myalgic encephalomyelitis/chronic fatigue syndrome (ME/CFS. Molecular Neurobiology 49 (2), 741–756. Available from: https://doi.org/10.1007/s12035-013-8553-0.

NCEH [National Center for Environmental Health, CDC (Centre for Disease Control and Prevention), Atlanta, Georgia], 2013. Investigating suspected cancer clusters and responding to community concerns: Guidelines from CDC and the Council of State and Territorial Epidemiologists. MMWR (Morbidity and Mortality Weekly Report) Recommendation Report 62 (RR-08), 1–24.

Ndiaye, F.C., Bourgeois, D., Leclercq, M.H., Berthe, O., 1999. Noma: public health problem in Senegal and epidemiological surveillance, Oral Diseases, 5. pp. 163–166.

Neal, A.P., Guilarte, T.R., 2013. Mechanisms of lead and manganese neurotoxicity. Toxicology Research 2 (2), 99–114.

Peres, T.V., Schettinger, M.R., Chen, P., Carvalho, F., Avila, D.S., Bowman, A.B., et al., 2016. Manganese-induced neurotoxicity: a review of its behavioral consequences and neuroprotective strategies". BMC Pharmacology and Toxicology 17 (1), 57. Available from: https://doi.org/10.1186/s40360-016-0099-0.

Pickering, G., Mazur, A., Trousselard, M., Bienkowski, P., Yaltsewa, N., Amessou, M., et al., 2020. Magnesium status and stress: the vicious circle concept revisited. Nutrients 12 (12), 3672. Available from: https://doi.org/10.3390/nu12123672.

Randolph, S., 2009. Tick-borne disease systems emerge from the shadows: the beauty lies in molecular detail, the message in epidemiology. Parasitology 136 (12), 1403–1413. Available from: https://doi.org/10.1017/S0031182009005782.

Reedman, A.J., 1973. Geochemical Atlas of Uganda. Geological Survey of Uganda, Entebbe, Uganda, 42 p.

Smith, T., 1934. Parasitism and Disease. Princeton University Press, Princeton, NJ.

Taylor, E.W., 1997. Selenium and viral diseases: facts and hypotheses. Journal of Orthomolecular Medicine 12 (4), 227–239.

US GCRG (US Global Change Research Group), 2016. Climate and Health Assessment: Temperature-Related Death and Illness. https://health2016.globalchange.gov/temperature-related-death-and-illness. (Accessed 22 January 2021).

US NIEHS (United States National Institute of Environmental Health Sciences), 2020. Autoimmune disease. https://www.niehs.nih.gov/health/topics/conditions/autoimmune/index.cfm. (Accessed 22 January 2021).

Whelan, E.A., 2010. Reproductive and ENDOCRINE TOXICOLOGY. In: McQueen, C.A. (Ed.), Comprehensive Toxicology, second ed. Elsevier Science.

WHO (World Health Organisation), 2020d. Climate Change. https://www.afro.who.int/health-topics/climate-change. (Accessed 28 November 2020).

Workshop, 1993. Papers from the Workshop on Statistics and Computing in Disease Clustering. Port Jefferson, New York, July 23–24, 1992. Statistics in Medicine 12 (19–20), 1751–1968. https://pubmed.ncbi.nlm.nih.gov/7903817/. (Accessed 01 February 2021).

Zermoglio, F., Scott, O., Said, M.,For USAID (United States Agency for International Development), 2019. Vulnerability and Adaptation in the Mara River Basin. doi: 10.13140/RG.2.2.20540.49282. https://www.researchgate.net/publication/332971520_Vulnerability_and_Adaptation_in_the_Mara_River_Basin/citation/download. (Accessed 22 January 2021).

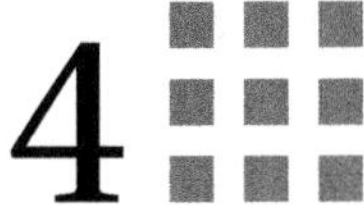

Environmental health impacts of geogenic pollution in Africa

"Environmental pollution is an incurable disease. It can only be prevented."

Ecologist, Barry Commoner (1917–2012).

Key chapter features

1. This Chapter summarizes the current status of environmental pollution research in Africa; and identifies and characterizes the main sources of pollutants that affect human health and the environment, with special attention paid to those pollutants from *geogenic sources* (i.e., those present in geological systems or processes involving geological materials).
2. Presents highlights of how the presence of toxic metal and other contaminants negatively affect soil, water, air, and environmental quality in Africa, with particular emphasis on threats to human health.
3. Puts forward cogent measures for addressing issues of soil, water, and air pollution on human health together with some recommendations of the best available techniques for assessing and remediating contaminated soils, water, and air.
4. Points out gaps in our knowledge of the sources, dispersal, and uptake of environmental contaminants, and the relationships with human health. Suggests areas of future research for bridging knowledge gaps on environmental health impacts of soil, water, and air pollution in Africa.
5. Explains why research to unravel these knowledge gaps would contribute greatly toward better diagnosis, therapeutic interventions, and management of associated health conditions.

Introduction

Soil, water, and air pollution has always been of worldwide concern. Through their discussions with public health practitioners and researchers, Lucero-Prisno et al. (2023) identified 10 key areas of public health research that would occupy center stage in 2023, among which, "environmental pollution" features prominently; and noted the desirability of implementing cogent measures to address these areas and facilitate proactive and innovative health interventions and practices. In 2023 the World Bank reported a total of 9 million premature deaths during 2016, or 16% of all deaths worldwide, with the most pollution-related mortality (about 92%) occurring in the less developed countries.

Medical Geology of Africa. DOI: https://doi.org/10.1016/B978-0-12-818748-7.00015-0

Concepts and definitions

It is important to attain unanimity among scientists in the understanding of the concepts and definitions in the study of *environmental pollution* since using a common and simplified language would lead to a better understanding of the issues and provide policymakers and other stakeholders with the basis for identification of strategies and techniques for assessing and addressing pollution.

The term *"geogenic sources of environmental pollution"* as used in this Chapter, refers to all processes and interactions, natural, or anthropogenic, that involve geological materials (minerals, rocks, soils, sediments, water, air, metals, and so on), including aspects of transport, deposition, accumulation, acid precipitation, atmospheric pollution, aquatic pollution, marine pollution and effects of pollutants on humans, other animals, vegetation, fish, aquatic species, and micro-organisms. The distinction between *"contaminants"* and *"pollutants"* is sometimes unclear, since the concentrations at which contaminants become pollutants is itself not always clear.

Contamination refers to the presence of an alien and potentially toxic substance in the environment or the presence of a substance at concentrations above *"background." Pollution* may be defined as contamination that has harmful effects on the environment (Stengel et al., 2006; Chapman, 2007). It is the harm that results from the presence of a substance or substances where they would not normally be found or because they are present in larger than normal quantities. A polluting substance may occur in the form of a solid, liquid, or gas. According to these definitions, all pollutants are considered contaminants, but the converse is not necessarily true.

When defining the extent of pollution or contamination, it is important to distinguish between *background values* and *baseline values.* Chapman (2007) has defined *background* (normal concentration) as "... a natural value for a given medium that has not been influenced by human activities," the determination of which, in the case of soil pollution or contamination, has to take account of parent material, climate, and weathering rate before *"thresholds"* (the upper limit of normal background fluctuations) can be established. *"Baseline values,"* on the other hand, refers to "... the actual content of an element in the superficial environment at any given point." See Carrillo et al. (2022) for a modern discussion on usage of the terms *background* and *baselines.*

Nature and sources of soil pollution in Africa

Soil pollution can be defined as the occurrence in the soil of a chemical or substance that is out of place and/or present at a higher than normal concentration (i.e., background) that has adverse effects on any nontargeted organism (Rodriguez-Eugenio et al., 2018). It is not always easy to assess soil pollution, nor can it be visually discerned, which makes it a hidden danger.

The Status of the World's Soil Resources Report (FAO, 2015a) referred to soil pollution as one of the foremost soil threats affecting world soils and the ecosystem services that they provide. About 95% of the food we consume come directly or indirectly from the soil (FAO and UNEP, 2021), underlining the threat posed by polluted soil on human health (Zhang et al., 2023a).

The African Continent bears the brunt of soil pollution from substances generated from diverse sources, both geogenic and anthropogenic. Contaminants such as toxic metals, radionuclides, asbestos, and a host of other contaminants may be present in soils due to geological and pedological processes with or without anthropogenic influence (FAO and UNEP, 2021; Xin et al., 2022). Soil pollution from purely geological processes (e.g., volcanic eruptions, earthquakes, weathering and erosion, flooding, and tsunamis) also occurs in many regions in Africa (Tindwa and Singh, 2023), but, according to Fayiga et al. (2018), the other sources (anthropogenic) of soil pollution in Africa including agricultural activities, roadside emissions, auto-mechanic workshops, refuse dumps, and e-waste are also important in disease causation.

Soil pollution from volcanic eruptions

Volcanic eruptions are one of the main geological processes that release toxic metals into soil. Gases from eruptions deposit toxic metals and metalloids such as Cd, Hg, Cr, Cu, Zn, Mn, Ni, and As on soils. Both ancient and modern volcanism in tectonically active regions of the African Continent leave a footprint on the lives of geographically marginal populations, for example, the Lake Nyos disaster of 1986 (Tchindjang, 2018) and the Mount Nyiragongo eruption of 2003 (Nicole, 2022), through consumption of food grown on polluted soils from volcanic emissions. Along the path of the African Rift Valley, associated volcanism has emitted toxic elements and ashes that have contaminated the root zone of soils as well as surface and groundwater bodies (Davies, 2008; Gao et al., 2023). Rift formation and separation of the African and Arabian tectonic plates, are other natural processes that cause pollution of soils on the eastern flank of the African Continent.

Of the elements released in volcanic exhalations, F and As bear special significance because of widespread maladies they cause among populations living in the neighborhood of certain ancient volcanoes, such as Mount Kenya (fluorosis) and the western sector of the Gregory Rift (arsenicosis) (See, e.g., Davies, 2008). Arsenic also occurs widely in the sulfide mineralized belts and coal producing areas of the Continent. Health conditions arising from exposure to metals and metalloids from volcanic areas or sulfide-mineralized or coal-producing areas are discussed further in Part II: "An exposition of some key geomedically important elements, with Africa case descriptions."

Point-source and diffuse source soil pollution

Pollution of soils can come about by processes often characterized as *"point source"* or *"diffuse (or nonpoint) pollution."*

Point source pollution comes from a specific source and enters a water body at a specific site from where the water can contaminate land and soil resources. Potential point sources of pollution include effluent discharges from a pipe, industrial sites, sewage treatment works, power stations, landfill sites, fish farms, and oil spillage via a pipeline from industrial sites. Point source pollution is in general, readily identified, can be monitored and controlled, and is in general, readily prevented. Anthropogenic activities represent the main sources of point-source pollution.

Diffuse pollution or *nonsource pollution* arises where polluting substances are distributed over wide areas. Identification of specific sources of such pollution sometimes proves difficult, obviating the possibility of taking immediate action to prevent it, since prevention often requires wholesale changes to land use and management practices. Diffuse pollution involves the transport of pollutants via soil-water-air systems. Examples of diffuse pollution include the leaching to surface water and groundwater of contaminants from roads, manures, nutrients, and pesticides used in agriculture and forestry, and atmospheric deposition of contaminants arising from industry. A special case arises where, for example, a power station emits sulfur dioxide and nitrous oxide to the air. Although this is a *point source*, the deposition (fallout) and hence, impact, will be over a wide area as *diffuse pollution.*

Soil contamination through mining and processing of ore

Mining and smelting are important economic activities in Africa, but also large sources of environmental pollution. The Continent ranks first or second in quantity of world reserves of bauxite, cobalt, industrial diamond, phosphate rock, platinum-group metals, vermiculite, and zirconium. Gold mining is Africa's main mining resource. The sheer volume of waste generated through mining and processing such a massive endowment of ore deposits, makes the Continent amenable to large scale pollution of the soil, water, and air environment through mining and ore processing activities. Raimi et al. (2022) place "mining" at the top of their list of contaminant sources of soil pollution. If as the United States Environmental Protection Agency (USEPA) noted in 1987—"*. . . problems related to mining waste may be rated as second only to global warming and stratospheric ozone depletion in terms of ecological risk. The release to the environment of mining waste can result in profound, generally irreversible destruction of ecosystems*" [Cited by Durand (2012)]. Then, surely, we are in very serious trouble.

The observation that mining and metal smelting to separate minerals have introduced huge quantities of toxic metals and other toxic elements to the soil environment has never been in doubt. These substances can persist for long periods, long after the end of these activities (Ogundele et al., 2017). Toxic mining wastes are stocked up in tailings, mainly formed of fine particles that can have different concentrations of toxic metals.

The erosion caused by rains, rivers, and winds, as well as over-use of soils for agriculture and low use of manures have resulted in soils becoming infertile, as for example, in the plains of the River Nile and the Orange River.

Trace element pollution of soils

Other major geogenic sources of trace elements in soils, besides mining and ore processing, include: (1) volcanism, which may yield fertile volcanic soils with high clay mineral content and cation exchange capacity; but may also emit a high toxic metal load; (2) geothermal input, yielding volatile substances, and gases, (F, As, Hg, CO_2, SO_2, etc.); (3) extreme tropical weathering; (4) element redistribution through leaching/eluviation, lateritization, podsolization, gleying, and so on, leading to depletion or enrichment of trace elements. Trace element input in soils also comes from human activities, such as mining, industry, agriculture, domestic activities, construction, and transportation.

Trace elements that are essential for the proper growth of animals (including humans), plants, and soil microorganisms are referred to as *micronutrients*. Examples are Mn, Fe, Zn, Se, Cu, Ni, Mo, and B (See, e.g., Mishra et al., 2023), while other elements have no known metabolic function, such as Hg, Cd, and Pb (Haidar et al., 2023). Geogenic processes, such as weathering and soil formation may lead to increased concentration of these elements in soils (Dror et al., 2023).

As the majority of Africa's population is rural, living close to the land and dependent on food and drinking water sources from their immediate vicinity, the region offers a unique opportunity for studying the health effects of varying concentration levels of nutritional and toxic elements in soils as well as in other environmental media in relation to the distribution of health conditions due to dietary intake of these elements. Soil supports diverse ecosystems and critical ecological services in various areas of the Continent. Soil pollutants wash into rivers causing water pollution as well.

Soil pollution due to oil and gas extraction

Spills of crude oil and brines are a major problem in oil-rich African countries such as Nigeria and Angola (Nkem et al., 2022); and represent a significant point-source soil pollution. Spills of crude oil occur from well sites and from pipelines. Brine spills are also widespread. Brines, by definition, have high salinity levels and can contain toxic trace elements and naturally occurring radioactive materials.

Geological aspects of waste disposal

Municipal waste disposal in *"landfills"* and *"incineration"* are the two most common waste disposal options in Africa. Landfills are controlled and engineered establishments, wherein the waste ought to abide with certain regulations regarding their quality and quantity. The three most popular "landfill" types are: (1) municipal solid waste landfill, (2) industrial waste landfill, and (3) hazardous waste landfill. There is also an emerging landfill type called *"green waste landfill"* (See "Glossary of Terms") that is occasionally used in some countries.

The form of *"landfilling"* in most African countries, however, is a crude version of a proper sanitary landfill (Abubakar et al., 2022; Siddiqua et al., 2022), and can be more accurately described as *open dumps*. In both cases (landfills and incineration systems), many pollutants, such as toxic metals, polyaromatic hydrocarbons (PAHs), and many other compounds can accumulate in the soil (El-Saadony et al., 2023) either directly as a result of migration of *landfill leachates* that may pollute soil and groundwater resources, or by ash fallout from incinerating plants (Siddiqua et al., 2022; Faragó et al., 2023).

Soil pollution by landmines

A relatively new development in modern African warfare is the use of nondegradable weapons of destruction and chemicals that can remain in the affected soils for years after the end of the conflict (FAO and ITPS, 2015). Soils can be considerably altered by warfare activities in both wartime and times of peace due to military activities, such as test-firing facilities. Recovery of the soils in these areas can take several years, and in some cases even centuries (See Certini et al., 2013; Broomandi et al., 2020).

Each of the countries of Chad and Angola has more than a 100 sq. km. of land still affected by the presence of landmines (ICBL-CMC, 2019). Several other African countries (e.g., Eritrea, Ethiopia, Somalia, South Sudan, and Zimbabwe) have smaller areas affected by landmines, none of which has undergone an assessment. Evidence from outside Africa has shown enrichment of forest soils in Croatia with Cd, Ni, and Zn as a result of historical explosion of landmines (Mesić Kiš et al., 2016). In Iraq, the deliberate detonation of landmines as part of demining programmes is reported to have caused soil contamination by trace metals (Hamad et al., 2019). Experiences and lessons learned from these problems on other continents can be applied to Africa today in their soil remediation programmes.

Interaction of pollutants with soil constituents

The filtering and buffering capacity of soil, as well as its ability to transform inorganic and organic contaminants are among the major ecological services provided by this resource.

These essential functions are responsible for maintaining good groundwater quality and safe food production (Kopittke et al., 2023). On entering the soil, pollutants undergo physical, physicochemical, microbiological, and biochemical processes whereby they are retained, reduced, or degraded.

Changes in chemical, biochemical, and microbial properties of soils and plant responses commonly reflect contamination of soils by metals (See Chibuike and Obiora, 2014). For risk assessments, the total content of toxic metals in soil is still the most widely used indicator, although, as Xin et al. pointed out in 2022, extractable amounts are considered to be more closely related to plant uptake. *Bioavailability* refers to the physical, chemical, and biological interactions that determine the exposure of organisms to chemicals associated with soils; or, in more general terms, the degree and rate at which the chemical is absorbed into a living system or is made available at the site of physiological activity. Of the several models that have been proposed for assessment of soil contamination by toxic metals, none of them, according to Xin et al. (2022), have gained consensus among scientists, with reference to their application to a wide range of soils.

Geogenic sources of water pollution

Water pollution is worsening in many regions of the world, not least in Africa, where the problem now represents a crisis of massive proportion. Mining is known to be one of the principal reasons for this crisis. Mining affects fresh water through heavy use of water in ore processing and through water pollution from discharged mine effluent and seepage from tailings and waste rock impoundments.

Through a systematic analysis of published literature, Ighalo et al. (2021) have typified the nature and regional distribution of pollution sources in Africa in the domain of surface water, groundwater, and rainwater quality by way of a case study of Nigeria, a West African country trying to cope with the enormous water pollution problems posed by petroleum resources development. It is well-established that leachates from landfills and dumpsites in Africa are significant to major sources of groundwater pollution with organics, salts, and toxic metals as pollutants (See, e.g., Igboama et al., 2022 and the Section: "Geological aspects of waste disposal," in this Chapter).

Water pollution through mining

While there have been improvements to mining practices in recent years, significant environmental risks remain. The environmental legacy of past mining activities that were undertaken with little concern for the environment continues to receive considerable attention.

Mining by its very nature, requires large amounts of water, which after use, should be disposed of in an environmentally friendly way, to obviate seriously pollution of water resources. One of the problems is that mining has become more mechanized and therefore able to handle more rock and ore material than ever before. Therefore mine waste has multiplied enormously. As mine technologies are developed to make it more profitable to mine low grade ore, even more waste will be generated in the future.

According to Evans et al. (2019), although water quality is currently receiving more attention now than in the past, much more attention to this problem is desirable. Factors to be considered, include: "… *source attribution, emerging contaminants, costs, and incentives for adoption of pollution reduction measures.*" Measures for addressing the problems of water pollution in Africa are considered in the Section: "Addressing prevention and control of water pollution," in this Chapter.

Types of water pollution from mining

There are four main types of water pollution from mining (SDWF, 2023)

1. Acid Mine Drainage

In acid rock drainage, which is a natural process, sulfuric acid is produced when exposed sulfides in rocks react with air and water. The term acid mine drainage (AMD) refers to a similar process, but greatly magnified. At a certain level of acidity, a naturally occurring type of bacteria called *Thiobacillus ferroxidans* enters the reaction mix, and speeds up the oxidation and acidification processes, leaching even more trace metals from the wastes. As long as source rock is exposed to air and water, the acid will leach from the rock until the sulfides are leached out—a process that can last hundreds, even thousands of years. Acid is carried off the mine site by rainwater or surface drainage and deposited into nearby streams, rivers, lakes, and groundwater. AMD severely degrades water quality and can kill aquatic life and make water virtually unusable.

AMD chemistry is very complex, a complexity that has considerably inhibited the design of effective treatment options. A number of chemical processes contribute to AMD formation, with pyrite oxidation being by far, the greatest contributor.

FIGURE 4–1 Postclosure coal mine AMD treatment on the East Rand, South Africa. *AMD*, Acid mine drainage. *Credit: AIDC. From https://aidc.org.za/acid-mine-drainage-environmental-social-risks-witwatersrand-gold-fields/post-closure-coal-mine-amd-treatment-east-rand-south-africa/. (Accessed 17 August 2023).*

A general equation for AMD formation is:

$$2FeS_2(s) + 7O_2(g) + 2H_2O(l) = 2Fe^{2+}(aq) + 4SO_4^{2-}(aq) + 4H^+(aq)$$

AMD formation in tailings storage facilities (TSF) can impact large areas due to redistribution of material from the facility by wind and water erosion (See Laker, 2023).

The problem of AMD is a particularly intractable one in South Africa (See Fig. 4–1), where, in the Bushveld Complex alone, such a wide variety of metal sulfide mining takes place, leading to lots of clamor by environmental activists.

2. Toxic metal and metalloid contamination and leaching

Toxic metal and metalloid pollution occurs when metals such as Co, Cu, Cd, or metalloids, such as As contained in excavated rock or exposed in an underground mine come into contact with water. Metals are leached out and taken downstream as water washes over the rock surface. Streams draining the tailings dumps are typically acidic and have high sulfate and heavy metal concentrations—(Pb, Hg, Zn, Pt, etc.) (See Bokar et al., 2020). Arsenic, a chalcophilic metalloid, is also abundant (See, e.g., Irunde et al., 2022).

3. Processing chemicals pollution

This kind of pollution occurs when spills, leaks, or leaches of chemical agents (such as cyanide or sulfuric acid used by mining companies to separate the sought after mineral from the ore) from the mine site gets into nearby waterways. These chemicals can be highly toxic to humans and wildlife.

4. Erosion and sedimentation

Soil and rock piles are disturbed in the process of constructing and maintaining roads, open pits, and waste impoundments during mineral development. If necessary prevention and control measures are not put in place, erosion of the exposed earth materials may carry substantial amounts of sediment into nearby waterways (See Section: "Addressing prevention and control of water pollution," in this Chapter). Excessive sediment can clog riverbeds and smother watershed vegetation, wildlife habitat, and aquatic organisms.

Other sources of water pollution

Aquatic ecosystems, so vital in Africa for their provision of requisite goods and services, are, at the same time, integral to a biosphere's survival; but are constantly under harmful anthropogenic impacts (Ouma et al., 2022) underlining the necessity to monitor and mitigate these detrimental impacts through established or developed scientifically feasible physical, chemical, or biological methods and integrated assessment systems.

Fayiga et al. (2018) consider urban and industrial discharges of untreated effluents as major pollution sources for surface waters. According to these authors, although some African countries have standards for effluent discharge into surface waters, the extent of enforcement of these standards is unknown. Though groundwater sources in wells and boreholes are the major sources of drinking water for the African populace, the biological water quality of these groundwater sources is quite poor in the region due to their close proximity to sanitary facilities (Fayiga et al., 2018).

Road runoff and urban stormwater discharges

Waters emanating from large runoff-producing areas such as roads and car parks in the urban environment of Africa are often contaminated with sediment, litter, oil and petrol, and with toxic metals from motor vehicles (See, e.g., Gupta, 2020). Water carrying these contaminants is washed off into drains and directly into nearby watercourses to which most surface water drains are directly connected, such that spillage of chemicals will tend to be washed into rivers.

Thankfully, *Sustainable Drainage Systems (SuDS)* (See "Glossary of terms," this Chapter) are now increasingly being adopted to ensure that urban areas behave more like natural catchments through the use of porous pavement surfaces and by diverting potentially polluted water from watercourses (See, e.g., GPG, 2020). This way, the potential for pollution caused by direct runoff as well as the volume of water flowing in the drainage network are reduced, thereby avoiding flooding and sewer overflows.

Hydraulic fracturing

Mining processes such as *"hydraulic fracturing"* (sometimes referred to as *"fracking"*), is a method for extracting gas and oil from shale. The process involves blasting huge amounts of water mixed with toxic chemicals and sand deep into the Earth at high pressure to fracture rock formations and release naturally occurring oil and natural gas into a well, and then be collected for market. This extreme form of energy production threatens the health of surrounding communities and wildlands. The environmental health impacts of fracking are discussed in the Section "Fracking and health," of this Chapter.

Flooding

In Africa, floods are the most frequent type of natural disaster and occur when an overflow of water submerges land that is usually dry. The main causes of flooding are related to heavy rainfall, rapid snowmelt or a storm surge from a tropical cyclone or tsunami in coastal areas.

There are three common types of floods:

1. *Flash floods* are caused by rapid and excessive rainfall that raises water heights quickly, and rivers, streams, channels or roads may be overtaken.
2. *River floods* are caused when consistent rain or snow melt forces a river to exceed capacity.
3. *Coastal floods* are caused by storm surges associated with tropical cyclones and tsunami.

Taking the region of West Africa, as an example, devastating flood events have been experienced since the year 2000, following the great drought that started in the 1970s (Al-Zu'bi et al., 2022). The mean occurrence of floods per year in the Region between 1966 and 1999 was 3, but this number rose to 12 per year between 2000 and 2017, with very high human and economic damages (Al-Zu'bi et al., 2022). Focusing on West Africa, the region has been experiencing devastating flood events since 2000, following the great drought that started in the 1970s. For instance, the mean occurrence of floods per year in West Africa between 1966 and 1999 was 3, but this number rose to 12 per year between 2000 and 2017, with very high human and economic damages. In this context, a question that would probably arise is about whether the current scientific literature on this region provides solutions for mitigation and flood prevention efforts.

The need to revise the hydrological standards for building flood mitigation infrastructures can never be overstated. Apart from a regional initiative launched by the World Meteorological Organization, research on the development of new hydrological standards is very rare in West Africa (which is also likely to be the case in other parts of the Continent). A call for further investigation on this aspect is therefore warranted.

In eastern Africa, the flood-drought pattern is not much different from that in West Africa. Bimodal rainfall changes consisting of long rains (March–May) and short rains (October–December), have severe impact on public health, socioeconomic and environmental conditions. Palmer et al. (2023) examine the drivers and corresponding impacts of Eastern Africa rainfall variability, assessing the extent to which remote teleconnections, *viz.,* the El Niño–Southern Oscillation and the Indian Ocean Dipole, influence interannual variability.

Flash flooding from heavy rains in the Horn of Africa (HA) killed dozens and affected 300,000 people in Ethiopia and Somalia in March 2023 (NASA Earth Observatory, 2023). The widespread flooding in the HA came after almost 3 years of extreme drought. Even with the unexpectedly early and heavy rains in March, predictive climate modeling suggests that the long rains in 2023 will be drier than normal and drought conditions are likely to continue (NASA Earth Observatory, 2023).

The nature and sources of air pollutants

Air pollutants comprising fine inhalable particles, with diameters generally 2.5 μm and smaller ($PM_{2.5}$) are generated from a range of both natural geogenic and anthropogenic sources (HEI, 2022). Air pollutants may be categorized as primary or secondary.

The sources responsible for $PM_{2.5}$ pollution vary within and between countries and regions across Africa (Tessum et al., 2022). Common natural (geogenic) sources include windblown dust, sea spray, and wildfires, while anthropogenic sources are exemplified by fossil fuel and biofuel combustion, transportation, industrial and semi-industrial (e.g., artisanal mining) processes, agriculture, and waste burning.

A massive amount of dust estimated at a million tons is released from the Sahara Desert each year into the atmosphere and travels over the North Atlantic Ocean; and commonly referred to as the *Saharan dust plume* (SDP). The SDP contributes to the high particulate matter concentrations in North Africa and the Middle East, as well as in some countries in western Sub-Saharan Africa, and further afield, for example, the Caribbean, North America, and Europe (WMO, 2023). Household solid fuel use is also an important contributor to ambient $PM_{2.5}$ levels in many African countries (HEI, 2022).

Other geogenic sources of dust

The greater percentage of particulate load in the African urban environment is considered to come from geogenic sources (e.g., Davies, 2008); and these may cause more toxicity problems in humans than dust components from other sources. The known sources of atmospheric dust in Africa include: (1) *harmattan dust*, which transports huge amounts of Saharan dust (the SDP) over the Sahelian region (from Niger Republic across northern Nigeria) during the months of December to March; and sometimes tinged with radioactive particles; (2) mines and quarries; (3) treatment plants; (4) solid waste disposal sites; (5) unpaved roads; (6) disturbed land; (7) construction sites; (8) sawdust; and (9) particulates from all types of combustion.

Exposed soils are eroded by the wind and windborne particles, wherever these are exposed for any length of time in dry conditions, giving rise to aerosols containing dust ranging from very fine particulates to silt-grade material. Once in the airstream, dust can be carried for very long distances before it is deposited on land or in water. For instance, it is now well-established that dust from North Africa (re.: the SDP) reaches the Americas and Europe depending on weather conditions (See, e.g., Gomes et al., 2022). Dust also arises from point sources, such as erupting volcanoes, fires, and events, such as atmospheric nuclear tests. Deposits of material that have not been stabilized, for instance, tips and lagoons containing mineral wastes, in dry conditions, also give rise to dust. Most roads in rural environments in Africa are made of "murram" (untarred or nontarmac); hence, dust admixed with toxic metal and metalloid particulates (e.g., Pb and As) as well as other pathogens are freely inhaled, causing asthma and other respiratory disorders.

Mine dusts are products of mining activities, which are formed when rocks are broken by impact, crushing or grinding. The composition of these dusts is determined by the source region, and closely reflects the composition of the soil cover, and are named according to the most abundant particulate mineral composition. Thus we have *silica dust, asbestos dust, coal dust,* and so on (See Section: "How big is the air pollution problem in Africa?," in this Chapter).

Environmental health impacts of soil pollution

Soil pollution is invisible to the human eye, but it compromises the quality of the food we eat, the water we drink, and the air we breathe. The results of scientific research demonstrate that soil pollution affects human health both indirectly through the consumption of contaminated food and drinking water, and directly through exposure to contaminated soil (FAO and UNEP, 2021). Soil pollution affects soil fertility, thus jeopardizing food security.

According to Tindwa and Singh (2023), the effects of pollution of soils on human health in the Sub-Saharan Africa region are inadequately reported, but they range from nonfatal, life-changing effects like skin damage due to acute, invariably fatal incidences of exposure to milt by chronic effects. Oliver and Gregory (2015) summarize six soil-related human health risks, three of which are related to soil pollution: risks from elemental contamination (e.g., As, Cd, and Pb); organic chemical contamination (e.g., PCBs, PAHs, POPs); and pharmaceutical contamination (e.g., estrogen and antibiotics). The three other risks according to Oliver and Gregory (2015) result from exposure to soil pathogens, such as anthrax and prions, micronutrient deficiencies, and under-nutrition due to degraded soils.

Rodriguez-Eugenio et al. (2018) point out two ways in which soil pollution can affect food security, *viz.*, reduction in crop yields, since toxic pollutants degrade soils over the long term; and second, soil pollution can make foods unsuitable for human consumption. Montanarella and Panagos (2021) also reiterated the value of healthy soils in providing nutritious food, clean drinking water, raw materials and carbon sequestration functions—ecosystem services that are essential for guaranteeing food security, tackling climate change, and safeguarding human health.

In the agricultural realm, Tindwa and Singh (2023) noted the significant reduction in the soil's ability to support crop growth and yield, in addition to jeopardizing safety and security of agricultural produce; and have reported a lack of coordination on development, enforcement, and implementation of legal and political instruments to tackle the growing risk of pollution to human health from soil contamination in Africa.

Environmental health impacts of water pollution

Several communities throughout Africa do not have access to safe, clean water for drinking, cooking, and hygiene. Water scarcity, pollution of water resources from both geogenic and anthropogenic sources, as well as poor sanitation practices have resulted in an abundance of illnesses, diseases, and deaths. Waterborne tropical diseases, such as typhoid fever, cholera, dysentery, and diarrheal diseases occur at different times of the year at various parts of the Continent (See Lin et al., 2022). Other conditions, such as plague, typhus, and trachoma (eye infection that can result in blindness), are also common.

As the population continues to grow and contributors like urbanization affect waterbodies throughout the Continent, the treats to water scarcity and pollution become bigger. Nor is this situation entirely unique to the African Continent. For example, Warren-Vega et al. (2023) report the presence of contaminants in the water as being a severe environmental and public health problem in the American Continent, leading to health risks encompassing skin damage, carcinogenic effects, nervous system damage, circulatory system issues, kidney damage, gastrointestinal damage, and impacts on the food chain.

Geogenic sources of water pollution, of which those processes and activities associated with mining are key (e.g., AMD and emission of Hg vapor in artisanal and small-scale Au mining operations), can degrade surface and groundwater supplies, causing water-related illnesses to users. Groundwater withdrawals may damage or destroy streamside habitat many miles from the actual mine site.

Environmental health impacts of saltwater intrusion

Salt in drinking water is generally found in low levels (in the vicinity of around 20 mg/L) and is considered a negligible contributor to daily salt intake. The World Health Organization, therefore, does not have a health-based standard, but an esthetic guideline value of 200 mg/L (WHO, 2017, 2022a). However, increasing salt intake can have substantial negative impacts on human health and well-being (Shammi et al., 2019; Rahaman et al., 2020), including the onset of conditions such as hypertension or high blood pressure in both male and female adults, leading to higher risks of stroke. Pregnant women have been found to be at a high risk of gestational hypertension, (*pre-*) *eclampsia* (See "Glossary of terms," this Chapter) and postpartum infant morbidity and mortality (Shammi et al., 2019; Talukder et al., 2023).

Environmental health impacts of floods

Floods can cause widespread devastation, resulting in loss of life, and damages to personal property and critical public health infrastructure.

The 2023 cholera outbreaks in Africa—a direct impact of flooding that year—were exacerbated by extreme climatic events and conflicts that increased vulnerabilities, as people were forced to flee their homes and grapple with precarious living conditions elsewhere. In the southern parts of the Continent where conditions were most severe, cholera outbreaks were ubiquitous, amid seasonal rains, and tropical storms that caused heavy flooding (WHO, 2023). The cholera response was supported by the WHO through the deployment of several experts to the affected countries, and shipment of 455 tons of critical cholera supplies to Malawi and Mozambique (WHO, 2023). The supplies were also delivered to countries in other parts of the Continent—Burundi, the Democratic Republic of the Congo (DRC), Ghana, Kenya, and Zambia to bolster outbreak preparedness and response.

While cholera is highly virulent and can be deadly, the disease is easy to treat. Most people can be treated successfully through prompt administration of oral rehydration solution or intravenous fluids (WHO, 2023). Sustainable and effective cholera control requires comprehensive measures such as improved detection and response, access to treatment and vaccination, safe water, and sanitation.

Suhr and Steinert's systematic 2022 review of the health outcomes of floods in Sub-Saharan Africa is instructive. According to these authors, most of the case reports on the after-effects of floods point to an increased risk of infection with cholera, scabies, taeniasis, Rhodesian sleeping sickness, malaria, alphaviruses, and flaviviruses.

Fracking and health

There are lots of environmental concerns regarding the impact of fracking on the environment and human health. The fracking procedure relies heavily on water and is one of the major causes of water pollution around localities where fracking is done. The process creates vast amounts of wastewater, which is stored in tanks where it often leaks out into groundwater, polluting it, and making it unfit for human consumption. The process also emits greenhouse gases, such as methane, releases toxic air pollutants, and generates noise. Studies have shown these gas and oil operations can lead to loss of animal and plant habitats, species decline, migratory disruptions and land degradation (See, e.g., Caldwell et al., 2022).

Sometimes, the polluted water will be injected back into the rock, where it causes additional harm. Water pollution can also be caused by underground fracking wells, if these are not properly sealed. Vogel (2017) identify 55% of fracking chemicals that could possibly be carcinogenic. Other fracking chemicals are known to harm the skin or reproductive system (CBD, n.d.). There is mounting evidence that these chemicals—as well as methane released by fracking—are making their way into aquifers through fractures and drinking water (Barth-Naftilan et al., 2018; URMC, 2022).

The release of dangerous petroleum hydrocarbons, including benzene and xylene is thought to occur during fracking, a possibility that is currently being investigated (CBD, n.d.). Fracking also leads to an elevation ground-level ozone levels, raising the risk of asthma, and other respiratory illnesses (CBD, n.d.). Wildlife, too, may also be endangered. Fish die when fracking fluid contaminates streams and rivers (CBD, n.d.).

How big a problem is air pollution in Africa?

Issues of air pollution are widely reported and documented worldwide. In 2022 air pollution was ranked topmost in the list of environmental cause of illness and premature death worldwide (Fuller et al., 2022; World Bank, 2022), with Africa experiencing some of the worst air pollution and some of the most severe health consequences relative to the rest of the world

Table 4–1 Top 10 countries with the highest number of deaths linked to PM$_{2.5}$ across Africa in 2019.

Country	Total number of PM$_{2.5}$-linked deaths
Egypt	90,600 (66,800−116,900)
Nigeria	68,500 (41,500−101,700)
Morocco	27,000 (20,300−34,000)
South Africa	25,800 (19,700−30,000)
Algeria	21,600 (15,300−29,000)
Sudan	16,600 (10,200−24,400)
Ghana	12,500 (8000−17,800)
Democratic Republic of the Congo	11,000 (4700−20,900)
Cameroon	10,200 (6100−14,800)
Ethiopia	9000 (4200−16,200)

Source: From HEI, 2022. The State of Air Quality and Health Impacts in Africa. A Report from the State of Global Air Initiative. Boston, MA: Health Effects Institute. ISSN 2578−6881. https://www. stateofglobalair.org/sites/default/files/documents/2022-10/soga-africa-report.pdf.

(Table 4−1). In 2019 air pollution was the second leading risk factor for death across the Continent after malnutrition (HEI, 2022), accounting for 1.1 million deaths, with the poor, elderly, and young children being the most vulnerable groups (Eze et al., 2024; Fisher et al., 2021; HEI, 2022).

Regular inhalation of fine air pollution particles or aerosols (PM$_{2.5}$) can lead to a wide range of diseases such as ischemic heart disease, chronic obstructive pulmonary disease (COPD), stroke, lung cancer, type 2 diabetes, and neonatal disorders (Wang et al., 2022; World Bank, 2022; GBD 2019 Chronic Respiratory Diseases Collaborators, 2023). Several studies have shown air pollution to be a contributor to brain disorders (e.g., neurodegenerative disease in adulthood and neurological effects in childhood and adolescence), as well as to a reduction of life expectancy (i.e., the number of years that a person might expect to live) (HEI, 2022).

Fisher et al. (2021) reported the loss in economic output in 2019 due to air pollution-related morbidity and mortality in some African countries [\$3 · 02 billion in Ethiopia (1 · 16% of GDP); \$1 · 63 billion in Ghana (0 · 95% of GDP); and \$349 million in Rwanda (1 · 19% of GDP)].

Health effects of geogenic emissions and emanations

From an environmental health perspective, the SDP degrades air quality, posing serious health threats to humans, especially for people with lung conditions. Recent literature documents health issues that include respiratory and cardiovascular diseases and even death in extreme cases (See, e.g., Kotsyfakis et al., 2019).

Explosive volcanic eruptions can emit noxious gases all the way into the stratosphere, the layer of the atmosphere that begins around 10,000 m (33,000 feet) overhead. The gases released by volcanoes can include sulfur dioxide (SO$_2$) (gas), which combines with water in the atmosphere to form minute particles that can go around the globe and stay in the air for years, depending on how large the plume is and how high it goes. The effect of these gases may be direct, such as asphyxiation, respiratory diseases, and skin burns; or indirect, such as regional famine caused by the cooling that results from the presence of sulfate aerosols injected into the stratosphere during explosive eruptions (Edmonds et al., 2015). Volcanic gases may often result in small-scale fatal events in diverse volcanic and geothermal regions. Volcanic ash nanoparticles can reach the upper troposphere and stratosphere, being able to remain in suspension for years affecting global climate and environment. The inhalation of natural nanoparticles released from volcanic activity may also pose a risk to human health (See, e.g., Yacobi et al., 2011; Sonwani et al., 2021; Schiavo et al., 2023).

The burning of fossil fuels releases sulfate particles and sulfur dioxide, which can reflect sunlight and make the atmosphere cooler. According to Kirk (2023), "Unlike volcanic aerosols, air pollution aerosols are produced continuously, but do not travel very high in the atmosphere." Inhalation of the tiny particles emitted during fossil fuel combustion causes asthma, respiratory infections, lung cancer, and heart disease.

Ozone

Ground-level (tropospheric) ozone is a highly reactive pollutant formed in the atmosphere from a chain of chemical reactions involving nitrogen oxides and volatile organic compounds in the presence of sunlight. Ozone exposure can cause adverse respiratory effects that can aggravate lung diseases like asthma, emphysema and chronic bronchitis (COPD), with premature death likely in certain cases (Zhang et al., 2019).

According to HEI (2022), the disease burden from ozone exposure is lower in Africa than the global average. Across Africa, ozone exposures vary from a low of about 32 ppb in Comoros to a high of 54 ppb in Algeria (HEI, 2022). Despite the protective role of stratospheric ozone against ultraviolet irradiation (UVR) [the potential disease burden attributable to solar UVR exposure of Africans is relatively higher (See Lucas et al., 2016)], its long-term exposure has been linked to a variety of health concerns in Africans. Harmful effects on the respiratory and cardiovascular system recorded include the development of COPD (See, e.g., Manisalidis et al., 2020; US EPA, 2023a). Ozone is also a greenhouse gas, contributing to the warming on which it thrives. Ozone threatens the production of major crops on the African continent. In an instructive case study, Borduas-Dedekind et al. (2023) gave the results of research on the missing chemical species, as well as primary emissions information that are needed for developing tropospheric ozone pollution mitigation strategies in the Johannesburg-Pretoria Megacity of South Africa.

Asbestos dust

The general industrial term *"asbestos"* is used to describe six different natural fibrous silicates. Amosite [grunerite, $(Fe, Mg)7Si_8O_{22}(OH)_2$], crocidolite [riebeckite, $Na_2(Fe, Mg)_3Fe_2Si_8O_{22}(OH)_2$], tremolite $[Ca_2Mg_5Si_8O_{22}O_{H2}]$, anthophyllite $[(Mg, Fe)7Si8O_{22}(OH)_2]$ and actinolite [Ca$_2$(Fe, Mg)$_5$ Si$_8$O$_{22}$(OH)$_2$], all belonging to the amphibole group, while chrysotile [(Mg$_6$(Si$_4$O$_{10}$)(OH)$_8$] is a serpentine. These minerals were exploited largely in the past century in industrial applications because of their versatile and unique properties. Asbestos refinement and use is being progressively restricted or banned by several countries, including African countries. However, because of some push back in recognition and control of the risk associated with silica exposure in modern day work practices, such as sandblasting, silicosis appears to be re-emerging around the world (Hoy and Chambers, 2020).

South Africa was the third largest exporter of asbestos in the world for more than a century. As a consequence of particularly exploitative social conditions, former workers and residents of mining regions suffered (and continue to suffer) from a serious, yet still largely, undocumented burden of asbestos-related disease. This epidemic has been invisible both internationally and inside South Africa.

Asbestos exposure has been known to cause diseases of the lungs and pleura, including mesothelioma, lung cancer, asbestosis, and pleural plaques (Markowitz, 2022; Caceres and Venkata, 2023).

In South Africa, "Asbestos Regulations" promulgated in 2002, prescribe the method for working with, and demolishing asbestos containing materials (Braun and Kisting, 2006). Despite the banning of asbestos in South Africa in 2008, as at 2017, there were over a million houses in South Africa with asbestos cement roofs (Rees and Phillips, 2017). Jardin (2023) has cautioned that "... nonuse of asbestos contributes to an increase of harm from fires, armed conflicts, and traffic accidents."

Silica

Silicosis, historically a disease of miners, is caused by a subset of the crystalline silica polymorphs, *viz.,* quartz, tridymite, and cristobalite. Silicosis and other respirable crystalline silica (RCS)-related diseases, most notably tuberculosis, have long been major causes of morbidity and mortality in South Africa (Brouwer and Rees, 2020). Despite the large body of work undertaken since the start of the 1950s on the pathogenicity of silica-related diseases, there is no unanimity among physicians in explaining the mechanism of action of crystalline silica particles at the molecular level (See, e.g., Skuland et al., 2020). Silicon is a component of diatomite (diatomaceous earth), a deposit mined and processed in Algeria, Ethiopia, Kenya, and South Africa for use in filters, as a filler and sometimes as a mild abrasive (Zuluaga-Astudillo et al., 2023). Exposure to dust from the processing and handling of diatomite poses serious danger to humans and can lead to grave and fatal illness, both in those with heavy occupational exposure, and also in people living in the vicinity of industrial installations, or even simply in the same household with an exposed worker. During calcining in particular, highly reactive, free crystalline silica (tridymite, cristobalite) is produced, leading to the onset of silicosis. The illnesses include nodular pulmonary fibrosis (a variant of silicosis), secondary heart disease, lung cancer, and bronchitis.

Other mineral dusts

The term *"pneumoconiosis"* refers to a kind of *lung disease* caused by inhalation of inorganic dusts; and is characterized by the accumulation of dust in the lungs and the tissue reactions to its presence (ILO, 1971). When the identity of the dust can be traced to the handling and processing of coal, the kind of silicosis contracted is referred to as *"coal workers' pneumoconiosis"* (CWP). Other mineral dusts that may cause adverse pulmonary responses, but have not been officially reported as human carcinogens include *kaolinite*, talc (*siderosis*), and iron oxides (*talcosis*) (See Nemery, 2022 and Borm, 2023). The etiology of diseases that result from inhalation of such minerals has been difficult to define, largely because of the complicated nature of epidemiological investigations engendering the coexposure to other minerals, such as quartz and asbestos (Huang et al., 2006; Fubini and Feneglio, 2007).

The intersection of climate change and air pollution

The two phenomena of air pollution and climate change are highly interconnected, because the emissions driving both occurrences largely derive from the same sources, *viz.*, fossil fuel or biofuel combustion (Fuller et al., 2022). Climate change affects air quality, which in turn can lead to adverse health outcomes. Disruptions to weather patterns influence air quality by increasing and distributing air pollutants, such as ground-level ozone, fine particulates, wildfire smoke, and dust. Changes to the weather during seasons also impact the production, distribution, and severity of airborne allergens.

Products of combustion, fine and ultrafine particulates (e.g., $PM_{2.5}$), long-lived greenhouse gases, and short-lived climate pollutants (SLCPs), are simultaneously air pollutants and climate warmers. For example, the formation of haze, viewed as an air pollution event, can be modulated by local weather conditions (Yang et al., 2022). The primary SLCPs are methane, black carbon (*viz.*, soot), and hydrofluorocarbons (IPCC, 2021); with methane emission being one of the main precursors to ground level ozone, and a major source of premature death (See Fuller et al., 2022). Increased air pollution can also alter the production of local and regional allergens, with the altered pattern depending on bioclimatic parameters (Singh and Kumar, 2022). These examples illustrate how adverse climate change scenarios can be directly or indirectly influenced by aerosol concentrations as well as by their optical properties; resulting not only in damage to Earth systems, but also to human health (Yang et al., 2022).

Addressing the issues

Environmental pollution is a complex problem, the solution of which would require a mutisectoral, multidisciplinary approach that empowers policy makers, partners, and other stakeholders to address pollution in a responsible and environmentally sound manner (See, e.g., Tindwa and Singh, 2023). Such an approach would provide different perspectives on the problems, create circumspective research questions, and develop consensus clinical definitions and guidelines for the establishment of comprehensive health services.

The UN Environment Program is collaborating with governments, business, and civil society to resolve pollution issues through the "Implementation Plan Towards a Pollution Free Planet" (UNEP, n.d.). This agenda was tabled at the "Fourth United Nations Environment Assembly," which identified five key action areas for addressing the gaps and challenges associated with pollution: Knowledge, Implementation, Infrastructure, Awareness, and Leadership.

Alternative solutions to tackle environmental pollution are leveraged on three fundamental aspects: (1) the detection of contaminants present in the environment (soil, water, and air) (Flores-Contreras et al., 2022; Banerjee et al., 2023) (2) assessment of risks to public and environmental health due to the presence of contaminants (e.g., Zhang et al., 2023a,b), and (3) a proposal for addressing the issues through: mitigation/remediation (soil); treatment technologies (water) or pollution control measures (air).

Remediation of affected soils

Remediation of polluted soils is essential, and research continues to develop novel, science-based remediation methods. However, a great problem impeding these efforts is the scarcity of official data and monitoring systems. The existing data on soil pollution levels come mostly from scientific research that is limited to small areas, or from nongovernmental agencies or international agency remediation projects that are, likewise, concentrated in specific areas. This sometimes means that governments are unaware of the risks and do not take the necessary prevention and remediation measures.

Risk assessment approaches consist of a series of steps to be taken to identify and evaluate whether natural or human-made substances are responsible for polluting the soil and the extent to which that pollution is posing a risk to the environment and to human health (US EPA, 2023b). Increasingly expensive physical remediation methods, such as chemical inactivation or sequestration in landfills are being replaced by science-based biological methods, such as enhanced microbial degradation or *phytoremediation,* with the latter methods taking sustainability factors into account. "*Sustainable remediation,*" as defined by Nathanail (2011) is: "... remediation that eliminates and/or controls unacceptable risks in a safe and timely manner, and which maximizes the overall environmental, social, and economic benefits of the remediation work." Here, the best available techniques are needed, both during the remediation process and for the entire process, including risk assessment and risk reduction.

Spatial distribution of soil pollution in Africa

Mapping the distribution of soil pollution is important if we are to design effective soil pollution abatement measures. International organizations have been at the forefront in the provision of support to countries in identifying stockpiles of pollutants in the soil that pose a high risk to human health (FAO and UNEP, 2021). "*Pure Earth,*" for example, has produced an inventory of polluted sites including several sites in Africa under the "Toxic Sites Investigation Programme" (See Caravanos et al., 2014; Williams et al., 2021). South Africa and Nigeria have national programs for monitoring soil (Tindwa and Singh, 2023). The monitoring system of Nigeria was specifically set up for spill reports related to the oil industry, while the South Africa system monitors all sources of a specific list of contaminants (FAO and UNEP, 2021). As there is not yet a harmonized soil pollution monitoring system available for the entire African region or for the different subregions of the Continent, a few examples are given in this Section of different approaches that have been used in the region for this lack.

Methods of soil remediation

Soil remediation involves the removal of contaminants from the soil through the application of a range of physical, chemical, and biological processes that can be applied to carry it out.

Many types of soil remediation technologies are being employed in Africa including mechanical methods, evaporation, burying, washing, and dispersion; and some, such as *phytoremediation* (See Odoh et al., 2019; Lee et al., 2021; Sharma et al., 2023) and *the use of artificial intelligence* (AI) (See Gautam et al., 2023), are in the developmental stage. Some of these methods such as electroremediation, soil replacement, and chemical leaching, are typically very expensive but have shown a strong potential in the partial breakdown of pollutants (See, e.g., Aparicio et al., 2022; Yadav and Joshi, 2023). Remediation is commonly done on a site-by-site basis, since for every combination of pollutant, soil property, land use, property and liability regimes, and technical and economic reality of the site or area, a different technique or combination of techniques may be more appropriate (Liu et al., 2018; Huang et al., 2023).

Ruppel and Ginzky (2022) recently advanced a set of recommendations that national governments should implement to regulate soil pollution and limit the accumulation of contaminants beyond established levels in order to guarantee human health and well-being, a healthy environment and safe food.

For soil management to be sustainable, the retention of contaminants at the point of contamination has to be given full consideration so that they can either be attenuated naturally or (in the case of more harmful contaminants) subjected to appropriate remediation techniques to render them harmless (FAO and UNEP, 2021).

Phytoremediation

Phytoremediation is a developing risk-based bioremediation process that uses plants to *stabilize*, transfer, and/or destroy contaminant metals in the soil and groundwater. The process is currently used in a number of African countries, especially the top metal mining countries, such as South Africa and Zambia (See, e.g., Lee et al., 2021).

The mechanisms by which different plants known as *hyperaccumulators* (e.g., *Berkheya coddii*; Fig. 4−2) with a high affinity of the contaminant, remove toxic metals from soil and water include *phytoextraction, rhizofiltration, phytovolatilization,* and *phytostabilization*

FIGURE 4–2 The nickel hyperaccumulator *Berkheya coddii*, native to South Africa (Croesus Projects Ltd., 2017). *From Croesus Projects Ltd., 2017. Could phytomining be coming of age? https://www.archimedesnz.com/single-post/ 2017/04/02/could-phytomining-be-coming-of-age. (Accessed 22 July 2023).*

(Yan et al., 2020; Tan et al., 2023). In some cases, these plants can transport the metals through their vascular tissue to the stems and leaves, which can then be treated or harvested from time to time and disposed of securely (Hemond and Fechner, 2023). *Phytostabilization* is particularly applicable for clean-up of soils contaminanted by noxious chemical elements at waste disposal sites.

The efficiency of the phytoremediation process is affected by several key factors, such as plant species, ambient temperature, and other environmental conditions (Tan et al., 2023). According to these authors (Tan et al., 2023), there are a number of inherent drawbacks in application of bioremediation and phytoremediation in contaminated soils and waters, one of the major ones being the potential of secondary pollution due to the rerelease of absorbed contaminants by plants.

The term *nanophytoremediation* is used to describe a blended approach between *nanotechnology* and *phytotechnology* for the remediation of contaminated habitats and restoring these tainted resources (Mocek-Płóciniak et al., 2023). Nanophytoremediation offers an environmentally friendly approach for cleanup and management of the soil environments without destroying the resource. *Nanomaterials* are characterized by a wide surface area, a large number of active sites, and high-adsorption capacity, making them quite amenable to soil remediation applications (Gul et al., 2022; Pandey et al., 2023; Yasmin et al., 2023). For example, the application of nanoparticles to soil remediation has been found to significantly enhance Pb, Ni, and Cd removal using the selected plant species (Jesitha et al., 2021).

Finally, as Tindwa and Singh noted in 2023, while advancement in technology has yielded a multiplicity of new alternative techniques for pollution control and remediation of affected soils, many of these techniques are currently inaccessible to some African countries. Several countries also lack the mechanisms to effectively coordinate technological developments, as well as enforce and implement the legal and political instruments needed to handle the growing risk of pollution to human health from soil contamination across the region. The lack of data on the status of soil pollution in most African countries also bedevils the countries' capacity to devise and plan policies that can help to allay the problems of soil pollution (Tindwa and Singh., 2023).

The Africa geochemical database program

Comprehensive geochemical maps are an important tool for charting the distribution pattern of potential health issues that may be caused by exposure to high levels of potentially toxic elements (PTEs), or for identifying regions where essential elements insufficiency (or, indeed, excess) in soil may have some bearing on the health of man and other animals (Medical Geography/Medical Geology). Other applications of baseline geochemical datasets include: Assessing the condition and health of soils and sediments for agricultural and ecosystem functions; identifying anomalous values resulting from, for example, mineralization and industrial contamination; providing a marker of the current state of the environment for the measurement and monitoring of future change (See Chapter 1, Part I).

To production of a robust baseline geochemical database for Africa; however, demands the possession of a strong geoanalytical capacity in African laboratories. Addressing the issue of the shortfall in intensity of Applied Geochemistry campaigns to effectively map chemical element distribution in soil and natural waters in Africa would require a concerted effort by respective governments and pertinent geoscientific institutions to train an adequate number of highly skilled applied geochemists, and concomitant massive investment in analytical geochemical infrastructure.

The call can never be louder for the attainment of complete geochemical maps of Africa for all nongaseous elements and other chemical parameters in the surface environment. Such maps should be based on the agreed guidelines (See Darnley et al., 1995), with pertinent modifications that take cognizance of the Continent's unique and complex surface geochemistry. It is suggested that the completion of such maps would, inter alia, constitute a massive contribution through correlation analysis and map overlays, toward improved therapeutic interventions and the broadening of the diagnostic spectrum of a whole array of environmental health problems in Africa.

Such a program has been designed for Africa, but its execution has been tardy due to technical reasons and unavailability of resources. The initiative was conceived during the early 1990s under the International Geological Correlation Program (IGCP) as Project 259: "Global Geochemical Baselines," and is referred to as: "The African Geochemical Database Program (AGDP)" (See Chapter 3, Part I). The surface area of each country is divided into a specific number of cells, and the grid generated referred to as a "Global Reference Network (GRN)" of cells (Fig. 4−3) in each of which, sampling is conducted at predetermined sampling intervals.

A number of African countries (Botswana, Nigeria, South Africa, Zimbabwe, and others) have made significant gains in their large-scale surface geochemical mapping programs, but most are not using the recommended sampling and analytical protocols in the Darnley et al. (1995) document, making integration of datasets problematic. AGDP activities for the period 2017−19 can be viewed at: https://www.earthobservations.org/documents/geo_xiii/GEO-XIII-5-3_2017-19_Work%20Programme.pdf (Accessed 06 August 2023); and for the period 2023−25, at: https://www.earthobservations.org/documents/geoweek2022/GEO-18-7.3_2023-2025%20Work%20Programme.pdf (Accessed 06 August 2023).

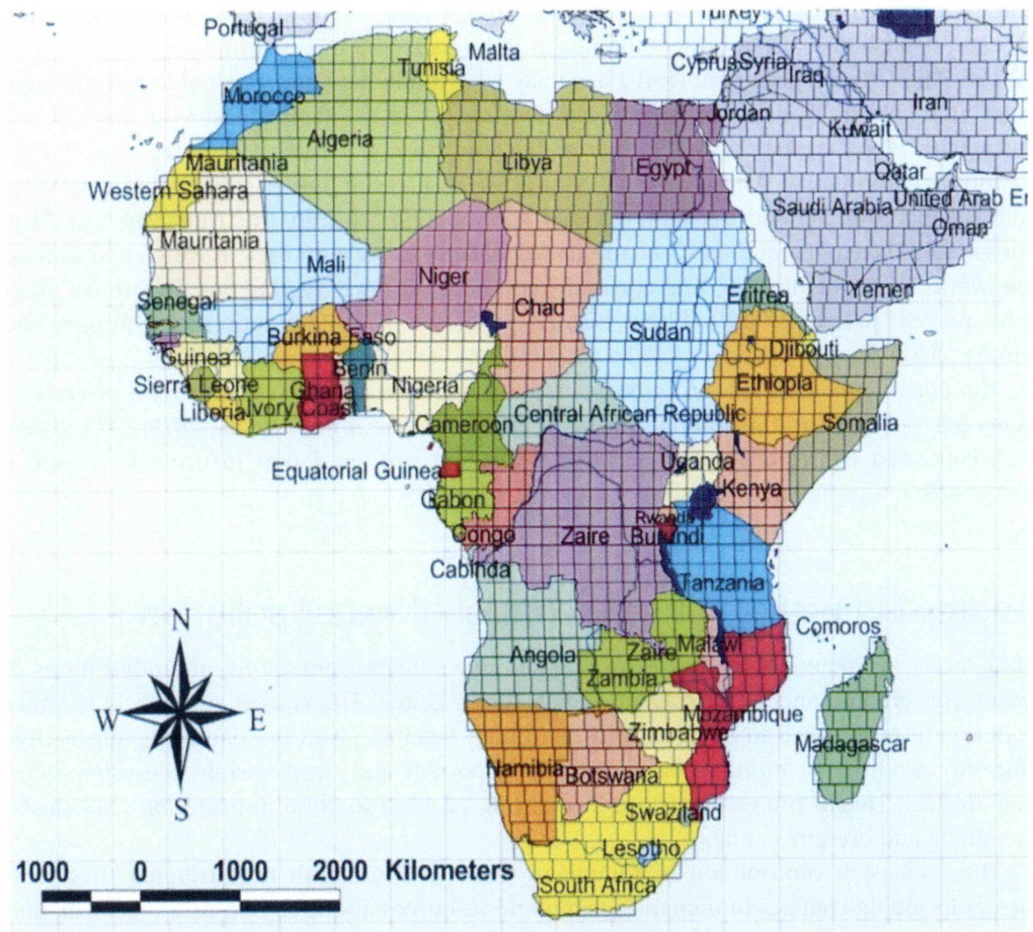

FIGURE 4–3 GRN cells in Africa. *GRN*, Global reference network. *From Darnley, A.G., Björklund, A., Bølviken, B., Gustavsson, N., Koval, P.V., Plant, J.A., et al., 1995. A Global Geochemical Database for Environmental and Resource Management: Final Report of IGCP Project 259. Earth Science Report 19, UNESCO Publishing, Paris. http:// globalgeochemicalbaselines.eu.176–31–41–129. hsservers.gr/datafiles/file/Blue_Book_GGD_IGCP259.pdf.*

Soil protection laws

The need for proper regulation of environmental pollutants in Africa can never be overstated. A number of legislations are being promulgated by governments and institutions around the world, not least in Africa to prevent soil pollution and ensure sustainable management.

Guidelines put forward *via* the FAO's "*Revised World Soil Charter*" (FAO, 2015b) on soil pollution require national governments to promulgate legislations limiting the accumulation of contaminants that are over and above established levels in order to obviate threats to human health and wellbeing; and guarantee a healthy environment and safe food.

The Chapter also implores governments to encourage remediation of contaminated soils that exceed established levels to protect the health of humans and the environment.

The Pan-African Parliament (PAP) has enacted a model law for sustainable soil management (PAP, 2020). Soil pollution was the foremost agenda item at the Fifth Global Soil Partnership Plenary Assembly (Rodríguez-Eugenio, 2021). Earlier (in 2018), the United Nations Environmental Assembly (UNEA-3) adopted a resolution eliciting rapid responses and collaboration to address and manage soil pollution within the framework of "The Sustainable Development Goals." At the national level, many countries in Africa and around the world have adopted or are currently adopting national regulations to protect their soils, prevent pollution, and address historic problems of contamination (See Ruppel and Ginzky, 2022).

The potential of soils for carbon sequestration has been the subject of several projects in many parts of the world including in African countries (See Ambaw et al., 2020; FAO, 2023), with concerted efforts made everywhere at updating soil legislation in order to ensure a sustainable management of soils.

Addressing prevention and control of water pollution

At all levels, awareness of the exposure risks of mine effluents, emissions, and other forms of waste from mining and processing of ores in Africa is low. The causes and effects of water pollution in Africa need to be dealt with in a timely fashion, given the damaging effects they currently produce on human health and the important and irreplaceable ecosystems that they destroy. Health professionals are in a strong position to communicate the risks posed by mining and ore processing.

The increasing contribution of mining to water pollution can be attributed to several factors including changes in demand for mineral resources and fuel, population growth, and climate change. Tackling issues of water pollution and minimizing risk from mining will, just as in the case of agriculture, according to Evans et al. (2019), require great understanding of:

"• Sources and fate of pollutants;
• The changing landscape of pollutants—emerging contaminants;
• The influence of drivers including globalization, trade and climate change;
• Impacts on people and the environment;
• The effectiveness of instruments and measures for pollution reduction and mitigation."

In 2017 Riemann et al. identified gaps in knowledge and understanding of water resource protection processes in South Africa (that can be applied elsewhere on the Continent); and put forward measures for implementation of associated regulations for different water resources. Later (in 2022), Ngeno et al. reviewed the reutilization of waste materials as adsorbents for water and wastewater treatment in Africa, pointed out research gaps, and proposed directions that future research should follow. A number of other research groups

FIGURE 4–4 The Anglo American eMalahleni Water Reclamation Plant. *From https://web.facebook.com/ AngloAmericanZA/photos/a.162541733936916/1325799947611083/?type = 3&theater. (Accessed 17 August 2023).*

(e.g., Baloyi et al., 2023; Nti et al., 2023) are working on the development of various water treatment technologies and bioremediation processes. Some of these water treatment techniques are now "tried and tested."

Improved mine wastewater control and treatment techniques

Despite the gains already noted in the development of water and wastewater treatment techniques in Africa, as recently as 2022, Onu et al. noted that relatively few studies have looked at possible new approaches that would ensure sustainable utilization of wastewater for the achievement of a circular economy in Africa.

It is worth noting that several mining companies in the region are known to be moving their focus away from conventional treatment technologies to advanced methods. In the last few years, *nanofiltration* (a membrane liquid-separation technology) for instance, has emerged as a promising mine wastewater treatment method owing to its inherent attributes such as high water recovery, low-energy consumption, high efficiency, simple operation processes, and the fact that it does not require the use of chemical reagents (See also Doucet et al., 2022).

Some mining companies are beginning to adopt Anglo American's model for transforming wastewater from its mines into drinking water, using *reverse osmosis* membranes to remove impurities from water at the eMalahleni water reclamation plant in Mpumalanga, South Africa. This plant (Fig. 4–4) operates at production efficiency of up to several million liters per day treatment capacity (UNCC, 2012).

Acid mine drainage treatment options

The postclosure decant of AMD is an enormous threat to human and ecosystem health, and this could become worse, if remedial measures are delayed further. The most commonly used treatment options for AMD are well-known and include (1) Lime neutralization;

(2) Calcium silicate neutralization; (3) Ion exchange; (4) Constructed wetland; (5) Precipitation of metal sulfides; (6) Technologies; and (7) Microbes and drug discovery.

Because of the serious public health consequences of AMD, as already highlighted (See Section "Types of water pollution from mining," in this Chapter). It is important for mining companies to be able to accurately predict, prevent, and effectively treat AMD using reliable and modern techniques (See Skousen et al., 2019; Ligate et al., 2021; Jiao et al., 2023). Certain important aspects regarding AMD generation, remediation, quality prediction using conventional, and AI techniques and their limitations are presented by Vadapalli et al. (2020).

Treatment of landfill leachates

The need for developing efficient methods of leachate purification and the use of environmentally friendly management systems can never be overemphasized (See, e.g., Podlasek, 2022). The determination of the leachate generation rate is crucial in the landfill design stage, as this is an essential basis for the correct selection of parameters of drainage systems or retention reservoirs for temporary storage of the leachate.

According to El-Saadony et al. (2023), "… the optimal treatment for leachate is often decided by its characteristics, technical applicability, potential limits, needed effluent limit, cost-effectiveness, regulatory requirements, and long-term ecological effects. ….." This is so, because leachate composition can be very complex and can vary considerably over time. Using *Technology Readiness Levels (TRL)* (See "Glossary of Terms," this Chapter) is essential for determining the recycling rate and net present value to undertake a comprehensive economic feasibility study of solid waste value addition and application (See Peng et al., 2023).

Bentonite (a very highly absorbent, granular clay formed from volcanic ash) is attached to synthetic materials in various ways to form the *"geosynthetic clay liners* (GCL)" system. GCLs represent a relatively new technology (developed in 1986) widely used today as a barrier system in municipal landfills (US EPA, 2001; AL-Soudany et al., 2023). In configurations using a *geomembrane* (impermeable barrier preventing the leakage of contaminants), the clay is affixed using an adhesive. In geotextile configurations, however, adhesives, *"stitchbonding"* (See "Glossary of Terms," this Chapter), *"needlepunching"* (See Glossary of Terms, in this Chapter), or a combination of the three, are used. *Stitchbonding* and *needlepunching* do create small holes in the geotextile; however, these holes are sealed when the installed GCL's clay layer takes up water.

Tailings should be usefully applied

Pollution emanating from *"tailings storage facilities"* (tailings dams, slimes dams) is a serious environmental concern in mining-active countries of Africa, for example, South Africa (See Laker, 2023). The most important forms of pollution from TSFs are from AMD and high levels of PTEs.

Mineral tailings produced by a number of South African mines are largely composed of silicates, oxides and sulfides (Amponsah-Dacosta, 2017; Okereafor et al., 2020). The application of engineering techniques to unravel the potential of recycled tailings in reducing carbon footprint and to evaluate their suitability for mineral carbonation (See, e.g., Amponsah-Dacosta, 2017) and for brick-making (e.g., Malatse and Ndlovu, 2015; Simão et al., 2022) are now subjects of intensive research in South Africa and worldwide.

Some mining companies are now also exploring the advantages of reprocessing of legacy tailings resources to extract minerals, such as uranium and gold (e.g., Singo and Kramers, 2021; Araujo et al., 2022). The technique is becoming attractive, because of the potential it has of maintaining margins and unlocking profit. Tailings reprocessing is also viewed favorably from the standpoint of environmental integrity, for it engenders additional production without increasing a mine's footprint.

Legislative issues of mine water pollution

- The general guidelines for ensuring that the quality of water ensuing from mine sites is not affecting water users downstream are embedded in the "Water Management Policy Documents" of Africa's top mining countries. In South Africa, for instance, various control techniques have been recommended for reducing the potential for water contamination and for minimizing the volume of water requiring treatment (See, e.g., Naidoo et al., 2023). New legislation in South Africa now makes it mandatory for mines to have a water-treatment plan after the mine closes (See, e.g., Grewar, 2019).

What are some of the interventions that African countries can implement to reduce air pollution?

There is a nexus between air pollution, population density, and industrialization. The population of Africa is heterogeneously distributed, with both large, overpopulated areas (*re*: megacities such as Cairo, Lagos, and Kinshasa) and those that are sparsely populated, such as some areas in the Sahel. In regions where there is little industrial development and low population density, air quality is still quite high; whereas in the densely populated and industrialized regions the air quality is low. Although the totality of air pollution production in Africa is generally low by international standards, air pollutants are causing a variety of health and environmental problems that have to be urgently addressed as these continue to be accentuated by the vagaries of climate change.

There is a lack of substantive capacity for air quality measurement and monitoring. Only 17 of the countries of Africa have adopted legislative instruments containing air quality standards. These countries are Algeria, Benin, Burkina Faso, Côte d'Ivoire, Egypt, Eswatini, Gambia, Ghana, Kenya, Mauritius, Morocco, Mozambique, Nigeria, Rwanda, Senegal, South Africa, and the United Republic of Tanzania (HEI, 2022).

Given that many African countries are still in the early stages of urbanization, there is still the chance to put in place the mechanisms for better air quality control, including accelerating the transition to wind and solar energy, and avoiding a reliance on fossil fuel-based economies (Fisher et al., 2021; Okello et al., 2023). Fisher et al. (2021) offer several other recommendations for prevention and control of air pollution in Africa.

Effective measures for improving air quality can also include transportation alternatives. Reducing the number of vehicle miles traveled while increasing carpooling, use of public transportation, and other alternative transportation options can lower the level of emissions of ozone, particulate matter, and other pollutants associated with respiratory and cardiovascular diseases. Reduced electricity use in areas that burn coal or other fossil fuels for energy production can lower exposure to toxic metals, such as Hg and Pb, and nitrogen oxides and sulfur oxides. Measures for improvng Africa's air quality are discussed further by Mead et al. (2023).

The document "Integrated Assessment of Air Pollution and Climate Change for Sustainable Development in Africa," an outcome of the 2022 COP (See "Glossary of Terms," in this Chapter) meeting in Egypt, lists five areas, *viz.*, transport, residential, energy, agriculture, and waste, where African leaders can act quickly to combat climate change and prevent air pollution, to protect human health (CCAC/UNEP/AUC, 2022).

Ground-based, real-time data help to capture fluctuations in air quality, which is important to improve public awareness and to help people alter their behaviors to reduce air pollution and exposure to it. Without high quality pollution data such as accurate measurements of $PM_{2.5}$, health risk assessment or epidemiological studies of air pollution are unfeasible; and respiratory diseases, difficult to track (Makoni, 2020). According to UNICEF (2019), only seven out of 54 African countries have reliable, real-time air pollution monitors.

A limited amount of recent data exists on health impacts of exposure to effluents and emissions from mining and ore processing activities in Africa; the majority of these works coming from southern Africa, and in particular, South Africa (e.g., Mwaanga et al., 2019; Mpanza et al., 2020; Dietler et al., 2021; Myllyvirta and Kelly, 2023). Cojunct actions put in place in the South African mining industry are meant to improve exposure assessment, such as the determination of the silica percentage of the sampled dust of each filter of the respirable dust sampler; and more importantly, to control exposure (See Brouwer and Rees, 2020; Ehrlich et al., 2023). Accuracy and reliability of the RCS measurements, especially in the underground mines, call for improvement.

Policy briefs

Innovative media strategies are needed to enhance awareness of health risks from environmental pollution and to improve public capacity to combat the challenges. One such strategy is through policy briefs. A *policy brief* is a much abridged version of a report that presents the findings and recommendations of a study or research project to a nonspecialized audience. It is a medium for exploring an issue and conveying lessons learnt from the study; or a vehicle for providing policy advice. "*Evidence-based policy briefs*" or "*evidence briefs for policy*" bring together data garnered from a systematic investigation or review or research evidence to inform deliberations (policy dialogs) about health policies and programs.

A policy brief would normally start with a description of a policy problem, then, present a summary of the best available evidence to elucidate the size and nature of the problem, bring out the likely impacts of the main alternatives for addressing the problem, taking cognizance of the potential barriers to implementing the alternatives and procedures for addressing these barriers.

Gaps in knowledge

Like in many other areas of Medical Geology, there are several gaps in our knowledge of the impact of geoenvironmental (soil, water, and air) pollution on the health of man and other animals. Progress in tackling geoenvironmental problems and health in Africa has been slower than anticipated (See Section: "Addressing the Issues," in this Chapter), and interventions continue to be very limited in their scale and impact (WHO, 2022b). According to Münzel et al. (2023), 70% of pollution-related diseases are noncommunicable diseases (NCDs) and more than 60% are cardiovascular diseases, but interventions against pollution are barely mentioned in the Global Action Plan for the Prevention and Control of NCD. Again, too, unlike the developed regions, such as Europe and the United States, where a vast amount of evidence on pollution-induced health effects exists, very limited quantitative data are available for the African Continent (Katoto et al., 2019). There is a dearth of quantitative epidemiological data; and differences in methological approaches for assessing exposure and outcomes underlie the need for more primary studies to fill knowledge and methodological gaps, in a bid to strengthen the current evidence to inform and support African policy makers. Such primary studies, according to Katoto et al. (2019), should preferably have prospective designs and rely on exposure measurements made with "Africa-friendly devices" (i.e., low cost and extended battery autonomy), for both large monitoring programs and comprehensive personalized exposure assessments.

Soil pollution and health: knowledge gaps

Despite a long history of soil pollution research in Africa, there remain notable and important knowledge gaps and great uncertainty regarding the quantity and scope of pollution in affected locations, a situation that is exacerbated by increasing mechanization of mining procedures and the emergence of new contaminants stemming from growth in industrialization processes. Here, examples of knowledge gaps in soil pollution research are given, where specific scientific evidence is needed for the formulation of appropriate policies on soil pollution management and prevention of related diseases in Africa.

- A flurry of research studies are on-going all around the African Continent, providing important insights on the harmful environmental effects of soil contamination. However, as noted by Akande et al. (2018), some of these studies lack detailed characterization of the contaminants present. These authors have gone on to illustrate the drastic reduction in biodiversity as a result of soil contamination by petroleum products in Nigeria, but without identifying the specific petroleum contaminants or their concentrations.

- According to Raimi et al. (2022), in the Global South, most contaminant releases to soil, other than pesticide inputs, remain difficult to quantify; nor are other pertinent attributes of these contaminants readily determinable. A widening gap also exists in our understanding of how soils impacted through *pollution diffusion* behave; and how they influence other environmental compartments (See Raimi et al., 2022).
- Sometime ago now (2015), Tully et al. called for more research focus on the subject of how soil degradation in Africa leads to changes in ecosystem services; as of today, efficient soil management is still unattainable in many regions.
- Because of the myriad of geogenic as well as anthropogenic activities concentrated on *urban soils* and their more complex exposure patterns, these soils (urban soils) are considered to merit special research attention, encompassing novel remediation techniques, and vulnerability-based risk assessment approaches (See WHO, 2013).
- By strengthening of remediation programs, African countries need to intensify efforts on researching ways to reverse the status of already polluted pieces of contaminated land so as to uncover innovative ways of gathering, maintaining, and complementing soil pollution data and actions that inform decision-making (See Tindwa and Singh, 2023).
- The capacity of countries or regions to adapt is largely a function of the level of vulnerability to the health consequences of natural disasters that bring about soil degradation. In this regard, Suhr and Steinert (2022) noted that low-income countries experience more than three-quarters of the global mortality burden caused by natural disasters. The impact of natural disasters on soil quality is an area that warrants intensive research.
- The nature and extent of soil pollution as a result of armed conflicts in Sub-Saharan Africa is largely unknown, despite the region's history of armed conflicts where landmines have widely been used (Demarest and Langer, 2018). Records of quantitative studies on the impact of landmine on the soil environment as either point or diffuse sources of pollution are yet to be documented.
- Few studies so far have analyzed the presence of emerging contaminants in landfill leachate in the African region (See Lindamulla et al., 2022); quantitative studies on the emerging contaminants, their mechanism of dispersion in the leachate system, and migration in surrounding soils, as well as the leachate treatment options in Africa and other tropical regions deserve more research (Lindamulla et al., 2022).
- "*Ventilation* is an efficient approach employed for accelerating stabilization and reducing aftercare of landfill, but its effect on leachate reduction is still elusive." (Jin et al., 2023). A deeper understanding of the leachate reduction mechanism in the microaerobic landfill is desirable. The mechanism of transfer of toxic metals in leachates from landfills into groundwater systems deserves further study, especially in cases where the landfills are not lined with a geomembrane (See, e.g., Nevondo et al., 2019; Makhadi et al., 2020; Amano et al., 2021).
- According to Tindwa and Singh (2023) the multitude of alternative techniques currently available for soil pollution control and remediation of affected soils, are not readily accessible to most Sub-Saharan countries (SSA) countries. These authors have also

pointed out the lack of coordination on development, enforcement and implementation of legal and political instruments to tackle the growing risk of pollution to human health from soil contamination across the SSA region. Coupled with this, is the lack of data on status of soil pollution in most SSA countries, a feature that grossly affects the countries' capacity to devise and plan policies that can help reduce soil pollution (Tindwa and Singh, 2023).

- Radioactivity maps of soils, rocks, and groundwater are urgently needed for African countries, especially the U producers (e.g., Gabon, Namibia, Niger, and South Africa) to be used as a basis for informing the development of policy recommendations regarding groundwater development and housing construction in areas where elevated U concentrations are predicted. In its U mining policy, South Africa, for example, is facing a potential human crisis, as well as an environmental one, because nuclear energy is seen as "clean energy" (See, e.g., Hanto et al., 2022).

- Despite the many possible sources of radon within South Africa (and possibly, other large uranium mining countries of SSA) and potential health hazards, there is a problem of data availability; and hence the need for direct quantification of radon concentrations and exposure.

- According to Tan et al. (2023), multiple drawbacks are inherent in the application of bioremediation and phytoremediation in contaminated soils and waters, one of the major constraints being the potential of secondary pollution due to the re-release of absorbed contaminants by plants. Another key mandate for phytoremediation today, is the *valorization* of phytoremediation biomass to offset remediation costs (See Mocek-Płóciniak et al., 2023).

Technical and intrinsic limitations of hyperaccumulator plants include deficiencies that are inherent in the postphytoremediation treatment, disposal and handling methods. Extensive research is still needed to overcome these drawbacks, as well as amplify the commercial value of phytoremediation process as a whole package.

- The creation of *transgenic plants* (See "Glossary of Terms," in this Chapter) having an outstanding capacity to chelate certain metals and prevent the harmful effects of these metals constitute an emerging subject of research with high potential for commercial application (Sharma et al., 2023). The identification of genes associated with metal tolerance through the use of *genome sequencing* might pave the way for the construction of *transgenics* with favorable characteristics that can be exploited in phytoextraction technology.

Gaps in knowledge: water pollution and health

- In Africa, the contribution of mining to water pollution is of great concern, because current mining practices have an unprecedented impact on water quality. Areas of mine water pollution that are likely to elicit the largest volume of future water pollution research would likely include: "emerging contaminants," "source attribution," and "costs and incentives for adoption of pollution reduction measures" (See Davies, 2017 and Evans et al., 2019).

- Evans et al. (2019) put forward other priorities for future water pollution research, including "identification and testing of locally appropriate markers"; "modeling the effects of aqueous contaminants on biota and pathways of microbial contaminants"; "harmonization of data collection and calculation of economic costs of mine water pollution across countries and projects"; and "how to better share relevant knowledge to incentivize improved mining practices."
- Point-of-use (POU) technologies are gaining in importance as a result of water pollution concerns and affordability. However, there is very little known about the efficiencies and mechanisms of POU technologies for the removal of regulated and emerging disinfection byproducts (DBPs) (See Chen et al., 2021). Chen et al. (2021) provided a summary of trends in performance of four technologies that are commonly in use (*viz.* boiling, adsorption, membrane filtration, and advanced oxidation) on mitigating preformed DBPs, and listed seven areas where research is woefully lacking *viz.* "... (1) data on DBP levels at the tap or cup in domestic applications; (2) certainty regarding the controls of DBPs by heating processes as DBPs may form and transform simultaneously; ...."
- Some of the major applications of micronano bubbles (MNB) technologies in water-treatment processes were reviewed by Temesgen et al. (2017), who also outlined various research areas and gaps in knowledge in these applications. Later on, in 2022, Sakr et al. performed a qualitative and quantitative *meta-analysis* of the effects of MNBs in the removal of target pollutants found in wastewater from different sources. Further work is needed on MNBs characterization techniques to identify aspects, such as bubble size, size distribution, as well as gas mass transfer rates, zeta potential, generation of free radicals, and bubble stability.
- Although the successful removal from wastewater by MNB of many contaminants, including toxic metals, organics, N_2 and total P has been proven by several studies, the removal efficiency of certain other contaminants, such as some emerging organic compounds and toxic metal species that pose great threat to human health remains uncertain (See, e.g., Sakr et al., 2022). Further work is also needed to examine the efficacy of combining physiochemical treatments, such as adsorption, ion exchange, filtration with MNBs, and aeration processes in wastewater treatment.
- According to Ouma et al. (2022), though rapid bioassessment schemes (RBS) have been developed, re., the South Africa SASS-5 method, uncertainties still exist in aquatic ecosystem monitoring due to the disaggregated approaches in bioassessments. A major knowledge gap, according to Ouma et al. (2022), is the exclusion of toxic-metal (TM) pollutant indicators in the RBS despite the TM constituting a significant source of water pollution in southern Africa.
- According to Kenfack et al. (2016), as a result of poor attention given to water quality and sanitation issues in West Africa, waterborne diseases such as cholera and diarrhea abound; and, recently, the Ebola fever, too. Many of the West African countries adopt WHO water quality guidelines, but testing and enforcement capacity is very weak. High standards of water quality assurance and adequate laboratory resources for water quality monitoring are among the essential requirements of the water sector in the region.

- Wagner et al. (2021) systematic review of research on flood risk management in West Africa reveals various dimensions of impacts from residual flood risk aside from material damage, in particular, health impacts, and economic losses. These authors have advocated for more work to be carried out in the area of local risk assessments and understanding of disaster risk in diverse aspects.
- According to Bayabil et al. (2021), as sea level rise continues, it is expected that salinity due to saltwater intrusion will impact human health, soil health, and agricultural production. As such, the significant threats of salinity strengthens the call for more work to be done to better understand the impact of salinity on soil health and associated functional ecosystem processes.
- As Riemann et al. (2017), stated climate change is not fully considered in water resource protection. Nor does the process of water resource classification and determination of reserve cater for climate change. The impacts of climate change on water flow needs to be modeled and to be incorporated into yield and planning models (Riemann et al., 2017). These authors conclude, however, that the implementation of these recommendations is beset by the lack of climate monitoring (rainfall gauges) to record trends of climate change. The cruciality of appropriate water resource management for climate change adaptation and mitigation cannot be overemphasized; so must be considered a priority area of research.

Air pollution and health: knowledge gaps

- In 2019 Katoto et al. called attention to the high levels of ambient air pollution currently affecting large populations in the Sub-Saharan Africa Region; and posited that the situation would get worse as the rate of urbanization rapidly increases. The dearth of data on aspects of air pollution extends to the whole of Africa, making it necessary to carry out primary studies covering the entire Continent (See Fomba et al., 2024).
- In 2021 Fisher et al. quantified air pollution effects on health, human capital, and the economy across Africa, looking particularly at Ethiopia, Ghana, and Rwanda. More studies along these lines, covering other countries in Africa are needed.
- According to Millar et al. (2022), the prevalence of adverse respiratory health outcomes among adolescents living in known air pollution hotspots in Africa is largely unknown; but Statistica (2023), placed *lower respiratory infections* in second place in the list of "The Ten Leading Causes of Death in Africa in 2019." Studies in the Highveld Air Pollution Priority Area (HPA) in South Africa by Millar et al. in 2022 found adolescents to be adversely affected by air pollution; in particular O_3 and PM, and reinstated the essentiality of working toward meeting the National Ambient Air Quality Standards (NAAQS) in this area to protect adolescent health. Future research should investigate long-term exposure and health outcomes among adolescents living in the HPA given the findings regarding risk factor outlined in the study by Millar et al. (2022).

- According to Stewart et al. (2022), much has been learnt in the last two decades, or so, in mitigating the effects of volcanic air pollution from both explosive and effusive events. However, there have been very few studies on the health effects of many of the major, and sometimes long-duration, eruptions near large African populations [e.g., Mount Nyiragongo, DRC, the most destructive effusive eruption in modern history (See, e.g., Allard et al., 2003; Michellier et al., 2020; Dhillon and Sasidharan, 2021; Nicole, 2022)]. Some eruptions have gone completely unmonitored. The question about whether chronic exposures to emissions and effluents from volcanoes can trigger development of potentially fatal diseases, remains to be fully elucidated (Stewart et al., 2022).
- Exposure to geogenic dust and the health consequences are a major problem in African countries rich in silicate ores and coal. RCS and respirable coal mine dust in Africa's underground and surface coal mines are an inherent byproduct of mining activities, and are of notable health and safety concerns. Cumulative inhalation of respirable mine dust can lead to obstructive lung diseases. According to Rahimi et al. (2023), despite the massive efforts to reduce dust exposure by decreasing the permissible exposure limits and improving monitoring techniques, the rate of mine workers with respiratory diseases in mines is still high.
- From an environmental health perspective, the SDP degrades air quality, posing serious health threats to humans, especially to people with lung conditions (Olarewaju et al., 2023). Despite this knowledge, large uncertainties exist in our understanding of the impact of the contained atmospheric particulates in SDP on health, ecosystems, and human activities. Such knowledge is vital for us to be able to formulate credible mitigation measures and unveil areas where further research is needed for improving on these formulations (Olarewaju et al., 2023).
- Although CWP has been known as an occupational disease for several centuries, we still do not fully understand the immunopathological mechanisms that underpin disease pathogenesis as a result of exposure to different PM forms generated during coal-mining operations (See Vanka et al., 2022).
- The work of Katoto et al. (2019) reveals the dearth of data on air quality monitoring in Sub-Saharan Africa, where cities are known to exhibit higher ambient air pollution levels than those given in the "WHO Guidelines." Most of the reported health studies of air quality in SSA, according to Katoto et al. (2019) were cross-sectional, and were done through the administration of questionnaires. This study reiterated the need for more Africa-specific longitudinal studies of ambient air pollution.
- In 2020 Makoni noted that the ability of satellite-based remote sensing to predict air quality in different regions can be harnessed to address the gaps in air quality monitoring of African cities without ground-based sensors; especially for tracking long-distance air pollutant transport. However, satellite-based remote sensing does not replace ground-based monitors, and is often highly dependent on reference grade ground-level measurements of air pollution to improve the accuracy of the estimates (UNICEF, 2019).

Further research is needed to investigate the efficacy of the current mass-concentration-based monitoring systems.

- South Africa was the world's largest *asbestos* producers in the 1970s. It was banned in 2008, but a relatively recent count shows that there are over a million houses with asbestos cement roofs in the Country (Rees and Phillips, 2017). Although asbestos has been banned in many other African countries, there are still lingering hazards from the millions of tons of asbestos currently in place (e.g., mine tailings) and the many manufactured goods (e.g., brakes, including brake linings, brake pads, and brake shoes) that currently contain asbestos.
- Much is known about the ability of asbestos to cause disease (e.g., mesothelioma); but the carcinogenic mechanism is not yet fully understood. Outstanding legitimate scientific questions, such as potential differential toxicity and carcinogenicity of different fiber types, remain to be answered. Illegitimate issues include the supposed "safe" use of asbestos, and the *"chrysotile hypothesis"* (carcinogenic potency). In 2006 Braun and Kisting suggested a way of addressing these issues, *viz.*, ". . . recast our understanding of disease causality by integrating historical knowledge about the social and political conditions that produce disease with biological understandings of the pathogenicity of asbestos fibers." [Sic.]
- The first pedestal of the "Climate Change-Related Data Gaps" comprises seven recommendations that seek to provide stakeholders with vital statistics that will form the basis for developing appropriate policy responses to help mitigate and adapt to the changing climate (See IMF, 2023).
- To step up cuts to carbon emissions, robust data on the spatiotemporal dynamics of carbon emissions are vital since policies must be based on a thorough understanding of the broad impacts of climate change, the green transition, and the associated risks.
- Detailed statistics on spaciotemporal dynamics of carbon dynamics are needed by policy makers to monitor the path of the energy transition as well as to assist them in formulating effective mitigation strategies that can deliver the fastest and least disruptive pathway toward net zero emissions (See Jiying et al., 2023).
- In 2023 Feng et al. pointed out the existence of a number of scientific knowledge gaps in the estimation of carbon emissions that are anathema to the effective formulation of climate policies. To formulate an effective national climate policy, Feng et al. (2023) reiterated the need to fill in the many gaps in our understanding of the impact of carbon emissions and sinks, and perform accurate computation of the amount of anthropogenic CO_2 released into the atmosphere and the amount of CO_2 sunk annually.
- In addition to their impacts on climate, atmospheric aerosols have tremendous impacts on health. Better understanding of aerosol dynamics—homogeneous and heterogeneous nucleation condensation, coagulation, and deposition loss—and the effects has on atmosphere dynamics is crucial in our effort to reduce air pollution, especially in West Africa, a region where populations are already struggling to adapt to climate change (See Al-Zu'bi et al., 2022).

- According to Al-Zu'bi et al. (2022), changes in hydroclimatic variables such as rainfall in response to greenhouse gas emissions have been detected with a high degree of confidence in many parts of the world, but the same cannot be said for Africa. Africa lacks a well-developed observation network, a problem exacerbated by a huge and constant decline in the number of observation stations fully meeting the standards set by the World Meteorological Organization (WMO). This calls for urgent actions for maintaining and densifying existing networks in addition to using new observation technologies offered by remote sensing.
- Another issue is the limited performance of climate models in reproducing observed hydroclimatic extremes, which has been widely demonstrated, even in data-rich areas (Al-Zu'bi et al., 2022). Improving both short-term forecasting and medium- to long-term future projections of hydroclimatic extremes is one of the greatest challenges facing researchers in this field in Africa.
- Research is also needed to better understand climate impacts on indoor air quality, including temperature, humidity, ventilation, fungi, mold, biological contaminants, and volatile organic compounds.
- In 2022 Shindell et al. listed a number of outcomes that would accrue from our effort in
- achieving climate change mitigation, and hence meeting some of the sustainable development goals; among which are the substantial reduction in the number of premature deaths due to air pollution.
- Results of the diagnosis by Ryan and Bustos (2019) of the main knowledge gaps facing climate adaptation policy in the Latin American countries can serve as input for a research and action agenda aiming to strength the interaction between science and policy on climate adaptation in African countries as well.
- The methods and chemicals used to frack are dependent on location. South Africa has about $0.4 \times 109 \text{ m}^3$ (13 Tcf) in technically recoverable, but not proven, reserves of natural gas in the Karoo Basin (de Kock et al., 2017). There is a paucity of information about the possible effects and risks of fracking in South Africa, and whether there are sufficient reserves that can be exploited. There are significant uncertainties about adverse health outcomes that may be associated with high-volume hydraulic fracturing, and that should give us cause for concern and open the doors for intensive research (See, e.g., Finkel and Hays, 2016; Black et al., 2021; Deziel et al., 2022).
- On the need for integrated effort to mitigate fracking while keeping an eye on environmental integrity, Deziel et al. (2022) recommend that all stakeholders should take a more integrated approach that embraces both public health and conservation interests, focusing more on the underprivileged, under-represented, historically disadvantaged, or poorly understood communities—initiatives that illustrate how a wide range of collaborations (multidisciplinary) can work.
- According to Fdez-Arroyabe et al. (2022), electrical charge becomes a determining factor in the *Brownian motion* of atmospheric nanoparticles (6−150 nm), and this may have relevant implications for human health. Future work should aim to confirm that nanoparticle electrical charge is dependent on size in different environments and geographic locations.

- Formal air quality standards are an essential component of policy actions to reduce air pollution and improve health. However, most countries in Africa lack national air quality standards.
- Evaluating pollution, air-borne allergens, aerosolized pathogens, and dust burdens through air quality monitoring will improve our ability to model future health impacts. Again too, the collection of data on climate-dependent variables, such as wildfires and land-use, will improve understanding of their impacts on health.
- Air quality impacts depend on complex interplay of the effects of temperature, weather variability, long-term climate change, and environmental exposures. According to US NIH (2022), only the use of transdisciplinary approaches would be capable of yielding tangible outcomes in this field of research.

Glossary of terms

- **Brownian motion** is the random movement of microscopic particles in a fluid due to their collisions with other fast-moving atoms or molecules in the surrounding medium.
- **Contaminant** is the substance, material, or agent that is unwanted in the environment.
- **COP**: Refers to "Conference of the Parties," which is the supreme decision-making body of the United Nations Framework Convention on Climate Change.
- **Eclampsia** refers to the onset of seizures (convulsions) in a woman with *preeclampsia,* which is itself is a hypertensive disorder of pregnancy.
- **Food security** is defined as the availability, access, utilization, and stability of food supply.
- **Green waste landfill** is a waste disposal site where organic waste is taken to naturally decompose and turn into compost.
- **Harm**: Physical injury or damage to the health of people or damage to property or the environment.
- **Hazard**: Potential source of harm.
- **Needle punching** is a textile manufacturing technique that involves interlocking fibers through the repeated insertion of barbed needles.
- **Persistent organic pollutant (POP)**: Synthesized carbon-based compounds from agrochemicals and industrial products that generally biodegrade very poorly and most of which will bioaccumulate in tissues of organisms.
- **Pollutant**: Contaminant present in the environment or which might enter the environment which, due to its properties or amount or concentration, causes harm.
- **Pathway**: The means by which harm is transmitted from the source to the receptor
- **Risk**: Combination of the probability of occurrence of harm and the severity of that harm.
- **Soil ecosystem services**: The capacity of natural processes and components to provide goods and services that satisfy human needs, directly or indirectly.
- **SuDS** are drainage solutions that work by slowing and holding back the water that runs off from a site, allowing natural processes to break down pollutants.
- **Stitchbonding** refers to a nonwoven construction where the fabric is formed by stitching or knitting the fibers to form a fabric with the appearance of a knit fabric.

- ***TRLs*** are a method for understanding the technical maturity of a technology during the acquisition phase of a program.
- ***Transgenic plants*** are those plants whose DNA has been modified using genetic engineering techniques, producing a new trait to the plant, which does not occur naturally in the species.

References

Abubakar, I.R., Maniruzzaman, K.M., Dano, U.L., AlShihri, F.S., AlShammari, M.S., Ahmed, S.M.S., et al., 2022. Environmental sustainability impacts of solid waste management practices in the Global South. International Journal of Environmental Research and Public Health 19 (19), 12717. Available from: https://doi.org/10.3390/ijerph191912717.

Akande, O., Ogunkunle, C., Ajayi, S., 2018. Contamination from petroleum products: impact on soil seed banks around an Oil Storage Facility in Ibadan, South-West Nigeria. Pollution 4 (3), 515−525. Available from: https://doi.org/10.22059/poll.2018.249913.375.

Allard, P., Baxter, P., Halbwachs, M., Kasareka, M., Komorowski, J.C., Joron, J.L., 2003. The most destructive effusive eruption in modern history: Nyiragongo (RD. Congo). EGS - AGU - EUG Joint Assembly, Abstracts from the meeting held in Nice, France, 6−11 April 2003, abstract id. 11970. https://ui.adsabs.harvard.edu/abs/2003 EAEJA. ...11970A/abstract. (Accessed 10 August 2023).

AL-Soudany, K.Y.H., Fattah, M.Y., Rahil, F.H., 2023. A comprehensive review of new trends in the use of geo-synthetics clay liner GCLs in landfill. International Journal of Intelligent Systems and Applications in Engineering 11 (5S), 124−130.

Al-Zu'bi, M., Dejene, S.W., Hounkpè, J., Kupika, O.L., Lwasa, S., Mbenge, M., et al., 2022. African perspectives on climate change research. Nature Climate Change 12, 1078−1084. Available from: https://doi.org/10.1038/s41558-022-01519-x.

Amano, K.O.A., Danso-Boateng, E., Adom, E., Kwame Nkansah, D., Amoamah, E.S., Appiah-Danquah, E., 2021. Effect of waste landfill site on surface and ground water drinking quality. Water Environment Journal 35, 715−729. Available from: https://doi.org/10.1111/wej.12664.

Ambaw, G., Recha, J.W., Nigussie, A., Solomon, D., Radeny, M., 2020. Soil carbon sequestration potential of climate-smart villages in East African countries. Climate 8, 124. Available from: https://doi.org/10.3390/cli8110124.

Amponsah-Dacosta, M., 2017. Mineralogical characterization of South African mine tailings with aim of evaluating their potential for the purposes of mineral carbonation (M.Sc. thesis). University of Cape Town. https://open.uct.ac.za/bitstream/item/28424/thesis_sci_2017_amponsah_dacosta_maxwell.pdf?sequence = 1. (Accessed 19 August 2023).

Aparicio, J.D., Raimondo, E.E., Saez, J.M., Costa-Gutierrez, S.B., Álvarez, A., Benimeli, C.S., et al., 2022. The current approach to soil remediation: a review of physicochemical and biological technologies, and the potential of their strategic combination. Journal of Environmental Chemical Engineering 10 (2), 107141. Available from: https://doi.org/10.1016/j.jece.2022.107141.

Araujo, F.S.M., Taborda-Llano, I., Nunes, E.B., Santos, R.M., 2022. Recycling and reuse of mine tailings: a review of advancements and their implications. Geosciences 12, 319. Available from: https://doi.org/10.3390/geosciences12090319. (Accessed 31.07.2023).

Baloyi, J., Ramdhani, N., Mbhele, R., Ramutshatsha-Makhwedzha, D., 2023. Recent progress on acid mine drainage technological trends in South Africa: Prevention, treatment, and resource recovery. Water 2023 15, 3453. Available from: https://doi.org/10.3390/w15193453 (Accessed 25 February 2024).

Banerjee, A., Singh, S., Ghosh, A., 2023. Detection and removal of emerging contaminants from water bodies: a statistical approach. Frontiers in Analytical Science 3, 1115540. Available from: https://doi.org/10.3389/frans.2023.1115540.

Barth-Naftilan, E., Sohng, J., Saiers, J.E., 2018. Methane in groundwater before, during, and after hydraulic fracturing of the Marcellus Shale. Proc. Natl. Acad. Sci. U. S. A. 115 (27), 6970−6975. Available from: https://doi.org/10.1073/pnas.1720898115.

Bayabil, H.K., Li, Y., Tong, Z., Gao, B., 2021. Potential management practices of saltwater intrusion impacts on soil health and water quality: a review. Journal of Water and Climate Change 12 (5), 1327−1343. Available from: https://doi.org/10.2166/wcc.2020.013.

Black, K.J., Boslett, A., Hill, E., Ma, L., McCoy, S., 2021. Economic, environmental and health impacts of the fracking boom. Annual Review of Resource Economics 13 (1), 311−334.

Bokar, H., Traoré, A.Z., Mariko, A., Diallo, T., Traoré, A., Sy, A., et al., 2020. Geogenic influence and impact of mining activities on water soil and plants in surrounding areas of Morila Mine, Mali. Journal of Geochemical Exploration 209, 106429. Available from: https://doi.org/10.1016/j.gexplo.2019.106429.

Borduas-Dedekind, N., Naidoo, M., Zhu, B., Geddes, J., Garland, R.M., 2023. Tropospheric ozone (O_3) pollution in Johannesburg, South Africa: exceedances, diurnal cycles, seasonality, O_x chemistry and O_3 production rates. Clean Air Journal 33 (1). Available from: https://doi.org/10.17159/caj/2023/33/1.15367.

Borm, P.J.A., 2023. Talc inhalation in rats and humans: a review and appraisal of available evidence. Journal of Occupational and Environmental Medicine 65 (2), 152−159. Available from: https://doi.org/10.1097/JOM.0000000000002702.

Braun, L., Kisting, S., 2006. Asbestos-related disease in South Africa: the social production of an invisible epidemic. American Journal of Public Health 96 (8), 1386−1396. Available from: https://doi.org/10.2105/AJPH.2005.064998.

Broomandi, P., Guney, M., Kim, J.R., Karaca, F., 2020. Soil contamination in areas impacted by military activities: a critical review. Sustainability 12, 9002. Available from: https://doi.org/10.3390/su12219002.

Brouwer, D.H., Rees, D., 2020. Can the South African milestones for reducing exposure to respirable crystalline silica and silicosis be achieved and reliably monitored? Frontiers in Public Health 8, 107. Available from: https://www.frontiersin.org/articles/10.3389/fpubh.2020.00107.

Caceres, J.D., Venkata, A.N., 2023. Asbestos-associated pulmonary disease. Current Opinion in Pulmonary Medicine 29 (2), 76−82. Available from: https://doi.org/10.1097/MCP.0000000000000939.

Caldwell, J.A., Williams, C.K., Brittingham, M.C., Maier, T.J., 2022. A consideration of wildlife in the benefit-costs of hydraulic fracturing: expanding to an E3 analysis. Sustainability 14, 4811. Available from: https://doi.org/10.3390/su14084811.

Caravanos, J., Gualtero, S., Dowling, R., Ericson, B., Keith, J., Hanrahan, D., et al., 2014. Simplified risk-ranking system for prioritizing toxic pollution sites in low- and middle-income countries. Annals of Global Health 80 (4), 278−285. Available from: https://doi.org/10.1016/j.aogh.2014.09.001.

Carrillo, K.C., Rodríguez-Romero, A., Tovar-Sánchez, Ruiz-Gutiérrez, G., Viguri Fuente, J.R., 2022. Geochemical baseline establishment, contamination level and ecological risk assessment of metals and As in the Limoncocha lagoon sediments, Ecuadorian Amazon region. Journal of Soils and Sediments 22, 293−315. Available from: https://doi.org/10.1007/s11368-021-03084-w.

CBD (Centre for Biological Diversity), n.d. Fracking. https://www.biologicaldiversity.org/campaigns/fracking/#:~:text = About%2025%20percent%20of%20fracking.into%20aquifers%20and%20drinking%20water. (Accessed 26 June 2020).

CCAC/UNEP/AUC (Climate and Clean Air Coalition/the United Nations Environment Programme/African Union Commission), 2022. Integrated assessment of air pollution and climate change for sustainable development in Africa: Summary for decision makers. https://www.ccacoalition.org/sites/default/files/resources//AFRICAN_ASSESMENT_SDM.pdf. (Accessed 08 August 2023).

Certini, G., Scalenghe, R., Woods, W.I., 2013. The impact of warfare on the soil environment. Earth Science Reviews 127, 1−15. Available from: https://doi.org/10.1016/j.earscirev.2013.08.009.

Chapman, P.M., 2007. Determining when contamination is pollution - weight of evidence determinations for sediments and effluents. Environment International 33 (4), 492−501. Available from: https://doi.org/10.1016/j.envint.2006.09.001.

Chen, B., Jiang, J., Yang, X., Zhang, X., Westerhoff, P., 2021. Roles and knowledge gaps of point-of-use technologies for mitigating health risks from disinfection byproducts in tap water: a critical review. Water Research 200, 117265. Available from: https://doi.org/10.1016/j.watres.2021.117265.

Chibuike, G.U., Obiora, S.C., 2014. Heavy metal polluted soils: effect on plants and bioremediation methods. Applied and Environmental Soil Science 2014, 752708. Available from: https://doi.org/10.1155/2014/752708.

Croesus Projects Ltd., 2017. Could phytomining be coming of age? https://www.archimedesnz.com/single-post/2017/04/02/could-phytomining-be-coming-of-age. (Accessed 22 July 2023).

Darnley, A.G., Björklund, A., Bølviken, B., Gustavsson, N., Koval, P.V., Plant, J.A., et al., 1995. A Global Geochemical Database for Environmental and Resource Management: Final Report of IGCP Project 259. Earth Science Report 19. UNESCO Publishing, Paris. Available from: http://globalgeochemicalbaselines.eu.176-31-41-129.hsservers.gr/datafiles/file/Blue_Book_GGD_IGCP259.pdf.

Davies, T.C., 2008. Environmental health impacts of East African rift volcanism. Environmental Geochemistry and Health 30 (4), 325−338. Available from: https://doi.org/10.1007/s10653-008-9168-7.

Davies, T.C., 2017. Environmental health impacts of mining in Africa. Academy of Science of South Africa (ASSAf). https://www.assaf.org.za/files/Science%20BUsiness%20Dialogue%202017/T.C.%20Davies%20ASSAF%20%20PRESENTATION.pdf. (Accessed 18 August 2023).

de Kock, M.O., Beukes, N.J., Adeniyi, E.O., Cole, D., Götz, A.E., Geel, C., et al., 2017. Deflating the shale gas potential of South Africa's Main Karoo Basin. South African Journal of Science 113 (9/10), 12. Available from: https://doi.org/10.17159/sajs.2017/20160331.

Demarest, L., Langer, A., 2018. The study of violence and social unrest in Africa: a comparative analysis of three conflict event datasets. African Affairs 117 (467), 310−325. Available from: https://doi.org/10.1093/afraf/ady003.

Deziel, N.C., Shamasunder, B., Pejchar, L., 2022. Synergies and trade-offs in reducing impacts of unconventional oil and gas development on wildlife and human health. Bioscience 72 (5), 472−480. Available from: https://doi.org/10.1093/biosci/biac014.

Dhillon, H.S., Sasidharan, S., 2021. Psychological impact of a volcano eruption - Mount Nyiragongo. Commentry: Egyptian Journal of Psychiatry 42 (3), 176−179.

Dietler, D., Loss, G., Farnham, A., de Hoogh, K., Fink, G., Utzinger, J., et al., 2021. Housing conditions and respiratory health in children in mining communities: an analysis of data from 27 countries in sub-Saharan Africa. Environmental Impact Assessment Review 89, 106591. Available from: https://doi.org/10.1016/J.EIAR.2021.106591.

Doucet, F.J., Dube, G., Mohamed, S., Gcasamba, S., Coetzee, H., Vadapalli, V.R.K., 2022. Chapter 4: Application of nanofiltration in mine-influenced water treatment: a review with a focus on South Africa. In: Fosso-Kankeu, E., Mkandawire, M., Mamba, B. (Eds.), Application of Nanotechnology in Mining Processes. https://doi.org/10.1002/9781119865360.ch4. (Accessed 06 August 2023).

Dror, I., Yaron, B., Berkowitz, B., 2023. The human impact on all soil-forming factors during the Anthropocene. ACS Environmental Au 2 (1), 11−19. Available from: https://doi.org/10.1021/acsenvironau.1c00010.

Durand, J.F., 2012. The impact of gold mining on the Witwatersrand on the rivers and karst system of Gauteng and North West Province, South Africa. Journal of African Earth Sciences 68, 24−43. Available from: https://doi.org/10.1016/j.jafrearsci.2012.03.013.

Edmonds, M., Grattan, J., Michnowicz, S., 2015. Volcanic gases: silent killers. In: Fearnley, C.J., Bird, D.K., Haynes, K., McGuire, W.J., Jolly, G. (Eds.), Observing the Volcano World. Advances in Volcanology. Springer, Cham, pp. 65−83. Available from: https://doi.org/10.1007/11157_2015_14.

Ehrlich, R., Barker, S., Montgomery, A., Lewis, P., Kistnasamy, B., Yassi, A., 2023. Mining migrant worker recruitment policy and the production of a silicosis epidemic in Late 20th-Century Southern Africa. Annals of Global Health 89 (1), 25. Available from: https://doi.org/10.5334/aogh.4059.

El-Saadony, M.T., Saad, A.M., El-Wafai, N.A., Abou-Aly, H.E., Salem, H.M., Soliman, S.M., et al., 2023. Hazardous wastes and management strategies of landfill leachates: a comprehensive review. Environmental Technology and Innovation 31, 103150. Available from: https://doi.org/10.1016/j.eti.2023.103150.

Evans, A.E.V., Mateo-Sagasta, J., Qadir, M., Boelee, E., Ippolito, A., 2019. Agricultural water pollution: key knowledge gaps and research needs. Current Opinion in Environmental Sustainability 36, 20–27. Available from: https://doi.org/10.1016/j.cosust.2018.10.003.

Eze, J.N., Vanker, A., Ozoh, O.B., 2024. Air pollution exposure among African school children in different microenvironments. The Lancet. Child and Adolescent Health 8 (1), 2–3. Available from: https://doi.org/10.1016/S2352-4642(23)00294-8 (Accessed 25 February 2024).

Fomba, K.W., Fankam, B.T., Mellouki, A., Westervelt, D.M., Giordano, M.R. (Eds.), 2024. Advances in Air Quality Research in Africa: Proceedings of the First International Conference on Air Quality in Africa (ICAQ'Africa 2022) (Advances in Science, Technology and Innovation). First Edition. Springer Nature Switzerland. 188 pages.

FAO (Food and Agricultural Organisation), 2015a. Status of the World's Soil Resources. Rome, Italy. https://www.fao.org/documents/card/en/c/c6814873-efc3-41db-b7d3-2081a10ede50/#:~:text = %23650%20p.&text = The%20SWSR%20is%20a%20reference.expert%20knowledge%20and%20project%20outputs. (Accessed 29 July 2023).

FAO (United Nations Food and Agricultural Organisation), 2015b. Revised World Soil Charter, Rome, Italy. https://www.fao.org/3/i4965e/i4965e.pdf. (Accessed 05 August 2023).

FAO (United Nations Food and Agricultural Organisation), 2023. FAO Soils Portal. Global Soil Sequestration Potential (GSOCseq) Map. https://www.fao.org/soils-portal/data-hub/soil-maps-and-databases/global-soil-organic-carbon-sequestration-potential-map-gsocseq/en/. (Accessed 06 August 2023).

FAO and ITPS (Food and Agriculture Organization of the United Nations and Intergovernmental Technical Panel on Soils), 2015. Status of the World's Soil Resources (SWSR) - Main Report. Rome, Italy. Food and Agriculture Organisation of the United Nations and Intergovernmental Technical Panel on Soils. http://www.fao.org/3/a-i5199e.pdf. (Accessed 16 August 2023).

FAO and UNEP (United Nations Food and Agricultural Organisation and United Nations Environment Programme), 2021. Global Assessment of Soil Pollution: Report, Rome. Available from: https://doi.org/10.4060/cb4894en. (Accessed 21 June 2023).

Faragó, T., Špirová, V., Blažeková, P., Lalinská-Voleková, B., Macek, J., Jurkovič, Ľ., et al., 2023. Environmental and health impacts assessment of long-term naturally-weathered municipal solid waste incineration ashes deposited in soil-old burden in Bratislava city, Slovakia. Heliyon 9 (3), e13605. Available from: https://doi.org/10.1016/j.heliyon.2023.e13605.

Fayiga, A.O., Ipinmoroti, M.O., Chirenje, T., 2018. Environmental pollution in Africa. Environmental Development and Sustainability 20, 41–73. Available from: https://doi.org/10.1007/s10668-016-9894-4.

Fdez-Arroyabe, P., Salcines, C., Kassomenos, P., Santurtún, A., Petäjä, T., 2022. Electric charge of atmospheric nanoparticles and its potential implications with human health. Science of the Total Environment 808, 152106. Available from: https://doi.org/10.1016/j.scitotenv.2021.152106.

Feng, R., Hu, L., Hu, X., Fang, X., 2023. Knowledge gaps are making it harder to formulate national climate policies. Proceedings of the National Academy of Sciences of the USA 120 (23). Available from: https://doi.org/10.1073/pnas.2218563120.

Finkel, M.L., Hays, J., 2016. Environmental and health impacts of 'fracking': Why epidemiological studies are necessary. Journal of Epidemiology and Community Health 70 (3), 221–222. Available from: https://doi.org/10.1136/jech-2015-205487.

Fisher, S., Bellinger, D.C., Cropper, M.L., Kumar, P., Binagwaho, A., Koudenoukpo, J.B., et al., 2021. Air pollution and development in Africa: Impacts on health, the economy, and human capital. The Lancet Planetary Health 5 (10), e681–e688. Available from: https://doi.org/10.1016/S2542-5196(21)00201-1 (Accessed 25 February 2024).

Flores-Contreras, E.A., González-González, R.B., González-González, E., Melchor-Martínez, E.M., Parra-Saldívar, R., Iqbal, H.M.N., 2022. Detection of emerging pollutants using aptamer-based biosensors: recent advances, challenges, and outlook. Biosensors 12 (12), 1078. Available from: https://doi.org/10.3390/bios12121078.

Fubini, B., Feneglio, I., 2007. The toxic potential of mineral dust. Elements 3, 407–414.

Fuller, R., Landrigan, P.J., Balakrishnan, K., Bathan, G., Bose-O'Reilly, S., Brauer, M., et al., 2022. Pollution and health: a progress update. Lancet Planet Health 6 (6), e535−e547. Available from: https://doi.org/10.1016/S2542-5196(22)00090-0. Erratum in: Lancet Planet Health. 2022 Jun 14, https://www.thelancet.com/journals/lanplh/article/PIIS2542-5196(22)00090-0/fulltext.

Gao, J., Gong, J., Yang, J., Wang, Z., Fu, Y., Tang, S., et al., 2023. Spatial distribution and ecological risk assessment of soil heavy metals in a typical volcanic area: Influence of parent materials. Heliyon 9 (1), e12993. Available from: https://doi.org/10.1016/j.heliyon.2023.e12993.

Gautam, K., Sharma, P., Dwivedi, S., Singh, A., Gaur, V.K., Varjani, S., et al., 2023. A review on control and abatement of soil pollution by heavy metals: emphasis on artificial intelligence in recovery of contaminated soil. Environmental Research 225, 115592. Available from: https://doi.org/10.1016/j.envres.2023.115592.

GBD 2019 Chronic Respiratory Diseases Collaborators, 2023. Global burden of chronic respiratory diseases and risk factors 1990 - 2019: an update from the global burden of disease study 2019. EClinicalMedicine (Part of the Lancet Discovery Science) 59, 101936. Available from: https://doi.org/10.1016/j.eclinm.2023.101936.

Gomes, J., Esteves, H., Rente, L., 2022. Influence of an extreme Saharan dust event on the air quality of the West Region of Portugal. Gases 2 (3), 74−84. Available from: https://doi.org/10.3390/gases2030005.

GPG (Gauteng Provincial Government), 2020. Literature Review on SuDS: Definitions, Science, Data and Policy and Legal Context in South Africa for Project 'Research on the Use of Sustainable Drainage Systems in Gauteng Province' Produced by Fourth Element, AquaLinks, Eco-Pulse, NM & Associates and GreenVision Consulting and Commissioned by Gauteng Department of Agriculture and Rural Development. https://futurewater.uct.ac.za/sites/default/files/content_migration/futurewater_uct_ac_za/792/files/Gauteng%2520PG%2520%25282020%2529%2520Del%25202%2520Literature%2520review_SuDS%2520in%2520Gauteng.pdf. (Accessed 22 October 2023).

Grewar, T., 2019. South Africa's options for mine-impacted water re-use: a review. Journal of the Southern African Institute of Mining and Metallurgy 119 (3), 321.

Gul, Z., Rupula, K., Beedu, S.R., 2022. Chapter 6: Nano-phytoremediation for soil contamination: an emerging approach for revitalizing the tarnished resource. Phytoremediation: Biotechnological Strategies for Promoting Invigorating Environs. Academic Press, pp. 115−138. Available from: https://doi.org/10.1016/B978-0-323-89874-4.00014-5.

Gupta, V., 2020. Vehicle-generated heavy metal pollution in an urban environment and its distribution into various environmental components. In: Shukla, V., Kumar, N. (Eds.), Environmental Concerns and Sustainable Development. Springer, Singapore. Available from: https://doi.org/10.1007/978-981-13-5889-0_5.

Haidar, Z., Fatema, K., Shoily, S.S., Sajib, A.A., 2023. Disease-associated metabolic pathways affected by heavy metals and metalloid. Toxicology Reports 10, 554−570. Available from: https://doi.org/10.1016/j.toxrep.2023.04.010.

Hamad, R., Balzter, H., Kolo, K., 2019. Assessment of heavy metal release into the soil after mine clearing in Halgurd-Sakran National Park, Kurdistan, Iraq. Environmental Science and Pollution Research International 26 (2), 1517−1536. Available from: https://doi.org/10.1007/s11356-018-3597-3.

Hanto, J., Schroth, A., Krawielicki, L., Oei, P.-Y., Burton, J., 2022. South Africa's energy transition - unraveling its political economy. Energy for Sustainable Development 69, 164−178. Available from: https://doi.org/10.1016/j.esd.2022.06.006.

HEI,, 2022. The State of Air Quality and Health Impacts in Africa. A Report from the State of Global Air Initiative. Health Effects Institute, Boston, MA. Available from: https://www.stateofglobalair.org/sites/default/files/documents/2022-10/soga-africa-report.pdf.

Hemond, H., Fechner, E., 2023. Chemical Fate and Transport in the Environment, fourth ed. https://shop.elsevier.com/books/chemical-fate-and-transport-in-the-environment/hemond/978-0-12-822252-2. (Accessed 03 July 2023).

Hoy, R.F., Chambers, D.C., 2020. Silica-related diseases in the modern world. Allergy 75 (11), 2805−2817. Available from: https://doi.org/10.1111/all.14202.

Huang, X., Rom, X., Gordon, T., Finkelman, R.B., 2006. Interaction of iron and calcium minerals in coals and their roles in coal dust-induced health and environmental problems. Reviews in Mineralogy and Geochemistry 64 (1), 153−178. Available from: https://doi.org/10.2138/rmg.2006.64.6.

Huang, K., Diao, Z., Lu, G., 2023. Advances in remediation of contaminated sites. Processes 11, 157. Available from: https://doi.org/10.3390/pr11010157.

ICBL-CMC (International Campaign to Ban Landmines - Cluster Munitions Coalition), 2019. Landmine and cluster munition monitor. Available from: the-monitor.org. http://www.the-monitor.org/en-gb/home.aspx. (Accessed 16 August 2023).

Igboama, W.N., Hammed, O.S., Fatoba, J.O., Aroyehun, M.T., Ehiabhili, 2022. Review article on impact of groundwater contamination due to dumpsites using geophysical and physiochemical methods. Applied Water Science 12, 130. Available from: https://doi.org/10.1007/s13201-022-01653-z.

Ighalo, J.O., Adeniyi, A.G., Adeniran, A.J., Ogunniyi, 2021. A systematic literature analysis of the nature and regional distribution of water pollution sources in Nigeria. Journal of Cleaner Production 283, 124566.

ILO (International Labour Office), 1971. Fourth International Pneumoconiosis Conference, 27 September−2 October, 1971, Bucharest. Report ILO-CIP/IV/1971/DEFINITION, Geneva, ILO, 1971. https://www.ilo.org/public/libdoc/ilo/1971/71B09_521.pdf. (Accessed 22 October 2023).

IMF (International Monetary Fund), 2023. G20 Data Gaps Initiative 3 - Workplan. https://www.imf.org/en/News/Seminars/Conferences/g20-data-gaps-initiative. (Accessed 25 June 2023).

IPCC (The Intergovernmental Panel on Climate Change), 2021. Climate Change 2021: The Physical Science Basis - Contribution of Working Group I to the Sixth Assessment Report of the Intergovernmental Panel on Climate Change. Cambridge University Press. https://www.ipcc.ch/report/ar6/wg1/#FullReport. (Accessed 02 August 2023).

Irunde, R., Ijumulana, J., Ligate, I.F., Maity, J.P., Ahmad, A.A., Mtamba, J., et al., 2022. Arsenic in Africa: potential sources, spatial variability, and the state of the art for arsenic removal using locally available materials. Groundwater for Sustainable Development 18, 100746. Available from: https://doi.org/10.1016/j.gsd.2022.100746.

Jardin, S.V., 2023. Asbestos-related cancer: exaggerated risk perception. Cancer Screening and Prevention 2 (1), 51−57. Available from: https://doi.org/10.14218/CSP.2022.00028.

Jesitha, K., Jaseela, C. and Harikumar, P.S., 2021. Chapter 22: Nanotechnology enhanced phytoremediation and photocatalytic degradation techniques for remediation of soil pollutants. In: Nanomaterials for Soil Remediation. pp. 463−499.

Jiao, Y., Zhang, C., Su, P., Tang, Y., Huang, Z., Ma, T., 2023. A review of acid mine drainage: Formation mechanism, treatment technology, typical engineering cases and resource utilization. Process Safety and Environmental Protection 170, 1240−1260. Available from: https://doi.org/10.1016/j.psep.2022.12.083.

Jin, P., Bian, S., Yu, W., Guo, S., Lai, C., Wu, L., et al., 2023. Insights into leachate reduction in landfill with different ventilation rates: balance of water, waste physicochemical properties, and microbial community. Waste Management 156, 118−129. Available from: https://doi.org/10.1016/j.wasman.2022.11.033.

Jiying, W., Beraud, J.D., Xicang, Z, 2023. Investigating the impact of air pollution in selected African developing countries. Environmental Science and Pollution Research International 30 (23), 64460−64471. Available from: https://doi.org/10.1007/s11356-023-26998-z (Accessed 26 February 2024).

Katoto, P.D.M.C., Byamungu, L., Brand, A.S., Mokaya, J., Strijdom, H., Goswami, N., et al., 2019. Ambient air pollution and health in Sub-Saharan Africa: current evidence, perspectives and a call to action. Environmental Research 173, 174−188. Available from: https://doi.org/10.1016/j.envres.2019.03.029.

Kenfack, S., Beguere, M., Boukerrou, L., 2016. Water, Climate, and Health Risks in West Africa: Perspectives From a Regional Water Quality Program, 2016. ISEE Conference Abstracts, Number:S-064. Environmental Health Perspectives. Available from: https://doi.org/10.1289/isee.2016.4802.

Kirk, K., 2023. Aerosols: Small Particles with Big Climate Effects. NASA Global Climate Change. https://climate.nasa.gov/explore/ask-nasa-climate/3271/aerosols-small-particles-with-big-climate-effects/. (Accessed 02 August 2023).

Kopittke, P.M., Minasny, B., Pendall, E., Rumpel, C., McKenna, B.A., 2023. Healthy soil for healthy humans and a healthy planet. Critical Reviews in Environmental Science and Technology . Available from: https://doi.org/10.1080/10643389.2023.2228651.

Kotsyfakis, M., Zarogiannis, S., Patelarou, E., 2019. The health impact of Saharan dust exposure. International Journal of Occupational Medicine and Environmental Health 32 (6), 749−760. Available from: https://doi.org/10.13075/ijomeh.1896.01466.

Laker, M.C., 2023. Environmental impacts of gold mining - with special reference to South Africa. Mining 3, 205−220. Available from: https://doi.org/10.3390/mining3020012.

Lee, J., Kaunda, R.B., Sinkala, T., Workman, C.F., Bazilian, M.D., Clough, G., 2021. Phytoremediation and phytoextraction in Sub-Saharan Africa: addressing economic and social challenges. Ecotoxicology and Environmental Safety 226, 112864. Available from: https://doi.org/10.1016/j.ecoenv.2021.112864.

Ligate, F., Ijumulana, J., Ahmad, A., Kimambo, V., Irunde, R., Mtamba, P.H.O., et al., 2021. Groundwater resources in the East African Rift Valley: understanding the geogenic contamination and water quality challenges in Tanzania. Scientific African 13, e00831. Available from: https://doi.org/10.1016/j.sciaf.2021.e00831.

Lin, L., Yang, H., Xu, X., 2022. Effects of water pollution on human health and disease heterogeneity: a review. Frontiers in Environmental Science 10, 880246. Available from: https://doi.org/10.3389/fenvs.2022.880246.

Lindamulla, L., Nanayakkara, N., Othman, M., Jinadasa, S., Herath, G., Jegathesan, V., 2022. Municipal solid waste landfill leachate characteristics and their treatment options in tropical countries. Current Pollution Reports 8, 273−287. Available from: https://doi.org/10.1007/s40726-022-00222-x.

Liu, L., Li, W., Song, W., Guo, M., 2018. Remediation techniques for heavy metal-contaminated soils: principles and applicability. Science of the Total Environment 15 (633), 206−219. Available from: https://doi.org/10.1016/j.scitotenv.2018.03.161.

Lucas, R.M., Norval, M., Wright, C.Y., 2016. Solar ultraviolet radiation in Africa: a systematic review and critical evaluation of the health risks and use of photoprotection. Photochemical & Photobiological Sciences: Official Journal of the European Photochemistry Association and the European Society for Photobiology 15 (1), 10−23. Available from: https://doi.org/10.1039/c5pp00419e.

Lucero-Prisno, D.E., Shomuyiwa, D.O., Kouwenhoven, M.B.N., et al., 2023. Top 10 public health challenges to track in 2023: shifting focus beyond a global pandemic. Public Health Challenges 2, e86. Available from: https://doi.org/10.1002/puh2.86.

Makhadi, R., Oke, S.A., Ololade, O.O., 2020. The influence of non-engineered municipal landfills on groundwater chemistry and quality in Bloemfontein, South Africa. Molecules (Basel, Switzerland) 25 (23), 5599. Available from: https://doi.org/10.3390/molecules25235599.

Makoni, M., 2020. Air pollution in Africa. Lancet Respiratory Medicine 8 (7), e60−e61. Available from: https://doi.org/10.1016/S2213-2600(20)30275-7.

Malatse, M., Ndlovu, S., 2015. The viability of using the Witwatersrand gold mine tailings for brickmaking. Journal of South African Institute of Mining and Metallurgy [online] 115 (4), 321−327.

Manisalidis, I., Stavropoulou, E., Stavropoulos, A., Bezirtzoglou, E., 2020. Environmental and health impacts of air pollution: a review. Frontiers in Public Health 8, 14. Available from: https://doi.org/10.3389/fpubh.2020.00014.

Markowitz, S.B., 2022. Lung cancer screening in asbestos-exposed populations. International Journal of Environmental Research and Public Health 19 (5), 2688. Available from: https://doi.org/10.3390/ijerph19052688.

Mead, M.I., Okello, G., Mbandi, A.M., Pope, F.D., 2023. Spotlight on air pollution in Africa. Nature Geoscience 16 (11), 930−931. Available from: https://doi.org/10.1038/s41561-023-01311-2.

Mesić Kiš, I., Karaica, B., Medunić, G., Romić, M., Šabarić, J., Balen, D., et al., 2016. Soil, bark and leaf trace metal loads related to the war legacy (the Prašnik Rainforest, Croatia. Rudarsko-geološko-naftni Zbornik 31 (2), 13–28. Available from: https://doi.org/10.17794/rgn.2016.2.2.

Michellier, C., Katoto, P.D.C., Dramaix, M., Nemery, B., Kervyn, F., 2020. Respiratory health and eruptions of the Nyiragongo and Nyamulagira volcanoes in the Democratic Republic of Congo: a time-series analysis. Environmental Health 19, 62. Available from: https://doi.org/10.1186/s12940-020-00615-9.

Millar, D.A., Kapwata, T., Kunene, Z., Mogotsi, M., Wernecke, B., Garland, R.M., et al., 2022. Respiratory health among adolescents living in the Highveld Air Pollution Priority Area in South Africa. BMC Public Health 22, 2136. Available from: https://doi.org/10.1186/s12889-022-14497-8.

Mishra, P., Mishra, J., Arora, N.K., 2023. Biofortification revisited: addressing the role of beneficial soil microbes for enhancing trace elements concentration in staple crops. Microbiological Research 275, 127442. Available from: https://doi.org/10.1016/j.micres.2023.127442.

Mocek-Płóciniak, A., Mencel, J., Zakrzewski, W., Roszkowski, S., 2023. Phytoremediation as an effective remedy for removing trace elements from ecosystems. Plants 12, 1653. Available from: https://doi.org/10.3390/plants12081653.

Montanarella, L., Panagos, P., 2021. The relevance of sustainable soil management within the European Green Deal. Land Use Policy 100, 104950. Available from: https://publications.jrc.ec.europa.eu/repository/handle/JRC119967.

Mpanza, M., Adam, E., Moolla, R., 2020. Perceptions of external costs of dust fallout from gold mine tailings: West Wits Basin. Clean Air Journal 30 (1), 1–12. Available from: https://doi.org/10.17159/caj/2020/30/1.7566.

Münzel, T., Hahad, O., Daiber, A., Landrigan, P.J., 2023. Soil and water pollution and human health: what should cardiologists worry about? Cardiovascular Research 119 (2), 440–449. Available from: https://doi.org/10.1093/cvr/cvac082.

Mwaanga, P., Silondwa, M., Kasali, G., Banda, P.M., 2019. Preliminary review of mine air pollution in Zambia. Heliyon 5 (9), e02485. Available from: https://doi.org/10.1016/j.heliyon.2019.e02485.

Myllyvirta, L., Kelly, J., 2023. Air quality, health, and economic impacts of a new coal mine and power plant in Lephalale. Centre for Research on Energy and Clean Air (CREA). https://energyandcleanair.org/wp/wp-content/uploads/2023/03/CREA_Air-Quality-Health-and-Economic-Impacts-of-a-New-Coal-Mine-and-Power-Plant-in-Lephalale.pdf. (Accessed 08 August 2023).

Naidoo, D., Roselt, M., Mpofu, A., 2023. '2023 Water Market Intelligence Report'. Friedrich Naumann Foundation/Greencape. https://greencape.co.za/wp-content/uploads/2023/04/WATER_MIR_2023_DIGITAL_SINGLES.pdf. (Accessed 07 August 2023).

NASA Earth Observatory, 2023. Heavy rains hit drought-stricken Horn of Africa. https://earthobservatory.nasa.gov/images/151208/heavy-rains-hit-drought-stricken-horn-of-africa. (Accessed 23 July 2023).

Nathanail, P., 2011. Sustainable remediation: Quo vadis? Remediation 21, 35–44. Available from: https://doi.org/10.1002/rem.20298.

Nemery, B., 2022. Metals and the respiratory tract. In: Nordberg, G.F., Costa, M. (Eds.), Handbook on the Toxicology of Metals, fifth ed., Volume 1, Chapter 19: General Considerations, pp. 421–443. https://doi.org/10.1016/B978-0-12-823292-7.00030-9. (Accessed 02 August 2023).

Nevondo, V., Malehase, T., Daso, A.P., Okonkwo, O.J., 2019. Leachate seepage from landfill: a source of groundwater mercury contamination in South Africa. Water SA 45 (2). Available from: https://doi.org/10.4314/wsa.v45i2.09.

Ngeno, E.C., Mbuci, K.E., Necibi, M.C., Shikuku, V.O., Olisah, C., Ongulu, R., et al., 2022. Sustainable re-utilization of waste materials as adsorbents for water and wastewater treatment in Africa: recent studies, research gaps, and way forward for emerging economies. Environmental Analyses 9, 10082. Available from: https://doi.org/10.1016/j.envadv.2022.100282.

Nicole, W., 2022. Clear and present dangers: the multiple health hazards of volcanic eruptions. Environmental Health Perspectives 130 (2), 22001. Available from: https://doi.org/10.1289/EHP10541.

Nkem, A.C., Topp, S.M., Devine, S., Li, W.W., Ogaji, D.S., 2022. The impact of oil industry-related social exclusion on community wellbeing and health in African countries. Frontiers in Public Health 10, 858512. Available from: https://doi.org/10.3389/fpubh.2022.858512.

Nti, E.K., Cobbina, S.J., Attafuah, E.E., Senanu, L.D., Amenyeku, G., Gyan, M.A., Forson, D., Safo, A.R., 2023. Water pollution control and revitalization using advanced technologies: Uncovering artificial intelligence options towards environmental health protection, sustainability and water security. Heliyon 9 (7), e18170. Available from: https://doi.org/10.1016/j.heliyon.2023.e18170 (Accessed 25 February 2024).

Odoh, C.K., Zabbey, N., Sam, K., Eze, C.N., 2019. Status, progress and challenges of phytoremediation - an African scenario. Journal of Environmental Management 237, 365−378. Available from: https://doi.org/10.1016/j.jenvman.2019.02.090.

Ogundele, L.T., Owoade, O.K., Hopke, P.K., Olise, F.S., 2017. Heavy metals in industrially emitted particulate matter in Ile-Ife, Nigeria. Environmental Research 156, 320−325. Available from: https://doi.org/10.1016/j.envres.2017.03.051.

Okello, G., Nantanda, R., Awokola, B., Thondoo, M., Okure, D., Tatah, L., et al., 2023. Air quality management strategies in Africa: a scoping review of the content, context, co-benefits and unintended consequences. Environment International 171, 107709. Available from: https://doi.org/10.1016/j.envint.2022.107709.

Okereafor, U., Makhatha, M., Mekuto, L., Mavumengwana, V., 2020. Gold mine tailings: a potential source of silica sand for glass making. Minerals 10, 448. Available from: https://doi.org/10.3390/min10050448.

Olarewaju, O., Fajinmi, O., Davies, T.C., Arthur, G., Naidoo, K., Coopoosamy, R., 2023. The Saharan dust plume: Current knowledge on the impact on health, human activities, and the ecosystem, with comments on research gaps. doi: 10.22541/essoar.167979167.72582095/v1.

Oliver, M.A., Gregory, P.J., 2015. Soil, food security and human health: a review: soil, food security and human health. European Journal of Soil Science 66 (2), 257−276. Available from: https://doi.org/10.1111/ejss.12216.

Ouma, K.O., Shane, A., Syampungani, S., 2022. Aquatic ecological risk of heavy-metal pollution associated with degraded mining landscapes of the Southern Africa River Basins: a review. Minerals 12, 225. Available from: https://doi.org/10.3390/min12020225.

Palmer, P.I., Wainwright, C.M., Dong, B., Maidment, R.I., Wheeler, K.G., Gedney, N., et al., 2023. Drivers and impacts of Eastern African rainfall variability. Nature Reviews Earth & Environment 4, 254−270. Available from: https://doi.org/10.1038/s43017-023-00397-x.

Pandey, G., Singh, P., Pant, B.N., Bajpai, S., 2023. Potentials and frontiers of nanotechnology for phytoremediation. In: Newman, L., Ansari, A.A., Gill, S.S., Naeem, M., Gill, R. (Eds.), Phytoremediation. Springer, Cham. Available from: https://doi.org/10.1007/978-3-031-17988-4_17.

PAP (The Pan African Parliament), 2020. Pan African Parliament maps first steps for model legislation on sustainable soil management in Africa (2020). https://www.africanparliamentarynews.com/2020/03/pap-maps-first-steps-for-model.html?m = 1. (Accessed 21 June 2023).

Peng, X., Jiang, Y., Chen, Z., Osman, A.I., Farghali, M., Rooney, D.W., et al., 2023. Recycling municipal, agricultural and industrial waste into energy, fertilizers, food and construction materials, and economic feasibility: a review. Environmental Chemistry Letters 21, 765−801. Available from: https://doi.org/10.1007/s10311-022-01551-5.

Podlasek, A., 2022. Modeling leachate generation: practical scenarios for municipal solid waste landfills in Poland. Environmental Science and Pollution Research 30, 13256−13269. Available from: https://doi.org/10.1007/s11356-022-23092-8.

Rahaman, M.A., Rahman, M.M., Nazimuzzaman, M., 2020. Impact of salinity on infectious disease outbreaks: experiences from the Global Coastal Region. In: Leal Filho, W., Wall, T., Azul, A.M., Brandli, L., Özuyar, P.G. (Eds.), Good Health and Well-Being. Encyclopaedia of the UN Sustainable Development Goals. Springer, Cham. Available from: https://doi.org/10.1007/978-3-319-95681-7_106.

Rahimi, E., Shekarian, Y., Shekarian, N., Roghanchi, 2023. Investigation of respirable coal mine dust (RCMD) and respirable crystalline silica (RCS) in the U.S. underground and surface coal mines. Scientific Reports 13, 1767. Available from: https://doi.org/10.1038/s41598-022-24745-x.

Raimi, M.O., Iyingiala, A.A., Sawyerr, O.H., Saliu, A.O., Ebuete, A.W., et al., 2022. Leaving no one behind: impact of soil pollution on biodiversity in the global south: a global call for action. In: Chibueze Izah, S. (Ed.), Biodiversity in Africa: Potentials, Threats and Conservation. Sustainable Development and Biodiversity 29. Springer, Singapore. Available from: https://doi.org/10.1007/978-981-19-3326-4_8.

Rees, D., Phillips, J.I., 2017. The legacy of *in situ* asbestos cement roofs in South Africa. Occupational and Environmental Medicine 74 (Suppl 1), A1−A166. Available from: https://www.nioh.ac.za/the-legacy-of-in-situ-asbestos-cement-roofs-in-south-africa/.

Riemann, K., McGibbon, D.C., Gerstner, K., Scheibert,S., Hoosain, M., Hay, E.R., 2017. Water Resource Protection: Research Report: A Review of the State-of-the-Art and Research and Development Needs for South Africa. Report to the Water Research Commission. WRC Report No. 2532/1/17. https://www.wrc.org.za/wp-content/uploads/mdocs/2532-1-17.pdf. (Accessed 22 June 2023).

Rodríguez-Eugenio, N., 2021. The global soil partnership: tackling global soil threats through collective action. In: Ginzky, H., Dooley, E., Heuser, I.L., Kasimbazi, E., Kibugi, R., Markus, T., et al.,International Yearbook of Soil Law and Policy, 2019. Springer, Cham, pp. 197−221. Available from: https://doi.org/10.1007/978-3-030-52317-6_11. (Accessed 06 August 2023).

Rodriguez-Eugenio, N., McLaughlin, M., Pennock, D., 2018. Soil Pollution: A Hidden Reality. FAO, Rome, p. 142. Available from: https://www.fao.org/3/I9183EN/i9183en.pdf. (Accessed 14 June 2023).

Ruppel, O.C., Ginzky, H. (Eds.), 2022. African Soil Protection Law: Mapping Out Options for a Model Legislation for Improved Sustainable Soil Management in Africa - A comparative Legal Analysis From Kenya, Cameroon and Zambia. Law and Constitution in Africa 41. https://d-nb.info/1233109731/34. (Accessed 20 June 2023).

Ryan, D., Bustos, E., 2019. Knowledge gaps and climate adaptation policy: a comparative analysis of six Latin American countries. Climate Policy 19 (10), 1297−1309. Available from: https://doi.org/10.1080/14693062.2019.1661819.

Sakr, M., Mohamed, M.M., Maraqa, M.A., Hamouda, M.A., Hassan, A.A., Ali, J., et al., 2022. A critical review of the recent developments in micro-nano bubbles applications for domestic and industrial wastewater treatment. Alexandria Engineering Journal 61 (8), 6591−6612. Available from: https://doi.org/10.1016/j.aej.2021.11.041.

Schiavo, B., Morton-Bermea, O., Meza-Figueroa, D., Acosta-Elías, M., González-Grijalva, B., Armienta-Hernández, M.A., et al., 2023. Characterization and polydispersity of volcanic ash nanoparticles in synthetic lung fluid. Toxics 11 (7), 624. Available from: https://doi.org/10.3390/toxics11070624.

SDWF (Safe Drinking Water Foundation), (Updated 2023). Mining and Water Pollution. Safe Drinking Water Foundation. https://www.safewater.org/fact-sheets-1/2017/1/23/miningandwaterpollution. (Accessed 30 June 2023).

Shammi, M., Rahman, M.M., Bondad, S.E., Bodrud-Doza, M., 2019. Impacts of salinity intrusion in community health: a review of experiences on drinking water sodium from coastal areas of Bangladesh. Healthcare (Basel, Switzerland) 7 (1), 50. Available from: https://doi.org/10.3390/healthcare7010050.

Sharma, J.K., Kumar, N., Singh, N.P., Santal, A.R., 2023. Phytoremediation technologies and their mechanism for removal of heavy metal from contaminated soil: an approach for a sustainable environment. Frontiers in Plant Science 14, 1076876. Available from: https://doi.org/10.3389/fpls.2023.1076876.

Shindell, D., Faluvegi, G., Parsons, L., Nagamoto, E., Chang, J., 2022. Premature deaths in Africa due to particulate matter under high and low warming scenarios. GeoHealth 6. Available from: https://doi.org/10.1029/2022GH000601.

Siddiqua, A., Hahladakis, J.N., Al-Attiya, W.A.K.A., 2022. An overview of the environmental pollution and health effects associated with waste landfilling and open dumping. Environmental Science of Pollution Research 29, 58514−58536. Available from: https://doi.org/10.1007/s11356-022-21578-z.

Simão, F.V., Chambart, H., Vandemeulebroeke, L., Nielsen, P., Adrianto, L.R., Pfister, S., et al., 2022. Mine waste as a sustainable resource for facing bricks. Journal of Cleaner Production 368, 133118. Available from: https://doi.org/10.1016/j.jclepro.2022.133118.

Singh, A.B., Kumar, P., 2022. Climate change and allergic diseases: an overview. Frontiers in Allergy 3, 964987. Available from: https://doi.org/10.3389/falgy.2022.964987.

Singo, N.K., Kramers, J.D., 2021. Feasibility of tailings retreatment to unlock value and create environmental sustainability of the Louis Moore tailings dump near Giyani, South Africa. Journal of the Southern African Institute of Mining and Metallurgy 121 (7), 361−368. Available from: http://www.scielo.org.za/pdf/jsaimm/v121n7/08.pdf.

Skousen, J.G., Ziemkiewicz, P.F., McDonald, L.M., 2019. Acid mine drainage formation, control and treatment: approaches and strategies. The Extractive Industries and Society 6, 241−249. Available from: https://doi.org/10.1016/j.exis.2018.09.008.

Skuland, T., Låg, M., Gutleb, A.C., Brinchmann, B.C., Serchi, T., Øvrevik, J., et al., 2020. Pro-inflammatory effects of crystalline- and nano-sized non-crystalline silica particles in a 3D alveolar model. Particulate and Fibre Toxicology 17, 13. Available from: https://doi.org/10.1186/s12989-020-00345-3.

Sonwani, S., Madaan, S., Arora, J., Suryanarayan, S., Rangra, D., Mongia, N., et al., 2021. Inhalation exposure to atmospheric nanoparticles and its associated impacts on human health: a review. Frontiers in Sustainable Cities 3, 690444. Available from: https://doi.org/10.3389/frsc.2021.690444.

Statistica, 2023. Leading 10 causes of death in Africa in 2019 (in deaths per 100,000 population). https://www.statista.com/statistics/1029287/top-ten-causes-of-death-in-africa/. (Accessed 03 September 2023).

Stengel, D., O'Reilly, S., O'Halloran, J., 2006. Contaminants and pollutants. In: Davenport, J., Davenport, J.L. (Eds.), The Ecology of Transportation: Managing Mobility for the Environment. Environmental Pollution, 10. Springer, Dordrecht. Available from: https://doi.org/10.1007/1-4020-4504-2_15.

Stewart, C., Damby, D.E., Horwell, C.J., Elias, T., Ilyaskaya, E., Tomašek, I., et al., 2022. Volcanic air pollution and human health: recent advances and future directions. Bulletin of Volcanology 84, 11. Available from: https://doi.org/10.1007/s00445-021-01513-9.

Suhr, F., Steinert, J.I., 2022. Epidemiology of floods in sub-Saharan Africa: a systematic review of health outcomes. BMC Public Health 22, 268. Available from: https://doi.org/10.1186/s12889-022-12584-4.

Talukder, B., Salim, R., Islam, S.T., Mondal, K.P., Hipel, K., Vanloon, G., 2023. Collective intelligence for addressing community planetary health resulting from salinity prompted by sea-level rise. The Journal of Climate Change and Health 10, 100203. Available from: https://doi.org/10.1016/j.joclim.2023.100203.

Tan, H.W., Pang, Y.L., Lim, S., Chong, W.C., 2023. A state-of-the-art of phytoremediation approach for sustainable management of heavy metals recovery. Environmental Technology and Innovation 30, 103043. Available from: https://doi.org/10.1016/j.eti.2023.103043.

Tchindjang, M., 2018. Lake Nyos, a multirisk and vulnerability appraisal. Geosciences 8, 312. Available from: https://doi.org/10.3390/geosciences8090312.

Temesgen, T., Bui, T.T., Han, M., Kim, T.I., Park, H., 2017. Micro and nanobubble technologies as a new horizon for water-treatment techniques: a review. Advances in Colloid and Interface Science 246, 40−51. Available from: https://doi.org/10.1016/j.cis.2017.06.011.

Tessum, M.W., Anenberg, S.C., Chafe, Z.A., Henze, D.K., Kleiman, G., Kheirbek, I., et al., 2022. Sources of ambient PM2.5 exposure in 96 global cities. Atmospheric Environment (Oxford, England, 1994) 286, 119234. Available from: https://doi.org/10.1016/j.atmosenv.2022.119234.

Tindwa, H.J., Singh, B.R., 2023. Soil pollution and agriculture in sub-Saharan Africa: state of the knowledge and remediation technologies. Frontiers in Soil Science 2, 1101944. Available from: https://doi.org/10.3389/fsoil.2022.1101944.

Tully, K., Sullivan, C., Weil, R., Sanchez, P., 2015. The State of soil degradation in Sub-Saharan Africa: baselines, trajectories, and solutions. Sustainability 7, 6523−6552. Available from: https://doi.org/10.3390/su7066523.

UNCC (United Nations Climate Change), 2012. eMalahleni: Water Reclamation Plant, South Africa. https://unfccc.int/climate-action/momentum-for-change/lighthouse-activities/emalahleni-water-reclamation-plant. (Accessed 15 August 2023).

UNEA-3 (UN Environment Assembly 3), 2018. Implementation Plan: "Towards a Pollution-Free Planet." Second Draft (23/11/2018). https://wedocs.unep.org/bitstream/handle/20.500.11822/26623/Second%20draft%20of%20the%20Implementation%20%20Plan%20with%20Annexes%20-%2023%20November%202018-%20for%20circulation%5B6%5D.pdf?sequence = 11&isAllowed = y. (Accessed 06 August 2023).

UNEP (United Nations Environment Programme), n.d. Pollution and Waste: World Environment Situation Room; Data, Information and Knowledge on the Environment. https://wesr.unep.org/article/pollution-and-waste. (Accessed 22 June 2023).

UNICEF (United Nations Childrens' Fund), 2019. Silent Suffocation in Africa, 2019. https://www.unicef.org/media/55081/file/Silent%20suffocation%20in%20africa%20air%20pollution%202019%20.pdf. (Accessed 08 August 2023).

URMC (University of Rochester Medical Centre), 2022. Study Links Fracking, Drinking Water Pollution and Infant Health. https://www.urmc.rochester.edu/news/story/study-links-fracking-drinking-water-pollution-and-infant-heath. (Accessed 31 June 2023).

US EPA (United States Environmental Protection Agency), 2001. Geosynthetic Clay Liners Used in Municipal Solid Waste Landfills. Solid Waste and Emergency Response (5306W). Document EPA530-F-97-002 (Revised December 2001). https://www.epa.gov/sites/default/files/2016-03/documents/geosyn.pdf. (Accessed 07 August 2023).

US EPA (United States Environmental Protection Agency), 2023a. Health Effects of Ozone Pollution. https://www.epa.gov/ground-level-ozone-pollution/health-effects-ozone-pollution. (Accessed 02 August 2023).

US EPA (United States Environmental Protection Agency), 2023b. Human Health Risk Assessment. https://www.epa.gov/risk/human-health-risk-assessment. (Accessed 05 August 2023).

US NIH (US National Institute of Environmental Health Sciences), (2022). Health Impacts of Air Quality: Climate Change and Human Health. https://www.niehs.nih.gov/research/programs/climatechange/health_impacts/asthma/index.cfm. (Accessed 10 August 2023).

Vadapalli, V.R.K., Sakala, E., Dube, G., Coetzee, H., 2020. Mine water treatment and the use of artificial intelligence in acid mine drainage prediction. In: Fosso-Kankeu, E., Wolkersdorfer, C., Burgess, J. (Eds.), Recovery of Byproducts from Acid Mine Drainage Treatment. https://doi.org/10.1002/9781119620204.ch2. (Accessed 05 August 2023).

Vanka, K.S., Shukla, S., Gomez, H.M., James, C., Palanisami, T., Williams, K., et al., 2022. Understanding the pathogenesis of occupational coal and silica dust-associated lung disease. European Respiratory Review: An Official Journal of the European Respiratory Society 31 (165), 210250. Available from: https://doi.org/10.1183/16000617.0250-2021.

Vogel, L., 2017. Fracking tied to cancer-causing chemicals. Canadian Medical Association Journal 189 (2), E94−E95. Available from: https://doi.org/10.1503/cmaj.109-5358.

Wagner, S., Souvignet, M., Walz, Y., Balogun, K., Komi, K., Kreft, S., et al., 2021. When does risk become residual? A systematic review of research on flood risk management in West Africa. Regional Environmental Change 21, 84. Available from: https://doi.org/10.1007/s10113-021-01826-7.

Wang, L., Xie, J., Hu, Y., Tian, Y., 2022. Air pollution and risk of chronic obstructed pulmonary disease: the modifying effect of genetic susceptibility and lifestyle. EBioMedicine, Part of the Lancet Discovery Science 79, 103994. Available from: https://doi.org/10.1016/j.ebiom.2022.103994.

Warren-Vega, W.M., Campos-Rodríguez, A., Zárate-Guzmán, A.I., Romero-Cano, L.A., 2023. A current review of water pollutants in American Continent: trends and perspectives in detection, health risks, and treatment technologies. International Journal of Environmental Research and Public Health 20 (5), 4499. Available from: https://doi.org/10.3390/ijerph20054499.

WHO (World Health Organisation), 2013. Contaminated Sites and Health. Copenhagen, Denmark. http://www.euro.who.int/__data/assets/pdf_file/0003/186240/e96843e.pdf. (Accessed 17 August 2023).

WHO, 2017. World Health OrganisationGuidelines for Drinking-Water Quality: Fourth Edition Incorporating the First Addendum. WHO, Geneva, Switzerland. Available from: https://www.who.int/publications/i/item/9789241549950.

WHO, 2022a. World Health OrganisationGuidelines For Drinking-Water Quality: Fourth edition Incorporating the First and Second Addenda. World Health Organisation. Available from: https://pubmed.ncbi.nlm.nih.gov/35417116/.

WHO (World Health Organisation), 2022b. Updated Regional Strategy for the Management of Environmental Determinants of Human Health in the African Region 2022–2032: Report of the Secretariat, AFR/RC72/10, Regional Committee for Africa, Seventy-Second Session Lomé, Republic of Togo, 22–26 August 2022. https://www.afro.who.int/sites/default/files/2022-07/AFR-RC72-10%20Updated%20Regional%20Strategy%20for%20the%20management%20of%20environmental%20determinants%20of%20human%20health%20in%20the%20African%20Region%202022-2023.pdf. (Accessed 30 June 2023).

WHO (World Health Organisation) Madagascar, 2023. Floods Raise Cholera Risk Even As Cases Decline in Africa. https://www.afro.who.int/countries/madagascar/news/floods-raise-cholera-risk-even-cases-decline-africa. (Accessed 23 July 2023).

WMO (The World Meteorological Organisation), 2023. WMO Highlights Efforts to Tackle Sand and Dust Storms. https://public.wmo.int/en/media/news/wmo-highlights-efforts-tackle-sand-and-dust-storms. (Accessed 17 August 2023).

World Bank, 2022. What You Need To Know About Climate. Change and Air Pollution. https://www.worldbank.org/en/news/feature/2022/09/01/what-you-need-to-know-about-climate-change-and-air-pollution. (Accessed 31 July 2023).

World Bank, 2023. Pollution Management and Environmental Health Programme. https://www.worldbank.org/en/programs/pollution-management-and-environmental-health-program. (Accessed 22 July 2023).

Xin, X., Shentu, J., Zhang, T., Yang, X., Baligar, V.C., He, Z., 2022. Sources, indicators, and assessment of soil contamination by potentially toxic metals. Sustainability 14 (23), 15878. Available from: https://doi.org/10.3390/su142315878.

Yacobi, N.R., Fazllolahi, F., Kim, Y.H., Sipos, A., Borok, Z., Kim, K.J., et al., 2011. Nanomaterial interactions with and trafficking across the lung alveolar epithelial barrier: implications for health effects of air-pollution particles. Air Quality, Atmosphere and Health 4 (1), 65–78. Available from: https://doi.org/10.1007/s11869-010-0098-z.

Yadav, S.K., Joshi, V., 2023. Phytoremediation: a sustainable approach to combat heavy metal contaminated soil - a review. NewBioWorld, A Journal of Alumni Association of Biotechnology 5 (1), 254. Available from: https://doi.org/10.52228/NBW-JAAB.2023-5-1-5.

Yan, A., Wang, Y., Tan, S.N., Mohd Yusof, M.L., Ghosh, S., Chen, Z., 2020. Phytoremediation: a promising approach for revegetation of heavy metal-polluted land. Frontiers in Plant Science 11, 359. Available from: https://doi.org/10.3389/fpls.2020.00359.

Yang, Y., Ho, H.C., Lolli, S.L., Yang, X., 2022. Editorial: Climate change, aerosol pollution and public health risk in an urban context. Frontiers in Climate. Climate Services 4. Available from: https://doi.org/10.3389/fclim.2022.862982.

Yasmin, T., Imadi, S.R., Gul, A., 2023. Nanotechnology in phytoremediation: application and future. In: Newman, L., Ansari, A.A., Gill, S.S., Naeem, M., Gill, R. (Eds.), Phytoremediation. Springer, Cham. Available from: https://doi.org/10.1007/978-3-031-17988-4_20.

Zhang, J.J., Wei, Y., Fang, Z., 2019. Ozone pollution: a major health hazard worldwide. Frontiers in Immunology 10, 2518. Available from: https://doi.org/10.3389/fimmu.2019.02518.

Zhang, S., Han, Y., Peng, J., Chen, Y., Zhan, L., Li, J., 2023a. Human health risk assessment for contaminated sites: a retrospective review. Environment International 171, 107700. Available from: https://doi.org/10.1016/j.envint.2022.107700.

Zhang, H., Du, P., Yuan, B., Zhang, Y., Chen, J., Liu, H., 2023b. Multifaceted insight into sensitivity analysis and environmental impact on human health of soil contamination risk assessment. Environmental Health 1 (3), 214−227. Available from: https://doi.org/10.1021/envhealth.3c00077.

Zuluaga-Astudillo, D., Ruge, J.C., Camacho-Tauta, J., Reyes-Ortiz, O., Caicedo-Hormaza, B., 2023. Diatomaceous soils and advances in geotechnical engineering - Part I. Applied Sciences 13, 549. Available from: https://doi.org/10.3390/app13010549.

Further reading

Frank, A.L., 2020. Global use of asbestos - legitimate and illegitimate issues. Journal of Occupational Medicine and Toxicology 15, 16. Available from: https://doi.org/10.1186/s12995-020-00267-y.

Onu, M.A., Ayeleru, O.O., Oboirien, B., Olubambi, P.A., 2023. Challenges of wastewater generation and management in sub-Saharan Africa: a review. Environmental Challenges 11, 100686. Available from: https://doi.org/10.1016/j.envc.2023.100686.

Texas Disposal Systems, 2023. Land Pollution: Causes, Effects and Prevention. https://www.texasdisposal.com/blog/land-pollution/. (Accessed 16 August 2023).

Williams, P.R.D., von Stackelberg, K., Guerra Lopez, M.G., Sanchez-Triana, E., 2021. Risk analysis approaches to evaluating health impacts from land-based pollution in low- and middle-income countries. Risk Analysis 41, 1971−1986. Available from: https://doi.org/10.1111/risa.13699.

5

Combatting effects of ionizing radiation exposure around uranium and gold mines in Sub-Saharan Africa

"The long-term risks to health of low radiation levels are still poorly known. A combination of more studies of exposed populations and basic research are needed."

David J. Brenner (Brenner, 2011).

"There is no safe level of exposure and there is no dose of (ionizing) radiation so low that the risk of a malignancy is zero"

Karl Z. Morgan, was dubbed the father of Health Physics (Morgan, 1978).

Key chapter features

1. A brief overview of natural background radiation, the different sources, and enhanced exposure to natural background radiation
2. A summary description of the requirements for radiological sampling and analysis required for different purposes during the different phases of mining and mineral processing operations
3. A summary of the regulatory framework pertaining to international requirements and guidelines, as well as national legislations and regulations in radioprotection
4. A bullet-point listing of urgently needed research imperatives on ionizing radiation exposure in Sub-Saharan Africa uranium and gold mining
5. The role of the Medical Geologist in understanding ionization radiation effects, and in contributing toward better diagnosis and therapy for associated maladies

(Box 5−1)

Introduction

Uranium and gold mining activities are increasing in Sub-Saharan Africa, where mining is not always strictly regulated and controlled. In this Chapter, after a brief overview of the

Medical Geology of Africa. DOI: https://doi.org/10.1016/B978-0-12-818748-7.00009-5

BOX 5–1   Learning objectives.

- Acquisition of a basic understanding of ionizing radiation, its sources, and pathways of exposure
- Understanding the range of potential health effects of exposure to ionizing radiation on human health and the environment
- Understanding the basic principles of radiation protection and their application
- Identifying and formulating urgently needed research themes in the area of ionizing radiation uptake and doses in Sub-Saharan Africa uranium and gold mining centers, especially with regard to miners, pregnant women, and young children
- Information packaging and presentation by Medical Geologists, of ionizing radiation health effects and their mitigation to government policy makers
- Demonstrating how science-based knowledge and understanding of ionizing radiation uptake and its subsequent fate in human physiological dynamics can help formulate more stringent policies regarding mitigation and prevention

fundamentals of ionizing radiation, its characteristics, and behavior, examples are given of the various impacts of its exposure, due to uranium and gold mining on the environment (air, soil, water, and biota) and human health, especially on miners and associated communities most notably on the developing fetus and young children; and recommendations provided for limiting or obviating these impacts.

The effects of ionizing radiation on the developing fetus depend on several factors, which can properly be evaluated only when jointly assessed, not by medical practitioners alone, but by a network or team comprising of Medical Geologists, geochemists, and allied specialists able to chart radiation source areas, migration dynamics, intake doses, and other geochemical characteristics. These effects include those directly related to the level and rates of exposure (as determined by a geochemist) and stage of fetal development (as determined by a pediatric clinician).

Finally, an assessment is made of mitigation and legislative measures that governments and industry in uranium and gold mining centers in Sub-Saharan African countries are taking; and these practices are compared with those in the Western World's two largest producers of uranium, *viz.*, Canada and Australia, where regulation is more robust and stringent. *This, it is hoped, would inform the development of novel methods for modeling ionizing radiation generation and intake schemes for improved prognosis, diagnoses, and therapy, for planning interventions as well as for formulating more appropriate legislative articles.*

A list of urgently needed research themes covering various aspects of ionizing radiation exposure in uranium and gold mining, milling, and waste disposal in Sub-Saharan Africa is presented at the end of the Chapter.

Background

We do know that uranium and gold mining operations in Sub-Saharan Africa have high impacts on the environment and society (e.g., Scheele et al., 2011; Dasnois, 2012; Chareyron et al., 2014; Hecht, 2012a; Winde et al., 2019; Zupunski et al., 2023), and can lead to deterioration of health of miners, other mine workers and surrounding communities. *However, the extent of exposure, the exposure pathways, rates of uptake and dose levels, as well as pathologies and therapy for ionizing radiation maladies are aspects that are still urgently in need of further research.*

In Sub-Saharan Africa, several studies have looked at the application of medical diagnostic and therapeutic ionizing radiation (e.g., Moifo et al., 2017; Rose and Rae, 2017; Chimangwa et al., 2018; Herbst/CANSA, 2019a; Kawooya et al., 2022), even though we still see the *paucity of stringent regulatory guidelines, training programs in radiation protection, consistent dose assessment protocols, quality control tests, and shielding* (See, e.g., Simo et al., 2018; Dauda et al., 2019; De-Kun et al., 2020; Tefera et al., 2020; Lewis et al., 2023). However, the scope of this Chapter does not include medical applications of ionizing radiation; and only a cursory mention of it is made in relevant the sections of the Chapter. Notwithstanding, a number of pertinent references are given in the Section: "Selected Further Reading" (this Chapter), *to aid the search process of Medical Geology researchers and coworkers wanting to launch further investigations on aspects of this pertinent and important subject.*

We know, from countries where good morbidity estimates exist, e.g., South Africa, that ionizing radiation exposure in major uranium and gold mining centers can lead, in particular, to a high percentage of neonatal deaths due to fetal abnormalities or congenital malformations (See, e.g., Shaw et al., 2011; Dauda et al., 2019). Although some recent studies indicate progress in knowledge of the effects of ionizing radiation on the embryo (e.g., Khadzhidekova et al., 2001; WHO, 2016; Simo et al., 2018; Saada et al., 2023) those often tasked with such studies are from a purely medical background, e.g., pediatric clinicians and radiologists. They go about their work, often taking little cognizance of uranium's geogenic characteristics and other geodynamic intricacies that bear so heavily on the mode of radioactivity uptake, the doses received, the dose rate, and dose distribution (See, e.g., Moges et al., 2023). These are physical exposure parameters that result in both qualitatively and quantitatively different cellular and molecular responses, thus demonstrating nonlinear responses concerning dose. Medical Geologists therefore, need to demonstrate unequivocally that their inclusion in teams investigating the effects of ionizing radiation exposure on pregnancy outcomes (neonatal and child radiation-related morbidity and mortality) in major uranium and gold mining countries of the Sub-Continent can lead to much-improved pathogenesis, diagnosis, and therapy, as well as prevention and other forms of intervention for these conditions. Such collaboration would constitute a leading example on the global public health stage on how the critical need for greater communication and coordination between the fields of radiation biology, geomedical science, and environmental toxicology can be met.

A brief history of radiation science and health

The relationship between radiation and health has been a subject of interest to scientists since the late 19th century. X-rays were discovered by Wilhelm Conrad Röntgen in 1895 (Babic et al., 2016) and associated dangers with exposure, were soon realized. In 1896 the first maladies due to X-ray exposure were documented, and in 1904 Thomas Edison's assistant, Clarence Dally, became the first person reported to have died as a result of X-ray exposure (Sansare et al., 2011). In spite of the injuries known at that time, the use of X-rays for a variety of applications rapidly gained traction. Portable X-ray instruments appeared in the early 1900s and were used on both warring sides during the First World War (1914−18) to locate shrapnel and to set broken bones (Matanoski et al., 2001). Radium was discovered in 1898 (Mould, 1998) and its use in medicine also spread very rapidly. In the 1920s, scientists and doctors who investigated cases of sickness and death in watch dial painters who ingested small quantities of radium in their work, quickly realized that internal exposure to radium could be harmful (US EPA, 2018). During that same period, the cancer risks from radiological applications became apparent. In 1928 a committee of experts established the first internationally recognized radiation safety principles and guidelines (See Section: "Ionizing radiation protection: legislatory and regulatory guidelines," in this Chapter).

The 1930s saw the emergence of World War II, and plans were already underway at that time for building a nuclear bomb. The discovery of plutonium took place in 1940 (Clark, 2016) and the Manhattan Project (Norris, 1996), with its goal of "make the bomb," was launched soon afterward. In 1945 the first nuclear weapon was detonated in New Mexico, USA (Widner and Flack, 2010). The resulting fallout from that test was at that time considered a revelation. Two nuclear bombs were dropped on Japan in Nagasaki and Hiroshima during August of that same year (1945). In the 1950s, the USA government put forth the concept of "civil nuclear power" signaling a campaign that engendered a friendlier attitude toward nuclear science (Young, 2003). At that time, the USA was operating hundreds of uranium mines across the country, serving both the nuclear power industry and the nuclear weapons industry. The Nevada Test Site, established in 1951, was intended to serve as a testing location that could ease logistical and security concerns (Fehner and Gosling, 2000). Before its inauguration, tests had already been conducted in the South Pacific. The fallout from the test sites and the occupational hazards of the workers in the mines and in the weapons factories were adding to the range of radiation exposures that were experienced.

Radiation fundamentals

We begin with a brief look at the characteristics (including the physics and chemistry) of *radiation*. Although the terminology is seemingly extensive and technical, an attempt is made here to express it in simple terms; and the information conveyed on the basics is kept to the minimum that Medical Geologists and allied researchers (the uninitiated, unfamiliar with the basics) may need in their investigations on the impacts of ionizing radiation exposure from mining and associated activities.

Radioactivity is a measure of the amount of radioactive decays that occurs in some form of matter during the spontaneous disintegration of atoms in that matter. The excess energy emitted is a form of *ionizing radiation* (our Chapter focus). Radioactivity cannot be destroyed, but could be converted into a stable form or extracted from the environment.

Radioactivity is a part of our Earth, having existed since the Earth was formed some 4.5 billion years ago. Naturally occurring radioactive materials are present in the Earth's crust, the floors and walls of our homes, classrooms, and offices. Our own bodies—our muscles, bones, and tissue—are naturally radioactive, and so also is the food chain, because we because we eat, drink, and breathe radioactive substances that are naturally present in the environment.

We will be better equipped to understand ionizing radiation and its effects, as we acquire more familiarization with its characteristics—causes, sources, physics, chemistry, exposure pathways, and so on.

Characteristics of radiation

Fast-moving energy that is emitted from a source and travels through material or space in the form of particles or electromagnetic waves, is called *radiation*. It occurs in atoms, when the unstable nuclei decay and release particles. The term covers many different forms of emission, e.g., visible light (VIS), ultraviolet (UV) light, heat, radio waves, microwaves from an oven, radar, X-rays, gamma rays from radioactive elements, and so on. Radiation is also the general name for the kind of energy given off by radioactive atoms, such as uranium and thorium.

Radioactive elements produce many types of radiation as they undergo decay. They can emit electrons, alpha particles, neutrons, neutrinos, and gamma rays. Measurement of these forms of radiation is given in units of Becquerel (Bq), which is defined as one decay per second (See Section: "Dosimetric quantities," in this Chapter, for more on definitions of units of radiation measurement).

Sources of radiation

"Natural sources account for most of the radiation we all receive each year."

WNA (2020a).

Natural background radiation

Natural background radiation pervades our entire environment (Fig. 5−1) and is present on Earth at all times. Most natural substances contain a quantity of radioactive material, but most background radiation comes naturally from minerals, with a small fraction coming from man-made elements. The human body, as already noted, also contains some of these naturally occurring radioactive minerals. There can be large variances in natural background radiation levels from place to place, as well as changes in the same location over time.

Natural radiation sources also include radon, cosmic radiation, and elements in the ground (Fig. 5−2). Radon is found in many places, but most prominently in the soil. The ground all around us has varying levels of naturally occurring radioactive elements that can produce *radon gas* (See Section on: "RADON," this Chapter). Radon is the leading source of radiation exposure and the second leading cause of lung cancer worldwide (ALA, 2020;

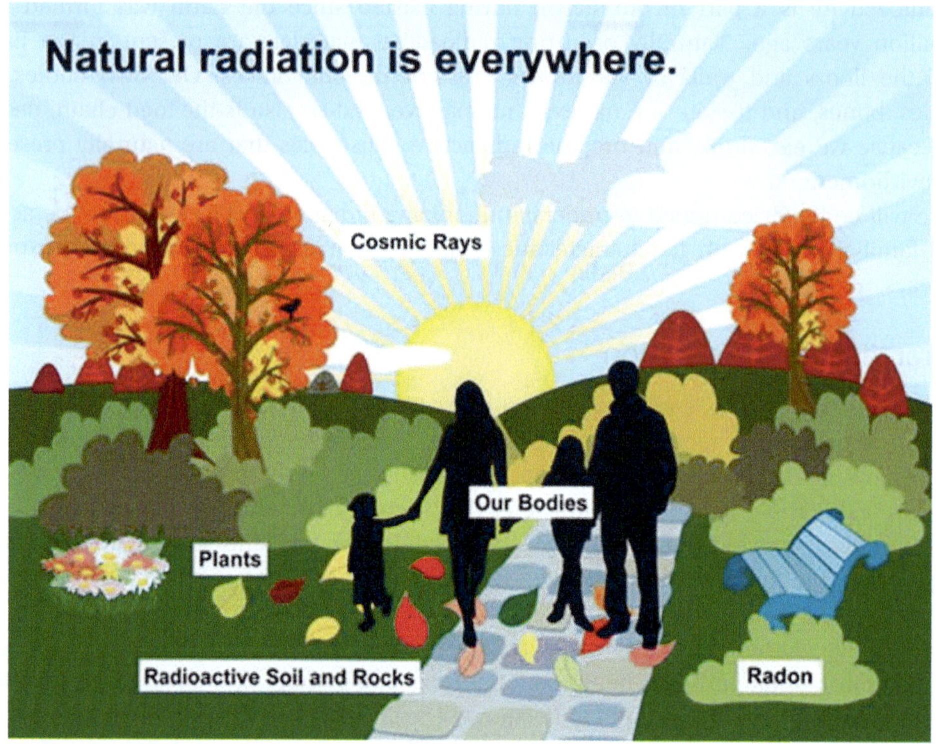

FIGURE 5–1 Natural background radiation (CNSC, 2013). *From CNSC (Canadian Nuclear Safety Commission), 2013. Natural Background Radiation. https://nuclearsafety.gc.ca/eng/resources/fact-sheets/natural-background-radiation. cfm. (Accessed 07.05. May 2020).*

Riudavets et al., 2022); *but very little is known about the incidence of lung cancer in Sub-Saharan Africa as a result of radon exposure* (Herbst/CANSA, 2019b).

The following *three* sources account for all the natural background radiation, we experience:

1. Cosmic Radiation
2. Terrestrial Radiation
3. Internal Radiation (US NRC, 2017).

Cosmic radiation

Cosmic radiation is comprised of heavily charged, extremely energetic particles, e.g., gamma rays, generated from outer space (the sun and stars), and certain celestial events such as supernova explosions, and contributes to the background radiation around us.

Differences in elevation, atmospheric conditions, and the Earth's magnetic field (MF) can change the amount (or *dose*) of cosmic radiation that we receive. The higher the altitude, the higher is the dose.

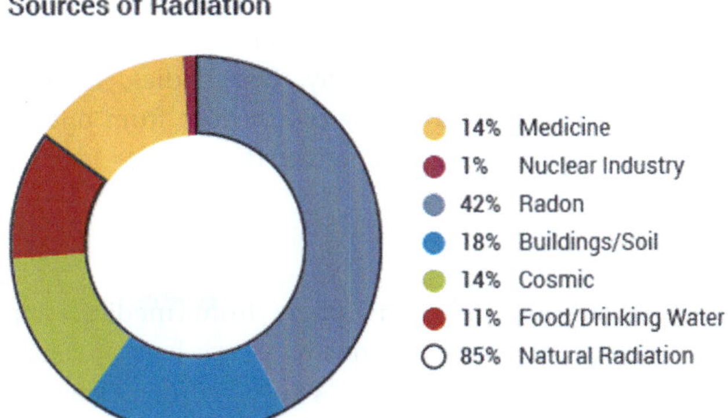

FIGURE 5–2 Sources of radiation. *From WNA (World NuclearAssociation), 2020a. Nuclear Radiation and Health Effects. https://www.world-nuclear.org/information-library/safety-and-security/radiation-and-health/nuclear-radiation-and-health-effects.aspx. (Accessed 07 May 2020).*

The Earth's atmosphere can absorb and filter some cosmic radiation, but some still make it through; so as the thickness of atmosphere above the receptor increases, we receive smaller and smaller doses (IAEA, 2004). Similarly, the intensity of cosmic rays at the altitude where aircraft fly is much greater than on the ground.

Terrestrial radiation

The Earth is a source of *terrestrial radiation*, long-wave electromagnetic radiation originating from the Planet and its atmosphere. When the Earth was formed, a number of radioactive elements were formed too. After the c. 4.5 billion years since the Earth's formation, all the shorter-lived *isotopes* have decayed. However, some of these isotopes having very long *half-lives* (See Section: "Breakdown of ionizing radiation," in this Chapter) in the order of billions of years, are still present. These *radionuclides* are known as primordial radionuclides and contribute to the annual dose that each of us receives.

As noted earlier, radioactive materials exist naturally in rock, soil, water, and vegetation. The major isotopes of concern for terrestrial radiation are potassium, uranium, and the decay products of uranium, such as thorium, radium, and radon. Terrestrial radiation engenders external exposure caused by these radionuclides, which are in trace amounts all around us. The dose from terrestrial sources varies in different parts of the world, but locations with higher soil concentrations of uranium and thorium generally have higher dose levels. Everyone of us is exposed, on average, to 2.4 mSv per year (See Section: "Dosimetric quantities" for definition of units) of ionizing radiation from natural sources (WNA, 2020a).

Internal radiation

All humans have small amounts of internal radiation, mainly from traces of radioactive potassium-40 and carbon-14 (but still lesser quantities of lead-210, and other isotopes) inside their bodies from birth. These are derived from the food, water and soil (*cf., geophagy, Chapter 6*) that we ingest. The body metabolizes the nonradioactive and radioactive forms of potassium and other elements in the same way. Our bodies, therefore, are sources of radiation exposure to others. However, the variation in dose from one person to another is not as great as that associated with cosmic and terrestrial sources.

Man-made radiation sources

The largest source of man-made radiation comes from medical applications (*c.* 14%; Fig. 5–2); in particular, from the use of diagnostic X-rays by physicians to determine the extent of disease or physical injury (See under: "X-rays," of this Chapter). Ever since radiation was discovered, people have benefited from this use. In the field of nuclear medicine, radioactive compounds called *radiopharmaceuticals* are also used to support diagnoses, while a further source of radiation exposure is radiation therapy.

Man-made radiation also comes from consumer products, such as construction materials, televisions, fluorescent lamp starters, smoke detectors, tobacco, and lantern mantles. A newer source of X-rays is the use of backscatter scanners used in security operations at airports.

Radiation from the various sources described is commonly divided into two broad categories, *viz.*: (1) *ionizing radiation* and (2) *nonionizing radiation*.

First, we take a brief look at "nonionizing radiation," before going on to "ionizing radiation" which is our Chapter focus.

Nonionizing radiation

Nonionizing radiation is low-frequency radiation that disperses energy through heat and increased molecular movement.

Radiations from sources such as power lines, near UV light, VIS, infrared light, low-frequency radio waves, cell phones, microwave, and traffic radars are all classified as *nonionizing* radiation because they are not capable of removing an electron, having sufficient energy only for excitation.

Tulchinsky and Varavikova (2014) have described two types of nonionizing radiation, *viz.*, (1) *optic*, and (2) certain *electromagnetic fields* (EMF's). *Optic radiation* includes the regions of UV radiation, VIS, and infrared radiation. *EMFs* are produced by moving electrical charges, such as in power lines and those charges induced by microwave or radio frequencies. They are described in terms of wavelengths or frequency.

There are three main kinds of harmful effects of nonionizing radiation: *photochemical (sunburn or snow blindness), thermal, and electrical.* UV radiation has been reported to result in morbidities that include increasing incidence of squamousa and basal cell carcinoma as well as melanoma of the skin, a highly malignant cancer. Long periods of exposure to the sun, can, in addition to these skin cancers, cause skin and eye burns, cataracts, reduced immunity, and injury to blood vessels.

Long-term exposure to mobile phone use, high-voltage power lines, and radio and radar transmitters is thought to be associated with increased risk of cancer, *but this notion remains controversial* (Tulchinsky and Varavikova, 2014; Lai and Levitt, 2023). Microwave exposures at high levels can cause injury to vulnerable tissues, *but the level of dangerous exposure has not yet been conclusively determined.*

Lasers (an acronym for light amplification by stimulated emission of radiation) are pulsed electromagnetic waves that find applications in many electrical devices and play an important role in legal administration. They are also used for military, industrial, medical, and surgical purposes, skin treatments, hair removal, and radiation science research. Lasers that are not designed for medical use (and misused medical lasers) can result in irreparable retinal damage and severe burns. The potential for formation of cataracts in the lens of the eye is the main issue behind laser application, standards, safety, and control measures.

Excessive use of magnetic resonance imaging, computed tomography (CT) scans (described in the Section: "X-rays," in this Chapter) and airport total body scanners may also contribute to excess radiation exposure, as may the use of cell phones by children and adults. *Safer technology with reduced radiation should be a high priority as these useful instruments are increasingly used in Sub-Saharan Africa and globally.*

A *taser* (Tom A. Swift Electric Rifle) is a gunlike device used for firing electrical probes that transmit a 50,000-V electric shock, temporarily stunning, debilitating, or incapacitating someone. Those exposed to tasers lose control of their muscles, as the electricity causes involuntary muscle contractions (Tulchinsky and Varavikova, 2014).

Ionizing radiation

Ionizing radiation is a type of energy released by atoms in radioactive substances in the form of electromagnetic waves (gamma or X-rays) or particles (neutrons, beta, or alpha). It occurs when radiation that has enough energy to remove an electron from a neutral atom or molecule (i.e., to ionize them), creates a free radical. Ionizing radiation differs from *nonionizing radiation* in that it is very energetic and has the capacity to ionize matter when it interacts with it.

Naturally occurring uranium in the Earth's crust is the main source of ionizing radiation to which humans are exposed. Other major contributors of ionizing radiation are natural deposits of uranium, potassium and thorium, which release small amounts of ionizing radiation during the process of natural decay.

When an atom loses an electron, it becomes unstable, and retains either excess energy or mass. The energy of the ionizing radiation is transferred to electrons which are consequently stripped from atoms in the tissue. These energetic electrons cause further stripping of electrons, thus forming many ionized atoms, which have high-chemical reactivity and can break chemical bonds in molecules, including DNA molecules, and can directly or indirectly alter their normal structure and function. DNA damage can lead to cancer (See Section: "Radiation and DNA," in this Chapter).

As the use of ionizing radiation increases, so does the potential for health hazards, if not properly used or contained. Such hazards may result from exposures that take place during radiotherapy and chemotherapy; but our greatest concern in this Chapter is with the exposure of miners and sensitive populations to ionizing radiation from uranium and gold mining centers in Sub-Saharan Africa.

Breakdown of ionizing radiation

A substance is called *radioactive* when the nucleus is unstable, i.e., when some atoms change by themselves and give out ionizing radiation in the process. The process of radioactive decay (*breakdown*) began at the time of formation of the universe, with uranium and thorium being two of the constituents.

Isotopes

The word *isotope* is built from the two Greek words "*iso*" (the same) and "*topos*" (place). Isotopes are atomic forms of the same element, all having the same chemical characteristics and only slightly differing physical properties from one another, such as nuclear masses and nuclear structures. Isotopes have the same number of protons but different number of neutrons. These different forms of an element may be stable or unstable (radioactive). For example, uranium-234, uranium-235, and uranium-238 are the three isotopes of uranium, all of which are radioactive, but each with a different nuclear mass (Fig. 5−3).

The activity

The term *activity* is used as a measure of the rate at which spontaneous transformations occur in a given amount of a radioactive material. Activity is related to mass because the greater the mass of radioactive material, the more atoms are present to undergo radioactive decay. Units used in measuring activity (the Becquerel or the Curie are described in the Section: "Dosimetric quantities," in this Chapter).

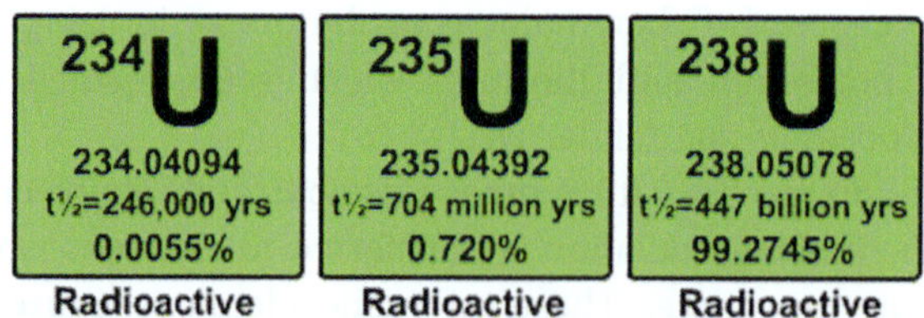

FIGURE 5–3 Uranium has three long-lived radioactive isotopes. *Credit: SAHARA Project, United States Geological Survey (USGS). From http://web.sahra.arizona.edu/programs/isotopes/uranium.html. (Accessed 15 July 2020).*

Half-lives and decay chains

Several radioactive isotopes (or *radionuclides*) occur naturally and are present in rocks, soils, river water, and seawater. Radionuclides undergo radioactive decay leading to the emission of ionizing radiation.

When it decays, a radionuclide transforms into a different atom—a decay product. The atoms keep transforming into new decay products until they reach a stable state and are no longer radioactive. The amount of a radionuclide decreases during the decay with its radioactivity also decreasing proportionately. These decreases cannot be brought about by any chemical or physical processes, such as cooling, heating, or melting of the radioactive substance. Only the passage of time brings about the reduction of the radioactivity of a substance.

The *half-life* is the time required for the activity of a radionuclide to decrease by decay to one half of its initial value. The half-life can range from a fraction of a second to billions of years (e.g., polonium-214, 0.00016 seconds; iodine-131, 8.04 days; carbon-14, 5730 years; uranium-238, 4.5 billion years). Radiation will continue to be emitted until finally a stable (nonradioactive) nuclide is formed.

The majority of radionuclides decay just once before becoming stable. Those that decay in more than one step are called *series radionuclides*. The series of decay products created to reach this balance, forming nuclides, is called the *decay chain*. Each series has its own unique decay chain, within which the decay products are always radioactive. Some decay products are different chemical element.

Many of the naturally occurring radionuclides belong to four distinct radioactive series (Fig. 5–4): the uranium (U-238), actinium (U-235), thorium (Th-232), and neptunium (Np-237) series, named according to the *progenitor* (or parent) to the series products. For Np-237, however, no primordial sources exist any longer, because its half-life is only 2.1 million years (Martin, 2006), which means that the level of decay of natural sources of Np-237 was insignificant since their creation some 4.5 billion years ago.

In the uranium-238 radioactive series, decay takes place very slowly, through a series of steps, its daughter products themselves being radioactive. There are fourteen radioactive nuclides, the last in the chain, polonium-210 transforming to lead-210, and eventually to the stable nuclide, lead-206 (Fig. 5–4; Table 5–1). Only the final, stable atom in the chain (lead-206) is not radioactive.

All the radioactive daughter products are metals (thorium-234, thorium-230, radium-226, lead-210, polonium-210, etc.) except one, radon-222, which is a radioactive gas.

Thorium, just like uranium, decays slowly, with its series having eleven radioactive elements (Fig. 5–4). The typical uranium-238 activity of the Earth' crust is about 40 Bq/kg, a level of radiation that has decreased by twofold since the creation of the Earth (Chareyron, 2008). This is so, because uranium-238 has a very long *half-life* equal to the age of Planet Earth (*c.* 4.5 billion years).

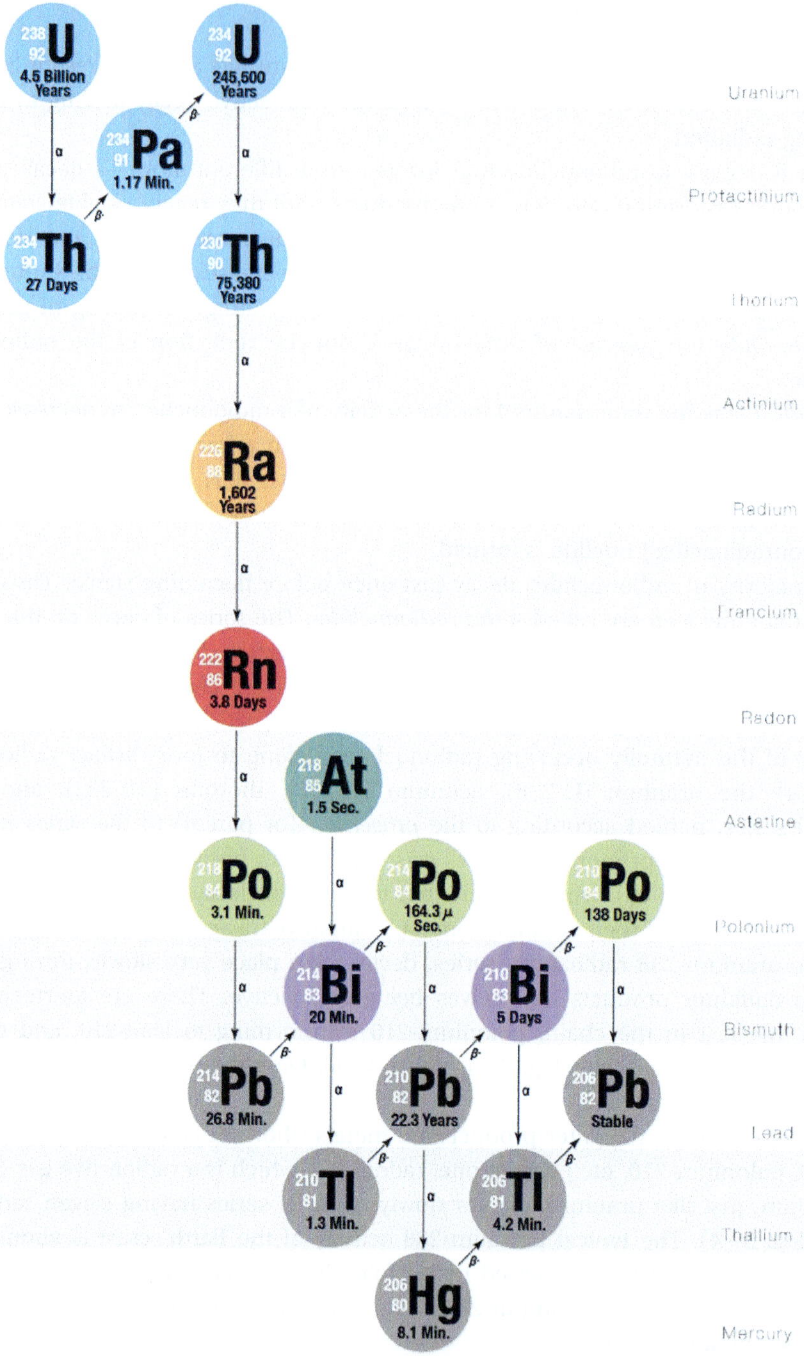

FIGURE 5–4 Uranium-238 decays through a series of steps to become a stable form of lead. Each step indicates a different nuclide. The numbers below each label indicate the length of a radionuclide's half-life. *Credit: InterNACHI, Copyright © 2008–2020. From https://www.nachi.org/gallery/radon/uranium-238-decay-chain. (Accessed 09 August 2020).*

Table 5–1   Main characteristics of Uranium-238 decay products (Browne and Firestone, 1986).

Radionuclide	Half life	Decay mode	Main X- or gamma emission	
			Energy (keV)	Intensity (%)
Uranium-238	4.47 billion years	Alpha	16.6	4.1
Thorium-234	24.1 days	Beta	63.3	3.8
Protactinium-234^m	1.17 min		1.001	0.65
Uranium-234	245.000 years	Alpha	53.2	0.12
Thorium-230	75.400 years	Alpha	67.7	0.38
Radium-226	1.600 years	Alpha	186.1	3.28
Radon-222 (gas)	3.8 days	Alpha	510	0.07
Polonium-218	3.1 min	Alpha	–	–
Lead-214	26.8 min	Beta		37.1
Bismuth-214	19.9 min	Beta	609.3	46.1
Polonium-214	0.16 ms	Alpha	798	0.01
Lead-210	22.3 years	Beta	46.5	4
Bismuth-210	5 days	Beta	–	–
Polonium-210	138.4 days	Alpha	803.1	0.001
Lead-206	Stable element (no decay)			

Source: Compiled from Browne, E., Firestone, R.B., 1986. Table of Radioactive Isotopes. John Wiley and Sons, pp. 237–243 and WUP (Wise Uranium Project), 2016. Uranium Radiation Properties. https://www.wise-uranium.org/rup.html. (Accessed 05 June 2020).

Radionuclides can be identified uniquely by the type of radiation they emit, the energy of the radiation, and their half-life (Table 5–1). Radioactive substances emit three principal types of ionizing radiation. These are: (1) *alpha particles*, (2) *beta particles*, and (3) *gamma rays*.

Alpha particles

Alpha particles (Fig. 5–5) consist of two protons and two neutrons, making them identical to the nucleus of a helium atom. Alpha particle radiation occurs when an unstable nucleus (the parent nucleus) releases a particle equivalent to the nucleus of a helium atom (two neutrons and two protons), thus leaving a nucleus with two less protons and neutrons (the daughter nucleus).

They result from the radioactive decay of the heaviest radioactive elements, such as uranium, radium and polonium. Alpha particles are positively charged, relatively large and heavy particles, which, though very energetic, cannot travel long distances, even in air; nor are they able to penetrate the skin or a sheet of paper.

Alpha particles can get inside the body through inhalation of radioactive dust, ingestion (swallowing) of alpha-emitting radioactive material, or through a cut. When once inside the body, alpha-emitting radioactive elements can pose a risk to living tissue. Sensitive body organs, such as the lungs and bones can be affected. The physical characteristics of alpha particles (large, heavy) and the nature of their interactions within the body (the ionizations they cause are very close together, resulting in the release of all their energy in a few cells)

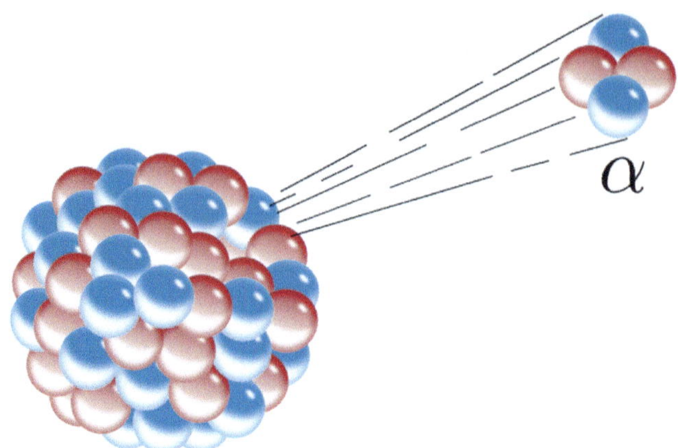

FIGURE 5–5 Alpha particle (Wikipedia, 2020a). *From Wikipedia, 2020a. Alpha particle. https://en.wikipedia.org/wiki/ Alpha_particle. (Accessed 24 July 2020).*

makes them more dangerous than other types of radiation. For example, they can inflict more severe damage to cells and DNA.

Thankfully, it is possible to reduce this risk to a minimum, simply by ensuring that the inhalation or ingestion of materials that emit alpha particles is obviated, either by installing dust control equipment or by the appropriate use of personal respiratory protection devices, such as dust masks.

Beta particles

Beta particles (β) (Fig. 5—6) are small, high-speed electrons (particles) with a negative electrical charge. They are emitted from the nucleus of certain unstable atoms in many radioactive elements during a decay process. Hydrogen-3 (tritium), carbon-14, and strontium-90 are examples of beta emitters. Since beta particles are smaller, they are more penetrating than alpha particles (roughly a centimeter into tissue), and are much less damaging to living tissue and DNA. This is so, because the ionizations they produce are more widely spaced. Some beta particles are capable of penetrating the skin and causing dermatological maladies like skin burns. However, as with alpha-emitters, beta-emitters are most hazardous when they are inhaled or swallowed.

Beta particles travel farther in air than alpha particles. They can pass through a piece of paper, but can easily be shielded by a few millimeters of most other materials, such as a layer of clothing or by a thin layer of a substance such as a sheet of aluminum foil.

Gamma rays

Gamma rays (γ) (Fig. 5—7) are photons, or weightless packets of light energy that are lost when the particles within the nucleus reorganize into more stable arrangements and accompany other forms of radiation. Gamma rays are pure energy, unlike alpha and beta particles,

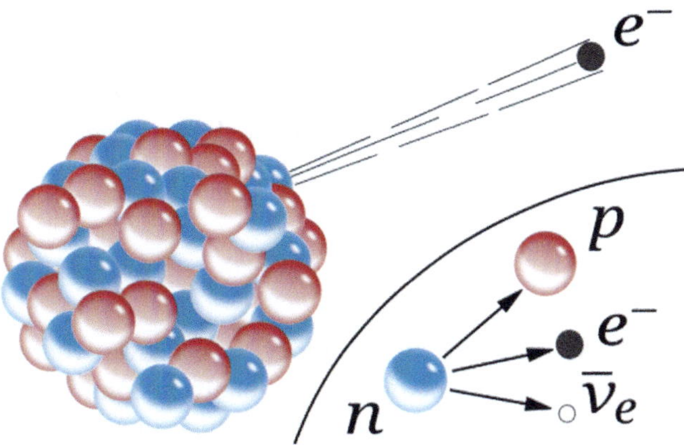

FIGURE 5–6 Beta decay. The inset shows beta decay of a free neutron (Wikipedia, 2020b). *From Wikipedia, 2020b. Beta decay. https://en.wikipedia.org/wiki/Beta_decay. (Accessed 09 May 2020).*

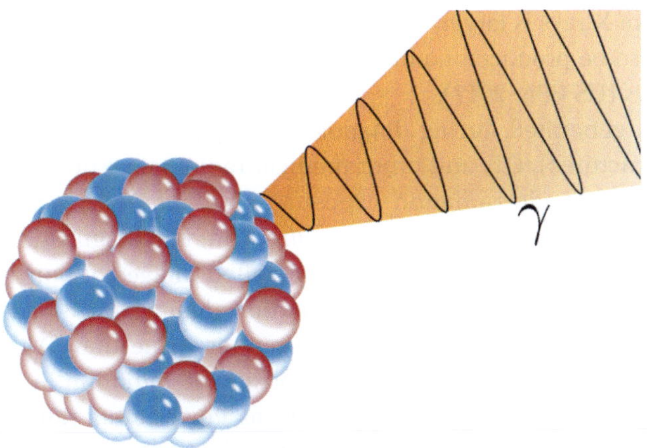

FIGURE 5–7 Illustration of an emission of a gamma ray (γ) from an atomic nucleus (Wikipedia, 2020c). *From Wikipedia, 2020c. Gamma ray. https://en.wikipedia.org/wiki/Gamma_ray. (Accessed 11 May 2020).*

which have both energy and mass. Gamma rays are similar to VIS, but their energy is much higher, enabling them to travel great distances through air.

Sources of gamma rays include emissions from nuclear power plants, radiation science research, military weapons testing, and nuclear medicine procedures, such as bone, thyroid, and lung scans (Brenner, 2011).

Gamma rays, just like X-rays, are very penetrating. They can pass through the body, as well as penetrate materials, such as clothing, paper, and aluminum foil, barriers that can stop alpha and beta particles. As they pass through the body, they can cause ionizations that can damage tissue and DNA. Gamma rays therefore require some form of shielding. They can be stopped by materials, such as thick shields of concrete or metal (lead) plating.

Gamma rays pose a health risk, regardless of whether the radioactive material is within or outside the body.

Since the naturally occurring radionuclides are present throughout the Earth's crust, background levels of gamma radiation are affected by the local geology. It is thus expected that the contribution from gamma radiation in different parts of the world, not least in Africa, may vary over several orders of magnitude.

X-rays

Due to their widespread use in diagnostic radiology, in a region where conditions such as tuberculosis and coal workers' pneumoconiosis are relatively common (See Section: "Other mineral dusts" in Chapter 4), many African people have heard of *X-rays*.

Just like gamma rays, X-rays are photons of pure energy. However, they differ from gamma rays in that they originate in electron fields around the nucleus. X-rays and gamma rays have sufficient energy to penetrate and damage body tissue below the surface of the skin; but, being generally lower in energy, X-rays are less penetrating than gamma rays.

The generation of X-rays is largely artificial and can be produced by machines using electricity; but it can also be produced naturally. X-rays are also used in industry for inspections and process controls (US EPA, 2017).

Medical X-rays, generated during diagnostic imaging, fluoroscopy (real-time imaging of moving body structures), CT, and other medical radiological procedures, are the single largest source of man-made radiation exposure. *Computerized tomography*, commonly known as a CT or *computerized* axial *tomography* scan, is a diagnostic imaging technique that uses special X-ray equipment to produce detailed images of bones and soft tissue in the body, thus providing more-detailed information than plain X-rays can. In 2007 concerned about the increasing use of CT scans, Brenner wrote that "Although the risks for any one person are not large, the increasing exposure to radiation in the population may be a public health issue in the future." However, as Power et al. (2016) noted, the use of automated exposure control, new image reconstruction techniques that reduce radiation dose, and other improvements in the technique, have been introduced in recent years, with promising results.

By 2005, Sub-Saharan Africa already had a large supply of X-ray equipment in teaching hospitals, though most of them remain inoperative because of lack of maintenance programs (Rabinowitz and Pretorius, 2005).

The damage potential of ionizing radiation

Different types of radiation have distinct levels of *damage potential*, described by the radiation's *linear energy transfer (LET)*. This is the energy the radiation deposits in the surrounding medium, and thus its potential damage per unit of distance traveled. Alpha radiation has a high LET because it deposits a relatively large amount of energy in a small area before it stops. Beta, gamma, and X-radiation have low in LET because they deposit energy in more diffuse patterns.

Naturally occurring radioactive materials and technologically enhanced naturally occurring radioactive materials

Exposure to naturally occurring radiation is responsible for a large proportion of an average person's yearly radiation dose, and is therefore not usually considered to be of substantial health or safety risk. However, certain occupational activities can give rise to significantly enhanced exposures that may have to be controlled by regulation (IAEA, 2018a).

The acronym "NORM" is generally used to describe radioactive materials which occur naturally, and where human activities increase the exposure of people to ionizing radiation. This potentially includes all radioactive elements found in the environment. From a radiation protection standpoint, the two most important NORMS are the long-lived radioactive elements, uranium and thorium and their decay products, such as radium and radon. These elements have always been present in the Earth's crust and in the atmosphere, and are concentrated in certain localities, such as uranium ore bodies, which may be mined.

Mining of uranium exposes those involved to radiation from radionuclides in the uranium-238 and thorium-232 decay series in the uranium ore body. NORM also results from activities, such as mining and burning of coal, making and using fertilizers (phosphate), oil and gas production, building industry, recycling and mineral sands (rare earth minerals, titanium, and zirconium) production.

In the Kamiesberg Project in Namaqualand, South Africa, for example, African Radiation Consultants (ARC) called attention in 2014, to potential harm caused by mining and mineral processing operations. According to ARC (2014), various wastes and rejects are generated during these operations especially during mineral separation and beneficiation processes. Due to the presence of naturally occurring radionuclides in the mineral sand deposit, many of these wastes and rejects are likely to contain NORMS. Consequently, ARC (2014) emphasized the need for great care and a huge sense of responsibility in handling of materials that generate NORM in mining operations, in the interest of worker and public safety.

Another example of the occurrence of NORM is radon infiltration into homes, giving rise to concern in many residential settings in the neighborhood of uranium mines and granite basements in Africa (See, e.g., Mathuthu et al., 2016; Ilori and Chetty, 2023). This calls for action to control it by applying one or the other of a number of available radon mitigation or prevention techniques, such as ventilation (See Section: "Radon migration in the home," in this Chapter for a comprehensive treatment of pertinent aspects of radon science).

Other NORM issues generally considered in the scientific literature include occupational health exposure to higher levels of cosmic radiation of flight crew, frequent flyers, and mountain dwellers, and to the exposure of tour guides to radon in caves and exposure of underground miners to radon (See Section: "Radon control in mines," in this Chapter).

Essentiality of the norm, potassium

Some NORMS are actually beneficial to human health. For example, potassium-40 (K-40), a major source of terrestrial NORM, has a long half-life of 1.251×10^9 years (WNA, 2020b), meaning that it still exists in measurable quantities today. It undergoes beta decay, mostly (89.28% of events) to calcium-40, and forms 0.012% (20 ppm) of natural potassium, which is otherwise made up of stable K-39 and K-41. Potassium is the seventh most abundant element in the Earth's crust, and K-40 averages 850 Bq/kg there. Potassium found in a range of foodstuffs, such as bananas, is an important dietary requirement, pertaining to the proper development of our bones [humans have about 65 Bq/kg of K-40 (Khan and Nakhabov, 2020)]. It is an essential ingredient for the proper functioning of nerve and muscles cells, particularly heart muscle cells. A 70 kg person has 4400 Bq of K-40 and 3000 Bq of carbon-14 (Khan and Nakhabov, 2020). Zhu et al. (2009) showed that potassium intake correlates positively with bone density in elderly women, suggesting that increasing consumption of food rich in potassium may play a role in osteoporosis prevention.

The acronym TENORM, or technologically enhanced NORM, is often used to refer to those materials where the amount of radioactivity, or radionuclide distributions, have actually been increased or concentrated, and exposed to the accessible environment as a result of human activities, technological or other industrial processes, such as manufacturing, mineral extraction, or water processing.

Pathways of occupational exposure to ionizing radiation

People are exposed to ionizing radiation under different circumstances, at home or in public places (public exposures), in a medical setting (as are patients, caregivers, and volunteers) or at their workplaces (*occupational exposures*).

Occupational exposure refers to the exposure of workers that takes place during the course of their work, regardless of the exposure situation. The various occupational activities that can lead to exposure to radiation include work associated with the different stages of the nuclear fuel cycle (WNA, 2016; Fig. 5–8); the use of radiation in medicine, scientific research, agriculture and industry; and occupations that involve exposure due to natural sources, such as the mining and milling of uranium and gold ores (See Section: "Uranium mining and processing," this Chapter).

The nuclear fuel cycle

The *nuclear fuel cycle* (Fig. 5–8) comprises the various steps involved in the production of electricity from nuclear reactions. Uranium mining, processing, decommissioning, and radioactive waste disposal are all the parts of this cycle and they present some of the more serious ionizing exposure situations likely to be encountered by proximal populations in the "Africa Uranium Story."

A number of theoretical considerations on pathways of human uptake of significant exposures in mining of uranium and gold ores are given in the literature, *but few data on directly measured human exposure are available.*

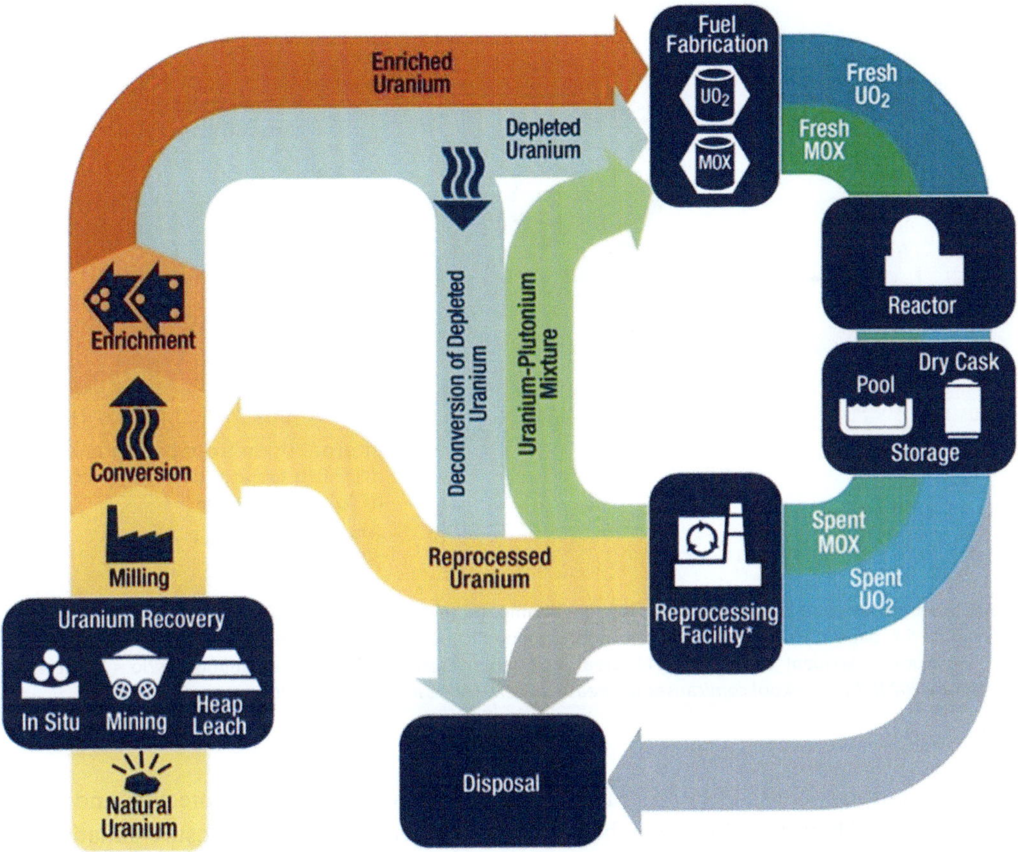

FIGURE 5–8 Key phases in the nuclear fuel cycle. *Credit: Wikipedia. From https://en.wikipedia.org/wiki/ Nuclear_fuel_cycle#/media/File:The_Nuclear_Fuel_Cycle_(44021369082)_(cropped).jpg. (Accessed 18 July 2020).*

Types of exposure

Understanding the type of radiation received, the radiation exposure pathway (external vs internal) (Fig. 5−9), and for how long a person is exposed are all important in estimating ionizing radiation health effects.

Internal exposure to ionizing radiation occurs through inhalation or ingestion of a radionuclide, or when it otherwise enters the bloodstream (for example, by direct injection or through wounds). Such exposure stops when the radionuclide is eliminated from the body, either spontaneously (such as through excreta) or as a result of a treatment.

External exposure refers to the deposition of airborne radioactive material, (such as dust, liquid, or aerosols) on skin or clothes. Airborne radioactive material can often be removed from the body by simply washing.

Irradiation from an external source can also occurs from exposure to ionizing radiation through diagnostic radiology (using X-rays). If the radiation source is shielded or when the person moves outside the radiation field, external irradiation stops.

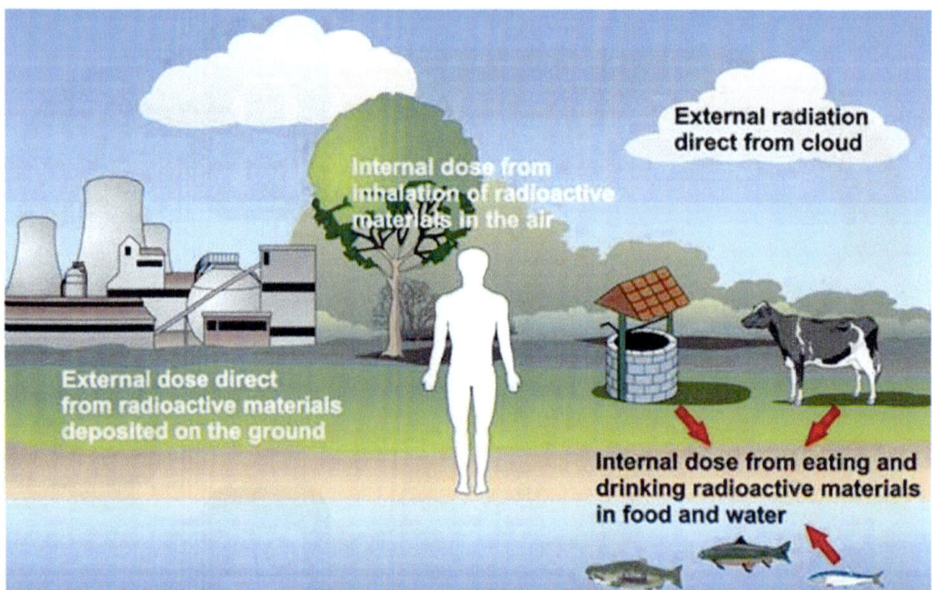

FIGURE 5–9 Exposure pathways to humans from environmental releases of radioactive material. *Source: IAEA, 2006; Figure 5.1, page 100.* https://www.researchgate.net/publication/271829775_Preliminary_Dose_Estimation_From_the_Nuclear_Accident_After_the_2011_Great_East_Japan_Earthquake_and_Tsunami/figures?lo = 1 *(Accessed 18 February 2024). https://vkool.com/causes-of-breast-cancerl. (Accessed 12 April 2020).*

For the purpose of establishing practical requirements for radiation protection and safety, GSG 8 (IAEA, 2018a,b) distinguishes between three different types of exposure situation: (1) planned exposure situations; (2) emergency exposure situations; and (3) existing exposure situations. According to IAEA (2018a,b):

"*(1)* A planned exposure situation *is a situation of exposure that arises from the planned operation of a source or from a planned activity that results in an exposure from a source.*

(2) An emergency exposure situation *is a situation of exposure that arises as a result of an accident, a malicious act, or any other unexpected event, and requires prompt action in order to avoid or reduce adverse consequences.*

(3) An existing exposure situation *is a situation of exposure, which already exists when a decision on the need for control needs to be taken.*"

Using examples from uranium mining in South Africa, Hecht (2012a) reviews the radiation hazards and occupational health in uranium production in South Africa, highlighting the major routes of entry of radionuclides (Fig. 5–10).

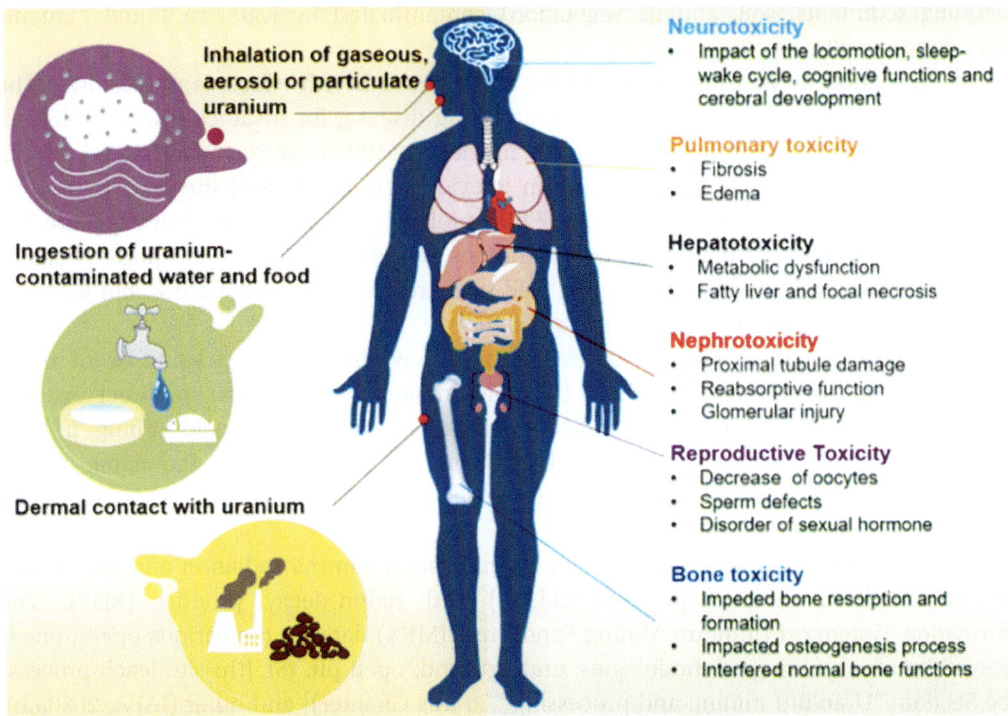

FIGURE 5–10 The primary exposure routes and health risks of uranium.Uranium can enter the human body through inhalation of gaseous and aerosoluranium, drinking water, food ingestion or dermal contact (Left), and induce health problems through impairing the kidneys, bones, liver, brain, lungs, and reproductive system (Right) (Ma et al., 2020).*From: Ma, M., Wang, R., Xu, L., Xu, M. and Liu, S., 2020. Emerging health risks and underlying toxicological mechanisms of uranium contamination: Lessons from the past two decades. Environment International 145, 106107. https://doi.org/10.1016/j.envint.2020.106107 (Accessed 18 February 2024). http://njmoldinspection.com/granite_table.html*

Occupational exposure to ionizing radiation during uranium and gold mining in Sub-Saharan Africa

Occupational exposure is the exposure of workers in the course of their work, whether full time or part time, as either a company employee or contract worker. Other characteristics of "occupational exposure" have been described in the Section: "Pathways of occupational exposure to ionizing radiation," in this Chapter.

Several types of radioactive materials from a uranium mine are encountered by workers and neighboring communities and can increase the ambient dose rates for these people. Materials that bring about external contamination include tailings from heap leaching activities and mills; uranium concentrate (*yellowcake*); contaminated equipment used in mines and mills (pipes, pumps, drums, etc.); vehicles; filters used for filtration of air or liquid effluents, etc.); material (including top soil and bioindicators) contaminated by the deposit of radioactive dust from the mine, mill, waste rock dumps or tailings dams, etc., and material (including sediments, soil, aquatic vegetation) contaminated by water or liquid effluents from mines or mills.

The association between uranium mining and adverse health effects is well-known. The association between the mine environment and lung disease, for instance, was established, even before the classification of lung cancer in the late 19th century (CDC/NIOSH , 2000; Eisenbud and Gesell, 1997). This connection is evidenced in many of the epidemiological studies of miners worldwide (See Section: "Epidemiology of ionizing radiation exposure," in this Chapter. A number of theoretical considerations on pathways of human uptake of significant exposures in mining of uranium and gold ores are given in the literature, *but few data on directly measured human exposure are available.*

According to the most recent studies (at time of writing, 2020), (e.g., those of IAEA, 2020b), the exposure of the workforce (*cf.* uranium and gold miners) to direct external gamma radiation is often the most significant pathway in uranium and gold mining. In Sub-Saharan Africa significant exposure to such radiation takes place during the actual mining and processing stages of uranium and gold ores (See, e.g., Dasnois, 2012; Matshusa and Makgae, 2017; Winde et al., 2019).

Exposure of mine workers arises mainly from external gamma radiation and the inhalation of long-lived radioisotope dust (LLRD) and radon-decay product (RDP). The Information System on Uranium Mining Exposure (UMEX) data for the various operations is resolved into four mining methodologies: underground, open pit, ISL [(in-situ leach process) (See Section: "Uranium mining and processing," in this Chapter)] and other (IAEA, 2020a,b). Both the underground and open Pit mining data are further separated into mining and processing personnel. The other category includes exposures from uranium recovery from rehabilitation, wastewater treatment, and toll milling (IAEA, 2020a).

Reducing the amount of radon ingress into the mine atmosphere is the first approach in combating health effects of ionizing radiation exposure of miners. An important part of this strategy is minimizing the level of mine development there is in areas of uranium mineralization.

Just as with all occupational exposure situations, a properly developed radiation monitoring program in the workplace is the only way to reliably assess the effective dose received by a worker exposed to ionizing radiation. The main reasons for conducting a monitoring program are summarized here, and an overview given of the techniques commonly used to monitor and subsequently assess doses to workers for different exposure situations.

Monitoring and assessment of occupational exposure

Objectives of a monitoring program

The major part of any monitoring program, entails the taking of measurements, but it also involves interpretation and assessment. As stated in Paragraph 3.98 of GSG-7 (IAEA, 2018a): "A program of monitoring may serve various purposes, depending on the nature and extent of the practice." This Document (GSG-7) also lists the purposes that are included in a holistic monitoring program, and adds that "... Furthermore, monitoring data may be used for the purpose of risk-benefit analysis and to supplement medical records."

IAEA Safety Standards Series No. RS-G-1.8 also states in its Section 4.1, that: "... The general objectives of any monitoring program for the protection of the public and the environment are: (1) To verify compliance with authorized discharge limits and any other regulatory requirements concerning the impact on the public and the environment due to the normal operation of a practice or a source within a practice; (2) To provide information and data for dose assessment purposes and to assess the exposure or potential exposure of critical groups and populations due to the presence of radioactive materials or radiation fields in the environment from the normal operation of a practice or a source within a practice and from accidents or past activities; (3) To check the conditions of operation and the adequacy of controls on discharges from the source and to provide a warning of unusual or unforeseen conditions and, where appropriate, to trigger a special environmental monitoring program." (IAEA, 2005). Some subsidiary objectives, which should usually be fulfilled by a monitoring program, are also given in this document (RS-G-1.8) (IAEA, 2005).

Monitoring in surface mining

The monitoring approach in surface mining is similar to that used in all other stages of uranium mining. For gamma exposures, normal practice is to use either individual personal dosimeters [e.g., thermoluminescent dosimeters (TLDs), optically stimulated luminescent dosimeters, electronic personal dosimeters, or workgroup (*similar exposure groups, SEG*) averaging of personal dosimeters]. For dust, personal (on a workgroup averaged basis) or work area sampling is normal way to proceed.

Process description
Surface (or open pit) mining extracts the ore via surface cutting. It has the largest surface signature because of the high ratio between waste rock and ore, the latter of which can be of radiological concern.

Control mechanisms
As the radiological risk associated with surface mining is relatively low, control mechanisms are usually incorporated into controls for common hazards. Radiological specific controls are generally only necessary for mines with very high ore grades or activities with very close and continual interaction with the material. For direct gamma exposure, the most common control mechanisms are the physical shielding and distance from the rock surface provided by the mining equipment and reducing the time spent in active areas.

Monitoring of engineered controls

The monitoring program needs to include a review schedule of the effectiveness of engineered controls. For example, how effective the ventilation system is needs to be reviewed when it is used to control RDP and LLRD. This includes reviewing the preventative maintenance program, downtime and repair times for the equipment.

Gamma radiation

Monitoring of gamma radiation is usually quite straightforward, and there is an array of both active and passive monitoring equipment used to perform area and personal monitoring. In developing a mine, gamma surveys (typically with hand held meters) are performed to determine the background radiation levels, as well as to identify any unexpected areas of mineralization. Periodic gamma surveys are also expected to be performed at main haul ways and high occupancy areas. In both cases, the primary objective is to determine whether the controls are working or whether any remedial actions need to be taken. Personal (electronic) direct reading dosimeters (See Section: "Radiation measuring instruments," in this Chapter) can be used for the continuous monitoring of doses, which can be high when working in the vicinity of gamma sources, such as in high grade mines.

Individual monitoring

Individual monitoring of workers should be performed for controlling individual exposure on a day-to-day basis, or during a particular task. For this, each worker should be provided with an integrating personal dosimeter. Direct reading dosimeters (See Section: "Radiation measuring instruments," in this Chapter) can provide estimates of an individual's dose with a frequency greater than that provided by typical routine dosimetry and can give information on dose rates. Such a dosimeter can be useful for optimization purposes.

Mining companies also use passive monitors or electronic dosimeters for the evaluation of individual doses of workers. Passive monitors have to be sent to a specialized laboratory for dose evaluation.

Dose assessment

Approaches in monitoring and methods used in dose calculation as well as assumptions used to estimate worker doses vary according to operations and regulations. Area monitoring can be used for dose assessment and estimates made of occupancy times. Alternatively, monitoring may be based directly on individual dose measurements. The different approaches and assumptions for dose assessment affect both the estimation of dose by pathway and the total dose; underlining the necessity to document any assumptions made in dose estimation and reporting, and the values of other key parameters used in calculating the dose.

Assessment of external exposure

The dose due to external irradiation can be monitored using different methods. The cumulative dose may be evaluated using electronic monitoring posts that perform continuous monitoring, or by passive detectors that will give an average value (See Section: "Radiation measuring instruments," in this Chapter).

Portable monitors like calibrated Geiger-Mueller counters or other radiation monitors like scintillometers may also be used to monitor dose rates (μSv/hour). These monitors are easily accessible and easy to use (See Section: "Radiation measuring instruments," in this Chapter).

Doses received by workers from external exposure can, in most cases, be readily done from the results of a systematic program of individual monitoring. Doses may also be assessed from the results of workplace monitoring.

For the requirements with regard to the use of individual monitoring and workplace monitoring for dose assessment purposes, see General Safety Guide No. GSG-7 (IAEA, 2018a) and Korir and Karam (2018).

The general aspiration to keep radiation exposure *"as low as reasonably achievable,"* concomitant with the growing number of workers whose exposure to radiation must be strictly controlled, requires strengthening of efforts in improvement of monitoring programs.

Proximal populations

Besides the miners themselves, populations around uranium and gold mining installations are also particularly at risk of ionizing radiation exposure, and regular monitoring of their absorbed dose rates should be imperative. For instance, Kneen, in 2015, estimated that in the Gauteng Province of South Africa, approximately 1.6 million people (including pregnant women and children) live in close proximity to uranium and gold tailings dams, and growing informal settlements are moving closer and closer to these dams. Kneen et al. (2015) also noted that inhalation of windblown tailings dust and radon, consumption of polluted river water and groundwater, ingestion of food produced with contaminated water, including home-grown vegetables, meat, and milk from domestic livestock, and fish from polluted waterways are some of the ionizing radiation exposure pathways for local residents. To these likely pathways, Winde et al. (2019) added the deliberate or sometimes inadvertent hand-to-mouth-consumption of contaminated soil, sediment, and tailings material (*geophagy*) by mostly pregnant women and children (See also Biira et al., 2021).

From a review of environmental uranium measurement surveys in the area (Gauteng) and predictions of people's exposure, performed during an expert workshop held in Johannesburg, South Africa in 2013, *it was concluded that "... direct measurements in human populations living around the tailings were needed to assess exposure distributions and conditions."* (See Schonfeldt et al., 2014). The aim of such measurements would be to confirm that theoretical considerations on the multiple exposure pathways and high environmental levels were indeed resulting in an increased exposure in humans.

Biokinetic models for assessing internal exposure

Although uranium toxicity has long been recognized in humans, we know far less about its pathways of entry into the food web, and in particular, into the bloodstream; nor have we as yet a firm grasp of the underlying mechanisms governing its uptake (See Bird, 2012; Croteau and Fuller, 2016; Croteau et al., 2016). To examine and model these processes, we need to understand the *chemical speciation* of uranium, which in turn, influences both its *bioavailability* (i.e., gut uptake) and *bioaccessibility* (i.e., fractional release), tasks that fall within the compass of the Medical Geologist and environmental geochemist.

There are various pathways through which radionuclides can enter the body—ingestion, injection, dermal absorption (through intact skin or through a wound), and so on (Fig. 5–10). However, the main route of intake in occupational exposure situations is by *inhalation*. A small amount of material may be deposited in the conventional respiratory tract and transferred to the throat by ciliary action. This fraction is subsequently swallowed, lending itself eventually to absorption in the gastrointestinal tract.

The entry of radionuclides into the body and their movement within the body are defined by biokinetic models of the alimentary tract and respiratory tract. Once in the body, radionuclides are absorbed into the blood and/or removed from the body.

Compartmental models describe intake, uptake, internal transfer and excretion of radionuclides. Specific systemic models for the behavior of radionuclides absorbed into the blood of occupationally exposed workers have been developed by the ICRP (ICRP, 1994, 2006; Paquet, 2019; Paquet et al., 2019). These models vary in complexity, and can be applied for the derivation of dose coefficients, for the interpretation of bioassay data, as well as for regulatory control of the workplace.

Routes of intake of radionuclides into the body, subsequent transfers within the body and excretion from the body are shown schematically in Fig. 5–11.

Methods of analysis

The detection and quantification of emissions from the radionuclides present are usually performed by the analysis of biological or physical samples by means of appropriate instrumentation.To achieve sensitive and reproducible detection, the radionuclides would normally have to be separated from the sample matrix. In certain instances, the detectors are unable to discriminate between radionuclides having similar emissions (e.g., some actinides). When this is the case, chemical separation of the elements in the samples should be done before counting.

Models for different routes of entry

Models for the different routes of entry of radionuclides are described in detail in IARC (2001), IAEA (2018a), and EU (2018). The following paragraphs briefly paraphrase these descriptions, but also incorporate some more recent references.

Inhalation

Particulates in windblown dust can contain radioactive materials from mining. These radionuclides when inhaled in dust produce ionizing radiation, which damages the cells and tissues in the body (Taylor, 2009; Dudu et al., 2018).

The behavior of radionuclides inhaled by workers was originally described in ICRP (1994) in a human respiratory tract model. Guidance on the use of the human respiratory tract models can be found in ICRP (2002).

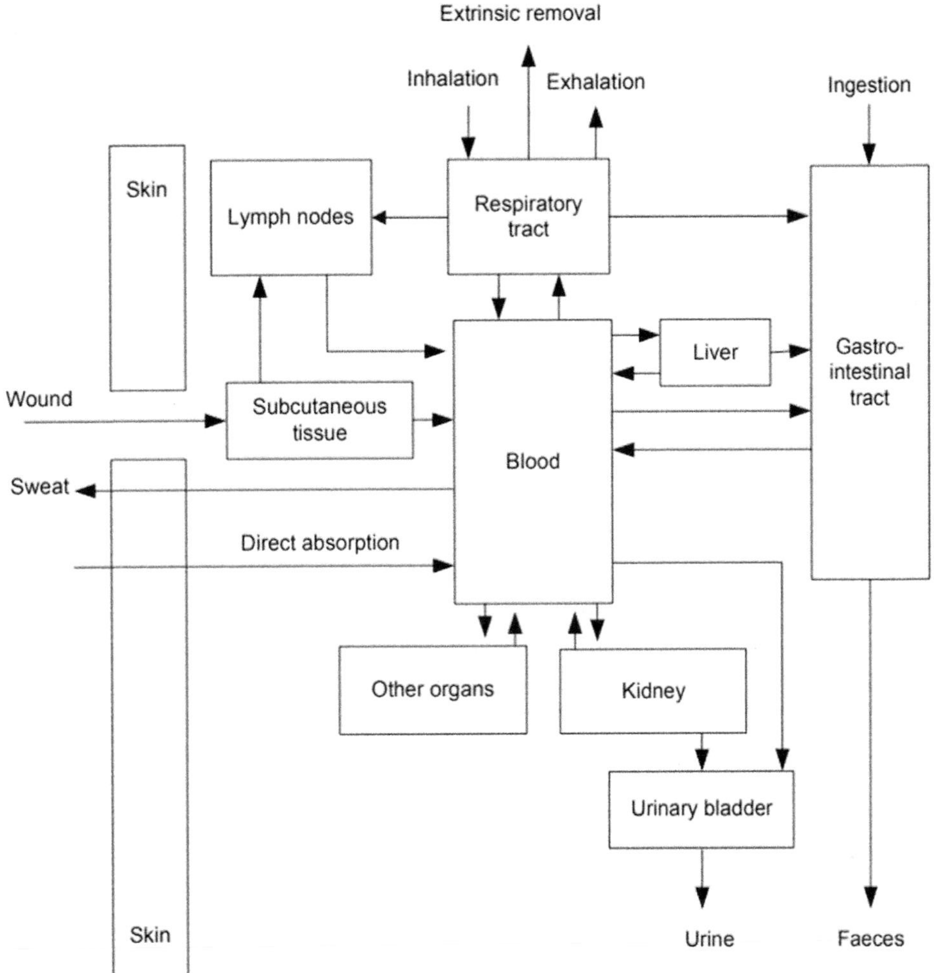

FIGURE 5–11 The main routes of intake, transfer, and excretion of radionuclides in the body (Ye, 2018). *From Ye, G., 2018. Thermodynamic and structural investigations on the interactions between actinides and phosphonate-based ligands. Radiochemistry. Université Paris-Saclay, 2018. English. ffNNT: 2018SACLS286ff.*

The human respiratory tract model treats deposition and clearance of inhaled radionuclides separately. It calculates doses to specific tissues of the respiratory tract and takes cognizance of differences in the radiosensitivity of tissues (IAEA, 2018a).

Inhalation of long lived radionuclide dust

The "dust" pathway is notably associated with the *yellowcake* drying and packaging areas, since up to the drying step, the *in situ* leaching(ISL) process is essentially aqueous and the risk of significant dust generation elsewhere in the process is low (See Section: "Uranium mining and processing," in this Chapter). However, there is a risk of exposure to

LLRD anywhere in the process if spills of process materials are not expeditiously cleaned up before they dry.

A number of different operations conducted during uranium and gold mining activities (e.g., blasting the rocks inside the mine, loading, and unloading the trucks with rocks, crushing the ore, and transporting various types of materials on unpaved roads) also give rise to considerable quantities of radioactive dust that can be inhaled by miners and proximal population. Radioactive elements present in this dust can be carried long distances by the wind.

The effective dose due to LLRD is calculated from:

$$H_{\text{LLRD}} = h_{\text{LLRD}} \times \text{RD} \times \text{BR} \times \text{IT} \times \text{PF} \tag{5-1}$$

Where, Alpha disintegrations per second is shortened to αdps. H_{LLRD} is the committed effective dose due to inhalation of LLRD (mSv); h_{LLRD} is the dose conversion factor (DCF) for relevant LLRD (mSv/αdps); RD is the concentration of LLRD in air (αdps/m^3); BR is the breathing rate for light activity (assumed to be 1.2 m^3/hour); IT is the inhalation time (hours); and PF is the protection factor for any respiratory protective equipment effectively worn (the default value being 1).

The LLRD is monitored for various occupational SEGs using personal dust samplers. Dose calculations are based on the average LLRD concentration for each SEG and the hours spent performing that task or in that SEG. Depending on the dose assessment being performed, the results are taken from the quarterly or annual summary LLRD records and inserted either in the rolling SEG or the annual dose assessment records.

The breathing rate used for dose assessments for inhaled LLRD for workers originates from the ICRP human respiratory tract model in which a worker who is occupationally exposed while doing light work (5.5 hours light exercise + 2.5 hours rest, sitting) breathes 9.6 m^3 of air, which is equivalent to a breathing rate of 1.2 m^3/hour (Bain et al., 1994; ICRP, 1994; IAEA, 2005; Menzel and ICRU, 2014).

Ingestion

The behavior of radionuclides ingested by workers was originally described in ICRP (1979) in a model based on four gastrointestinal tract compartments representing the stomach, the small intestine, the upper large intestine, and the lower large intestine. The mean residence times in the four compartments are: 1, 4, 13, and 24 hours, respectively. The uptake to blood takes place from the small intestine and is specified by the fractional uptake (f1) values (IAEA , 2018a).

Entry through wounds

Although much of the radioactive material can be retained at the wound site, soluble material can be taken to the blood and then distributed to other parts of the body (IAEA, 2018a). Insoluble material is then slowly translocated to regional lymphatic tissue, where gradually dissolution takes place before eventual transfer to the blood.

Models for systemic radionuclides

A *systemic biokinetic model* describes the time dependent distribution and retention of a radionuclide in the body after it gets to the systemic circulation, and its excretion from the body. The systemic biokinetic models used in ICRP (1979) model engendered a relatively simple structure that included the passage of material from the systemic circulation to selected tissues and organs, and eventually to excretion.

Further development of the systemic biokinetic models has been carried out (Skrable et al., 1994; IARC, 2001; Paquet, 2019), leading to the definition of model structures with an increased physiological realism compared with those described in previous publications of the ICRP (e.g., ICRP, 1994).

Indirect methods

Indirect monitoring is based on the determination of activity concentrations in biological materials separated from the body—usually urine, feces, breath or blood—or in physical samples taken from the work environment, such as samples of air or samples of contamination from surfaces.

According to IAEA (2018a), indirect methods should be used for those radionuclides that do not emit strongly penetrating radiation to any significant degree.

Uranium: the world scenario

The element uranium was first discovered in 1789. It is a naturally occurring silvery gray radioactive heavy metal, relatively abundant in the Earth's crust (much more than in the mantle).

Uranium has both military and civilian applications, its earliest use being as a coloring agent in glass and ceramics industry. In 1898 Marie Curie discovered two decay products of uranium, *viz.*, radium and polonium and used them for therapeutic purposes. During World War 1, X-rays were used in the medical field (See Section: "A brief history of radiation science," in this Chapter). Today, technology developed from this early discovery of uranium, the process of fission, and nuclear weapons, continues to provide us with many life-saving diagnostic tools in the field of Nuclear Medicine.

Uranium resources and demand

Worldwide, uranium occurs in ores, in soils, and even in seawater (Gupta and Walther, 2019).

Average concentrations of uranium in the crust of the Earth are around 0.0003% (Herring, 2012). Uranium-238 comprises the bulk (about 99.28% by weight) of all natural uranium. In contrast, uranium-235, the isotope required for the production of nuclear fission leading to a chain reaction, constitutes only 0.7% of natural uranium (Breeze, 2017).

Table 5–2 Production from mines (tons U) up to 2019.

Country	2010	2011	2012	2013	2014	2015	2016	2017	2018	2019
Kazakhstan	17,803	19,451	21,317	22,451	23,127	23,607	24,586	23,321	21,705	22,808
Canada	9783	9145	8999	9331	9134	13,325	14,039	13,116	7001	6938
Australia	5900	5983	6991	6350	5001	5654	6315	5882	6517	6613
Namibia	4496	3258	4495	4323	3255	2993	3654	4224	5525	5476
Niger	4198	4351	4667	4518	4057	4116	3479	3449	2911	2983
Russia	3562	2993	2872	3135	2990	3055	3004	2917	2904	2911
Uzbekistan (est.)	2400	2500	2400	2400	2400	2385	2404	2404	2404	2404
China (est.)	827	885	1500	1500	1500	1616	1616	1885	1885	1885
Ukraine	850	890	960	922	926	1200	1005	550	1180	801
USA	1660	1537	1596	1792	1919	1256	1125	940	582	67
India (est.)	400	400	385	385	385	385	385	421	423	308
South Africa (est.)	583	582	465	531	573	393	490	308	346	346

Source: From WNA (World Nuclear Association), 2020d. World Uranium Mining Production. https://www.world-nuclear.org/information-library/nuclear-fuel-cycle/mining-of-uranium/world-uranium-mining-production.aspx. (Accessed 26 June 2020).

Following the tsunami-induced Fukushima disaster of 2011, there has been a slowdown in nuclear reactor development. Demand for uranium, however, is forecast to rise again as there are reactors currently under construction, mainly in China, Russia, India, S. Korea, and the United States, while 173 are planned that are likely to be operational within 10 years (Kobayashi, 2017; Schneider and Froggatt, 2019; WNA, 2020c). According to the World Nuclear Association, this growing market can be provided by a rising supply from current mines, but by 2025–30, new mines will be needed (WNA, 2020d). Currently, Kazakhstan (25%), Canada (13%), and Australia (9%), supply almost 50% of the current world supply with Africa contributing approximately 18% (Table 5–2; Fig. 5–12).

Comprehensive accounts on world uranium production and mining are given in Winde et al. (2017), Garside (2019), WNA (2020d,e).

Uranium geology and mineralogy

Geology

Fifteen major types of uranium deposits are recognized worldwide, based on their geological setting (WNA, 2020f). Some categories, though, have several subtypes. The 2013 IAEA classification (Bruneton, 2014) is the most recent to date (2020) and has been adopted in the *Red Book* since 2014 (IAEA , 2016a). The "Red Book" published jointly by the Organization for Economic Co-operation and Development (OECD), the Nuclear Agency (NEA) and the International Atomic Energy Agency (IAEA), was first appeared in 1965, and has since been recognized as a world reference on uranium.

The classification in the "Red Book" is based on fundamental characteristics and recognition criteria of the deposit types. Although named essentially by host rock, the types are really empirical models, based on observable characteristics. For more details on types of

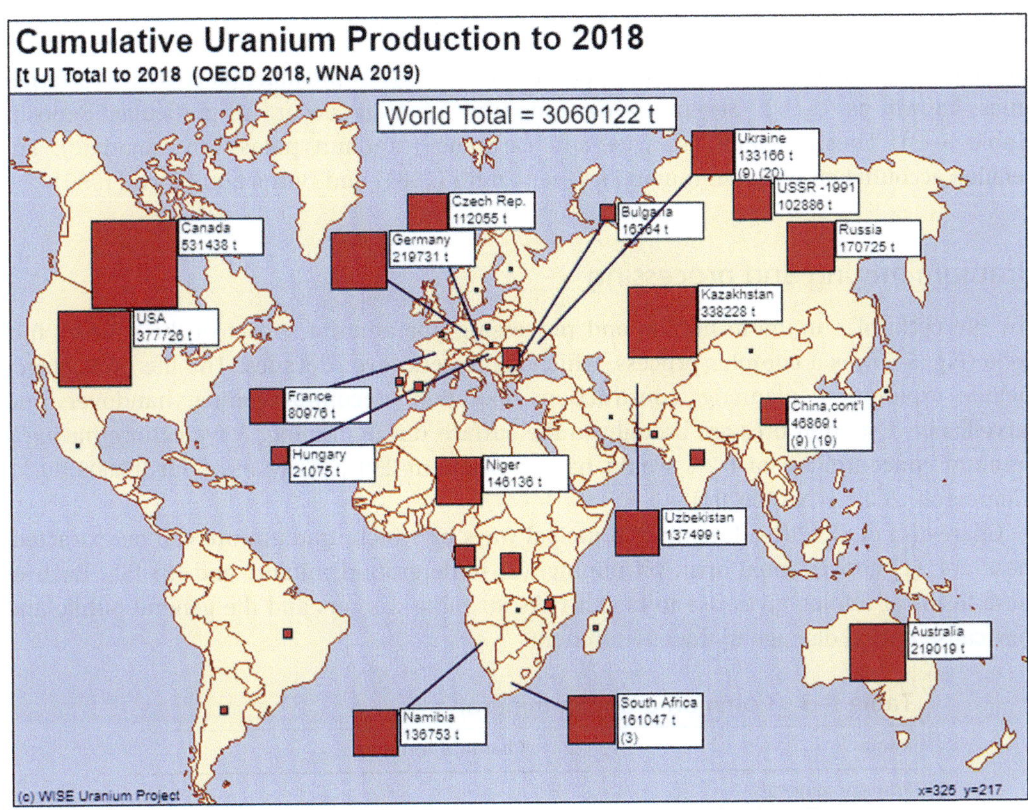

FIGURE 5–12 World cumulative uranium production [tU] per country up until 2018. *From http://wise-uranium.org/ umaps.html?set = cumu. (Accessed 10 April 2018).*

uranium deposits, see, e.g., Bruneton, 2014; Dahlkamp, 1993; IAEA, 2008; Dahlkamp, 2009; WNA, 2020f and other references on the "Selected further reading" list.

Distribution of uranium

Natural uranium is present in the Earth's crust and different rock types, the uranium content of which varies widely depending on the type of rock and the geology of the area; and therefore in a lot of building material made out of Earth materials, it is the main source of exposure of mankind to ionizing radiation. Of the various rock types in which uranium occurs, its concentration is usually higher in granite rocks (typically c. 200 Bq/kg) (Dżaluk et al., 2018).

The diffusion of *radon gas* from underlying geological materials (rock and soil) containing uranium, and its accumulation in the air inside homes and other buildings and (See Section: "Radon," this Chapter) is well-known. This radiological hazard is now extensively documented and international regulations crafted for determining action levels and recommendations in order to lower radon concentration inside homes and other buildings and reduce cancer risks (See Section: "Radon as a carcinogen—management issues," in this Chapter).

Mineralogy

The major primary ore mineral is uraninite (basically UO_2) or pitchblende ($U_2O_5.UO_3$, sometimes known as U_3O_8). Several other uranium minerals occur in particular deposits (Table 5–3). These include tardavite and carnotite (uranium potassium vanadate). For detailed accounts on uranium mineralogy, see Smith (1984) and Burns and Finch (1999).

Uranium mining and processing

The life cycle of a uranium mining and processing operation, a subset of the nuclear fuel cycle (Fig. 5–9), is a complex process, which can extend over decades. The life cycle stages include exploration, planning, construction, operation, decommissioning, handover, and surveillance. Uranium ores are brought to the surface during mining. A typical ore having a uranium concentration of 0.2% would have a uranium-238 activity of about 25,000 Bq/kg (Chareyron, 2008; WNA, 2020g).

Chareyron et al. (2014) lists three principal ways by which uranium ore can be extracted, these are: (1) conventional open-pit mining, (2) underground mining, and (3) ISL. Each of these forms of mining gives rise to health risks for mine workers and the general public and may cause untold damage to the environment.

Table 5–3 Common uranium minerals.

Name	Chemical formula
Primary minerals	
Uraninite or pitchblende	UO_2
Coffinite	$U(SiO_4)_{1-x}(OH)_{4x}$
Brannerite	UTi_2O_6
Tadavidite	$(REE)(Y,U)(Ti,Fe^{3+})_{20}O_{38}$
Tthucholite	Uranium-bearing pyrobitumen
Secondary minerals	
Name	Chemical formula
Autunite	$Ca(UO_2)_2(PO_4)_2 \times 8-12\ H_2O$
Carnotite	$K_2(UO_2)_2(VO_4)_2 \times 1-3\ H_2O$
Gummite	Gum like mixture of various uranium minerals
Saleeite	$Mg(UO_2)_2(PO_4)_2 \times 10\ H_2O$
Torbernite	$Cu(UO_2)_2(PO_4)_2 \times 12\ H_2O$
Tyuyamunite	$Ca(UO_2)_2(VO_4)_2 \times 5-8\ H_2O$
Uranocircite	$Ba(UO_2)_2(PO_4)_2 \times 8-10\ H_2O$
Uranophane	$Ca(UO_2)_2(HSiO_4)_2 \times 5\ H_2O$
Zeunerite	$Cu(UO_2)_2(AsO_4)_2 \times 8-10\ H_2O$

Source: Compiled from Smith, D.K., 1984. Uranium mineralogy. In: De Vivo, B., Ippolito, F., Capaldi, G., Simpson, P.R. (Eds.), Uranium Geochemistry, Mineralogy, Geology, Exploration and Resources. Springer, Dordrecht and Burns, P.C., Finch, R.J. (Eds.), 1999. Uranium: Mineralogy, Geochemistry and the Environment. Reviews in Mineralogy 38. Mineralogical Society of America. 679 p. ISBN: 0–9399 50-50-2.

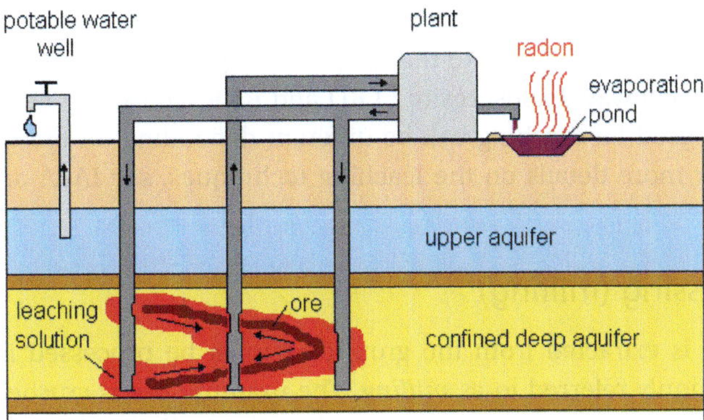

FIGURE 5–13 In situ leaching technology (Diehl, 2011). *From Diehl, P., 2011. Uranium Mining and Milling Wastes: An Introduction. https://www.wise-uranium.org/uwai.html. (Accessed 04 August 2020).*

1. Open-pit mining is used to remove near-surface deposits and requires the removal of rock and soil to access the uranium ore. Open pit mining generates huge quantities of waste per ton of ore. Seepage from waste rock may contain traces of uranium, uranium by-products, heavy metals, and acids. Rainwater runoffs from open pit mines require the development of large evaporation ponds for storage and expensive treatment facilities for processing. Open-pit mining also releases dust and emits radon gas. These radioactive and toxic particulates can end-up in waterways.
2. Underground mines are created using a series of shafts and tunnels. Miners must go underground to build machinery and access the uranium ore. This exposes workers to high levels of radon. When water is present in large quantities, the release of radon can be exacerbated, and surrounding rock can become unstable. Underground mining may also lead to soil subsidence and erosion that may affect neighboring properties.
3. ISL is a combined mining and processing technology (Fig. 5–13). A leaching liquid comprising a mix of chemicals (*lixivants*) is pumped into underground uranium deposits through a series of patterned drill-holes. The chemicals used are typically ammonium carbonate or sulfuric acid. These chemicals separate the uranium ore from surrounding rock, and the uranium-bearing liquid is pumped out from below for further processing. The ISL technology can only be used for uranium deposits located in an aquifer in permeable rock, confined in nonpermeable rock (Fig. 5–13). Once underground, this chemical solution can leach into surrounding groundwater; and it would require a long and expensive process to restore the aquifer back to its original state.

Heap leaching

Heap leaching is an alternative method of extracting uranium-rich liquor from extracted ore. In this technology, the ore is placed on surface pads where extractive liquors (acid or alkaline) are pumped over and through the material. This process can be repeated until liquor of sufficient uranium content is transferred for further processing to extract the uranium. Most of the occupational exposures associated with this process will occur in individuals who spend a great deal of their time near to the heap leaching pads. Gamma exposure will be the dominant exposure pathway. Exposures to LLRD and RDP are usually not significant.

The techniques used for leaching will be different depending on the uranium concentration in the ore. For more details on the leaching techniques, see IAEA and Gaspar (2018d), WNA (2020g).

Uranium processing (milling)

Once uranium ore is extracted from the ground, it must be processed into a usable form. Processing is commonly referred to as *milling*. The uranium ore is crushed and infused with sulfuric acid or sometimes a strong alkaline solution, to allow the separation of uranium from the unusable rock. It is then recovered from solution and precipitated as uranium oxide (U_3O_8) concentrate, a dry powder, sometimes called *yellowcake* (though it is usually khaki in color) (WNA, 2020g). After drying and usually heating, it is packed in 200-L drum and is the uranium product that is sold.

Yellowcake typically contains more than 80% uranium (Hausen, 1998). The original ore, by comparison, may contain as little as 0.1% uranium (WNA, 2020g). With a typical purity of 70%–90% of U_3O_8, it is possible to calculate the uranium-238 activity in the yellowcake at between 7.4and 9.5 million Bq/kg (WUP, 2016).

Conventional mines usually have a mill located close to the mine where the ore is crushed, ground and then leached to dissolve the uranium oxides.

The choice of mining and processing technique is largely dependent on the ore grade and the characteristics of the ore body. Other important factors include topography, hydrogeology, geotechnical aspects, logistics, and the perspectives of interested parties (e.g., the public, indigenous people, regulatory bodies). Therefore awareness of the impact of the design approach on the control of occupational exposures is a critical aspect.

The works of Scheele et al. (2011), Dasnois (2012), Chareyron et al. (2014), Winde et al. (2017), Winde et al. (2019) incorporate comprehensive accounts of uranium mining and milling, characteristics of mine tailings, and a number of other pertinent aspects of uranium mining in Africa, in a way that makes clearly evident, areas of the subject in need of further research.

Decommissioning

The final stage of the uranium mine life cycle is decommissioning, which involves the demolition or removal of plant structures, the rehabilitation of tailings and waste rock structures, and other longer term activities to process wastes arising from the decommissioned facility (mainly contaminated surface water and groundwater). Occupational exposures from the three main exposure pathways will occur and a radiation protection program would need to be put in place (See Section: "Ionizing radiation protection: legislatory and regulatory guidelines," in this Chapter). The highest level of exposures will take place around the contaminated plant and land, during dusty operations, entry into plant vessels, and the decontamination, cutting, and grinding of contaminated objects.

Tailings generate radon (Rn) gas that can travel up to thousands of kilometers in a light breeze. As it drifts, it deposits radioactive byproducts on vegetation. It is transmitted to animals, fish, and plants miles away from the uranium (or gold) mine.

Radioactivity of uranium ore and waste rock (tailings)

The uranium concentration in the material to be mined varies widely and so does the activity of the waste rock and the ore to be processed. The biggest impacts come from waste rock and tailings. The activity of uranium-238 in waste rocks can reach 1250−5000 Bq/kg (Chareyron et al., 2014).

After the processing of uranium ore, the residual material or tailings still contain about 85% of the original radioactivity with a half-life of 80,000 years (Robinson et al., 1979; Diehl, 2011). This material has to be stored and isolated from the environment. The options available include the use of purpose-built impoundments or natural features, or disposal back into the mine pit or underground workings. Occupational exposures at tailings facilities are usually low due to the low grade of many uranium ores. The dominant exposure pathway is generally gamma exposure, although both RDP and LLRD exposures can become significant in specific circumstances involving high-grade ore tailings.

Radioactive waste is produced by uranium mines in many different forms, increasing the ambient dose rates for workers and people living in the vicinity of these materials. These include solid waste (waste rocks, tailings, contaminated equipment, etc.) and liquid effluents, which are not properly managed and are usually disposed of without proper confinement, allowing for airborne and water contamination. The residual radioactivity of uranium tailings will depend on the initial activity of the ore and the efficiency of the leaching process.

Remediation of legacy sites

Uranium mining creates enormous amounts of radioactive waste and, according to Chareyron et al. (2014), neither the companies nor the states in Sub-Saharan Africa know how to successfully manage this problem in the long term. When mines are shut down, radioactive waste remains, and the costs for managing this radioactive legacy including the maintenance of water treatment systems, are passed on to society instead of being taken care of by mining companies. Nor are these costs integrated into the calculation of the overall cost of nuclear electricity.

Africa's uranium story

Africa has been on the global stage of uranium production since 1945 when it was discovered in the DRC (then the Belgian Congo) (Kinnaird and Nex, 2016). Namibia, Niger, and South Africa were the African Continent's largest uranium producers as at 2019 (WNA, 2020d; Table 5−2).

Readers are referred to the recent studies that cover several pertinent aspects of uranium in Africa—mining production, milling, tailings handling, and environment and health impacts on workers and proximal communities. These include works by Hecht (2009), Scheele et al. (2011), Dasnois (2012), Hecht (2012a,b), Chareyron et al. (2014), Kinnaird and Nex (2016), Winde et al. (2017), Winde et al. (2019), Postar (2017), WNA (2020a) and other pertinent references in the Section: "Further Reading," this Chapter. In addition, case histories that cover many aspects of uranium mining are given for Namibia, Niger, and South Africa, Africa's three largest producers of uranium in 2019 (See Section on "Case histories," this Chapter).

Nonconventional uranium extraction

Most of the world's uranium production comes from facilities dedicated to uranium extraction. However, a small percentage is as a by-product of mining for gold, copper, nickel, phosphate, silver, vanadium, and rare earth elements, but also from water-treatment facilities. Occupational exposures from these facilities are heavily influenced by the specifics of the extraction processes and the quantity of the uranium being produced.

Association of uranium with gold in South African Mining

This association is best illustrated using the classic Witwatersrand gold deposits of South Africa. Here, gold and uranium commonly occur together, a realization that dates back to 1915 (Wendel, 1998). So, with gold mining, uranium is often brought to the surface. According to Stassen (2015), mine tailings in the Witwatersrand contain an estimated 600,000 tons (544,000 tons) of uranium.

Wymer (2001) reports that in South Africa, uranium has also been produced as a by-product of copper since 1971, at an average U_3O_8 grade of 0.004%. In 1997 the CMSA reported an average U_3O_8 grade for all goldfields in the Witwatersrand Basin to be of the order of 0.013% with an average recovery grade of 0.0184% (CMSA, 1997).

(Box 5—2)

BOX 5—2   Association of uranium with gold in South African Mining.

"At many South African gold mines, the ore extracted contains not only gold but also considerable amounts of uranium (U), which is brought to the surface inadvertently. Mine waste in hydraulically generated deposits (locally called "tailings dams") cover about 400 km^2, mostly in regions that, through mining, developed into densely populated urban agglomerations, like Johannesburg. The gold tailings of the Witwatersrand basin have an average U concentration of about 100 mg/kg U_3O_8 (ranging from about 10 mg/kg to several hundreds of mg/kg); these gold tailings thus contain as much U as, or more U than, tailings from dedicated U mines, for example, those in Germany or Namibia." Excerpted from Winde et al. (2019).

Wendel (1998) lists the following three geological processes to account for the presence of uranium in ores that are mined for their gold content in the Witwatersrand Basin: "… the concentration of uranium into almost pure blocks of uranium minerals, the weathering of these blocks into a granular detrital form, and finally the gravity concentration of these weathering products into the reefs in the Basin."

Dosimetric quantities

Dose—what amount of ionizing radiation is safe?

The question on "safe dose" is very well-embodied in the following excerpt from Mothersill et al. (2019): "When people discuss the risks associated with low doses of ionizing radiation, central to the discussion is the definition of a low dose and the nature of harm. Standard answers such as "doses below 0.1 Gy are low" or "cancer is the most sensitive measure of harm" obscure the complexity within these seemingly simple questions." Other well-known quotes regarding the uncertainty in radiation dose estimates are given as an epigraph at the beginning of the Chapter (Morgan, 1978; Brenner, 2011).

The above quotations reflect the intractability of researching the effects of very low levels of ionizing radiation, since they are well below the levels of natural background radiation. Consequently, there is hardly any conclusive evidence for setting a "safe" level of exposure above background. It is widely recognized that any exposure carries some risk and the risk increases as the exposure increases.

The field of radiation protection (See Section: "Ionizing radiation protection: legislatory and regulatory guidelines," in this Chapter) is concerned with the protection of humans and the environment from the harmful effects of ionizing radiation after external or internal exposures. Such studies entail both a quantitative description of radiation fields, and of the exposure of the human body. Our principal concern in this Section would be with the metrology of ionizing radiation due to external and internal exposures (of the human body). However, we would start with a brief mention of dosimetry of radiation fields.

Dosimetry of radiation fields

A specific type of *radiation field* is fully described by the number (N) of particles, their distributions in energy and direction, as well as their spatial and temporal distribution. This description requires a clear definition of *scalar* and *vectorial* quantities. Definitions of radiation field quantities are given in detail in ICRU Report 60 (ICRU, 1998). *Vectorial* quantities are mainly applied in radiation transport theory and calculations, and they inform on directional distributions; *scalar* quantities, on the other hand, such as *particle fluence* or *kerma*, are often used in dosimetric applications.

Fluence (of particle radiation) refers to the number of radiant-energy particles emitted from or incident on a surface in a given period of time, divided by the area of the surface. Although fluence is an important quantity in describing radiation fields it is not generally considered appropriate and simple enough for use in radiological protection and the definition of limits. This is so, mainly because of complications in particle specification and particle energy correlation with detriments.

Kerma (K) is an acronym for "kinetic energy released per unit mass" (or "kinetic energy released in matter" "kinetic energy released in material" (Sherer et al., 2014). It is defined as the sum of the initial kinetic energies of all the charged particles liberated by uncharged ionizing radiation (i.e., indirectly ionizing radiation, such as photons and neutrons) in a sample of matter, divided by the mass of the sample:

$$K = dE_{tr}/dm \qquad\qquad (5-2)$$

where dE_{tr} is the sum of the initial kinetic energies of all the charged ionizing particles (electrons and positrons) liberated by uncharged particles (photons) in a material of mass dm (ICRU, 1980, 2005).

The SI unit of Kerma is the gray (Gy) (or Joule per kilogram), the same as the unit of *absorbed dose* (defined in the Section: "External or internal dosimetry," in this Chapter).

The *air Kerma* is used to describe the kinetic energy released per unit mass of air.

Body related dose quantities

In describing the exposure of humans to ionizing radiation, we need, in addition to physical quantities, information about the biokinetics of radionuclides and other parameters of the human body. This precept is engendered in the definition of *dosimetry* within the realm of health physics and radiation protection, *viz.*, that *dosimetry* refers to the measurement, calculation and assessment of the ionizing radiation dose absorbed by an object, usually the human body. It is applied on a routine basis for monitoring occupational radiation workers. The biological effect is related to the dose and depends on the nature of the radiation.

Within the public realm, dose uptake is measured and calculated from a variety of indicators, such as ambient measurements of gamma radiation, radioactive particulate monitoring, and the measurement of levels of radioactive contamination.

To evaluate the *total effective dose,* it is necessary to take into consideration *external irradiation* and *internal contamination.*

External or internal dosimetry

Dosimetry can be applied internally, when radioactive substances are ingested or inhaled, or externally, when the ionization dose absorbed is due to irradiation by sources of radiation.

Assessment of the effect of ionizing radiation on tissue is influenced by the "damage potential" (See Section: "The damage potential of ionizing radiation," in this Chapter) of the radiation type (e.g., energy, size, charge, half-life, etc.), the administered dose, and the dose rate (Kummar and Takimoto, 2018). The quantities obtained from calculations in dosimetry are what we use to determine issues on radiation protection, risk assessment, diagnostic dose estimates, and treatment planning (Fisher and Fahey, 2017).

Internal dosimetry estimates rely on a variety of monitoring, bioassay, or radiation imaging techniques. *External dosimetry* is predicated on measurements that employ a dosimeter, or inferred from measurements that are done using some other radiological protection equipment.

Clear distinctions need to be made with respect to terminology, and units used to describe or define the various aspects of radiation and radiation dose. We have (1) units of radioactivity (2) units of energy deposited in matter, and (3) units of biologically relevant dose.

We now take a look at the relevant units for measuring dosimetric quantities.

1. *The absorbed dose*

Radiation damage to tissue and/or organs depends on the dose of radiation received, or the *absorbed dose.*

1. *The Gray:* The "absorbed dose" is expressed in the international system unit called the *Gray (Gy)*, where 1 Gy = 1 J/kg or 100 rad). The potential damage from an absorbed dose depends on the type of radiation and the sensitivity of the different tissues and organs. The Gray thus relates to the amount of energy actually deposited in some material (tissue and/or organs), and is used for any type of radiation and any material through which they pass (e.g., water, tissue, air). An absorbed dose of 1 rad means that 1 g of material absorbed 100 ergs of energy (a small but measurable amount) as a result of exposure to radiation.

2. *The Rad*: The *Rad* (radiation absorbed dose) is the older unit of *absorbed dose.*

The *radiation-absorbed dose (Rad)* is the amount of energy (from any type of ionizing radiation) deposited in any medium (e.g., water, tissue, air). One Rad is equal to 0.01 Gy.

Formally, absorbed dose at a point is defined as:

$$D = \Delta\varepsilon / \Delta m \qquad (5.3)$$

where $\Delta\varepsilon$ is the mean energy transferred by the radiation to a mass Δm.

2. *Activity* (See Section: "The activity," this Chapter, for definition).

Activity is expressed in Becquerel (Bq) or in Curie (Ci).

1. *Becquerel (Bq):* The Becquerel, named after the French physicist Henri Becquerel, is a unit that describes one radioactive disintegration or transformation of 1 g of radium per second. It is therefore a much smaller version of the Curie. There are 37,000,000,000 Bq in one Ci.

As the unit (Bq) is so small, multiples of the Bq are often used, such as the *megabecquerel, MBq*, which is 1 million becquerels. One gram of radium-226, for instance, has an activity of approximately 37,000 MBq: it emits about 37,000 million alpha particles each second.

2. *The Curie (Ci)*: The *curie* (Ci) is an older unit that is also sometimes used to measure radioactivity. Named in honor of the Polish-born French scientist, Marie Curie, one Curie is a quantity of a radioactive material that will have 37,000,000,000 transformations, or nuclear decays, in one second. Often radioactivity is expressed in smaller units like a thousandth of a Curie (mCi), a millionth of a curie (uCi) or even a billionth of a curie (nCi).

Note: Bq and Ci are units of measurement of radioactivity, and not of radiation or radiation dose.

3. *Specific Activity/Activity Concentration* refers *to* radionuclide activity per mass (volume) unit of a matter. Measurement unit: Bq/kg (specific activity) and Bq/m^3 (activity concentration).

4. *The Effective Dose*

The Effective Dose is a measure of the biological damage potential of some amount of ionizing radiation. It is a measure of the risk of long-term effects (in terms of the potential for causing harm) of ionizing radiation on the human body, organs and tissues.

1. *Sievert (Sv): Effective dose* is expressed in *Sievert*. It is the unit of measurement used for the biological effect of radiation on the human body [recorded in milliSievert (mSv)] since it takes into account the type of radiation and sensitivity of tissues and organs. The reason for recording this measure (radiation on the human body) in milliSievert is because the Sv is a very large unit; so it is more practical to use smaller units such as milliSieverts (mSv) or microSieverts (μSv). There are 1000 μSv in 1 mSv, and 1000 mSv in 1 Sv. 1 milliSievert (1 mSv) = 1000 microSievert (1000 μSv).

   In addition to the amount of radiation (dose), it is often useful to express the rate at which this dose is delivered (dose rate), such as microSieverts per hour (μSv/hour) or milliSievert per year (mSv/year) (Table 5–4).

   The average global exposure to natural radiation is 2.4 mSv per year (Table 5–4). The annual average total dose from ionizing radiation, over the population of the world, is about 2.8 mSv (Hung et al., 2013). It is known that very large doses of over 5000 mSv, received by the entire body over a short time, result in death within a few days (Gupta et al., 2019). Doses over 100 mSv can have a harmful effect on humans, such as a higher incidence of developing cancer. At even lower doses of radiation, below 100 mSv, *we still are uncertain about the overall effects* (Table 5–5).

   What we do know is that the risk of adverse effects in this dose range is very low. To be on the safe side, we assume that there is a risk even in this low dose range and this risk is proportional to the dose by the same amount as in the high dose ranges.

Table 5–4   Summary of the annual effective doses to humans from different types of natural radiation (IAEA, 2004).

Source	Worldwide average dose mSv	Typical dose range
Cosmic radiation	0.4	0.3–1.0
Gamma radiation	0.5	0.3–0.6
Radon inhalation	1.2	0.2–10
Internal radiation	0.3	0.2–0.8
Total (rounded)	2.4	1.0–10

Table 5–5 Dose limits (IAEA, 2018c).

Occupational public		
Effective dose	20 mSv per year averaged over 5 consecutive calendar years	1 mSv in a year
Annual equivalent dose in:		
Skin	500 mSv	50 mSv
Hands and feet	500 mSv	—
Lens of the eye	20 mSv averaged over 5 consecutive years, 50 mSv in any 1 year	15 mSv

Source: Compiled from IAEA (International Atomic Energy Agency), 2018c. IAEA Safety Standards for Protecting People and the Environment: General Safety Guide No. GSG-8 - Radiation Protection of the Public and the Environment. http://regelwerk.grs.de/sites/default/files/cc/dokumente/dokumente/PUB_GSG-8_web.pdf. (Accessed 16 July 2020). Data on dose limits.

2. *Rem.* The *rem* (roentgen equivalent in man) is the older unit of effective dose. It is still in use in some of the leading nuclear countries such as the United States and Russia. One rem is equal to 0.01 Sv.

   In addition, radiation scientists today, appear to favor the use of the two larger units of *Gray* and *Sievert* in place of the more traditional, common units of *Rad* and *Rem* for the same things.

   Effective dose is typically calculated by multiplying the absorbed dose by a factor specific to the type of radiation. This is usually called a *quality factor* or *relative biological effectiveness factor.*

   For low-LET radiations this factor is typically close to or equal to one, so that one Sv is approximately equal to one Gy. For high-LET radiations like alpha particles the factor might be as high as twenty.

5. The Collective Effective Dose

   The Collective Effective Dose is a measure of collective risk of *stochastic effects* (See Section: "Types of radiation effects," in this Chapter) of ionizing radiation, equal to the total of individual effective doses. The measurement unit is the *man-Sieverts* (man-Sv).

6. The Dose Limit

   The Dose Limit (DL) is the annual effective or equivalent dose of human-induced radiation that must not be exceeded in normal operation conditions. DL compliance helps prevent *deterministic effects* (See Section: "Types of radiation effects," in this Chapter) and keep potential *stochastic effects* at an acceptable level.

7. Equivalent Dose

   The Equivalent Dose is a dose quantity (H) used to assess the extent of biological damage that is expected from the absorbed dose. It is calculated by multiplying the absorbed dose in an organ or tissue by a proper coefficient depending on the radiation type. The unit of measurement is also the Sievert.

   For all these units, appropriate conversions can always be made to suite the measurement context.

Acute respiratory syndrome

Over and above certain thresholds (with doses higher than 0.7−1.0 Gy, short duration), whole body exposure to high amounts of ionizing radiation can produce acute health effects due to cellular degradation resulting from damage to DNA and other key molecular structures within the cells in various tissues (Barabanova et al., 2019). The effects an include nausea, vomiting, and diarrhea, developing within hours or minutes of an exposure.

The collection of health effects is known as *acute radiation syndrome (ARS)*; also referred to *as radiation poisoning, radiation sickness*, or *radiation toxicity*. These effects are more severe at the higher dose levels. For instance, the dose threshold for ARS is about 1 Sv (1000 mSv). Doses of radiation below 0.15 Gy (15 Rad) produce no observable symptoms or signs. According to US EPA (2017), ARS is rare, and comes from extreme events like a nuclear explosion or accidental handling or rupture of a highly radioactive source.

These conditions are discussed further in Section: "Summary of health impacts," in this Chapter.

The radiation dose-response curve

An apparently intractable problem with the genetic studies on humans is one that engenders the many remaining *uncertainties* about dosimetry (See, e.g., Grandjean, 2016). The ability of epidemiological studies to demonstrate (or exclude) small health effects or the effects of very low radiation doses, say below 100−200 mSv, are very nebulous, since these studies are fraught with a number of uncertainties, sources of error and confounders.

The health effects of ionizing radiation will depend on many factors. First, if the radiation dose is low and/or it is delivered over a long period of time (low-dose rate), the risk is substantially lower because there is a greater chance of repairing the damage. Low doses of ionizing radiation produce effects that include the increase of various types of cancers, genomic instability, and life-shortening and negative impacts on all bodily functions.

The clinical signs of toxicity from high doses of radiation follow the classic dose-response curve, with some organs more severely affected at each dose level than others. High doses of ionizing radiation can lead to various effects, such as cancer, skin burns, hair loss, birth defects, and even death. For cancer induction, increases in the radiation dose do not increase the severity of the cancer; but rather, it increases the chance of cancer induction.

In general, the higher the radiation dose, the sooner the effects will appear, and the higher the probability of death. The risk of long-term effects for cancer may appear years or even decades later (the latent period). Effects of this type will not always occur, but their likelihood is proportional to the radiation dose. This risk is higher for children and adolescents, as they are significantly more sensitive to radiation exposure than adults (See Section: "Sensitive populations," in this Chapter).

The basic toxicology dictum: "*... the dose makes the poison...*" vividly applies to the toxicology of ionizing radiation much as it does to all other branches of toxicology.

Radiation measuring equipment

The use of calibrated radiation protection instruments for practical radiation measurements is essential, for monitoring an area, in evaluating the effectiveness of protection measures, and also for individual monitoring, in assessing the radiation dose likely to be received by individuals. The ability of radiation to penetrate matter has already been noted. This attribute means that measurements can be made without direct physical contact of the sensor with the material being measured.

Two types of instrument can be distinguished: (1) those used for measuring radiation levels, referred to as *area survey meters* (or *area monitors*), which are either gas-filled detectors or solid-state detectors (e.g., scintillator or semiconductor detectors); and (2) those instruments used for recording the equivalent doses received by individuals working with radiation, referred to as *personal dosimeters* (or *individual dosimeters*).

Radiation measuring equipment for radiation protection can be: (1) installed (in a fixed position), or (2) portable (hand-held or transportable).

Installed instruments

Installed instruments are fixed in positions, which are known to be important in assessing the general radiation hazard in an area. Examples are installed "area" radiation monitors, gamma interlock monitors, personnel exit monitors (PEM), and airborne particulate monitors.

The area radiation monitor will measure the ambient radiation, usually X-Ray, Gamma, or neutrons; these are radiations, which can have significant radiation levels over a range in excess of tens of meters from their source, and thereby cover a wide area.

Gamma radiation "interlock monitors" are used in applications to prevent inadvertent exposure of workers to an excess dose by preventing personnel access to an area when a high-radiation level is present. These interlock the process access directly.

Airborne contamination monitors measure the concentration of radioactive particles in the ambient air to guard against radioactive particles being ingested, or deposited in the lungs of personnel. These instruments will normally give a local alarm, but are often connected to an integrated safety system so that areas of plant can be evacuated and personnel are prevented from entering an air of high airborne contamination.

PEM are used to monitor workers who are exiting a "contamination controlled" or potentially contaminated area. These can be in the form of hand monitors, clothing frisk probes, or whole body monitors. These monitor the surface of the worker's body and clothing to check if any radioactive contamination has been deposited. They generally measure alpha, beta, or gamma radiation, or combinations of these.

Portable instruments

Portable instruments are hand-held or transportable. The hand-held instrument is generally used as a survey meter to check an object or person in detail or assess an area where no installed instrumentation exists. They can also be used for personnel exit monitoring or personnel contamination checks in the field. These generally measure alpha, beta, or gamma, or combinations of these.

Transportable instruments are generally instruments that would have been permanently installed, but are temporarily placed in an area to provide continuous monitoring where it is likely there will be a hazard. Transportable instruments are often mounted on trolleys for ease of deployment, and are associated with temporary operational situations.

The following detection instrument types are the ones commonly used for both *fixed* and *survey* monitoring:

1. *Ionization chambers.* An *ionization chamber* measures the charge from the number of ion pairs created within a gas caused by incident radiation. It consists of a gas-filled chamber with two electrodes, an anode and a cathode.
2. *Proportional counters.* The *proportional counter* is a type of gaseous ionization detector device used in a regime where gas amplification takes place.
3. *Geiger counters.* A *Geiger Counter* detects ionizing radiation, such as alpha particles, beta particles, and gamma rays using the ionization effect produced in a low-pressure gas in a Geiger-Müller tube (commonly referred to as "GM tube"). The Geiger Counter is a very prominent hand-held radiation survey instrument, and is used widely around the world. It is considered by many as the best-known radiation detection instrument. However, these instruments detect the presence of radiation quantity but not magnitude/energy.

   Digital Geiger Counters are extremely sensitive and will detect and measure background radiation in addition to detecting and measuring radioactivity above background radiation.
4. *Semiconductor detectors.* A *semiconductor detector* device uses a semiconductor (usually silicon or germanium) to measure the effect of incident charged particles or photons.
5. *Scintillation detectors.* A *scintillation (light emission) detector* or *scintillation counter* detects and measures ionizing radiation by using the excitation effect of incident radiation on a scintillating material, and detecting the resultant light pulses.
6. *Airborne particulate radioactivity (APR) monitoring. Continuous particulate air monitors* are used in nuclear power plants for measuring releases of APR from the facility, monitoring levels of APR for protection of plant personnel, monitoring the air in the reactor containment structure to detect leakage from the reactor systems, and to control ventilation fans, when the APR level has exceeded a defined threshold in the ventilation system.

A fuller description of each of these instruments is given in Rajan and Izewska (2006).

All instruments must be calibrated in terms of the appropriate quantities used in radiation protection.

Calibration standards for measuring instruments

A network of secondary calibration laboratories has been set up by WHO/IAEA in many countries (IAEA, 2000). The sources of radiation and associated apparatus and calibration techniques are presented in IAEA (2000). Examples are provided of what established calibration laboratories have deemed adequate. However, there are other methods and instruments that have been found to be suitable for proper calibration. In addition to presenting a description of calibration facilities and procedures, the IAEA (2000) Safety Report gives appropriate definitions and describes appropriate methods for the statement of uncertainties in measurements.

As the human body is constituted of approximately 70% water by weight and has an overall density close to 1 g/cm^3 (Liberman and Marks, 2013), dose measurements are usually calculated and calibrated as dose to water.

Dosimeters in radiotherapy (linear particle accelerator in external beam therapy) are routinely calibrated using ionization chambers (Hill et al., 2009), diode technology, or gel dosimeters. National standards laboratories would normally provide calibration factors for ionization chambers and other measurement devices to convert from the instruments read-out to absorbed dose.

Estimation of uncertainties

The assessment of *uncertainty* in measurement is the basis for quantifying the accuracy of the measurements. International guidance on the metrological aspects of dosimetry can be found in publications developed by the Joint Committee for Guides in Metrology (JCGM).

The fundamental reference document is the International Vocabulary of Metrology: Basic and General Concepts and Associated Terms (VIM) (JCMG, 2017). In the evaluation of uncertainty, all knowledge of the dosimeter and its associated evaluation system (e.g., TLD reader, densitometer and track counting system) or of the workplace monitoring instrument, both from experience and from type testing should be used, possibly in combination with information from the client or customer, such as information on local exposure and storage conditions.

The evaluation of the uncertainty should use a mathematical model of the dosimetry system. This mathematical model can be given as:

$$Y = f(X_1, X_2 \ldots) = f(X) \tag{5-4}$$

Where, Y is the output quantity or measured, for instance, $H_p(10)$, and X is an array containing the input and influence quantities of the measurement system.

The evaluation of the uncertainty then consists of two stages: The formulation stage and the calculation stage.

Measuring radioactivity of materials in a uranium mine or mill

The activity (See Section: "Activity," in this Chapter) may be given in Bq/kg for samples of soil, sediments, or food; Bq/m^3 for air samples; Bq/L for water samples; Bq/cm^2 for contaminated surface samples, etc. Each radionuclide has a specific level of radioactivity for a given number of atoms or weight. This number is called the *specific activity*, which for uranium-238 is about 12,500 Bq/g (12.5 Bq/kg) (Chareyron et al., 2014).

Chareyron et al. (2014) caution that when calculating the radioactivity of solid materials in a uranium mine or mill, it is important to distinguish "… between the mining process, in which uranium-238 is still in equilibrium with its daughter products (natural ore or waste rocks) and the milling process in which ore is processed and leads to various disequilibrium in waste (tailings) or in the commercial product (yellowcake)."

A number of other pertinent aspects regarding the measurement of radioactivity of materials in an African uranium mine or mill are given comprehensive coverage in Chareyron et al. (2014) and Obasi et al. (2020).

Types of radiation effects

The harmful effect on human tissue that could result from exposure to ionizing radiation can be divided into two broad types: (1) *Chance (Stochastic) effects*, and (2) *Threshold/nonstochastic* (deterministic) effects (Rose and Rae, 2017).

Stochastic effects

Stochastic (Chance) effects are chance events whose probability of occurring is proportional to the dose. However, the severity of the effect is independent of the dose. These effects can be caused by both chronic and acute exposures and can occur among both exposed and unexposed individuals. In the realm of radiation protection, the main stochastic effect is *cancer*, but hereditary disorders are stochastic as well, with a combined detriment of $\sim 5\%$/Sv [ICRP publication 103 (ICRP, 2007)].

Threshold (deterministic or nonstochastic) effects

Threshold (Deterministic or Nonstochastic) effects are based on tissue reactions/tissue damage. They occur only if the *threshold* [i.e., the minimum dose that produces an effect; typically, of the order of magnitude of 0.1 Gy or higher (ICRP, 2007)] is exceeded. The degree of severity beyond the threshold dose increases as the dose increases. Threshold effects are caused by *acute* (high level, short duration) exposures (See Section: "Acute exposures," in this Chapter).

Effects of chronic exposures

Uranium and gold miners may be *chronically exposed*, i.e., they repeatedly receive small doses over long periods (years). The human body is better able to tolerate a chronic dose than an acute dose. The body has time to replace dead or nonfunctioning cells with new, healthy cells. Chronic external or internal exposure causes only an increased possibility of some stochastic effects such as cancer and hereditary abnormalities in children or grandchildren, with the note that, cancer and hereditary diseases occur naturally among people and may be caused by many other etiological factors apart from radiation.

Somatic effects

Somatic effects are those that are observable in the exposed individual. The formation of cataracts in the lens of the eye or blood pH changes could be taken as examples. *Teratogenic effects* are those that occur in the developing fetus during gestation.

Mutagenic effects are those effects, which engender the disruption of the base pair sequence in DNA molecules, which may result in a genetic defect or mutation. Genetic effects are similar to mutagenic effects but the genetic aberration occurs in one or both of the cells, which may affect the offspring.

Summary of health effects of ionizing radiation in uranium and gold mining

The research literature is awash with in-depth information and reviews on the biological, toxicological, and toxicokinetic effects of ionizing radiation exposure in humans; and it would be difficult, if not unfeasible, to summarize all of this knowledge in one place. *Despite the enormity of this database, however, much remains to be learned about certain aspects, such as the specific mechanisms by which ionizing radiation produces its effects, how these effects can be minimized in living tissues, what the long-term effects of very low doses of ionizing radiation are, over the normal human lifespan, and in what ways the contribution of Medical Geologists would help improve ionizing radiation diagnosis and therapy.*

Much of what we know has been derived from data obtained from human epidemiological studies and from diagnostic radiology, but a good portion has come from laboratory animal studies and then extrapolated to humans.

We know from recent studies that uranium exposure is of dire health consequences to miners and surrounding communities, especially sensitive groups, such as pregnant women and young children. Based on the new dosimetry system (Urban and Skubacz, 2015), there are sufficient data from these large-scale human exposures to derive some conclusions about the cancer-inducing effects of external radiation.

The various health impacts of radioactivity and ionizing radiation released during mining, processing, and related activities involving uranium on miners and proximal communities around uranium and gold mining centers of Sub-Saharan Africa have been presented in relevant sections of the Chapter. Here, a systematic summary of what we know of the more important aspects is given with a special focus on uranium and gold miners, and sensitive populations (the developing fetus and young children).

Recent findings by Winde et al. (2019) from investigations of health effects of uranium, (including depleted uranium used in military conflicts), suggest a much wider range of toxicity, beyond carcinogenicity, including teratogenicity, disruption of the endocrine system by uranium mimicking estrogen's effects, and genetic damage, as well as neurotoxic effects.

Several excellent references and review articles are also available in the open literature that provide the salient background material in radiation toxicology (e.g., Harley, 2011;

Scheele et al., 2011; Hecht, 2012a; Chareyron et al., 2014; OECD/NEA, 2014; Mathuthu et al., 2016; Winde et al., 2017; Rose and Rae, 2017; Ngetu et al., 2019; Winde et al., 2019; and more are given under: "Selected further reading," this Chapter).

In the following summary, the end points produced by ionizing radiation have been divided into (1) *carcinogenic effects* and (2) effects that were (at least initially) *noncarcinogenic*. The noncarcinogenic effects are discussed in relation to major organ systems affected plus teratogenic effects. This is done primarily to help the reader understand the broad scope of adverse health effects that can be produced by ionizing radiation.

Carcinogenic effects from ionizing radiation exposure

Cancer is the major latent harmful effect produced by ionizing radiation and the malady of greatest concern for most people exposed to radiation. Exposure to radiation can produce cancer at any site within the body, but some sites appear to be more common than others (e.g., Raabe, 2012). *The mechanism by which cancer is induced in living cells is a subject of great complexity and one that is being intensively researched at present.* The current notion is that the radiation damage acts in concert with a number of other factors, such as chemicals and genetic make-up, which eventually lead to the uncontrollable multiplication of cells.

Among the large volume of studies in the oncology literature involving external exposures to ionizing radiation and resulting in cancer, perhaps the best known available study to date is the *Life Span Study* (Socol and Dobrzyński, 2015) conducted by the "Radiation Effects Research Foundation (RERF)" with the survivors of the atomic bombings of Japan in 1945 (See Section: "Epidemiology of cancer due to ionizing radiation exposure," in this Chapter, for more details).

In addition, with regard to the effects of cancer induced by the chemical and radiological effects of uranium mining, readers are referred to the Agency for Toxic Substances and Disease Registry (ATSDR) Toxicologic Profile for Uranium (ATSDR/US DHHS, 2013).

In humans, exposure to low-levels of radiation does not cause immediate health effects, but can cause a small increase in the risk of cancer after some minimum *latent period* (defined in this Section: "the time from exposure to the time of observation"). Radiation exposure increases the chance of getting cancer, and the risk increases as the dose increases, *viz.*, the higher the dose, the greater the risk. Conversely, cancer risk from radiation exposure reduces as the dose falls: the lower the dose, the lower the risk (See Section: "Radon inhalation and carcinogenesis," this Chapter).

Radiation and DNA

One of the characteristic features of ionizing radiation (high-frequency, high-energy waves) as opposed to nonionizing radiation, is their content of sufficient energy to displace an electron from its orbit around a nucleus. The most important consequence of this displaced electron on human tissue is the potential damage it can cause on their genetic material (DNA), which may occur directly or indirectly.

Direct damage occurs when the displaced electron strikes and breaks a DNA strand. *Indirect damage* occurs when the electron reacts with a water molecule, creating a strong hydroxyl radical, which then damages the cell's DNA (Fig. 5—14). Damage to a cell's DNA in either of these ways can have several consequences. A single-strand DNA break is usually repaired appropriately by the cell with no subsequent deleterious sequelae. However, a break affecting both strands of DNA allows the potential for abnormal reconnection of the strands, which likely accounts for all the adverse biological effects ionizing radiation has on humans.

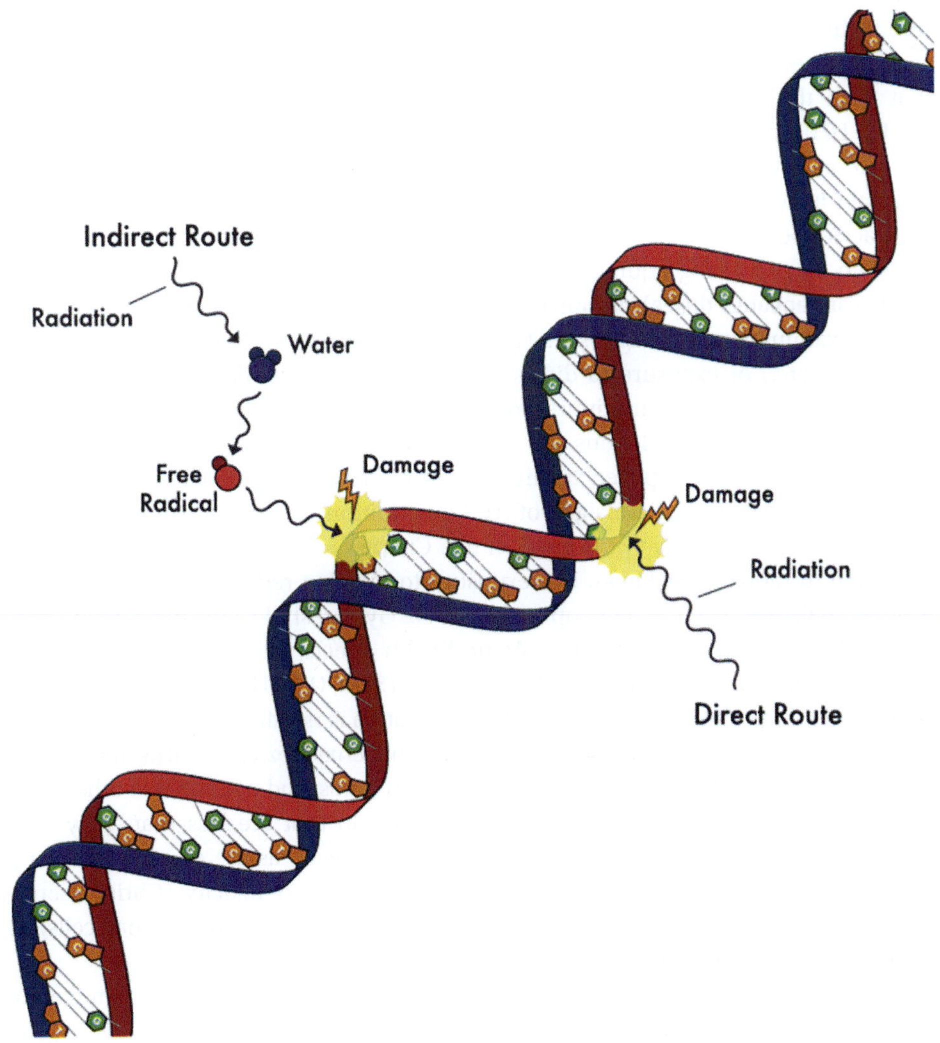

FIGURE 5–14 DNA damage due to radiation. *Credit: Duke University Word Press. https://sites.duke.edu/ missiontomars/the-mission/cancer/what-is-cancer/ (Accessed 12 April 2020).*

Thankfully, the cells in our bodies are extremely efficient at repairing this damage with no subsequent deleterious sequelae; though if not repaired correctly, a cell may die or eventually become cancerous.

Acute (immediate and noncarcinogenic) effects from ionizing radiation exposure

Acute exposure occurs when one receives a large radiation dose within a short period of time (few minutes or hours). If acute radiation dose exposure is large, it may result in effects, which are observable within a period of hours to weeks. The over-riding effect of acute exposure is the threshold effect.

There are three exposure situations in respect of acute high-level doses of ionizing radiation that from the outset would result in one or many immediate *noncarcinogenic effects* (ATSDR/US DHHS, 1999). The first situation can be likened to what one experiences in the vicinity of an atomic blast; this can be exemplified by the Japanese populations of Hiroshima and Nagasaki in August 1945. A second situation is akin to involvement in a laboratory or industrial accident, where only those present at the site receive high-intensity radiation from sources, such as generating equipment. The third and most likely scenario for exposure to high levels (or repeated doses) of ionizing radiation would involve medical sources such as radiological examinations or diagnostics (protracted exposures to X-rays, fluoroscopy, radio-iodine therapy, etc.) or exposure to displaced medical or industrial radiation sources, such as mining and processing of uranium and gold ores. Workers at uranium and gold mines and associated proximal populations fall into the third category. People who have a large-surface area of their body exposed to high doses (>100 rad) of radiation in any of these situations may exhibit immediate signs of ARS a condition that has already been treated in the Section: "Acute respiratory syndrome," in this Chapter.

In addition to radiation sickness, overexposure to ionizing radiation can result in a number of other maladies, such as lens opacities (~ 0.2 Gy threshold and protracted exposure) and fetal and developmental anomalies. Acute health effects, such as skin burns can occur when doses of radiation exceed certain thresholds (ATSDR/US DHHS, 1999).

The sensitive to the biological effects of ionizing radiation is not the same for all cells that comprise the body's tissues and organ systems. According to Hernández et al. (2015) the sensitivity of cells is affected by age at the time of exposure, sex, health status, and a number of other factors.

Individuals exposed to single acute doses of radiation that are less than 1 Gy (100 rad) hardly experience significant clinical signs of toxicity; however, as the dose is doubled (2 Gy, 200 rad), some cells and tissues begin to show signs of overt toxicity (Christensen et al., 2014). At this dose, the cells that multiply the most rapidly (gastrointestinal cells, blood-forming cells) are only being mildly affected (nausea/vomiting, leukopenia).

Cells that proliferate more slowly (e.g., the cells of the central nervous system, connective tissues, etc.) are largely unaffected (ASTDR, 1999).

Over and above a dose limit of 8 Gy (800 rad) of ionizing radiation, a dose-dependent increase in the severity of the hematological, gastrointestinal, and central nervous system toxicity occurs (Bolus, 2017), and death will likely follow due to disastrous multiorgan failure (ASTDR, 1999). Death is probably the result of deleterious effects that radiation has on the cell functionality within these organ systems.

Gastrointestinal effects

Gastrointestinal effects can occur after inhalation exposure to radionuclides and uptake of high acute doses of radiation. Localized, very large doses of external radiation [(> 1000 rad (10 Gy)] can lead to inflammation and swelling of the oral cavity, including the cheeks, soft and hard palate, tongue, and throat (Otterson, 2007). These effects are probably the result of the lodgment of inhaled particles in the nasopharyngeal mucus and in the tracheobronchial mucus layers of the conducting airways of the lungs (Shadad et al., 2013).

Sensitive populations

(Box 5–3)

A number of authors have reported a link between ionizing radiation exposure from uranium and gold mining in Sub-Saharan Africa, and acute maternal and neonatal morbidity and mortality [(See, e.g., IAEA, 2018b; Scheele et al., 2011; Govender et al., 2013; Clement et al., 2017)]. *However, many questions related to this problem and ways of intervention still remain unanswered; which underlines the need for more research in the area.*

Effects of radiation on the developing fetus

Pregnant women worldwide are at risk of exposure to both ionizing and nonionizing radiation resulting from necessary medical procedures, such as diagnostic or therapeutic interventions, and workplace exposure; and a number of features regarding irradiation of the developing

---

BOX 5–3   Neonatal, maternal and child exposure to ionizing radiation.

The percentage of neonatal, maternal, and child deaths in Sub-Saharan Africa due to fetal abnormalities or congenital malformations is relatively high. These conditions are due largely to the overexposure of pregnant women to ionizing radiation, which is so prevalent around uranium and gold mining centers in countries, such as Namibia, Niger, and South Africa. Similar clinical conditions are observed around these mining centers, *but mortality and morbidity data are woefully lacking.* It is considered that these conditions, so tenuous and so ill-defined, can be more effectively tackled if geogenic factors of uranium release during mining, and its distribution, are taken into account. These are tasks that can be far more effectively carried out by Medical Geologists who *must* be incorporated in teams studying environmental disease distribution and etiology in Sub-Saharan Africa.

fetus are now well-established. However, according to the CDC (2014), most radiation exposure events will not expose the fetus to levels likely to cause health effects; and, according to Williams and Fletcher (2010), *in utero* exposure to nonionizing radiation is not associated with significant risks, and ultrasonographic procedures are considered a safe undertaking during pregnancy.

However, the health impacts on the fetus of ionizing radiation released during mining, processing, and related activities involving uranium and gold ores, are less well understood and have to date, been given very little research attention. This may be because knowledge of the effects of ionizing radiation exposure on fetuses is mostly based on observations rather than on scientific research (Yoon and Slesinger, 2020). *Ethical issues too, limit our research effort in this direction.*

Impact of ionizing radiation on the developing fetus, maternal, and child health

In recent years, a revaluation has been made of the possible health effects of prenatal exposure to ionizing radiation (Williams and Fletcher, 2010; Shaw et al., 2011; Gök et al., 2015; Schüz et al., 2017; Sreetharan et al., 2017; WHO, 2023; Fig. 5–15). Ionizing radiation can injure the developing embryo due to cell death or chromosomal disorder. The severity of damage to the embryo depends on the dose absorbed and the stage of development at which the exposure occurs (See Isaksson and Rääf, 2017).

We now also know from Sub-Saharan Africa countries where good morbidity and mortality data exist (such as South Africa), that *in utero* exposure to ionizing radiation can have severe consequences (See example, Dauda et al., 2019). Effects can be teratogenic, carcinogenic, or mutagenic (Williams and Fletcher, 2010; Gök et al., 2015). Radiation risk to the fetus is most significant during *organogenesis* (2–7 weeks after conception) and in the early fetal period (8–15 weeks after conception) (De Santis et al., 2007; Williams and Fletcher, 2010; Table 5–6).

Some other fetal effects, which have been observed in humans include mental retardation, IQ reduction, and microcephaly (Goel and Carachi, 2014). Brain damage may also be induced in fetuses through prenatal exposure that results in an acute dose uptake exceeding 100 mSv between weeks 8 and 15 of pregnancy and 200 mSv between weeks 16 and 25 of pregnancy (Verreet et al., 2016; WHO, 2016). Before week 8 or after week 25 of pregnancy, however, no radiation risk to fetal brain development has been found from studies based on human subjects (WHO, 2016).

Recently, the use of animal models has been shown to be very useful in further characterizing defects in brain development and consequent functional deficits (Verreet et al., 2016). *However, the establishment of a definite dose threshold is far from being accomplished, and a credible link between early defects and persistent anomalies has not yet been drawn.*

Ionizing radiation can also injure the developing fetus as a result of cell death or chromosome injury. The sensitive of cells that comprise the body's tissues and organ systems to the biological effects of ionizing radiation is not the same for all cells (US NRC, 2003). This sensitivity is affected by age at the time of exposure, sex, health status, and several other variables. The level of damage to the embryo depends on the dose absorbed, the type of radiation that is delivered to it, the stage of fetal development at which the exposure occurs, as well as the

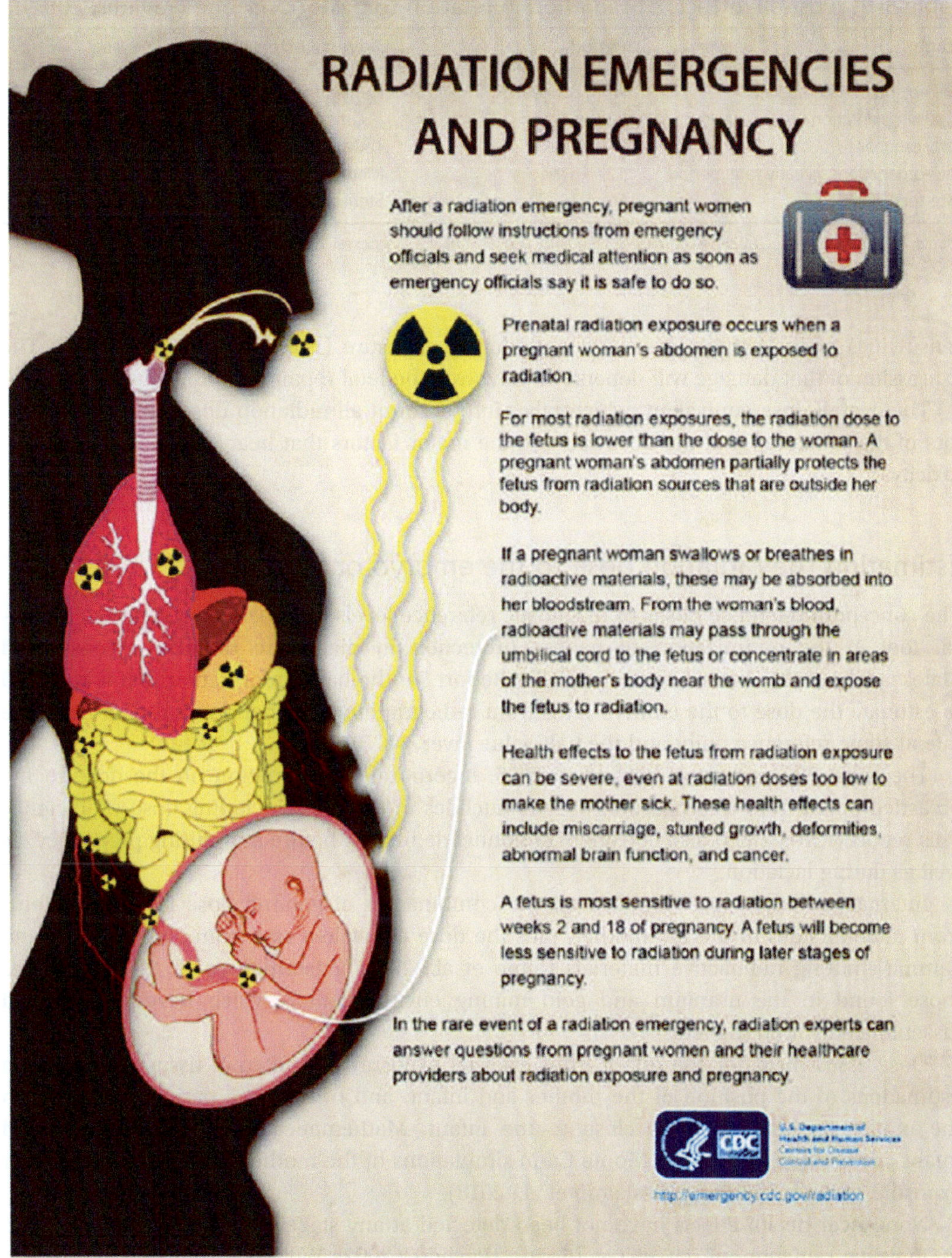

FIGURE 5–15 Radiation and the developing fetus (CDC, 2018). *From CDC (Centres for Disease Control and Prevention), 2018. Radiation and your health. https://www.cdc.gov/nceh/radiation/emergencies/resourcelibrary/all. htm. (Accessed 12 April 2020).*

Table 5–6 Possible effects of radiation in relation to gestational age (Pavlidis, 2002).

Stage	Period	Adverse effect
Preimplantation/immediate postimplantation	From conception to 9–10 days	Lethal; if not lethal, the fetus will recover completely
Early orogenesis	2–6 weeks	Teratogenesis, growth retardation
Late-organogenesis/early fetal period	12–16 weeks	Mental and growth retardation, microcephaly
Late-fetal stage	From 20–25 weeks to birth	Sterility, malignancies, genetic defects

Source: From Pavlidis, N.A., 2002. Coexistence of pregnancy and malignancy. Oncologist 7, 279–287.

sensitivities of the various organ systems during exposure (Isaksson and Rääf, 2017). The expression of that damage will depend on how well the fetal repair system works.

The overall dose equivalent, which takes into account all radiation doses, the quality factors of the radiation, and the dose rate are the major factors that bear on prediction of fetal toxicity.

Estimating the radiation dose to the embryo or fetus

The concept of reference doses or diagnostic reference levels has been developed as a practical tool in the optimization of patient protection in diagnostic radiology (Toossi and Malekzadeh, 2012). According to the NCRP Report 54, the basic information that is required to estimate the dose to the embryo-fetus from radiographic examinations is the air exposure rate at some reference point and the half-value layer (NCRP, 1977).

The ICRP 2004 report published in 2005, incorporates information on the dose to the breastfed infant from the ingestion of radionuclides in the milk, including dose coefficients. This report (ICRP, 2005) also considers radionuclide intakes before and during pregnancy, as well as during lactation.

Internal radiation dose depends on the combination of gamma dose rate to the fetus from radionuclides inside the mother, plus the dose equivalent rate from alpha-, beta-, and gamma-emitting radioactive materials (Simo et al., 2018; Vaiserman et al., 2018) (such as those found in the uranium and gold mining environments), which have entered and disseminated within the fetus.

Dose assessment for the infant from external exposure of maternal tissues is based on estimations of the position of the mother and infant, and of the time period during which the mother is holding, or is close to the infant. Mathematical models of mother and infant can be used to perform Monte Carlo simulations of the mother's tissues as sources for exposure of the infant (e.g., Kieskamp et al., 2018).

Noncancer health effects have not been detected at any stage of gestation after exposure to ionizing radiation of <0.05 Gy (5 rad) (Pavlidis, 2002; Williams and Fletcher, 2010). Spontaneous abortion, growth restriction, and mental retardation may occur at higher exposure levels (Table 5–7).

Table 5–7   Radiation dose effect on fetal life.

Dose (mGy)	Effects on fetus
< 100	No deterministic effects, only minor risk of stochastic effects
100–150	Increased risk of teratogenicity
2500	Malformation in most cases
> 30,000	Abortion

Source: From Pavlidis, N.A., 2002. Coexistence of pregnancy and malignancy.
Oncologist 7, 279–287.

Because fetal sensitivity to radiation exposure depends largely on the radiation dose to the fetus, the dose needs to be determined accurately before potential health effects can be assessed (CDC, 2020).

However, observers like Marci et al. (2018) have noted significant drawbacks in present methods for quantifying radiation doses to the testes, ovaries, uterus and pituitary glands such as lack of refinement, an observation that strongly underlines the need for Medical Geologists and geochemists who are better equipped with an understanding of the tools required for perfecting such measurements, to work closely with dosimetrists. Similarly, the risk of stillbirths or neonatal deaths can be more precisely determined when physicians work with geochemists and geostatisticians employing their hallmark modeling techniques and statistical procedures, such as Poisson regression analysis.

In the Gauteng Province of South Africa, there is some documentation of the harmful effects of ionizing radiation on neonates (e.g., Govender et al., 2013), including death, due to exposure to radioactive tailings material produced from uranium and gold mining activities. In Namibia and Niger, the other major uranium mining countries of Sub-Saharan Africa at present (2020), the situation is much more critical; *and although there is very little by way of documented studies that provide evidence to confirm this*, it is seemingly obvious that these problems are pervasive in the uranium mining centers of these countries.

Monitoring and management of female workers

Insofar as monitoring and occupational radiation protection of female workers are concerned, there is no justification to make any general distinction between male and female workers. However, additional protection measures would have to be considered for a female worker during and after pregnancy, in order to protect the embryo or fetus or the breastfed infant.

Monitoring of female workers during and after pregnancy should be predicated on the more stringent restrictions on dose. Assessment of doses should be based on a consideration of all relevant pathways of external exposure and internal exposure (See, e.g., Yoon and Slesinger, 2020). Once pregnancy is revealed, the monitoring program should be re-evaluated to ensure that the estimated dose to the embryo or fetus or the breastfed infant (including the dose from intakes by the female worker prior to conception) that could be due to occupational exposure, will not be in excess of the 1 mSv threshold (ICRP, 2007; IAEA

(International Atomic Energy Agency), 2018a,c). Information on the dose to the embryo or fetus from maternal intakes of radionuclides has been published by the ICRP (ICRP, 2000).

Finally, in line with Requirement 28 of GSR Part 3 (IAEA, 2014) and ICRP recommendations in ICRP Publication 84 (ICRP, 2000), the mine should have a process whereby: "…the employer should carefully review the exposure conditions of pregnant women. In particular, if required, their working conditions should be changed such that, during pregnancy, the probability of accidental doses and radionuclide intakes is extremely low."

Children's susceptibility

Although Sub-Saharan Africa and the Continent as a whole have made good progress towards reducing under-5 mortality, many children continue to die from preventable causes. Lloyd (2013) reported that neonatal deaths in South Africa accounted for approximately 40% of all deaths in children <5 years of age.

Children differ from adults. Their unique physiology and behavior can influence the extent of their exposure to ionizing radiation as well as their susceptibility to it; although their vulnerability may change with developmental age (NRC Committee on Pesticides in the Diets of Infants and Children, 1993; Hines et al., 2010; WHO, 2009a,b). There are also differences in pharmacokinetics and metabolism between children and adults. Absorption may be different in neonates because of the immaturity of their gastrointestinal tract and their larger skin surface area in proportion to body weight (NRC Committee on Pesticides in the Diets of Infants and Children, 1993; Keith et al., 1999; WHO, 2009a).

Again, too, children are exposed through pathways that differ from those of adults due to their size and developmental stage. Young children engage in normal exploratory behaviors, such as hand-to-mouth and object-to-mouth behaviors, and the deliberate or sometimes inadvertent intake of nonfood items, as they play on soil or mine tailings *(cf. geophagy)* which may dramatically increase exposure to radionuclides over that in adults.

Ionizing radiation carcinogenesis in children

Children are particularly vulnerable to the carcinogenic effects of ionizing radiation. The biological plausibility of children to radiation-induced cancer arises from the fact that their tissues are still growing and therefore the dividing cells are more prone to somatic genetic damage. Epidemiological studies indicate that the cancer risk after fetal exposure to radiation is similar to the risk after exposure in early childhood (WHO, 2016).

Radiological impacts on nonhuman biota

ICRP publications 103 (ICRP, 2007) and 108 (ICRP, 2008) deal, respectively, with the development of dose limits for nonhuman biota, but no specific guidance or assessment framework for practical application have so far been developed. Radiological impacts on nonhuman biota have always been excluded from the scope of the safety assessment, internationally, since it is assumed that if individual humans are shown to be adequately protected, then nonhuman biota are also protected, at least at the species level (ICRP, 2017). This assumption is buttressed by a comprehensive international study, which confirmed that both people and nature have been adequately protected from radioactive releases from all kinds of nuclear sites through the application of dose limits (Saint-Pierre, 2010).

Uranium mining in Sub-Saharan Africa—influence on nonhuman biota

The environmental consequences of uranium mining in Sub-Sahara African countries on nonhuman biota are not known with certainty. What we do know is that uranium, separated from its ore and released into the nonhuman biotic environment becomes more radioactive. Some of the radionuclides involved (e.g., radium-226 with a 1600-year half-life) can last for thousands of years. Namibia's "Uranium Rush" environmental assessment' (Dasnois, 2012) lists three cumulative impacts of uranium mining on nonhuman biota that appear to be valid for the rest of the Sub-Continent. These are deterioration of water quantity and quality for biodiversity and ecosystem functioning; habitat loss, degradation and fragmentation caused by mines and infrastructure; and threats to specific plants and animals. However, present legislative framework in Sub-Saharan Africa does not require the conduct of radiological the impact assessment for nonhuman biota.

Radon

A good knowledge of *radon* concentration levels in the vicinity of uranium mining zones, both in the outdoor environment and within buildings, is fundamental to assessing the health risks posed to associated populations.

What is radon?

Radon is a naturally occurring radioactive gas. It is chemically inert, colorless, odorless, and tasteless.

Radon results from the radioactive decay of Ra-226, which is a decay product of U-238 and radium-226. Three natural isotopes of radon are widely distributed, *viz.*, actinon (Rn-219), thoron (Rn-220), radon (Rn-222).

Both uranium and radium are present in nearly all soils and rocks, and traces of radon that they produce are in the air we all breathe. However, at high concentrations radon is a health hazard since its short half-life (3.8 days) means that disintegrations give of alpha particles, which can be discharged in the lungs and can later give rise to lung cancer *(and perhaps, on a more conjectural basis to date, also to some other cancers)* (See a later Section: "Health impacts of radon," in this Chapter).

Exposure to radon in the home and workplace is one of the main risks of ionizing radiation causing tens of thousands of deaths from lung cancer each year, globally. To reduce this burden in Sub-Saharan Africa, it is incumbent on policy makers from all nations to come up with methods and tools based on solid scientific evidence and sound public health policies formulated by teams of health professionals that should, ideally include Medical Geologists. The public needs to be aware of radon risks and the means to reduce and prevent them.

Recent case-control studies in many countries on lung cancer and exposure to radon in homes have led to substantial improvement in risk estimates and for further consolidation of knowledge by pooling data from these studies (Gillmore et al., 2018).

Unlike thorium and radium, radon gas reaches the environment (atmosphere) mainly by seepage through fractures and overlying soils. Radon gas escapes easily from rocks and soils into the air and tends to concentrate in enclosed spaces, such as underground mines, houses, and other buildings.

This is especially due to the diffusion of radon gas or Rn-222 from the soil and materials containing uranium. Accumulation of radon takes place in the air inside buildings and dwellings; and represents the main source of exposure of mankind to ionizing radiation, with the level of exposure affected by factors, such as local geography, building construction, and lifestyle.

This radiological hazard is now well known globally, *but in Sub-Saharan Africa it is little researched*. The ICRP and European Atomic Energy Community (EURATOM) have put forward recommendations that specify action levels for lowering radon concentration inside buildings and reduce cancer risks. The ICRP recommends that workplace radon levels be kept below 300 Bq/m^3, equivalent to about 10 mSv/year (WNA, 2020a).

Links have been made between radon concentrations in mines and the incidence of lung cancer in mine workers (Lubin et al., 1995; see also Gillmore et al., 2001, 2002) (See Sections: "Occupational exposure to ionizing radiation during uranium and gold mining in Sub-Saharan Africa" and "Radon in underground mines," respectively, in this Chapter). Other research has demonstrated a link between ocean and earth tides and indoor radon levels in some locations (Yong and Wei, 1995; Crockett et al., 2006). *While such studies represent some progress toward understanding the drivers behind indoor atmospheric radon concentrations (and other aspects of crustal deformation), a lot still remains to be uncovered.*

Similarly, radon is being used as a potential indicator in the monitoring of earthquake and volcanic activity (See, e.g., Alam et al., 2020), though these studies are riddled with controversies, and radon in groundwater wells has been used to monitor such activity in Japan (e.g., Igarashi et al., 1995).

Radon hazard assessment is a multi- and interdisciplinary undertaking as it requires inputs from a battery of professionals, such as geologists, geographers (physical and human), mathematicians/statisticians, physicists, chemists, and physicians.

Geology and migration characteristics of radon

The higher the uranium level is in an area, the greater the chances that houses in the area will have high levels of indoor radon. The underlying geology is one of the key factors that determine the levels of uranium, and hence, of indoor radon; but how the house or building is constructed, and how well it is ventilated are important determinants, too.

A geologic map would show the types of rocks and their structural orientations in a specific area. Because different types of rocks have different amounts of uranium, a geologic map would indicate to the geologist the general level of uranium or radium to be expected in the rocks and soils of the area. Higher than normal radon contamination potential can also be expected on soils of moderate to high permeability (Godish, 2001, 2018). Such soils can be identified from soil maps.

Some granites, ironstones, phosphatic rocks, and shales rich in organic material emit relatively high levels of radon (Appleton, 2004). These rocks and their soils may contain as much as 100 ppm uranium. High radon emission potential is also expected in houses constructed on glacially deposited features, such as eskers and kames (Godish, 2018).

Unlike thorium and radium, radon gas reaches the environment (atmosphere) mainly by seepage through fractures and overlying soils (Appleton, 2004). Radon is quickly diluted in the atmosphere, so that concentrations in the open air are normally very low, and probably do not present much hazard. However, radon that enters poorly ventilated buildings, caves, mines, and tunnels can reach high concentrations (Appleton, 2004).

Radon migration in the home

Exposure to airborne radon progeny in the domestic environment, but also in an underground mine setting, yields the largest source of exposure to ionizing radiation for the population. When radon gas enters the atmosphere from the ground, it disperses in the air, so outdoor concentrations are low. However, radon can move up from the ground into buildings predominantly through openings in floors or walls that are in contact with the ground (IAEA, 2013, 2015). When the gas enters a building or gets trapped within an underground mine, the concentration of activity builds up within the enclosed space. Radon accumulating in buildings or underground mine over time, can pose a serious health hazard (See Section: "Radon and health," in this Chapter).

The effective dose due to inhalation of radon decay products (RDPs) whether in a dwelling or in an underground mine setting is calculated from:

$$H_{\mathrm{RDP}} = h_{\mathrm{RDP}} \times \mathrm{RDP} \times \mathrm{IT} \qquad (5-5)$$

where H_{RDP} is the committed effective dose due to inhalation of RDP (mSv); h_{RDP} is the DCF for relevant RDP [mSv $\cdot$ (μJ hour)$^{-1}$ m^{-3}]; RDP is the RDP concentration (μJ/m^3); and IT is the inhalation time (hour).

Compare Eq. (4) with the general equation for calculating the effective dose due to inhalation of LLRD [Eq. (1)] in the Section: "Inhalation of long lived radionuclide dust," this Chapter.

The DCFs for radon and decay products are derived from epidemiological studies (See Section: "Epidemiology of ionizing radiation," in this Chapter) by the ICRP, whereas the LLRD DCFs are derived from the ICRP human respiratory tract model and physical data on the interaction of radionuclides on human tissue (ICRP, 1994). RDP concentrations are measured in various areas around the building or mine site. The inhalation time is given as the number of hours spent/worked in each area. The periodic (quarterly or annual) dose assessment for an individual with respect to the inhalation of RDP will be the average RDP concentration in an area multiplied by the occupancy time in that area for the individual, with this then multiplied by the individual RDP DCF (IAEA, n.d.).

Modes of entry into homes/buildings

Radon gas can get into homes and other buildings through cracks in solid floors and walls, building materials, construction joints, gaps in foundations around pipes, wires or pumps, and so on.

Site selection: to build or not to build?

High levels of radon can accumulate in any building, including new and old homes, well-sealed and drafty homes, office buildings and schools, and homes with or without basements. The Environmental Protection Agency of the United States (US EPA) recommends fixing your home/building when the radon level is at or above 148 Bq/m^3 (4 pCi/L) (US EPA, 2012).

One might choose to build (or not to build) on a site, based on knowledge of geology or soil characteristics, or detection of unacceptably high levels of radon emission from the underlying strata. The geology of an area is the ultimate predictor for high radon potential in a building (See Section: "Geology and migration characteristics of radon," in this Chapter). But in instances where radioactive waste rock or mine tailings have been used for the construction of buildings, the amount of radon in buildings may be drastically increased as well.

Before selecting a house or a site for building construction, prospective new home owners and their constructors should have the site or house and surrounding soil tested for levels of radon and other radioactive gases. Testing procedures (and testing kits), thankfully, can be inexpensive and easy; and only requires a few minutes to carry out (See Section: "Radon measurements," in this Chapter).

If unacceptably high levels of radon are found, decisions can be made on whether to build on a particular site, construct the house using techniques that minimize radon entry, or install an authentic radon mitigation system during the construction phase. Radon reduction systems are now readily available, and they are cost effective (See Section: "Radon measurements," in this Chapter).

In the case of proposed new (public) building construction, it is the responsibility of the regulatory authority or other relevant body to establish in advance, the basis for identifying those buildings for which preventive measures should be included in the design and construction phases; and, after construction, should ascertain the effectiveness of the specified preventive measures. Should there be indications that fill material on made-up ground might have elevated concentrations of Ra-226, appropriate adjustments should be made to the building development plan. It is important to note that some preventive measures may necessitate major changes to the original design and construction of the foundations.

Exposure of underground miners to radon

The mining of uranium ore (and sometimes also, of certain other minerals) can involve external exposure and internal exposure of workers to radon. Concentrations of radon in underground mines depend on the ventilation conditions and can reach high levels in some locations. Particular attention should be given to the impact of the air vents, which can emit large amounts of radon into the atmosphere. In order to lower the amount of radon and radon daughter products inhaled by miners, underground mines have to be vented. Fresh air is pumped down to the mine tunnels while extracting air from the shafts whose radon concentration will increase as it is flowing close to the uranium bearing rocks.

Radon and its decay products are a well-recognized cause of lung cancer in miners working underground (NRC Committee on Health Risks of Exposure to Radon (BEIR VI), 1999; Gillmore et al., 2001). The increased incidence of lung cancer among uranium miners has been known for decades (See Section: "Occupational exposure to ionizing radiation during uranium and gold mining in Sub-Saharan Africa.") Lead-210 in the bones of miners can be derived from the decay of radon gas and radon progeny and also from uranium dust. In vivo gamma spectrometric measurements of lead-210 in skull bones have been made in miners exposed to very high radon levels (WHO, 2001). However, *WHO has warned that a cautionary approach should be taken in interpreting these results, because of insensitivity of the method and the influence of other sources of lead-210 intake (water, food, etc.). For the same reasons, such measurements have not been feasible for occupants exposed to radon in dwellings; nor are they for in vitro determinations in bone. Add to all of that, are the practical and ethical problems involved in sampling for in vitro measurements.*

The *Biological Effects of Ionizing Radiation Committee* (BEIR VI), is a committee of the National Research Council of the USA, which publishes a series of reports informing the US government on the effects of ionizing radiation. Sometime ago now (1999), BEIR VI considered the entire body of evidence on radon and lung cancer, integrating findings from epidemiological studies with evidence from animal experiments and other lines of laboratory investigation (NRC Committee on Health Risks of Exposure to Radon (BEIR VI), 1999). The conclusions will be important to policymakers and environmental advocates, while the technical findings will be of interest to environmental scientists and engineers.

Radon control in mines

Since underground radon and radon progeny concentrations can vary rapidly both temporally and spatially, it is important that real time monitoring and control approaches be put in place.

The IAEA (2018a) publication: *Occupational Radiation Protection—General Safety Guide* gives the following hierarchy of control measures for radon and radon progeny emitted from uranium or thorium bearing ores into the working environment of operational uranium mines and mills: "(1) Adequate and effective ventilation systems; (2) Management of the source of the radiation; (3) Management of water sources and process liquors containing dissolved radium and radon; (4) Working in an enclosed and filtered operating environment (e.g., ventilated driver's cab or static plant control room); (5) Administrative controls establishing action levels for airborne contaminants."

Workplace monitoring of the short lived progeny of Rn-222 is carried out by drawing air through a filter to capture the progeny radionuclides. As with the monitoring of Rn-222 gas concentrations, the monitoring of Rn-222 progeny can be carried out either by instantaneous measurements or by measurements over a given time period. Owing to the short half-lives of the Rn-222 progeny, counting of the alpha or beta activity on the filter should be performed during, or shortly after, sampling.

Through the development of automated sampling and analytical techniques, instruments have become available for semicontinuous monitoring using integrated measurements (See, e.g., Masok et al., 2016; Masok et al., 2018) and for continuous monitoring (e.g., Van Schalkwyk, 2005).

In some instruments that perform alpha or beta spectroscopy, raw data can be stored on a continuous basis within the instrument and downloaded later for processing to determine the individual radionuclide concentrations over time.

Radon in mines is often measured in the historical units of *working level months* (WLMs), already defined in the later Section on: "Dosimetry of inhaled radon," this Chapter.

Radon research in Africa

There is a dearth of studies on radon concentrations and its fate in the African geosphere. This situation is well-reflected in the data-deficient radon map produced during the World Health Organization project of 2017 (Zielinsky, 2014).

Among the other studies on "radon in Africa" encountered in the literature are those of Botha and Human (2001), Nemangwele (2005), Lindsay et al. (2008), Botha et al. (2016), Botha et al. (2017), Botha et al. (2018), Pule and Speelman (2016), Masevhe et al. (2017), Bezuidenhout (2019), Herbst/CANSA (2019b), Moshupya et al. (2019).

A few references specifically on "health effects of indoor radon" in Sub-Saharan Africa exist, mainly reporting studies in South Africa. See, e.g., Leuschner et al. (1989), Lindsay et al. (2008), IAEA (2013), Pule and Speelman (2016), Kamunda et al. (2017), Herbst/CANSA (2019b), Le Roux et al. (2019).

Radon in Sub-Saharan Africa underground mines

In underground mining in Sub-Saharan Africa, the exposure of workers to ionizing radiation (especially radon and its short-time progeny) is of major concern for unions, workers, and government regulators. Such exposure has led to the death of hundreds of miners in the Sub-region's uranium mining industry (See Scheele et al., 2011; Hecht, 2012a,b). Other noteworthy accounts of health impacts of radon due to underground mining of uranium, as exemplified in Sub-Saharan Africa are given in: Axelson (1991), NRC (US) Committee on Health Risks of Exposure to Radon (BEIR VI) (1999), Chareyron (2008), Mudd (2008), Hecht

(2009), Scheele et al. (2011), Hecht (2012a), Chareyron et al. (2014), Cothern and Smith Jnr. (2013); and other references in the "Selected further reading" list at the end this Chapter.

Exposure to radiation from radon is a health threat not limited only to uranium miners. It is a continuing concern in the gold mining industry as well, especially in South Africa (See the earlier Section on: "Association of uranium with gold in South African mining," this Chapter). It is also a significant concern for coal miners as well.

Toxicokinetics of radon

Radon progeny in the environment can be taken up directly or be created within the body after consumption of radon. A unique feature of radon progeny is the occurrence of both aerosol bound radionuclides of 100—400 nm diameter and ultra-fine clusters in the diameter range of 1—4 nm, called the "unattached fraction" of radon progeny (Butterweck et al., 2005). *The biokinetic behavior of aerosol particles has been well studied, whereas experimental studies on the biokinetic behavior of unattached radon progeny are scarce.*

Case control studies of indoor radon have been carried out around the world and these have had mixed results (See, e.g., Sheen et al., 2016; Carmine et al., 2017; Gillmore et al., 2017a,b; Dobrzyński et al., 2018).

Radon inhalation and carcinogenesis

Since radon gas is pervasive in the environment, inhalation is the primary source of radiation to the body from radon exposure. After inhalation, a proportion of radon and its radioactive progeny enter the bloodstream and may reach other organs apart from the lungs (Keith et al., 2012). Most of the radon that gets into the lungs, though, is exhaled. Radon progeny that is not exhaled is deposited and attaches to the bronchial and bronchiolar epithelium (WHO, 2001). The basal and secretory cells in the bronchial epithelium are thought to be the most critical cells.

The effective half-life for irradiation of the bronchial epithelium by radon progeny following inhalation exposure is about 30 minutes (WHO, 2001). From the lymph and blood, radon, and other nuclides in the uranium chain are distributed to other organs in the body. The cells are irradiated with alpha particles by deposited activity.

A number of leading international institutions involved with oncology research, notably, the American Cancer Society, the International Agency for Research on Cancer and the World Health Organization, classify radon as a carcinogen. According to the United States Environmental Protection Agency (US EPA), exposure to radon gas is the leading cause of lung cancer in nonsmokers (US EPA, 2012).

The increased incidence of lung cancer among uranium miners has been known for decades (See, e.g., US DHEW, 1967). More recently, epidemiological studies have confirmed that inhalation of radon increases the risk of lung cancer even at very low doses, and even in the context of exposure at home (Darby et al., 2005).

The global literature on health effects of radon inhalation, whether in indoor environments or in underground mine setup, is voluminous. Reference to the more recent peer reviewed articles on the subject (See Section on: "Selected further reading," this Chapter) is instructive for Medical Geology researchers and coworkers from allied disciplines; for this would provide valuable data and insight on proven methodologies that could provide the bases for expanding research in this area in Sub-Saharan Africa.

A small proportion of deposited activity is circulated by the body fluids to other organs. Unattached progeny are largely filtered by the nose and, to some degree, in the larynx region. Ingested radon and the other radionuclides in the uranium chain may also be taken up by the gastrointestinal tract and distributed to other organs (WHO, 2001).

Some researchers believe that other pathologies, such as childhood leukemia may result from exposure to radon (See, e.g., WHO, 2009a).

Dosimetry of inhaled radon

The frequency of lung cancer deaths in the workforce in underground uranium mines is associated with the duration of work underground, and the cumulative exposure to radon.

The highest doses of radiation are produced by alpha particles emitted from short-lived radon progeny. These are deposited on bronchial airway surfaces, from where they are taken up by the bronchial region of the human lung (Kreuzer and McLaughlin, 2010; Hofmann et al., 2012a). The primary health effect due to radon inhalation, therefore, is the formation of bronchogenic carcinomas mainly in bronchial airways.

According to Kluszczyński (2005), due to the large number of exposed miners in underground mining situations, and the very difficult conditions under which they carry out their work, it is often difficult to cover the entire population by dosimetric measurements. This, together with the high variability of radon progeny concentrations, often result in the wrong assessment of the annual dose to the miners. Kluszczyński (2005) also noted that the formulae to determine the number of measurements that need to be taken (for miners) should necessarily include extreme values; and went on further to present the theoretical and empirical considerations, which may be useful in dose assessment for such exposed populations.

Because of the impracticability of measuring the activities of individual radon progeny in the environment, and while they decay inside the body, inhaled radon progeny in mines is measured in a unit termed the *Working Level (WL)*. This unit quantifies the cumulative radiation, which may not be in secular equilibrium (Keith et al., 2012). One WL was originally defined as the concentration of potential alpha energy associated with the short-lived radon progeny in equilibrium with 100 pCi/L (3700 Bq/m^3) (WHO, 2001). Exposure to the equivalent of one WL for 1 month of work is called one *WLM*. The conversion of WLM to Sieverts is fraught with difficulty, since it depends on breathing rate and the size of the radon daughter aerosols. UNSCEAR (2000) suggests that the average value of roughly 5.7 mSv per WLM be adopted.

Since the quantity (WLM) was introduced for specifying occupational exposure, 1 month was taken to be 170 hours. See UNSCEAR (2000), WHO (2001)and IAEA (2018a) for more details regarding the WLM measure and other essentials of radon metrology.

Radon as a diagnostic tool

The use of radon as a precursor for diagnosing volcanic eruptions and earthquakes has always been strongly debated. An underlying factor engenders the necessity to fully understand the behavior of fluids in the geological substrate (Otton, 1992; Appleton, 2005; Gillmore et al., 2018). This is well illustrated by Cigolini et al. (2016), who stress that understanding the behavior of fluids in hydrothermal systems is a key factor in volcano monitoring. Better knowledge of the factors of variability and heterogeneity of the radon sources has led to more cautious approaches. *This, in turn, will lead to better quantitative models better able to account for the complexity of the problems and, in due course, lead to more reliable forecasts even if precise predictions might remain unattainable owing the degree of complexity.*

Radon as a therapy? The radiation hormesis theory

The use of radon in therapeutics has been known for over a century (e.g., Przylibski, 2016; Eckert, 2024) based on the radiation hormesis (or radiation homeostasis) theory, [the (beneficial) simulation from low doses of ionizing radiation] (See Feinendegen, 2005). Radon's radioactive nature and the fact that its alpha-radioactive decay is the source of other radionuclides, are the reasons for elicitation of claims and controversies regarding its therapeutic application (See, e.g., Zdrojewicz and Strzelczyk, 2006; Kojima et al., 2019).

Granite tabletops and countertops and radiation

A variety of geomaterials are commonly used in residential settings in a number of African countries such as Egypt, Kenya, and South Africa, to make tabletops and countertops in kitchens, dining rooms and bathrooms. These materials include minerals, such as quartz, and rocks, such as granite, slate, and marble.

The durability of granite in particular, and its exquisite decorative properties make it a popular building material in many homes and buildings in the aforementioned countries. *However, there is huge concern as to whether there is a link between granite used to make kitchen tabletops and bathroom countertops and cases of death and irreversible lung injury in workers who cut, grind, and polish this increasingly popular material (See, e.g., Myatt et al., 2010; Turhan, 2012).*

The propensity of granite to emit radon has already been described (See Section: "Geology and migration characteristics of radon," in this Chapter). Some granites may contain veins of naturally occurring radioactive elements like uranium [around 10−20 parts per million (ppm) (Maurice, 1982)], thorium, and their radioactive decay products. These trace concentrations may vary depending on the type of granite (type, being defined by minor but significant compositional variation reflecting its radon emission potential), or even within a single slab of granite. So, we know that most granite tabletops and countertops may emit some level of radiation, an observation backed by studies that have reported on the concentrations of natural radionuclides (i.e., activity concentrations) in Egypt, Kenya, and South Africa for granite samples obtained from these countries, e.g., for Egypt: Arafa (2004), Ahmed

(2005), El-Taher et al. (2007); for Kenya: Mustapha et al. (1997), and for South Africa: Bezuidenhout (2019) and Le Roux et al. (2019).

Identifying radiation in granite countertops

Identifying the presence and concentration of radioactive elements in each specific granite tabletop or countertop requires sophisticated instruments. These instruments require proper calibration and a knowledgeable and trained user to interpret the results. Techniques for measuring radon either in the built environment or in underground mines is given in the Section: "Radon measurements," in this Chapter.

Risk reduction in indoor radon and health studies

Detailed radon distribution maps of rocks, soils and groundwater are urgently needed for Africa, to provide information on radon migration and be used as a basis for informing the development of policy recommendations regarding groundwater development and housing construction in areas where elevated uranium concentrations are predicted. It will permit radon hazard zones to be reliably isolated in a short time, as well as the evaluation of the radon hazard within building sites and prospective building zones (Komov, 2001, 2003).

International radon management initiatives

Research on "radon in Africa" can benefit from consideration of results from recent and on-going international projects on radon. These projects are generally aimed at mapping the distribution of radon in the natural environment, so that these maps can be used by government bodies, policy makers, and other stakeholders, to inform decisions as to where best to deploy their inevitably finite resources (Gillmore et al., 2018).

To address the serious health issue of radon hazard in homes, at an international level, the World Health Organization (WHO) established the International Radon Project (IRP) in 2005 (Zielinsky et al., 2006). The project was launched in January 2005 with its first meeting attended by 36 experts representing 17 countries. The project's scope and key activities during the period 2005–2008 included: "(1) a worldwide database on national residential radon levels, radon action levels, regulations, research institutions, and authorities; (2) public health guidance for awareness-raising and mitigation; and (3) an estimation of the global burden of disease (GDB) associated with radon exposure." (Zielinsky et al., 2006).

The Radon Hotline (http://www.radonhotline.org) Internet site is owned by the nonprofit making Hotline Limited, and operated as a partnership between The Radon Council and University College Northampton (UCN) (Gillmore, 2001). According to Gillmore (2001), "This site has been constructed to disseminate radon gas-related information to the public and professionals. The web pages aim to answer initial queries concerning radon gas and remediation methods. Information on radon and the associated health and construction issues is also included."

WHO has formed a network of key partner agencies from over 40 Member States as the basis for the WHO IRP, which was launched in 2005. The WHO handbook on indoor radon (WHO, 2009b) is a key product of this WHO IRP. This handbook focuses on residential radon exposure from a public health point of view and provides detailed recommendations on reducing health risks from radon and sound policy options for preventing and mitigating radon exposure.

In 2014 the Geological Survey of Austria (Bundesministerium für Land- und Forstwirtschaft, Umwelt und Wasserwirtschaft) undertook a detailed survey of radionuclides in groundwater, rocks, and stream sediments in Austria, and published the results in the form of an overview map with notes. The work of Schubert et al. (2018) summarizes the entire content of this survey.

Gillmore et al. (2018) reported on a 5-year project on radon, done under the auspices of UNESCO/IUGS/IGCP, on the realization that in recent years there has been an increasing interest in radon from a range of different aspects. The project (named UNESCO/IUGS/IGCP Project 571) focused on a variety of impacts and hazard-associated manifestations of radon. Gillmore et al. (2018) brought together the final outputs of this project and presented new data on radon in the built and natural environments, radon as a diagnostic tool of geophysical phenomena, reflections, and recommendations on the future of radon research and a critique of radon's asserted use as a therapy. *The* Gillmore et al. (2018) *Report incorporated a suggestion that radon science has the potential to be a useful tool in understanding our environment as well as its impacts on human health.*

Methods for reducing radon concentrations in homes and buildings

It has long been recognized that knowledge of prevention and mitigation methods related to indoor radon and natural radionuclides in building materials is an important public health requirement, worldwide (King, 1993). The latest awareness creation exercise to underline the necessity for organizing radon prevention and mitigation programs was IAEA's 2019 Webinar (IAEA, 2019a). *The radiation doses attributable to the use of building materials containing radionuclides is not known with certainty* (Valentin, 2006); however, the worldwide exposure for this cause has been estimated to be in the region of 0.4 mSv/year, with a typical range of 0.3−0.6 mSv/year (UNSCEAR, 2000).

Radon mitigation is the process or any system or steps designed to reduce radon gas concentrations in the occupied spaces of buildings to reduce occupants' exposure to elevated levels of radon. The goal of a radon mitigation system is to reduce indoor radon to a level as low as reasonably achievable.

A few African countries, such as South Africa (See, e.g., Pule and Speelman, 2016) have developed regulations on indoor radon and have verified these regulations, but remedial and preventative measures are not sufficiently used. This in part, could be related to an absence of a trained workforce on remedial and preventative techniques (IAEA, 2020c).

Most of the available literature on prevention and mitigation of radon in dwellings focus on South Africa (See: Leuschner et al., 1992, 2002, n.d.; Karam and Venter, 2007; Lindsay et al., 2008; Eilers et al., 2015; Pule and Speelman, 2016; Kamunda et al., 2017; Botha et al.,

2016, 2018; Masevhe et al., 2017; Bezuidenhout, 2019; Moshupya et al., 2019; Le Roux et al., 2019; Newman et al., 2019). As at 2016, according to Pule and Speelman (2016), regulatory framework did not provide for regulatory control of radon in dwellings (in South Africa).

Established preventative or mitigation methods for indoor radon

There are many steps that could be taken to reduce elevated radon levels in a home, and so minimize the amount of exposure to radon. A variety of source control techniques are now available, some of which can be applied during the construction of new houses, while others can be used on a retrofit basis.

Recent authoritative works on radon prevention and mitigation can be found in WHO (2009a,b), Zhou et al. (2018), Godish (2018), Khan et al. (2019) and other references in the "Selected further reading" list at the end of this Chapter. Here, key features of the main techniques that are available are briefly summarized.

- Subfloor depressurization

    This technique is used when foundations and basements are in contact with the soil. The method comprises of a reduction of the soil gas pressure in the vicinity of the foundation, relative to the pressure in the structure. This is normally done by installing a system of pipes leading from the soil under the foundation that maintains a negative pressure gradient between the soil and the foundation.
- Subfloor ventilation

    The amount of radon entering the structure can be reduced by ventilating the space beneath the floor, provided the ground floor is not in contact with the soil. This can be achieved by increasing the natural ventilation or by installing a fan that removes air from under the floor and replacing it with outdoor air (Dixon, 2005). However, this method is considered to be generally less effective than depressurization and ventilation, since it is difficult to seal all entry routes adequately; and seals tend to deteriorate over time.
- Increased ventilation

    Increasing the ventilation by use of a fan or by ingress of outside air, is the method often preferred when radon levels are very high, getting toward, say, 1000 Bq/m^3. At this level or greater, one may need to use more than one fan. However, in houses with full cellars or basements, this solution is likely to be inappropriate for use.
- Removal of subsoil

    Sometimes high indoor radon concentrations are caused by elevated Ra-226 values in the soil beneath or surrounding the building. Indoor radon concentrations can be reduced in such cases by removing the subsoil and putting in its place, uncontaminated soil. This is usually done as a last resort, when there are no feasible alternatives.
- Water treatment

    In some situations, the water used in the building could be a significant source of indoor radon. Prior treatment of the water by aeration (moving the radon from the water to the air), can be effective. This procedure can be carried out at a treatment plant, point of entry, or point of use. However, aeration can sometimes worsen the problem in the municipal water treatment plants where it is carried out; so the operation must be done in a meticulous fashion (See: Kinner et al., 1987; Dixon, 2005; Dixon et al., 1991; Drago, 2013; Ghernaout, 2019).

Filtration with granular-activated charcoal (GAR) can be used, though this is less likely to be effective (See, e.g., Zhou et al., 2018). The air spaces of frequently accessed areas in water treatment plants should be well-ventilated to avoid buildup of radon in high concentrations. It is necessary to state that the periods of occupancy of plant workers should be restricted in areas of high radon concentrations in radon treatment plants.

Radon as a carcinogen—management issues

Radiation therapy is one of the most important present methods for treating cancer (NCI, 2019). However, alpha radiation damages healthy body cells as well as the diseased ones. In the area of cancer therapy and treatment, although much progress has been made in recent years (e.g., Chaves-Pérez et al., 2019; Martinez Marignac et al., 2019; Yang et al., 2020), *there still remain significant challenges in these aspects for many types of cancer.* As I write (2020), Bayer AG researchers are working on the development of targeted radiation therapies using biomolecules, which specifically deliver alpha radiation to the tumor cells (See Cuthbertson, 2020). *Additional research along these lines are desirable.*

Policy statement: radon in homes and underground mines

Appropriate construction codes and best practices for construction methods should be developed for indoor radon. To do this, decision makers need a characterization of the risk of radon exposure across the range of exposures people actually receive. In response to this need, the BEIR VI committee has developed a mathematical model for the lung cancer risk associated with radon, incorporating the latest information from epidemiology and scientific studies (NRC Committee on Health Risks of Exposure to Radon (BEIR VI), 1999).

In the case of radon accumulation in underground mines, the BEIR committee has also provided a revised assessment of exposure-dose relationships, and discussed key issues, such as the weight of biological evidence and extrapolation from radon-exposed miners to the larger population, in evaluating the risk posed by indoor radon. Such uncertainties as the combined effects of smoking and radon, and the impact of exposure rate are also addressed.

Measurement of radon

Measurements of indoor radon are usually made subannually and then seasonally corrected, using measurement protocols and widely observed seasonal variations in indoor radon concentrations for the statistical derivation of the seasonal correction factors.

While radon is generally known to be a naturally occurring radiological hazard, Crockett and Gillmore (2016) report measurements of significant, hazardous radon (and potentially also thoron, Rn-220) concentrations that arise from anthropogenic sources, such as radium-dial watches and uranium glass artifacts. Modern indoor radon-measuring devices must therefore, have the capacity to accurately determine radiation from these sources as well.

Instruments for measuring radon

The instruments and counting methods used for measuring the concentrations of Rn-222 and its progeny can be adapted for measuring the concentrations of Rn-220 and its progeny, with certain limitations.

The range of techniques and available instruments is wide, with each technique or instrument having its advantages and disadvantages. There is, however, no single technique that can meet all the requirements of the different types of radon measurements and of its decay products (El-Taher, 2018; Elzain, 2017).

Radon measurements are often discussed in terms of either a short-term or a long-term test (Quindos et al., 1991).

An activated charcoal detector or some other type of detector, such as an electret ion chamber can be used to perform a short-term test for radon. Such a test can provide a first indication of the mean long-term radon concentration in a home. However, when performing short-term radon measurements, one should also take cognizance of diurnal and seasonal radon variations.

In many cases, though, what we want to know are the long-term exposures; even though for diagnostic purposes, a continuous measurement may be more appropriate (Miles, 2001, 2004).

Other techniques available for instantaneous Rn-222 measurement include the pulse counting ionization chamber technique and the double filter sampler technique, the latter of which can be used for measuring both Rn-222 and Rn-220 (WHO, 2009b).

Most registered radon measuring instruments are accurate and practical for screening potentially high-radon concentrations (Rolle, 2004); so, the least expensive instrument is the one usually chosen. When subsequent significant remediation decisions hinge on these measurements, it is important to remember that sensitive measurement of RDP concentrations would be the best for evaluating key source and sink dynamic features (Rolle, 2004).

Excellent accounts on radon measurements, as well as the different instruments used for measuring radon and its decay products are given in Miles (2001, 2004), Rolle (2004), WHO (2009b), Hofmann et al., 2012b, Elzain (2017), Kozempel (2017), and other relevant references in the "Selected further reading" list at the end of the Chapter.

Epidemiological approach in studying effects of ionizing radiation exposure

Epidemiology is the statistical study of the incidence of disease in groups of people. Epidemiologists compare the rate of occurrence of the health effects under consideration, and attempt to identify and analyze relationships between these effects and their possible causes. This is difficult, in general, because there are many *confounding factors* in any study that make a simple cause-and-effect relationship hard to separate.

The most important question to be asked in the case of health effects of ionizing radiation is "Is there a relationship between this agent/risk factor (ionizing radiation) and the effect?"

The most practical and informative approach to assessing the toxicity of ionizing radiation in humans is provided by the use of human data pools (Keith et al., 1999). This is a major strength of this kind of epidemiological study, for the approach obviates the need for extrapolation from species, dose levels, or outcomes to another. Direct inference outweighs the uncertainties usually inherent in other kinds of epidemiological studies; especially so, in the case of nonrandomized studies. Laboratories using controlled exposure scenarios usually employ this kind of approach.

Despite the perceived advantages, however, epidemiological studies may be said to be fraught with a number of potentially critical limitations and sources of error when used to demonstrate (or exclude) small health effects. The main issues (in the case of ionizing radiation as a causative factor for cancer), according to McCormick (2011), include"... causality versus association, limitations in statistical power, timing, and duration of exposure prior to hazard identification, and dosimetry and exposure assessment." Of these issues, dosimetry and exposure assessment pose especially important questions in evaluating the possible risks of human exposure to ionizing radiation. Although average exposures can be estimated, we never know exactly the amount of radiation one has been exposed to, nor from what sources the radiation came. For example, uranium miners, have always been exposed to some background radiation (and exposure to other carcinogens), and there has always been a background cancer rate. The effects of small additional doses from the mining environment can be very difficult to detect.

Epidemiology of cancer due to ionizing radiation exposure

In recent years, a dose-dependent increase in cancer risk has been repeatedly demonstrated in population-level studies in humans with consistent findings in different populations (See: Hayashi et al., 2003; Shah et al., 2012; Brooks et al., 2016; Clewell et al., 2019). According to SCENIHR (2012), the latency from exposure to the occurrence of excess cancer is typically approximately 10 years, but shorter (2−5 years) for leukemia and thyroid cancer have also been observed.

The cancer risk for prenatal exposure was judged in the ICRP publication 103 (ICRP, 2007), to be similar to that following irradiation in early childhood. However, recent evidence on *in utero* exposure to ionizing radiation (See Preston et al., 2008) indicates that the lifetime risks of solid cancers (but not leukemia) are considerably lower than those for exposure in childhood, i.e., at most, about three times that of the population as a whole.

The disease risk caused by radiation can be expressed in terms of relative risk (RR), i.e., as a multiple of the underlying disease risk (SCENIHR, 2012). A RR of 1 indicates a similar occurrence in the study population as in the reference group, while for instance, an RR of 1.5 implies a relative increase by 50% (i.e., 1.5-fold occurrence). Such a RR model inherently assumes that the health effect is proportional to the baseline risk of the population, i.e., whether the disease risk due to other factors, such as age is low or high, the exposure would always increase the risk as a multiple of the baseline.

The life-span study

The Report of the Scientific Committee on Emerging and Newly Identified Health Risks (SCENIHR) on an epidemiological study of the effects of radiation on the atomic bomb survivors of Hiroshima and Nagasaki, known as the *Life-Span Study*, is perhaps the most well-known source of epidemiological knowledge on health effects of radiation. In this Report, SCENIHR (2012) writes: "... A wide range of doses (from several Sv down to 5 mSv), good quality dose estimates, a large study population covering a wide age span and long follow-up with information on both cancer incidence and mortality, increase the amount and quality of evidence from the study. Among atomic bomb survivors, a significant dose-response relationship is seen in the dose range 0−150 mGy for solid cancers and the existence of a threshold (below which no effect is seen) can be excluded at 85 mGy or higher (but not below)."

Awareness/education

According to Rose and Rae (2017), "... formalized training in radiation safety and protection would improve knowledge on the subject and facilitate greater compliance in safety practices." Rose and Rae (2017) also consider that such training should be included as a basic knowledge requirement in nursing curricula, and as radiation-safety guidelines needed in radiography suites in African Medical Physics curricula (See also Thambura et al., 2019).

In Sub-Saharan Africa, there is currently a huge campaign on awareness and education of radiation workers on the need to adhere to the principles of radiation protection and radiation safety; but this effort has been directed largely at radiation health workers, and not at miners and proximate communities at mining centers (See, e.g., Nyabanda and AFROSAFE, 2015; Dauda et al., 2019; Thambura et al., 2019; Ribeiro et al., 2020).

Although many of these studies had limitations on training protocols, *viz.*, that they were conducted as single-center studies, which prevented researchers from making comparisons, general recommendations from them, [e.g., Dauda et al., 2019] embody the typical conclusion, *viz.*, emphasizes be put on the proper conduct of diagnostic radiation courses and education programs, as well as justification of referral for imaging among clinicians of all specialties at graduate and postgraduate levels.

Some noteworthy efforts on awareness raising and education on reducing public doses at uranium mining and milling legacy sites are those of IAEA, which has recently organized a number of workshops and training courses (IAEA, 2020c). For example, from June 2013 to June 2017, the IAEA conducted a project that aimed to reduce the radiation doses of people living near uranium mining and milling legacy sites. Three training events targeting African countries having legacy sites, were also held as part of the project. The project resulted in a draft IAEA Technical Document (TECDOC) titled "Short-Term Measures for Reducing Risks from Legacy Sites Associated with Uranium Production," and related training materials. The TECDOC and training materials targeted Member States that lack resources or experience to implement a remediation program that would provide for long-term solutions. The document's principles and approaches can be used to address legacies from the mining and processing of (NORM) residues (IAEA, 2020c).

The first Regional Training Workshop on: "Practical Intervention Techniques to Reduce Public Doses at Uranium Mining and Milling Legacy Sites," took place in Centurion, South Africa June 15–19, 2015, with 20 participants from nine African states.

The Second Regional Training Workshop on: "Practical Intervention Techniques to Reduce Public Doses at Uranium Mining and Milling Legacy Sites" for Portuguese-speaking Member States (of IAEA) took place from 28 September to 2 October 2015 in Vienna with nine participants from Angola, Brazil, and Mozambique.

A Regional Meeting in conjunction with Project B.2 on "The Prevention of Future Legacy Sites in Uranium Mining and Processing" was held December 14–15, 2015 in Vienna with the participation of 12 senior management representatives from nuclear regulatory bodies or other competent authorities in nine African Member States. An IAEA Technical Meeting to present and review the draft TECDOC and training materials developed under Project B1 was held in Vienna February 23–27, 2015 (IAEA, 2020c). The participants, from countries with planned or operating uranium mining and milling facilities enhanced their ability to establish sound policies, regulatory frameworks, and infrastructure to achieve sustainable levels of safety in line with IAEA safety standards.

Outcomes of the IAEA meetings

The IAEA 2013 to 2017 Project culminated in some very important outcomes, which can be listed as "Improved ability by regulators in African countries with uranium legacy sites to develop and implement site-specific measures to protect the public; increased awareness among legacy site managers of low-cost, mitigation techniques that reduce public exposure now and in the future; Improved safety of legacy sites, which will contribute to increased protection of the public and a reduction in future legacy issues in the long term." (IAEA, 2020c). The project enhanced the regulatory frameworks and national capacities for countries that have legacy uranium production sites, so that they can mitigate impacts and improve the safety of populations living near legacy sites.

Tanzania has set regional precedence by establishing an Atomic Energy Commission and gone ahead to start construction of a state-of-the-art laboratory designed for the management and use of radiation materials. In Sub-Saharan Africa, the country is only second to South Africa in this frontier and has already begun the first phase of construction (Mdoe, 2020).

Awareness of radon infiltration in homes

To develop appropriate public policy for indoor radon, decision makers need a characterization of the risk of radon exposure across the range of exposures people actually receive. In response to this realization, as far back as 1994, an international group of experts published a comprehensive re-analysis of data from 11 existing studies of radon and lung cancer risk, which covered underground miners in Australia, Canada, China, the Czechoslovakia, France, Sweden, and the United States. According to Hecht (2012b) "… African exposures could not be reanalyzed, because they had never existed *as data* in the first place (Lubin, 1994)." *Today (2020), such data for Africa are at best still only patchy, exposing the urgent need for their acquisition from more comprehensive studies that would no doubt serve as basis for formulation of much more meaningful radiation protection guidelines by public health officials in Africa.*

Finally, governments in Africa are becoming increasingly concerned about emission of radioactive materials in the environment and the contamination of food. The establishment of an appropriate radiation protection infrastructure is now, thankfully, regarded by all countries as an important obligation and a governmental responsibility to the health and safety of the citizenry (Bakr, 1989; Hecht, 2012a; Chareyron et al., 2014). Despite the increasing awareness, however, there is still (until recently; 2020) the lack of definite steps for institution of adequate radiation protection measures by African governments (See, e.g., Bakr, 1989). Until the countries with legacy sites and active uranium and gold mines as well as nuclear processing facilities are able to implement well-funded and resourced remediation efforts, short-term intervention techniques must be adopted in order to reduce public doses of ionizing radiation.

Case histories

Excellent accounts of uranium occurrence, mining, processing, mine decommissioning, and health effects in Africa are given in Koos and Basedau (2012), Scheele et al. (2011), Dasnois (2012), Kinnaird and Nex (2016), Winde et al. (2017), WNA (2020g), and other pertinent references in the "Selected further reading" list. Here, a brief summary of these accounts is presented, citing the three leading African uranium producers in 2019, Namibia, Niger, and South Africa (WNA, 2020d).

Types of uranium occurrences in Sub-Saharan Africa

The main types of uranium deposits in Africa can be classified into four broad groups (Kinnaird and Nex, 2016). These are: the South Africa quartz-conglomerate hosted gold-uranium deposits of Archean age; the sheeted leucogranites and small stocks of Namibia that mark the Neoproterozoic end-orogeny; the sandstone-hosted roll-front deposits (Mesozoic) of Niger and Malawi; and the recent channel-hosted calcrete and alluvial deposits in Namibia, which locally are of huge economic potential. Minor unconformity-style deposits are found in countries, such as Algeria and Malawi. Phosphate-hosted deposits occur in Morocco and offshore South Africa. Vein-hosted deposits are found at Shinkolobwe in the DRC and Democratic Republic of the Congo at Lumwana in Zambia. In Namibia, uraniferous chert veins cross-cut Damaran metasediments at Rössing Mine. Uranium is also found in fluorite veins at Husab (Jacob, 1974) and calcite-boltwoodite veins near Goanikontes (Palfi et al., 2001).

Kinnaird and Nex (2016) consider primary deposits (Fig. 5—16) to be those where mines extract uranium as the principal commodity or the sole product; whereas secondary deposits (e.g., the Witwatersrand Basin), produce another commodity as the prime metal with uranium as a by-product.

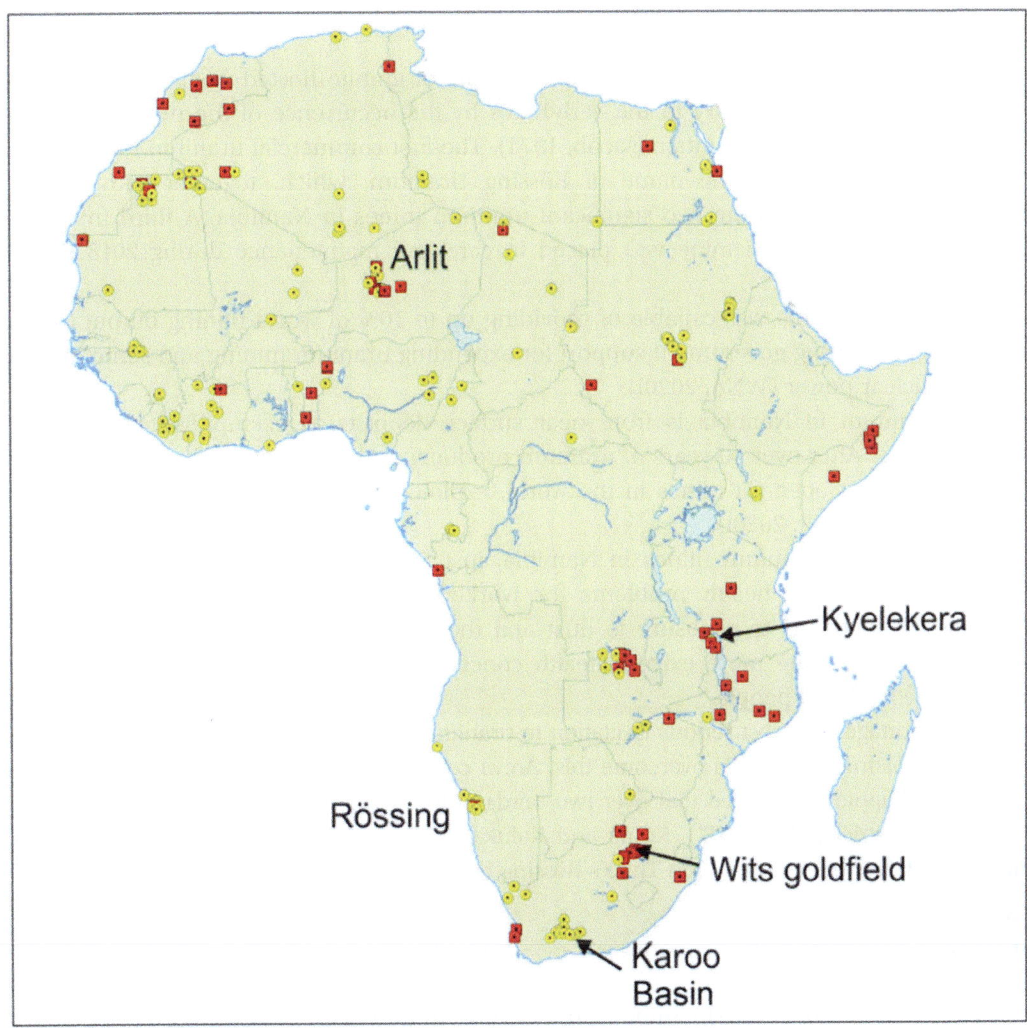

FIGURE 5–16 Primary deposits shown in yellow circles are those where uranium occurs as the main, or sole commodity, whereas those shown as secondary deposits in red squares, such as the Witwatersrand Basin, produce another commodity as the prime metal with uranium as a byproduct. *From Kinnaird, J., Nex, P., 2016. Uranium in Africa. Episodes 39 (2), 335. https://doi.org/10.18814/epiiugs/2016/v39i2/95782.*

Africa's three leading uranium producing countries as at 2019 were Namibia (5476 tU/annum), Niger (2983 tU/annum), and South Africa [346 tU/annum (est.)] (WNA, 2020d). The World Nuclear Association has described exploration and mine development in 13 other African countries Algeria, Botswana, Central African Republic, the DRC, Guinea, Equatorial Guinea, Mali, Mauritania, Morocco, Nigeria, Tanzania, Zambia, and Zimbabwe, which have not hitherto supplied uranium, as well as Malawi and Gabon, which have been significant uranium suppliers in the past (WNA, 2020e).

Namibia

The anorogenic granites of Namibia are an example of granite-hosted uranium mineralization, first noted in the country in the early1920s by the occurrence of the mineral davidite in the vicinity of Rössing Mountain (Jacob, 1974). The first commercial uranium mine started operating in 1976 under the name of Rössing Uranium, which, together with Swakop Uranium, now constitute the two significant uranium mines in Namibia. A third mine, the Langer Heinrich Uranium mine was placed in care and maintenance during 2018 (WNA, 2020i).

As of 2020, Namibia was capable of providing up to 10% of world mining output (WNA, 2020i). There is strong government support for expanding uranium mining and some interest in using nuclear power (WNA, 2020i).

Most uranium in Namibia is from near surface deposits as exemplified by Rossing's Uranium mine. After over 40 years of uranium production, this operation developed to form one of the largest open-pit mines in the world, exploiting low-grade hard rock with below 0.03% uranium (WNA, 2020i).

Working in the uranium mines in Namibia, just like in other uranium mines in Sub-Saharan Africa, poses health problems for workers and the surrounding communities. Fleischer (2018) points to exposure to dust and the respiratory complications that accompany it as some of the more exigent health concerns. Radon has also been considered a threat to the health of people.

Water shortage is a most serious limitation to uranium mining operations in the arid environment of the Namib Desert. To overcome this, Areva constructed a desalination plant for its new mine at Trejkoppie and agreed that over two-thirds of the water will be supplied to residents of the Erondo district (Njini, 2007). Kohrs and Kafuka (2014) consider the environmental health effects of low-level radiation on Rio Tinto's Rössing Uranium Mine workers.

Niger

Excellent accounts of uranium mining, processing, mine decommissioning, health effects, and sustainability issues in Niger are given in Chamaret et al. (2007), Larsen and Mamosso (2013), Alhacen (2017), Elliott (2019) and WNA (2020j). Here, a brief synthesis of these accounts is presented.

The first occurrence of Uranium in Niger in Azelik in 1957, was reported by the French Bureau de Recherches Geologiques et Minières (BRGM), who were actually prospecting for copper at the time (WNA, 2020j). The main uranium deposits occur as roll-fronts or in tabular bodies of various ages (Kinnaird and Nex, 2016). According to Cazoulat (1985), these deposits were the result of the interaction of palaeogeographic, tectonic and lithostratigraphic factors, and the movement of meteoric water that deposited uranium in fluviatile or deltaic sediments.

The first commercial uranium mine in Niger started operations in 1971. Today (2020), Niger provides about 5% of world mining output from Africa's highest-grade uranium ores

(WNA, 2020j). As a francophone country much of the uranium mining in Niger is controlled by the French company Areva.

Uranium is mined close to the twin mining towns of Arlit and Akokan, 900 km northeast of the capital Niamey.

The serious environmental and human health impacts of exploitation of uranium in Niger during the last 50 years or so by Areva are still very much in evidence (Alhacen, 2017). Workers were unaware of the risks of ionizing radiation, and the surrounding population, similarly, was not informed. Thankfully, this situation has changed somewhat, with the mobilization of environmental protection organizations, and in particular, after the publication of alarming reports following visits to mining centers in Niger by CRIIRAD, SHERPA, and Greenpeace (Valentina, 2015). Several steps are now being put in place, in partially solving the environmental and public health crisis (Larsen and Mamosso, 2013; Ejolt, 2015).

According to the United Nations Human Development Report of 2019, Niger is the least (in terms of human development index) of 189 countries assessed around the world (UNDP, 2019). Uranium mining is a major contributor to the economy. By 2019, Niger was the world's fifth largest producer of uranium after Kazakhstan, Canada, Australia, and Namibia, producing 2983 tU in 2019 (WNA, 2020e, 2020j). Uranium mining is, therefore, strongly supported by the government, including the will for expansion of the industry.

South Africa

The history of uranium production in South Africa dates back to 1944 (Brynard et al., 1987; Fordt, 1993; WNA, 2020k), when the mining industry was involved in evaluating the low-grade uranium mineralization found in association with gold in the Witwatersrand Basin, as part of the bid to acquire uranium for the Manhattan Project (Wymer, 2001). This Basin hosts the classic type of "quartz-pebble conglomerate reef deposit" (Kinnaird and Nex, 2016). The Basin lies south and southwest of Johannesburg, and has an aerial extent of about 330 km $\times$ 150 km. Uranium production is largely as a by-product of gold ores from this Basin (Frimmel, 2019; WNA, 2020k), which comprises a 2500−4500 m thick Witwatersrand Supergroup sediments of shale, sand, and gravel, deposited 3.0−2.7 billion years ago (Kinnaird and Nex, 2016).

Gold and uraninite particles, together with rounded pebbles of pyrite and other heavy minerals are preserved in thin quartz pebble beds, referred to as *reefs*, that individually extend for several kilometers (Kinnaird and Nex, 2016). Here, gold is more abundant than uranium, but, a total resource of >200 kt U_3O_8 is still present within some of the gold-bearing conglomerates, though most of the uranium, where it has not been extracted, is now concentrated in voluminous tailings dams close to 1 Mt U_3O_8 (Frimmel, 2019). This large amount of uranium accounts for close to 10%, and the world's third largest uranium resource, in terms of deposit type (Frimmel, 2010).

In 1951 a company was formed to exploit the uranium-rich slurries from gold mining (Koos and Basedau, 2012). In 1967 this function was taken over by Nuclear Fuels

Table 5–8   Uranium production in South Africa (tons U).

	2011	2012	2013	2014	2015	2016	2017
Ezulwini-Cooke	34	0	0	69	47	67	0
Vaal River	548	465	531	504	346	423	308
Total	582	465	531	573	393	490	308

Source: From WNA (World Nuclear Association), 2020k. Nuclear Power in South Africa. https://www.world-nuclear.org/information-library/country-profiles/countries-o-s/south-africa.aspx. (Accessed 09 April 2020).

Corporation (Nufcor), which in 1998, became a subsidiary of AngloGold Limited, now Anglo Gold Ashanti (Hore-Lacy, 2016), one of the major players in African uranium mining.

The two main producers of uranium in the Witwatersrand Basin are the Vaal River Mine and the Ezulwini Mine (Table 5–8). The AngloGold Ashanti Vaal River mine near Klerksdorp is the dominant producer.

Lacustrine deposits

In South Africa, some relatively small deposits hosting uranium mineralization are found within inland basins lacking an outward drainage. One example in this category of deposits is the Henkries Uranium Project in the Northern Cape Province, 80 km north-northeast of the town of Springbok. Here, uraninite and urano-organic complexes occur within a buried lake of Neogene age (Kinnaird and Nex, 2016). On its northern end, the deposit is bound by a rocky impediment of the drainage. Mineralization occurs in a 5.8 m thick, soft peat-like organic-rich layer overlain by whitish diatomite, alluvium, calcrete, and aeolian sand and has a grade of approximately 570 ppm. In 2009 Niger Uranium defined a measured and indicated resource of 1.35 million tons grading at 501 ppm U_3O_8 (Kinnaird and Nex, 2016).

Mining and processing of gold and uranium in South Africa pose one of the greatest mining-induced environmental health risks in the country (See, e.g., Winde and Sandham, 2004; Utembe et al., 2015; Winde et al., 2019, and so on). There are about 400 tailings dams and dumps arising from gold mining in the Witwatersrand area of Gauteng province alone, where much of the available uranium occurs today (Hore-Lacy, 2016). Several tailings dams and dumps near Klerksdorp, close to the Vaal River are being reprocessed to recover gold and sometimes uranium. Radionuclide and heavy metal pollution arising from these reprocessing operations include the dreaded *acid mine drainage* (See Section on: Acid Mine Drainage, Chapter 4).

Canada and Australia—summary

As of 2020, Canada and Australia were the second and third largest uranium-producing countries in the World (after Kazakhstan) (WNA, 2020d). Both countries host several mining operations, with a number of these companies, such as AREVA having operations in Africa, and are also active in either Canada or Australia, or both (Scheele et al., 2011).

There is little doubt about the considerable experience in Canada and Australia on uranium mining and its consequences. As these countries have healthy economies (OECD, 2019; World Atlas, 2020), sound legislation, proper law enforcement, stable and democratic political systems, a sense of social and environmental consciousness, and strong civil societies (Macdonald et al., 2012; GoC, 2019; IDEA, 2019) it can be expected that the mitigation processes are functioning better than in the African states where some or all of these factors are lacking (Scheele et al., 2011). Indeed, ample legislations and guidelines exist in these countries (Canada and Australia) that cover all aspects of uranium mining (Dasnois, 2012).

The Canadian Nuclear Safety Commission (CNSC) is responsible for regulating and licensing all existing and future uranium mining and milling operations in Canada (CNSC, 2020). The CNSC's work is undertaken in accordance with the comprehensive requirements of the *Nuclear Safety and Control Act* (NSCA) (GoC, 1997) and its related regulations, which reflect Canadian and international safety standards. The CNSC and its staff focus on health, safety, security, and the environment, to ensure Canada implements its international obligations on the safe use of nuclear materials (CNSC, 2020).

The pertinence of environmental concerns and their abatement in uranium mining in Australia were discussed by Kay (2007). Later, Waggitt (2014) looked at the regulatory processes that are mandated via the "Mining Management Plan" of the Northern Territory Government of Australia.

Federal and provincial authorities in Canada and Australia have the capacity to implement uranium mining-related policies and legislation, as well as to monitor the health and environmental impact of mining activity. Indeed, there are documented examples of good practice with regard to uranium mining in both of these countries (Dasnois, 2012).

However, just as in the African countries, environmental legislation breaches or offences, such as spillages and leaks do occur regularly in Canadian and Australian mines (See Scheele et al., 2011), and opposition to uranium mining is widespread among stakeholders in both countries. According to Scheele et al. (2011), a difference may be that chances are higher that authorities will actually discover and prosecute the responsible company; so, these environmental concerns should not be seen as an example to African uranium mining countries, where the situation might be much worse.

Ionizing radiation protection: legislatory and regulatory guidelines

Radioprotection, also referred to as radiation protection or radiological protection, is defined by the International Atomic Energy Agency as: "The protection of people from harmful effects of exposure to ionizing radiation, and the means for achieving this" (IAEA, 2019a,b). The source of exposure can be from radiation that is external to the human body or from internal irradiation caused by the ingestion of radionuclides.

Exposure to natural background radiation is ostensibly not amenable to regulatory guidelines. Of widespread concern is the need to regulate exposure to radiation in the nuclear, medical, and mining industries. This concern gave rise to the development of the international framework for radiation protection and the formation of a number of expert working groups tasked with the design of tangible recommendations on what needs to be done. One of the earliest (groups) to be established (in 1928) was an independent nongovernmental body of specialists in radiation science, the International X-ray and Radium Protection Committee (Kang, 2016). It was later (1950) renamed the International Commission on Radiological Protection (ICRP) to better take account of uses of radiation outside the medical area (Clarke and Valentin, 2008). The Commission's mandate was the establishment of basic principles for, and issue recommendations on, radiological protection (Fig. 5−17).

Other intergovernmental bodies that play a key role in formulating radiation protection principles include the United Nations Scientific Committee on the Effects of Atomic Radiation (UNSCEAR), and the International Atomic Energy Agency (IAEA) (IAEA, 2004). UNSCEAR was set up in 1955 by the General Assembly of the United Nations (UNGA, 2013) with a mandate of the General Assembly of the United Nations in 1955, "... to assemble, study and disseminate information on observed levels of ionizing radiation and radioactivity (natural and man-made) in the environment; and on the effects of such radiation on man and the environment ..." (UNGA, 2013). Consequently, UNSCEAR has published several informative documents, the articles therein, serving governments and organizations throughout the world as the scientific basis for evaluating radiation risk and for establishing protective measures and national regulations governing the exposure of radiation workers and members of the public. These articles have also been incorporated by the IAEA into its *Basic Safety Standards for Radiation Protection* published jointly with the World Health Organization (WHO), International Labor Organization (ILO), and the Nuclear Energy Agency (NEA) of the Organisation for Economic Co-operation and Development. These standards are used worldwide to ensure safety and protection of radiation workers and the general public (IAEA, 2018a).

The technical terms used in the *IAEA Safety Standards* and other safety related IAEA publications are defined and explained in a series of *glossaries*, the first of which was issued in 2007 (IAEA, 2007). The glossaries also provide information on how the terms in the IAEA Safety Standards should be used (https://www.iaea.org/resources/safety-standards/search) (Accessed 19 May 2020). The 2018 Edition of the IAEA Safety Glossary (IAEA, 2019b) is a new edition (revised and updated) of the IAEA (2007) Safety Glossary. The 2018 version takes into account new terminology and usage in safety standards issued between 2007 and 2018. The revisions and updates reflect developments in the technical areas of application of the safety standards and changes in regulatory approaches in IAEA Member States.

Because of the huge strategic importance for energy production and potential use in nuclear weapons, uranium trade is also closely superintended by international bodies, such as the NEA of OECD and the United Nation's IAEA. These two agencies jointly publish a biannual survey of uranium resources, production, and demand, known as The *Red Book*. This Book lists the supply and demand development for each country, and quantifies their available resources using various categories of extraction costs (Karpas, 2014).

International Policy Relationships for Radiological Protection

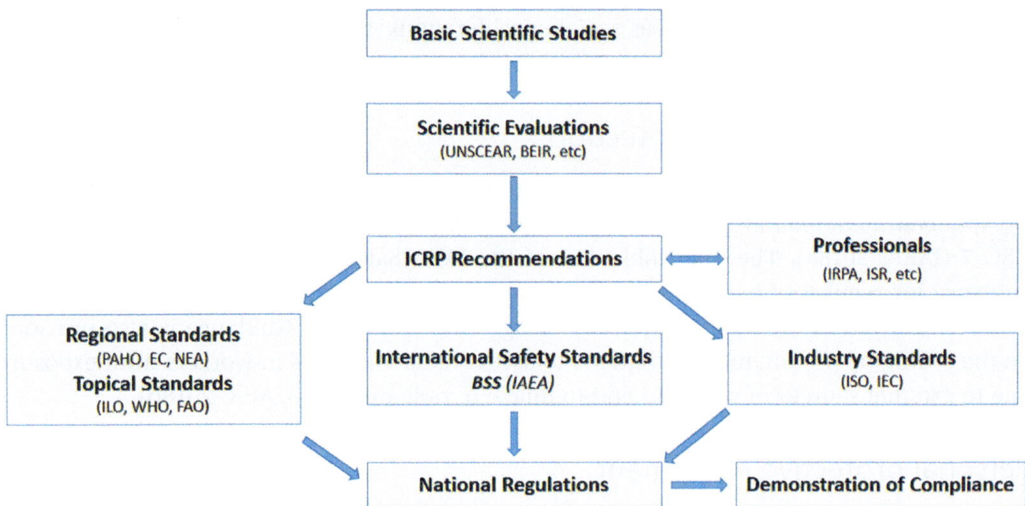

FIGURE 5–17 International policy relationships in radiological protection. *Credit: Doug Sim: Based upon Fig. 5–3, ICRP Publication 109, by Clarke and Valentin (2008).*

At the regional and national levels, there are a number of other institutional bodies that control the possession, handling, and related actions of radioactive material. These bodies include the Forum of Nuclear Regulatory Bodies in Africa (FNRBA), which was: "… established on March 26, 2009, to strengthen and harmonize radiation and nuclear safety and security regulatory infrastructures in its member countries, and serve as an effective platform for the exchange of regulatory experiences and practices among the nuclear regulatory bodies." (Fisher, 2018; IAEA, 2018c; FNRBA, 2020)

Another example of a national regulatory body for radioprotection is the Atomic Energy Board (AEB) of South Africa. AEB functions in terms of the (South African) Atomic Energy Act 90 of 1967 (Rabie, 1973; AEB, 1979; Herbst and Fick, 2012), which was itself amended by the Atomic Energy Amendment Act 46 of 1979 (Welsh, 1979).

System of radiological protection

The system of radiological protection used worldwide is based on the revised and updated recommendations of the ICRP (ICRP, 2007). Similarly, measures to be taken for protection against ionizing radiation are, notably, consistent throughout the world, thanks to the existence of a well established and internationally recognized frame of reference (See Section: "Ionizing radiation protection: Legislatory and regulatory guidelines," in this Chapter.

The system of radiological protection is based on three principle requirements, i.e., *justification*, *optimization*, and *dose or risk limitation* (ICRP (International Commission on Radiological Protection), 2007). The limitation of radiation doses refers to more than just the dose limits. Rather, it applies to an entire system of radiation protection, i.e., all steps that need to be taken to keep doses as low as possible (See Allisy-Roberts and Williams, 2008). Each of the three principles involves social considerations, and inherent in them, is the great need for the principles to be articulated judiciously.

Occupational radiation protection

The most widely accepted recent (at the time of writing, 2020) recommendations on *Occupational Radiation Protection* are those given in the *new* IAEA General Safety Guide No. GSG-7 (IAEA, 2018a). The main objective of the IAEA Safety Guides is to provide general guidance for regulatory bodies on meeting the requirements for the radiation protection of workers involved in the mining and processing of raw materials. Guidance is also provided on the setting up of monitoring programs to assess radiation doses to workers from exposure due to external sources of radiation and to intake of radionuclides (IAEA, 2018b).

Personal protective equipment

To protect workers from radiation in uranium mining and milling, *personal protective equipment (PPE)* in the form of respiratory protection is generally used as a protective measure against RDP and LLRD.

PPE is often considered an effective prescription for minimizing exposure where the application of higher level controls cannot be achieved timeously. The selection and use of PPE has to be done in a way that takes cognizance of the relevant regulatory requirements. However, PPE is also known to be the lowest control on the hierarchy of control measures, and can cause discomfort and interference with other safety equipment and work efficiency.

Legislations for protection against ionizing radiation in uranium mining

Distinct from other resources is the fact that uranium is subject to strict national regulations regarding its extraction, trade, and export, especially in Canada, Australia, and the USA. Some parts of these countries have banned uranium mining or adopted temporary moratoria (Winde et al., 2017).

In 2011 the IAEA initiated the *Information System on Uranium Mining Exposure* to enhance radiation protection of workers in the uranium mining and processing industry (IAEA, 2020b). In a first step, IAEA, in 2012, developed a questionnaire, which was distributed to uranium-producing countries in a global survey to evaluate worldwide occupational radiation protection (Okar et al., 2018). Following an analysis of the results, IAEA was able to identify both good practices and opportunities for improvements. The results of the questionnaire survey are embodied in the most recent (at the time of writing, 2020) IAEA Publication: "Safety Reports Series No. 100" (IAEA, 2020a), which publication also identifies actions to assist industry, workers and regulatory bodies in implementing the principle of optimization of protection. Also presented in the (IAEA, 2020a) publication, is information on uranium mining and processing methods, radiation risk to workers and safety considerations, monitoring, dose assessment, and radiation protection programs for the range of commonly used mining and processing techniques.

Radiation protection programs

Scheele et al. in 2011, noted with regard to uranium mining and associated activities in Africa, that: "Dealing with a type of mining which is more hazardous than other mining types, and which has very specific and extremely long-term effects, requires at the least, excellent laws, excellent law enforcement, disciplined, knowledgeable and dedicated governments and institutions, a strong civil society, and a healthy civil society."

IAEA (2020a) has pointed out the integrality of the radiation protection program with respect to all stages in the mining and processing of uranium; and how the level and type of radiation protection practices required would vary depending on the different stages of the operation and the specific needs of each individual operation.

Factors that should be emphasized in any radiation protection program include the radiation risks associated with handling radioactive ore and concentrates, and a way of reducing these risks to the lowest possible level. By reducing the radiation risks, the radiation protection program would also provide a framework for ensuring compliance with regulatory legislation, corporate policies, standards, and guidelines. Another important program attribute would be the ability to assist all personnel to meet their *duty of care* regarding maintenance of their obligations with respect to radiation protection (IAEA, 2018a).

In the African scenario, following several decades of protests by NGOs, against environmental pollution and the sometimes unacceptable behavior toward communities, workers, and indigenous peoples, some uranium mining companies in Africa are now coming up with somewhat credible corporate social and environmental responsibility programs (Hecht, 2012a,b). However, others seem to disregard these issues altogether, or simply make a minimum effort to address them. Mitigation of uranium mining impacts on society and environment in Africa is fully discussed in Scheele et al. (2011), Hecht (2012a), Nyanda (2014) and Winde et al. (2017).

Conclusion

Radioactive substances and ionizing radiation have many important applications in Sub-Saharan Africa, ranging from nuclear power generation to uses in medicine, manufacturing, construction, and agriculture. However, there are certain concerns with these applications, the greatest of which is the risk of ionization radiation exposure of workers, the public and the environment linked with certain key phases of the nuclear fuel cycle, such as uranium mining, processing and tailings management. This is of especial concern in the case of sensitive populations, such as pregnant women and young children. Medical geologists have an important role to play in their contribution toward controlling or obviating these risks.

This Chapter has provided an overview of the basics of ionizing radiation science and the many challenges related to ionizing radiation exposure in miners and residents around uranium and gold mining centers in Sub-Saharan Africa. By way of case histories, examples are given of negative health and environmental effects of on-going open-pit uranium mining activities by foreign companies in countries such as Namibia, Niger, and South Africa, which have prompted strong media and public critical reaction.

Health hazards from ionizing radiation exposure can be both carcinogenic and noncarcinogenic in nature, as well as result in long-term contamination of the environment (water, soil, and biota).

Carcinogenic effects may be produced in various organ systems; but symptoms may not present for several years after the initial exposure. Various types of cancers may already become evident even at low doses; so also may life-shortening and negative impacts on all the body functions as well as genomic instability.

Noncarcinogenic effects are immediately seen in organs with rapidly dividing cells, including the hematopoietic system, gastrointestinal tract, and skin, or delayed effects, such as embryo/fetal development problems and cataracts.

However, the dose-response relationships for both carcinogenic and noncarcinogenic effects is an area where there is currently a dearth of knowledge, despite the huge amount of data already garnered from studies on both humans and animals.

It is considered that the most fruitful approach to gaining a full understanding of the risk from exposure to ionizing radiation exposure could be through molecular studies that engender identification of unique biomarkers and pathogenic pathways at both the cellular and tissue levels.

In Sub-Saharan Africa, there is at the political sphere, a lack of proper regulations, poor awareness of the radiological hazards associated with uranium and its daughter-products, insufficient monitoring practices and a lack of control from local and national administrative authorities. Existing mining legislations are not adapted to uranium mining, and institutions are often bereft of the requisite tools for ensuring adequate monitoring and compliance by mining companies. Uranium miners and communities should receive proper education on radiological characteristics of materials and waste from the mines, principles of radiation protection, and methods of dose evaluation. More stringent legislative articles that would promote and assure confidence in safety are needed.

It has been shown that if these negative impacts on societies and ecosystems remain poorly addressed and mismanaged, uranium and gold mining in Sub-Saharan Africa would be seen more as a bane than a blessing. The need for maximum control, proper government action, stringent laws and their enforcement, as well as responsible corporate behavior can, therefore, never be overemphasized.

In general, the medical follow-up of pregnant women exposed to ionizing radiation in mining centers, is a delicate subject awaiting much research; and requires the collaborative effort of Medical Geologists, Medical Physicists, and other coworkers in a public health investigative team (PHIT).

The two goals of radiation monitoring are to confirm that the radiation levels remain within the expected range and to assess the dose to workers. Monitoring needs to be in place for all of the exposure pathways and needs to be carried out by well trained personnel. Where specific engineered controls are in place, routine monitoring should be carried out to review their effectiveness.

A global radiation safety regime has been developed by the IAEA, and is being continuously improved. The IAEA safety standards support the implementation of binding international instruments and national safety infrastructures, and are a cornerstone of this global regime. *However, according to* Monken-Fernandes et al. (2013). *". . . current efforts of the IAEA and other international organizations to assist developing countries with remediation of uranium legacy sites are commendable, but appear to be insufficient to address the problem on the required scale. More stringent legislative articles are therefore needed to promote and assure confidence in safety and in assessment of governments' capacities to monitor uranium mining."*

The discourse on "Case histories" ends by evaluating today's mining practices in the uranium mining sector in Sub-Saharan Africa, and compares these to the industry's practices as they are carried out in Canada and Australia, where best practices exist, and where institutions are stronger and laws and regulations match the issues. Also, as both countries have strict laws and proper monitoring systems, they might provide less experienced African countries with a good example of how to manage uranium mining operations. Notwithstanding, it should be noted that despite good laws, a strong judicial system, powerful NGOs, and democratic governments, uranium mining practices still threaten indigenous societies and natural protected areas in Canada and Australia (See, e.g., Scheele et al., 2011).

In summary, synthesizing the collective amount of information gathered, it can be clearly seen that substantial problems and negligence exist in Sub-Saharan countries with respect to the operation of the uranium and gold mines and the safety of miners and local citizens. There is also a general and persistent lack of (accessible) monitoring data on past and current levels of radioactivity in and near the mining sites, data on health, and environmental effects associated with ionizing radiation, and the safe management and disposal of mining material. *These are areas in desperate need of urgent research.*

To buttress such research, a comparison is made between the nature and articulation of legislations to limit or prevent ionizing radiation exposure in uranium and gold mining in Sub-Saharan Africa with those existing in two of the world's leading uranium producing countries, Canada and Australia, where legislative articles are considered to be more strict. The rationale is that, since the health problems outlined in this Chapter (DNA damage, cancer, fetal abnormalities, and so on) can be seen as evidence of a global environmental problem, not just specific to Sub-Saharan Africa, experiences and lessons learnt from these countries can be applied, perhaps with appropriate modifications, in reviewing and updating existing legislative and regulatory guidelines.

Our knowledge of the health hazards posed by ionizing radiation will continue to expand as we design better and better in vitro and in vivo biokinetic models and laboratory experiments using animals. As additional new information concerning the potential public health impact of ionizing radiation becomes available, IAEA, WNA, and the other international radioprotection and regulatory bodies will continue to evaluate and update their databases to ensure an up-to-date assessment of all relevant biomedical data to protect miners and the public from exposure to harmful levels of ionizing radiation.

Finally, in concordance with the aim of the European Commission concerted action CURE (Concerted Uranium Research in Europe), in order to better characterize the effects of occupational ionizing radiation exposure, future collaborative research projects would need to develop protocols that integrate dosimetry, epidemiology, and biology (See, e.g., Laurent et al., 2016). To this list should be added Medical Geology.

A listing of topics under which these various interrelated problems can be most effectively researched, for the purpose of finding effective solutions, is given in the Section on: "Suggestions for future research," at the end of the Chapter. The listing could also form the basis on which a network of investigative teams of Medical Geologists in ionizing radiation research, particularly in universities and other research establishments in Sub-Saharan Africa could perform experimental, educational and outreach activities.

Interdisciplinary networks in researching ionizing radiation exposure: effects, mitigation and management

The role of the Medical Geologist

It is important to recognize the important role of Medical Geologists in teams researching ionizing radiation exposure, the environmental and health impacts, and the value of their (Medical Geologists) input in improving diagnostic therapy and etiology.

Interdisciplinary networks should provide to stakeholders, a clear demonstration of the important role of Medical Geologists and geochemists in complementing medical practitioners, radiologists, and epidemiologists in diagnosis, therapy, and prognosis of radiation-related mortality and morbidity, including neonatal malformations and congenital conditions. These networks should reflect the value of interdisciplinary linkage between Medical Geologists/geochemists and medical staff of existing public health clinics in uranium and gold mining centers in the countries concerned (Namibia, Niger, South Africa, Malawi, and others) through the institution of maternal and health care programs and the strengthening of existing ones. Through these links postpartum and pregnant women and children would be educated in their homes, mentored, and provided with simple, user-friendly radiation detection devices.

It is realized that the sources, intake pathways, dynamics of uptake of radioactive contaminants (*sensu lato*), as well as the dose taken up by the human body are best explained by the geochemical specialist—geochemical evolution (chemistry) of the tailings; distribution and migration characteristics; speciation/forms of occurrence; bioavailability/bioaccessibility; factors determining enrichment/depletion, horizons where we have accumulations—a knowledge of all of these factors is crucial in determining the fate of ionizing radiation that enters the body, and in diagnoses and therapy, as well as in planning interventions. However, those tasked with formulating solutions to radiation exposure of humans are traditionally medics (pediatric clinicians, radiologists, and epidemiologists) and even legal practitioners (e.g., Rabie, 1973), who go about their business with little cognizance taken of the element's geogenic characteristics and other geodynamic intricacies. These factors bear heavily on the mode of radioactivity uptake by humans, further underlining the importance of the Medical Geologist's contribution in investigative teams studying ionizing radiation exposure and effects.

Again too, consideration should be given to the possible disturbance by geochemical processes of secular equilibrium reached between U-238 and U-235 and their decay products in a uranium ore deposit (Kamunda et al., 2016). After uranium extraction from the ore, the mill tailings may contain virtually all of the radionuclides in the uranium decay series, especially those of U-238 (Kamunda et al., 2016). It is important to determine the geochemistry and mineralogy of these radionuclides, in particular, Th-230 and Ra-226, which have the longest half-lives, which determinations normally fall within the compass of the geoscientific investigator.

The exposure, and more importantly, the bioavailability (the amount or fraction of the element which is actually available for uptake) depends on several factors, such as concentration in the source, particle size, and the specific physical and chemical properties of the contaminant itself (Gomes and Pereira de Silva, 2007).

Previous studies on ionizing radiation dose uptake have re-examined existing paradigms and provided the results that support the development of new, biological paradigms concerning exposure to a low versus high dose of radiation resulting in both qualitatively as well as quantitatively different cellular and molecular responses, thus demonstrating nonlinear response with respect to dose. These biological processes are shaped by physical exposure parameters that include dose, dose rate, and dose distribution. These researches have underscored the importance of studying the physical ionizing exposure parameters as stressors, and how they influence intact-tissue biological response, rather than studying only the initial events within an individual cell. Such studies, too, are within the sphere of the Medical Geologist.

The fetal radiation dose estimate is extremely important, and is a necessity for assessing the potential health effects on the neonate. The risk of pregnancy outcomes and damage to the offspring of women who have been exposed to a certain level of radiation doses are still unclear.

Medical Geologists, hospital medical physicists, and health physicists are good resources for expertise in radiation dose estimation. Physicians should therefore contact these specialists for assistance in estimating fetal radiation dose.

The Medical Geologist would determine geogenic factors, such as what controls the quantities of uranium's release during mining, its migration pathways and its accumulation in environmental compartments, geochemical determinants that modulate the element's uptake into living systems, factors that are seldom considered in routine radiometric surveys. The geochemistry of U-238, Th-232, and K-40 and their activity concentrations is important in the evaluation of absorbed doses that can lead to the estimation of their radiological hazard to the population. A thorough knowledge of these is necessary in order to devise ways of reducing incorporation into the human body.

Finally, the Medical Geologists' input in radiological site assessment must be emphasized. These contributions include: the identification of contaminated areas and the characterization and quantification of radionuclide contamination, as well as the identification of pollution resulting from human activities (Coetzee and Lakin, 2011; Isinkaye et al., 2023).

A nonexhaustive list of mandates or *"job description"* reflecting a Medical Geologist's cumulative interlinked tasks in an investigative team on ionizing radiation exposure impacts and management, would include the following:

1. Establishing and maintaining a strong liaison with multidisciplinary collaborating partners in the countries concerned, for: (1) Executing fieldwork involving geochemical sample collection and on-site analyses around uranium and gold mining centers for the purpose of determining radiation source areas, characterization of radioactive waste materials, charting migration pathways, bioavailability and bioaccessibility relations and uptake mechanisms; (2) Performing outreach and educational activities.

2. Participation in conducting airborne surveys to support a radon potential mapping process in uranium and gold mining centers of the countries/regions concerned. An aircraft can provide a continuous quantitative profile of radioelement abundance over any type of land surface. Flight-line spacing can be varied according to the sampling density required.

3. Evaluating the effectiveness of airborne radiometric surveys (noting the strengths and weaknesses) flown over areas contaminated by uranium and gold mining and other activities, in aiding in the identification and mapping of contamination throughput in the life cycle of a mine.

4. Researching the surface remediation of most of the uranium and gold mill tailings sites in the study localities, providing geochemical and Medical Geology expert knowledge needed by structural engineers in their procedures for designing and constructing special "repositories" to contain uranium mill tailings.

5. Examining the effectiveness and economics of existing dosimeters in the countries/regions concerned, and contributing towards their refinement and usability by the study populations.

6. Performing hazard estimations using data from routinely and universally registered birth and death certificates for developing statistical methods of ionizing radiation hazard estimation where populations are subject to predictable or measurable radiation in the study areas.

7. Estimating death rates by examining the relationship between neonatal death rate for specific periods (statistically determined) and using revised estimates of natural or background radiation exposure (taken as an indication of a higher than normal environmental radiation level) in the countries concerned.

8. The altitude variable—Since there has been so much talk for decades about the (possible) relationship between mortality and the presence of uranium ore reserves in the absence of the altitude variable (e.g., Landau, 1972), the Medical Geologist would participate in determining whether there is any evidence to support a possible role of reduced partial pressure of oxygen in causing a reduced fetal growth rate and increased neonatal death rate in the study areas.

9. Monitoring—Write relevant and reliable radiation monitoring programs for uranium and gold mines in the study areas capable of reducing ionizing radiation exposure levels to below 50 mSv/a. Development of a database that contains the results of personal monitoring for statistical analysis and engineering control purposes.

10. Instrumentation—Contribute toward the design of simple, efficient, cheap, and affordable dosimeters for use by residents around uranium tailings disposal facilities in the study areas.
11. Risk assessment—Maintain a continuous and progressive risk assessment of radiation exposure after the study is completed; and seek to further reduce exposure below 20 mSv over a specific number of consecutive years.
12. Generation of a critical mass of postgraduate students and other budding researchers with interest in Medical Geology for deployment in the various phases of well-designed research programs on ionizing radiation exposure sources, pathways, effects, and management.

The role of Medical Physicists in enhancing radiological applications of ionizing radiation

It must be pointed out that this list of tasks in the Medical Geologist's compass can never be holistically accomplished without the help of the Medical Physicists whose role in these teams must therefore also be emphasized. These are professionals trained in the concepts and techniques of applying physics in medicine and competent to practice in subfields, such as radiation oncology, nuclear medicine, and diagnostic radiology (See, e.g., Trauernicht et al., 2022). They are well-versed in methodologies to ensure that patients are imaged or treated safely and effectively with radiation technologies; which underlines the need for them to work with Medical Geologists and allied specialists in radiation therapy and management of associated conditions.

Suggestions for further research

The literature is replete with research information on reducing radiation risks from medical X-ray applications in Sub-Saharan Africa, even though certain knowledge gaps still exist in this domain. For completeness in this Chapter, some comments on research gaps in diagnostic radiology have been included. However, the main focus of the Chapter has been on the effects of exposure to ionizing radiation from uranium and gold mining, processing and tailings management, *areas where a substantial dearth of research has been revealed.*

Following is a bullet point listing that summarizes some of the more important areas awaiting research, musing particularly on the potential data needs regarding health effects that may be associated with exposure to ionizing radiation:

- *Today, very little site and operational specific ionizing radiation data are available at several uranium and gold mining locations in Sub-Saharan Africa.* The few available data are qualitative in nature. Comprehensive site-specific data and information are required to perform a radiation impact assessment on a quantitative basis. It is important to address such shortcoming and provide information required to perform a quantitative assessment.

Concomitant with the generation of site specific radiation data is the need to urgently produce radioactivity maps of rocks, soils and groundwater for uranium producing Sub-Sahara African countries (Namibia, Niger, South Africa, Gabon, Tanzania, etc.), to be used as a basis for informing the development of policy recommendations regarding housing construction, groundwater development and other land uses in areas where elevated uranium concentrations are predicted. We need to generate sufficient data for the preparation of an appropriate grid, suitable for the production of regional or national atlases and ultimately, radioelement maps.

- Those governments considering use of nuclear energy in their uranium mining policy, e.g., South Africa, are facing a potential human crisis, as well as an environmental one, because they see nuclear energy as "clean energy." It is necessary to urge these governments to take steps in ensuring that pertinent environmental impact assessment processes be conducted around power station sites.

- *Dose measurements and calculations*—Calculation of the potential dose to man via different pathways using mathematical models (See, e.g., Bain et al., 1994; SAHPRA, 2022).

We need to produce maps of air dose rate (Gy/s) or effective dose equivalent (mSv/a) for use in environmental health radiation monitoring purposes. We need to ensure that knowledge and instrumentation acuity forms the basis for awareness in the invisible world of ionizing radiation in Sub-Sahara African countries, by investing in the design of simple, efficient, and affordable dosimeters for use by residents around uranium mining and tailings disposal facilities.

Ultimately, what we want to achieve is a limitation of the extra and unnecessary exposures that are indicated in these maps, and that are preventable. It is essential that authorities ensure that miners wear protective clothing at all times and are afforded monthly health checks, adequate medical aid cover and regular checks of their DNA.

- *Studies on the potential mobility, bioavailability and bioaccessibility of uranium in soil and tailings in the neighborhood of abandoned uranium and gold mining sites would greatly aid in effective management and environmental restoration of these sites.*

- *Radon progeny in the indoor environment*: Despite the many possible sources of radon in Sub-Sahara African countries (Namibia, Niger, South Africa, Gabon, Tanzania, etc.), and potential health hazards, there is a problem of data availability and hence the need for direct quantification of radon concentrations and exposure. In Sub-Saharan Africa, there are a number of nonuranium producing countries having an undercover geology that is significantly granitic (See, e.g., Bezuidenhout, 2019). Thus more radon measurements of the indoor air environment are required, not only around uranium and gold mining areas, but also in homes situated in the neighborhood of granitic terrain. Further modeling of the indoor air environment is needed to assess potential health consequences of indoor radon exposure. In addition, adding a thoron (Rn-220 and Rn-220 plus decay products) study to the radon studies would be of benefit, because thoron is hard to get into the home. If it does, the decay products are long-lived; there are limited data on in-home thoron and thoron decay product levels.

- *Radon in underground mines*—More epidemiological studies need to be conducted in order to firmly describe the risks of lung cancer in African underground miners (See, e.g., Agius et al., 2023). Data from well designed underground-miner surveys should be analyzed for the interaction between radon and smoking. Such studies should also provide more information on radon dosimetry and narrow some uncertainties in applying the lung-cancer risk data derived from the miner data sets to estimation of risk of radon exposure in the general population. The role of Po-210 in tobacco smoke and lung cancer should continue to be evaluated; this includes bronchial and lung dosimetry, identification and characterization of target cells, and the role of cofactors in the carcinogenic response. The deterministic acute and delayed health effects from Polonium-210, particularly those affecting the renal, cardiovascular, and reproductive systems, should continue to be investigated.

  More research on controlled recirculation in underground mines (e.g., in South Africa) is needed. Indeed, as Butterworth (1999) advocated in the case of South Africa, "To implement controlled recirculation … a method to reduce the radiation level in the recirculated fraction of the ventilation air to a level at or near that of the fresh intake air, will need to be investigated."

  As Gillmore et al. (2018) noted, in contemplating the future of radon research: "… In particular, we hope to see arguably neglected aspects of radon research such as indoor sources in the built environment and, controversially perhaps, research into the radiation hormesis theory, can be taken forward …."

- *Construction of predictive models for epidemiological studies*—According to Winde et al. (2019), further methodological work is needed on modeling in epidemiological and health studies (based on human tissue sampling, such as *hair*) of the likelihood of uranium exposure, taking into account measured environmental levels, distance to mine tailings, wind directions, water flow, consumption of local water and food, and how common personal habits, such as geophagy, which can influence exposure variables. Such prediction models, according to Winde et al. (2019), "… would need to be validated with a new series of individual (rather than composite) human measurements, with a choice of barbers *(in the case of human hair sampling)* serving fewer commuting customers while serving the most highly exposed indigent population."

- *Prenatal radiation exposure studies*—The possible effects associated with prenatal radiation exposure include immediate effects (such as fetal death or malformations) or increased risk for cancer later in life. However, there is still a dearth of reported research on reduction of ionizing radiation risks that confront pregnant women living in the neighborhood of uranium and gold mine tailings, the risk of pregnancy outcomes and damage to the offspring of women who have been exposed to a certain level of radiation doses still remain unclear.

  De-Kun et al. (2020) have reported an association between maternal exposure to MF nonionizing radiation during pregnancy and the risk of attention-deficit/hyperactivity disorder in both animal and human studies. Further studies (e.g., by Medical Physicists) on such associations are desirable.

Since somatic and reproductive cell chromosomes are radiosensitive tissues that can sustain damage after exposure to ionizing radiation, damage to the chromosomes, and the genes on them in exposed populations has potentially serious implications. Better methods are needed by which to estimate the levels of exposure to ionizing radiation that may result in an increased risk of hereditary disease. Important gaps in knowledge should be filled to permit more reliable estimation of genetic and hereditary risks.

- *Disposal of radioactive tailings and site remediation:* Even decades after the shutdown of uranium mines and mills, the radioactive contamination of the environment will remain. This is due to the fact that uranium-238 half life is very long (*c.* 4.5 billion years). But even the tailings, whose uranium content is lower than the initial uranium concentration in the ore—will remain radioactive on the long term, because of the presence of thorium-230 and radium-226, whose half-lives are 75,000 years and 1600 years, respectively.

  The situation in Niger, for instance, *shows that the disposal of radioactive tailings and their control on the long term, has not yet received satisfactory solutions.* And, according to Winde et al. (2019): "... if high uranium levels are confirmed in human populations exposed to these tailings, investigations of subsequent adverse health effects, including cancer, would be justified."

- *Assessing the risk of cancer:* The mechanisms by which cancer is induced in living cells are complex and an area under intense study. The design of studies on background radiation has been largely of ecologic nature and this type of study is limited in its informative ability. Based on the linear no-threshold model we can expect some cancer risk to be associated with background radiation. Detecting this risk is very challenging because of uncertainty and variability in individual exposure to both radiation and the variety of other factors contributing to background cancer risk.

  More research is therefore required to better understand the mechanisms by which cancer is induced after exposure to chemical carcinogens and to ionizing radiation. This should include the identification of unique biomarkers and biochemical pathways at the cellular and tissue level, and the study of radiological and chemical mixtures.

- However, the magnitude of health risks at low doses and dose-rates (below 100 mSv and/ or 0.1 mSv/minute) remains controversial due to a lack of direct human evidence. Some observers, e.g., Pernot et al. (2012) anticipate the emergence of significant insights from the integration of epidemiological and biological research, made possible by molecular epidemiology studies incorporating biomarkers and bioassays.

  Studies like the meta-analysis of Lubin and Boice (1997) are useful in demonstrating that background radiation risk estimates are consistent with other estimates, in this case, miner data. Knox et al. (1988) and Hatch and Susser (1990) have generated childhood cancer risk estimates that are larger than what we might expect, based on studies of *in utero* X-ray exposures or risk estimates from larger exposures. Although the exact magnitude of the background risk from radiation exposure is unclear we can see that there is likely to be some risk; natural radiation is not harmless. This is an important context in which to consider the additional exposures that are discussed in relevant sections of this Chapter.

- *Cancer therapy and management*—Radiation therapy is one of the most important methods for treating cancer. However, alpha radiation damages healthy body cells as well as the diseased ones. In the area of cancer therapy and treatment, although much progress has been made in recent years (e.g., Chaves-Pérez et al., 2019; MartinezMarignac et al., 2019; Cuthbertson, 2020; Licciulli, 2023), there still remain significant challenges in these aspects for many types of cancer. Bayer AG researchers are (as at 2020) working on the development of targeted radiation therapies using biomolecules, which specifically deliver alpha radiation to the tumor cells (Cuthbertson, 2020). Additional researches along these lines are desirable.
- More effort is required toward a comprehensive estimation of uranium from *unconventional sources* that will provide important inputs to long-term nuclear fuel cycle supply (See Bangoto et al., 2018). More education, training, and human resource development programs in unconventional uranium exploration and the uranium production cycle (See Bangoto et al., 2018) should be mounted.
- *Uranium contaminated building materials*—The study by Veit and Srebotnjak (2010) is specifically concerned with the utilization and associated risks of using radioactively polluted construction materials that originate from uranium mines in Gabon and Niger. In the study framework, a desk study and a fact-finding field mission to Gabon were undertaken to collect information on the use of uranium-contaminated materials in the construction of residential homes in Gabon and Niger and general safety practices surrounding uranium mining. This particular study concludes with "... a summary of the findings, their limitations, and a description of the remaining data and information gaps that would be needed to conduct an exhaustive analysis of the issue." (See Veit and Srebotnjak, 2010).
- *Monitoring dust fallout in uranium mines*—An assessment of dust fallout radionuclides in uranium and gold mining areas of Sub-Saharan Africa is urgently needed. More dust fallout monitoring using, say, the method (ASTM) D-1739 of 1998 (ASTM, 2017), employing multidirectional buckets, must be undertaken in different seasons; and results compared with dust fallout rates within stipulated residential limit according to International Dust Control Regulations (See Dudu et al., 2018).
- *Assessing levels of radionuclides in biota*—Determination of the levels of radionuclides in water, soil, and agricultural products, such as vegetables, milk, and meat from farms potentially affected by contaminated effluents from uranium mine tailings and calculation of relevant concentration ratios, must be made. Sustained monitoring of these radionuclides from the point of release and, where possible, via aquatic to terrestrial ecosystems (See Bain et al., 1994) must be undertaken.

  Experimental studies to determine with different irrigation schemes the amount of radium-226 uptake through the leaves and roots of vegetables are desirable.
- *Use of human tissues as sampling media in measuring and predicting uranium uptake levels*—Winde et al.'s (2019) measurements of uranium concentrations in the hair of the resident population of a South African gold mining area indicate elevated U levels that merit research on possible adverse health consequences. These authors (Winde et al.,

2019) mentioned the need for further methodological work on whether the exposure likelihood can be better modeled taking into account measured environmental levels, distance to mine tailings, wind directions, water flow, consumption of local water and food; and how common personal habits, such as geophagy influence these measurements. According to Winde et al. (2019), the need for validating such prediction models with a new series of individual (rather than composite) human measurements is considered to be of extreme importance.

- *Awareness creation and education*—Shindondola-Mote, in 2008, wrote, in the case of Namibia: "In Namibia general knowledge and awareness about the nuclear industry and its complex of impact on humans and the environment is negligible. To make use of their democratic rights and influence development toward sustainability, citizens need to understand issues and problems related to the nuclear industry. Earthlife Namibia has started taking the lead in filling this gap through education and awareness drives, targeting the general public. However, this activity needs to be intensified. The public campaigns are not meant to discredit any stakeholder, but to find ways to achieve the best possible practices for uranium mining."

  To improve awareness and knowledge acquisition by the public, Chareyron et al. (2014) noted "... the importance of promoting independent and collaborative work that synergizes the contributions of local communities and independent scientists; and that if local citizens are provided with information and tools, they are in a better position to make their own assessments and radiological checks." According to these authors: "Communities will also gain confidence in their ability to understand what radiation is, how to monitor and limit its impacts. ... It is important to develop the critical capacities of communities, so that they might have more information with which to face conflicts with states or companies in relation to uranium mining projects."

- *Revision of legislative articles*—In 2001 Wymer called attention to the need for revisiting nuclear legislations of the South Africa, and the nuclear regulatory systems that they support, not least, to enable them to reflect modern democratic principles of accountability, transparency and public participation. Wymer (2001) noted that: "... there is no longer any need for the legislation to be framed around secrecy requirements and the strategic uses of uranium. In proceeding with such revision, the opportunity should be taken to ensure that the application of nuclear regulation, and in particular, the system of nuclear licensing, ..." be done in a more transparent and holistic fashion.

- Although the following research gaps regarding radium and thorium isotopes were listed by ATSDR way back in 1998, some of these gaps still exist today, and largely duplicate some of the bullet point listings given earlier:

  - *More quantitative information regarding the Radium-224, Radium-226, and Radium-228 human exposures is needed to more adequately evaluate the magnitude of some dosimetric uncertainties and what impact these uncertainties have on quantitative risk estimation.*

  - *The bone cancer information from all of the human Ra-224, Ra-226, and Ra-228 exposures should be integrated and more adequately analyzed.*

• *Research should continue on identifying the cells at risk after exposure to radium. This should include cell behavior over time, changes in cell behavior, location of cells in relation to the microenvironment of the radiation field, responses of the cell to the radiation, and the time course and distribution of radioactivity in the bone.*

• *The dosimetry of the mastoids should be examined in order to calculate the risk per unit of epithelial tissue and per unit of cell dose.*

• *Data should be obtained from the five major epidemiological studies of Thorotrast-exposed patients and the data analyzed to develop risk models for liver and other cancers.*

• *The dosimetry of the thorium isotopes at the cellular level in target organs should be closely examined.*

• *The mechanism of uranium deposition and redistribution in bone should be further investigated in order to more accurately define the potential carcinogenic effect of natural uranium based on results obtained from enriched uranium or other alpha particle emitters.*

• *The current epidemiological studies of worker populations exposed to transuranic elements should be continued.*

• *Efforts to assess the carcinogenic risks of exposure to low levels of ionizing radiation, for both single dose and protracted and fractioned doses, should continue.*

• *The carcinogenicity of neutron radiation exposure in human populations should continue to be examined. Similarly, the mutagenicity of low doses of neutron radiation should continue to be investigated in order to more comfortably predict the potential genetic risks observed in laboratory animals and extrapolate those findings to human populations.*

References

AEB (The Atomic Energy Board of South Africa), 1979. The Atomic Energy Board of South Africa. https://inis.iaea.org/collection/NCLCollectionStore/_Public/10/463/10463837.pdf?r = 1&r = 1. (Accessed 20 June 2020).

Agius, R., Batistatou, E., Gittins, M., Jones, S., McNamee, R., Liu, H., et al., 2023. An epidemiological study of lung cancer and selected other cancers among Namibian uranium workers. Radiation Research 20 (4), 340−348. Available from: https://doi.org/10.1667/RADE-23-00051.1 (Accessed 19 February 2024).

Ahmed, N.K., 2005. Measurement of natural radioactivity in building materials in Qena city, Upper Egypt. Journal of Environmental Radioactivity 83 (1), 91−99.

ALA (American Lung Association), 2020. Radon. https://www.lung.org/clean-air/at-home/indoor-air-pollutants/radon. (Accessed 03 June 2020).

Alam, A., Wang, N., Zhao, G., Barkat, A., 2020. Implication of radon monitoring for earthquake surveillance using statistical techniques: a case study of Wenchuan Earthquake. Geofluids 2020, 2429165. Available from: https://doi.org/10.1155/2020/2429165.

Alhacen, A., 2017. History and consequences of uranium mining in Niger from 1969 to 2017. https://static1. squarespace.com/static/58bd8808e3df28ba498d7569/t/59bd250780bd5e7ca76585f3/1505568010268/ Almoushapha_20170910_ + English_HN.pdf. (Accessed 29 July 2020).

Allisy-Roberts, P., Williams, J., 2008. Radiation hazards and protection. In: Farr's Physics for Medical Imaging, second ed. https://www.sciencedirect.com/topics/pharmacology-toxicology-and-pharmaceutical-science/ dose-limit. (Accessed 20 May 2020).

Appleton, D., 2004. Geology and Radon Protection. Earthwise 21, 20−21. British Geological Survey (BGS)/ National Environmental Research Council (NERC), UK.

Appleton, J.D., 2005. Radon in air and water. In: Selinus, O., et al., (Eds.), Essentials of Medical Geology: Impacts of the Natural Environment on Public Health. Elsevier, London, UK, pp. 227−262.

Arafa, W., 2004. Specific activity and hazards of granite samples collected from the Eastern Desert of Egypt. Journal of Environmental Radioactivity 75 (3), 315−327.

ARC (African Radiation Consultants), 2014. Preliminary Radiation Impact Assessment of the Kamiesberg. Technical Report; Project Report No. ASC-1031L-1. file:///C:/Users/admin/Documents/Radiation% 20Assessment.pdf. (Accessed 06 April 2020).

ASTM (The American Society for Testing and Materials), 2017. ASTM D1739-98 (2017): Standard Test Method for Collection and Measurement of Dustfall (Settleable Particulate Matter), ASTM International, West Conshohocken, PA. https://www.astm.org/Standards/D1739.htm. (Accessed 29 July 2020).

ATSDR/US DHHS (Agency for Toxic Substances and Disease Registry)/(United States Department of Health and Human Services), 1999. Toxicological Profile for Ionizing Radiation. file:///C:/Users/admin/ Downloads/cdc_6393_DS1.pdf. (Accessed 31 May 2020).

ATSD/US DHHS (Agency for Toxic Substances and Disease Registry)/(United States Department of Health and Human Services), 2013

Axelson, O., 1991. Occupational and environmental exposures to radon: cancer risks. Annual Reviews of Public Health 12, 235−255.

Babic, R.R., Stankovic-Babic, G., Babic, S.R., Babic, N.R., 2016. 120 years since the discovery of X-rays. Medicinski Pregled 69 (9 - 10), 323−330.

Bain, C.A.R., Schoonbee, H.R., de Wet, L.P.D., Hancke, J.J., 1994. Investigations Into the Concentration Ratios of Selected Radionuclides in Aquatic Ecosystems Affected by Mine Drainage Effluents With Reference to the Study of Potential Pathways to Man. Report to the Water Research Commission, South Africa. http:// www.wrc.org.za/wp-content/uploads/mdocs/313-1-94.pdf. (Accessed 07 April 2020).

Bakr, A.A., 1989. Radiation Protection Services for the International Atomic Energy Agency. https://www.iaea. org/sites/default/files/publications/magazines/bulletin/bull31-1/31105381315.pdf. (Accessed 24 March 2020).

Bangoto, R.R., Cuney, M., Tulsidas, H., Biandja, J., 2018. The Uranium exploration history of the Central African Republic. Journal of Geological Resource and Engineering 6, 65−79. Available from: https://doi. org/10.17265/2328-2193/2018.02.003.

Barabanova, A.V., Bushmanov, A.J., Kotenko, K.V., 2019. In: Reference Module in Earth Systems and Environmental Sciences. https://www.sciencedirect.com/topics/medicine-and-dentistry/acute-radiation-syndrome. (Accessed 13 June 2020).

Bezuidenhout, J., 2019. Estimation of radon potential through measurement of uranium in granite geology. South African Journal of Science 115 (7 - 8). Available from: https://doi.org/10.17159/sajs.2019/5768.

Biira, S., Ochom, P., Oryema, B., 2021. Evaluation of radionuclide concentrations and average annual committed effective dose due to medicinal plants and soils commonly consumed by pregnant women in Osukuru, Tororo (Uganda). Journal of Environmental Radioactivity 227, 106460.. Available from: https:// doi.org/10.1016/j.jenvrad.2020.106460 (Accessed 20 February 2024).

Bird, G.A., 2012. Uranium in the environment: behavior and toxicity. In: Meyers, R.A. (Ed.), Encyclopaedia of Sustainability Science and Technology. Springer, New York, NY. Available from: https://doi.org/10.1007/ 978-1-4419-0851-3_294.

Bolus, N., 2017. Basic review of radiation biology and terminology. Journal of Nuclear Medicine Technology 45, 259–264. Available from: https://doi.org/10.2967/jnmt.117.195230.

Botha, J.C., Human, C.F., 2001. The radiological impact of radon and dust from gold mining operations, on the public. In: Conference on Environmentally Responsible Mining in Southern Africa, September 25–28, 2001, Muldersdrift, Johannesburg. Papers, vol. 1, 2C11–2C18.

Botha, R., Newman, R.T., Maleka, P.P., 2016. Radon levels measured at a touristic Thermal Spa Resort in Montagu (South Africa) and associated effective doses. Health Physics 111 (3), 281–289. Available from: https://doi.org/10.1097/HP.0000000000000527.

Botha, R., Newman, R.T., Lindsay, R., Maleka, P.P., 2017. Radon and thoron in-air occupational exposure study within selected wine cellars of the Western Cape (South Africa) and associated annual effective doses. Health Physics 112 (1), 98–107. Available from: https://doi.org/10.1097/HP.0000000000000574.

Botha, R., Labuschagne, C., Williams, A.G., Bosman, G., Brunke, E.-G., Rossouw, A., et al., 2018. Radon-222 measurements at Cape Point: a characterization of a 15 year time series. Clean Air 28 (2). Available from: https://doi.org/10.17159/2410-972X/2018/v28n2a11.

Breeze, P., 2017. Nuclear fission. Nuclear Power. Elsevier BV, pp. 75–83, 10.1016/b978-0-08-101043-3.00008-0. ISBN: 978-0-08--101043-3.

Brenner, D.J., 2011. We don't know enough about low-dose radiation risk. Nature: International Weekly Journal of Science . Available from: https://doi.org/10.1038/news.2011.206.

Brewer, R., Nebbe, C. (Talk of Iowa), 2019. Estimated 400 Deaths Per Year In Iowa Are Caused By Radon-Induced Lung Cancer. https://www.iowapublicradio.org/post/estimated-400-deaths-year-iowa-are-caused-radon-induced-lung-cancer#stream/0. (Accessed 12 April 2020).

Brooks, A.L., Hoel, D.G., Preston, R.J., 2016. The role of dose rate in radiation cancer risk: Evaluating the effect of dose rate at the molecular, cellular and tissue levels using key events in critical pathways following exposure to low LET radiation. International Journal of Radiation Biology 92 (8), 405–426. Available from: https://doi.org/10.1080/09553002.2016.1186301.

Browne, E., Firestone, R.B., 1986. Table of Radioactive Isotopes. John Wiley and Sons, pp. 237–243.

Bruneton, P., 2014. IAEA Classification of Uranium Deposits. International Atomic Energy Agency - URAM 2014, Vienna, Austria. Available from: https://www-pub.iaea.org/iaeameetings/cn216pn/Monday/Session1/187-Bruneton.pdf.

Brynard, H.J., Ainslie, L.C., van der Merwe, P.J., 1987. Uranium in South Africa. ISBN: 0–86960-857-6. https://inis.iaea.org/collection/NCLCollectionStore/_Public/20/033/20033347.pdf. (Accessed 14 July 2020).

Burns, P.C., Finch, R.J. (Eds.), 1999. Uranium: Mineralogy, Geochemistry and the Environment. Reviews in Mineralogy 38. Mineralogical Society of America. 679 p. ISBN: 0–9399 50-50-2.

Butterweck, G., Schuler, C., Vezzù, G., Marsh, J.W., Birchall, A., 2005. *In vivo* measurement of deposition and absorption of unattached radon progeny. In: McLaughlin, J.P., Simopoulos, S.E., Steinhäusler, F. (Eds.), The Natural Radiation Environment VII. The Seventh International Symposium on the Natural Radiation Environment (RRE - VII), Rhodes, Greece, 20–24 May, 2002. Elsevier.

Butterworth, M., 1999. Controlled recirculation in deep South African Gold Mines. In: 8th U.S. Mine Ventilation Symposium, Rolla, Missouri, vol. 5, Section XVIII, Miscellaneous Topics. https://scholarsmine.mst.edu/usmvs/8usmvs/8usmvs-theme18/5/. (Accessed 20 April 2020).

Carmine, G., Pugliese, M., Loffredo, F., Quarto, M., 2017. Indoor radon exposure and lung cancer risk: a meta-analysis of case-control studies. Translational Cancer Research 6 (5), S934–S943.

Cazoulat, M., 1985. Geologic environment of the uranium deposits in the Carboniferous and Jurassic sandstones of the western margin of the Air Mountains in the Republic of Niger: Vienna, International Atomic Energy Agency, Geological environments of sandstone-type uranium deposits, Tecdoc-328, pp. 247–603.

CDC (Centres for Disease Control and Prevention), 2014. Radiation and Pregnancy: A Fact Sheet for Clinicians. https://emergency.cdc.gov/radiation/prenatalphysician.asp. (Accessed 12 February 2017).

CDC (Centres for Disease Control and Prevention), 2020. Radiation and Pregnancy: A Fact Sheet for Clinicians. https://www.cdc.gov/nceh/radiation/emergencies/prenatalphysician.htm. (Accessed 25 June 2020).

CDC/NIOSH (Centres for Disease Control and Prevention)/(National Institute for Occupational Safety and Health), 2000. Worker Health Studies Summaries: Research on Long-Term Exposure - Uranium Miners (1). https://www.cdc.gov/niosh/pgms/worknotify/uranium.html. (Accessed 25 July 2020).

Chamaret, A., O'Connor, M., Récoché, G., 2007. Top-down/bottom-up approach for developing sustainable development indicators for mining: application to the Arlit uranium mines (Niger). International Journal of Sustainable Development 10 (1/2), 161–174.

Chareyron, B., 2008. Radiological hazards from uranium mining. In: Merkel, B.J., Hasche-Berger, A. (Eds.), Uranium, Mining and Hydrogeology. Springer, Berlin, Heidelberg. Available from: https://doi.org/10.1007/978-3-540–87746-2_56.

Chareyron, B., Živčič, L., Tkalec, T., Conde, M../CRIIRAD, 2014. Uranium mining. Unveiling the impacts of the nuclear industry. EJOLT Report No. 15, 116 p. http://www.ejolt.org/wordpress/wp-content/uploads/2014/11/141115_U-mining.pdf. (Accessed 07 April 2020).

Chaves-Pérez, A., Yilmaz, M., Perna, C., de la Rosa, S., Djoude, N., 2019. URI is required to maintain intestinal architecture during ionizing radiation. Science 364 (6443). Available from: https://doi.org/10.1126/science.aaq1165.

Chimangwa, G., Amoako, J.K., Fletcher, J.J., 2018. Investigation of the status of occupational radiation protection in Malawian hospitals. Malawi Medical Journal: The Journal of Medical Association of Malawi 30 (1), 22–24. Available from: https://doi.org/10.4314/mmj.v30i1.5.

Christensen, D.M., Iddins, C.J., Parrillo, S.J., Glassman, E.S., Goans, R.E., 2014. Management of ionizing radiation injuries and illnesses, Part 4. Acute radiation syndrome. Journal of the American Osteopathic Association 114 (9), 702–711. Available from: https://doi.org/10.7556/jaoa.2014.138.

Cigolini, C., Laiolo, M., Coppola, D., Trovato, C., Borgogno, G., 2016. Radon surveys and monitoring at active volcanoes: learning from Vesuvius, Stromboli, La Soufrière and Villarrica. In: Gillmore, G.K., Perrier, F.E., Crockett, R.G.M. (Eds.), Radon, Health and Natural Hazards. Geological Society, London, Special Publications, 451.

Clewell, R.A., Thompson, C.M., Clewell III, H.J., 2019. Dose-dependence of chemical carcinogenicity: biological mechanisms for thresholds and implications for risk assessment. Chemico-Biological Interactions 301, 112–127. Available from: https://doi.org/10.1016/j.cbi.2019.01.025.

Clark, D.L., 2016. Historical aspects of the discovery of plutonium. [(Los Alamos National Laboratory (United States)]. In: Proceedings of the DAE-BRNS Theme Meeting Commemorating Seventy Five Years of the Discovery of Plutonium 239, Kalpakkam (India), 23–25 May, 2016.

Clarke, R.H., Valentin, J., 2008. The History of ICRP and the Evolution of Its Policies. ICRP Publication 109; Annals of the ICRP. Elsevier. Available from: https://www.icrp.org/docs/The%20History%20of%20ICRP%20and%20the%20Evolution%20of%20its%20Policies.pdf.

Clement, C., Valentin, J., Ogino, H., Foote, D., Reyjal, J., Omar-Nazir, L. (Eds.), 2017. In: Proceedings of the 14th International Congress of the International Radiation Protection Association Cape Town, South Africa 9–13 May, 2016, vols. 1–5. ISBN 978-0-9989666-4-9, 1781. http://www.irpa.net/docs/IRPA14%20Proceedings%20Volume%204.pdf. (Accessed 04 April 2020).

CMSA (Chamber of Mines of South Africa), 1997. South African Mining Industry Statistical Tables 1996. Chamber of Mines of South Africa, Johannesburg.

CNSC (Canadian Nuclear Safety Commission), 2013. Natural Background Radiation. https://nuclearsafety.gc.ca/eng/resources/fact-sheets/natural-background-radiation.cfm. (Accessed 07 May 2020).

CNSC (Canadian Nuclear Safety Commission), 2020. Uranium Mines and Mills. http://nuclearsafety.gc.ca/eng/uranium/mines-and-mills/index.cfm. (Accessed 21 May 2020).

Coetzee, H., Lakin, J., 2011. Radiometric surveying for the management of health, safety and the environment in the uranium mining industry: Potential applications and limitations. In: Merkel, B., Schipek, M. (Eds.), The New Uranium Mining Boom, pp. 483−492.

Cothern, C.R., Smith (Jnr.), J.E., 2013. Environmental Radon. Springer Science and Business Media, p. 363.

Crockett, R.G.M., Gillmore, G.K., 2016. Radon as an anthropogenic indoor air pollutant as exemplified by radium-dial watches and other uranium- and radium containing artefacts. In: Gillmore, G.K., Perrier, F. E., Crockett, R.G.M. (Eds.), Radon, Health and Natural Hazards. Geological Society, LondonSpecial Publications, 451. Available from: https://doi.org/10.1144/SP4510.4.

Crockett, R.G.M., Gillmore, G.K., Phillips, P.S., Denman, A.R., Groves-Kirkby, C.J., 2006. Tidal synchronicity of built-environment radon levels in the UK. Geophysical Research Letters 33, L05308. Available from: https://doi.org/10.1029/2005GL024950.

Croteau, M.-N., Fuller, C., 2016. Processes that control uranium bioavailability in a freshwater snail. https://www.usgs.gov/mission-areas/environmental-health/science/study-reveals-processes-control-uranium-bioavailability-a?qt-science_center_objects = 0#qt-science_center_objects. (Accessed 14 August 2020).

Croteau, M.-N., Fuller, C.C., Cain, D.J., Campbell, K.M., Aiken, G.R., 2016. Biogeochemical controls of uranium bioavailability from the dissolved phase in natural freshwaters. Environmental Science and Technology 50 (15), 8120−8127. Available from: https://doi.org/10.1021/acs.est.6b02406.

Cuthbertson, A. (for Bayer AG), 2020. Deploying antibodies to deliver targeted radiation energy. The Bayer Scientific Magazine: Research. https://www.research.bayer.com/en/deploying-antibodies-to-deliver-targeted-radiation-energy.aspx. (Accessed 24 July 2020).

Dahlkamp, F.J., 1993. Uranium Ore Deposits. Springer-Verlag, Berlin, p. 460, ISBN: 978-3-662−02892-6.

Dahlkamp, F.J., 2009. Uranium Deposits of the World. Springer-Verlag, Berlin, Heidelberg, p. 494, Print ISBN: 978-3-540−78557-6.

Darby, S., Hill, D., Aluvinen, A., Barros-Dios, J.M., Baysson, H., et al., 2005. Radon in homes and risk of lung cancer: collaborative analysis of individual data from 13 European case-control studies. British Medical Journal 330 (7485), 223. Available from: https://doi.org/10.1136/bmj.38308.477650.63.

Dasnois, N., 2012. Uranium Mining in Africa: A Continent at the Centre of a Global Nuclear Renaissance. Governance of Africa's Resources Programme; South African Institute of International Affairs Occasional Paper No. 122. African perspectives. Global insights. https://www.files.ethz.ch/isn/154040/saia_sop_122%20_dasnois_20121005.pdf. (Accessed 03 April 2020).

Dauda, A.M., Ozoh, J.O., Towobola, O.A., 2019. Medical doctors' awareness of radiation exposure in diagnostic radiology investigations in a South African academic institution. South African journal of Radiology 23 (1), 1707.

De-Kun, L., Chen, H., Ferber, J.R., Hirst, A.K., Odouli, R., 2020. Association between maternal exposure to magnetic field nonionizing radiation during pregnancy and risk of Attention-Deficit/Hyperactivity disorder in offspring in a Longitudinal Birth Cohort. JAMA Network Open 3 (3), e201417. Available from: https://doi.org/10.1001/jamanetworkopen.2020.1417.

De Santis, M., Cesari, E., Nobili, E., Straface, G., Cavaliere, A.F., Caruso, A., 2007. Radiation effects on development. Birth Defects Research C Embryo Today 81 (3), 177−182.

Diehl, P., 2011. Uranium Mining and Milling Wastes: An Introduction. https://www.wise-uranium.org/uwai.html. (Accessed 04 August 2020).

Dixon, D., 2005. Understanding radon sources and mitigation in buildings. Journal of Building Appraisal 1 (2), 164−176. Available from: https://doi.org/10.1057/palgrave.jba.2940015.

Dixon, J.L., Lee, R.G., Smth, J., Zielinsky, P., 1991. Evaluating aeration technology for radon removal. Journal AWWA, Research and Technology 83 (4), 141−148.

Dobrzyński, L., Fornalski, K.W., Reszczynska, J., 2018. Meta-analysis of thirty-two case-control and two ecological radon studies of lung cancer. Journal of Radiation Research 59 (2), 149−163. Available from: https://doi.org/10.1093/jrr/rrx061.

Drago, J., 2013. Chapter 22: Radon removal technologies for small communities. In: Advances in Water and Wastewater Treatment. https://ascelibrary.org/doi/pdf/10.1061/9780784407417.ch22. (Accessed 29 May 2020).

Dudu, V.P., Mathuthu, M., Manjoro, M., 2018. Assessment of heavy metals and radionuclides in dust fallout in the West Rand mining area of South Africa. Clean Air Journal 28 (2), 42−52. Available from: https://doi.org/10.17159/2410-972x/2018/v28n2a17.

Dżaluk, A., Malczewski, D., Żaba, J., Dziurowicz, M., 2018. Natural radioactivity in granites and gneisses of the Opava Mountains (Poland): a comparison between laboratory *and in situ* measurements. Journal of Radioanalytical and Nuclear Chemistry 316 (1), 101−109. Available from: https://doi.org/10.1007/s10967-018-5726-3.

Eckert, D., Evic, M., Schang, J., Isbruch, M., Er, M., Dörrschuck, L., et al., 2024. Osteo-immunological impact of radon spa treatment: Due to radon or spa alone? Results from the prospective, thermal bath placebo-controlled RAD-ON02 trial. Frontiers in Immunology 14, 1284609. Available from: https://doi.org/10.3389/fimmu.2023.1284609 (Accessed 19 February 2024).

Eilers, A., Miller, A., Swana, K., Botha, R., Talma, S., Newman, R., et al., 2015. Characterisation of radon concencentrations in Karoo groundwater, South Africa, as a prelude to potential shale-gas development. Procedia Earth and Planetary Science 13, 269−272. Available from: https://doi.org/10.1016/j.proeps.2015.07.063.

Eisenbud, M., Gesell, T.F., 1997. Environmental radioactivity from natural, industrial and military sources, From Natural, Industrial and Military Sources, Fourth Edition Academic Press, ISBN: 9780122351549.

Ejolt, 2015. Mapping Environmental Justice: Areva's Uranium Mines. http://www.ejolt.org/2015/07/arevas-uranium-mines-niger/. (Accessed 11 April 2020).

Elliott, O., 2019. Uranium mining and exploitation in Niger. https://storymaps.arcgis.com/stories/de9dff2-f59964ab1a1bb46c2ea905f0c. (Accessed 12 April 2020).

El-Taher, A., 2018. An overview of instrumentation for measuring radon in environmental studies. Journal of Radiation and Nuclear Applications 3 (3), 135. Available from: https://doi.org/10.18576/jrna/030302.

El-Taher, A., Uosif, M.A.M., Orabi, A.A., 2007. Natural radioactivity levels and radiation hazard indices in granite from Aswan to Wadi El-Allaqi southeastern desert, Egypt. Radiation Protection Dosimetry 124 (2), 148−154.

Elzain, A.E.A., 2017. Radon Monitoring in the Environment. DOI: 10 5772/interchopen.69902. https://www.intechopen.com/books/radon/radon-monitoring-in-the-environment. (Accessed 12 July 2020).

EU (The European Union), 2018. Radiation Protection: No. 188; Technical Recommendations for Monitoring Individuals for Occupational Intakes of Radionuclides. ISBN: 978-92-79-86304. https://ec.europa.eu/energy/sites/ener/files/rp_188.pdf. (Accessed 31 July 2020).

Fehner, T.R., Gosling, F.G. 2000. Origins of the Nevada Test Site. Report No. DOE/MA-0518. United States Department of Energy. https://www.nnss.gov/docs/docs_LibraryPublications/DOE_MA0518.pdf. (Accessed 04 May 2020).

Feinendegen, L.E., 2005. Evidence for beneficial low level radiation effects and radiation hormesis. British Journal of Radiology 78 (925), 3−7. Available from: https://doi.org/10.1259/bjr/63353075.

Fisher, M., 2018. IAEA safeguards at uranium mines provide more complete picture of countries s' nuclear activities. In: IAEA/Gaspar, M., et al. (Eds.), Uranium: From Exploration to Remediation. IAEA Bulletin. vol. 59, No. 2, pp. 14−117. https://www.iaea.org/publications/magazines/bulletin/59-2. (Accessed 01 May 2020).

Fisher, D.R., Fahey, F.H., 2017. Appropriate use of effective dose in radiation protection and risk assessment. Health Physics 113 (2), 102−109.

Fleischer, B., 2018. Working in a uranium mine in Namibia. http://large.stanford.edu/courses/2018/ph241/fleischer2/#:~:text = The%20Health%20Implications%20of%20Working%20in%20R%C3%B6ssing%20Uranium%20Mine&text = Some%20by%20products%20of%20uranium,companies%20in%20Namibia%20is%20R%C3%B6ssing. (Accessed 26 June 2020).

FNRBA (Forum of Nuclear Regulatory Bodies in Africa), 2020. FNRBA. https://www.irdp-online.org/organization/participants/fnrba. (Accessed 04 April 2020).

Fordt, M.A., 1993. Uranium in South Africa. Journal of the South African Institute of Mining and Metallurgy 93 (2), 37−58.

Frimmel, H.E., 2010. Verfügbarkeit von natürlich vorkommendem Uran. Unpublished Report, Office of Technology Assessment at the German Bundestag (TAB), Berlin, 123 p.

Frimmel, H.E., 2019. The Witwatersrand Basin and its gold deposits. In: Kröner, A., Hofmann, A. (Eds.), The Archaean Geology of the Kaapvaal Craton, Southern Africa. Regional Geology Reviews. Springer, Cham.

Garside, M., 2019. Uranium Statistics and Facts. https://www.statista.com/topics/1553/uranium/. (Accessed 29 June 2020).

Ghernaout, D., 2019. Aeration process for removing radon from drinking water - a review. https://www.researchgate.net/publication/333748175_Aeration_Process_for_Removing_Radon_from_Drinking_Water_-A_Review. (Accessed 29 May 2020).

Gillmore, G.K., 2001. The radon hotline Internet site. The Journal of the Royal Society for the Promotion of Health 121 (4), 212.

Gillmore, G.K., Phillips, P., Denman, A., Sperrin, M., Pearce, G., 2001. Radon levels in abandoned metalliferous mines, Devon, Southwest England. Ecotoxicology and Environmental Safety 49, 281−292.

Gillmore, G.K., Grattan, J., Pyatt, F.B., Phillips, P.S., Pearce, G. 2002. Radon, water and abandoned metalliferous mines in the UK: Environmental and human health implications. In: Merkel, B.J., PlanerFriedrich, B., Woldersdorfer, C. (Eds.), Uranium in the Aquatic Environment. Proceedings of the International Conference Uranium Mining and Hydrogeology III and the International Mine Water Association Symposium, Freiberg, Germany, 15−21 September, 2002. Springer-Verlag, Berlin, Heidelberg, pp. 65−76.

Gillmore, G.K., Crockett, R.G.M., Phillips, P.S., 2017a. Radon as a carcinogenic built-environmental pollutant. In: Gillmore, G.K., Perrier, F.E., Crockett, R.G.M. (Eds.), Radon, Health and Natural Hazards. Geological Society, London Special Publications, 451. Available from: https://doi.org/10.1144/SP451.5.

Gillmore, G.K., Crockett, R.G.M., Phillips, P.S., 2017b. Radon as a carcinogenic built-environmental pollutant. Journal of Radiology 78, 3−7. Available from: https://doi.org/10.1259/bjr/63353075.

Gillmore, G.K., Perrier, F.E., Crockett, R.G.M., 2018. Radon, Health and Natural Hazards: A Signpost for Assessment and Protection in the 21st Century. Geological Society of London, Special Publications 451, pp. 1−5. In: Gillmore, G.K., Perrier, F.E., Crockett, R.G.M. (Eds.), Radon, Health and Natural Hazards. Geological Society, London, Special Publications, 451. doi: https://doi.org/10.1144/SP451.5.

GoC (Government of Canada), 1997. The Nuclear Safety and Control Act (S.C. 1997, c.9). https://laws-lois.justice.gc.ca/eng/acts/N-28.3/FullText.html. (Accessed 21 May 2020).

GoC (Government of Canada), 2019. Canada Occupational Health and Safety Regulations (SOR/86-304). https://laws-lois.justice.gc.ca/eng/regulations/SOR-86-304/index.html. (Accessed 21 May 2020).

Godish, T., 2001. Indoor Environmental Quality. CRC Press LLC/Lewis Publishers, Boca Raton.

Godish, T., 2018. Indoor Air Pollution Control, Third Edition CRC Press, p. 416. Available from: https://books.google.de/books?id = NBalDwAAQBAJ&pg = PT112&lpg = PT112&dq = Removal + of + subsoil + radon + mitigation&source = bl&ots = SEzUzH4IiT&sig = ACfU3U2PhsLTUSwhPhccCRI (Accessed 25 July 2020).

Goel, K.M., Carachi, R., 2014. Atlas of Paediatric Physical Diagnosis. Jaypee Brothers Medical Publishers (P) Ltd, New Delhi.

Gök, M., Bozkurt, M., Guneyli, S., Kara Bozkurt, D., Korkmaz, M., Peker, N., 2015. Prenatal radiation exposure. Proceedings in Obstetrics and Gynecology 5 (1), 2.

Gomes, C.de Sousa F., P. de Silva, J.B.P., 2007. Minerals and clay minerals in medical geology. Applied Clay Science 36, 4–21.

Govender, V., Ntholeng, T., Vawda, F., Larson, C.O., 2013. The profile of indications for radiography in the Neonatal Intensive Care Unit at Universitas Academic Hospital, Bloemfontein. South African Journal of Child Health 7 (1).

Grandjean, P., 2016. Paracelsus revisited: the dose concept in a complex world. Basic and Clinical Pharmacology and Toxicology 119 (2), 126–132. Available from: https://doi.org/10.1111/bcpt.12622.

Gupta, D.K., Walther, C., 2019. Uranium in Plants and the Environment. Springer, p. 246. Available from: https://books.google.de/books? id = yx6VDwAAQBAJ&dq = Average + concentrations + of + uranium + in + the + crust + of + the + earth + are + around + 0.0003%25.&source = gbs_navlinks_s (Accessed 10 June 2020).

Gupta, A.K., Garg, A., Sandhu, M.S. (Eds.), 2019. Diagnostic Radiology: Advances in Imaging Technology. AIIMS-MAMC-PGI Imaging Series. Jaypee Brothers Medical Publishers (P) Ltd.

Hatch, M., Susser, M., 1990. Background gamma radiation and childhood cancers within ten miles of a US nuclear plant. Journal of Epidemiology 19, 546–552.

Harley, N., 2011. Radiation Toxicology: Toxicology of Specific Groups of Substances. Wiley Online Library. Available from: https://doi.org/10.1002/9780470744307.gat117.

Hausen, D.M., 1998. Characterizing and classifying uranium yellow cakes: a background. JOM 50, 45–47. Available from: https://doi.org/10.1007/s11837-998-0307-5.

Hayashi, T., Kusunoki, Y., Hakoda, M., Morishita, Y., Kubo, Y., Maki, M., et al., 2003. Radiation dose-dependent increases in inflammatory response markers in A-bomb survivors. International Journal of Radiation Biology 79 (2), 129–136. Available from: https://doi.org/10.1080/0955300021000038662.

Hecht, G., 2009. Africa and the nuclear world: labor, occupational health, and the transnational production of uranium. Comparative Studies in Society and History 51 (4), 896–926.

Hecht, G., 2012a. The work of invisibility: radiation hazards and occupational health in South African uranium production. International Labor and Working-Class History 81, 94–113. Available from: https://doi.org/10.1017/S0147547912000051.

Hecht, G., 2012b. Being Nuclear: Africans and the Global Uranium Trade. The MIT Press, Cambridge. Available from: https://www.academia.edu/2459111/Being_Nuclear_Africans_and_the_Global_Uranium_Trade, ISSN: 978-0-262-01726-8.

Herbst, C.P., Fick, G.H., 2012. Radiation protection and the safe use of X-ray equipment: laws, regulations and responsibilities. South African Journal of Radiology 16 (2).

Herbst, M.C., CANSA (Cancer Association of South Africa), 2019a. Fact Sheet on the Amount of Ionizing Radiation Received from Routine Imaging. file:///C:/Users/admin/Documents/Fact-Sheet-Amount-of-Ionizing-Radiation-Received-from-Routine-Imaging-Oct-2018.pdf. (Accessed 18 March 2020).

Herbst, M.C., CANSA, 2019b. Fact Sheet and Position Statement on Radon as a Cause of Lung Cancer in South Africa. https://www.cansa.org.za/files/2019/11/Fact-Sheet-Position-Statement-Radon-as-Cause-of-Lung-Cancer-in-South-Africa-Nov-2019.pdf. (Accessed 24 May 2020).

Hernández, L., Terradas, M., Camps, J., Martín, M., Tusell, L., Genescà, A., 2015. Aging and radiation: bad companions. Aging cell 14 (2), 153–161. Available from: https://doi.org/10.1111/acel.12306.

Herring, J.S., 2012. Uranium and thorium resources. In: Meyers, R.A. (Ed.), Encyclopedia of Sustainability Science and Technology. Springer, New York, NY.

Hill, R., Mo, Z., Haque, M., Baldock, C., 2009. An evaluation of ionization chambers for the relative dosimetry of kilovoltage X-ray beams. Medical Physics 36 (9), 3971–3981. Available from: https://doi.org/10.1118/1.3183820.

Hines, R.N., Sargent, D., Autrup, H., Birnbaum, L.S., et al., 2010. Approaches for assessing risks to sensitive populations: lessons learned from evaluating risks in the pediatric population. Toxicological Sciences 113 (1), 4–26. Available from: https://doi.org/10.1093/toxsci/kfp217.

Hofmann, W., Arvela, H.S., Harley, N.H., Marsh, J.W., McLaughlin, J., Röttger, A., et al., 2012a. Radon and radon progeny inhalation and resultant doses. Journal of the International Commission on Radiation Units and Measurements 12 (2), 29−53. Available from: https://doi.org/10.1093/jicru/ndv010.

Hofmann, W., Arvela, H.S., Harley, N.H., Marsh, J.W., McLaughlin, J., Röttger, A., et al., 2012b. Glossary: definitions, quantities, and units. Journal of the International Commission on Radiation Units and Measurements 12 (2), 3−15. Available from: https://doi.org/10.1093/jicru/ndv002.

Hore-Lacy, I., 2016. Production of byproduct uranium and uranium from unconventional resources. Uranium for Nuclear Power - Resources, Mining and Transformation to Fuel 239−251. Available from: https://doi.org/10.1016/B978-0-08-100307-7.00009-0.

Hung, Y.-T., Wang, L.K., Shammas, N.K. (Eds.), 2013. Handbook of Environmental and Waste Management: Land and Groundwater Pollution Control, vol. 2. World Scientific, p. 1116.

IAEA, n.d., Monitoring and Dose Assessment: Training Package on Occupational Radiation Protection in Uranium Mining and Processing Industry. https://nucleus.iaea.org/sites/orpnet/training/uranium/Shared%20Documents/Module%206%20Monitoring%20and%20Dose%20Assessment.pdf. (Accessed 31 July 2020).

IAEA (International Atomic Energy Agency), 2000. Calibration of Radiation Protection Monitoring Instruments. Safety Reports Series No. 16. Vienna, Austria. 153 p. https://www-pub.iaea.org/MTCD/Publications/PDF/P074_scr.pdf. (Accessed 14 June 2020).

IAEA (International Atomic Energy Agency), 2004. Occupational Radiation Protection in the Mining and Processing of Raw Materials. Safety Standards Series No. RS-G-1.6. https://www-pub.iaea.org/MTCD/Publications/PDF/Pub1183_web.pdf.

IAEA (International Atomic Energy Agency), 2005. Environmental and Source Monitoring for Purposes of Radiation Protection. IAEA Safety Standards Series No. RS-G-1.8. https://www-pub.iaea.org/MTCD/Publications/PDF/Pub1216_web.pdf. (Accessed 15 June 2020).

IAEA (International Atomic Energy Agency), 2007. Terminology Used in Nuclear Safety and Radiation Protection. 2007 Edition. IAEA Vienna. https://www-pub.iaea.org/MTCD/Publications/PDF/Pub1290_web.pdf. (Accessed 19 May 2020).

IAEA (International Atomic Energy Agency), 2008. Uranium 2007 - Resources, Production and Demand, OECD Nuclear Energy Agency and International Atomic Energy Agency, Paris 2004.

IAEA (International Atomic Energy Agency), 2013. National and Regional Surveys of Radon Concentration in Dwellings Review of Methodology and Measurement Techniques. Analytical Quality in Nuclear Applications; Series No. 33. Report IAEA/AQ/33 IAEA. https://www-pub.iaea.org/MTCD/Publications/PDF/IAEA-AQ-33_web.pdf. (Accessed 09 July 2020).

IAEA (International Atomic Energy Agency) - Jointly sponsored by: EC (European Commission), FAO (Food and Agriculture Organization of the United Nations), IAEA (International Atomic Energy Agency), ILO (International Labour Organization), OECD NEA (Organisation for Economic Co-operation and Development; Nuclear Energy Agency), PAHO (Pan American Health Organization), UNEP (United Nations Environment Programme), WHO (World Health Organization, 2014. Safety Standards for Protecting People and the Environment: Radiation Protection and Safety of Radiation Sources: International Basic Safety Standards - General Safety Requirements, No. GSR Part 3. IAEA, Vienna. ISBN: 978-92-0-135310-8. ISSN: 1020−525X, No. GSR Part 3. https://www-pub.iaea.org/MTCD/publications/PDF/Pub1578_web-57265295.pdf. (Accessed 07 July 2020).

IAEA (International Atomic Energy Agency), 2015. Protection of the Public Against Exposure Indoors due to Radon and Other Natural Sources of Radiation: Specific Safety Guide; IAEA Safety Standards Series No. SSG-32. https://www-pub.iaea.org/MTCD/publications/PDF/Pub1651Web-62473672.pdf. (Accessed 30 July 2020).

IAEA (International Atomic Energy Agency), 2016a. New Edition of "Red Book" Uranium Report Is Published. https://www.iaea.org/newscenter/pressreleases/new-edition-of-red-book-uranium-report-is-published. (Accessed 31 July 2020).

IAEA (International Atomic Energy Agency), 2018a. Safety Standards for Protecting People and the Environment: Occupational Radiation Protection - General Safety Guide No.GSG-7. International Atomic Energy Agency, Vienna. https://www-pub.iaea.org/MTCD/Publications/PDF/PUB1785_web.pdf. (Accessed 27 MArch2020).

IAEA (International Atomic Energy Agency), 2018b. Radiation Exposure of the Public and the Environment. https://www-pub.iaea.org/MTCD/Publications/PDF/PUB1781_web.pdf. (Accessed 10 June 2020).

IAEA (International Atomic Energy Agency), 2018c. IAEA Safety Standards for Protecting People and the Environment: General Safety Guide No. GSG-8 - Radiation Protection of the Public and the Environment. http://regelwerk.grs.de/sites/default/files/cc/dokumente/dokumente/PUB_GSG-8_web.pdf. (Accessed 16 July 2020).

IAEA (International Atomic Energy Agency), Gaspar, M., et al., (Eds.), 2018d. Uranium: from exploration to remediation. IAEA Bulletin 59-2. https://www.iaea.org/publications/magazines/bulletin/59-2. (Accessed 10 October 2020).

IAEA (International Atomic Energy Agency), 2019a. Prevention and mitigation methods related to indoor radon and natural radionuclides in building materials. A Webinar. https://www.iaea.org/resources/webinar/prevention-and-mitigation-methods-related-to-indoor-radon-and-natural-radionuclides-in-building-materials. (Accessed 26 May 2020).

IAEA (International Atomic Energy Agency), 2019b. IAEA Safety Glossary: 2018 Edition. ISBN: 978−92−0−104718−2, 261 p. https://www-pub.iaea.org/MTCD/Publications/PDF/PUB1830_web.pdf. (Accessed 19 May 2020).

IAEA (International Atomic Energy Agency), 2020a. Occupational Radiation Protection in the Uranium Mining and Processing Industry. Safety Reports Series No. 100. file:///D:/IAEA%20%20MOST%20RECENT%20%20RADIATION%20%20PROTECTION.pdf. (Accessed 02 May 2020).

IAEA, 2020b. Information System on Uranium Mining Exposures (UMEX). https://nucleus.iaea.org/sites/orpnet/worldwide/umex/SitePages/Home.aspx. (Accessed 31 July 2020).

IAEA (International Atomic Energy Agency), 2020c. B.1. Practical intervention techniques to reduce public doses at uranium mining and milling legacy sites. https://www.iaea.org/about/partnerships/european-union/european-commission-funded-projects/b1-practical-intervention-techniques-to-reduce-public-doses-at-uranium-mining-and-milling-legacy-sites. (Accessed 05 April 2020).

IARC (International Agency for Research on Cancer), 2001. Ionizing Radiation, Part 2: Some Internally Deposited Radionuclides. IARC Monographs on the Evaluation of Carcinogenic Risks to Humans, No. 78. ISBN: 92 832 1278 9. https://www.ncbi.nlm.nih.gov/books/NBK396552/. (Accessed 10 August 2020).

ICRP (International Commission on Radiological Protection), 1979. Limits for Intakes of Radionuclides by Workers. ICRP Publication 30, Part 1. Annals of the ICRP 2 (3/4). Pergamon Press, Oxford.

ICRP (International Commission on Radiological Protection), 1994. Human Respiratory Tract Model for Radiological Protection. Annals of the ICRP. ICRP Publication 66, vol. 24, Nos. 1−3. Pergamon Press, Oxford and New York. https://www.icrp.org/publication.asp?id = ICRP%20Publication%2066. (Accessed 15 June 2020).

ICRP (International Commission on Radiological Protection), 2000. Pregnancy and Medical Radiation. Annals of the ICRP 30 (1), iii−viii, 1−43.

ICRP (International Commission on Radiological Protection), 2002. Guide for the Practical Application of the ICRP Human Respiratory Tract Model, ICRP Supporting Guidance 3. Pergamon Press, Oxford and New York.

ICRP (International Commission on Radiological Protection), 2005. Doses to Infants from Ingestion of Radionuclides in Mothers' Milk. The 2004 Annual Report of the International Commission on Radiological Protection (ICRP). International Commission on Radiological Protection 95 (3−4), . Available from: https://www.icrp.org/docs/2004_Annual_Rep_52_150_05.pdf (Accessed 07 June 2020).

ICRP (International Commission on Radiological Protection), 2006. Human Alimentary Tract Model for Radiological Protection. ICRP Publication 100. A report of The International Commission on Radiological

Protection. Annals of the ICRP 36 (1-2), 25–327. Available from: https://doi.org/10.1016/j. icrp.2006.03.004. iii.

ICRP (International Commission on Radiological Protection), 2007. The 2007 recommendations of the International Commission on Radiological Protection. ICRP Publication 103. Annals of the ICRP 37 (2 - 4), 1–332.

ICRP (International Commission on Radiological Protection), 2008. Environmental protection - the concept and use of reference animals and plants. Annals of the ICRP 38 (4–6), ICRP Publication 108.

ICRP (International Commission on Radiological Protection), 2017. Dose coefficients for nonhuman biota environmentally exposed to radiation. Annals of the ICRP 46 (2), ICRP Publication 136.

ICRU (International Commission on Radiation Units and Measurements), 1980. Radiation Quantities and Units. Report No. 33. Washington DC. International Commission on Radiation Units and Measurements.

ICRU (International Commission on Radiation Units and Measurements), 1998. Fundamental Quantities and Units for Ionizing Radiation. ICRU Report 60, Volume os-31, Issue 1. https://journals.sagepub.com/toc/ crub/os-31/1. (Accessed 02 July 2020).

ICRU (International Commission on Radiation Units and Measurements), 2005. Patient dosimetry for X-rays used in medical imaging. ICRU Report 74, ICRU, Bethesda, Maryland, USA.

IDEA (International Institute for Democracy and Electoral Assistance), 2019. The Global State of Democracy 2019: Addressing the Ills, Reviving the Promise. Reprinted 2019. https://www.idea.int/sites/default/files/ publications/the-global-state-of-democracy-2019.pdf. (Accessed 21 May 2020).

Igarashi, G., Saeki, S., Takahata, N., Sumikawa, K., 1995. Groundwater radon anomaly before the Kobe earthquake in Japan. Science 269, 60–61. Available from: https://doi.org/10.1126/science.269.5220.60.

Ilori, A.O., Chetty, N., 2023. A review of the occurrence of naturally occurring radioactive materials and radiological risk assessment in South African soils. International Journal of Environmental Health Research . Available from: https://doi.org/10.1080/09603123.2023.2280661 (Accessed 19 February 2024).

Isaksson, M., Rääf, C.L., 2017. Environmental Radioactivity and Emergency Preparedness. CRC Press, Taylor and Francis Group. Available from: https://books.google.de/books?id = oLTZDQAAQBAJ&pg = SA2-PA31&lpg = SA2-
PA31&dq = The + severity + of + damage + to + the + embryo + depends + on + the + dose + absorbed- + and + th (Accessed 05 July 2020).

Jacob, R.E. 1974. Geology and Metamorphic Petrology of part of the Damara orogen along the Lower Swakop River, South West Africa. Cape Town. Thesis, University of Cape Town, Faculty of Sciences. https://open. uct.ac.za/handle/11427/23253. (Accessed 05 August 2020).

Jaiswal, I., 2014. Radiation carcinogenesis. https://pt.slideshare.net/ishajaiswal169/radiation-carcinogenesis? smtNoRedir = 1. (Accessed 05 July 2020).

JCMG (Joint Committee for Guides in Metrology), 2017. International Vocabulary of Metrology - Basic and General Concepts and Associated Terms (VIM, third ed.). JCGM 200: 2012 (JCGM 200: 2008 With Minor Corrections).

Isinkaye, M.O., OlaOlorun, O.A., Chandrasekaran, A., Adekeye, A.S., Dada, T.E., Tamilarasi, A., et al., 2023. Quantification of radiological hazards associated with natural radionuclides in soil, granite and charnockite rocks at selected fields in Ekiti State, Nigeria. Heliyon 9 (11), e22451. Available from: https://doi.org/ 10.1016/j.heliyon.2023.e22451 (Accessed 20 February 2024).

Kamunda, C., Mathuthu, M., Madhuku, M., 2016. An assessment of radiological hazards from gold mine tailings in the Province of Gauteng in South Africa. International Journal of Environmental Health 13 (1), 138. Available from: https://doi.org/10.3390/ijerph13010138.

Kamunda, C., Mathuthu, M., Madhuku, M., 2017. Determination of radon in mine dwellings of Gauteng Province of South Africa using AlphaGUARD Radon Professional Monitor. Journal of Environmental Toxicology Studies 1 (1), 1–4. Available from: https://doi.org/10.16966/2576-6430.107.

Kang, K.W., 2016. History and organizations for radiological protection. Journal of Korean Medical Science 31 (Suppl 1), S4–S5. Available from: https://doi.org/10.3346/jkms.2016.31.S1.S4.

Karam, A., Venter, N., 2007. Affordable housing on contaminated land in Johannesburg. Acta Structilia 14 (2), 35–57. Available from: https://journals.ufs.ac.za/index.php/as/article/view/359.

Karpas, Z., 2014. Analytical chemistry of uranium. Environmental, Forensic, Nuclear and Toxicological Applications. CRC Press, Boca Raton. London. New York, p. 308, ISBN: 13d978-1-4822-2058-2.

Kay, P., 2007. Australia's Uranium Mines: Past and Present. https://www.aph.gov.au/Parliamentary_Business/Committees/Senate/Former_Committees/uranium/report/c07. (Accessed 21 May 2020).

Keith, S., et al., 1999. Toxicological Profile for Ionizing Radiation U.S. Department of Health and Human Services Public Health Service Agency for Toxic Substances and Disease Registry. https://www.atsdr.cdc.gov/toxprofiles/tp149.pdf. (Accessed 26 July 2020).

Kawooya, M.G., Kisembo, H.N., Remedios, D., Malumba, R., del Rosario Perez, M., Ige, T., et al., 2022. An Africa point of view on quality and safety in imaging. Insights Imaging 13 (58), . Available from: https://doi.org/10.1186/s13244-022-01203-w (Accessed 19 February 2024).

Keith, S., Doyle, J.R., Harper, C., Mumtaz, M., Tarrago, O., Wohlers, D.W., et al., 2012. Toxicological Profile for Radon. Agency for Toxic Substances and Disease Registry (US), Atlanta (GA). Available from: https://www.ncbi.nlm.nih.gov/books/NBK158784/ (Accessed 05 August 2020).

Khadzhidekova, V., Shishkova, R., Khadzhidekov, V., 2001. Effects of ionizing radiation on the embryo and fetus. Akush Ginekol (Sofiia) 41 (1), 24–28.

Khan, S.U., Nakhabov, A. (Eds.), 2020. Nuclear Reactor Technology Development. Woodhead Publishing Series in Energy. Elsevier Ltd. Available from: https://books.google.de/books?id = pJrgDwAAQBAJ&pg = PA78&lpg = PA78&dq = humans + have + about + 65 + Bq/kg + of + K-40). &source = bl&ots = vmyGMxV9zQ&sig = ACfU3U3b7zvFDNKASQjgEJf8EMwV6R9xfw&hl = en&sa = X&ved = 2ahUKEwjVosL8ianqAhUR7KYKHYdoAQkQ6AEwAHoECAgQAQ#v = onepage&q = humans%20have%20about%2065%20Bq%2Fkg%20of%20K-40).&f = false.

Khan, S.M., Gomes, J., Krewski, D.R., 2019. Radon interventions around the globe: a systematic review. Heliyon 5 (5), e01737. Available from: https://doi.org/10.1016/j.heliyon.2019.e01737.

Kieskamp, K.K., Worley, R.R., McLanahan, E.D., Verner, M.A., 2018. Incorporation of fetal and child PFOA dosimetry in the derivation of health-based toxicity values. Environment International 11, 260–267. Available from: https://doi.org/10.1016/j.envint.2017.12.019.

King, R.D., 1993. The legal implications of residential radon contamination: The first decade, 18 William and Mary Journal of Environmental Law and Policy Review 107/18 (1/4). https://scholarship.law.wm.edu/cgi/viewcontent.cgi?referer = &httpsredir = 1&article = 1354&context = wmelpr. (Accessed 05 August 2020).

Kinnaird, J., Nex, P., 2016. Uranium in Africa. Episodes 39 (2), 335. Available from: https://doi.org/10.18814/epiiugs/2016/v39i2/95782.

Kinner, N.E., Lassard, C.E., Schnell, G.S., Fox, K.R., 1987. Low cost/low technology aeration techniques for drinking water. EPA Environmental Research Brief EPA/600/M-87/031. https://nepis.epa.gov/Exe/tiff2png.cgi/300024G5.PNG?-r + 75 + -g + 7 + D%3A%5CZYFILES%5CINDEX%20DATA%5C86THRU90%5CTIFF%5C00000240%5C300024G5.TIF. (Accessed 29 May 2020).

Kluszczyński, D., 2005. Idea of assessment of the annual dose when non-continuous data exist. Radioactivity in the Environment 7, 788–793. Available from: https://doi.org/10.1016/S1569-4860(04)07098-6.

Kneen, M.A., Ojelede, M.E., Annagarn, H.J., 2015. Housing and population sprawl near tailings storage facilities in the Witwatersrand: 1952 to current. South African Journal of Science 111, 142–150.

Knox, E.G., Stewart, A.M., Gilman, E.A., Kneale, G.W., 1988. Background radiation and childhood cancers. Journal of Radiological Protection 8 (1), 9–18.

Kobayashi, M., 2017. Nuclear development status in the world (4) Four new emerging countries (China, Russia, India, and South Korea) leading global nuclear development. Nippon Genshiryoku Gakkai-Shi (Atomos) 59 (11), 638–642.

Kohrs, B., Kafuka, P., 2014. Study on low-level radiation of Rio Tinto's Rössing Uranium mine workers. EJOLT and Earthlife Namibia Report. http://www.criirad.org/mines-uranium/namibie/riotinto-rossing-workers-EARTHLIFE-LARRI-EJOLT.pdf. (Accessed 26 June 2020).

Kojima, S., Cuttler, J.M., Inoguchi, K., Yorozu, K., Horii, T., Shimura, N., et al., 2019. Radon therapy is very promising as a primary or an adjuvant treatment for different types of cancers: 4 case reports. Dose-Response: A Publication of International Hormesis Society 17 (2), . Available from: https://doi.org/10.1177/15593258198531631559325819853163.

Komov, I.L., 2001. Health effects and indoor radon risk reduction. Medical Geology Newsletter 4, 5–6.

Komov, I.L., 2003. Monitoring of radon in Ukraine. In: Proceedings of the 2003 International Radon Symposium - Volume I1 American Association of Radon Scientists and Technologists, Inc. 5–8, October, 2003. aarst-nrpp.com/proceedings/2003/2003_05_Monitoring_Of_Radon_In_Ukraine.pdf. (Accessed 02 April 2020).

Koos, C., Basedau, M., 2012. Does Uranium Mining Increase Civil Conflict Risk?: Evidence from a Spatiotemporal Analysis of Africa from 1945 to 2010. German Institute for Global and Area Studies (GIGA). http://www.jstor.org/stable/resrep07671. (Accessed 27 June 2020).

Korir, G., Karam, P.A., 2018. A novel method for quick assessment of internal and external radiation exposure in the aftermath of a large radiological incident. Health Physics 115 (2), 235–251. Available from: https://doi.org/10.1097/HP.0000000000000858.

Kozempel, J., 2017. Radon nuclides and radon generators. doi: 10.5772/intechopen.69901. https://www.inte-chopen.com/books/radon/radon-nuclides-and-radon-generators. (Accessed 12 July 2020).

Kreuzer, M., McLaughlin, J., 2010. Radon. WHO Guidelines for Indoor Air Quality: Selected Pollutants. World Health Organization, Geneva.

Kummar, S., Takimoto, C. (Eds.), 2018. Novel Designs of Early Phase Trials for Cancer Therapeutics. Academic Press. Available from: https://books.google.de/books?id = e_JcDwAAQBAJ&pg = PA140&lpg = PA140&dq = effect + of + ionising + , Accessed 13 June 2020.

Landau, E., 1972. Radiation and infant mortality: Some hazards of methodology. In: Le Cam, L.M., Neyman, J., Scott, E.L. (Eds.), Proceedings of the Sixth Berkeley Symposium on Mathematical Statistics and Probability, vols. 1–3. University of California Press, Berkeley, California. http://projecteuclid.org/euclid.bsmsp/1200514240. (Accessed 14 February 2017).

Larsen, R.K., Mamosso, C.A., 2013. Environmental Governance of Uranium Mining in Niger - A Blind Spot for Development Cooperation? Danish Institute for International Studies. http://www.jstor.org/stable/resrep13351. (Accessed 27 June 2020).

Lai, H., Levitt, B.B., 2023. Cellular and molecular effects of non-ionizing electromagnetic fields. Reviews on Environmental Health Advance online publication, . Available from: https://doi.org/10.1515/reveh-2023-0023 (Accessed 19 February 2024).

Laurent, O., Gomolka, M., Haylock, R., Blanchardon, E., Giussani, A., Atkinson, W., et al., 2016. Concerted Uranium Research in Europe (CURE): toward a collaborative project integrating dosimetry, epidemiology and radiobiology to study the effects of occupational uranium exposure. Journal of Radiological Protection 36 (2), 319–345.

Le Roux, R.R., Bezuidenhout, J., Smit, H.A.P., 2019. The influence of different types of granite on indoor radon concentrations of dwellings in the South African West Coast Peninsula. Journal of Radiation Research and Applied Sciences 12 (1), 375–382. Available from: https://doi.org/10.1080/16878507.2019.1680043.

Leuschner, A.H., van As, D., Grundling, A., Steyn, A., 1989. A survey of indoor exposure to radon in South Africa. The Clean Air Journal 7 (7). Available from: https://doi.org/10.17159/caj/1989/7/7.7292.

Leuschner, A.H. Steyn, A. Strydom, R., De Beer, G.P., 1992. Indoor radon concentrations in South African homes. In: International Congress of the International Radiation Protection Association (IRPA8), International Radiation Protection Association, May 17–22, 1992, Montreal, Quebec (Canada), vol. 2, pp. 1347–1350. ISBN: 1-55048-657-8.

Leuschner, A.H., Steyn, A., Strydom, R., de Beer, G.P. 2002. Indoor radon concentrations in South African homes. Atomic Energy Corporation of South Africa Limited. http://www.irpa.net/irpa8/cdrom/VOL.2/M2_82.PDF. (Accessed 26 July 2020).

Lewis, S., Downing, C., Hayre, C.M., 2023. Radiation protection among South African diagnostic radiographers - A mixed method study. Health Physics 124 (3), 208−216. Available from: https://doi.org/10.1097/HP.0000000000001655 (Accessed 19 February 2024).

Liberman, M., Marks, A.D., 2013. Basic Biochemistry: A Clinical Approach, 4th Edition Lippincott Williams and Wilkins, A Wolters Kluver Business.

Licciulli, S., 2023. Experts Forecast Cancer Research and Treatment Advances in 2023. American Association of Cancer Research . Available from: https://www.aacr.org/blog/2023/01/13/experts-forecast-cancer-research-and-treatment-advances-in-2023/ (Accessed 19 February 2024).

Lindsay, R., Newman, R., Speelman, W.J., 2008. A study of airborne radon levels in Paarl houses (South Africa) and associated source terms, using electret ion chambers and gamma-ray spectrometry. Applied Radiation and Isotopes 66 (11), 1611−1614. Available from: https://doi.org/10.1016/j.apradiso.2008.01.022.

Lloyd, L.G., 2013. Neonatal mortality in South Africa: how are we doing and can we do better? The South African Medical Journal 103 (8), 518−519. Available from: https://doi.org/10.7196/samj.7200.

Lubin, J.H., 1994. Radon and lung cancer risk: A joint analysis of 11 underground miners studies. NIH Report 94-3644, U.S. Department of Health and Human Services, Public Health Service, National Institutes of Health.

Lubin, J.H., Boice Jr., J.D., 1997. Lung cancer risk from residential radon: meta-analysis of eight epidemiologic studies. Journal of the National Cancer Institute (JNCI) 89 (1), 49−57. Available from: https://doi.org/10.1093/jnci/89.1.49.

Lubin, J.H., Boice Jr., J.D., Eding, C., Hornung, R.W., Howe, G.R., et al., 1995. Lung cancer in radon-exposed miners and estimation of risk from indoor exposure. Journal of the National Cancer Institute 87, 817−827. Available from: https://doi.org/10.1093/jnci/87.11.817.

Macdonald, W., Driscoll, T., Stuckey, R., Oakman, J., 2012. Occupational health and safety in Australia. Industrial Health 50 (3), 172−179. Available from: https://doi.org/10.2486/indhealth.ms1374.

McCormick, D., 2011. Chapter 8: Animal studies in carcinogen identification: the example of power frequency (50/60 Hz) magnetic fields. In: Obe, G., Jandrig, B., Marchant, G.E., Schütz, H., Wiedemann, P.M. (Eds.), Cancer Risk Evaluation: Methods and Trends. Wiley-Blackwell. Available from: https://books.google.de/books?id = 5hLZEuLth7cC&pg = PT135&lpg = PT135&dq = The + strength + of + the + epi.

Masevhe, L., Mavunda, R.D., Connell, S., 2017. A general survey of radon concentration in water from rivers in Gauteng, South Africa using a solid-state α-detector. Journal of Environment and Analytical Toxicology 7 (4). Available from: https://doi.org/10.4172/2161-0525.1000472.

Masok, F.B., Masiteng, P.L., Mavunda, R.D., Maleka, P.P., 2016. Health effects due to radionuclides content of solid minerals within Port of Richards Bay, South Africa. International Journal of Environmental Research and Public Health 13, 1180.

Masok, F.B., Masiteng, P.L., Mavunda, R.D., Maleka, P.P., Winkler, H., 2018. Measurement of radioactivity concentration in soil samples around phosphate rock storage facility in Richards Bay, South Africa. Journal of Radiation Research and Applied Sciences 11 (1), 29−36. Available from: https://doi.org/10.1016/j.jrras.2017.10.006.

Matanoski, G.M., Boice Jr., J.D., Brown, S.L., Gilbert, E.S., Puskin, J.S., O'Toole, T., 2001. Radiation exposure and cancer: case study. American Journal of Epidemiology 154 (12), S91−S98.

Marci, R., Mallozzi, M., Di Benedetto, L., Schimberni, M., Mossa, S., Soave, I., et al., 2018. Radiations and female fertility. Reproductive Biology and Endocrinology 16, 112. Available from: https://doi.org/10.1186/s12958-018-0432-0.

Martin, J.E., 2006. Physics for Radiation Protection: A Handbook, Second Edition Wiley-VCH, Weinheim.

Martinez Marignac, V.L., Mondragon, L., Favant, J.L., 2019. Sources of ionizing radiation and their biological effects: an interdisciplinary view, from the physics to cell and molecular biology. Clinical Cancer Investigation Journal 8 (4), 129–138.

Mathuthu, M., Kamunda, C., Madhuku, M., 2016. Modelling of radiological health risks from gold mine tailings in Wonderfonteinspruit Catchment Area, South Africa. International Journal of Environmental Research and Public Health 13 (6), 570. Available from: https://doi.org/10.3390/ijerph13060570.

Matshusa, K., Makgae, M., 2017. Prevention of future legacy sites in uranium mining and processing: The South African perspective. Ore Geology Reviews 86, 70–78.

Maurice, Y.T. (Ed.), 1982. Uranium in Granites. Paper 81–23. In: Proceedings of a Workshop Held in Ottawa, Ontario, November 25–26, 1980. Ministry of Supply and Services, Canada. https://inis.iaea.org/collection/NCLCollectionStore/_Public/15/020/15020917.pdf. (Accessed 11 July 2020).

Mdoe, G., 2020. Tanzania: Pros and cons of Africa's atomic age. The Exchange. https://theexchange.africa/industry-and-trade/tanzania-pros-and-cons-of-africas-atomic-age/. (Accessed 04 April 2020).

MEGJ (Ministry of the Environment, Government of Japan), 2008. Chapter 2: Internal exposure to radon and thoron through inhalation. In: Booklet to Provide Basic Information Regarding the Health Effects of Radiation, first ed., 'Radiation Exposure', Section 2.5. https://www.env.go.jp/en/chemi/rhm/basic-info/1st/02-05-08.html. (Accessed 12 April 2020).

Menzel, H.G., ICRU (International Commission on Radiation Units and Measurements), 2014. Radiation Monitoring and Dose Assessment - Key Requirements for the Implementation of the Principles of Limitation and Optimization in Occupational Exposure: International Conference on Occupational Radiation Protection, IAEA, December 2014. www-ns.iaea.org/tech-areas/communication-networks/orpnet/documents/cn223/2-menzel-radiation-monitoring-and-dose-assessment.pdf. (Accessed 01 August 2020).

Miles, J.C.H., 2001. Temporal variation of radon levels in houses and implications for radon measurement strategies. Radiation Protection Dosimetry 93 (4), 369–375.

Miles, J., 2004. Methods of Radon Measurement and Devices. In: Proceedings of the 4th European Conference on Protection Against Radon at Home and at Work. Conference Programme and Session Presentations. Czech Republic, p. 377. https://inis.iaea.org/collection/NCLCollectionStore/_Public/36/010/36010924.pdf?r = 1. (Accessed 12 July 2020).

Moges, N., Anley, D.T., Zemene, M.A., Adella, G.A., Solomon, Y., Bantie, B., et al., 2023. Congenital anomalies and risk factors in Africa: a systematic review and meta-analysis. BMJ Paediatrics Open 7 (1), e002022. Available from: https://doi.org/10.1136/bmjpo-2023-002022 (Accessed 19 February 2024).

Moifo, B., Tene, U., Tapouh, J.R.M., Ngano, O.S., Youta, J.T., Simo, A., et al., 2017. Knowledge on irradiation, medical imaging prescriptions, and clinical imaging referral guidelines among physicians in a Sub-Saharan African Country (Cameroon). Radiology Research and Practice 7, 1245236. Available from: https://doi.org/10.1155/2017/1245236.

Monken-Fernandes, H., Santamaria Recio, M., Forsstrom, H., Carson, P.M., 2013. International cooperation and support in environmental remediation - is there any room for improvement? Journal of Environmental Radioactivity 119, 13e20.

Morgan, K.Z., 1978. Cancer and low-level ionizing radiation. The Bulletin of the Atomic Scientists 34 (7), 30–41.

Mothersill, C., Rusin, A., Seymour, C., 2019. Towards a new concept of low dose. Health Physics 117 (3), 330–336. Available from: https://doi.org/10.1097/HP.0000000000001074.

Moshupya, P., Abiye, T., Mouri, H., Levin, M., Strauss, M., Strydom, R., 2019. Assessment of radon concentration and impact on human health in a region dominated by abandoned gold mine tailings dams: a case from the West Rand Region, South Africa. Geosciences 9, 466. Available from: https://doi.org/10.3390/geosciences9110466.

Mould, R.F., 1998. The discovery of radium in 1898 by Maria Sklodowska-Curie (1867-1934) and Pierre Curie (1859-1906) with commentary on their life and times. The British Journal of Radiology 71 (852), 1229–1254. Available from: https://doi.org/10.1259/bjr.71.852.10318996.

Mudd, G.M., 2008. Radon sources and impacts: a review of mining and non-mining issues. Reviews in Environmental Science and Biotechnology 7, 325−353. Available from: https://doi.org/10.1007/s11157-008-9141-z.

Mustapha, A.O., Narayana, D.G.S., Patel, J.P., Otwoma, D., 1997. Natural radioactivity in some building materials in Kenya and the contributions to the indoor external doses. Radiation Protection Dosimetry 71 (1), 65−69.

Myatt, T.A., Allen, J.G., Minegishi, T., Mccarthy, W.B., Stewart, J.H., Macintosh, D.L., et al., 2010. Assessing exposure to granite countertops-part 1: Radiation. Journal of Exposure Science and Environmental Epidemiology 20 (3), 273−280. Available from: https://doi.org/10.1038/jes.2009.44.

NCI (National Cancer Institute of the US), 2019. Radiation Therapy to Treat Cancer https://www.cancer.gov/about-cancer/treatment/types/radiation-therapy. (Accessed 03 March 2020).

NCRP (National Council on Radiation Protection and Measurements), 1977. Report No. 54 Medical radiation exposure of pregnant and potentially pregnant women. Bethesda, Maryland. https://ncrponline.org/shop/reports/report-no-054-medical-radiation-exposure-of-pregnant-and-potentially-pregnant-women-1977/. (Accessed 26 July 2020).

Nemangwele, F., 2005. Radon in the Cango Caves (M.Sc. thesis). University of the Western Cape, Cape Town, South Africa. https://core.ac.uk/download/pdf/58912552.pdf. (Accessd 24 May 2020).

Newman, R., Lindsay, R., Botha, R., Bezuidenhout, J., Maleka, P., Ocwelwang, A., et al., 2019. Design of a National Indoor-Radon Survey for South Africa: Overview of Existing Indoor-Radon Concentration Data and Proposed Features of the Mapping Program. https://app.oxfordabstracts.com/events/543/program-app/submission/125565. (Accessed 26 July 2020).

Ngetu, L., Marais, W., Rose, A., Rae, I.D., 2019. Ophthalmic manifestations of ionising radiation among interventionalists. African Vision and Eye Health 78 (1).

Njini, F., 2007. Desalination Plan for Namibia's Third Uranium Mine, December 14. http://www.miningweekly.co.za.

Norris, M., 1996. Dividing the Indivisible: The fissured story of the Manhattan Project. Cultural Critique 35, 5−38. Available from: https://doi.org/10.2307/1354570.

NRC [National Research Council (US)] Committee on Pesticides in the Diets of Infants and Children, 1993. Pesticides in the Diets of Infants and Children, vol. 3, Perinatal and Pediatric Toxicity. National Academies Press (US), Washington (DC).

NRC [(National Research Council (US)] Committee on Health Risks of Exposure to Radon (BEIR VI), 1999. Health Effects of Exposure to Radon: BEIR VI. National Academies Press (US), Washington (DC). https://www.ncbi.nlm.nih.gov/books/NBK233259/. (Accessed 02 April 2020). Appendix E, Exposures of Miners to Radon Progeny. https://www.ncbi.nlm.nih.gov/books/NBK233269 (Accessed 09 July 2020).

Nyabanda, R., AFROSAFE, 2015. Africa-wide campaign on radiation safety. https://inis.iaea.org/collection/NCLCollectionStore/_Public/47/035/47035637.pdf. (Accessed 05 August 2020).

Nyanda, P., 2014. Uranium mining and milling in Tanzania. In: RLS. Uranium Mining Impact on Health and Environment. http://www.cesopetz.org/wp-content/uploads/2013/06/Uranium_Mining_Impact.pdf. (Accessed 06 April 2020).

Obasi, I.A., Ogwah, C., Selemo, A.O.I., Afiukwa, J.N., Chukwu, C.G., 2020. *In situ* measurement of radionuclide concentrations (^{238}U, ^{40}K, ^{232}Th) in middle Cretaceous rocks in Abakaliki-Ishiagu areas, southeastern Nigeria. Arabian Journal of Geosciences 13, 374. Available from: https://doi.org/10.1007/s12517-020-05360-4.

OECD (Organisation for Economic Co-operation and Development), 2019. Australia Economic Snapshot. http://www.oecd.org/economy/australia-economic-snapshot/. (Accessed 03 June 2020).

OECD/NEA (Organisation for Economic Co-operation and Development/Nuclear Energy Agency), 2014. Managing Environmental and Health Impacts of Uranium Mining. https://inis.iaea.org/collection/NCLCollectionStore/_Public/45/073/45073527.pdf. (Accessed 04 June 2020).

Okar, H.B., Ma, J., Pinak, M., 2018. UMEX Project: an IAEA Survey of Global Uranium Mining and Processing - Occupational Doses. International Symposium on Uranium Raw Material for the Nuclear Fuel Cycle: Exploration, Mining, Production, Supply and Demand, Economics and Environmental Issues (URAM-2018). Report No. IAEA-CN-261. Vienna. Book of Abstracts and Extended Abstracts. https://inis.iaea.org/search/search.aspx?orig_q = RN:49097470. (Accessed 19 May 2020).

Otterson, M.F., 2007. Effects of radiation upon gastrointestinal motility. World Journal of Gastroenterology 13 (19), 2684−2692. Available from: https://doi.org/10.3748/wjg.v13.i19.2684.

Otton, J.K., 1992. The Geology of Radon. US Department of the Interior/US Geological Survey, Open Report 30. https://pubs.usgs.gov/gip/7000018/report.pdf. (Accessed 07 August 2020).

Palfi, A.G., Wartha, R., Niedermayr, G., Jahn, S., 2001. Das Boltwoodite-vorkommen von Goanikontes in der Namib. In: Jahn, S., Medenbach, O., Niedermayr, G., Schneider, G. (Eds.), Namibia Zauberwelt edler stein und Kristalle. Reiner Bode, Borkum, Germany, pp. 126−135.

Paquet, F., 2019. The new ICRP biokinetic and dosimetric models. BIO Web of Conferences, vol. 14, 02001. In: The 12th International Conference on the Health Effects of Incorporated Radionuclides (HEIR 2018). Fontenay-aux-Roses, France, 8−11 October, 2018. https://www.bio-conferences.org/articles/bioconf/full_html/2019/03/bioconf_heir2018_02001/bioconf_heir2018_02001.html. (Accessed 26 July 2020).

Paquet, F., Bailey, M.R., Leggett, R.W., et al., 2019. Occupational Intakes of Radionuclides, Part 4. ICRP Publication 141. https://journals.sagepub.com/doi/full/10.1177/0146645319834139. (Accessed 08 August 2020). A correction has been published: https://journals.sagepub.com/doi/full/10.1177/0146645320916587. (Accessed 08 August 2020).

Pavlidis, N.A., 2002. Coexistence of pregnancy and malignancy. Oncologist 7, 279−287.

Pernot, E., Hall, J., Baatout, S., Benotmane, M.A., Blanchardon, E., Bouffler, S., et al., 2012. Ionising radiation biomarkers for potential use in epidemiological studies. Mutation Research/Reviews in Mutation Research 751 (2), 258−286. Available from: https://doi.org/10.1016/j.mrrev.2012.05.003.

Postar, S., 2017. The half-lives of African uranium: a historical review. The Extractive Industries and Society 4 (2), 398−409.

Power, S.P., Moloney, F., Twomey, M., James, K., O'Connor, O.J., Maher, M.M., 2016. Computed tomography and patient risk: facts, perceptions and uncertainties. World Journal of Radiology 8 (12), 902−915. Available from: https://doi.org/10.4329/wjr.v8.i12.902.

Preston, D.L., Cullings, H., Suyama, A., Funamoto, S., Nishi, N., Soda, M., et al., 2008. Solid cancer incidence in atomic bomb survivors exposed *in utero* or as young children. Journal of the National Cancer Institute 100, 428−436.

Przylibski, T.A., 2016. Radon: a radioactive therapeutic element. In: Gillmore, G.K., Perrier, F.E., Crockett, R. G.M. (Eds.), Radon, Health and Natural Hazards. Geological Society, LondonSpecial Publications, 451. Available from: https://doi.org/10.1144/SP451.7.

Pule, J., Speelman, W., 2016. South African perspective for radon in dwellings and the anticipated regulatory control measures. In: NNR Regulatory Information Conference, October 05−07, 2016. Pretoria, South Africa. http://www.nnr.co.za/wp-content/uploads/2016/10/4.-Pres-SA-Perspective-for-Radon.pdf. (Accessed 24 May 2020).

Quindos, L.S., Fernandez, P.L., Soto, J., 1991. National survey on indoor radon in Spain. Environment International 17, 449−453.

Raabe, O.G., 2012. Chapter 15: Ionizing radiation carcinogenesis. In: Nenoi, M. (Ed.), Current Topics in Ionizing Radiation Research. InTech, Rijeka, Croatia, pp. 299−348.

Rabie, M.A., 1973. South African legislation for protection against ionizing radiation. The Comparative and International Law Journal of Southern Africa 6 (3), 403−412.

Rabinowitz, D.A., Pretorius, E.S., 2005. Postgraduate radiology in sub-Saharan Africa: A review of currenteducational resources. Radiologic Education 12 (2), 224−231. Available from: https://doi.org/10.1016/j.acra.2004.11.014.

Rajan, G., Izewska, J., 2006. Chapter 4: Radiation Monitoring Instruments. IAEA Review of Radiology Oncology Physics: A Handbook for Teachers and Students, pp. 101−121. http://www-naweb.iaea.org/nahu/DMRP/documents/Chapter4.pdf. (Accessed 14 June 2020).

Ribeiro, A., Husson, N., Drev, I., Murray, K., May, K., Thurston, J., et al., 2020. Ionising radiation exposure from medical imaging − a review of patient's (un) awareness. Radiography 26 (2), e25−e30. Available from: https://doi.org/10.1016/j.radi.2019.10.002.

Riudavets, M., Garcia de Herreros, M., Besse, B., Mezquita, L., 2022. Radon and lung cancer: current trends and future perspectives. Cancers 14 (13), 3142. Available from: https://doi.org/10.3390/cancers14133142 (Accessed 19 February 2024).

Robinson, P., Hector, A., Luis, J., Benavides, D., Hancock, D., 1979. Uranium Mining and Milling: A Primer. The Workbook, vol. 4. Southwest Research and Information Center, Albuquerque, New Mexico, pp. 6−7. http://www.sric.org/uranium/docs/1979_SRIC_URANIUM_PRIMER.pdf. (Accessed 10 August 2020).

Rolle, R., 2004. A comparison of 'radon' measurement devices. In: Proceedings of the Fourth European Conference on Protection Against Radon at Home and at Work Conference Programme and Session Presentations, Czech Republic, p. 377. https://inis.iaea.org/search/search.aspx?orig_q = RN:36010929. (Accessed 12 July 2020).

Rose, A., Rae, W.I., 2017. Perceptions of radiation safety training among interventionalists in South Africa. Cardiovascular Journal of Africa 28 (3), 196−200. Available from: https://doi.org/10.5830/CVJA-2017-028.

Rosenman, K.D., 2020. Pneumoconioses. British Medical Journal Best Practice .

Rössing Uranium, 2020. History and location of Rössing. https://www.rossing.com/images/page_pics/history_page/map.jpg. (Accessed 12 April 2020).

Saint-Pierre, S., 2010. Towards Greater Harmonization of the System of Radiological Protection - Views from the Global Nuclear Industry. IAEA: N. p. Web. https://www.osti.gov/etdeweb/servlets/purl/21190229. (Accessed 07 July 2020).

Saada, M., Sanchez-Jimenez, E., Roguin, A., 2023. Risk of ionizing radiation in pregnancy: just a myth or a real concern?, EP Europace. EP Europace 25 (2), 270−276. Available from: https://doi.org/10.1093/europace/euac158 (Accessed 19 February 2024).

SAHPRA, (South African Health Products Regulatory Authority), 2022. Ionizing Radiation Dose Limits and Annual Limits on Intake of Radioactive Material. Pretoria . Available from: https://www.sahpra.org.za/wp-content/uploads/2022/09/SAHPGL-RDN-XR-15_v1-Guideline-for-Ionising-Radiation-Dose-Limits-and-Annual-Limits-on-Intake-of-Radioactive-Material.pdf (Accessed 19 February 2024).

Sansare, K., Khanna, V., Karjodkar, F., 2011. Early victims of X-rays: a tribute and current perception. Dentomaxillofacial Radiology 40 (2), 123−125. Available from: https://doi.org/10.1259/dmfr/73488299.

SCENIHR (The Scientific Committee on Emerging and Newly Identified Health Risks), 2012. Health effects of security scanners for passenger screening. The European Union, Brussels. https://ec.europa.eu/health/scientific_committees/emerging/docs/scenihr_o_036.pdf. (Accessed 12 July 2020).

Scheele, F., Wilde-Ramsing, J., de Haan, E. (for WISE and SOMO, Amsterdam), 2011. Uranium from Africa: mitigation of uranium mining impacts on society and environment by industry and governments. ISBN: 978-90-71284-82-3. https://www.somo.nl/wp-content/uploads/2011/07/Uranium-from-Africa.pdf. (Accessed 01 April 2020).

Schneider, M., Froggatt, A., 2019. The Current Status of the World Nuclear Industry. 2018 Status Report, (WNISR2018). https://link.springer.com/content/pdf/10.1007%2F978-3-658-25987-7_3.pdf. (Accessed 29 June 2020).

Schonfeldt, S.J., Winde, F., Albrecht, C., Kielkowski, D., Liefferink, M., Patel, M., et al., 2014. Health effects in populations living around the uraniferous gold mine tailings in South Africa: gaps and opportunities for research. Cancer Epidemiol. 38, 628e632.

Schubert, G., Berka, R., Katzlberger, C., Motschka, K., Denner, M., Grath, J., et al., 2018. Radionuclides in groundwater, rocks and stream sediments in Austria - results from a recent survey. In: Gillmore, G.K., Perrier, F.E., Crockett, R.G.M. (Eds.), Radon, Health and Natural Hazards.

Schüz, J., Deltour, I., Krestinina, L.Y., Tsareva, Y.V., Tolstykh, E., Sokolnikov, M.E., et al., 2017. *In utero* exposure to radiation and haematological malignancies: pooled analysis of Southern Urals cohorts. British Journal of Cancer 116, 126−133. Available from: https://doi.org/10.1038/bjc.2016.373.

Shadad, A.K., Sullivan, F.J., Martin, J.D., Egan, L.J., 2013. Gastrointestinal radiation injury: symptoms, risk factors and mechanisms. World Journal of Gastroenterology 19 (2), 185−198. Available from: https://doi.org/10.3748/wjg.v19.i2.185.

Shah, D.J., Sachs, R.K., Wilson, D.J., 2012. Radiation-induced cancer: a modern view. The British Journal of Radiology 85 (1020), e1166−e1173. Available from: https://doi.org/10.1259/bjr/25026140.

Shaw, P., Duncan, A., Vouyouka, A., Ozsvath, K., 2011. Radiation exposure and pregnancy. Journal of Vascular Surgery 53 (15S), 28S−34S.

Sheen, S., Lee, K.S., Chung, W.Y., Nam, S., Kang, D.R., 2016. An updated review of case-control studies of lung cancer and indoor radon - is indoor radon the risk factor for lung cancer? Annals of Occupational and Environmental Medicine 28, 9. Available from: https://doi.org/10.1186/s40557-016-0094-3.

Sherer, M.A.S., Visconti, P.J., Ritenour, E.R., Hayes, K.W., 2014. Radiation Protection in Medical Radiography, 7th Edition Elsevier/Mosby.

Shindondola-Mote, H., 2008. Uranium mining in Namibia: The mystery behind 'low level radiation. https://www.issuelab.org/resources/20335/20335.pdf. (Accessed 07 April 2020).

Simo, A., Ateba, J.F.B., Manyol, E.N., Ndah, T.N., Ndontchueng, M.M., et al., 2018. Medical radiation exposure of pregnant patients: case study of the gynaeco-obstetric and paediatric hospital of Douala, Cameroon. Radiological and Diagnostic Imaging 2. Available from: https://doi.org/10.15761/RDI.1000126.

Skrable, K.W., Chabot, G.E., French, C.S., Labone, T.R., 1994. Use of multicompartment models of retention for internally deposited radionuclides. In: Raabe, O.G. (Ed.), Internal Radiation Dosimetry. Medical Physics Publishing, Madison, WI, pp. 271−354.

Smith, D.K., 1984. Uranium mineralogy. In: De Vivo, B., Ippolito, F., Capaldi, G., Simpson, P.R. (Eds.), Uranium Geochemistry, Mineralogy, Geology, Exploration and Resources. Springer, Dordrecht.

Socol, Y., Dobrzyński, L., 2015. Atomic Bomb Survivors Life-Span Study: Insufficient Statistical Power to Select Radiation Carcinogenesis Model. Dose-Response: A Publication of International Hormesis Society, 13(1), dose-response.14-034.Socol. https://doi.org/10.2203/dose-response.14-034.Socol.

Sreetharan, S., Thome, C., Tharmalingam, S., Jones, D.E., Kulesza, A.V., Khaper, N., et al., 2017. Ionizing radiation exposure during pregnancy: effects on postnatal development and life. Radiation Research 187 (6), 647−658. Available from: https://doi.org/10.1667/RR14657.1.

Stassen, W., 2015. Gauteng's mine dumps brimming with radioactive uranium. Environmental Health. https://www.health-e.org.za/2015/10/15/gautengs-mine-dumps-brimming-with-radioactive-uranium/. (Accessed 04 January 2017).

Taylor, D.M., 2009. Human respiratory tract model for radiological protection. Journal of Radiological Protection 16 (1). Available from: https://doi.org/10.1088/0952-4746/16/1/013.

Tefera, E., Qureshi, S.A., Gezmu, A.M., Mazhani, L., 2020. Radiation protection knowledge and practices in interventional cardiologists practicing in Africa: a cross sectional survey. Journal of Radiological Protection 40, 311. Available from: https://doi.org/10.1088/1361-6498/ab5840.

Thambura, M.J., Conradie, D.R., Vinette, I., 2019. Nurses' knowledge of ionizing radiation in Northern Gauteng State Hospitals in South Africa. Journal of Radiology Nursing 38 (1), 56−60.

Toossi, M.T.B., Malekzadeh, M., 2012. Radiation dose to newborns in neonatal intensive care units. Iranian Journal of Radiology 9 (3), 145−149. Available from: https://doi.org/10.5812/iranjradiol.8065.

Trauernicht, C., Hasford, F., Khelassi-Toutaoui, N., Bentouhami, I., Knoll, P., Tsapaki, V., 2022. Medical physics services in radiology and nuclear medicine in Africa: challenges and opportunities identified through workforce and infrastructure surveys. Health Technolology 12, 729−737. Available from: https://doi.org/10.1007/s12553-022-00663-w (Accessed 19 February 2024).

Tulchinsky, T.H., Varavikova, E.A., 2014. Environmental and occupational health. In: Tulchinsky, T.H., Varavikova, E.A. (Eds.), The New Public Health, 3rd Edition Elsevier, Academic Press, San Diego, pp. 471–532.

Turhan, Ş., 2012. Estimation of possible radiological hazards from natural radioactivity in commercially-utilized ornamental and countertops granite tiles. Annals of Nuclear Energy 44, 34–39. Available from: https://doi.org/10.1016/j.anucene.2012.01.018.

UNDP (United Nations Development Programme), 2019. Human Development Report 2019: Beyond income, beyond averages, beyond today: Inequalities in human development in the 21st century. UNDP, New York. http://hdr.undp.org/sites/default/files/hdr2019.pdf. (Accessed 27 June 2020).

UNGA (United Nations General Assembly), 2013. Report of the United Nations Scientific Committee on the Effects of Atomic Radiation, Sixtieth session (May 27–31, 2013), General Assembly Official Records Sixty-Eighth Session. Supplement No. 46. https://www.unscear.org/docs/GAreports/A-68-46_e_V1385727.pdf. (Accessed 19 May 2020).

UNSCEAR (United Nations Scientific Committee on the Effects of Atomic Radiation), 2000. Sources and Effects of Ionizing Radiation. https://www.unscear.org/docs/publications/2000/UNSCEAR_2000_Report_Vol.I.pdf. (Accessed 27 July 2020).

Urban, P., Skubacz, K., 2015. New dosimetry system based on the thermoluminescence method for evaluation of ionising radiation doses to workers of the health centers. Wiad Lek 68 (1), 71–78.

US DHEW (United States Department of Health, Education and Welfare), 1967. A Strategy for a Liveable Environment: A Report to the Secretary of Health, Education and Welfare, Washington DC, 90 p. https://books.google.de/books?id = LxdSAAAAMAAJ&printsec = frontcover&hl = de&source = gbs_ge_summary_r&cad = 0#v = onepage&q&f = false. (Accessed 10 July 2020).

US EPA (United States Environmental Protection Agency), 1999–2012. Radiation emission from granite table tops. http://njmoldinspection.com/granite_table.html. (Accessed 12 April 2020).

US EPA (United States Environmental Protection Agency), 2012. A Citizen's Guide to Radon. Document EPA402/K-12/002. https://www.epa.gov/sites/production/files/2016-02/documents/2012_a_citizens_guide_to_radon.pdf. (Accessed 09 July 2020).

US EPA (US Environmental Protection Agency), 2017. Radiation Protection: Radiation Basics. https://www.epa.gov/radiation/radiation-basics. (Accessed 27 July 2020).

US EPA (United States Environmental Protection Agency), 2018. Significant Discoveries and the History of Radiation Protection. RadTown Radiation Protection Activity Set US EPA 402-B-19-015. https://www.epa.gov/sites/production/files/2018-12/documents/significant_discoveries_history_radiation_protection-worksheet_rp_1.pdf. (Accessed 04 May 2020).

US NRC (United States Nuclear Regultory Commission), 2003. Reactor Concepts Manual: Biological Effects of Radiation. US NRC Technical Training Center, vol. 0603. https://www.nrc.gov/reading-rm/basic-ref/students/for-educators/09.pdf. (Accessed 27 July 2020).

US NRC, 2017. Sources of Radiation: Natural Background Sources. https://www.nrc.gov/about-nrc/radiation/around-us/sources/nat-bg-sources.html. (Accessed 12 August 2020).

Utembe, W., Faustman., E.M., Matatiele, P., 2015. Hazards identified and the need for health risk assessment in the South African mining industry. Human & Experimental Toxicology 34 (12), 1212–1221. Available from: https://doi.org/10.1177/0960327115600370.

Vaiserman, A., Koliada, A., Zabuga, O., Socol, Y., 2018. Health impacts of low-dose ionizing radiation: current scientific debates and regulatory issues. Dose-response: A Publication of International Hormesis Society 16 (3), . Available from: https://doi.org/10.1177/1559325818796331 1559325818796331.

Valentin, J., 2006. The scope of radiological protection regulations. Annals of the International Commission on Radiological Protection (ICRP). Elsevier. Available from: https://www.icrp.org/docs/Scope_of_rad_prot_draft_02_258_05v06.pdf (Accessed 04 July 2020).

Valentina, B., 2015. Areva's Uranium Mines, Niger. EJOLT Factsheet No. 30, 4 p. http://www.envjustice.org/wp-content/uploads/2015/07/FS-301.pdf. (Accessed 27 June 2020).

Van Schalkwyk, A., 2005. A radiation monitoring program in a south African gold mine Mini dissertation submitted in partial fulfilment of the requirements for the degree MSc in Physiology at the North-West University, South Africa. C:/Users/admin/Downloads/vanschalkwyk_adelle.pdf. (Accessed 03 March 2020).

Veit, S., Srebotnjak, T., 2010. Potential Use of Radioactively Contaminated Mining Materials in the Construction of Residential Homes From Open Pit Uranium Mines in Gabon and Niger. Published on Behalf of the Directorate-General for External Policies of the Directorate of the European Parliament, Policy Department. Report: EXPO/B/DEVE/FWC/2009-01/Lot05-07 November/ 2010 PE 433.662. https://www.ecologic.eu/sites/files/project/2013/Potential%20use%20of%20radioactively%20contaminated%20mining%20materials%20in%20Gabon%20and%20Niger_published.pdf. (Accessed 03 April 2020).

Verreet, T., Verslegers, M., Quintens, R., Baatout, S., Benotmane, M.A., 2016. Current evidence for developmental, structural, and functional brain defects following prenatal radiation exposure. Neural Plasticity 2016, 1243527. Available from: https://doi.org/10.1155/2016/1243527.

Waggitt, P., 2014. Uranium mine regulation and remediation in Australia's Northern Territory. https://www-pub.iaea.org/iaeameetings/cn216pn/Tuesday/Session4/045-Waggitt.pdf. (Accessed 21 May 2020).

Welsh, R.S., 1979. Mining Law. The 1979 Annual Survey of South African Law. Johannesburg, South Africa. HEINONLINE, p. 258.

Wendel, G., 1998. Radioactivity in mines and mine water - sources and mechanisms. The South African Institute of Mining and Metallurgy. https://www.saimm.co.za/Journal/v098n02p087.pdf. (Accessed 09 April 2020).

WHO (World Health Organisation), 2001. Chapter 8.3: Radon. Air Quality Guidelines, Second Edition WHO Regional Office for Europe, Copenhagen, Denmark. Available from: http://www.euro.who.int/__data/assets/pdf_file/0005/123089/AQG2ndEd_8_3Radon.pdf?ua = 1.

WHO (World Health Organisation), 2009a. Children's Health and the Environment: WHO Training Package for the Health Sector World Health Organization. https://www.who.int/ceh/capacity/radiation.pdf. (Accessed 25 July 2020).

WHO (World Health Organisation), 2009b. Handbook on Indoor Radon: A Public Health Perspective. NCBI Resources, 94 p. https://www.ncbi.nlm.nih.gov/books/NBK143222/. (Accessed 29 May 2020).

WHO (World Health Organisation), 2016. Ionising radiation, health effects and protective measures. https://www.who.int/news-room/fact-sheets/detail/ionizing-radiation-health-effects-and-protective-measures. (Accessed 22 April 2020).

WHO, (World Health Organization)., 2023. Ionizing radiation and health effects. WHO, Geneva . Available from: https://www.who.int/news-room/fact-sheets/detail/ionizing-radiation-and-health-effects (Accessed 19 February 2024).

Widner, T.E., Flack, S.M., 2010. Characterization of the world's first nuclear explosion, the Trinity test, as a source of public radiation exposure. Health Physics 98 (3), 480−497. Available from: https://doi.org/10.1097/HP.0b013e3181c18168.

Wikipedia, 2020a. Alpha particle. https://en.wikipedia.org/wiki/Alpha_particle. (Accessed 24 July 2020).

Wikipedia, 2020b. Beta decay. https://en.wikipedia.org/wiki/Beta_decay. (Accessed 09 May 2020).

Wikipedia, 2020c. Gamma ray. https://en.wikipedia.org/wiki/Gamma_ray. (Accessed 11 May 2020).

Williams, P.A., Fletcher, S., 2010. Health effects of prenatal radiation exposure. American Family Physician 82 (5), 488−493.

Winde, F., Sandham, L.A., 2004. Uranium pollution of South African streams: an overview of the situation in gold mining areas of the Witwatersrand. GeoJournal 61 (3), 131−149. Available from: https://doi.org/10.1007/s10708-004-2867-4.

Winde, F., Brugge, D., Nidecker, A., Ruegg, U., 2017. Uranium from Africa: an overview on past and current mining activities: re-appraising associated risks and chances in a global context. Journal of African Earth Sciences 129, 1–20. Available from: https://doi.org/10.1016/j.jafrearsci.2016.12.004.

Winde, F., Geipel, G., Espina, C., Schüz, J., 2019. Human exposure to uranium in South African gold mining areas using barber-based hair sampling. PLoS One 14 (6), e0219059. Available from: https://doi.org/10.1371/journal.pone.0219059.

WNA (World NuclearAssociation), 2016. The Nuclear Fuel Cycle Overview. https://www.world-nuclear.org/information-library/nuclear-fuel-cycle/introduction/nuclear-fuel-cycle-. (Accessed 18 July 2020).

WNA (World NuclearAssociation), 2020a. Nuclear Radiation and Health Effects. https://www.world-nuclear.org/information-library/safety-and-security/radiation-and-health/nuclear-radiation-and-health-effects.aspx. (Accessed 07 May 2020).

WNA (World Nuclear Association), 2020b. Naturally-Occurring Radioactive Materials (NORM). https://www.world-nuclear.org/information-library/safety-and-security/radiation-and-health/naturally-occurring-radioactive-materials-norm.aspx. (Accessed 30 June 2020).

WNA (World Nuclear Association), 2020c. Plans for New Reactors Worldwide. https://www.world-nuclear.org/information-library/current-and-future-generation/plans-for-new-reactors-worldwide.aspx. (Accessed 29 June 2020).

WNA (World Nuclear Association), 2020d. World Uranium Mining Production. https://www.world-nuclear.org/information-library/nuclear-fuel-cycle/mining-of-uranium/world-uranium-mining-production.aspx. (Accessed 26 June 2020).

WNA (World Nuclear Association). 2020e. Supply of Uranium. https://www.world-nuclear.org/information-library/nuclear-fuel-cycle/uranium-resources/supply-of-uranium.aspx. (Accessed 29 June 2020).

WNA (World Nuclear Association), 2020f. Geology of Uranium Deposits. https://www.world-nuclear.org/information-library/nuclear-fuel-cycle/uranium-resources/geology-of-uranium-deposits.aspx. (Accessed 11 June 2020).

WNA (World Nuclear Association), 2020g. Uranium Mining Overview. https://www.world-nuclear.org/information-library/nuclear-fuel-cycle/mining-of-uranium/uranium-mining-overview.aspx. (Accessed 02 August 2020).

WNA (World Nuclear Association), 2020i. Uranium in Namibia. https://www.world-nuclear.org/information-library/country-profiles/countries-g-n/namibia.aspx. (Accessed 09 April 2020).

WNA (World Nuclear Association), 2020j. Uranium in Niger. https://www.world-nuclear.org/information-library/country-profiles/countries-g-n/niger.aspx. (Accessed 09 April 2020).

WNA (World Nuclear Association), 2020k. Nuclear Power in South Africa. https://www.world-nuclear.org/information-library/country-profiles/countries-o-s/south-africa.aspx. (Accessed 09 April 2020).

World Atlas, 2020. The economy of Canada. https://www.worldatlas.com/articles/important-facts-related-to-the-economy-of-canada.html. (Accessed 03 June 2020).

WUP (Wise Uranium Project), 2016. Uranium Radiation Properties. https://www.wise-uranium.org/rup.html. (Accessed 05 June 2020).

Wymer, D.G., 2001. Nuclear Regulation of South African Mines: An Industry perspective. In: IAEA - Impact of New Environmental and Safety Regulations on Uranium Exploration, Mining, Milling and Management of Its Waste. Report IAEA-TECDOC-1244. Proceedings of a Technical Committee Meeting held in Vienna, September 14–17, 1998, pp. 73–88. https://www-pub.iaea.org/MTCD/publications/PDF/te_1244_prn.pdf. (Accessed 04 April 2020).

Yang, N., Xiao, W., Song, X., Wang, W., Dong, X., 2020. Recent advances in tumor microenvironment hydrogen peroxide-responsive materials for cancer photodynamic therapy. Nano-Micro Letters 12, 15. Available from: https://doi.org/10.1007/s40820-019-0347-0.

Ye, G., 2018. Thermodynamic and structural investigations on the interactions between actinides and phosphonate-based ligands. Radiochemistry. Université Paris-Saclay, 2018. English. ffNNT: 2018SACLS286ff.

Yong, S., Wei, Z., 1995. The correlation between radon variation and Earth solid tidchange in rock-groundwater system- the mechanical foundation for using radon change to predict earthquake. Journal of Earthquake Prediction Research 4, 423−430.

Yoon, I., Slesinger, T.L., 2020. Radiation exposure in pregnancy. StatPearls [Internet]. StatPearls Publishing, Treasure Island (FL). Available from: https://www.ncbi.nlm.nih.gov/books/NBK551690/.

Young, W., 2003. Atomic energy: from "public" to "private" power - the US, UK and Japan in comparative perspective. Annales Historiques de l'Électricité 1 (1), 133−153. Available from: https://doi.org/10.3917/ahe.001.0133.

Zdrojewicz, Z., Strzelczyk, J.J., 2006. Radon treatment controversy. Dose-Response: A Publication of International Hormesis Society 4 (2), 106−118. Available from: https://doi.org/10.2203/dose-response.05-025.Zdrojewicz.

Zhou, Q., Zhao, G., Xiao, D., Qui, S., Lei, Q., Kearfott, K.J., 2018. Prediction of radon removal efficiency for a flow-through activated charcoal system and radon mitigation characteristics. Radiation Measurements 119, 112−120.

Zhu, K., Devine, A., Prince, R.L., 2009. The effects of high potassium consumption on bone mineral density in a prospective cohort study of elderly postmenopausal women. Osteoporosis International 20 (2), 335−340. Available from: https://doi.org/10.1007/s00198-008-0666-3.

Zielinsky, J.M., 2014. Mapping of residential Radon in the World. https://inis.iaea.org/collection/NCLCollectionStore/_Public/45/079/45079288.pdf. (Accessed 24 May 2020).

Zielinsky, J.M., Carr, Z., Krewski, D., Repacholi, M., 2006. World Health Organization's International Radon Project. Journal of Toxicology and Environmental Health, Part A 69 (7 - 8), 759−769. Available from: https://doi.org/10.1080/15287390500261299.

Further readings

Abdel-Wahab, M., Bourque, J.-M., Pynda, Y., Iżewska, J., Van der Merwe, D., Zubizarreta, E. et al., 2003. Status of radiotherapy resources in Afrsica: An International Atomic Energy Agency analysis. Cancer Control in Africa 4. http://globalrt.org/wp-content/uploads/2014/09/adal-wahab-et-al-status-of-RT-in-africa-Lancet-oncol-2013.pdf. (Accessed 08 April 2020).

Abdel-Wahab, M., Rosenblatt, E., Holmberg, O., Meghzifene, A., 2011. Safety in radiation oncology: The role of international initiatives by the International Atomic Energy Agency. The Lancet Oncology 14 (4), e168−e175. Available from: https://doi.org/10.1016/S1470-2045(12)70532-6.

Akpanowo, M.A., Umaru, I., Iyakwari, S., 2019. Assessment of radiological risk from the soils of artisanal mining areas of Anka, North West Nigeria. African Journal of Environmental Science and Technology 13 (8), 303−309.

Andrews, A.H., Humphreys Jr., R.L., DeMartini, E.E., Nichols, R.S., Brodziak, J., 2011. Bomb Radiocarbon and Lead-Radium Dating of Opakapaka (Pristipomoides filamentosus). Affiliation: NOAA Fisheries; Administrative Report H-11-07, 58 p. + Appendices.

ATSD/US DHHS (Agency for Toxic Substances and Disease Registry)/(United States Department of Health and Human Services), 2013. Toxic Substances Portal: Toxicological Profile for Uranium. CAS#: 7440-61-1. https://www.atsdr.cdc.gov/ToxProfiles/tp150.pdf. (Accessed 18 June 2020).

ATSDR/US DHHS (Agency for Toxic Substances and Disease Registry)/(United States Department of Health and Human Services), 1998. Toxcological Profile for Ionizing Radiation. https://books.google.de/books?id = JUDvcSRsUyUC&pg = PA175&lpg = PA175&dq = The + dosimetry + of + the + mastoids + should + - be + examined + in + order + to + calculate + the + risk&source = bl&ots = lRF6EDPYfr&sig = ACfU3U0x-XEwsqkYCSPpeiskiCZvYS4_RmQ&hl = en&sa = X&ved = 2ahUKEwjB5c3XwY3pAhWD-6QKHVuEB2sQ6 AEwAHoECAYQAQ#v = onepage&q = The%20dosimetry%20of%20the%20mastoids%20should%20be% 20examined%20in%20order%20to%20calculate%20the%20risk&f = false. (Accessed 29 April 2020).

Basson, I.J., Greenway, G., 2004. The Rössing uranium deposit: a product of late-kinematic localization of uraniferous granites in the Central Zone of the Damara Orogen, Namibia. Journal of African Earth Sciences 38, 413–435.

Bayang, D., Kurz, M., Lerche, U., Lyamunda, A., Maro, F.L., Mbogoro, D.K., et al., 2014. Radiating Africa: The Menace of Uranium Mining; Case Studies on Cameroon, Mali and Tanzania. World Information Service on Energy (WISE), 76 p. https://www.sortirdunucleaire.org/IMG/pdf/wise-2014-radiating_africa-the_menace_of_uranium_mining-case_studies_on_cameroon_mali_and_tanzania.pdf. (Accessed 04 April 2020).

Bigotte, G., Molinas, E., 1973. How French Geologists discovered Niger uranium deposits. World Mining 34–39.

Bigotte, G., Obellianne, J.M., 1968. Découverte de minéralisations uranifères au Niger. Mineralium Deposita 3, 317–333.

Bowden, P., Bennett, J.N., Kinnaird, J.A., Whitley, J.E., Abaa, S.I., Hadzigeorgiou-Stavrakis, P.K., 1981. Uranium in the Niger-Nigeria younger granite province. Mineralogical Magazine 44 (336), 379–389.

Bowden, R.A., Shaw, R.P., 2007. The Kayelekera uranium deposit, Northern Malawi: past exploration activities, economic geology and decay series disequilibrium. Applied Earth Science (Transactions of the Institute of Mining and Metallurgy, Section B) 116, 55–67.

Brenner, D.J., 2007. Computed tomography - an increasing source of radiation exposure. New England Journal of Medicine 357, 2277–2284.

Briqueu, L., Lancelot, J.R., Valois, J.-P., Walgenwitz, F., 1980. Géochronologie U-Pb et genèse d'un type de mineralisation uranifère. Les alaskites de Goanikontès (Namibie) et leur encaissant. Bulletin Centres Recherches Exploration Production Elf, Aquitaine 4, 759–811.

Brugge, D., Buchner, 2011. Health effects of uranium: new research findings. Reviews of Environmental Health 26, 231–249.

Brugge, D., Buchner, V., 2012. Radium in the environment: exposure pathways and health effects. Reviews of Environmental Health 27, 1e17.

Buglova, E., 2009. Radiation Health Effects. IAEA. https://www-pub.iaea.org/mtcd/meetings/PDFplus/2009/36489/p36489/Top%201.1%20E.%20Buglova.pdf. (Accessed 06 May 2020).

Bundesministerium für Land- und Forstwirtschaft, Umwelt und Wasserwirtschaft, Abteilung Wasserhaushalt (HZB), 2014. Hydrologischer Atlas von Österreich. ISBN: 3-85437-250-7. Österreichischer Kunst- und Kulturverlag, Seite 250–259.

CDC (Centre for Disease Control and Prevention), 2014. Radiation and Pregnancy: A Fact Sheet for Clinicians. https://emergency.cdc.gov/radiation/prenatalphysician.asp. (Accessed 12 February 2017).

CDC (Centres for Disease Control and Prevention), 2018. Radiation and your health. https://www.cdc.gov/nceh/radiation/emergencies/resourcelibrary/all.htm. (Accessed 12 April 2020).

Chareyron, B., 2007. The radiological impact of impact of uranium mining (France and Africa). CRIIRAD Laboratory Studies, Stockholm. http://nonuclear.se/waste2007/03-chareyron20070427.pdf. (Accessed 07 August 2020).

Cole, D. 1998. Uranium. In: Wilson, M.G.C., Anhaeusser, C.R. (Eds.), The Mineral Resources of South Africa: Handbook. Council for Geoscience, vol. 16, pp. 642–658.

Cole, D.J., Labuschagne, L.S., Sohnge, A.P.G. 1991. Aeroradiometric survey for uranium and ground follow-up in the main Karoo Basin. Geological Survey of South Africa, Memoir, vol. 76, 145 pp.

Crockett, R., Gillmore, G., Perrier, F., Guzzetti, F. (Eds.), 2012. Radon, health and natural hazards II. Natural Hazards and Earth System Science, Special Issue, 12, 130.

Crockett, R.G.M., Groves-Kirkby, C.J., Denman, A.R., Phillips, P.S., 2016. Significant annual and subannual cycles in indoor radon concentrations: seasonal variation and correction. In: Gillmore, G.K., Perrier, F.E., Crockett, R.G.M. (Eds.), Radon, Health and Natural Hazards. Geological Society, London Special Publications, 451. Available from: https://doi.org/10.1144/SP451.2..

Dendy, P., Ringertz, H., 2002. Serious deficiencies in numbers of medical physics experts in diagnostic radiology. European Radiology 12 (8), 2125.

Derricks, J.J., Oosterbosch, R., 1958. The Swambo and Kalongwe deposits compared to Shinkolobwe: Contribution to the study of Katanga Uranium. In: Proceedings of the Second United Nations International conference on the peaceful uses of Atomic Energy, Geneva, September 1–13, 1958. Survey of raw material resources, vol. 2, pp. 663–695.

Derricks, J.J., Vaes, J.F., 1956. The Shinkolobwe uranium deposit: current status of our geological and metallogenic knowledge. In: Geology of uranium deposits – International Conference on Peaceful Uses of Atomic Energy. United Nations, Geneva, vol. 6, pp. 468–472.

Djamel Ghernaout, 2019. Aeration process for removing radon from drinking water – a review. Applied Engineering 3 (1), 32–45. Available from: https://doi.org/10.11648/j.ae.20190301.15.

Eberle, D., Andritzky, G., Wackerle, R., 1995. The new magnetic data set of Namibia: its contributions to the understanding of crustal evolution and regional distribution of mineralization. Communications of the Geological Survey of Namibia 10, 141–150.

Eberly, S., Deloule, E., De Putter, T., Dewaele, S., Mees, F., Yans, J., et al., 2011. SIMS U-Pb dating of uranium mineralization in the Katanga Copperbelt: constraints for the geodynamic context. Ore Geology Reviews 40, 81–89.

Fleurance, S., Cuney, M., Kinnaird, J.A., 2011. Mineralogical, Geochemical and Isotopic Study of 6 Samples from the Langer Heinrich Calcrete Deposit (Namibia). Internal Report to Langer Heinrich Mine.

Freemantle, G.G., 2015. Primary Uranium Mineralisation of the Central Damara Orogen, Namibia (Ph.D. thesis). University of the Witwatersrand, South Africa.

Frimmel, H.E., Groves, D., Kirk, J., Ruiz, J., Chesley, J., Minter, W.E.L., 2005. The formation and preservation of the Witwatersrand Goldfields, the world's largest gold province. In: Hedenquist, J.W., Thompson, J.F.H., Goldfarb, R.J., Richards, J.P. (Eds.), Economic Geology One Hundredth Anniversary Volume. Society of Economic Geologists, pp. 769–797.

George, A.C., 1990. An overview of instrumentation for measuring environmental radon and radon progeny. IEEE Transactions on Nuclear Science 37 (2), 892–901. Available from: https://doi.org/10.1109/23.106733.

GIBC (Global Investment and Business Centre, USA), 2014. South Africa: Mining Laws and Regulations Handbook. International Business Publications, Washington DC, USA, p. 330, ISBN: 1-4 330-7822-8 https://books.google.de/books?id = IAkABwAAQBAJ&pg = PA85&lpg = PA85&dq = Africa+ + ionizing + radiation+ +mining+ +Africa&source = bl&ots = -rT2dKzUYq&sig = ACfU3U398Y_Gigmu_5X7T09epf8ctmPD7g&hl = en&sa = X&ved = 2ahUKEwi6vdTStc7oAhWy0KYKHT2ACQM4FBDoA-TAIegQICxA_#v = onepage&q = Africa%20%20ionizing%20radiation%20%20mining%20%20Africa&f = false.

Gillmore, G., Crockett, R., Przylibski, T., Guzzetti, F. (Eds.), 2010. Radon Health and Natural Hazards. Natural Hazards and Earth System Science, Special Issue, 108.

Gillmore, G.K., Grattan, J., Brian Pyatt, F., Phillips, P.S., Pearce, G., Radon, water and abandoned metalliferous mines in the UK: environmental and human health implications. In: Merkel, B.J., Planer-Freidrich, B., Woldersdorfer, C. (Eds.), Uranium in the Aquatic Environment. Proceedings of the International Conference Uranium Mining and Hydrogeology III and the International Mine Water Association Symposium, Freiberg, Germany, 15–21 September, 2002. Springer-Verlag, Berlin, Heidelberg, pp. 65–76.

Girault, F., Koirala, B.P., Bhattarai, M., Perrier, F., 2016b. Radon and carbon dioxide around remote Himalayan thermal springs. In: Gillmore, G.K., Perrier, F.E., Crockett, R.G.M. (Eds.), Radon, Health and Natural Hazards. Geological Society, London Special Publications, 451. Available from: https://doi.org/10.1144/SP451.6.

Girault, F., Perrier, F., Przylibski, T.A., 2016a. Radon-222 and radium-226 occurrence in water: a review. In: Gillmore, G.K., Perrier, F.E., Crockett, R.G.M. (Eds.), Radon, Health and Natural Hazards. Geological Society, London Special Publications, 451. Available from: https://doi.org/10.1144/SP451.3.

Gray, T. 2015. The Geological Setting of Uranium Mineralisation in the Rössing Area, Namibia. PhD thesis, University of the Witwatersrand, South Africa. 198 pp.

Hambleton-Jones, B.B., Toens, P.D., 1978. The geology and geochemistry of calcrete/gypcrete uranium deposits in duricrust: Namib Desert, South West Africa. Economic Geology 73, 1407−1408.

Herd, D.A., 1996. Geochemistry and Mineralisation of Alaskite in Selected Areas of The Rössing Uranium Mine, Namibia (Unpublished M.Sc. thesis). University of St. Andrews, Scotland, 144 p.

Hunter, A., 2012. Radiation biology - an important science for an advanced nuclear nation like South Africa. South African Journal of Science 108 (7/8). Available from: https://doi.org/10.4102/sajs.v108i5/6.972 Art. #972.

IAEA (International Atomic Energy Agency), 1996. Radiation Protection and the Safety of Radiation Sources. Safety Series No. 120. IAEA, Vienna. https://www.iaea.org/publications/5125/radiation-protection-and-the-safety-of-radiation-sources-a-safety-fundamental. (Accessed 31 July 2020).

IAEA (International Atomic Energy Agency), 2003, Radiation Protection and the Management of Radioactive Waste in the Oil and Gas Industry, Safety Report Series No. 419, STI/PUB/1171 (ISBN: 9201140037).

IAEA (International Atomic Energy Agency), 2003. Extent of Environmental Contamination by Naturally Occurring Radioactive Material (NORM) and Technological Options for Mitigation, Technical Reports Series No. 419, STI/DOC/010/419, ISBN: 9201125038.

IAEA (International Atomic Energy Agency), 2014. Radiation Protection and Safety of Radiation Sources: International Basic Safety Standards, STI/PUB/1578 (July 2014).

IAEA (International Atomic Energy Agency), 2015. Naturally Occurring Radioactive Material (NORM VII). In: Proceedings of an International Symposium Beijing, China, April 22−26, 2013, STI/PUB/1664, ISBN: 9789201040145.

IAEA (International Atomic Energy Agency), 2016. Leadership and Management for Safety, IAEA Safety Standards Series No. GSR Part 2, IAEA, Vienna, 26 p. https://www.iaea.org/publications/11070/leadership-and-management-for-safety. (Accessed 05 August 2020).

IAEA (International Atomic Energy Agency), 2016b. IAEA Brief: Human Health - Enhancing Patient Care in Africa Through Safe Medical Imaging. https://www.iaea.org/sites/default/files/16/11/enhancing-patiient-care-in-africa-through-safe-medical-imaging.pdf. (Accessed 27 March 2020).

IAEA/ILO (International Atomic Energy Agency/International Labour Office), 1999. Assessment of Occupational Exposure Due to Intakes of Radionuclides, Safety Standards Series No. RS-G-1.2. IAEA, Vienna.

IARC (International Agency for Research on Cancer)/WHO, 2020. Environment and Radiation − Section of Environment and Radiation. https://www.iarc.fr/research-sections-env/. (Accessed 04 April 2020).

ICMM (International Council on Mining and Metals), 2014. The Role of Mining in National Economies e, Second Edition Mining's Contribution to Sustainable Development, UK, London.

ICRP (International Commission on Radiological Protection), 1977. Recommendations of the International Commission on Radiological Protection. ICRP Publication 26. Annals of the ICRP, 1(3). Reprinted (with additions) in 1977.

ICRP (International Commission on Radiological Protection), 1991. 1990 Recommendations of the ICRP. ICRP Publication 60. Annals of the ICRP, 21 (1−3). Pergamon Press, Oxford.

ICRP (International Commission on Radiological Protection), 2010. ICRP 2009 Annual Report ICRP reference 4836-3026-995. http://www.icrp.org/docs/ICRP%20Annual%20Report%202009.pdf. (Accessed 07 July 2020).

Iilende, A., 2012. The Source of Uranium and Vanadium at the Langer Heinrich and Klein Trekkopje Uranium Deposits - Genesis and Controlling Factors (M.Sc. thesis). University of Namibia, 266 p.

ILO (International Labour Organization), 2005. Zambia: Ionising Radiation Protection Act, 2005 [No. 16 of 2005]. https://www.ilo.org/dyn/natlex/natlex4.detail?p_lang = en&p_isn = 86273. (Accessed 03 April 2020).

Jacob, R.E., Corner, B., Brynard, H.J., 1986. The regional and structural setting of the uraniferous granite provinces of southern Africa. In: Anhaeusser, C.R., Maske, S. (Eds.), Mineral Deposits of Southern Africa, Vol. 2. Geological Society of South Africa, Johannesburg, pp. 1807–1818.

Kamiya, K., Ozasa, K., Akiba, S., Niwa, S., Kodama, K., Takamura, N., et al., 2015. Long-term effects of radiation exposure on health; from the series: "From Hiroshima and Nagasaki to Fukushima". Lancet 386 (9992), 469–478. Available from: https://doi.org/10.1016/S0140-6736(15)61167-9.

Kinnaird, J.A., Nex, P.A.M., 2007. A review of geological controls on uranium mineralisation in sheeted leucogranites within the Damara Orogen, Namibia. Applied Earth Science, Transactions of the Institute of Mining and Metallurgy 116, 68–85.

Kinnaird, J.A., Nex, P.A.M., Freemantle, G., 2009. Uranium deposits in Central Namibia. Uranium field guide, June 22–26, 2009.

Kinnaird, J.A., Youlton, B., Theron, S., Martin, G., Freemantle, G., Nex, P., 2011. Geometallurgy of two major uranium deposit types. In: 23rd Colloquium of African Geology, University of Johannesburg, South Africa, 8–14 January, 2011.

Kirk, J., Ruiz, J., Chesley, J., Walshe, J., England, G., 2002. A major Archean, gold and crust-forming event in the Kaapvaal Craton, South Africa. Science 297, 1856–1858.

Lamas, Maria del Carmen, 2005. Factors affecting the availability of uranium in soils Landbauforschung Völkenrode Sonderheft 278; Braunschweig Federal Agricultural Research Centre (FAL). https://literatur.thuenen.de/digbib_extern/bitv/zi036810.pdf. (Accessed 14 August 2020).

Le Roux, J.P., 1993. Genesis of stratiform U-Mo deposits in the Karoo Basin of South Africa. Ore Geology Reviews 7, 485–509.

Lottering, M.J., Lorenzen, L., Phala, N.S., Smit, J.T., Schalkwyk, G.A.C., 2008. Mineralogy and uranium leaching response of low grade South African ores. Minerals Engineering 21, 16–22.

Marlow, A.G., 1983. Geology and geochronology of mineralised and anomalous granites and alaskites, Namibia. Geological Society of South Africa Special Publication 11, 289–298.

Marlow, A.G.M. 1981. Remobilisation and Primary Uranium Genesis in the Damaran Orogenic Belt (Ph.D. thesis). University of Leeds, p. 247.

McCarthy, T.S., 2006. The witwatersrand supergroup. In: Johnson, M.R., Anhaeusser, C.R., Thomas, R.J. (Eds.), The Geology of South Africa. Geological Society of South Africa, Johannesburg, pp. 155–186. , Council for Geoscience, Pretoria.

Meshik, A.P., 2005. The Workings of an Ancient Nuclear Reactor. Scientific American, November 2005. https://www.scientificamerican.com/article/ancient-nuclear-reactor/. (Accessed 07 August 2020).

Miller, R.Mc.G.. 2008. The geology of Namibia. Ministry of Mines and Energy, 3 volumes, Namibia.

Mischo, H., Ellmies, R., 2015. Uranium boom in Namibia — hausse or baisse. In: Merkel, B., Arab, A. (Eds.), Uranium - Past and Future Challenges. Springer, Cham, https://doi.org/10.1007/978-3-319-11059-2_1. (Accessed 29 April 2020).

Mishra, U.C., 2004. Environmental impact of coal industry and thermal power plants in India. Journal of Environmental Radioactivity 72 (1-2), 35–40.

Mokotso, E.A., Mogoru, W., Sikakana, I., Neeson, H., 2012. Safety practices of industrial radiography license holders in South Africa. In: 18th World Conference on Nondestructive Testing, April 16–20, 2012, Durban, South Africa. https://www.ndt.net/article/wcndt2012/papers/629_wcndtfinal00629.pdf. (Accessed 07 August 2020).

Morris, P., Perkins, A., 2012. Diagnostic imaging. The Lancet 379, 1525–1533.

Mostert, J.C., Undated. An overview of the facilities of the ionizing radiation laboratory, South Africa. https://inis.iaea.org/collection/NCLCollectionStore/_Public/33/020/33020680.pdf. (Accessed 07 August 2020).

Mouillac, J.L., Valois, J.-P., Walgenwitz, F., 1986. The Goanikontes uranium occurrence in South West Africa/Namibia. In: Anhaeusser, C.R., Maske, S. (Eds.), Mineral Deposits of Southern Africa. Geological Society of South Africa, vol. II, pp. 1833–1843. NEA.IAEA, 1994. Uranium—Resources, Production and Demand. OECD, Paris.

Naino, N., 2013. Environmental and sanitation conditions of uranium mining in Niger. In: Rosa Luxemburg Stiftung and Legal and Human Rights Centre (2014): Uranium Mining: Impact on Health and Environment. Dar es Salam, pp. 16–18.

Nelson, A.W., Eitrheim, E.S., Knight, A.W., May, D., 2015. Understanding the radioactive ingrowth and decay of naturally occurring radioactive materials in the environment: an analysis of produced fluids from the Marcellus Shale. Environmental Health Perspectives 123 (7), 689–696. Available from: https://doi.org/10.1289/ehp.1408855.

Nex, P.A.M., Kinnaird, J.A., Oliver, G.J.H., 2001. Petrology, geochemistry and uranium mineralisation of post-collisional magmatism around Goanikontes, southern Central Zone, Damaran Orogen, Namibia. Journal of African Earth Sciences 33 (3-4), 481–502.

Nex, P.A.M., 1997. Tectono-metamorphic Setting and Evolution of Granitic Sheets in the Goanikontes Area, Namibia (Unpublished Ph.D. thesis). National University of Ireland, University College, Cork.

Nex, P.A.M., Herd, D., Kinnaird, J.A., 2002. Fluid extraction from quartz in sheeted leucogranites as a monitor to styles of uranium mineralization: an example from the Rössing area, Namibia. Geochemistry: Exploration, Environment, Analysis 2 (1), 83–96.

Niepraschk, A., 2015. Uranium Mining Companies in Africa: The Case of Paladin Energy in Malawi. Nuclear Monitor 807; Nuclear Information and Resource Service (NIRS). https://www.nirs.org/monitor/issue/nuclear-monitor-807-uranium-mining-companies-africa-case-paladin-energy-malawi/#. (Accessed 05 April 2020).

NIOSH (National Institute for Occupational Safety and Health), 2002. NIOSH Hazard Review: Health Effects of Occupational Exposure to Respirable Crystalline Silica. Department of Health and Human Services/Centers for Disease Control and Prevention. Publication No. 2002-129. https://www.cdc.gov/niosh/docs/2002-129/pdfs/2002-129.pdf?id = 10.26616/NIOSHPUB2002129. (Accessed 07 August 2020).

Njinga, R.L., Tshivhase, V.M., 2017. Radio-epidemiological evaluation and remediation in water sources from two mines in South Africa. International Journal of Radiation Research 15 (3), 307–315.

NNR SA (National Nuclear Regulator of the Republic of South Africa), 2014. South African National Report on the Compliance to Obligations Under the Joint Convention on Safety of Spent Fuel Management and on the Safety of Radioactive Waste Management. Third Report September 2014. http://www.nnr.co.za/wp-content/uploads/2015/02/NNR%20Joint%20Report_2014.pdf. (Accessed 04 April 2020).

OECD (Organisation for Economic Co-operation and Development), 2010. Uranium 2009: Resources, Production and Demand. OECD Publishing. NEA No. 6891, p. 305. https://www.oecd-nea.org/ndd/pubs/2010/6891-uranium-2009.pdf. (Accessed 07 August 2020).

OECD iLibrary, 2020. National Accounts of OECD Countries. https://www.oecd-ilibrary.org/economics/national-accounts-of-oecd-countries_2221433x. (Accessed 20 April 2020).

Oliver, G.J.H., Kinnaird, J.A., 1996. The Rössing-SJ Dome, Central Zone, Damara Belt, Namibia: an example of mid-crustal extensional ramping. Communications of the Geological Survey of Namibia 11, 53–64.

Osseo-Asare, A., 2019. Radiation within: monitoring particles in bodies. Atomic Junction - Nuclear Power in Africa After Independence. Cambridge University Press, pp. 107–138. Available from: https://www.cambridge.org/core/books/atomic-junction/radiation-within-monitoring-particles-in-bodies/C8393560A63EA82672701446476B80B8 (Accessed 07 August 2020).

Pedley, A., 2007. The Trekkopje Project. Conference abstract: Uranium Exploration and Exploitation in Africa. DPP Course Geological Society of South Africa.

Perrier, F., Girault, F., Bouquerel, H., 2016. Effective radium-226 concentration in rocks, soils, plants and bones. In: Gillmore, G.K., Perrier, F.E., Crockett, R.G.M. (Eds.), Radon, Health and Natural Hazards. Geological Society, London Special Publications, 451. Available from: https://doi.org/10.1144/SP451.8.

Phillips, G.N., Powell, R., 2012. Origin of Witwatersrand gold: A metamorphic devolatilisation hydrothermal replacement model. Applied Earth Science (Transactions of the Institution of Mining and Metallurgy, Section B) 120, 112−129.

Robb, L.J., Meyer, F.M., 1995. The Witwatersrand Basin, South Africa: geological framework and mineralization processes. Ore Geology Reviews 10, 67−94.

RSA (Republic of South Africa), 1999. National Nuclear Regulator Act of 1999 (Act 47 of 1999), and Regulations. https://www.gov.za/documents/national-nuclear-regulator-act?gclid = EAIaIQobChMIs JPXt6GL6wIVBKp3Ch0RiQoYEAAYASAAEgIauvD_BwE. (Accessed 08 August 2020).

RSIC (Radiation Safety Institute of Canada), 2009. Radiation Safety and the Mine Worker. International Workshop, May 27, 2009, Johannesburg. https://radiationsafety.ca/wp-content/uploads/2011/10/ Radiation-Safety-for-Mine-Workers-Africa-fn-final.pdf. (Accessed 08 August 2020).

Russ, A., Burns, C., Tuler, S., Taylor, O., 2006. Health Risks of Ionizing Radiation: An Overview of Epidemiological Studies. A Report by the Community-Based Hazard Management, The George Perkins Marsh Institute of Clark University, Worcester, USA. https://www2.clarku.edu/mtafund/prodlib/clark/ round6/Ionizing_Radiation.pdf. (Accessed 08 August 2020).

Saager, R., Utter, T., Meyer, M., 1982. Pre-Witwatersrand and Witwatersrand conglomerates in South Africa: a mineralogical comparison and bearings on the genesis of gold-uranium placers. In: Amstutz, G.C., El Goresy, A., Frenzel, G., Kluth, C., Moh, G., Wauschkuhn, A., Zimmermann, R.A. (Eds.), Ore Genesis, the State of the Art. Springer-Verlag, Berlin, pp. 38−56.

Sadiq, A.A., Liman, M.S., Agba, E.H., Abdullahi, E., 2010. Assessment of exposure to ionizing radiation at selected mining sites in Nasarawa state, Nigeria. International Journal of Natural and Applied Sciences 6 (4), 478−481.

Schoonhoven, M., 2012. Occupational Exposure to Radon in a South African Platinum Mine (M.S. Mini-dissertation). Northwest University, South Africa. https://pdfs.semanticscholar.org/bbdb/ 880ec81fe963225a645ffbda69c14ab8abeb.pdf. (Accessed 05 April 2020).

South Africa Atomic Energy Board (SAAEB), 1971. Lung Cancer and Exposure to Radon Daughters in South African Gold/Uranium Mines (J.K. Basson, AEB Project Leader, 1971).

Spivey, M., Penkethman, A., Culpan, N., 2010. Geology and mineralization of the recently discovered Rössing South uranium deposit, Namibia. Special Publications of the Society of Economic Geologists 15, 729−746.

Srivastava, R.R., Pathak, P., Perween, M., 2020. Environmental and health impact due to uranium mining. In: Gupta, D., Walther, C. (Eds.), Uranium in Plants and the Environment - Part of the 'Radionuclides and Heavy Metals in the Environment' Book Series. Springer, Cham. Available from: https://doi.org/10.1007/ 978-3-030-14961-1_3 (Accessed 29 April 2020).

Terblanche, P.W.Le. R., Murray, A.P.S., 1991. Indoor exposure to radon: what are the health effects in non-occupational environments? South African Journal of Science 87 (11-12), 548−550.

Thomadsen, B., 2004. The shortage of radiotherapy physicists. Journal of the American College of Radiology 1 (4), 280−282. Available from: https://doi.org/10.1016/j.jacr.2003.12.036.

Turner, B.R., 1985. Uranium mineralization in the Karoo Basin, South Africa. Economic Geology 80, 256−269.

Turpina, L., Clauer, N., Forbes, P., Pagel, M., 1991. U-Pb, Sm-Nd and K-Ar systematics of the Akouta uranium deposit, Niger. Chemical Geology 87 (3 - 4), 217−230.

UNSCEAR (United Nations Scientific Committee on the Effects of Atomic Radiation), 2000. Exposures from natural radiation sources, Annex B to Volume I of the Report to the General Assembly, Sources and Effects of Ionizing Radiation, available on the UNSCEAR 2000 Report, Vol. I Webpage: https://www. unscear.org/docs/publications/2000/UNSCEAR_2000_Annex-B.pdf. (Accessed 08 August 2020).

UNSCEAR (United Nations Scientific Committee on the Effects of Atomic Radiation), 2009. Sources-to-effects assessment for radon in homes and workplaces, Annex E to Volume II of the Report to the General Assembly, Effects of Ionizing Radiation. Available on the UNSCEAR 2006 Report, Vol. II Webpage: https:// www.unscear.org/docs/reports/2006/09-81160_Report_Annex_E_2006_Web.pdf. (Accessed 08 August 2020).

UNSCEAR (United Nations Scientific Committee on the Effects of Atomic Radiation), 2010. Exposures of the Public and Workers from Various Sources of Radiation, Annex B to Volume I, 2008 Report to the General Assembly, Sources and Effects of Ionizing Radiation, available on the UNSCEAR 2008 Report Vol. I webpage: https://www.unscear.org/docs/reports/2008/09-86753_Report_2008_Annex_B.pdf. (Accessed 08 August 2020).

US EPA (United States Environmental Protection Agency), 2017. Radiation Protection: Radiation Health Effects. https://www.epa.gov/radiation/radiation-health-effects. (Accessed 31 March 2020).

USGS (United States Geological Survey), 1997. Radioactive Elements in Coal and Fly Ash: Abundance, Forms, and Environmental Significance. USGS Fact Sheet FS-163-97. https://pubs.usgs.gov/fs/1997/fs163-97/FS-163-97.pdf. (Accessed 08 August 2020).

US-NRC (United States Nuclear Regulatory Commission), 2011. Fact Sheet on Uranium Recovery. https://www.nrc.gov/reading-rm/doc-collections/fact-sheets/fs-uranium-recovery.html. (Accessed 08 August 2020).

Vaupotič, J., Brodar, A., Gregorič, A., Kobal, I., 2017. Radon dynamics in a dwelling with high radon levels in a karst area. In: Gillmore, G.K., Perrier, F.E., Crockett, R.G.M. (Eds.), Radon, Health and Natural Hazards. Geological Society, London Special Publications, 451. Available from: https://doi.org/10.1144/SP451.9.

Venter, R., Boylett, M. 2009. The evaluation of various oxidants used in acid leaching of uranium. In: Hydrometallurgy Conference 2009. The Southern African Institute of Mining and Metallurgy, 2009.

von Backstrom, J.W. 1970. The Rössing uranium deposit near Swakopmund, South West Africa. A preliminary report, Uranium Exploration Geology. IAEA Report IAEA-PL-391/12, Vienna, pp. 143–150.

von Backstrom, J.W., 1975. Uranium and the generation of power, a South African perspective. Transactions, Geological Society South Africa 78, 275–292.

Whiteside, H.C.M., 1970. Uraniferous Precambrain conglomerates of South Africa. Uranium Exploration Geology. International Atomic Press Agency, Vienna, pp. 49–75.

WHO (World Health Organisation), 2009. Handbook on Indoor Radon: A Public Health Perspective. World Health Organization, Geneva. Available from: https://apps.who.int/iris/bitstream/handle/10665/44149/9789241547673_eng.pdf;jsessionid = 23FEA54920E895E300D5EDA2057795E8?sequence = 1, Accessed 08 August 2020.

Winde, F., Villiers, A.B., 2002. The nature and extent of uranium contamination from tailings dams in the Witwatersrand gold mining area (South Africa). In: Uranium in the Aquatic Environment. ISBN: 978-3-642-62877-1.

WNA (World Nuclear Association), 2020h. Uranium in Africa. https://www.world-nuclear.org/information-library/country-profiles/others/uranium-in-africa.abon.aspx. (Accessed 09 April 2020).

Youlton, B., 2007. Controls on Uranium Mineralization at the Klein Trekkopje Prospect, Namibia (B.Sc. Honours thesis). University of the Witwatersrand, 63 p.

Youlton, B. 2014. The Process Mineralogy of Selected Southern African Uranium Ores (Unpublished Ph.D. thesis). University of the Witwatersrand, 370 p.

Zupunski, L., Street, R., Ostroumova, E., Winde, F., Sachs, S., Geipel, G., et al., 2023. Environmental exposure to uranium in a population living in close proximity to gold mine tailings in South Africa. Journal of Trace Elements in Medicine and Biology. 77, 127141. Available from: https://doi.org/10.1016/j.jtemb.2023.127141 (Accessed 18 February 2024).

6 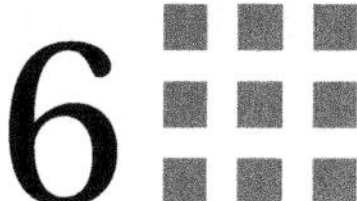

Geophagic practices in Africa

"The oldest evidence of geophagy practiced by humans comes from the prehistoric site at Kalambo Falls on the border between Zambia and Tanzania."

Root-Bernstein and Root-Bernstein (2000)

Key chapter features

1. Prevalence of geophagy in Africa
2. Etiological theories of geophagy
3. Geophagy in pregnancy
4. Predominance of the *nutrient supplementation theory* of geophagy
5. Geophagy in the animal kingdom
6. Suggested areas for further research

(Box 6−1)

BOX 6−1   Learning objectives.

After going through this Chapter, students should be better able to:

1. Identify research gaps in various aspects of geophagy
2. Analyze the prevalence of geophagy in African communities
3. Identify the causal factors for geophagy in different groups of practitioners
4. Assess the effects of geophagy on pregnancy outcomes
5. Able to design a research study on geophagy with a high community impact potential
6. Provide a more unified approach to the design of in vitro studies in bioaccessibility and bioaccessibility in geophagy
7. Able to propose remedial interventions for (reducing) harmful effects of the practice

---

Introduction

Geophagy (or *geophagia*), the habit of eating Earth materials (soil, clay, soft stone, wall scraping, sand, termite mound, anthill, dried-up stream sediment, etc.) is a practice that is common throughout the world, among members of the animal kingdom, including humans.

Medical Geology of Africa. DOI: https://doi.org/10.1016/B978-0-12-818748-7.00003-4

Geophagy has been described as long ago as in the ancient world. Pliny (Gaius Plinius Secundus, CE 23–79), for example, recorded the ingestion of soil on the Greek island of Lemnos, north of the Aegean Sea. The consumption of soil from this island was noted up until the 14th century (Woywodt and Kiss, 2002; Abrahams, 2013). The Greek physician, Hippocrates (460–377 BCE), commonly regarded as the father of medicine, made reference to geophagy in his medical textbook. Reference to geophagy was also made in another famous medical textbook edited by A. Cornelius Celsus (CE 14–37), and titled: *De Medicina*, (Woywodt and Kiss, 2002).

The term "geophagy" is sometimes used interchangeably with the term *pica*. However, according to Young (2011), pica refers to a *craving* and deliberate ingestion of high amounts of nonfood items, such as ice, chalk, ash, and soil over significant periods. Reid (1992) added to this definition: "… the compulsive consumption of otherwise normal food items." Thus according to Reid (1992) and Huebl et al. (2016), geophagy can be considered as a form of pica.

In Africa, geophagy remains a largely misunderstood phenomenon, and terms such as *"confusing," "strange," "mysterious," "aberrant," "puzzling," "curious," "odd,"* and *"perverted,"* have been used at various times when referring to human geophagy. The use of such words to describe the practice, obviously underlines the extent of such misunderstanding. Abrahams (2013) comments that: "This is perhaps understandable for members of a developed urban society that is educated, has ready access to modern pharmaceuticals, and which has increasingly, in both a physical and mental sense, become more remote from soils."

The clinical effects of geophagy are thought to include both the beneficial and the deleterious. There are archeological, biological, cultural, linguistic, religious, symbolic, and other dimensions to the phenomenon and according to Henry and Cring (2013), thorough research on these aspects "…may offer a valid paradigm to better understand this sporadic, puzzling, yet human, behavior," as well as explain factors related to its prevalence, diversity, and distribution (See also, Engberg, 1995). The ensuing sections of this chapter are geared toward the identification of some of these research gaps based on syntheses of the more pertinent information we have to date on various aspects of the subject.

Historical prevalence, diversity, and distribution of geophagy in Africa

A number of authors affirm that humans first ate soil in Africa. Root-Bernstein and Root-Bernstein (2000), for instance, report that "The oldest evidence of geophagy practiced by humans comes from the prehistoric site at Kalambo Falls on the border between Zambia and Tanzania." At this site, a calcium-rich white clay was found to exist alongside the bones of *Homo habilis* (the immediate predecessor of *Homo sapiens*).

Early accounts of geophagy, including its historical prevalence, spatial distribution, and possible etiological explanation are given in several early works, including those of Laufer (1930), Cooper (1957), Halstead (1968), and Hunter (1973).

In Africa, geophagy is now a well-established phenomenon, practiced by members of the animal kingdom, including people, especially those from the abounding tribal and traditional rural

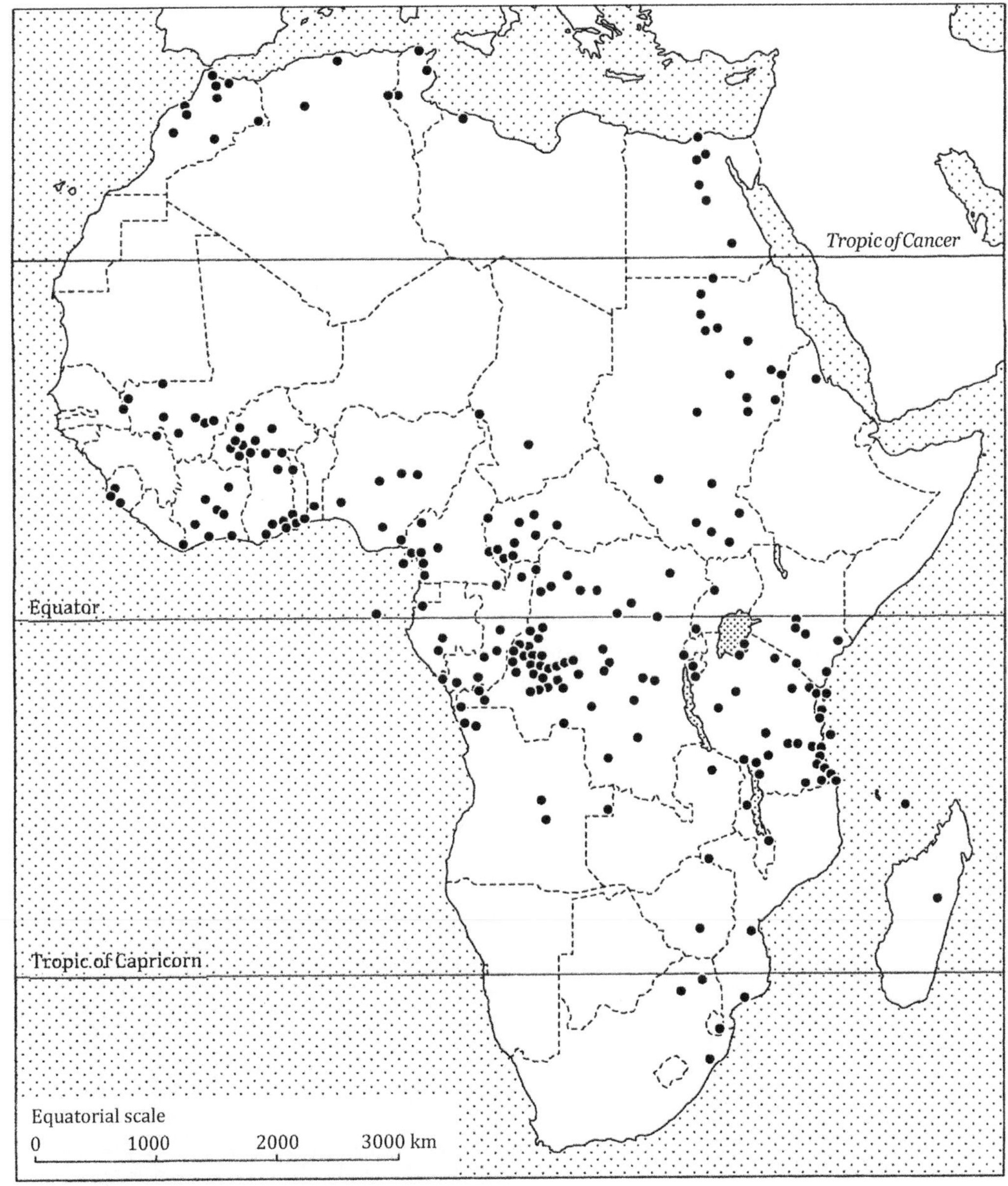

FIGURE 6–1 The distribution of geophagy in Africa (shown by the dots) is widespread, though not all countries of the Continent (e.g., some countries in the Sahara) practice the phenomenon. *Henry, J.M., Cring, F.D., 2013. Chapter 8: Geophagy: an anthropological perspective. In: Brevik, E.C., Burgess, L.C. (Eds.), Soils and Human Health. CRC Press/Taylor and Francis Group, pp. 179–198. https://doi.org/10.1201/b13683-12. From Anell, B., Lagercrantz, S., 1958. Geophagical customs. Studia Ethnographica Upsaliensia 17. Humanistiska Fonden, Stockholm, Sweden and Abrahams, P.W., Parsons, J.A., 1996. Geophagy in the tropics: a literature review. Geographical Journal 162, 63–72.*

societies (Abrahams, 2013; Fig. 6–1). In many other societies outside the Continent, geophagy is generally seen as an unhealthy aberration (See, e.g., Engberg, 1995). Among African societies, however, geophagy is generally considered a normally prescribed behavior. The MUV (2016) gives a figure of between 30% and 80% for the probable number of geophagy practitioners in Africa.

In nearly all societies (worldwide), the highest prevalence of geophagy is recorded among pregnant women [commonly referred to as *geophagy in pregnancy* (GiP) in the literature] and children; but men seldom engage in the practice. In sub-Saharan Africa, for instance, up to 84% of practitioners in some regions are pregnant women (Njiru et al., 2011; Fawcett et al., 2016; Huebl et al., 2016; MUV, 2016; Odongo et al., 2016; Kambunga et al., 2019) (Fig. 6–2A). In Nigeria, the most populous country in Africa, the prevalence of GiP is estimated to be as high as 50% (Njiru et al., 2011). Working in a gold mining area of northwestern Tanzania, Nyanza et al., 2014 reported that 45.6% of pregnant women practiced geophagy, with 54.8% initiating the practice in the first trimester. Njiru et al.'s (2011) study also showed that a total of 101 (65%) of pregnant women ate soil two to three times per day while 20 (13%) ate soil more than three times per day. The amount of Earth material consumed daily varies among geophagists, but is typically in the range of between 5 and 219 g (Huebl et al., 2016) (Fig. 6–2B).

In 2013 Abrahams remarked that "The migration of people from societies where geophagy is especially prevalent results in a cultural transfer of the practice to countries that many would consider to be not typically associated with this deliberate consumption." For example, in the United Kingdom, geophagy is known to be associated with immigrants from south Asia (Hunter, 1973; Decaudin et al., 2018) and West Africa (Abrahams et al., 2013), with the latter (from West Africa) consuming *Calabash chalk* that has been imported from Nigeria and sold in ethnic shops. Similarly, MUV in 2016 noted the habit of eating soil to be prevalent among some migrants from Africa to Europe, in particular, in Vienna, Austria, where that particular study was conducted. They buy portions of geophagic material from exotic supermarkets and health food stores that also offer pharmaceutical

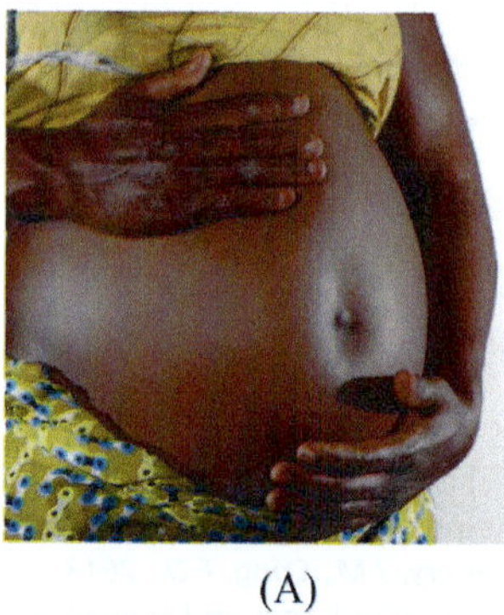
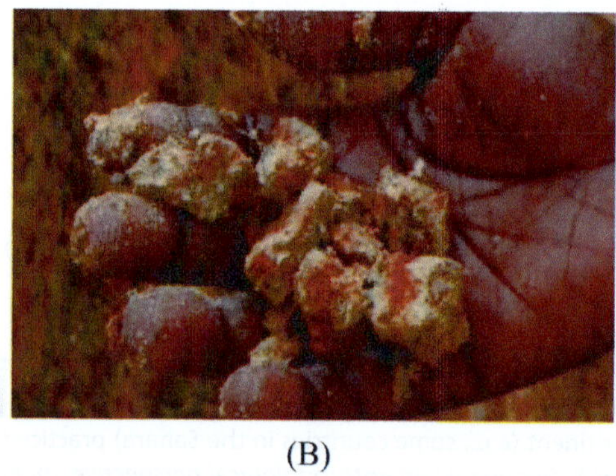

(A)   (B)

FIGURE 6–2 What are the effects of geophagic practices on the developing fetus? (A) In sub-Saharan Africa, up to 84% of geophagy practitioners in some regions are thought to be pregnant women. (B) Geophagic soil consumed by pregnant women on Pemba Island, Zanzibar, a Tanzanian archipelago. *From (A) MUV (Medical University of Vienna), 2016. Geophagy: 'Soil-eating' as an addictive behavior. https://www.sciencedaily.com/releases/2016/12/161205085943.htm. (Accessed 21 January 2020). (B) Credit: Sera Young, Columbia University Press.*

additives such as bentonite clay for internal use (See Section: "Detoxification hypothesis" and Fig. 5, in this Chapter; Abrahams, 2013; MUV, 2016).

In researching the distribution of geophagy worldwide, it has to be noted that there is a large degree of underreporting of the phenomenon (See, e.g., Young, 2011; Nyanza et al., 2014; Huebl et al., 2016). Reasons advanced for underreporting include embarrassment regarding the behavior, lack of knowledge regarding craving, and sensitive questioning on the part of certain investigators inquiring about geophagy, as well as the differing perceptions, beliefs, and cultural norms associated with the phenomenon (Nyanza et al., 2014).

Also, clinicians do not usually ask patients about their craving for Earth materials (Huebl et al., 2016). Pregnant women might not report their geophagic tendencies, because eating soil does not augur with the hygiene concept associated with western medicine, and might feel ashamed or fear castigation from family members or medics (Huebl et al., 2016.) who fear that the practice would harm them or their developing fetus (Young et al., 2011; Huebl et al., 2016). Furthermore, many observers in Africa hold the practice as normal during gestation and therefore might not be seen as necessary to mention (Huebl et al., 2016).

Contemporary practices

The nature of consumed Earth material

The nature of geophagic materials varies markedly with reference to their types, mode of formation, geochemical composition, and so on. The influence of these variables on consumption patterns and health consequences is the subject of numerous researches (e.g., Ngole et al., 2010; Young et al., 2010a,b; Jumbam, 2013; Lar et al., 2015; Bisi-Johnson et al., 2016; Ekosse et al., 2017; Phakoago et al., 2019), which have in turn, significantly influenced the etiological debate.

Selectivity of Earth material in human geophagy

The consumed Earth material is carefully selected (See Geissler, 2000; Henry and Cring, 2013; Nyanza et al., 2014; Ogomaka, 2015; Huebl et al., 2016), and can be gathered from a variety of local sources including specific soil horizons, riverbanks and swamps, clay pits, anthills, termite mounds, and wall scrapings; or can be bought at local markets, from where it can be transported over long distances (See Section: "Historical prevalence, diversity, and distribution of geophagy in Africa," in this Chapter).

Henry and Cring (2013) write: "Not all soil is considered good to eat; it is carefully selected on the basis of appearance, texture, and taste." Earlier research by Geissler (2000) showed that "… among children of the Luo tribe in western Kenya, the preferred material for consumption is collected from termite mounds, with a particular liking for the dark red clay collected from the inside walls where it is considered to be purer."

Ogomaka (2015) writes "Different types of Earth materials from these sources are consumed, the material typically containing a high content of clay." Huebl et al. (2016), working in northern Uganda noted that the consumed material needs to have special qualities engendering color, odor, flavor, softness, and plasticity. Nyanza et al. (2014) report that in Tanzania, pregnant women commonly eat soil sticks sold in the market (called *pemba* in Kiswahili), soil from walls of houses, termite mounds, and ground soil (*kichuguu*).

Processing of consumed Earth material

The literature gives several methods used by geophagists to prepare Earth material before consumption. Henry and Cring (2013) discuss some of these methods.

The processing of some soils may start with cleaning. The material is mostly air dried, but can also be baked, smoked, salted (Shinondo and Mwikuma, 2009; Huebl et al., 2016), mixed with herbs or water (Engel, 2007) or flavored with spices, such as *black pepper* and *cardamom* (Gabbatiss, 2016). Processing of geophagic materials is generally thought to improve their (food) quality in terms of appearance, texture, palatability (Ekosse, 2010; Ngole et al., 2010), and freedom from harmful organisms, such as helminth ova (Shinondo and Mwikuma, 2009).

Geophagy among animals

Geophagy is widespread in the animal kingdom, both small and large creatures alike consuming some form of Earth material and for some purpose, which, to date, in many cases, remain unclear. Galen (CE 130−200), the Greek philosopher and physician, was the first to observe how sick or injured animals used clay for healing purposes in the second century AD (Ghisalberti, 2007; Nuton, 2020).

The practice has been recorded in all the chordate orders, being particularly common in birds, mammals, reptiles, and fish (Henry and Cring, 2013; Pebsworth et al., 2019; Panichev et al., 2023). Some invertebrates, such as earthworms and termites also indulge in the practice, but much of the recent research has focused on mammals, from bats to zebras, and on primates, especially monkeys, macaques, and chimpanzees, as well as a variety of ungulates (Krishnamani and Mahaney, 2000). There is, however, a huge gap still existing in our understanding of different aspects of geophagy in animals, especially the nonhuman primates (See Section: "Suggestions for further research," in this Chapter). The factors driving the process, for instance are still not yet firmly established (Holdø et al., 2002; Panichev et al., 2018). In the case of avian geophagy, some evidence suggests that sodium is the most important driver (See, e.g., Brightsmith et al., 2008).

According to Engel (2007), it appears that geophagy is far more common in animals that subsist on a diet predominantly consisting of plant food. The presumption is that they eat Earth materials for the purpose of gaining minerals, such as salt (sodium chloride), lime (calcium carbonate), copper, iron, or zinc. The original explanation for geophagy in animals was that although wild animals do seek minerals from natural deposits, a need for minerals could not by any means provide a universally accepted explanation for geophagy (Engel, 2007) in all practitioners in the animal kingdom.

Similarly, in bats, the debate continues over whether geophagy is primarily for nutritional supplementation or for detoxification. Some researchers do believe that certain species of bats regularly visit *mineral- or salt licks* (places where animals can go to *lick* essential *mineral* nutrients from a deposit of *salts* and other minerals) to increase mineral consumption (See, e.g., Mills and Milewski, 2007). However, Voigt et al. (2008) demonstrate that both mineral-deficient and healthy bats visit salt licks at the same rate. In the absence of incontrovertible scientific evidence to date, therefore, mineral supplementation is unlikely to be the primary reason for geophagy in bats.

Parrots are known to eat toxic foods globally, but geophagy is concentrated in very specific regions (Lee et al., 2010; See Section on: "Selectivity of Earth material in animal geophagy," in this Chapter). Lee et al. (2010) further showed that parrots in South America practice their geophagy at sites with a significantly positively correlation with distance from the ocean. This correlation can be interpreted as the parrot's predilection for particular geophagic sites being based on an overall lack of sodium in the ecosystem, rather than variation in food toxicity, in accounting for the spatial distribution of geophagy.

Also, it has been observed that presence at salt licks increases during periods of high energy demand (See, e.g., Klein et al., 2008). This is a feature especially evident in lactating and pregnant bats, as their food intake increases to meet higher energy demands. Voigt et al. (2008) concluded that "... the primary purpose for bat presence at salt licks is for detoxification purposes, compensating for the increased consumption of toxic fruit and seeds."

Selectivity of Earth material in animal geophagy

As is the case with geophagy in humans, the type of Earth material consumed by animals is carefully selected. Regular visitation of salt licks has been reported among African forest elephants (Brevik and Burgess, 2012), besides many other animals engaged in geophagy. Wild birds show that they often prefer clayey soils and clayey sediments consisting largely of minerals of the smectite family of clay minerals that includes montmorillonite and bentonite (Brightsmith et al., 2008).

The preference for certain types of clay or soil can lead to peculiar feeding behavior. In Africa, avian species showing geophagy can be broadly divided into those congregating and feeding on grit, and those feeding on clay (Downs, 2006; Bentley, 2018; Downs et al., 2019). Parrots, for instance, avoid eating the substrate in layers one meter above or below the preferred layer (Bentley, 2018; Downs et al., 2019; Fig. 6−3).

The preferred soil bands are shown to have much higher levels of sodium than those that are not chosen.

The etiological debate

Up until now, many theories have been advanced to account for the etiology of Earth material consumption, many of which are still largely unsubstantiated. The intensity of this

FIGURE 6–3 Meyer's parrot eating clay at a bird hide in Kafue National Park, Zambia. *From Credit: Fons Buts, (2017). https://www.flickr.com/photos/cirdan-travels/38709238452. (Accessed 11 March 2020).*

long-standing debate on causative factors is still gathering pace and will only gradually diminish as present-day researchers continue to focus their effort on the remaining knowledge gaps, a number of which are tabulated in the Section: "Suggested areas for further research," this Chapter.

The abounding theories about why people practice geophagy include as a nutritional supplement; as food detoxifier; as diarrheal pharmaceutical; soothing gastrointestinal (GI) or gastroesophageal reflux disorders, such as hypersalivation, heartburn, spitting, and vomiting during pregnancy; as famine food; as a natural stimulant; boosting appetite; for psychological (comforting reasons), neuropsychiatric, cultural and religious reasons; and many more (See Laufer, 1930; Engberg, 1995; Callahan, 2003; Abrahams, 2005; Njiru et al., 2011; Huebl et al., 2016; Ekosse et al., 2017 and others in the "Further Reading List," this Chapter). The symbolic dimension of geophagy cannot be overlooked. Henry and Cring (2013) describe "... how eating soils can mean more than simply fulfilling a need or a craving," and give evidence of how the mode of practice varies according to social class, sex, and age.

Knudsen (2001) observes that, for the Chaggas of Tanzania, geophagy appears to be sacred to women, and, according to Woywodt and Kiss (2002), South African urban women, ingest soils for enhancement of their beauty.

The association between geophagy and spiritual and religious beliefs, commented on by Frate back in 1984, has also more recently been revisited (See, e.g., Engberg, 1995; Geissler, 2000; MUV, 2016). In 2012 Brevik and Burgess noted geophagy's early historic relationship with religion, which "... ranges from the use of antique lozenges of *terra sigilatta*, extracted by a priestess and mixed with goat's blood, to the clay tablets marked with Roman Catholic symbols and images of the cult of Esquipulas in Guatemala." Consumption of0 Earth material from sacred sites for its expected healing properties has also been noted in India (Laufer, 1930) and New Mexico (Callahan, 2003) (Box 6–2).

BOX 6–2   Motivation for consumption.

The initial motivation for consuming Earth materials and the intricacies of pica and addiction are well epitomized by Huebl et al. (2016) study of geophagy practitioners in northern Uganda. In this study, most of the geophagists interviewed (25 out of 35) attributed their geophagic practice to a kind of craving. Nonpregnant women and men were less likely to use the term craving and 10 out of 18 of them (nonpregnant women and men) rather saw themselves as "addicted" to soil. Huebl et al. (2016) findings also indicated that pregnant women ate soil before, during, and beyond pregnancy.

Source: Paraphrased from Huebl, L., Leick, S., Guettl, L., Akello, G, Kutalek, R., 2016. Geophagy in northern Uganda: perspectives from consumers and clinicians. American Journal of Tropical Medicine and Hygiene 95, 1440–1449. https://doi.org/10.4269/ajtmh.15-0579.

Health perspectives

Some of the foregoing theories and causal explanations seem particularly robust, especially those that center around geophagy's medicinal efficacy, which is predicated on the idea that geophagy, [in the words of Engberg (1995)] "… is a rational behavior for people living in environments and social situations that do not otherwise adequately accommodate their vitamin and mineral requirements." This thesis gathers strength, especially when considered in the context of pregnancy (e.g., Njiru et al., 2011), in which case, many practitioners believe that the soil or clay affords nutrients and minerals, such as iron, calcium, and potassium or vitamins such as cobalamin, that may otherwise be absent from the diet, but whose transfer to the developing fetus is quintessential for its optimal development.

How far established is the micronutrient supplementation theory as a causal factor for geophagy?

> *"It remains a matter of conjecture whether ingestion of soil actually satisfies a nutritional deficiency."*
>
> *Abrahams (2005).*

Present-day researchers on the subject of geophagy generally agree that one of the principal values is the curbing of micronutrient deficiencies, especially in GiP; and this remains the most pervasive and perhaps the most credulent explanation of human geophagy so far (See, e.g., Abrahams, 2005; Yanai et al., 2009; Njiru et al., 2011; Adehossi et al., 2017; and other references given in the "Further Reading List," this Chapter).

The strength of the nutrient supplementation theory (Table 6–1) is predicated on the importance of the direct soil-animal pathway of mineral nutrients that complements the soil-plant-animal route in agricultural systems; and the fact that soils do have the potential to supply mineral nutrients especially iron, and vitamins, such as cobalamin where the ingestion of soil (Fig. 6–4) can account for a major proportion of the recommended daily intake.

Table 6–1 Major hypotheses (*micronutrient supplementation and detoxification attribution*) on geophagy.

Sources	Nutrient hypotheses	Detoxification hypotheses
Abrahams and Parsons (1996)	Minerals—especially iron	Detoxification (bimodal)
Brevik (2009)	Enhanced mineral nutrition	
Hooda and Henry (2007)	Calcium deficiency	Detoxification
Johns and Duquette (1991)	Calcium, sodium	Acorn tannins
Johns (1999)		Detoxification
Vermeer (1966)	Calcium deficiency	Plant toxins
Young et al. (2007)	Calcium deficiency	Detoxification
Wiley and Katz (1998)	Calcium and other minerals	Detoxification (bimodal)

Source: From Henry, J.M., Cring, F.D., 2013. Chapter 8: Geophagy: an anthropological perspective. In: Brevik, E.C., Burgess, L.C. (Eds.), Soils and Human Health. CRC Press/Taylor and Francis Group, pp. 179–198. https://doi.org/10.1201/b13683-12: Table 8.1.

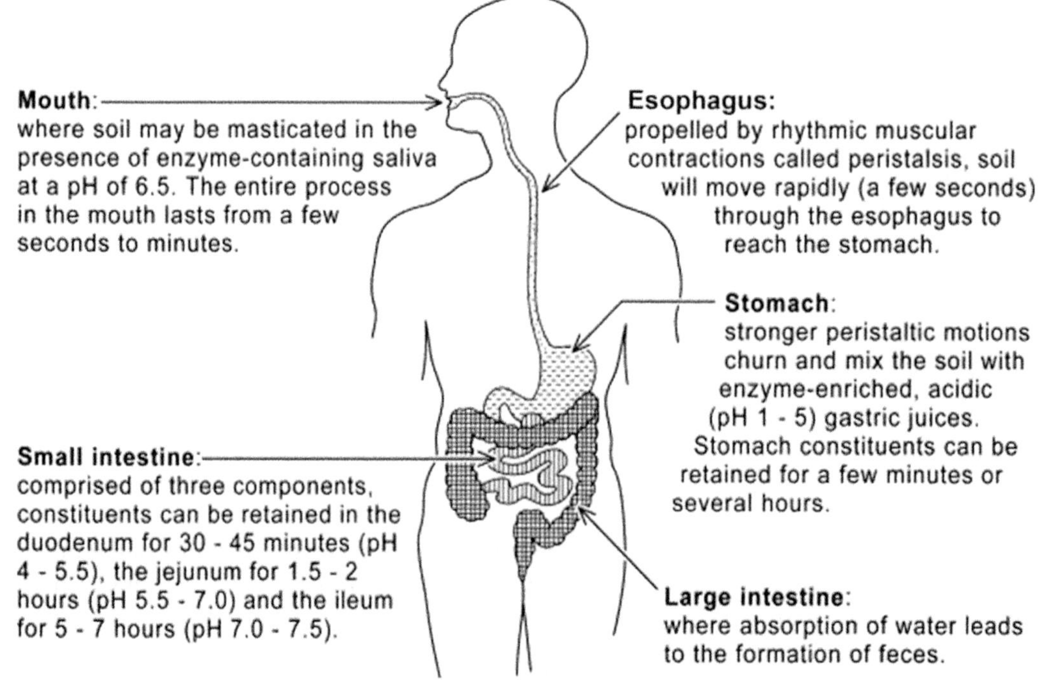

FIGURE 6–4 The human digestion system: Conversion of soil into absorbable substances in the gastrointestinal tract starts in the mouth and continues in the stomach and intestines. *From Abrahams, P.W., 2012. Involuntary soil ingestion and geophagia: a source and sink of mineral nutrients and potentially harmful elements to consumers of earth materials. Applied Geochemistry 27, 954–968.*

The consumption of soil for supplementation of iron and calcium would depend on the concentrations of these elements in the soil, which to a large extent depends on the type of soil and the degree of weathering the soil has undergone (Mahaney et al., 2000; Wilson, 2003; Hooda et al., 2004); chemical weathering being one of the main processes by which weathered material is altered prior to deposition.

Cobalamin supplementation

According to Rosenthal et al. (2015), women of childbearing age from low-resource settings and those with low intake of animal products are the ones often at risk of cobalamin (vitamin B12) deficiency. An increased store of cobalamin is quintessential during pregnancy and lactation to meet the demands of the mother, the fetus, and the infant (Obeid et al., 2017; Samuel et al., 2013; Duggan et al., 2014).

Humans cannot synthesize cobalamin (e.g., See Obeid et al., 2017; Rowley and Kendall, 2019); and the only way it is obtained is through dietary intake (Froese and Gravel, 2010). Geophagy may therefore be a behavioral adaptation to obtain cobalamin produced by bacteria and archea in the soil. More research on the role of geophagy in supplementing cobalamin (*and possibly inducing pica*) in pregnancy is warranted (See Section: "Suggestions for further research," in this Chapter).

The detoxification hypothesis

The microbiological effects of clay consumption to animal health have been known for a long time (See, e.g., Williams and Haydel, 2010). These effects include binding of mycotoxins (fungal toxins), bacterial endotoxins (internal toxins), manmade toxic chemicals, parasites, and pathogens (Box 6—3).

BOX 6—3   Geophagy: soil-eating as an addictive behavior.

". . . However, the reason behind it could be quite different—and also quite multilayered: Soil contains clay, which binds toxins, in the same way as charcoal tablets combat diarrhea. This clay could influence the pH of the stomach acid and help to combat heartburn—many women in Africa predominantly eat maize, cassava, and pulses—but, as Kutalek *(Ruth Kutalek, Medical University of Vienna's Center for Public Health, Institute of Social Medicine)* explains, there are also indications that the soil helps with morning sickness. Many African ethnic groups therefore regard soil-eating as "womanly" and the increased consumption of soil is seen as a sign that a woman is pregnant. However, according to the Medical University of Vienna experts, men are also starting to eat soil more frequently, especially since clay is also regarded as a natural stimulant."

Source: Excerpted from MUV (Medical University of Vienna), 2016. Geophagy: 'Soil-eating' as an addictive behavior. https://www.sciencedaily.com/releases/2016/12/161205085943.htm. (Accessed 21 January 2020).

Detoxification of harmful substances present in the diet of individuals by soils and the relief from GI disorders depend on the soil sorption capacity, which is determined by its cation exchange capacity (CEC), underlining the necessity of understanding the mineralogy and geochemistry of geophagic soils (See under Section: "Suggestions for further research," in this Chapter).

Unbaked soil, which is commonly consumed in northern Uganda, for instance, may be microbially contaminated and cause GI upsets (Huebl et al., 2016). Clay protects the gut lining from corrosion, acts as an antacid and curbs diarrhea.

According to Kreulen (1985), addition of bentonite clay, which is sold worldwide as a digestive aid, can improve food intake, feed conversion efficiency, and absorption patterns in domestic cattle by 10%−20%. Veterinarians therefore find bentonite clay an effective antacid that can bring relief to clay-fed cattle having some form of GI malaise (Engel, 2007).

The effectiveness of bentonite clay as an antacid derives from its special properties (hydration, swelling, water absorption, viscosity, and thixotropy), making it a valuable material not only as the base for some medicines, but also for a wide range of other uses and applications.

Similarly, kaolin [mainly comprising the clay mineral kaolinite $(Al_2O_3(SiO_2)2(H_2O)_2)$], is widely used as a digestive aid and is the base for some medicines, such as Kaopectate, for suppressing diarrhea and reducing toxic effects and inflammation in the digestive system.

Attapulgite (sepiolite and palygorskite), another type of clay, is structurally different from bentonite and kaolin, and is an active ingredient in many antidiarrheal medicines (See, e.g., Huggett, 2015).

Demerits of geophagy

Ingestion of potentially harmful elements

Despite the potential to supply micronutrients, there are a number of apparent risks associated with the practice of geophagy. Some earlier studies suggest that the nutrient value of the soil is overestimated (e.g., EVM, 2003; Wilson, 2003), and contrary to the micronutrient supplementation theory, some researchers (e.g., Anell and Lagercrantz, 1958; Hooda et al., 2004 and Mylonas, 2013) believe that excessive consumption of Earth materials can interfere with bioavailability of micronutrients (See Box 6−4). Such interference, according to a number of researchers (e.g., Rieuwerts et al.,1998; Lohn et al., 2000; Abrahams, 2003; Comerford, 2005; Olaniran et al., 2013; Nyanza et al., 2014; Ekosse et al., 2017; Ngole-Jeme et al., 2018) can lead to or exacerbate, micronutrient and vitamin deficiencies that could cause infectious disease, lead poisoning, bowel impaction, and so on. These conditions can put pregnant woman and the developing fetus at risk.

More recently, Pebsworth et al. (2019) reaffirmed that the high content of iron in consumed Earth materials does not necessarily translate into high bioavailability (Box 6−5), again putting the supplementation theory for iron into question. Hooda et al. (2004) also indicate that the sorption capacity of some geophagical soils (e.g., those with a high clay content) does account for a lowering of the bioaccessibility of copper, iron, and zinc; although

> **BOX 6–4   Bioavailability and bioaccessibility experiments.**
>
> After initial contact with digestive fluids, micronutrient elements can be solubilized from soils and this *bioaccessible soil content* made available for absorption (Fig. 6–4). However, concern has often been expressed about the high concentration levels of some of these elements, generally referred to as potentially harmful elements (PHEs) in some consumed Earth materials (e.g., Abrahams et al., 2013). Although these *total* concentrations may be significantly higher than World Health Organization guideline limits (See Otten et al., 2006; Frisbie et al., 2015), it is important to take account of the *bioavailability* (defined as the fraction that reaches the human systemic circulation from the GI tract) of soil-PHE (Fig. 6–4). The bioavailability of this PHE is strongly dependent on bioaccessibility, since if an element is not bioaccessible it will not be available for absorption (Cave et al., 2011), and both bioavailability and bioaccessibility are influenced by a number of soil variables (mineralogy, particle size, and morphology) as well as factors associated with the human individual, such as age, sex, genetics, and socioeconomic status (WHO, 1996; Abrahams, 2012).
>
> A number of research groups (e.g., USEPA, 2007; Wragg et al., 2011; Denys et al., 2012) have, in the last two decades, been working on the use of in vitro bioaccessibility (IVBA) tests that mimic the conditions of the human GI environment, and determine the bioaccessibility of ingested soil chemical elements. However, a number of problems are evident with the use of these IVBA procedures. For instance, there have (until recently) been a lack of Certified Reference Materials needed for the evaluation of the accuracy of these analyses; there is often insufficient in vivo information against which the bioaccessible concentrations can be compared; and the various models employed tend to produce different results—though the BioAccessibility Research Group of Europe has relatively recently developed and published information about a fasted state IVBA method that begins to address some of these issues (Wragg et al., 2011), and which has been correlated against in vivo data for arsenic, cadmium, and lead (Denys et al., 2012).
>
> Despite advances made in the development of these IVBA procedures, there has been only a limited application in understanding uptake dynamics of geophagic materials, largely due to aforementioned criticisms regarding the efficacy of the experimental techniques applied (See Yanai et al., 2009; Kutalek et al., 2010).

other materials in the soils can be identified as sources of calcium, magnesium, and manganese that humans could potentially utilize.

Moreover, the possible excessive presence of useful components may have negative consequences; for instance, too much iron may result in health problems such as hemosiderosis (Brevik and Burgess, 2012; Henry and Cring, 2013). Brouillard and Rateau (1989), Hooda et al. (2002) and Severance et al. (1988) have shown that ingesting soils with high CEC could also cause iron deficiency.

The presence of very high levels of PHEs in geophagic materials is particularly likely in mineralized areas, associated with mineral extraction, or polluted urban environments (See Davies, 2010; Jovine et al., 2023). By virtue of its rich mineral endowment, several such geochemically anomalous areas exist in Africa from where even low levels of trace metal exposure, such as cadmium and lead can be linked to numerous negative health outcomes, including cognitive deficits and other delayed developmental milestones (Wigle et al., 2007; Wigle et al., 2008; Rosado et al., 2007).

Ingestion of radioactive materials derived from mining of gold, uranium, or other minerals associated with ionizing radiation emitting substances can occur in localities exemplified by the tin-mining areas of the Jos Plateau in Nigeria and the gold and uranium mining sectors of Gauteng Province in South Africa.

The Jos Plateau in Nigeria is underlain largely by the extensively mineralized Younger Granite formation with high-radiation levels and yields geophagic materials with probably significant levels of radionuclides, which is a source of danger to consumers. We simply do not yet have enough information on possible maladies caused by ingestion of soil with high radiation counts, such as were recorded for surficial materials from the Jos Plateau by Solomon in 2005. This is one gap that urgently awaits further research.

Microbiological infections

External elements such as helminth ova and feces may sometimes be present in surface soils and, when consumed, can lead to helminth infections and diarrhea through interaction with the human intestinal biome (Shigova and Moturi, 2009; Blum et al., 2019; Fig. 6−5). Soil contaminated by industrial pollutants pose considerable threat to geophagists, including infections from various pathogenic soil organisms (Sumbele et al., 2014; Ivoke et al., 2017; Box 6−5). Helminth infection associated with geophagy has

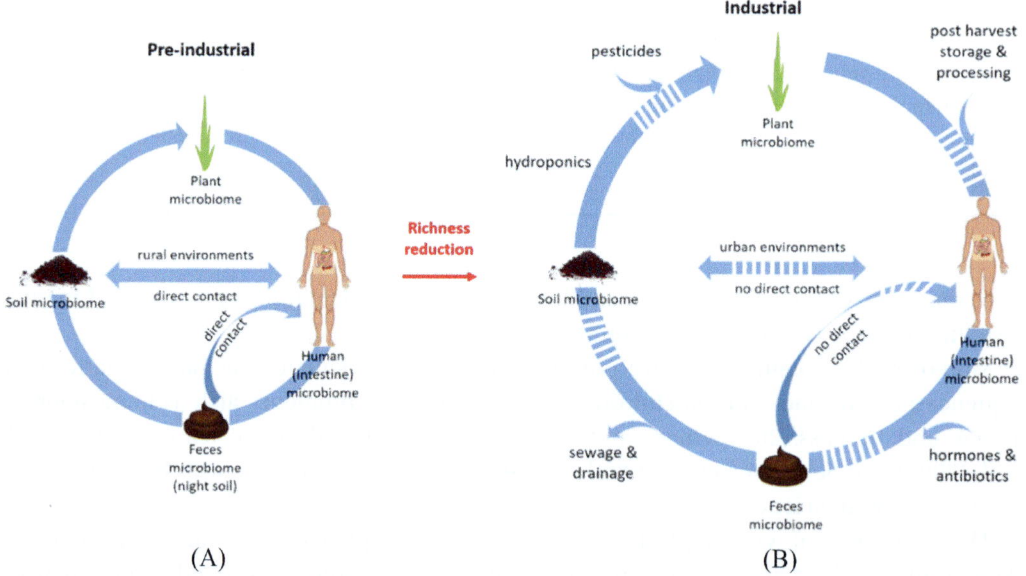

FIGURE 6–5 Illustration of how the microbiota in our environment can influence the human intestine microbiome, through direct contact with soil and feces as well as through food (quality). (A) A cycle for preindustrial microbiota. (B) A cycle for industrial microbiota. *From Blum, W.E.H., Zechmeister-Boltenstern, S., Keiblinger, K.N., 2019. Does soil contribute to the human gut microbiome? Microorganisms 7 (9), 287. https://doi.org/10.3390/microorganisms7090287.*

BOX 6–5   Key facts—soil transmitted helminth infections.

- Soil-transmitted helminth infections are caused by different species of parasitic worms.
- They are transmitted by eggs present in human feces, which contaminate the soil in areas where sanitation is poor.
- Approximately 1.5 billion people are infected with soil-transmitted helminths worldwide.
- Infected children are nutritionally and physically impaired.
- Control is based on periodical deworming to eliminate infecting worms, health education to prevent reinfection, and improved sanitation to reduce soil contamination with infective eggs.
- Safe and effective medicines are available to control infection.

Source: WHO (World Health Organization), 2019. Key facts: soil transmitted helminth infections. https://www.who.int/news-room/fact-sheets/detail/soil-transmitted-helminth-infections. (Accessed 12 February 2020).

been linked with the frequency of inflammatory bowel diseases (Callahan, 2003) as well as an important unrecognized risk factor for environmental enteropathy and stunting (George et al., 2015).

A number of studies (e.g., Saathoff et al., 2002; Saathoff et al., 2004; Ivoke et al., 2017) have shown ingested soil to be of particular concern as a risk factor for geohelminth infection among children in Africa; but the species involved, their epidemiology and the kinds of infection they produce remain unclear (Ozumba and Ozumba, 2002), except in certain cases. For example, Callahan (2003) noted that among children in Nigeria, the most common parasitic infection associated with eating dirt is *ascariasis*.

In adults, however, geophagy is thought to be an unlikely cause for adult infestation (Young et al., 2007; Kutalek et al., 2010), though Ozumba and Ozumba (2002) had earlier noted that Hookworm, Ascaris, Trichuris, and Strongyloides were common helminths in a mixed-population sample [adults (18 years +), adolescents, and children] in Enugu State, Nigeria. We also now know that helminth eggs, such as Ascaris, which can stay viable in the soil for years, can lead to helminth infections (Bisi-Johnson et al., 2010; UCPJ, 2011).

However, accurate knowledge of the inherent biological dangers of soils contaminated with untreated human or animal waste, has uptil now, proved difficult to assess; but, through well-designed microbiological investigations, it is possible to successfully address this aspect of the problem.

Other banes of geophagy

There exist a number of other demerits of the geophagic practice in Africa that have been given very little attention in research circles (See, e.g., Abbey and Lombard, 1973; King et al.,1999). Geophagy's possible association with lead poisoning, blockage of the large intestines, hyperkalemia, phosphorous intoxication, and dental injury (Tayie, 2004), have been little researched.

Abrahams et al. (2006) call for attention to the risk of soil-lead (Pb) toxicity affecting pregnant women and their fetus; whereas excessive tooth wear and dental enamel damage as a consequence of human geophagy was very recently commented on by Ekosse et al. (2017). Ingesting soils high in coarse particles could no doubt affect dental enamel (Anderson et al., 1991; King et al.,1999), and provoke obstruction and rupturing of the sigmoid colon (Key et al., 1982; Engel, 2007; Narayanan et al., 2008; Solaini et al., 2012). However, the physiological intricacies of these processes remain largely unclear.

As of 2016, hardly any research data on appendicitis caused by geophagy existed in the literature (Huebl et al., 2016). Tetanus, peritonitis, and eclampsia and iron-deficiency anemia are also considered to pose a further risk (UCPJ, 2011; Odangowei and Okiemute, 2015), but very few detailed studies on these aspects exist in the contemporary literature.

Some attention has recently been drawn on the internal accumulation of soil that can lead to constipation (Ivoke et al., 2017), intestinal occlusion (Kutalek et al., 2010), the reduction of the power of absorption of food materials by the body (Brand et al., 2009), and severe abdominal pain (Abrahams, 2013), but more data on these aspects are required for a holistic appraisal of these processes.

Social, cultural, psychological, and religious perspectives on geophagy

In Africa, the eating of soil has come to be seen as a socially accepted practice in many quarters, and a common habit in pregnant women (See, e.g., Kortei et al., 2020; Kimassoum et al., 2023). In studying geophagy as practiced among the Luo community of western Kenya, Geissler (2000), describes "... how the practice is associated with social and cultural motives, related to position in the family and community on the one hand and aspects of the meaning of life and one's place in the world, on the other."

Geissler (2000) further describes how: "... beyond the significance of earth-eating in relation to age, gender and power, it [*geophagy practice*] relates to several larger cultural themes, namely, fertility, belonging to a place, and the continuity of the lineage. Earth symbolizes female, life-bringing forces."

The psychological hypothesis centers around the craving ideas wrought by feelings of misery, homesickness, depression, and alienation (Henry and Kwong, 2003).

Comments on the association between geophagy and spiritual and religious beliefs are given in the Section: "The Etiological Debate" (this Chapter).

Other important perspectives on the social, cultural, religious, and other seemingly extraneous motivations for the practice of geophagy in Africa are explored in the under-listed works, in which knowledge gaps are also exposed and major directions for future research indicated. These works include those of: de Garine (1972), Hunter (1973), Prince (1989), Engberg (1995), Geissler (2000), Izugbara (2003), Pérez-Rodríguez (2003), Wilson (2003), Dominy et al. (2004), Young et al. (2008), Minami (2009), Perridge et al. (2011), Young et al. (2011), Brevik and Burgess (2012), Henry and Cring (2013), Bhatia and Kaur (2014), Msibi

(2014), Nyanza et al. (2014), Pemunta (2014), Ogomaka (2015), Okereke et al. (2015), Huebl et al. (2016), Gabbatiss (2016), Macheka et al. (2016), Fairhead (2017), Ekosse et al. (2017), Kambunga et al. (2019), Ravuluvulu (2018), Chung et al. (2019), Psychology Wiki (2019).

Suggested areas for further research

Despite the large volume of recently recorded research on the subject of geophagy in Africa, there still exist substantial knowledge gaps in certain aspects of a phenomenon that is still considered largely misunderstood (See: Davies, 2023). Some of the more important areas (of potentially high-societal impact) of needed research include:

1. *Accurate determination of bioavailability and bioaccessibility of PHEs in consumed Earth material from a particular contaminative source.* This is deemed necessary, as these characteristics need to be taken into account during site-specific risk assessments (See Section: "Bioavailability and bioaccessibility experiments, Box 6−4," in this Chapter).

   Despite the progress made in the development of IVBA tests that mimic the conditions of the human GI environment, and determination of the bioaccessibility of ingested soil chemical elements, some methodological constraints still remain (See, e.g., Deshommes et al., 2012; Abrahams et al., 2013). Refinements in the methodology of systematic in vivo studies are therefore recommended on geophagic soils (George and Abiodun, 2012/2013). There is also the need for more validation studies in which in vivo results are compared with in vitro results (See, e.g., Etcheverry et al., 2012).

2. (i) *During pregnancy, excessive amounts of soil are often consumed and may have an implication on health, both for pregnant women and the infant* (See, e.g., Cham et al., 2023). Despite its association with anemia, pregnancy, and micronutrients, many antenatal clinics (ANC) or national guidelines on micronutrient deficiency control are silent on the subject of GiP. The guidelines generally recommend iron supplementation and deworming of pregnant women as anemia control measures. However, not all women seek antenatal services; hence, there is need for more innovative ways of addressing micronutrient deficiencies in pregnancy (See, e.g., Njiru et al., 2011).

   Despite the full awareness by health care workers of the intricacies of the practice of geophagy, especially during pregnancy, the phenomenon has received only limited attention in ANC in many African countries (e.g., Njiru et al., 2011; George et al., 2015). Health care workers lack critical information on etiology and therapy upon which to create guidelines for safer practices, which could be included in medical training and practice.

   There is, therefore, an urgent need for intensive and extensive health education regarding the detrimental consequences of this common practice and health promotion in the community. Although geophagy can be part of the topics of "nutrition" and "native medicine" in health education protocols at ANC in many African countries, no local or national guidelines or uniform recommendations on geophagy exist.

From the ongoing observations, it is necessary to create guidelines for geophagy or modify clinical guidelines and strengthen or expand the existing supplementation programmes. Sensitization and mass education of pregnant women on the dangers of geophagy is needed.

(ii) The role of geophagy in cobalamin (vitamin B) supplementation during pregnancy and lactation has received little attention (See, e.g., Obeid et al., 2017). The intake requirements for pregnant and lactating women as well as in children need to be re-evaluated in light of their desperation to regularly consume soil (See, e.g., EVM, 2003).

(iii) The enhancement of intestinal triacylglycerol hydrolysis and nonesterified fatty acid absorption by clay ingestion (See, e.g., Habold et al., 2009), as a cause of *pica* among pregnant women needs further investigation.

3. *Examination of geophagic materials for presence of intestinal parasites and their effect on consumers (especially children) of Earth materials* (See, e.g., Ozumba and Ozumba, 2002; Araújo et al., 2015). In 2003 Callahan noted that, among children in Nigeria, the most common parasitic infection associated with eating Earth materials was *ascariasis*. Today, we still do not have accurate knowledge on which childhood infections are most important, although several studies (e.g., Kutalek et al., 2010; Ngure et al., 2013; Nyanza et al., 2014; Ogomaka, 2015) have implicated different types of mycobacterial infection. The extent to which helminth infection associated with geophagy affects the frequency of inflammatory bowel diseases (Callahan, 2003) still needs further attention. A determination of the role of geophagy in the transmission of geohelminth infections among pregnant women attending ANC should also be undertaken. Furthermore, deworming of pregnant women should be integrated into the healthcare delivery system (See Ivoke et al., 2017).

4. *Despite a long period of residence in western countries, geophagy is still a current practice among a significant group of western travelers, who are poorly informed of its harmful effects* (See, e.g., Decaudin et al., 2018). Therefore specific information about the risks of geophagy should be transmitted in western countries, and the international importance of geophagy brought out in migration studies and global public health protocols. Preventive education should be integrated into care of HIV adults, not only in Africa, but also in countries outside the Continent into which this category of geophagy practitioners (HIV-infected) have migrated (See Kmiec et al., 2017).

5. *A structural analysis of the distribution channels of consumed Earth materials should be determined by identifying the participants and their relationships (i.e., equal, collaborative, and exploitative).* It is also necessary to establish the extent to which stakeholders in the distribution chain participate in the exploitation, preparation, and marketing of the soil products.

6. *Soil mineralogy also plays an important role in the behavior of soils, along with the soil texture and organic matter content (Palm et al., 2007; Ekosse et al., 2017), and contributes to the effect of geophagy among humans through its influence on soil CEC (Sanchez, 2019).* For these reasons, it is important for researchers studying the direct

soil-animal pathway of mineral nutrients to have a thorough understanding of the mineralogy, geochemistry and source characterization of geophagic materials. A lot of progress has been made on these kinds of studies (e.g., Aufreiter et al., 2009; Young et al., 2010a,b; Gichumbi et al., 2012; Ekosse and Ngole, 2012; Ekosse et al., 2017; Kambunga et al., 2019; Bernardo et al., 2022; Bonglaisin et al., 2022; Buthelezi-Dube et al., 2022; Malepe et al., 2023; Mouri et al., 2023; Olisa et al., 2023; and other references in the "Further Reference List," this Chapter), but much more needs to be done.

7. *The association between human geophagy and age, gender, and educational status appears to be still conjectural* [See, e.g., Vermeer and Frate, 1975; Engberg, 1995; Abrahams and Parsons, 1996; Geissler, 2000; Brevik and Burgess, 2012; Golden et al., 2012]. In the animal kingdom, the association between geophagy and age and gender appears to be clearer (See, e.g., Holdø et al., 2002; Pebsworth et al., 2011; Young et al., 2011; Myers, 2019). More research on these associations need to be undertaken.

8. *The role of clay minerals (e.g., bentonite clay, attapulgite) as a base for certain pharmaceuticals (digestive aid; antidiarrheal medicines) should be further researched.* As Moosavi (2017) writes "As traditional remedies seem to have a deep root in maintaining body health, it merits doing more research works on bentonite clay and its impacts on body function."

9. *Although as recently as 2013, Henry and Cring noted that estimates of geophagy in large populations are few and quite imprecise, it is well known that the proportion of male practitioners of geophagy is far less than that of women and children (Huebl et al., 2016).* The extent of geophagy among children is not much better documented because studies are fraught with a number of methodological impairments (Brevik and Burgess, 2012). Given that the causative reasons for male geophagy appear to differ from those of female geophagists [e.g., "Many men believed that eating clay increased sexual prowess, and some females claimed that eating clay helped pregnant women to have an easy delivery." (Wayne, 2004)], geophagy in men should be included in further studies and form part of every health education protocol.

10. *Underlain largely by the extensively mineralized Younger Granite formation with high radiation levels, the Jos Plateau (Nigeria) yields geophagic materials with probably significant levels of radionuclides, a source of danger to consumers.* We simply do not yet have enough information on possible maladies caused by ingestion of soil with high radiation counts, such as were recorded for surficial materials from the Jos Plateau (Solomon, 2005). This is one gap in radiation health research that still needs to be filled.

11. (i). *The role of geophagy among nonhuman primates is still far from well understood.* More data are required on the behavioral and dietary characteristics, as well as preferential soil types consumed, to more rigorously investigate the hypotheses of *protection* and *mineral supplementation* across representative species of all taxonomic groups, geographical regions, and dietary classification (See, e.g., Pebsworth et al., 2019).

(ii). *The role of geophagy in species conservation and biodiversity for many endangered species in Africa needs further attention.*

For example, despite baboons' widespread distribution across Africa, geophagy among all subspecies of these primates has been poorly documented (See, e.g., Pebsworth et al., 2011).

Also, the adaptive functions of geophagy, found in a number of bird species in Africa remain unclear (See, e.g., Downs, 2006). Indeed, as recently as 2014, Lee and Marsden wrote: "… we still do not know how common and widespread geophagy is in birds—largely because it is difficult to observe in the vast and little-known tropical forests."

Factors driving avian geophagy is therefore an another area awaiting further investigation (See, e.g., Brightsmith et al., 2008; Lee et al., 2010).

12. *Finally, we need to answer the question of whether humans can practice geophagy safely, in cases where the practice cannot be discontinued (cf. the merits of processing the material before consumption, such as drying, baking, smoking, salting it)?*

Writing on "geohelminth infection" in geophagic practices, Saathoff et al. (2002) noted: "Therefore it does not seem justified to convey the message that people should stop this practice completely. Instead, health education should aim to explain the risks involved and make suggestions regarding how to avoid them. People should be advised to try to find "safe" soil, which is most probably not contaminated with feces or other organic matter. … Alternatively, people could be told that they should cook, fry or bake the soil, as is practiced in some parts of the world (Anell and Lagercrantz, 1958, cited in Saathoff, et al., 2002). Especially in urban areas, health education should also explain where unpolluted soils can be found, which areas should be avoided. This approach would accept geophagy as a normal habit, and thus work within local culture and traditions instead of trying to counteract them. It would allow people to continue practicing a habit that is obviously important to them and that might in addition serve essential nutritional functions. At the same time it would enable people to protect themselves from the negative consequences of geophagy."

Conclusion

The practice of geophagy is still prevalent in many contemporary societies, even the most exclusive of them. But, it is not a widely discussed topic, largely because it still suffers stigmatization from many quarters. In Africa, however, the eating of soil has come to be seen largely as a socially accepted practice, and one that is particularly common with pregnant women.

The etiology of geophagy remains elusive. Both physiological (e.g., mineral deficiency or hunger) and psychological (e.g., craving, obsessive-compulsive spectrum disorder) models have been proposed (See Knudsen, 2001; Allen et al., 2003; Hergüner et al., 2008). Cultural and socioeconomic factors have also been identified as influencing the practice of geophagy, thereby highlighting its complex and little understood nature (Young et al., 2011).

According to Brevik and Burgess (2012), "The physical condition of individual geophagists, framed in part by their ecological environment, their culturally sanctioned diet, their access to nutrients, and their nutritional needs, is clearly related to the urge and propensity to consume soil." This is increasingly confirmed by the physical and chemical diversity of geophagic materials and their differential treatment by the human digestive system, the analyses of which now comprise the most advanced and informative research on the topic.

Though several of the reasons so far advanced for the practice of geophagy still remain conjectural, there are certain characteristics of the phenomenon that are now well established. For instance, geophagy has been shown to be multicausal, and, according to Pebsworth et al. (2019) it has physiological, social, cultural, religious and symbolic underpinnings. However, not all of these functions are adaptive or positive (Henry and Cring, 2013), since there are serious demerits with some, not least, the negative health ramifications.

The most pervasive, and perhaps most credulent explanation of both human and animal geophagy to date is the value of Earth materials consumed as a source of micronutrients (e.g., Johns and Duquette, 1991; Wilson, 2003; Yanai et al., 2009; Mills and Milewski, 2007; Nyanza et al., 2014.) such as for iron deficiency (Adehossi et al., 2017). However, a number of methological impairments are inherent in experiments on micronutrient intake dynamics, such as in determining bioavailable and bioaccessible PHEs consumed from Earth materials (e.g., Abrahams et al., 2013). This is an area, among others, that is intensively being researched (e.g., Wragg et al., 2011; Denys et al., 2012; See Section: "Suggested areas for future research," in this Chapter).

The health impacts of geophagy also remain controversial and inconclusive, with reports in the literature showing the practice to have both health benefits and harmful effects; as well the absence of effects (Al-Rmalli et al., 2010; Kutalek et al., 2010; Njiru et al., 2011; Young, 2010; Khan and Tisman, 2010; Crawford and Bodkin, 2011; Vermeer and Ferrell, 1985; Habold et al., 2009). Substances with clay constituents have long been used (e.g., Kaopectate) for treating gastroenteritis, nausea, diarrhea, and vomiting (Njiru et al., 2011; Vermeer and Ferrell, 1985; Szajewska et al., 2006).

The risk associated with the ingestion of contaminated soil depends on the element of interest, how much is consumed (dose), how often (frequency) and the bioavailability, defined broadly by Chou et al. (1998), as the dose of an unchanged substance that is absorbed and consequently distributed throughout the body. This can depend upon the form or state of a chemical element. Nutrients can also interact with each other. Calcium, for example, can interact with iron and zinc to reduce the absorption of the two latter elements (Anderson, 2004).

Among the demerits of the practice of geophagy, a major concern, especially among children, is helminth infection, acquired through ingestion of soil contaminated with fecal matter. This can lead to anemia due to blood loss from the intestine (See, e.g., Osazuwa et al., 2011).

Our literature review also clearly indicates that geophagy is not limited to any particular age group, race, sex, geographic region, or time period; though most researchers (e.g., Abrahams and Parsons, 1996) consider the practice to be most common among the world's poorer or more tribally oriented people (largely pregnant women and children), and is therefore particularly extensive in the tropics.

Geophagy has also taken new dimensions, especially in the proliferation of nutritional mineral supplements and additives (Brevik and Burgess, 2012) and commercialization of geophagic materials is of some importance (Henry and Cring, 2013). While most consumers of Earth materials, especially in Africa, continue to indulge in the practice, evidence from the United States and Great Britain suggests that store-bought material is becoming the norm, although there are currently efforts in Great Britain to stop the sale of geophagic materials (Brevik and Burgess, 2012; Gabbatiss, 2016).

When packaged, advertised, and sold as scientifically balanced, and doctor recommended, the ingestion of clay components, a form of clay eating, becomes acceptable. This is certainly demonstrated by the wide use of over-the-counter drugs, such as Kaopectate and multivitamins (Brevik and Burgess, 2012). Pills containing mineral supplements, digestive aids, high doses of calcium added to milk and orange juice, and mineral-loaded vegetables can be viewed as acceptable forms of geophagic regimens.

The existence of ethnic grocery stores in many Western cities and the growth of Internet shopping can counter, if not eliminate, the effect of displacement. Thus immigrants and internal migrants are still able to maintain their habits even when removed from their preferred sites (Brevik and Burgess, 2012).

Research on various aspects of geophagy, once conducted in separate, respective fields and shaped by different paradigms (Reid, 1992), is now generally pluridisciplinary, an approach predicated on the evolution of our current understanding (nonunderstanding) of the phenomenon.

In conclusion, this Chapter has highlighted the biological, physical, cultural, religious, symbolic, and other dimensions of geophagy. Several findings have emerged about the basis and effects of the behavior and about the research conducted till now, to shed light on the many aspects that are still unclear or unknown. The avenues for further research on geophagy are legion and a number of them are listed under "Suggestions for further research," in this Chapter.

References

Abbey, L.M., Lombard, J.A., 1973. The etiological factors and clinical implications of pica: report of case. Journal of the American Dental Association 87, 885–887.

Abrahams, P.W., 2003. Human geophagy: a review of its distribution, causes, and implications. In: Skinner, H.C., Berger, A.R. (Eds.), Geology and Health: Closing the Gap. Oxford University Press, USA.

Abrahams, P.W., 2005. Chapter 17: Geophagy and the involuntary ingestion of soil. In: Selinus, O., Alloway, B., Centeno, J.A., Finkelman, R.B., Fuge, R., Lindh, U., et al.,Essentials of Medical Geology. Elsevier Academic Press, Amsterdam, pp. 435–458.

Abrahams, P.W., 2012. Involuntary soil ingestion and geophagia: a source and sink of mineral nutrients and potentially harmful elements to consumers of earth materials. Applied Geochemistry 27, 954–968.

Abrahams, P.W., 2013. Geophagy and the involuntary ingestion of soil. In: Selinus, O., Alloway, B., Centeno, J.A., Finkelman, R.B., Fuge, R., Lindh, U., et al.,Essentials of Medical Geology, Revised Edition. Springer, pp. 433–454. Available from: https://doi.org/10.1007/978-94-007–4375-5_18.

Abrahams, P.W., Parsons, J.A., 1996. Geophagy in the tropics: a literature review. Geographical Journal 162, 63–72.

Abrahams, P.W., Follansbee, M.H., Hunt, A., Smith, B., Wragg, J., 2006. Iron nutrition and possible lead toxicity: an appraisal of geophagy undertaken by pregnant women of UK Asian communities. Applied Geochemistry 21 (1), 98–108.

Abrahams, P.W., Davies, T.C., Solomon, A.,O., Trow, A.J., Wragg, J., 2013. Human geophagia, calabash chalk and undongo: mineral element nutritional implications. PLoS One 8 (1), e53304. Available from: https://doi.org/10.1371/journal.pone.0053304.

Adehossi, E., Malam-Abdou, B., Andia, A., Djibrilla, A., Sani Beydou, S., Brah, S., et al., 2017. Geophagy associated with severe anemia in non-pregnant women: a case series of 12 patients. La Revue de Médecine Interne 38 (1), 53—55. Available from: https://doi.org/10.1016/j.revmed.2016.02.019.

Allen, A., King, A., Hollander, E., 2003. Obsessive-compulsive spectrum disorders. Dialogues in Clinical Neuroscience 5 (3), 259—271.

Al-Rmalli, S.W., Jenkins, R.O., Watts, M.J., Haris, P., 2010. Risk of human exposure to arsenic and other toxic elements from geophagy: trace element analysis of baked clay using inductively coupled plasma mass spectrometry. Environment and Health 9, 79. Available from: https://doi.org/10.1186/1476-069X-9-79.

Anderson, J.J.B., 2004. In: Mahan, L.K., Escott-Stump, S. (Eds.), Minerals. Krause's Food, Nutrition, and Diet Therapy, tenth ed. W.B. Saunders, 2000, Philadelphia, USA, pp. 120—163.

Anderson, J.E., Akmal, M., Kittur, D.S., 1991. Surgical complications of pica: report of a case of intestinal obstruction and a review of the literature. American Surgery 57, 663—667.

Anell, B., Lagercrantz, S., 1958. Geophagical customs. Studia Ethnographica Upsaliensia. Humanistiska Fonden, Stockholm, Sweden, p. 17.

Araújo, A., Luiz, K.R., Ferreira, F., 2015. Palaeoparasitology - human parasites in ancient material. In: De Baets, K., Timothy, D., Littlewood, J. (Eds.), Fossil Parasites: Advances in Parasitology, 90. pp. 349—387.

Aufreiter, S., Hancock, R.G.V., Mahaney, A., Stambolic-Robb, Sanmugadas, K., 2009. Geochemistry and mineralogy of soils eaten by humans. International Journal of Food Sciences and Nutrition 48 (5), 293—305.

Bentley, C.S., 2018. Soil-eating by grey parrots in Cameroon: an answer to mineral deficiencies or toxins in the diet? A Thesis submitted to the Department of Agriculture, in partial fulfillment of the requirements for the award of the Bachelor's Degree in Biochemistry of the University of Arizona; May 1999. https://medium.com/@agreenmoment/soil-eating-by-grey-parrots-in-cameroon-an-answer-to-mineral-deficiencies-or-toxins-in-the-diet-b2ffc60e2fb7. (Accessed 09 February 2020).

Bernardo, B., Candeias, C., Rocha, F., 2022. Geophagic materials characterization and potential impact on human health: The case study of Maputo City (Mozambique). Applied Sciences 12 (10), 4832. Available from: https://doi.org/10.3390/app12104832 (Accessed 22 February 2024).

Bhatia, M.S., Kaur, J., 2014. Pica as a cultural bound syndrome. Delhi Psychiatry Journal 17 (1), 143—147.

Bisi-Johnson, M.A., Obi, C.L., Ekosse, G.E., 2010. Microbiological and health related perspectives of geophagia: an overview. African Journal of Biotechnology 9 (36), 5784—5791.

Bisi-Johnson, M.A., Adebayo, O.H., Abraham, A.A., Olarinde, A.S., 2016. Physical and chemical evaluation of geophagic and cosmetic clays from southern and western Nigeria: the health implications. International Journal of Science and Engineering Investigations 5 (59), 45—51.

Blum, W.E.H., Zechmeister-Boltenstern, S., Keiblinger, K.N., 2019. Does soil contribute to the human gut microbiome. Microorganisms 7 (9), 287. Available from: https://doi.org/10.3390/microorganisms7090287.

Bonglaisin, J.N., Kunsoan, N.B., Bonny, P., Matchawe, C., Tata, B.N., Nkeunen, G., Mbofung, C.M., 2022. Geophagia: Benefits and potential toxicity to human - A review. Frontiers in Public Health 10893831. Available from: https://doi.org/10.3389/fpubh.2022.893831.

Brand, C.E., de Jager, L., Ekosse, G.-I.E., 2009. Possible health effects associated with human geophagic practise: an overview. Medical Technology SA 23 (1), 11—13.

Brevik, E.C., 2009. Soil, food security and human health. In: Verheye, W. (Ed.), Soils, Plant Growth and Crop Production. Encyclopaedia of Life Support Systems (EOLSS). Developed under the Auspices of the UNESCO, EOLSS Publishers, Oxford. Available from: http://www.eolss.net.

Brevik, E.C., Burgess, L.C. (Eds.), 2012. Soils and Human Health. CRC Press/Taylor and Francis Group. Available from: http://doi.org/10.1201/b13683-12.

Brightsmith, D.J., Taylor, J., Phillips, T.D., 2008. The roles of soil characteristics and toxin adsorption in avian geophagy. Biotropica 40 (6), 766−774. Available from: https://doi.org/10.1111/j.1744-7429.2008.00429.x.

Brouillard, M.Y., Rateau, J.G., 1989. Smectite and kaolin on bacterial enterotoxins. Gastroentrology and Clinical Biology 13, 18−24.

Buthelezi-Dube, N.N., Muchaonyerwa, P., Hughes, J., Modi, A., Caister, K., 2022. Properties and indigenous knowledge of soil materials used for consumption, healing and cosmetics in KwaZulu-Natal, South Africa. Soil Science Annual 73 (4), 157408. Available from: https://doi.org/10.37501/soilsa/157408 (Accessd 23 February 2024).

Callahan, G.N., 2003. Eating dirt. Emerging Infectious Diseases 9 (8), 1016−1021. Available from: https://doi.org/10.3201/eid0908.030033.

Cave, M.R., Wragg, J., Denys, S., Jondreville, C., Feidt, C., 2011. Oral bioavailability. In: Swartjes, F.A. (Ed.), Dealing with contaminated sites (from theory towards practical application). Springer, Dordrecht, pp. 287−384.

Cham, L.C., Kayeme, Z., Bokanya, I., Tambwe, M.A., Bito, V., Kakoma, S.Z., 2023. Contributing factors to the awareness of health risks of geophagy among pregnant women in Lubumbashi. Fortune Journal of Health Sciences 6, 325−331. Available from: https://www.fortunejournals.com/articles/contributing-factors-to-the-awareness-of-health-risks-of-geophagy-among-pregnant-women-in-lubumbashi.html (Accessed 23 February 2024).

Chou, C.H.S.J., Holler, J., De Rosa, C., 1998. Minimal risk levels (MRLs) for hazardous substances. Journal of Clean Technology, Environmental Toxicology and Occupational Medicine 7 (1), 1−24.

Chung, E.O., Mattah, B., Hickey, M.D., Salmen, C.R., Milner, E.M., Bukusi, E.A., et al., 2019. Characteristics of pica behavior among mothers around Lake Victoria, Kenya: a cross-sectional study. International Journal of Environmental Research and Public Health 16, 2510. Available from: https://doi.org/10.3390/ijerph16142510.

Comerford, N., 2005. Soil factors affecting nutrient bioavailability. In: BassiriRad, H. (Ed.), Nutrient Acquisition by Plants. Ecological Studies (Analysis and Synthesis), vol. 181. Springer, Berlin, Heidelberg.

Cooper, M., 1957. Pica. Charles C. Thomas Publisher, Springfield, IL.

Crawford, L., Bodkin, K., 2011. Health and social impacts of geophagy in Panama. MSURJ 6 (1), 31−37.

Davies, T.C., 2010. Chapter 8: Medical geology in Africa. In: Selinus, O., Finkelman, R.B., Centeno, J. (Eds.), Medical Geology - A Regional Synthesis, first ed. Springer Verlag, Amsterdam, The Netherlands, pp. 199−219.

Davies, T.C., 2023. Current status of research and gaps in knowledge of geophagic practices in Africa. Frontiers in Nutrition 9, 1084589. Available from: https://doi.org/10.3389/fnut.2022.1084589 (Accessed 23 February 2024).

de Garine, I., 1972. The socio-cultural aspects of nutrition. Ecology of Food and Nutrition 1 (2), 143−163. Available from: https://doi.org/10.1080/03670244.1972.9990282.

Decaudin, P., Kanagaratnam, L., Kmiec, I., Nguyen, Y., Migault, C., Lebrun, D., et al., 2018. Prevalence of geophagy and knowledge about its health effects among native Sub-Saharan Africa, Caribbean and South America healthy adults living in France. Eating and Weight Disorders. Studies on Anorexia, Bulimia and Obesity. Springer. Available from: https://doi.org/10.1007/s40519-018-0624-9.

Denys, S., Caboche, J., Tack, K., Rychen, G., Wragg, J., et al., 2012. *In vivo* validation of the Unified BARGE method to assess the bioaccessibility of arsenic, antimony, cadmium, and lead in soils. Environmental Science and Technology 46, 6252−6260.

Deshommes, E., Tardif, R., Edwards, M., Sauvé, S., Prévost, M., 2012. Experimental determination of the oral bioavailability and bioaccessibility of lead particles. Chemistry Central Journal 6, 138. Available from: https://doi.org/10.1186/1752-153X-6-138.

Dominy, N.J., Davoust, E., Minekus, M., 2004. Adaptive function of soil consumption: an *in vitro* study modeling the human stomach and small intestine. Journal of Experimental Biology 207, 319−324. Available from: https://doi.org/10.1242/jeb.00758.

Downs, C.T., 2006. Geophagy in the African Olive Pigeon, *Columba aquatrix*. Ostrich Journal of African Ornithology 77 (1 - 2), 40−44.

Downs, C.T., Bredin, I.P., Wragg, P.D., 2019. More than eating dirt: a review of avian geophagy. African Zoology 54 (1), 1−19.

Duggan, C., Srinivasan, K., Thomas, T., Samuel, T., Rajendran, R., Muthayya, S., et al., 2014. Vitamin B-12 supplementation during pregnancy and early lactation increases maternal, breast milk, and infant measures of vitamin B-12 status. Journal of Nutrition 144, 758−764.

Ekosse, G.-I.E., 2010. Kaolin deposits and occurrences in Africa. Applied Clay Science 50, 212−236.

Ekosse, G.-I.E., Ngole, V.M., 2012. Mineralogy, geochemistry and provenance of geophagic soils from Swaziland. Applied Clay Science 57, 25−31.

Ekosse, G.-I.E., Ngole-Jeme, V.M., Diko, M.L., 2017. Environmental geochemistry of geophagic materials from Free State Province in South Africa. Open Geosciences 9 (1), 114−125. Available from: https://doi.org/ 10.1515/geo-2017-0009.

Engberg, D.E., 1995. Geophagy: Adaptive or aberrant behavior. Nebraska Anthropologist 84. https:// digitalcommons.unl.edu/cgi/viewcontent.cgi?article = 1083&context = nebanthro. (Accessed 25 January 2020).

Engel, C., 2007. Chapter 2: Zoopharmacognosy. In: Wynn, S.G., Fougére, B.J. (Eds.), Veterinary Herbal Medicine. Mosby, Elsevier Press, pp. 7−16. Available from: https://doi.org/10.1016/B978-0-323-02998-8.X5001-X.

Etcheverry, P., Grusak, M.A., Fleige, L.E., 2012. Application of *in vitro* bioaccessibility and bioavailability methods for calcium, carotenoids, folate, iron, magnesium, polyphenols, zinc, and vitamins B_6, B_{12}, D, and E. Frontiers in Physiology 3, 317. Available from: https://doi.org/10.3389/fphys.2012.00317.n.

EVM (Expert Group on Vitamins and Minerals), 2003. Safe upper levels for vitamins and minerals. http:// food.gov.uk/multimedia/pdfs/vitmin2003.pdf. (Accessed 22 March 2020).

Fairhead, J.R., 2017. Termites, mud daubers and their earths: a multispecies approach to fertility and power in West Africa. Conservation and Society 14 (4), 359−367.

Fawcett, E.J., Fawcett, J.M., Mazmanian, D., 2016. A meta-analysis of the worldwide prevalence of pica during pregnancy and the postpartum period. International Journal of Gynaecology and Obstetrics 133, 277−283. Available from: https://doi.org/10.1016/j.ijgo.2015.10.012.

Fons Buts, 2017. Meyer's parrot eating clay at dawn. Flickr. Available from: https://www.flickr.com/photos/ cirdan-travels/38709238452 (Accessed 22 February 2024).

Frate, D.A., 1984. Last of the earth eaters. Sciences 24 (6), 34−38.

Frisbie, S.H., Mitchell, E., Sarkar, B., 2015. Urgent need to reevaluate the latest World Health Organization guidelines for toxic inorganic substances in drinking water. Environmental Health 14, 63. Available from: https://doi.org/10.1186/s12940-015-0050-7.

Froese, D.S., Gravel, R.A., 2010. Genetic disorders of vitamin B_{12} metabolism: eight complementation groups - eight genes. Expert Reviews on Molecular Medicine 12, e37. Available from: https://doi.org/10.1017/ S1462399410001651.

Gabbatiss, J., 2016. The people who can't stop eating dirt. https://www.bbc.com/future/article/20160615-the-people-who-cant-stop-eating-dirt. (Accessed 09 February 2020).

Geissler, P., 2000. The significance of earth-eating: social and cultural aspects of geophagy among Luo children. Africa (Lond) 70 (4), 653−682. Available from: https://doi.org/10.3366/afr.2000.70.4.653.

George, G., Abiodun, A., 2013. Physiological effects of geophagy (soil eating) with reference to iron nutritional status in pregnant women: - A study in selected antenatal clinics in KSD Municipal Area of the Eastern

Cape, South Africa. 3rd International Conference on Biology, Environment and Chemistry, IPCBEE, vol. 46. IACSIT Press, Singapore. doi: 10.7763/IPCBEE.

George, C.M., Oldja, L., Biswas, S., Perin, J., Lee, G.O., Kosek, M., et al., 2015. Geophagy is associated with environmental enteropathy and stunting in children in rural Bangladesh. American Journal of Tropical Medicine and Hygiene 92, 1117–1124.

Ghisalberti, E.L., 2007. Detection and isolation of bioactive natural products. In: Colegate, S.M., Molyneux, R.J. (Eds.), Bioactive Natural Products: Detection, Isolation and Structural Determination, second ed. CRC Press, pp. 11–76.

Gichumbi, J.M., Ombaka, O., Gichuki, 2012. Geochemical and mineralogical characteristics of geophagic materials from Kiambu, Kenya. International Journal of Modern Chemistry 2 (3), 108–116.

Golden, C.D., Rasolofoniaina, B.J.R., Benjamin, R., Young, S.L., 2012. Pica and amylophagy are common among Malagasy men, women and children. PLoS One 7 (10), e47129. Available from: https://doi.org/10.1371/journal.pone.0047129.

Habold, C., Reichardt, F., Le Maho, Y., Angel, F., Liewig, N., Lignot, J.H., et al., 2009. Clay ingestion enhances intestinal triacylglycerol hydrolysis and non-esterified fatty acid absorption. British Journal of Nutrition 102, 249–257. Available from: https://doi.org/10.1017/S0007114508190274.

Halstead, J.A., 1968. Geophagia in man: its nature and nutritional effects. American Journal of Clinical Nutrition 21, 1384–1393.

Henry, J., Kwong, A.M., 2003. Why is geophagy treated like dirt. Deviant Behavior 24 (4), 353–371. Available from: https://doi.org/10.1080/713840222.

Henry, J.M., Cring, F.D., 2013. Chapter 8: Geophagy: an anthropological perspective. In: Brevik, E.C., Burgess, L.C. (Eds.), Soils and Human Health. CRC Press/Taylor and Francis Group, pp. 179–198. Available from: http://doi.org/10.1201/b13683-12.

Hergüner, S., Ozyildirim, I., Tanidir, C., 2008. Is pica an eating disorder or an obsessive-compulsive spectrum disorder? Progress in Neuropsychopharmacology and Biological Psychiatry 32 (8), 210–211. Available from: https://doi.org/10.1016/j.pnpbp.2008.09.011.

Holdø, R.M., Dudley, J.P., McDowell, L.R., 2002. Geophagy in the African elephant in relation to availability of dietary sodium. Journal of Mammalogy 83 (3), 652–664. Available from: https://doi.org/10.1644/1545-1542 (2002)083 < 0652:GITAEI > 2.0.CO.

Hooda, P., Henry, J., 2007. Geophagia and human nutrition. In: MacClancy, J., Henry, J., MacBeth, H. (Eds.), Consuming the Inedible. Neglected Dimensions of Food Choice. Berghahn Books, New York, NY, pp. 89–98.

Hooda, P.S., Henry, C.J.K., Seyoum, T.A., Armstrong, L.D.M., Fowler, M.B., 2002. The potential impact of geophagia on the bioavailability of iron, zinc and calcium in human nutrition. Environmental Geochemistry and Health 24, 305–319.

Hooda, P.S., Henry, C.J.K., Seyoum, T.A., Armstrong, L.D.M., Fowler, M.B., 2004. The potential impact of soil on human mineral nutrition. Science of the Total Environment 333, 75–87. https//doi.org/1016/j.scitotenv.2004.04.023.

Huebl, L., Leick, S., Guettl, L., Akello, G., Kutalek, R., 2016. Geophagy in northern Uganda: perspectives from consumers and clinicians. American Journal of Tropical Medicine and Hygiene 95, 1440–1449. Available from: https://doi.org/10.4269/ajtmh.15-0579.

Huggett, J.M., 2015. Clay minerals. Reference Module in Earth Systems and Environmental Sciences. Encyclopaedia of Geology. Elsevier. Available from: http://doi.org/10.1016/B978-0-12-409548-9.09519-1.

Hunter, J.M., 1973. Geophagy in Africa and in the United States: a culture-nutrition hypothesis. Geographical Review 63, 170–195.

Ivoke, N., Ikpor, N., Ivoke, O., Ekeh, F., Ezenwaji, N., Odo, G., et al., 2017. Geophagy as risk behaviour for gastrointestinal nematode infections among pregnant women attending antenatal clinics in a humid tropical zone of Nigeria. African Health Sciences 17 (1), 24–31. Available from: https://doi.org/10.4314/ahs.v17i1.5.

Izugbara, C.O., 2003. The cultural context of geophagy among pregnant and lactating Ngwa women of southeastern Nigeria. The African Anthropologist 10 (2), 180−199.

Johns, T., 1999. The chemical ecology of human ingestive behaviors. Annual Review of Anthropology 28, 27−50.

Johns, T., Duquette, M., 1991. Detoxification and mineral supplementation as functions of geophagy. The American Journal of Clinical Nutrition 53, 448−456.

Jumbam, N.D., 2013. Geophagic materials: the possible effects of their chemical composition on human health. Transactions of the Royal Society of South Africa 68 (3), 177−182. Available from: https://doi.org/10.1080/0035919X.2013.843606.

Jovine, J., Nyanza, E.C., Asori, M., Thomas, D.S.K., 2023. Prenatal arsenic and mercury levels among women practicing geophagy in areas with artisanal and small-scale gold mining activities, Northwestern Tanzania. BMC Pregnancy Childbirth 23 (854). Available from: https://doi.org/10.1186/s12884-023-06174-4 (Accessed 23 February 2024).

Kambunga, S.N., Candeias, C., Hasheela, I., Mouri, H., 2019. Review of the nature of some geophagic materials and their potential health effects on pregnant women: some examples from Africa. Environmental Geochemistry and Health 41, 2949−2975. Available from: https://doi.org/10.1007/s10653-019-00288-5.

Key, T.C., Horger 3rd., E.O., Miller, J.M., 1982. Geophagia as a cause of maternal death. Obstetrics and Gynaecology 60, 525−526.

Khan, Y., Tisman, G., 2010. Pica in iron deficiency: a case series. Journal of Medical Case Reports? 4, 86. Available from: https://doi.org/10.1186/1752-1947-4-86.

Kimassoum, D, Ngum, N.L., Bechir, M., Haroun, A., Tidjani, A., Frazzoli, C., 2023. Geophagy: A survey on the practice of soil consumption in N'Djamena. Journal of Global Health Reports 7e2023010. 10.29392/001c.74955.

King, T., Andrews, P., Boz, B., 1999. Effect of taphonomic processes on dental microwear. American Journal of Physiology and Anthropology 108, 359−373.

Klein, N., Fröhlich, F., Krief, S., 2008. Geophagy: soil consumption enhances the bioactivities of plants eaten by chimpanzees. Die Naturwissenschaften 95 (4), 325−331. Available from: https://doi.org/10.1007/s00114-007-0333-0.

Kmiec, I., Nguyen, Y., Rouger, C., et al., 2017. Factors associated with geophagy and knowledge about its harmful effects among native Sub-Saharan African, Caribbean and French Guiana HIV patients living in Northern France. AIDS Behaviour 21, 3630−3635. Available from: https://doi.org/10.1007/s10461-016-1661-x.

Knudsen, J.W., 2001. Akula udongo (earth eating habit): a soil and cultural practice among Chagga women on the slopes of the Mount Kilimanjaro. African Jounal of Indigenous Knowledge Systems 1, 19−26.

Kortei, N.K., Koryo-Dabrah, A., Akonor, P.T., Manaphraim, N.Y.B., Ayim-Akonor, M., Boadi, N.O., et al., 2020. Potential health risk assessment of toxic metals contamination in clay eaten as pica (geophagia) among pregnant women of Ho in the Volta Region of Ghana. BMC Pregnancy and Childbirth 20 (1), 160. Available from: https://doi.org/10.1186/s12884-020-02857-4 (Accessed 22 February 2024).

Kreulen, D.A., 1985. Lick use by large herbivores: a review of benefits and banes of soil consumption. Mammal Review 15 (3), 107−150. Available from: https://doi.org/10.1111/j.1365-2907.1985.tb00391.x.

Krishnamani, R., Mahaney, W.C., 2000. Geophagy among primates: adaptive significance and ecological consequences. Animal Behaviour 59, 899−915.

Kutalek, R., Wewalka, G., Gundacker, C., Auer, H., Wilson, J., et al., 2010. Geophagy and potential health implications: geohelminths, microbes and heavy metals. Transactions of the Royal Society of Tropical Medcine and Hygiene 104, 787−795.

Lar, U.A., Agene, J.I., Umar, A.I., 2015. Geophagic clay materials from Nigeria: a potential source of heavy metals and human health implications in mostly women and children who practice it. Environmental Geochemistry and Health 37 (2), 363−375. Available from: https://doi.org/10.1007/s10653-014-9653-0. Epub 2014 Nov 23.

Laufer, B., 1930. Geophagy. Fieldiana Anthropology, 18. Field Museum Press, Chicago, IL, p. 104.

Lee, A., Marsden, S., 2014. Dishing the dirt on Peru's parrots. http://stuartmarsden.blogspot.com/2014/02/dishing-dirt-on-perus-parrots.html. (Accessed 09 March 2020).

Lee, A.T.K., Kumar, S., Brightsmith, D.J., Marsden, S.J., 2010. Parrot claylick distribution in South America: do patterns of "where" help answer the question "why"? Ecography 33 (3), 503−513. Available from: https://doi.org/10.1111/j.1600-0587.2009.05878.x.

Lohn, J.W.G., Austin, R.C., Winslet, M.C., 2000. Unusual causes of small-bowel obstruction. Journal of the Royal Society of Medicine 93 (7), 365−368. https://doi.org/10.177/014107/680009300707.

Macheka, L.R., Olowoyo, J.O., Matsela, L., Khine, A.A., 2016. Prevalence of geophagia and its contributing factors among pregnant women at Dr. George Mukhari Academic Hospital, Pretoria. African Health Sciences 16 (4), 972−978. Available from: https://doi.org/10.4314/ahs.v16i4.13.

Mahaney, W.C., Milner, M.W., Hs, M., Hancock, R.G.V., Aufreiter, S., Reich, M., et al., 2000. Mineral and chemical analyses of soils eaten by humans in Indonesia. International Journal of Environmental Health Research 10, 93−109.

Malepe, R.E., Candeias, C., Mouri, H., 2023. Geophagy and its potential human health implications - A review of some cases from South Africa. Journal of African Earth Sciences 200, 104848. Available from: https://doi.org/10.1016/j.jafrearsci.2023.104848 (Accessed 22 February 2024).

Mills, A., Milewski, A., 2007. Geophagy and nutrient supplementation in the Ngorongoro Conservation Area, Tanzania, with particular reference to selenium, cobalt and molybdenum. Journal of Zoology 271, 110−118. Available from: https://doi.org/10.1111/j.1469-7998.2006.00241.x.

Minami, K., 2009. Soil and humanity: culture, civilization, livelihood and health. Soil Science and Plant Nutrition 55 (5), 603−615. Available from: https://doi.org/10.1111/j.1747-0765.2009.00401.x.

Moosavi, M., 2017. Bentonite clay as a natural remedy: a brief review. Iranian Journal of Public Health 46 (9), 1176−1183.

Mouri, H., Malepe, R.E., Candeias, C., 2023. Geochemical composition and potential health risks of geophagic materials: An example from a rural area in the Limpopo Province of South Africa. Environmental Geochemistry and Health 45 (8), 6305−6322. Available from: https://doi.org/10.1007/s10653-023-01551-6 (Accessed 22 February 2024).

Msibi, A.T., 2014. The Prevalence and Practice of Geophagia in Mkhanyakude District of KwaZulu-Natal, South Africa (Master of Science Thesis in Human Nutrition). University of KwaZulu-Natal, South Africa. http://ukzndspace.ukzn.ac.za/bitstream/handle/10413/12567/Msibi_Agnes_Thembisile_2014.pdf?sequence = 1&isAllowed = y. (Accessed 25 January 2020).

MUV (Medical University of Vienna), 2016. Geophagy: 'Soil-eating' as an addictive behavior. https://www.sciencedaily.com/releases/2016/12/161205085943.htm. (Accessed 21 January 2020).

Myers, B., 2019. Exploring Adaptive Functions of Geophagy Across Non-human Primates: New Evidence for Sexual Selection (Honours Thesis (HONR 499)). Ball State University Muncie, Indiana. http://cardinalscholar.bsu.edu/bitstream/handle/123456789/201922/2019MyersBre-combined.pdf?sequence = 1. (Accessed 10 February 2020).

Mylonas, M.M., 2013. Geophagy. Encyclopaedia of Arkansas. https://encyclopediaofarkansas.net/entries/geophagy-7022/. (Accessed 25 January 2020).

Narayanan, S.K., Akbar Sherif, V.S., Babu, P.R., Nandakumar, T.K., 2008. Intestinal obstruction secondary to a colonic lithobezoar. Journal of Pediatric Surgery 43 (7), e9−e10. Available from: https://doi.org/10.1016/j.jpedsurg.2008.02.065 (Accessed 22 February 2024).

Ngole, V.M., Ekosse, G.E., de Jager, L., Songca, S.P., 2010. Physicochemical characteristics of geophagic clayey soils from South Africa and Swaziland. African Journal of Biotechnology 9 (36), 5929−5937.

Ngole-Jeme, V., Ekosse, G., Songca, S., 2018. An analysis of human exposure to trace elements from deliberate soil ingestion and associated health risks. Journal of Exposure Science and Environmental Epidemiology 28, 55−63. Available from: https://doi.org/10.1038/jes.2016.67.

Ngure, F.M., Humphrey, J.H., Mbuya, M.N.N., Majo, F., Mutasa, K., et al., 2013. Formative research on hygiene behaviours and geophagy among infants and young children and implications of exposure to faecal bacteria. American Journal of Tropical Medicine and Hygiene 89 (4), 709−716. Available from: https://doi.org/10.4269/ajtmh.12-0568.

Njiru, H., Elchalal, U., Paltiel, O., 2011. Geophagy during pregnancy in Africa. A literature review. Obstetrical and Gynecological Survey 66 (7), 452−459.

Nuton, V., 2020. Galen: Greek physician. Encyclopaedia Britannica. https://www.britannica.com/biography/Galen. (Accesssed 09 February 2020).

Nyanza, E.C., Joseph, M., Premji, S.S., Thomas, D.S.K., Mannion, C., 2014. Geophagy practices and the content of chemical elements in the soil eaten by pregnant women in artisanal and small scale gold mining communities in Tanzania. BMC Pregnancy and Childbirth 14, 144. Available from: https://doi.org/10.1186/1471-2393-14-144.

Obeid, R., Murphy, M., Solé-Navais, P., Yajnik, C., 2017. Cobalamin status from pregnancy to early childhood: Lessons from global experience. Advances in Nutrition 8 (6), 971−979. Available from: https://doi.org/10.3945/an.117.015628.

Odangowei, O.I., Okiemute, O., 2015. Geophagic practice and its possible health implications - a review. Journal of Sciences and Multidisciplinary Research 7 (2), 100−110.

Odongo, A.O., Moturi, W.N., Wangari, S.N., 2016. Geophagic behaviour and factors influencing it among pregnant women: a case study of Nakuru Municipality, Kenya'. International Journal of Behavioural and Healthcare Research 6 (1), 28−41.

Ogomaka, I.A., 2015. Microorganisms associated with clay (NZU) consumption (Geophagy) in some parts of Imo State, Nigeria. International Journal of Current and Microbiology and Applied Sciences 4 (1), 55−557.

Okereke, J.N., Obasi, K.O., Nwadike, P.O., Ezeji, E.U., Udebuani, A.C., 2015. Geo-helminths associated with geophagic pupils in selected primary schools in Oyi, Anambra State. Science Journal of Public Health 3 (5-1), 45−50. Available from: https://doi.org/10.11648/j.sjph.s.2015030501.19.

Olaniran, A.O., Adhika, B., Pillay, B., 2013. Bioavailability of heavy metals in soil: impact on microbial biodegradation of organic compounds and possible improvement strategies. International Journal of Molecular Sciences 14, 10197−10228. Available from: https://doi.org/10.3390/ijms140510197.

Olisa, Olajide-Kayode, J.O., Adebayo, B.O., Ajayi, O.A., Odukoya, K., Olalemi, A.A., Uyakunmor, T.B-M., 2023. Mineralogy and geochemical characterization of geophagic clays consumed in parts of southern Nigeria. Journal of Trace Elements and Minerals. Journal of Trace Elements and Minerals 4. Available from: https://doi.org/10.1016/j.jtemin.2023.100063 (Accessed 22 February 2024).

Osazuwa, F., Ayo, O.M., Imade, P., 2011. A significant association between intestinal helminth infection and anaemia burden in children in rural communities of Edo state, Nigeria. North American Journal of Medical Sciences 3 (1), 30−34. Available from: https://doi.org/10.4297/najms.2011.330.

Otten, J.J., Hellwig, J.P., Meyers, L.D., 2006. Dietary DRI Reference Intakes: The Essential Guide to Nutrient Requirements. Institute of Medicine of the National Academies. The National Academies Press, Washington, DC.

Ozumba, U.C., Ozumba, N., 2002. Patterns of helminth infection in the human gut at the University of Nigeria Teaching Hospital, Enugu, Nigeria. Journal of Health Science 48, 263−268.

Palm, C., Sanchez, P., Ahamed, S., Awiti, A., 2007. Soils: a contemporary perspective. Annual Review of Environment and Resources 32 (1), 99−129.

Panichev, A.M., Baranovskaya, N.Y., Seryodkin, I.V., Chekryzhov, I.Y., Soktoev, B.R., Ivanov, V.V., Vakh, E.A., Desyatova, T.V., Lutsenko, T.N., et al., 2023. The man cause of geophagy according to extensive studies on Olkhon Island, Lake Baikal. Geosciences 13 (7), 211. Available from: https://doi.org/10.3390/geosciences13070211 (Accessed 22 February 2024).

Panichev, A.M., Seryodkin, I.V., Kalinkin, Y.N., et al., 2018. Development of the "rare-earth" hypothesis to explain the reasons for geophagy in Teletskoye Lake are kudurs (Gorny Altai, Russia). Environmental Geochemistry and Health 40, 1299—1316. Available from: https://doi.org/10.1007/s10653-017-0056-x.

Pebsworth, P.A., Bardi, M., Huffman, M.A., 2011. Geophagy in chacma baboons: patterns of soil consumption by age class, sex, and reproductive state. American Journal of Primatology 74 (1), 48—57. Available from: https://doi.org/10.1002/ajp.21008.

Pebsworth, P.A., Huffman, M.A., Lambert, J.E., Young, S.L., 2019. Geophagy among nonhuman primates: a systematic review of current knowledge and suggestions for future directions. https://doi.org/10.1002/ajpa.23724.

Pemunta, N., 2014. The 'gendered field' of kaolinite clay production: performance characteristics among the Balengou. Social Analysis: Journal of Cultural and Social Practice 58 (2), 21—41. Available from: https://doi.org/10.3167/sa.2014.580202.

Pérez-Rodríguez, J.L. (Ed.), 2003. Applied Study of Cultural Heritage and Clays. Conselo Superior de Investigaciones Scientifícas, Madrid.

Perridge, A., Fossey, A., Olivier, D., Ekosse, G.E., de Jager, L., 2011. Geophagic soil colour and nematode content in the District of Thabo Mofutsanyane, Free State, South Africa. In: Ekosse, G.E., de Jager, L., Ngole, V.M. (Eds.), An innovative perspective on the role of clays and clay minerals, and geophagia on eco-economic development. Conference Proceedings of the First International Conference on Clays and Clay Minerals in Africa, and Secondnd International Conference on Geophagia in Southern Africa, Bloemfontein, South Africa, pp. 314—320.

Phakoago, M.V., Ekosse, G.E., Odiyo, J.O., 2019. The prevalence of geophagic practices and causative reasons for geophagia in Sekhukhune area, Limpopo Province, SouthAfrica. Transactions of the Royal Society of South Africa 74 (1), 19—26. Available from: https://doi.org/10.1080/0035919X.2019.1572669.

Prince, I., 1989. Pica and geophagia in cross-cultural perspective. Transcultural Psychiatric Research Review 26 (3), 167—197. Available from: https://doi.org/10.1177/136346158902600301.

Psychology Wiki, 2019. Geophagy. https://psychology.wikia.org/wiki/Geophagy. (Accessed 15 February 2020).

Ravuluvulu, F.R., 2018. Effects of Open Defecation on Geophagic Soils and Water Resources: A Case Study of Siloam Village in Limpopo Province, South Africa (Masters Degree thesis). Department of Hydrology and Water Resources, University of Venda, South Africa.

Reid, R.M., 1992. Cultural and medical perspectives on geophagia. Medical Anthropology 13, 1337—1351.

Rieuwerts, J.S., Thornton, I., Farago, M.E., Ashmore, M.R., 1998. Factors influencing metal bioavailability in soils: preliminary investigations for the development of a critical loads approach for metals. Chemical Speciation and Bioavailability 10 (2), 61—75. Available from: https://doi.org/10.3184/095422998782775835.

Root-Bernstein, R., 2000. Honey, Mud, Maggots and Other Medical Marvels: The Science Behind Folk Remedies and Old Wives' Tales. Pan Books, London, p. 280.

Rosado, J.L., Ronquillo, D., Kordas, K., Rojas, O., Alatorre, J., Lopez, P., et al., 2007. Arsenic exposure and cognitive performance in Mexican schoolchildren. Environmental Health Perspectives 115 (9), 1371—1375. Available from: https://doi.org/10.1289/ehp.9961.

Rosenthal, J., Lopez-Pazos, E., Dowling, N.F., Pfeiffer, C.M., Mulinare, J., Vellozzi, C., et al., 2015. Folate and vitamin B12 deficiency among non-pregnant women of childbearing-age in Guatemala 2009-2010: prevalence and identification of vulnerable populations. Journal of Maternal and Child Health 19, 2272—2285.

Rowley, C.A., Kendall, M.M., 2019. To B12 or not to B12: five questions on the role of cobalamin in host-microbial interactions. PLoS Pathogens 15 (1), e1007479. Available from: https://doi.org/10.1371/journal.ppat.1007479.

Saathoff, E., Olsen, A., Kvalsvig, J.D., Geissler, P.W., 2002. Geophagy and its association with geohelminth infection in rural schoolchildren from northern KwaZulu-Natal, South Africa. Transactions of the Royal Society of Tropical Medicine and Hygiene 96, 485—490. Available from: https://doi.org/10.1016/S0035-9203(02)90413-X.

Saathoff, E., Olsen, A., Kvalsvig, J.D., et al., 2004. Patterns of geohelminth infection, impact of albendazole treatment and re-infection after treatment in schoolchildren from rural KwaZulu-Natal/South-Africa. British Medical C Infectious Diseases 4, 27. Available from: https://doi.org/10.1186/1471-2334-4-27.

Samuel, T.M., Duggan, C., Thomas, T., Bosch, R., Rajendran, R., Virtanen, S.M., et al., 2013. Vitamin B(12) intake and status in early pregnancy among urban South Indian women. Annals of Nutrition and Metabolism 62, 113−122.

Sanchez, P.A., 2019. Chapter 8: Mineralogy. In: Sanchez, P.Z. (Ed.), Properties and Management of Soils in the Tropics, Part II - Pedology, Physics, Chemistry and Biology. Cambridge University Press, pp. 196−209. Available from: https://doi.org/10.1017/9781316809785.010.

Severance, H.W., Holt, T., Patrone, N.A., Chapman, L., 1988. Profound muscle weakness and hypokalemia due to clay ingestion. Southern Medical Journal 18, 272−274.

Shigova, W., Moturi, W., 2009. Geophagia as a risk factor for diarrhoea. Journal of Infection in Developing Countries 3, 94−98.

Shinondo, C.J., Mwikuma, G., 2009. Geophagy as a risk factor for helminth infections in pregnant women in Lusaka, Zambia. Medical Journal of Zambia 35 (2), 48−52.

Solaini, L., Gardani, M., Ragni, F., 2012. Geophagia: an extraordinary cause of perforation of the sigmoid colon. Surgery 152, 136−137.

Solomon, A.O., 2005. Radiation levels in rocks from the Jos Plateau, Nigeria (Ph.D. thesis). University of Jos, Plateau State, Nigeria.

Sumbele, I., Ngole, V.M., Ekosse, G., 2014. Influence of physico-chemistry and mineralogy on the occurrence of geohelminths in geophagic soils from Eastern Cape, South Africa, and their possible implication on human health. International Journal of Environmental Health Research 24 (1), 18−30.

Szajewska, H., Dziechciarz, P., Mrukowicz, J., 2006. Meta-analysis: smectite in the treatment of acute infectious diarrhoea in children. Alimentary Pharmacology and Therapeutics 23 (2), 217−227. Available from: https://doi.org/10.1111/j.1365-2036.2006.02760.x.

Tayie, F., 2004. Pica: motivating factors and health issues. African Journal of Food, Agriculture, Nutrition and Development 4 (1), 1684−5374.

UCPJ (University of Chicago Press Journals), 2011. Eating dirt can be good for the belly, researchers find. Science Daily https/…. (Accessed 09 August 2015).

USEPA (United States Environmental Protection Agency), 2007. Guidance for evaluating the oral bioavailability of metals in soils for use in human health risk assessment. OSWER 9285.7-80. https://nepis.epa.gov/Exe/ZyNET. exe/93001C3I.TXT?ZyActionD = ZyDocument&Client = EPA&Index = 2006 + . (Accessed 24 February 2020).

Vermeer, D.E., 1966. Geophagy among the Tiv of Nigeria. Annals of the Association of American Geographers 56 (2), 197−204.

Vermeer, D.E., Frate, D.A., 1975. Geophagy in a Mississippi County. Annals of the Association of American Geographers 65 (3), 414−424.

Vermeer, D.E., Ferrell, R.E., 1985. Nigerian geophagical clay: a traditional antidiarrheal pharmaceutical. Science (New York, NY) 227, 634−636. Available from: https://doi.org/10.1126/science.3969552.

Voigt, C.C., Capps, K.A., Dechmann, D.K., Michener, R.H., Kunz, T.H., 2008. Nutrition or detoxification: why bats visit mineral licks of the Amazonian Rainforest. PLoS One 3 (4), e2011. Available from: https://doi. org/10.1371/journal.pone.0002011.

Wayne, F., 2004. Dixie's forgotten people: the South's poor whites. *Indiana University Press*, p. 40.

WHO (World Health Organization), 1996. Trace Elements in Human Nutrition and Health. World Health Organization, Geneva, p. 343.

WHO (World Health Organization), 2019. Key facts: soil transmitted helminth infections. https://www.who. int/news-room/fact-sheets/detail/soil-transmitted-helminth-infections. (Accessed 12 February 2020).

Wigle, D.T., Arbuckle, T.E., Walker, M., Wade, M.G., Liu, S., Krewski, D., 2007. Environmental hazards: evidence for effects on child health. Journal of Toxicology and Environmental Health Part B; Critical Reviews 10 (1 - 2), 3–39. Available from: https://doi.org/10.1080/10937400601034563.

Wigle, D.T., Arbuckle, T.E., Turner, M.C., Bérubé, A., Yang, Q., Liu, S., et al., 2008. Epidemiologic evidence of relationships between reproductive and child health outcomes and environmental chemical contaminants. Journal of Toxicology and Environmental Health Part B; Critical Reviews 11 (5 - 6), 373–517. Available from: https://doi.org/10.1080/10937400801921320.

Wiley, A.S., Katz, S.H., 1998. Geophagy in pregnancy: a test of a hypothesis. Current Anthropology 39 (4), 532–545.

Williams, L.B., Haydel, S.E., 2010. Evaluation of the medicinal use of clay minerals as antibacterial agents. International Geology Reviews 52 (7/8), 745–770. Available from: https://doi.org/10.1080/00206811003679737.

Wilson, M.J., 2003. Clay mineralogical and related characteristics of geophagic material. Journal of Chemical Ecology 29 (7), 1525–1547.

Woywodt, A., Kiss, A., 2002. Geophagia: the history of earth-eating. Journal of the Royal Society of Medicine 95 (3), 143–146. Available from: https://doi.org/10.1258/jrsm.95.3.143.

Wragg, J., Cave, M.R., Basta, N., Brandon, E., Casteel, S., et al., 2011. An inter-laboratory trial of the unified BARGE bioaccessibility method for arsenic, cadmium and lead in soil. Science of the Total Environment 409, 4016–4030.

Yanai, J., Noguchi, J., Yamada, H., Sugihara, S., Kilasara, M., Kosaki, T., 2009. Functions of geophagy as supplementation of micronutrients in Tanzania. Soil Science and Plant Nutrition 55 (1), 215–223. Available from: https://doi.org/10.1111/j.1747-0765.2008.00346.x.

Young, S.L., 2010. Pica in pregnancy: New ideas about old condition. Annual Reviews of Nutrition 30, 403–422. Available from: https://doi.org/10.1146/annurev.nutr.012809.104713.

Young, S., 2011. Craving Earth: Understanding Pica. Columbia University Press, New York, NY.

Young, S.L., Goodman, D., Farag, T.H., Ali, S.M., Khatib, M.R., Khalfan, S.S., et al., 2007. Geophagia is not associated with Trichuris or hookworm transmission in Zanzibar, Tanzania. Transactions of the Royal Society of Tropical Medicine and Hygiene 101, 766–772. Available from: https://doi.org/10.1016/j.trstmh.2007.04.016.

Young, S., Wilson, M.J., Miller, D., Hillier, S., 2008. Toward a comprehensive approach to the collection and analysis of pica substances, with emphasis on geophagic material. PLoS One 3 (9), e3147. Available from: https://doi.org/10.1371/journal.pone.0003147.

Young, S.L., Khalfan, S.S., Farag, T.H., et al., 2010a. Association of pica with anemia and gastrointestinal distress among pregnant women in Zanzibar, Tanzania. American Journal of Tropical Medicine and Hygiene 83, 144–151. Available from: https://doi.org/10.4269/ajtmh.2010.09-0442.

Young, S., Wilson, M.J., Hillier, S., Delbos, E., Ali, S.M., Stoltzfus, R.J., 2010b. Differences and commonalities in physical, chemical and mineralogical properties of Zanzibari geophagic soils. Journal of Chemical Ecology 36, 129–140.

Young, S., Sherman, P., Lucks, J., Pelto, G., Rowe, L., 2011. Why on Earth?: Evaluating hypotheses about the physiological functions of human geophagy. The Quarterly Review of Biology 86 (2), 97–120. Available from: https://doi.org/10.1086/659884.

Further reading

Abrahams, P.W., 1997. Geophagy (soil consumption) and iron supplementation in Uganda. Tropical Medicine and International Health 2, 617–623.

Abrahams, P.W., 1999. The chemistry and mineralogy of three Savanna lick soils. Journal of Chemical Ecology 25 (10), 2215–2228.

Abu, B.A., van den Berg, V.L., Raubenheimer, J.E., Louw, V.J., 2017. Pica practices among apparently healthy women and their young children in Ghana. Physiology and Behaviour 177, 297–304. Available from: https://doi.org/10.1016/j.physbeh.2017.04.012.

Agomuo, E.N., Amadi, P.U., Adumekwe, C., 2019. Gestational geophagia affects nephrocardiac integrity, ATP-driven proton pumps, the Renin-Angiotensin-Aldosterone system, and F2-Isoprostane Status. Medical Sciences 7 (2), 13. Available from: https://doi.org/10.3390/medsci7020013.

Antelman, G., Msamanga, G.I., Spiegelman, D., Urassa, E.J., Narh, R., Hunter, D.J., et al., 2000. Nutritional factors and infectious disease contribute to anemia among pregnant women with human immunodeficiency virus in Tanzania. Journal of Nutrition 130, 1950–1957.

ATSDR (Agency for Toxic Substances and Disease Registry), 2001. Summary Report for the ATSDR Soil-Pica Workshop, Atlanta, GA. Prepared by Eastern Research Group, Lexington, MA.

ATSDR (Agency of Toxic Substances and Disease Registry), 1999. Toxicological Profile for mercury. US Department of Public Health and Human Services. Public Health Services, Atlanta, GA.

ATSDR (Agency of Toxic Substances and Disease Registry), 2004. Toxicological Profile of Copper. US Department of Public Health and Human Services, Public Health Services, Atlanta, GA.

ATSDR (Agency of Toxic Substances and Disease Registry), 2005. Toxicological Profile of Zinc. US Department of Public Health and Human Services, Public Health Services, Atlanta, GA.

ATSDR (Agency of Toxic Substances and Disease Registry), 2007a. Toxicological Profile for Arsenic. US Department of Public Health and Human Services, Public Health Services, Atlanta, GA.

ATSDR (Agency of Toxic Substances and Disease Registry), 2007b. Toxicological Profile of Lead. US Department of Public Health and Human Services, Public Health Services, Atlanta, GA.

ATSDR (Agency of Toxic Substances and Disease Registry), 2009. Toxicological Profile of Nickel. US Department of Public Health and Human Services, Public Health Services, Atlanta, GA.

ATSDR (Agency for Toxic Substances and Disease Registry), 2012. Toxicological Profile for Manganese. U.S. Department of Public Health and Human Services, Public Health Services, Atlanta, GA.

ATSDR (Agency of Toxic Substances and Disease Registry), 2013. Minimal risk levels for hazardous substances (MRLs). http://www.atsdr.cdc.gov/mrls/mrllist.asp.

Baidoo, S.E., Tay, S.C.K., Obiri-Danso, K., Abruquah, H.H., 2010. Intestinal helminth infection and anaemia during pregnancy: a community based study in Ghana. Journal of Bacteriological Research 2, 9–13.

Baptista, S.L., Pinto, P.V., da Conceição Freitas, M., Cruz, C., Palmeirim, J.M., 2012. Geophagy by African ungulates: the case of the critically endangered giant sable antelope of Angola (*Hippotragus niger variani*). African Journal of Ecology 51 (1), 139–146. Available from: https://doi.org/10.1111/aje.12020.

Basden, G.T., 1938. Cited in: Young, S.L., 2011. Craving Earth: Understanding Pica: The Urge to Eat Clay, Starch, Ice, and Chalk. Columbia University Press, New York, p. 228.

Bonglaisin, J.N., 2015. Induced geophagy with local kaolin from Cameroon Market and heavy metals (lead, cadmium and mercury) profile of rat blood, liver, placentas and litters. Journal of Medical Sciences 15, 10–17. Available from: https://doi.org/10.3923/jms.2015.10.17.

Bonglaisin, J.N., Mbofung, C.M.F., Lantum, D.N., 2011. Intake of lead, cadmium and mercury in kaolin-eating: a qualitative assessment. Journal of Medical Sciences 1, 267–273. Available from: https://doi.org/10.3923/jms.2011.267.273.

Bonglaisin, J.N., Mbofung, C.M.F., Lantum, D.N., 2015. Geophagy and heavy metals (Pb, Cd and Hg) content of local kaolin varieties in the Cameroon market: assessment indices for contamination and risk of consumption or toxicity to the population. Journal of Medical Sciences 15, 1–9. Available from: https://doi.org/10.3923/jms.2015.1.9.

Brooker, S.J., Bundy, D.A.P., 2014. 55 - Soil-transmitted helminths (geohelminths). In: Farrar, J., et al., (Eds.), Manson's Tropical Infectious Diseases, twenty-third ed. W.B. Saunders, London, pp. 766–794. Available from: https://doi.org/10.1016/B978-0-7020-5101-2.00056-X.

Cassenote, A.J.F., de Lima, A.R.A., Neto, J.M., Rubinsky-Elefant, G., 2014. Seroprevalence and modifiable risk factors for *Toxocara* spp. in Brazilian schoolchildren. PLoS Neglected Tropical Diseases 8, e2830. Available from: https://doi.org/10.1371/journal.pntd.0002830.

Casteel, S.W., Weis, C.P., Henningsen, G.M., Brattin, W.J., 2006. Estimation of relative bioavailability of lead in soil and soil-like materials using young swine. Environmental Health Perspectves 114, 1162−1171.

Cave, M.R., Wragg, J., Harrison, I., Vane, H., Van de Wiele, T., et al., 2010. Comparison of batch mode and dynamic physiologically based bioaccessibility tests for PAHs in soil samples. Environmental Science and Technology 44, 2654−2660.

Corbett, R.W., Ryan, C., Weinrich, S.P., 2003. Pica in pregnancy: does it affect pregnancy outcomes? MCN American Journal of Maternal and Child Nursing 28, 183−189. Available from: https://doi.org/10.1097/00005721-200305000-00009.

Danford, D., 1982. Pica and nutrition. Annual Review of Nutrition 2, 303−322.

de Reims, C.H.U., 2019. Prevalence of geophagy and knowledge about its health effects among native Sub-Saharan Africa, Caribbean and South America healthy adults living in France (Prevalgeophagy). US National Library of Medicine; ClinicalTrials.Gov. https://clinicaltrials.gov/ct2/show/NCT03948126. (Accessed 26 January 2020).

Dean, J.R., Deary, M.E., Gbefa, B.K., Scott, W.C., 2004. Characterisation and analysis of persistent organic pollutants and major, minor and trace elements in Calabash chalk. Chemosphere 57, 21−25.

Diamond, J.M., 1999. Evolutionary biology: dirty eating for healthy living. Nature 400 (6740), 120−121. Available from: https://doi.org/10.1038/22014.

Diko, M.L., Siewe épse Diko, C.N., 2014. Physico-chemistry of geophagic soils ingested to relief nausea and vomiting during pregnancy. African Journal of Traditional, Complementary and Alternative Medicine 11 (3), 21−24. Available from: https://doi.org/10.4314/ajtcam.v11i3.4.

Dissanayake, C.B., Chandrajith, R., 2009. Introduction to Medical Geology. Springer Science and Business Media, Dordrecht, The Netherlands, p. 297.

Dooyema, C.A., Neri, A., Lo, Y., Durant, J., Dargan, P.I., Swarthout, T., et al., 2001. Oral bioavailability of lead and arsenic from a NIST standard reference soil material. Archives of Environmental Contamination and Toxicology 40, 128−135.

Enviromedica, 2018. Eating clay: lessons on medicine from worldwide cultures. https://www.enviromedica.com/wellness/eating-clay-lessons-on-medicine-from-worldwide-cultures/. (Accessed 15 February 2020).

FDA (Food and Drug Administration), 2001. Dietary Reference Intakes for Vitamin A, Vitamin K, Arsenic, Boron, Chromium, Copper, Iodine, Iron, Manganese, Nickel, Silicon, Vanadium, and Zinc. Report of the Panel on Micronutrients, Food and Drug Administration. Dietary Supplements. Center for Food Safety and Applied Nutrition, National Academy Press, Washington DC.

FNB (Food and Nutrition Board), 2001. Dietary Reference Intakes: Vitamin A, Vitamin K, Arsenic, Boron, Chromium, Copper, Iodide, Iron, Manganese, Molybdenum, Nickel, Silicon, Vanadium and Zinc. Institute of Medicine, National Academy Press, Washington DC.

Fosso-Kankeu, E., Waanders, F., Netshitanini, T., Ubomba-Jaswa, E., Abia, K., 2015. Identification of metals in geophagic clays: investigation of their behaviour in simulated gastric fluid, Seventh International Conference on Latest Trends in Engineering and Technology (ICLTET'2015), November 26−27, 2015, Irene, Pretoria, South Africa.

Geissler, P.W., Mwaniki, D.L., Thiong'o, F., Michaelsen, K.F., Friis, H., 1998a. Geophagy, iron status and anaemia among primary school children in Western Kenya. Tropical Medicine and International Health 3, 529−534.

Geissler, P.W., Shulman, C.E., Prince, R.J., Mutemi, W., Mnazi, C., Friis, H., et al., 1998b. Geophagy, iron status and anaemia among pregnant women on the coast of Kenya. Transactions of the Royal Society of Tropical Medicine and Hygiene 5, 549−553.

Glickman, L.T., Camara, A.O., Glickman, N.W., McCabe, G.P., 1999. Nematode intestinal parasites of children in rural Guinea, Africa: prevalence and relationship to geophagia. International Journal of Epidemiology 28, 169−174.

Gomes, C.S.F., Silva, J.B.P., 2007. Minerals and clay minerals in medical geology. Applied Clay Science 36 (1 - 3), 4−21.

Grøn, C., Andersen, L., 2003. Human bioaccessibility of heavy metals and PAH from soil. Danish Environmental Protection Agency. Environmental Project 840, 1−113.

Haladu, S., Sani-Gwarzo, N., Nguku, P.M., Akpan, H., Idris, S., Bashir, A.M., et al., 2012. Outbreak of fatal childhood lead poisoning related to artisanal gold mining in northwestern Nigeria. Environmental Health Perspectives 120 (4), 601−607. Available from: https://doi.org/10.1289/ehp.1103965.

Hertz, H., 1947. Notes on clay and starch eating among Negroes in a southern urban community. Social Forces 25, 343−344.

Highfield, R., 2008. The medicinal monkey; Roots of modern healing found in our hairy relatives. Daily Telegraph, 3 April, AL9.

Horner, R.D., Lackey, C.J., Kolasa, K., Warren, K., 1991. Pica practices of pregnant women. Journal of the American Dietetic Association 91, 34−38.

Hunter, J.M., de Kleine, R., 1984. Geophagy in Central America. Geographical Review 74 (2), 157−169.

Hunter, J.M., 1993. Macroterm geophagy and pregnancy clays in southern Africa. Journal of Cultural Geography 14, 69−92.

Huyck, K.L., Kile, M.L., Mahiuddin, G., Quamruzzaman, Q., Rahman, M., Breton, C.V., et al., 2007. Maternal exposure associated with low birth weight in Bangladesh. Journal of Occupational Environmental Medicine 49 (10), 1097−1104. Available from: https://doi.org/10.1097/JOM.0b013e3181566ba0.

Intawongse, M., Dean, J.R., 2006. In-vitro testing for assessing oral bioaccessibility of trace metals in soil and food samples. Trends in Analytical Chemistry 25, 876−886.

Johns, T., 1996. The Origins of Human Diet and Medicine. University of Arizona Press, Tucson, AZ, p. 176.

Johns, T., 2000. Well-grounded diet. In: Goodman, A., Dufour, D., Pelto, G. (Eds.), Nutritional Anthropology: Biocultural Perspectives on Food and Nutrition. McGraw-Hill, San Francisco, CA, pp. 122−126.

Kalguem, E., 2019. Geophagic clayey materials of Sabga Locality (northwest Cameroon): genesis and medical interest. Earth Sciences 8, 45. Available from: https://doi.org/10.11648/j.earth.2019081.14.

Kalumanga, E., Mpanduji, D.G., Cousins, S.A.O., 2016. Geophagic termite mounds as one of the resources for African elephants in Ugalla Game Reserve, Western Tanzania. African Journal of Ecology 55 (1), 91−100.

Katz, S., Hediger, M.L., Valleroy, L.A., 1974. Traditional maize processing techniques in the New World. Science (New York, NY) 184, 765−773.

Kawai, K., Saathoff, E., Antelman, G., Msamanga, G., Fawzi, W.W., 2009. Geophagy (soil-eating) in relation to Anemia and Helminth infection among HIV-infected pregnant women in Tanzania. American Journal of Tropical Medicine and Hygiene 80, 36−43.

Kenne, K.E.D., Armand, S.L.W., Njopwouo, D., Ekosse, G.-I., 2018. Physico-chemical characterization of clayey materials consumed by geopagists in locality of Sabga (northwestern Cameroon): health implications. International Journal of Applied Science and Technology 8. Available from: https://doi.org/10.30845/ijast.v8n3p6.

Kennedy, K., 2008. Dirt poor: eating mud to survive. Maclean's 121 (32), 30.

Klaus, G., Klaus-Hügia, C., Schmid, B., 1998. Geophagy by large mammals at natural licks in the rain forest of the Dzanga National Park, Central African Republic. Journal of Tropical Ecology 14, 829−839.

Korfali, S.I., Hawi, T., Mroueh, M., 2013. Evaluation of heavy metals content in dietary supplements in Lebanon. Chemical Central Journal 7, 10. Available from: https://doi.org/10.1186/1752-153X-7-10.

Kortei, N.K., Annor, I.A., Aboagye, G., Yaw, N., et al., 2019. Elemental minerals and microbial compositions as well as knowledge and perceptions regarding kaolin (clay) consumption by pregnant women in the Ho municipality of Ghana. Pan African Medical Journal 113 (34), 1−17. Available from: https://doi.org/10.11604/pamj.2019.34.113.17394.

Krief, S., Huffman, M.A., Sévenet, T., Hladik, C.-M., Grellier, P., Loiseau, P.M., et al., 2006. Bioactive properties of plant species ingested by chimpanzees in the Kibale National Park, Uganda. American Journal of Primatology 68 (1), 51−71.

La Voliere Parrot Boutique, 2020. Clay-Cal. https://www.lavoliere.ca/product_info.php?products_id = 1439&language = en. (Accessed 09 March 2020).

Lacey, E.P., 1990. Broadening the perspective of pica: literature review. Public Health Reports 105, 29−35.

Lallanilla, M. (Ed.), 2005. Eating Dirt: It Might Be Good for You. ABC News, Retrieved 9 August 2015.

Lambert, V., Boukhari, R., Nacher, M., et al., 2010. Plasma and urinary aluminum concentrations in severely anemic geophagous pregnant women in the Bas Maroni region of French Guiana: a case-control study. American Journal of Tropical Medicine and Hygiene 83, 1100−1105. Available from: https://doi.org/10.4269/ajtmh.2010.10-0370.

Lander, E.R., Feller, C. (Eds.), 2010. Soil and Culture. Springer, Dordrecht. Available from: https://doi.org/10.1007/978-90-481-2960-7_1.

Lin, J.W., Temple, L., Trujillo, C., et al., 2015. Pica during pregnancy among Mexican-born women: a formative study. Maternal and Child Nutrition 11, 550−558. Available from: https://doi.org/10.1111/mcn.12120.

MacClancy, J., Henry, J., MacBeth, H. (Eds.), 2007. Consuming the Inedible: Neglected Dimensions of Food Choice. Berghahn Books, New York, NY.

Maddaloni, M., Lolacono, N., Manton, W., Blum, C., Drexler, J., et al., 1998. Bioavailability of soilborne lead in adults, by stable isotope dilution. Environmental Health Perspectives 106 (Supplement 6), 1589−1594.

Mascolo, N., Summa, V., Tateo, F., 2004. *In vivo* experimental data on the mobility of hazardous chemical elements from clays. Applied Clay Science 25, 23−28.

Mathee, A., Naicker, N., Kootbodien, T., et al., 2014. A cross-sectional analytical study of geophagia practices and blood metal concentrations in pregnant women in Johannesburg, South Africa. South African Medical Journal of S Afr Tydskr Vir Geneeskd 104, 568−573. Available from: https://doi.org/10.7196/samj.7466.

Matsubayashi, H., Lagan, P., Majalap, N., Tangah, J., Sukor, J.R.A., Kitayama, K., 2007. Importance of natural licks for the mammals in Bornean inland tropical rain forests. Ecological Research 22, 742−748.

Meel, B.L., 2012. Geophagia in Transkei region of South Africa: case reports. African Health Sciences 12 (4), 566−568. Available from: https://doi.org/10.4314/ahs.v12i4.27.

Middleton, J.D., 1989. Sikor: an unquantified hazard. British Medical Journal 298, 407−408.

Mireku, M.O., Davidson, L.L., Zoumenou, R., Massougbodji, A., Cot, M., Bodeau-Livinec, F., 2018. Consequences of prenatal geophagy for maternal prenatal health, risk of childhood geophagy and child psychomotor development. Tropical Medicine and International Health 23, 841−849. Available from: https://doi.org/10.1111/tmi.13088.

Momoh, A., Davies, T.C., Akinsola, H., Akinyemi, S.A., Mhlongo, S., Gitari, W., et al., 2013. Human bioaccessibility of Fe, Mn, Zn and Cu from consumed earth materials in Vhembe District, South Africa. Transactions of the Royal Society of South Africa 68 (1), 33−39. Available from: https://doi.org/10.1080/0035919X.2012.755715.

Momoh, A., Akinsola, H.A., Nengovhela, M., Akinyemi, S.A., Ojo, O.J., 2015. Geophagic practice in Vhembe District, Limpopo Province, South Africa. Journal of Human Ecology 51 (3), 273−278. Available from: https://doi.org/10.1080/09709274.2015.11906922.

Msoffe, C.U., Nyanza, E.C., Thomas, D.S.K., Jahanpour, O., Dewey, D., 2018. The sources and chemical content of edible soil sticks sold in markets in Tanzania: a cross-sectional analytical study. Environmental Geochemistry and Health 41, 893−906. Available from: https://doi.org/10.1007/s10653-018-0185-x.

Mutura, A.W., Wesogah, J.O., Mkoji, G.M., 2015. Geophagy and parasitic infections in pregnant women attending an ante-natal care clinic in Thika Hospital, Kiambu County, Kenya. East African Medical Journal 92 (7).

Ngozi, P.O., 2008. Pica practices of pregnant women in Nairobi, Kenya. East African Medical Journal 85 (2), 72−79.

Nielsen, H., 1984. Fluoride, vanadium, nickel, arsenic and silicon in total parenteral-nutrition. Bulletin of the New York Academy of Medicine 60 (2), 177−195.

Nwafor, A.O., 2008. Reasons Pregnant Women Who Attend Antenatal Care in Mecklenburg Hospital Eat Soil (Ph.D. Dissertation). University of Limpopo (Medunsa Campus), South Africa.

Omoniwa, B.P., Lenka, J.L., Soji-Omoniwa, O., 2017. Hepatotoxicity evaluation of geophagic clay soil from Uzola, Edo State, Nigeria, in albino rats. Journal of Pharmacy and Bioresources 14 (2), 260. Available from: https://doi.org/10.4314/jpb.v14i2.21.

Oomen, A.G., Hack, A., Minekus, M., Zeijdner, E., Cornelis, C., et al., 2002. Comparison of five *in vitro* digestion models to study the bioaccessibility of soil contaminants. Environmental Science and Technology 36, 3326−3334.

Pain, S., Fauconneau, B., Bouquet, E., Vasse-Terrier, L., Pérault-Pochat, M.-C., 2018. Severe craving associated with kaolin consumption. Eating and Weight Disorders - Studies on Anorexia, Bulimia and Obesity 24 (2), 379−381. Available from: https://doi.org/10.1007/s40519-018-0583-1.

Parry-Jones, B., Parry-Jones, W.L., 1992. Pica: symptom or eating disorder? A historical assessment. British Journal of Psychiatry 160, 341−354.

Reeder, R.J., Schoonen, M.A.A., Lanzirotti, A., 2006. Metal speciation and its role in bioaccessibility and bioavailability. In: Sahai, N., Schoonen, M.A.A. (Eds.), Medical Mineralogy and Geochemistry; Reviews in Mineralogy and Geochemistry, 64. Geochemical Society of America, Washington, DC, pp. 71−77. (3).

Reid, R.M., 1992. Cultural and medical perspectives on geophagia. Medical Anthropology 13, 1337−1351.

Reilly, C., Henry, J., 2000. Geophagia: why do humans consume soil? Nutrition Bulletin 25, 141−144.

Reinbacher, W.R., 2003. Healing Earths: The Third Leg of Medicine. 1st Books Library, Bloomington, IN.

Rutherford, D., 1881. Notes on the people of Batanga. The Journal of Anthropological Institute of Great Britain and Ireland 10, 463−470.

Sayetta, R., 1986. Pica: an overview. American Family Physician 33 (5), 181−185.

Sienne, J.M., Buchwald, R., Wittemyer, G., 2014. Differentiation in mineral constituents in elephant selected versus unselected water and soil resources at Central African bais (forest clearings). European Journal of Wildlife Research 60, 377−382.

Sileshi, G.W., Nyeko, P., Nkunika, P.O.Y., Sekematte, B.M., Akinnifesi, F.K., Ajayi, O.C., 2009. Integrating ethno-ecological and scientific knowledge of termites for sustainable termite management and human welfare in Africa. Ecology and Society 14 (1), 48.

Sing, D., Sing, C.F., 2010. Review impact of direct soil exposures from airborne dust and geophagy on human health. International Journal of Environmental Research and Public Health 7, 1205−1223. Available from: https://doi.org/10.3390/ijerph7031205.

Smith, B., Rawlins, B.G., Cordeiro, M.J.A.R., Hutchins, M.G., Tiberindwa, J.V., et al., 2000. The bioaccessibility of essential and potentially toxic trace elements in tropical soils from Mukono District, Uganda. Journal of the Geological Society of London 157, 885−891.

Smulian, J., Motiwala, S., 1995. Pica in a rural obstetric population. Southern Medical Journal 88, 1236−1240.

Songca, P., Ngole, V.M., Ekosse, G., de Jager, L., 2010. Demographic characteristics associated with consumption of geophagic clays among ethnic groups in the Free State and Limpopo provinces. Indilinga African Journal of Indigenous Knowledge Systems 9 (1), 110−123.

Steinbaum, L., Kwong, L.H., Ercumen, A., Negash, M.S., Lovely, A.J., Njenga, S.M., et al., 2017a. Detecting and enumerating soil-transmitted helminth eggs in soil: new method development and results from field testing in Kenya and Bangladesh. PLoS Neglected Tropical Diseases 11 (4), e0005522. Available from: https://doi.org/10.1371/journal.pntd.0005522.

Steinbaum, L., Swarthout, J., Mboya, J., Pickering, A.J., 2017b. Following the worms: detection of soil-transmitted helminth eggs on mothers' hands and household produce in rural Kenya. American Journal of Tropical Medicine and Hygiene 97 (5), 1616–1618. Available from: https://doi.org/10.4269/ajtmh.17-0072.

Stokes, T., 2006. The earth eaters. Nature 444 (30), 543–544.

Sturmey, P., Hersen, M., 2012. Handbook of Evidence-Based Practice in Clinical Psychology, Child and Adolescent Disorders. John Wiley and Sons, p. 304.

Sule, S., Madugu, H.N., 2001. Cited in: Njiru, H., Elchalal, U. and Paltiel, O., 2011. Geophagy during pregnancy in Africa: A literature review, Obstetrical and Gynecolological Survey, 66. pp. 452–459.

Trow, A.J., 2008. Geophagy and Human Nutrition: The bioaccessibility of Essential and Potentially Toxic Elements in Nigerian and Other Soils (M.Sc. thesis). Aberystwyth University.

Turpin, J., 2004. Baked clay balls and earth ovens in south central Texas. Current Archeology in Texas 6 (2), 9–14.

Vermeer, D., 1971. Geophagy among the Ewe of Ghana. Ethnology 10, 56–72.

Vermeer, D., 1979. A note on geophagy among the Afenmai and adjacent peoples of Bendel State. Nigeria. African Notes 8, 13–14.

Vermeer, D., Frate, D.A., 1979. Geophagia in rural Mississippi: environmental and cultural context and nutritional implications. American Journal of Clinical Nutrition 32, 2129–2135.

Warshall, P. 1999. Eating Earth. Whole Earth. http://www.wholeearth.com/issue/2096/article/75/eating.earth. (Accessed 22 March 2020).

Wedeen, R.P., Mallik, D.K., Batuman, V., Bogden, J.D., 1978. Geophagic lead nephropathy: case report. Environmental Research 17, 409–415.

Wendler, R., Glahn, R., Ta, C.A.K., Arnason, J.T., et al., 2019. Geophagy among East African Chimpanzees: Consumed soils provide protection from plant secondary compounds and bioavailable iron. Environmental Geochemistry and Health 41 (6), 2911–2927. Available from: https://doi.org/10.1007/s10653-019-00366-8.

Ziegler, J., 1997. Geophagia: a vestige of paleonutrition? Tropical Medicine and International Health 2 (7), 609–611. Available from: https://doi.org/10.1046/j.1365-3156.1997.d01-359.x.

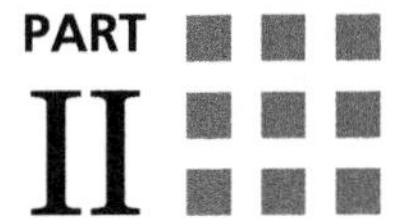

An exposition of some key geomedically important elements, with africa case descriptions

7

Introduction

A major objective of research in Medical Geology is the generation of geoscientific information that would assist the medical profession broaden the diagnostic and therapeutic spectrum of environmental diseases, particularly those whose precise etiologies are still uncertain. Most environmental diseases of concern to the Medical Geologist are *geochemical diseases* wrought by various positive or negative systemic interactions involving trace elements/metals/metalloids/isotopes (e.g., with the immune system) after different types of exposure episode. Thus important contributions the Medical Geologist can make engender the depiction of possible causative geochemical cofactors of disease, which are themselves predicated on pertinent geochemical characteristics of the individual chemical elements (trace elements/metals/metalloids) or their isotopes.

In the same vein, the principal innovations that have been made in unraveling the link between geochemistry and disease have been in the understanding of interactions of trace elements/metals/metalloids and isotopes in the immune system and how these interactions produce nutritional benefits or cause disease. The goal in Part II of these compilations is to combine new observations (up to the year 2023) with existing data to give fresh insights into the main sources, entry, distribution, and metabolic interactions of some of the more geomedically important chemical elements, and to provide a wider latitude in the search for diagnostic and therapeutic certainties for geochemical diseases.

In the ensuing chapters, the characteristics of some of the more geomedically important chemical elements (*cf.*, the Periodic Table, Table 7−1) are presented to illustrate their role in metabolic processes as essential elements (micronutrients) or as potentially toxic elements (PTEs) (disease-producing agents), or both; a role, which also depends on the bioavailable levels that are taken up mainly through the diet, and their involvement in humoral and cellular interactions in the body.

Of the elements for which there are proven or suspected direct causal relationships with man's health, significant data now exist for a host of them, both nutritional and potentially toxic ones. The need to address the types of data needed for drawing up meaningful correlations between patterns of trace element distribution and environmental diseases in humid tropical settings is strongly emphasized throughout this Primer. Seven of the geomedically important elements (As, I, F, Mn, Pb, Se, and Zn) have been selected for characterization in terms of their position in nutrition and disease, constituting the subject matter of Part II. Although not treated in detail in this discourse, there are other trace elements, particularly metals, such as Cd and Cr that can cause localized pollution and pose serious health risks.

Medical Geology of Africa. DOI: https://doi.org/10.1016/B978-0-12-818748-7.00012-5

Table 7–1 Periodic table illustrating major elements (pink), minor elements (blue), trace elements (yellow), and noble gases (gray) in the biosphere (Selinus et al., 2013).

$_1$H																	$_2$He
$_3$Li	$_4$Be											$_5$B	$_6$C	$_7$N	$_8$O	$_9$F	$_{10}$Ne
$_{11}$Na	$_{12}$Mg											$_{13}$Al	$_{14}$Si	$_{15}$P	$_{16}$S	$_{17}$Cl	$_{18}$Ar
$_{19}$K	$_{20}$Ca	$_{21}$Sc	$_{22}$Ti	$_{23}$V	$_{24}$Cr	$_{25}$Mn	$_{26}$Fe	$_{27}$Co	$_{28}$Ni	$_{29}$Cu	$_{30}$Zn	$_{32}$Ga	$_{32}$Ge	$_{33}$As	$_{34}$Se	$_{36}$Br	$_{36}$Kr
$_{37}$Rb	$_{38}$Sr	$_{39}$Y	$_{40}$Zr	$_{41}$Nb	$_{42}$Mo	$_{43}$Tc	$_{44}$Ru	$_{45}$Rh	$_{46}$Pd	$_{47}$Ag	$_{48}$Cd	$_{49}$In	$_{50}$Sn	$_{51}$Sb	$_{52}$Te	$_{53}$I	$_{54}$Xe
$_{55}$Cs	$_{56}$Ba	57-71	$_{72}$Hf	$_{73}$Ta	$_{74}$W	$_{75}$Re	$_{76}$Os	$_{77}$Ir	$_{78}$Pt	$_{79}$Au	$_{80}$Hg	$_{81}$Tl	$_{82}$Pb	$_{83}$Bi	$_{84}$Po	$_{85}$At	$_{86}$Rn
$_{87}$Fr	$_{88}$Ra	89-103	$_{104}$Db	$_{105}$Jo	$_{106}$Rf	$_{107}$Bh	$_{108}$Hn	$_{109}$Mt	110	111							

		$_{57}$La	$_{58}$Ce	$_{59}$Pr	$_{60}$Nd	$_{61}$Pm	$_{62}$Sm	$_{63}$Eu	$_{64}$Gd	$_{65}$Tb	$_{66}$Dy	$_{67}$Ho	$_{68}$Er	$_{69}$Tm	$_{70}$Yb	$_{71}$Lu
		$_{89}$Ac	$_{90}$Th	$_{91}$Pa	$_{92}$U	$_{93}$Np	$_{94}$Pu	$_{95}$Am	$_{96}$Cm	$_{97}$Bk	$_{98}$Cf	$_{99}$Es	$_{100}$Fm	$_{101}$Md	$_{102}$No	$_{103}$Lr

Those in green are essential trace elements. Known established toxic elements are shown in red.
From Selinus, O., Alloway, B., Centeno, J., Finkelman, R., Fuge, R., Lindh, U., Smedley, P. (Eds.), 2013. Essentials of Medical Geology, New Revised Edition. Springer Dordrecht, Netherlands. https://doi.org/10.1007/978-94-007-4375-5_1.

Important advances over the last two decades of research on geomedically important elements have engendered an improvement in our mechanistic understanding of their cycling and bioavailability, as well as their role in metabolic interactions; in turn, fostering our ability to more precisely assess their nutritional efficacy or disease-causing potential. These attributes are briefly summarized, the current status of geomedical research on each of these selected elements briefly reviewed with particular reference to Africa, and a list of suggested areas of further research, presented.

Sources

The main sources (unless specifically stipulated otherwise) of the elements are given as average values or range, largely obtained from the most recent and authoritative reference sources.

- Igneous rocks: Felsic or acid; Mafic or basic; Ultramafic rock.
- Sedimentary rocks: Limestone or dolomite; sandstone; shale; coal.
- Minerals: The mineralogical composition of rocks, ores, and weathering products in which the element is normally found in significant quantities, is given, with a note on the

stability of the minerals in the surface environment. This list is not restricted to minerals containing the element as a major constituent.

- Soils: The known constituents of soils in which each element is enriched are given.
- Plants: The known categories of plants in which each element is enriched are given.
- Food crops and other major dietary sources: The richest sources of the element together with the element content in these sources, are given.
- Natural waters: The mode of entry into the food chain of elements that may occur in the aqueous phase (natural waters) as known or unknown organic and inorganic complexes, are given.
- Drinking water: The range and maximum concentration levels in drinking water are given based on recommended values provided by authoritative regulatory bodies such as the World Health Organization (WHO) and the United States Environmental Protection Agency (US EPA).
- Air: The level of contribution of the element to the atmospheric budget is given, together with any links between atmospheric fluxes and uptake by humans.

Associations (synergism/antagonism)

Several versions of the Periodic Table now indicate elements that are geomedically relevant, either as nutritional elements or as PTEs (Table 7−1). Elements within these groupings that can act in concert, either synergistically or antagonistically, are indicated.

Chemical form or speciation in soil and aqueous phases

The chemical form or speciation in which each element occurs determines, in part, the level of health risk that it poses. Chemical forms of an element affect its absorption, retention, and utilization from diets. Interactions within the transporting medium can alter the "form" of the element from low-toxicity form to high-toxicity form and vice versa.

Environmental circulation and impact of climate change

The cycling of the element through the atmosphere, marine, and terrestrial systems is given, as well as an indication of how the element's environmental circulation is affected by climate change.

Mobility, mode of entry into food chains, and bioavailability

The relative mobility of each element in the surficial environment is indicated, as well as its mode of entry into the food chain and the principal limiting factors controlling its bioavailability.

Absorption and distribution

The mode of transportation and distribution in the body after absorption, is given.

Metabolic function (essentiality/clinical toxicity)

Interactions of the element in metabolic processes and the clinical outcomes in living organisms are reviewed.

Supplementation

The most common supplementation regimes that are currently in use are given, along with the richest food sources.

Reduction of body burden

Available clinical protocols for treatment of element toxicities are given.

Analytical determination

Modern analytical methods are listed for determination of the individual elements in environmental and biological materials, (including plant ash and tissue samples—serum, blood, urine, scalp hair, and toenail). The strengths and weaknesses of the listed analytical protocols are indicated.

Current status of geomedical research on the element in Africa

A brief review is made of recent research on geomedical aspects of each element, with particular reference to Africa, largely focusing on work done in the last two decades.

Other suggested areas for further research in Africa

A listing is made of topics that warrant consideration by graduate students for their thesis work, or by allied researchers, for future investigations.

Glossary of terms

Definition is given of terms not commonly in use in the geological sciences and of medical terms that are unfamiliar to geoscientists.

Bibliography

The bibliography comprises a list of cited references (Over 90% of the references postdate the year 2000), and a list of other pertinent references for further reading.

Reference

Selinus, O., Alloway, B., Centeno, J., Finkelman, R., Fuge, R., Lindh, U., Smedley, P., 2013. Essentials of Medical Geology, New Revised Edition. Springer, Dordrecht, Netherlands. https://doi.org/10.1007/978-94-007-4375-5_1.

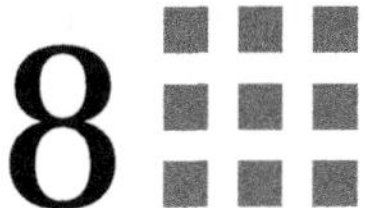

8

Medical geology of arsenic

Key chapter features

1. Brief overviews of the most relevant and important characteristics of As that are relevant in Medical Geology.
2. Attributes that make As one of the most common and potent naturally occurring environmental contaminants.
3. Latest advances in As research in Medical Geology ($>90\%$ of the references postdate the year 2000), and how they apply to the African situation.
4. Suggested areas of future As research in Africa.

Sources

"Arsenic is an elusive element, with a mysterious ability to change color, behavior, reactivity and toxicity." - [Sic] - Peggy A. O'Day (2006).

Arsenic occurs naturally in the Earth's crust and is widely distributed in the environment (air, water, soil, and biota). The terrestrial abundance of As is approximately 5 ppm, with higher concentrations being associated with sulfide deposits, sedimentary Fe and Mn ores as well as phosphate rock deposits, which occasionally contain levels of As of up to 2900 ppm (IARC, 2012; WHO, 2001).

According to Masuda (2018), the three most important natural sources of As contamination are hydrothermal activity, ore deposits, and Cenozoic sediments. Coal and coal combustion constitute other important sources.

Igneous rocks

Silicic volcanic rocks: 0.2–12.2 ppm (range); 3.0 ppm (average) [Estimate of US NRS US National Research Council, Committee on Medical and Biological Effects of Environmental Pollutants (1977) from Onishi (1969) and Boyle and Jonasson (1973)].

Granitic rocks: 0.2–13.8 ppm (range); 1.5 ppm (average) [Estimate of US NRS US National Research Council, Committee on Medical and Biological Effects of Environmental Pollutants (1977) from Onishi (1969) and Boyle and Jonasson (1973)].

Andesites, dacites: 0.5–5.8 ppm (range); 2.0 ppm (average) [Estimate of US NRS US National Research Council, Committee on Medical and Biological Effects of Environmental Pollutants (1977) from Onishi (1969) and Boyle and Jonasson (1973)].

Medical Geology of Africa. DOI: https://doi.org/10.1016/B978-0-12-818748-7.00006-X

Basalts, gabbros: 0.06−113 ppm (range); 2.0 ppm (average) [Estimate of US NRS US National Research Council, Committee on Medical and Biological Effects of Environmental Pollutants (1977) from Onishi (1969) and Boyle and Jonasson (1973)].

Ultrabasic rocks: 0.3−16 ppm (range); 3.0 ppm (average) [Estimate of US NRS US National Research Council, Committee on Medical and Biological Effects of Environmental Pollutants (1977) from Onishi (1969) and Boyle and Jonasson (1973)].

Sedimentary rocks

Limestones: 0.1−20 ppm (range); 1.7 ppm (average) [Estimate of US NRS US National Research Council, Committee on Medical and Biological Effects of Environmental Pollutants (1977) from Onishi (1969) and Boyle and Jonasson (1973)].

Sandstones: 0.6−120 ppm (range); 2.0 ppm (average) [Estimate of US NRS US National Research Council, Committee on Medical and Biological Effects of Environmental Pollutants (1977) from Onishi (1969) and Boyle and Jonasson (1973)].

Shales and clays: 0.3−490 ppm (range); 14.5 ppm (average) [Estimate of US NRS US National Research Council, Committee on Medical and Biological Effects of Environmental Pollutants (1977) from Onishi (1969) and Boyle and Jonasson (1973)].

Coal: 0.2−2000 ppm (range); 13 ppm (average) [Estimate of US NRS US National Research Council, Committee on Medical and Biological Effects of Environmental Pollutants (1977) from Onishi (1969) and Boyle and Jonasson (1973)].

Minerals

In nature, As occurs primarily in its sulfide form in complex minerals containing other *chalcophile elements, viz.*, those elements having an affinity for a sulfide phase, such as Ag, Pb, Cu, Ni, Sb, Co, and Fe. Arsenic is present in well over 200 mineral species, including elemental As, arsenides, sulfides, oxides, arsenates, and arsenites. Arsenate minerals comprise a large class with extensive substitution and solid solution, and new minerals continue to be identified. The most common As mineral is arsenopyrite, FeAsS.

Soils

Weathering of As-bearing minerals and the subsequent migration of As via geologic pathways account for the naturally occurring soil As concentrations (Aide et al., 2016). Typical range of natural soil As concentrations is of the order of 0.1−40 ppm, with an average of 5−6 ppm (Ning, 2005; IARC, 2012). Arsenic concentrations tend to increase during weathering processes (soil formation), as natural and anthropogenic As inputs continue during soil evolution (Reimann et al., 2009). Anthropogenic inputs (e.g., from mine/smelter wastes, arsenical pesticide use) can raise soil As concentrations as far as to thousands of ppm.

Plants

Arsenic is found everywhere in the plant kingdom; but it is not known to be essential for plant growth (Farooq et al., 2016). In plants, As concentrations can vary from <0.01 to c. 5 ppm (dry weight basis) (US NRS US National Research Council, Committee on Medical and Biological Effects of Environmental Pollutants, 1977). Plants are usually well-protected against As uptake, but there are plant species that can accumulate unusually high amounts of As (Reimann et al., 2009), at levels that may prove toxic to them, and may also enter the food chain, posing health risk to humans. Multiple mechanisms are involved in the uptake and metabolism of As in plants. Methylated As and arsenite (as As III) move through the *noduline 26-like intrinsic protein* (NIP) *aquaporin channels* while arsenate (as As V) is taken up through the *phosphate transporters* (*PiTs*) (Farooq et al., 2016; Kumar et al., 2019).

Food crops and other major dietary sources

The main sources of dietary intake of As are fish, shellfish, meat, poultry, dairy products, and cereals (WHO, 2019). The intake of total As per day from food and beverages is known to be in the range of 20−300 µg/day (IARC, 2012). The majority of foodstuffs generally contain low levels of As, normally less than 0.25 ppm. However, certain seafoods can contain higher quantities of As, such that the amount of As ingested daily by humans via food is greatly influenced by the amount of seafood in the diet (WHO, 2000).

Natural waters

Transport of As in the environment is mainly by water, deriving As through weathering, erosion, and dissolution. Arsenic content in natural waters is typically low (<10 ppb in surface freshwater sources, such as rivers and lakes); but it can be as high as 5 ppm near areas of certain human activities (IARC, 2012; WHO, 2001). In surface waters of the ocean and in groundwater, average As contents average 1−2 ppb, but groundwater concentrations can reach 3 ppm in areas where volcanic rocks are found, as well as in sulfide-mineralized terrains (WHO, 2001). Waters from geothermal activity can also be sources of As in groundwaters (Dinwiddie and Liu, 2018).

An important addition into the hydrological system is As released from acid mine drainage (AMD) generated through inorganic and microbially mediated sulfide oxidation. Sadly, As contamination of surface drainage and groundwater continues in many African countries having large base metal sulfide and coal mines (e.g., Ghana, South Africa, Zimbabwe), and the overall significance of this hazard remains difficult to quantify due to a paucity of data.

Modern innovative techniques for removal of As from mine drainage waters now involve the use of buffering methods (e.g., Paikaray, 2015), sequential extraction (e.g., Gault et al., 2005; Torres and Auleda, 2013; Hwang et al., 2019), polymer inclusion membrane (e.g., Zawierucha et al., 2020), and geochemical modeling (Bortnikova et al., 2020).

Drinking water

Worldwide, concentrations in groundwater can be as high as 3200 ppb, putting populations dependent on such waters for domestic purposes at risk of excessive As exposure. Such populations are estimated to be in the region of 57 million (Smedley and Kinniburgh, 2002; Yunus et al., 2016). More common drinking water sources generally contain As at concentrations of less than 10 ppb.

Contaminated groundwater poses the greatest threat to public health of the African people, since majority of the Continent's population still live in rural settings and depend on groundwater for drinking purposes, crop irrigation, and food preparation.

The original guideline value for As in drinking water set by the WHO was 50 ppb (WHO, 1993). The current WHO guideline value is put five times lower at 10 ppb, a value that WHO stipuates, is only *provisional*, in the light of foreseen practical difficulties in removing As in drinking water (WHO, 2018; WHO, 2017). The WHO further asserts that "Every effort should therefore be made to keep concentrations as low as reasonably possible and below the guideline when resources are available." However, many developing countries, including those in Africa still apply the older standard of 50 ppb (See Section "Regulatory implications: Tolerable intake levels," in this Chapter).

Air

Arsenic in the atmosphere is derived from wind erosion, emissions from volcanic activities, low temperature volatilization from soils, sea sprays and pollution; and is taken back to the Earth's surface through wet and dry deposition (Smedley and Kinniburgh, 2001; WHO, 2019). Substantial amounts of As are derived from industrial smelter plants and combustion of fossil fuel (Ciarrocca et al., 2012).

In air, As is present mainly in particulate forms as inorganic As (mixture of both arsenite and arsenate) with usually very small amounts of organic As species (Chung et al., 2014). The WHO (2000) gives the range of As in airborne concentrations of: $1-10$ ng/m^3 in rural areas and from a few nanograms per cubic meter to c. 30 ng/m^3 in urban areas bereft of contamination. By 2000, the WHO was still unable to establish a safe level for inhalation exposure of As.

Associations (additive/synergistic/antagonistic)

The affinity of As for bonding with other elements and species underlines its (As) rarity of occurrence as a native element, with the occasional exception of its occurrence in hydrothermal ore deposits. Arsenic belongs to the group of chalcophile elements, which are the essential components of sulfide deposits, and show association with epigenetic gold ore. Such deposits are common in the widespread greenstone belts of Africa.

In the biological realm, As surprisingly exhibits antagonistic relationships with some of the very elements (chacophiles) with which it is geochemically associated. For example, Se, in its role as an antitoxin, is shown to be antagonistic to As, promoting the excretion of As from the body (See Krohn et al., 2016; Skalny et al., 2016); and As stress removal from rice seedlings (Moulick et al., 2023; Pandey et al., 2015).

Data presented by Guzman-Rangel et al. (2018) revealed that Zn-As antagonisms (total concentrations) were partially related to the increased Zn immobilization of As in soil.

Islam et al. (2005) concluded that supplementation of Se, Fe, and Zn reduces the As concentration in tissues like liver, kidney, spleen, heart, intestine, stomach, muscle, and dermis of rats as well as lowers the tissue damage caused by As.

Early documented accounts of relationship between As and I (goiter) were presented by Polya and Watts (2017) and Liebhafsky (1931), way back in 1931, and by Sharpless and Metzger (1941). Another example illustrating the association between As exposure and I (thyroid function) is given in the work of Jain (2016), who found for I deficient male subjects, a positive association between *thyroid-stimulating hormone* and *urine dimethylarsinic acid* (UDMA) ($P < .01$); and a negative association between the levels of total serum thyroxine (TT4) and UDMA ($P < .01$). Results from these studies appear to corroborate those from earlier studies on the association between As exposure and hypothyroidism by Ciarrocca et al. (2012) and Gong et al. (2015). See also the work of Henjum et al. (2023).

Other elements having antagonistic relationship with As include: Pb and Cd (See Li et al., 2019; Vellinger et al., 2012).

A review of the recent literature tends to suggest a rarity of synergistic effects between As and other chemical elements; but synergism is rife between As and several compounds as well as a whole array of other toxic agents [e.g., dithiothreitol (Tsai et al., 2010); smoking, exposure to sunlight, fertilizer use, and pesticide use (Melkonian et al., 2011); *erlotinib* (Kryeziu et al., 2013); radiation (Moloudi et al., 2017); effect of mitochondrial instability, oxidative stress, and hormonal-neurotransmitter impairment (Medda et al., 2020)].

Chemical form or speciation in soil and aqueous phases

The strong propensity of As to bond with other elements and species makes it unlikely to occur as a native element, save occasionally in hydrothermal ore deposits (O'Day, 2006; Patel et al., 2023).

In nature, As bonds mainly to O and S forming a whole range of aqueous species and minerals. Four main oxidation states, *viz.*: -3, 0, $+3$, and $+5$, are found.

In soils, As exists in both organic and inorganic forms. However, the inorganic forms, As (III) (arsenite), and As(V) (arsenate) are the ones most often encountered (Moreno-Jiménez et al., 2012). In oxygenated soil, inorganic As is present in the pentavalent form. Under reducing conditions, it is in the trivalent form.

In aqueous solutions, As is found as the oxo-anions arsenite, $H_3As^{3+}O_3$ [sometimes written as $As^{3+}(OH)_3$], and arsenate, $H_3As^{5+}O_4$ (O'Day, 2006). Arsenic also forms stable bonds with methyl groups. Biological activities in surface waters produce organic forms of As, but these are of limited quantitative importance, except where substantial amounts of industrial wastewaters are inputted in these waters (See Smedley and Kinniburgh, 2001).

Given such dexterous chemistry, it is perhaps unsurprising, from a toxicological standpoint, that the behavior of As is somehow contradictory, acting both as a toxin and as a curative.

Just as in the case of N and P, the organic chemistry of As is quite rich and varied. It can bind to different ligands, each having its own distinct coordination geometry. *This structural attribute, together with its greater redox potential (than say, P), may help elucidate the mechanisms of As toxicity in living organisms, which till now remain incompletely known* (See, e.g., Carter et al., 2003).

Environmental circulation and impact of climate change

According to Masuda (2018), the release of As in the environment is strongly influenced by the four phenomena of: active plate tectonism, magmatism, hydrothermal activity, and rapid rates of erosion (Fig. 8–1). Some of the activities that engender these phenomena involve crustal As degassing, considered by Reimann et al. (2009) as an important process that leads

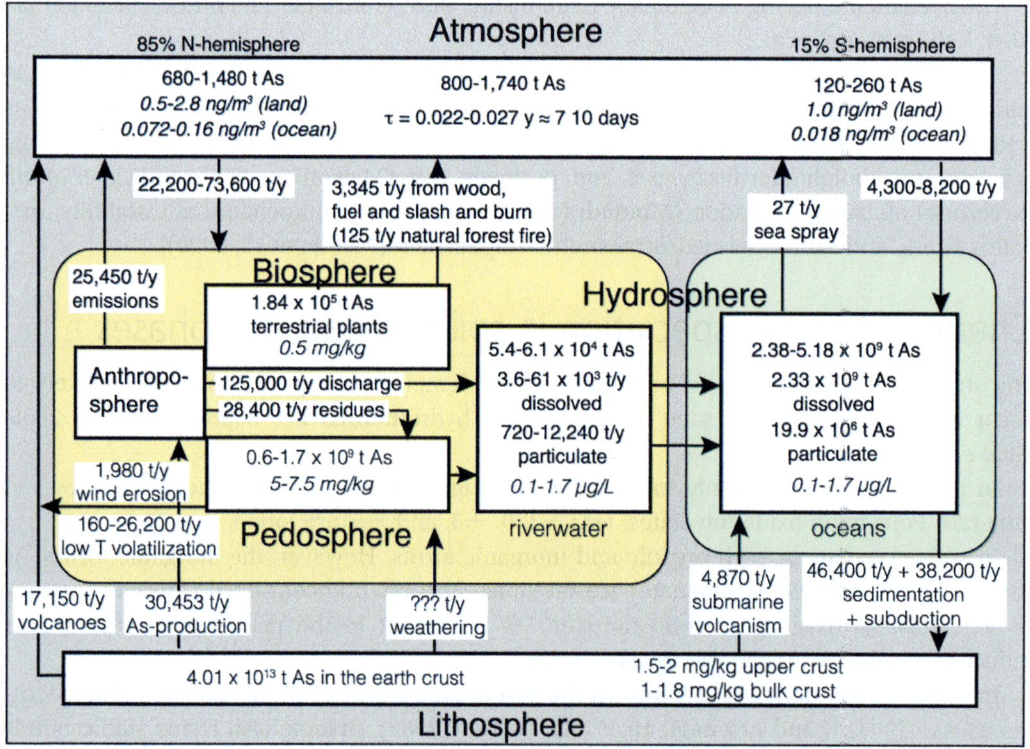

FIGURE 8–1 Global arsenic cycle. Fluxes and reservoir sizes in different parts of the geosphere (Matschullat, 2000). *Adapted from Matschullat, J., 2000. Arsenic in geosphere - a review. Science of the Total Environment 249, 297–312. https://doi.org/10.1016/S0048-9697(99)00524-0.*

to As enrichment in the surface environment. Natural mineralization and activities of microorganisms enhance the mobility of As in the environment.

Once brought to the surface environment, As mobility continues via a combination of natural processes, such as weathering reactions and biological activity. Environmental input occurs through As release in operations in the mining, metal smelting (e.g., ironworks), electronic, and glassmaking industries. Other human activities, such as geothermal operations, combustion of fossil fuels, as well as excess pumping of shallow groundwater, disperse As in the environment, thereby expanding areas of As contamination. Although the use of pesticides, herbicides, and other As-based products in the farm has decreased significantly in the last few decades, their use in wood preservation is still quite prevalent. Increased concentrations of As in urban areas have also been noted (e.g., Reimann et al., 2009).

According to Bondu et al. (2016), climate change is expected to affect groundwater quality by altering recharge and flow dynamics. Increased As mobilization is brought about by changes in land use patterns associated with climate change, particularly those causing increased groundwater withdrawals and pollution (Bondu et al., 2016; Aribam et al., 2021). One notable land use effect is seen in the area of rice production, where climate change is projected to cause both a decrease in rice yields and an increase of grain inorganic As concentrations (Wichelns, 2016; Muehe et al., 2019; UW, 2019).

Mobility, routes of exposure/mode of entry into food chains, and bioavailability

Environmental migration of As is mainly through water. For many African populations, drinking water is sourced from underground sources, where water As content can be high depending on the nature of the aquifer (See Section "Igneous rocks" and "Sedimentary rocks," in this Chapter). Since drinking water is a major source of oral intake, in regions where there are high-As concentrations in aquifers or mine drainage areas, human exposure to elevated levels of inorganic As is likely to occur (WHO, 2019). Human exposure to elevated levels of inorganic As also occurs through consumption of food prepared with this water, and food crops irrigated with high-As water sources [See Section "Drinking water," in this Chapter].

Arsenate is the common form of As in oxygenated water, whereas under reducing conditions, arsenites seem to be the most important chemical form (Smedley and Kinniburgh, 2002). Oxidation conditions result from the development of high pH (>8.5) values in semiarid or arid environments, wrought by the combined effects of mineral weathering and high evaporation rates. The development of strongly reducing conditions at near-neutral pH values leads to the desorption of As from mineral oxides and to the reductive dissolution of Fe and Mn oxides (Smedley and Kinniburgh, 2002).

Factors that determine As availability in soils are many, but the more important ones are pH, organic matter, clay content, amount of rainfall, the presence of Fe-oxides, and/or phosphorus and (co)precipitation in salts (Moreno-Jiménez et al., 2012).

Absorption and distribution

The main routes of As absorption in the general population are ingestion and inhalation, though inhalation of As from ambient air is usually a minor exposure route. Arsenic compounds inhaled in particulate form are deposited in the respiratory tract and absorbed into the bloodstream. Human and animal data indicate that over 90% of the ingested dose of dissolved inorganic trivalent or pentavalent As is absorbed from the gastrointestinal tract (IARC, 2004; US NRC, 1999; Wang et al., 2023). Following absorption, blood is the main vehicle for the transport of As to other parts of the body and its primarily distribution in the liver, kidneys, spleen, and lungs (WHO, 2000).

Organic As compounds in seafood are also readily absorbed (75%−85%) (WHO, 2000). Soil and dust in the vicinity of copper smelters can contain As in high concentrations, too (WHO, 2000), and are likely to engender high rates of entry of As into the food chain.

Metabolic function (essentiality/clinical toxicity)

Arsenic metabolism may comprise both detoxification as well as bioactivation processes (Fig. 8−2). Repetitive reduction and oxidative methylation are generally accepted as constituting the chemical metabolic pathway of As (Kobayashi and Agusa, 2019). Metabolism (methylation) of As (both arsenite and arsenate) occurs mainly in the liver and excreted in the urine as methylated arsenicals and accumulated in the hair and nails. The primary detoxification mechanism is considered to be biomethylation, since the highly reactive species of inorganic As are potentially more toxic to living organisms, including humans (Roy and Saha, 2002).

Arsenic does not appear to be biomagnified in the food chain (Roy and Saha, 2002). Both methylated species, methylarsonic acid, and dimethylarsonic acid are considered to be less toxic and show less affinity for binding to tissues. These characteristics make for their (methylated species) more rapid elimination than the unmethylated form (WHO, 2000).

Essentiality

Arsenic has several industrial uses (glass processing, pigments, textiles, paper, metal adhesives, wood preservatives, ammunition, and so on). To a limited extent, As is used in pesticides, feed additives, and pharmaceuticals. The contribution to water contamination by As resulting from these applications has already been briefly indicated in Section "Environmental Circulation and Impact of Climate Change," in this Chapter.

The notion that dietary As may have some beneficial effects in humans is quite conjectural. As Uthus (2003): "That is, there is an amount of dietary arsenic that is not only not harmful, but beneficial." [*Sic*]. Other studies have suggested an essential nutrient role for As, with a function in *methionine* metabolism (e.g., Uthus, 1992; Pal and Chatterjee, 2004), but some researchers (e.g., Anke, 1986; Kruger et al., 2013) have ruled out any role for As in human metabolism.

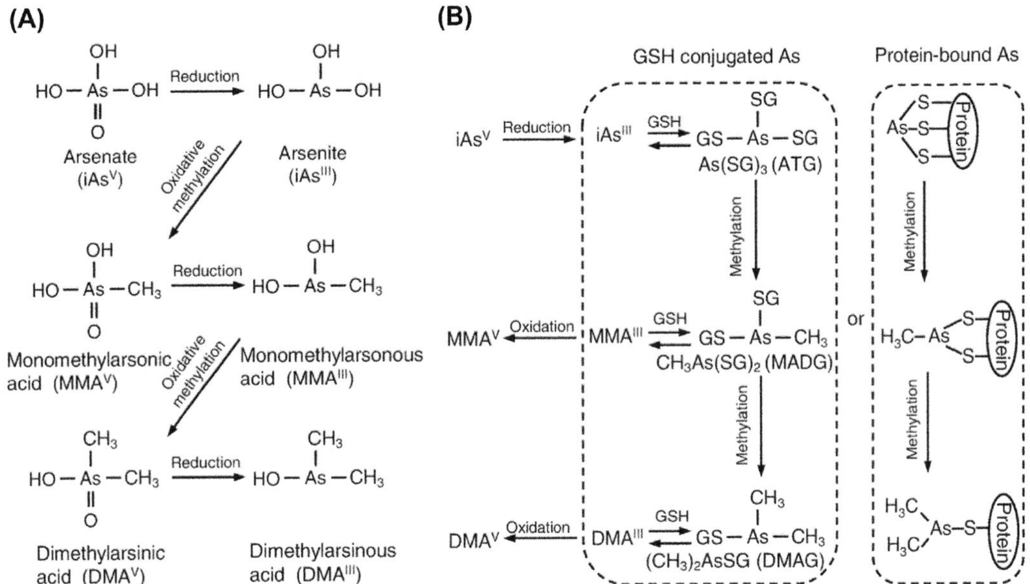

FIGURE 8–2 Metabolic pathways for arsenic in mammals. (A) Generally accepted pathway as proposed by Challenger (1945). (B) Proposed pathways for arsenic methylation via GSH-conjugated arsenic by Hayakawa et al. (2005) and via protein-bound arsenic by Naranmandura et al. (2006). *Adapted from Kobayashi, Y., Agusa, T., 2019. Chapter 2: Arsenic metabolism and toxicity in humans and animals: racial and species differences. In: Yamauchi, H., Sun, G. (Eds.), Arsenic Contamination in Asia: Biological Effects and Preventive Measures. Current Topics in Environmental Health and Preventive Medicine, pp. 13–28. https://doi.org/10.1007/978-981-13-2565-6_2.*

Despite serious concerns regarding safety, As has been shown to have some therapeutic effect, too (US NRC, 1999; Chakraborti et al., 2003; Liu et al., 2008). Its use as a part of extremely diluted *homeopathic* remedies for digestive disorders, food poisoning, sleep problems (insomnia), allergies, anxiety, depression, and obsessive-compulsive disorder, has been documented (US NRC, 1999; Chakraborti et al., 2003). Low doses of arsenite (As^{3+} administered as AsO_3) have been shown to hold promise as an effective treatment for some forms of leukemia and some types of tumors (Miller et al., 2002).

Certain bacteria species are able to use As compounds as respiratory metabolites (Omreland et al., 2001; Stolz et al., 2006; Drewniak and Sklodowska, 2013.). Trace amounts of As are reported to be an essential dietary element in rats, hamsters, goats, chickens, and probably other species as well (Anke, 1986; Uthus, 1992).

Toxicological effects

It is now well-established that long-term exposure to inorganic As, mainly through drinking water and food, can lead to chronic As poisoning. To date, the most notable regions in the world where excessive contents of As have been measured in drinking water, and the problems of *arsenicosis* deemed most serious, are Bangladesh, the People's Republic of China, Taiwan, West Bengal (India), and smaller areas of a number of other countries.

A wide spectrum of toxic effects of As can occur on a variety of organisms, including humans. Chronic exposure symptoms can range from skin disorders, commonly including hyperkeratosis, hyperpigmentation, skin malignancies, peripheral arteriosclerosis (Blackfoot Disease), vascular and heart disease, diabetes, and skin and lung cancer to disruption of immune system function, and eventually death; all of which are known in populations consuming water with 100−1000 ppb As (e.g., Brown and Ross, 2002; Tseng, 2005; Kim et al., 2011; McCarty et al., 2011; Abhyankar et al., 2012; Naujokas et al., 2013; Chen et al., 2017, and other references in the "List of selected references for further reading," this Chapter). Of these conditions, skin lesions and skin cancer show the most characteristic effects of As toxicity (WHO, 2018).

Back in 1987, the IARC (1987) found sufficient evidence for the carcinogenicity of As in humans, which led the Agency to classify it (As) and its compounds as Group I *carcinogens* (Farzan et al., 2013; Tolins et al., 2014). In 2007 Navarro Silvera and Rohan (2007) adduced compelling evidence in support of positive associations between As and risk of both lung and bladder cancers; and Pearce et al. (2012) showed that As may act as a carcinogen in conjunction with ultraviolet radiation, resulting in nonmelanocytic skin cancers.

The details of the biochemical reaction mechanisms driving the effect of As behavior both as a toxicant and a curative, are complex and incomplete (See, e.g., Muzaffar et al., 2023; Wysocki et al., 2023). *Further intensive research in unraveling these mechanisms will benefit the medical profession in their efforts at improving arsenicosis therapy* (See section "Suggested Areas for Further Research in Africa," in this chapter).

Supplementation

The possible therapeutic applications of "As supplements" have been highlighted under the heading "Essentiality" in Section G of this Chapter. However, the use of certain pharmaceutical products containing As is known to cause severe and sometimes life-threatening health problems in some individuals (e.g., Chen et al., 2013). Ma et al. (2021) recently found that dietary As supplementation can induce oxidative stress by suppressing nuclear factor erythroid 2-related factor 2 in the livers and kidneys of laying hens. Fowler's solution, which contains 1% potassium arsenite is an example of a pharmaceutical that should be administered in a highly judicious way (Bates et al., 1992) to prevent counter reactions.

For these reasons, many national and international regulatory bodies, for example, the Food and Drug Administration (FDA) of the United States, the US Centers for Disease Control (CDC), and the WHO monitor and regulate levels of As in dietary supplements, certain foods, and cosmetics (See Section "Regulatory Implications: Tolerable Intake levels," in this Chapter).

Reduction of body burden

Wang et al. (1997) have suggested that both *glutathione peroxidase* and *catalase* are involved in defense against arsenite toxicity, and have indicated that increasing the intracellular antioxidant level of *glutathione peroxidase* may have preventive or therapeutic effects in As poisoning.

The WHO (2019) has recommended the inclusion of routine screening of drinking water supplies in an effort to reduce human exposure to As and the clear identification of those delivering water that exceeds the WHO provisional guideline value for As of 10 ppb or national guideline values, along with awareness-raising campaigns.

Bretzler et al. (2020) recently showed that nail-based filters have the potential to achieve As removal to the tune of over 90% in a field context if certain conditions, such as filter bed saturation, flow rate, pauses between filtrations, are well-controlled.

Other mitigation options recommended by WHO (2019) include "… use of alternative groundwater sources, use of microbiologically safe sources, such as rainwater and treated surface water, use of arsenic removal technologies, or dilution of high-arsenic-content source water with lower-arsenic-content source water that is microbiologically safe." [*Sic.*].

Analytical determination

Monitoring of recent environmental exposure of As is commonly carried out by determination of total As in urine, since urine is the main excretion route for most As species (See, e.g., Byun et al., 2013; James et al., 2013). Assessment of recent As exposure by measuring As in blood samples is considered "inadequate," since As is unlikely to be detected in samples drawn more than 2 days (the halflife of inorganic As in blood is 4–6 hours, and the halflife of the methylated metabolites, 20–30 hours) (See, e.g., Hall et al., 2006). Periodic blood levels though, can be determined to follow the effectiveness of the therapy (See MFMER Mayo Foundation for Medical and Educational Research, 2021).

Katz (2019) argues that biomonitoring of As using hair samples could be a better alternative (to blood or urine), since hair is considered by some to be a metabolic end product providing a more permanent record of effects of assimilated environmental As (See also US ATSDR, 2007).

Modern techniques for determination of total As and As speciation (Ardini et al., 2020; Francesconi and Kuehnelt, 2004; Leybourne et al., 2014; Menezes and Silva, 2013) involve ICP-MS, ICP-AES, HG-AAS, XRF, INNA, or field-based separations and techniques, as well as a variety of data reduction protocols. Both logistical considerations (e.g., convenience, availability, turnaround time, cost, and documentation/reporting standards) and technical suitability (e.g., precision, accuracy, detection limits, limits of quantification, sensitivity, selectivity, and traceability) are involved in selection of the most appropriate methods (See Bhat et al., 2022; Polya and Watts, 2017). Similar instrumental techniques are applied to the determination of As in waters; but, according to Dominguez-Gonzalez et al. (2014) [quoted in Polya and Watts (2017)], the attainment of a better detection limit (2.5 ppb) is realizable by visible spectrometry using functionalized gold nanoparticles.

Current status of geomedical research on arsenic in Africa

Arsenic is one of the first of toxic elements known to spread globally in surficial environments, with associated health hazards, especially in some Asian countries. However, in Africa, the scarcity of pertinent analytical data on the element's abundance in the different environmental compartments (rocks, waters, soils, and air), as well as its mode of transport into biological systems and health effects, does not imply that As-associated health problems are nonexistent on the Continent. It is well worth noting that to date, many African laboratories still lack the technical capacity to determine As at the very low concentration levels (sub-ppm) needed for making tangible correlations with public health. Nor are multidisciplinary efforts of epidemiologists and Medical Geologists reached a level at which credible diagnostic correlations can be rapidly established. This is evident from the absence of data in many of the countries in the distribution map of "As in African waters" (Fig. 8−3) by Ahoulé et al. (2015). The situation is hardly different for the distribution of As in the soil, air, and foodstuff of Africa.

The association of As with metal sulfide ores, such as those containing Au, in the extensive African greenstone belt terrains [e.g., Birimian (West Africa); Barberton (South Africa); Ngayu (DRC); Belingwe (Zimbabwe); and others] warrants an assessment of possible links to public health, especially to miners and residents in the vicinity of sulfide mines, as well as workers in the metal smelting industries. The release of arsine gas into the atmosphere from metal smelting plants, also poses a hazard to nearby residents.

The risk is also high among populations that depend on shallow or deep groundwaters for domestic purposes as evidence presented by Norman et al. (n.d.), Smedley and Kinniburgh (2013) and Ahoulé et al. (2015) clearly indicates. *Detailed epidemiological studies should be carried out to establish the possibility that high As concentrations in groundwaters (borehole waters, dug wells) in these terrains bear links to the vast array of skin disorders seen in these areas* (See, e.g., Smedley et al., 2007).

In spite of these observations, there are very few documented studies to date, on As abundance and its circulation in the water, soil, and air environments in Africa as a whole (See, e.g., Hay et al., 2006; Kortatsi et al., 2008; Fatoki et al., 2013; Ahoulé et al., 2015). Similarly, very little is known regarding possible linkage between exposure of African populations to As and prevalence of the myriad skin disorders observed on the Continent. Hardly any of the documented studies on skin diseases, such as *noma* and *infantile acropustulosis*, make mention of a possible connection with As exposure. Other commonly occurring skin diseases in Africa (general) are described by Kennedy, 1965; van Hees and Naafs, 2014; for Burkina Faso (See, e.g., Bretzler et al., 2017); for Cameroon (See, e.g., Bissek et al., 2012); for Co\it{o\ucentre\char195}te de Voire (See, e.g., Yotsu et al., 2018); for Ghana (See, e.g., Norman et al., n.d.; Doe et al., 2001; Rosenbaum et al., 2017); for Nigeria (See, e.g., Henshaw and Olasode, 2015), and for South Africa (See, e.g., Marshall, 1963; Hartshorne, 2003; Horn and Ramudzuli, 2020). Aspects of As research in Africa that need urgent attention are given in the Section "Suggested Areas of Further Arsenic Research in Africa," in this Chapter.

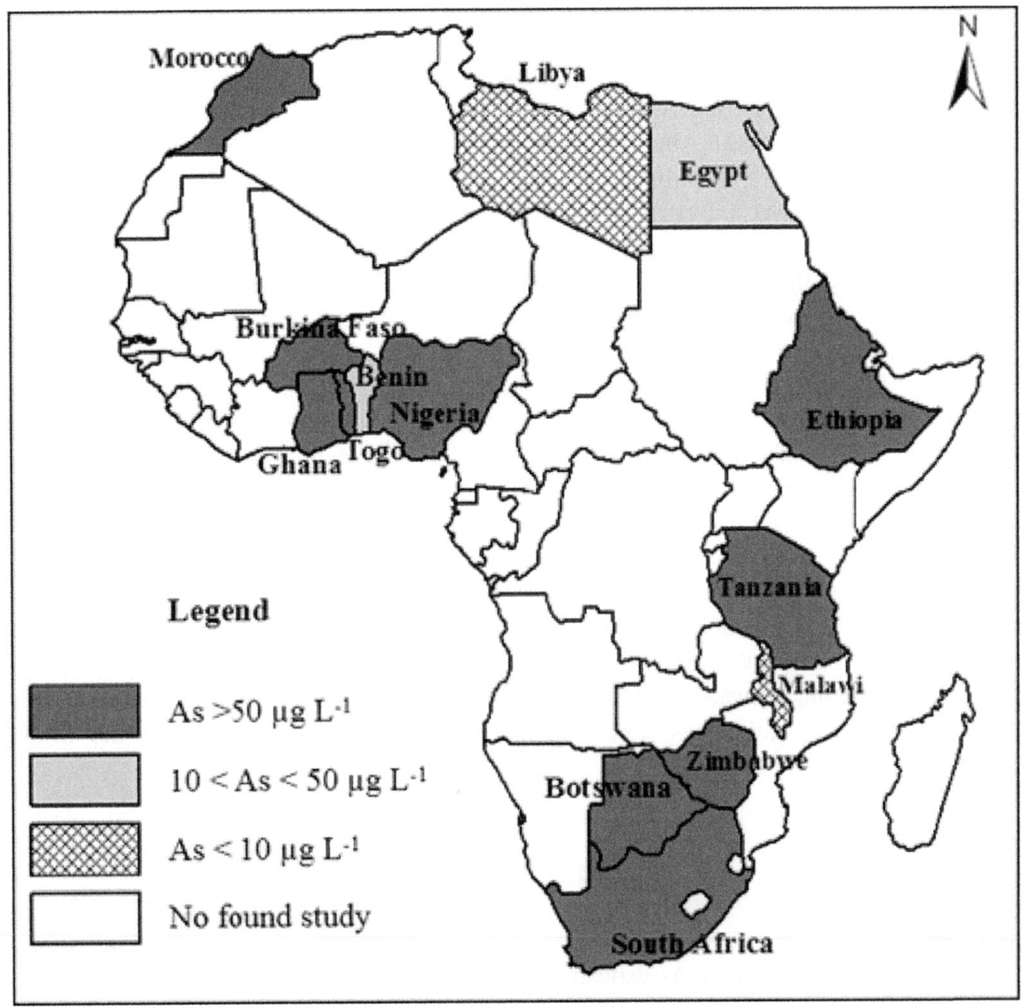

FIGURE 8–3 Distribution of arsenic in African waters (Ahoulé et al., 2015). *From Ahoulé, D.G., Lalanne, F., Mendret, J., Brosillon, S., Maïga, A.H., 2015. Arsenic in African waters: a review. Water, Air, and Soil Pollution 226, 302. https://doi.org/10.1007/s11270-015-2558-4.*

Regulatory implications: tolerable intake levels

Following the accumulation of evidence for the chronic toxicological effects of As in drinking water, recommended and regulatory limits of many authorities are being reduced. The WHO guideline value for As in drinking water was provisionally reduced in 1993 from 50 to 10 ppb. The new recommended value was based on the increasing awareness of the toxicity of As, particularly its carcinogenicity, and on the ability to measure it quantitatively (WHO, 1993).

Because of the varying hydrological (groundwater) and pedological conditions existing in different regions of Africa, as well as extensive (As) sulfide-mineralized terrains, it is incumbent on each country to define the maximum amount of As that it permits in drinking water. However, only a few African countries have developed drinking water standards [e.g., South Africa (SA DWAF, 1996); Nigeria (SON, 2007)], with most countries now in the process of developing and publishing their own guidelines [e.g., Ghana (GSA, 2009; Rossiter et al., 2010; GWCL, 2018)]. Those that have done so, have, to varying extent, used criteria, referrals, derivations, or supplementary material published by international regulatory bodies, mainly the WHO. The WHO guideline value and current national standards in Africa are quite frequently exceeded in drinking water sources, and often unexpectedly so (See Lapworth et al., 2020). Many national authorities in African countries are seeking to reduce their limits in line with the WHO guideline value (10 ppb); but, in the meantime, they still operate at the 50 ppb WHO standard, in part because of a lack of adequate testing facilities and analytical capacity for determining As at the very low concentration levels (sub-ppm) required for setting safety limits.

Because of certain food sources and dietary mineral supplements that are commonly used in Africa are often derived from natural sources that might contain high As (e.g., Ahoulé et al., 2015), it is important for clinicians to be aware of the potential for As toxicity in patients coming from high As areas (metal sulfide mining areas, areas with high As groundwaters or areas in the vicinity of metal smelting plants). As of 2020, only seven countries in Africa (Benin, Kenya, Namibia, Nigeria, Seychelles, Sierra Leone, and South Africa) report having dietary guidelines (FAO Food and Agricultural Organization of the United Nations, 2021).

Suggested areas for further arsenic research in Africa

Carlin et al. (2016), Tchounwou et al. (2019) and Sanyal et al. (2020) have identified several areas of research on inorganic As exposure (including mechanisms of action, systemic and carcinogenic effects, risk characterization, nutritional supplementation for reducing health impacts, a multidisciplinary approach to counter the health risks, and regulatory guidelines) that should be prioritized, globally. The following list, while incorporating many of these global research priorities, also gives priority areas in As research that are of particular relevance to Africa; but is however, nonexhaustive.

- The virtual "absence" of documentation of conditions akin to arsenicosis in African countries, does not in any way signify that the problems of As poisoning do not exist. As geospacial modeling results of Podgorski and Berg (2020) and Podgorski et al. (2020) clearly show, these conditions might well be present, but as yet unrecognized as such, due to analytical and diagnostic limitations in their characterization. Urgent multidisciplinary efforts between Medical Geologists, epidemiologists, and allied scientists should be mounted to improve diagnostic protocols for skin diseases in Africa.

- The mechanisms of chronic As poisoning vary between countries and regions; so also does the pathological difference in occurrence in different countries. Though keenly studied in parts of Asia, where large-scale chronic As poisoning is concentrated, the symptomatic presentation of arsenicosis is still only poorly understood for other regions, not least for Africa.
- Recent work has emphasized the inadequacy of mechanisms that have been advanced so far for explaining all of the observed toxicity of As, as well as its therapeutic effects (Carter et al., 2003; Ghosh and Shi, 2015; Hu et al., 2020; Miller et al., 2002; Obinaju, 2009; Shi et al., 2004; Singh et al., 2011). This is a situation that must be put right, as: "A better understanding of the mechanism(s) of action of arsenic will lead to a more confident determination of the risks associated with exposure to this chemical." Hughes (2002).
- As Smedley and Kinniburgh put it in 2002: "Even so, important uncertainties still remain in relation to the interactions of As(III) and As(V) at environmental concentrations and in the presence of other interacting ions." And Feldmann (2009): "... However, the importance of distinguishing among arsenic species in guidelines is an area that requires further progress."
- High concentrations of As occur in AMD, but the mechanisms governing its distribution along the flow of AMD are not fully understood (e.g., See Zhu et al., 2020).
- The organic chemistry of As is complex. This element is able to bind to myriads of organic ligands, having different coordination geometries. Again too, As has a greater redox potential than P. Further studies on the (structural) organic chemistry of As may help throw light on the reasons why mechanisms of As toxicity in living organisms are still incompletely understood, despite the volume of research that has gone into this study (Carter et al., 2003). As O'Day suggested in 2006, "Perhaps increased understanding of the subtleties of bonding and rates of exchange between As and its ligands in cells and biological fluids will further elucidate mechanisms associated with both the detrimental and beneficial effects of arsenic on living systems." [*Sic*].
- The etiology of many of the myriad skin diseases seen in Africa, some of which are analogous (in terms of presentation) to those observed in arsenicosis-endemic areas of Asia and elsewhere, remains unknown. A massive effort in collecting As abundance and geospatial distribution data in waters, soils, and air in Africa is therefore warranted; and correlation tests made between these data and those portrayed in epidemiological maps of these diseases.
- The amount of As ingested daily by Africans via food (information urgently needed for formulating regulatory guidelines), remains largely undocumented. The use of Se supplement pills to counteract As toxicity warrants further investigation (See Krohn et al., 2016).
- The use of mineral arsenicals in traditional medicines need to be further evaluated. Arsenolite/arsenic trioxide has been used as a cure for a subset of human leukemias, having shown to have a strong antitumor effect with little toxicity (Fang and Zhang, 2020). It is also used as a mineral arsenical in traditional medicines. Although pharmacological data indicate that the use of orpiment and realgar in traditional medicines may be desired in some cases, the therapeutic basis for such applications

remains to be fully validated (See, e.g., Liu et al., 2008); nor do we know much about any possible side effects such as secondary cancers that may result from their long-term use.

- As Bondu et al. remarked in 2016, "… the link between climate change and arsenic concentration in groundwater remains sparsely documented and not fully understood …
- Climate change could significantly affect the activity of microorganisms by changing the conditions in the aquifers, and these changes need to be evaluated. Future research focusing on the effects of climate change on groundwater quality is critically needed."
- A number of challenges confront the analytical geochemist of As in biological and environmental samples. For example, insufficient attention has been given to date to the field measurements of organoarsenic compounds, since they are less acutely toxic than inorganic As. However, they constitute an important fraction of the total environmental As, and so, their quantitative estimation should be considered important (See, e.g., Luong et al., 2007).
- Designing remediation strategies that incorporate clay should be further researched to determine its (clay) potential as a low-cost strategy for As remediation in mining-impacted areas (See Lefticariu et al., 2019; Bretzler et al., 2020). "… However, the implementation of successful mitigation requires this knowledge to be applied more effectively. This can only be achieved by enhancing the understanding of the interactions between socio-economic and scientific issues, thus enabling practitioners to develop effective and appropriate mitigation programs." Garelick and Jones (2008).
- Since it is becoming increasingly evident that undocumented areas of high As concentrations do exist in Africa, as well as areas where data on environmental As distribution are nonexistent (See, e.g., Ahoulé et al., 2015; Podgorski and Berg, 2020), it is suggested that the creation of prediction maps of areas of As greater than 10 μg/L with the aid of available customized modeling techniques would help greatly, efforts in diagnosis and therapy of some of the various skin conditions seen in Africa.
- Finally, as noted by Bowell et al. (2014), "In understanding the dispersion of As in the environment, geological materials provide a basic framework for characterizing As concentrations in ecosystems. Large variations can be observed on all spatial scales influenced by a variety of natural processes including nongeological influences, such as climate and vegetation. These natural variations must be documented and the processes causing them understood before reliable estimates of the anthropogenic impact on the natural environment can be made." [*Sic*]. And, as Carlin et al. observed in 2016, "Furthermore, integration of *omics data* with mechanistic and epidemiological data is a key step toward the goal of linking biomarkers of exposure and susceptibility to disease mechanisms and outcomes."

Glossary of terms

- *Aquaporins*, also called *water channels*, are *channel* proteins from a larger family of major intrinsic proteins that form pores in the membrane of biological cells, mainly facilitating transport of water between the cells.

- *Arsenicosis* is defined by the WHO Working Group as a "chronic health condition arising from prolonged ingestion (not less than 6 months) of arsenic above a safe dose, usually manifested by characteristic skin lesions, with or without the involvement of internal organs" (WHO, 2003).
- *Catalase* is a common enzyme found in nearly all living organisms exposed to oxygen, which catalyzes the decomposition of hydrogen peroxide to water and oxygen.
- *Chalcophile elements* are those elements having an affinity for a sulfide phase (e.g., As, Cd, Cu, Hg, Ni, Pb, Se, Zn). They occur predominantly in sulfide ores, such as found in the extensive greenstone belts of Africa. Chalcophile elements are found to pose greater health risks than elements belonging to other groupings in the Periodic Table.
- *Erlotinib*, molecular formula: $C_{22}H_{23}N_3O_4$, is a medication used alone to treat *nonsmall cell lung cancer*, or in combination to treat pancreatic cancer.
- *Glutathione peroxidase (Gpx)* is the general name of an enzyme family with peroxidase activity whose main biological role is to protect the organism from oxidative damage.
- *Homeopathy* (*homeopathy or homeopathic medicine*) is a pseudoscientific form of alternative medicine—a philosophy and practice based on the idea that the body has the ability to heal itself.
- *Methionine* is a sulfur-containing amino acid, which is a constituent of most proteins and is essential to the diet of vertebrates.
- The *NIP* family is a group of plant specific major intrinsic proteins, which transport a variety of uncharged solutes ranging from water to ammonia to glycerol.
- The word *omics* refers to a field of study in biological sciences that ends with -omics, such as genomics, transcriptomics, proteomics, or metabolomics. The ending -ome is used to address the objects of study of such fields, such as the genome, proteome, transcriptome, or metabolome, respectively.
- The inorganic *PiT* family is a group of carrier proteins derived from Gram-negative and Gram-positive bacteria, archaea, and eukaryotes.

References

Abhyankar, L.N., Jones, M.R., Guallar, E., Navas-Acien, A., 2012. Arsenic exposure and hypertension: a systematic review. Environmental Health Perspectives 120 (4), 494−500. Available from: https://doi.org/10.1289/ehp.1103988.

Ahoulé, D.G., Lalanne, F., Mendret, J., Brosillon, S., Maïga, A.H., 2015. Arsenic in African waters: a review. Water, Air, and Soil Pollution 226, 302. Available from: https://doi.org/10.1007/s11270-015-2558-4.

Aide, M.T., Beighley, D., Dunn, D., 2016. Arsenic in the soil environment: a soil chemistry review. International Journal of Applied Agricultural Research 11 (1), 1−28.

Anke, M., 1986. Arsenic. In: Mertz, W. (Ed.), Trace Elements in Human and Animal Nutrition, 5th ed. Vol. 2. Academic Press, Orlando, FL, pp. 347−372.

Ardini, F., Dan, G., Grotti, M., 2020. Arsenic speciation analysis of environmental samples. Journal of Analytical Atomic Spectrometry (Royal Society of Chemistry) (RSC) 35 (2), 215−237. Available from: https://doi.org/10.1039/c9ja00333a, https://www.researchgate.net/publication/337299536_Arsenic_speciation_analysis_of_environmental_samples.

Aribam, B., Alam, W., Thokchom, B., 2021. Chapter 8: Water, arsenic and climate. In: Thokchom, B., Qui, P., Singh, P., Iyer, P.K. (Eds.), Water Conservation in the Era of Climate Change. Elsevier, pp. 167−190.

Bates, M.N., Smith, A.H., Hopenhayn-Rich, C., 1992. Arsenic ingestion and internal cancers: a review. American Journal of Epidemiology 135 (5), 462−476. Available from: https://doi.org/10.1093/oxfordjournals.aje.a116313.

Bhat, A.O., Hara, T., Tian, F., 2022. Review of analytical techniques for arsenic detection and determination in drinking water. Environmental Science: Advances 2 (2), 175−195. Available from: https://doi.org/10.1039/d2va00218c.

Bissek, A.C., Tabah, E.N., Kouotou, E., Sini, V., Yepnjio, F.N., Nditanchou, R., et al., 2012. The spectrum of skin diseases in a rural setting in Cameroon (sub-Saharan Africa). BMC Dermatology 12, 7. Available from: https://doi.org/10.1186/1471-5945-12-7.

Bondu, R., Cloutier, V., Rosa, E., Benzaazoua, M., 2016. A review and evaluation of the impacts of climate change on geogenic arsenic in groundwater from fractured bedrock aquifers. Water, Air, and Soil Pollution 227 (9), 296. Available from: https://doi.org/10.1007/s11270-016-2936-6.

Bortnikova, S., Gaskova, O., Yurkevich, N., Saeva, O., Abrosimova, N., 2020. Chemical treatment of highly toxic acid mine drainage at a gold mining site in southwestern Siberia, Russia. Minerals 10 (10), 867. Available from: https://doi.org/10.3390/min10100867.

Bowell, R.J., Alpers, C.N., Jamieson, H.E., Nordstrom, D.K., Majzlan, J., 2014. The environmental geochemistry of arsenic - an overview. Reviews in Mineralogy and Geochemistry 79 (1), 1−16. Available from: https://doi.org/10.2138/rmg.2014.79.1.

Boyle, R.W., Jonasson, I.R., 1973. The geochemistry of arsenic and its use as an indicator element in geochemical prospecting. Journal of Geochemical Exploration 2, 251−296.

Bretzler, A., Lalanne, F., Nikiema, J., Podgorski, J., Pfenninger, N., Berg, M., et al., 2017. Groundwater arsenic contamination in Burkina Faso, West Africa: predicting and verifying regions at risk. Science of the Total Environment 584−585, 958−970.

Bretzler, A., Nikiema, J., Lalanne, F., Hoffmann, L., Biswakarma, J., Siebenaller, L., et al., 2020. Arsenic removal with zero-valent iron filters in Burkina Faso: field and laboratory insights. Science of the Total Environment 737, 139466. Available from: https://doi.org/10.1016/j.scitotenv.2020.139466.

Brown, K.G., Ross, G.L., 2002. Arsenic, drinking water, and health: a position paper of the American Council on Science and Health. Regulatory Toxicology and Pharmacology 36, 162−174. Available from: https://doi.org/10.1006/rtph.2002.1573.

Byun, K., Won, Y.L., Hwang, Y.I., Koh, D.H., Im, H., Kim, E.A., 2013. Assessment of arsenic exposure by measurement of urinary speciated inorganic arsenic metabolites in workers in a semiconductor manufacturing plant. Annals of Occupational and Environmental Medicine 25 (1), 21. Available from: https://doi.org/10.1186/2052-4374-25-21.

Carlin, D.J., Naujokas, M.F., Bradham, K.D., Cowden, J., Heacock, M., Henry, H.F., et al., 2016. Arsenic and environmental health: state of the science and future research opportunities. Environmental Health Perspectives 124 (7), 890−899. Available from: https://doi.org/10.1289/ehp.1510209.

Carter, D.E., Aposhian, H.V., Gandolfi, A.J., 2003. The metabolism of inorganic arsenic oxides, gallium arsenide, and arsine: a toxicochemical review. Toxicology and Applied Pharmacology 193, 309−334.

Chakraborti, D., Mukherjee, S.C., Saha, K.C., Chowdhury, U.K., Rahman, M.M., Sengupta, M.K., 2003. Arsenic toxicity from homeopathic treatment. Journal of Toxicology and Clinical Toxicology 41 (7), 963−967. Available from: https://doi.org/10.1081/clt-120026518.

Challenger, F., 1945. Biological methylation. Chemical Reviews 36 (3), 315−361. Available from: https://doi.org/10.1021/cr60115a003.

Chen, C.-E., Wang, Y.-C., Ma, H., 2017. Multiple Bowen's disease and epithelioid malignant peripheral nerve sheath tumor in a patient who experienced chronic arsenic poisoning. Formosan Journal of Surgery 50 (3), 110−113. Available from: https://doi.org/10.4103/fjs.fjs_37_17.

Chen, S.J., Yan, X.J., Chen, Z., 2013. Arsenic in therapy. In: Kretsinger, R.H., Uversky, V.N., Permyakov, E.A. (Eds.), Encyclopaedia of Metalloproteins. Springer, New York, NY. Available from: https://doi.org/10.1007/978-1-4614-1533-6_492.

Chung, J.Y., Yu, S.D., Hong, Y.S., 2014. Environmental source of arsenic exposure. Journal of Preventive Medicine and Public Health 47 (5), 253−257. Available from: https://doi.org/10.3961/jpmph.14.036.

Ciarrocca, M., Tomei, F., Caciari, T., Cetica, C., Andrè, J.C., Fiaschetti, M., et al., 2012. Exposure to arsenic in urban and rural areas and effects on thyroid hormones. Inhalation Toxicology 24 (9), 589−598. Available from: https://doi.org/10.3109/08958378.2012.703251.

Dinwiddie, E., Liu, X.-M., 2018. Examining the geologic link of arsenic contamination in groundwater in Orange County, North Carolina. Frontiers in Earth Science 6, 111.

Doe, P.T., Asiedu, A., Acheampong, J.W., Rowland, P.C.M., 2001. Skin diseases in Ghana and the UK. International Journal of Dermatology 40 (5), 323−326. Available from: https://doi.org/10.1046/j.1365-4362.2001.01229.x.

Dominguez-Gonzalez, R., Varela, L.G., Bermejo-Barrera, P., 2014. Functionalised gold nanoparticles for the detection of arsenic in water. Talanta 118, 262−269. Available from: https://doi.org/10.1016/j.talanta.2013.10.029.

Drewniak, L., Sklodowska, A., 2013. Arsenic-transforming microbes and their role in biomining processes. Environmental Science and Pollution Research 20, 7728−7739. Available from: https://doi.org/10.1007/s11356-012-1449-0.

Fang, Y., Zhang, Z., 2020. Arsenic trioxide as a novel anti-glioma drug: a review. Cell and Molecular Biology Letters 25, 44. Available from: https://doi.org/10.1186/s11658-020-00236-7.

FAO (Food and Agricultural Organisation of the United Nations), 2021. Food-based Dietary Guidelines: Africa. Available from: http://www.fao.org/nutrition/education/food-dietary-guidelines/regions/africa/en/. (Accessed 09.03.21).

Farooq, M.A., Islam, F., Ali, B., Najeeb, U., Mao, B., Gill, R.A., et al., 2016. Arsenic toxicity in plants: cellular and molecular mechanisms of its transport and metabolism. Environmental and Experimental Botany 132, 42−52. Available from: https://doi.org/10.1016/j.envexpbot.2016.08.004.

Farzan, S.F., Karagas, M.R., Chen, Y., 2013. *In utero* and early life arsenic exposure in relation to long-term health and disease. Toxicology and Applied Pharmacology 272 (2), 384−390.

Fatoki, O.S., Akinsoji, O.S., Ximba, B.J., Olujimi, O., Ayanda, O.S., 2013. Arsenic contamination: Africa the missing gap. Asian Journal of Chemistry 25 (16), 9263−9268.

Feldmann, J., 2009. Onsite testing for arsenic: field test kits, Reviews of Environmental Contamination and Toxicology (Continuation of Residue Reviews), vol. 197. Springer, New York, NY. Available from: https://doi.org/10.1007/978-0-387-79284-2_3. (Accessed 10 March 2021).

Francesconi, K., Kuehnelt, D., 2004. Determination of arsenic species: a critical review of methods and applications, 2000−2003. Analyst 129, 373−395. Available from: https://www.semanticscholar.org/paper/Determination-of-arsenic-species%3A-a-critical-review-Francesconi-Kuehnelt/534b6c8961d423fb445a04aa2d365ed31025f350. (Accessed 3 March 2021).

Garelick, H., Jones, H., 2008. Mitigating arsenic pollution: bridging the gap between knowledge and practice. Chemistry International 30 (4).

Gault, A.G., Cooke, D.R., Townsend, A.T., Charnock, J.M., Polya, D.A., 2005. Mechanisms of arsenic attenuation in acid mine drainage from Mount Bischoff, western Tasmania. Science of The Total Environment 345 (1-3), 219−228. Available from: https://doi.org/10.1016/j.scitotenv.2004.10.030.

Ghosh, J., Shi, P., 2015. Chapter 8: Mechanism for arsenic-induced toxic effects. In: Flora, S.J.S. (Ed.), Handbook of Arsenic Toxicology. Academic Press, pp. 203−231. Available from: https://doi.org/10.1016/B978-0-12-418688-0.00008-3.

Gong, G., Basom, J., Mattevada, S., Onger, F., 2015. Association of hypothyroidism with low-level arsenic exposure in rural West Texas. Environmental Research 138, 154–160. Available from: https://doi.org/10.1016/j.envres.2015.02.001.

GSA (Ghana Standards Authority), 2009. Water Quality Specification for Drinking Water. Document GS 175-1, third ed. Accra, Ghana. Available from: https://www.gsa.gov.gh/wp-content/uploads/2018/08/2018-Catalogue-of-Ghana-Standards.pdf (Accessed 27.01.2024).

Guzman-Rangel, G., Montalvo, D., Smolders, E., 2018. Pronounced antagonism of zinc and arsenate on toxicity to barley root elongation in soil. Environments 5 (7), 83. Available from: https://doi.org/10.3390/environments5070083.

GWCL (Ghana Water Company Limited), 2018. Laboratory Quality Assurance Manual Quality Assurance Manual for Ghana Water Company Limited, 2018. Document No.: GWCL-WQA-001. Available from: https://www.epa.gov/sites/production/files/2018-10/documents/quality_assurance_manual_for_ghana_-water_company_limited.pdf. (Accessed 09 March 2021).

Hall, M., Chen, Y., Ahsan, H., Slavkovich, V., van Geen, A., Parvez, F., et al., 2006. Blood arsenic as a biomarker of arsenic exposure: results from a prospective study. Toxicology 225 (2–3), 225–233.

Hartshorne, S.T., 2003. Dermatological disorders in Johannesburg, South Africa. Clinical and Experimental Dermatology 28 (6), 661–665. Available from: https://doi.org/10.1046/j.1365-2230.2003.01417.x.

Hayakawa, T., Kobayashi, Y., Cui, X., Hirano, S., 2005. A new metabolic pathway of arsenite: arsenic-glutathione complexes are substrates for human arsenic methyltransferase Cyt19. Archives of Toxicology 79 (4), 183–191. Available from: https://doi.org/10.1007/s00204-004-0620-x.

Hay, R., Bendeck, S.E., Chen, S., Estrada, R., Haddix, A., McLeod, T., Mahé, A., 2006. Chapter 37: Skin diseases. In: Jamison, D.T., Breman, J.G., Measham, A.R., et al.,Disease Control Priorities in Developing Countries, second ed. The International Bank for Reconstruction and Development/The World Bank, Washington (DC), Co-published by Oxford University Press, New York.

van Hees, C., Naafs, B., 2014. Common Skin Diseases in Africa: An Illustrated Guide, third ed. Health Books International. Available from: https://healthbooksinternational.org/product/common-skin-diseases-in-africa-an-illustrated-guide-3rd-revised-edition/. (Accessed 08 March 2021).

Henjum, S., Groufh-Jacobsen, S., Aakre, I., Gjengedal, E.L.F., Langfjord, M.M., Heen, E., et al., 2023. Thyroid function and urinary concentrations of iodine, selenium, and arsenic in vegans, lacto-ovo vegetarians and pescatarians. European Journal of Nutrition 3329–3338. Available from: https://doi.org/10.1007/s00394-023-03218-5.

Henshaw, E.B., Olasode, O.A., 2015. Skin diseases in Nigeria: the Calabar experience. International Journal of Dermatology 54 (3), 319–326. Available from: https://doi.org/10.1111/ijd.12752.

Horn, A.C., Ramudzuli, M.R., 2020. Arsenic contamination of soil in relation to water in northeastern South Africa. In: Fares, A., Singh, S. (Eds.), Arsenic Water Resources Contamination. Advances in Water Security. Springer, Cham. Available from: https://doi.org/10.1007/978-3-030-21258-2_7 (Accessed 15 March 2021).

Hughes, M.F., 2002. Arsenic toxicity and potential mechanisms of action. Toxicology Letters 133 (1), 1–16. Available from: https://doi.org/10.1016/s0378-4274(02)00084-x.

Hu, Y., Li, J., Lou, B., Wu, R., Wang, G., Lu, C., et al., 2020. The role of reactive oxygen species in arsenic toxicity. Biomolecules 10, 240. Available from: https://doi.org/10.3390/biom10020240.

Hwang, S., Her, Y., Jun, S.M., Song, J.-H., Lee, G., Kang, M., 2019. Characteristics of arsenic leached from sediments: agricultural implications of abandoned mines. Applied Sciences 9, 4628. Available from: https://doi.org/10.3390/app9214628.

IARC (International Agency for Research on Cancer), 2004. Some drinking-water disinfectants and contaminants, including arsenic. IARC Monographs on Evaluation of Carcinogenic Risks to Humans 84, 1–477.

IARC (International Agency for Research on Cancer), 2012. Arsenic, Metals, Fibres, and Dusts - Arsenic and Arsenic Compounds. IARC Monographs on Evaluation of Carcinogenic Risks to Humans, vol. 100C. IARC/WHO, Lyon, France.

IARC (The International Agency for Research on Cancer), 1987. Summaries and Evaluations: Arsenic and Arsenic Compounds (Group 1). International Agency for Research on Cancer. IARC Monographs on the Evaluation of Carcinogenic Risks to Humans, Supplement 7. Lyon, France, p. 100. Available from: http://www.inchem.org/documents/iarc/suppl7/arsenic.html. (Accessed 05 March 2021).

Islam, M.N., Awal, M.A., Kabir, S., Islam, M., 2005. Effects of selenium, iron and zinc supplementation on arsenic concentration in tissues with gross pathology in long Evans rats. Bangladesh Journal of Veterinary Medicine 3, 83−85.

Jain, R.B., 2016. Association between arsenic exposure and thyroid function: Data from NHANES 2007-2010. International Journal of Environmental Health Research 26 (1), 101−129. Available from: https://doi.org/10.1080/09603123.2015.1061111.

James, K.A., Meliker, J.R., Marshall, J.A., Hokanson, J.E., Zerbe, G.O., Byers, T.E., 2013. Validation of estimates of past exposure to arsenic in drinking water using historical urinary arsenic concentrations. Journal of Exposure Science and Environmental Epidemiology 23, 450−454. Available from: https://doi.org/10.1038/jes.2013.8.

Katz, S.A., 2019. On the use of hair analysis for assessing arsenic intoxication. International Journal of Environmental Research and Public Health 16 (6), 977. Available from: https://doi.org/10.3390/ijerph16060977.

Kennedy, B., 1965. Skin diseases in Africa: an essay in epidemiology. Archives of Dermatology 92 (3), 348. Available from: https://doi.org/10.1001/archderm.1965.01600150138037.

Kim, D., Miranda, M.L., Tootoo, J., Bradley, P., Gelfand, A.E., 2011. Spatial modeling for groundwater arsenic levels in North Carolina. Environmental Science and Technology 45, 4824−4831. Available from: https://doi.org/10.1021/es103336s.

Kobayashi, Y., Agusa, T., 2019. Chapter 2: Arsenic metabolism and toxicity in humans and animals: racial and species differences. In: Yamauchi, H., Sun, G. (Eds.), Arsenic Contamination in Asia: Biological Effects and Preventive Measures. Current Topics in Environmental Health and Preventive Medicine, pp. 13−28. Available from: https://doi.org/10.1007/978-981-13-2565-6_2.

Kortatsi, B.K., Asigbe, J., Dartey, G.A., Tay, C., Anornu, G.K., Hayford, E., 2008. Reconnaissance survey of arsenic concentration in groundwater in south-eastern Ghana. West African Journal of Applied Ecology 13 (1), 16−26. Available from: https://doi.org/10.4314/wajae.v13i1.40586.

Krohn, R.M., Raqib, R., Akhtar, E., Vandenberg, A., Smits, J.E.G., 2016. A high-selenium lentil dietary intervention in Bangladesh to counteract arsenic toxicity: study protocol for a randomized controlled trial. Trials 17, 218. Available from: https://doi.org/10.1186/s13063-016-1344-y.

Kruger, M.C., Bertin, P.N., Heipieper, H.J., Arsène-Ploetze, F., 2013. Bacterial metabolism of environmental arsenic: mechanisms and biotechnological applications. Applied Microbiology and Biotechnology 97 (9), 3827−3841. Available from: https://doi.org/10.1007/s00253-013-4838-5.

Kryeziu, K., Jungwirth, U., Hoda, M.A., Ferk, F., Knasmüller, S., Karnthaler-Benbakka, C., et al., 2013. Synergistic anticancer activity of arsenic trioxide with erlotinib is based on inhibition of EGFR-mediated DNA double-strand break repair. Molecular Cancer Therapeutics 12 (6), 1073−1084. Available from: https://doi.org/10.1158/1535-7163.MCT-13-0065.

Kumar, K., Gupta, D., Mosa, K.A., Ramamoorthy, K., Sharma, P., 2019. Arsenic transport, metabolism, and possible mitigation strategies in plants. In: Srivastava, S., Srivastava, A., Suprasanna, P. (Eds.), Plant-Metal Interactions. Springer, Cham. Available from: https://doi.org/10.1007/978-3-030-20732-8_8.

Lapworth, D., MacDonald, A., Kebede, S., Owor, M., Chavula, G., Fallas, H., et al., 2020. Drinking water quality from rural handpump-boreholes in Africa. Environmental Research Letters 15, 064020. Available from: https://doi.org/10.1088/1748-9326/ab8031.

Lefticariu, L., Sutton, S.R., Lanzirotti, A., Flynn, T.M., 2019. Enhanced immobilization of arsenic from acid mine drainage by detrital clay minerals. ACS Earth and Space Chemistry 2019 3 (11), 2525−2538. Available from: https://doi.org/10.1021/acsearthspacechem.9b00203.

Leybourne, M.I., Johannesson, K.H., Asfaw, A., 2014. Measuring arsenic speciation in environmental media: sampling, preservation, and analysis. Reviews in Mineralogy and Geochemistry 79 (1), 371−390. Available from: https://doi.org/10.2138/rmg.2014.79.6.

Liebhafsky, H.A., 1931. The reaction between arsenious acid and iodine. Journal of Physics and Chemistry 35 (6), 1648−1654. Available from: https://doi.org/10.1021/j150324a006.

Liu, J., Lu, Y., Wu, Q., Goyer, R.A., Waalkes, M.P., 2008. Mineral arsenicals in traditional medicines: orpiment, realgar, and arsenolite. The Journal of Pharmacology and Experimental Therapeutics 326 (2), 363−368. Available from: https://doi.org/10.1124/jpet.108.139543.

Li, H.-B., Chen, X.-Q., Wang, J.Y., L, M.-Y., Zhao, D., Luo, X.-S., et al., 2019. Mouse gastrointestinal tract and their influences on metal relative bioavailability in contaminated soils. Environmental Science and Technology 53 (24), 14264−14272. Available from: https://doi.org/10.1021/acs.est.9b03656.

Luong, J.H.T., Majid, E., Male, K.B., 2007. Analytical tools for monitoring arsenic in the environment. The Open Analytical Chemistry Journal 1, 7−14. Available from: https://doi.org/10.2174/1874065000701010007.

Marshall, J., 1963. Skin diseases in South Africa. Archives of Dermatology 87 (4), 419−427. Available from: https://doi.org/10.1001/archderm.1963.01590160011003.

Masuda, H., 2018. Arsenic cycling in the Earth's crust and hydrosphere: interaction between naturally occurring arsenic and human activities. Progress in Earth and Planetary Science 5, 68. Available from: https://doi.org/10.1186/s40645-018-0224-3.

Matschullat, J., 2000. Arsenic in geosphere - a review. Science of the Total Environment 249, 297−312. Available from: https://doi.org/10.1016/S0048-9697(99)00524-0.

Ma, Y., Shi, Y., Wu, Q., Ma, W., 2021. Dietary arsenic supplementation induces oxidative stress by suppressing nuclear factor erythroid 2-related factor 2 in the livers and kidneys of laying hens. Poultry Science 100 (2), 982−992. Available from: https://doi.org/10.1016/j.psj.2020.11.061.

McCarty, K.M., Hanh, H.T., Kim, K., 2011. Arsenic geochemistry and human health in South East Asia. Reviews on Environmental Health 26, 71−78. Available from: https://doi.org/10.1515/reveh.2011.010.

Medda, N., Patra, R., Ghosh, T.K., Maiti, S., 2020. Neurotoxic mechanism of arsenic: synergistic effect of mitochondrial instability, oxidative stress, and hormonal-neurotransmitter impairment. Biological Trace Element Research 198 (1), 8−15. Available from: https://doi.org/10.1007/s12011-020-02044-8.

Melkonian, S., Argos, M., Pierce, B.L., Chen, Y., Islamm, T., Ahmed, A., et al., 2011. A prospective study of the synergistic effects of arsenic exposure and smoking, sun exposure, fertilizer use, and pesticide use on risk of premalignant skin lesions in Bangladeshi men. American Journal of Epidemiology 173 (2), 183−191. Available from: https://doi.org/10.1093/aje/kwq357.

Menezes, M.A. de B.C., Silva, M.A., 2013. Neutron activation analysis on determination of arsenic in biological matrixes. Associação Brasileira de Energia Nuclear (ABEN). International Nuclear Atlantic Conference (INAC), 24−29 November, 2013, Recife, PE, Brazil. Available from: https://inis.iaea.org/collection/NCLCollectionStore/_Public/45/086/45086024.pdf. (Accessed 06 March 2021).

MFMER (Mayo Foundation for Medical and Educational Research), 2021. Arsenic: Blood; ID:ASB. Available from: https://neurology.testcatalog.org/show/ASB#:∼:text = Arsenic%20is%20not%20likely%20to,follow%20the%20effectiveness%20of%20therapy. (Accessed 16 March 2021).

Miller Jr., W.H., Schipper, H.M., Lee, J.S., Singer, J., Waxman, S., 2002. Mechanisms of action of arsenic trioxide. Cancer Research 62 (14), 3893−3903. Available from: https://aacrjournals.org/cancerres/article/62/14/3893/508952/Mechanisms-of-Action-of-Arsenic-Trioxide1. (Accessed 27 January 2024).

Moloudi, K., Neshasteriz, A., Hosseini, A., Eyvazzadeh, N., Shomali, M., Eynali, S., et al., 2017. Synergistic effects of arsenic trioxide and radiation: triggering the intrinsic pathway of apoptosis. Iranian Biomedical Journal 21 (5), 330−337. Available from: http://ibj.pasteur.ac.ir/article-1-2029-en.html.

Moreno-Jiménez, E., Esteban, E., Peñalosa, J.M., 2012. The fate of arsenic in soil-plant systems. Reviews in Environmental Contamination and Toxicology 215, 1—37. Available from: https://doi.org/10.1007/978-1-4614-1463-6_1.

Moulick, D., Ghosh, D., Mandal, D., Bhowmick, S., Mondal, D., Choudhury, S., et al., 2023. A cumulative assessment of plant growth stages and selenium supplementation on arsenic and micronutrients accumulation in rice grains. Journal of Cleaner Production 386, 135764. Available from: https://doi.org/10.1016/j.jclepro.2022.135764.

Muehe, E.M., Wang, T., Kerl, C.F., Planer-Friedrich, B., Fendorf, S., 2019. Rice production threatened by coupled stresses of climate and soil arsenic. Nature Communications 10, 4985. Available from: https://doi.org/10.1038/s41467-019-12946-4.

Muzaffar, S, Khan, J., Srivastava, R., Gorbatyuk, M.S., Athar, M., 2023. Mechanistic understanding of the toxic effects of arsenic and warfare arsenicals on human health and environment. Cell Biology and Toxicology 39 (1), 85—110. Available from: https://doi.org/10.1007/s10565-022-09710-8.

Naranmandura, H., Suzuki, N., Suzuki, K.T., 2006. Trivalent arsenicals are bound to proteins during reductive methylation. Chemical Research in Toxicology 19, 1010—1018. Available from: https://doi.org/10.1021/tx060053f.

Naujokas, M.F., Anderson, B., Ahsan, H., Aposhian, H.V., Graziano, J.H., Thompson, C., et al., 2013. The broad scope of health effects from chronic arsenic exposure: update on a worldwide public health problem. Environmental Health Perspectives 121, 295—302. Available from: https://doi.org/10.1289/ehp.1205875.

Navarro Silvera, S.A., Rohan, T.E., 2007. Trace elements and cancer risk: a review of the epidemiologic evidence. Cancer Causes and Control 18 (1), 7—27. Available from: https://doi.org/10.1007/s10552-006-0057-z.

Ning, R.Y., 2005. Arsenic in natural waters. In: Lehr, J.H., Keeley, J. (Eds.), Water Encyclopaedia. Available from: https://doi.org/10.1002/047147844X.mw13. (Accessed 02 March 2021).

Norman, D.I., Miller, G.P., Branvold, L., Thomas, T., Appiah, H., Ayamsegna, J., et al., (n.d.). Arsenic in Ghana, West Africa, Ground waters. Available from: https://wwwbrr.cr.usgs.gov/projects/GWC_chemtherm/FinalAbsPDF/norman.pdf. (Accessed 07 March 2021).

Obinaju, B.E., 2009. Mechanisms of arsenic toxicity and carcinogenesis. African Journal of Biochemistry Research 3 (5), 232—237. Available from: https://academicjournals.org/article/article1380102206_Obinaju%20%20pdf.pdf. (Accessed 12 March 2021).

O'Day, P., 2006. Chemistry and mineralogy of arsenic. Elements 2, 77—83.

Omreland, R., Newman, D., Kail, B., Stolz, J., 2001. Bacterial respiration of arsenate and its significance in the environment. In: Frankenberger Jr, W.T. (Ed.), Environmental Chemistry of Arsenic. Publisher, Marcel Dekker, Inc., pp. 273—295.

Onishi, H., 1969. Chapter 33: Arsenic. In: Wedepohl, K.H. (Ed.), Handbook of Geochemistry. Springer-Verlag, Berlin.

Paikaray, S., 2015. Arsenic geochemistry of acid mine drainage. Mine Water and the Environment 34, 181—196. Available from: https://doi.org/10.1007/s10230-014-0286-4.

Pal, S., Chatterjee, A.K., 2004. Protective effect of methionine supplementation on arsenic-induced alteration of glucose homeostasis. Food Chemistry and Toxicology 42 (5), 737—742. Available from: https://doi.org/10.1016/j.fct.2003.12.009.

Pandey, C., Raghuram, B., Sinha, A.K., Gupta, M., 2015. miRNA plays a role in the antagonistic effect of selenium on arsenic stress in rice seedlings. Metallomics 7 (5), 857—866. Available from: https://doi.org/10.1039/c5mt00013k.

Patel, K.S., Pandey, P.K., Martín-Ramos, P., Corns, W.T., Varol, S., Bhattacharya, P., Zhu, Y., 2023. A review on arsenic in the environment: contamination, mobility, sources, and exposure. Royal Society of Chemistry Advances 13, 8803—8821. Available from: https://doi.org/10.1039/D3RA00789H.

Pearce, D., Dowling, K., Sim, M., 2012. Cancer incidence and soil arsenic exposure in a historical gold mining area in Victoria, Australia: a geospatial analysis. Journal of Exposure Science and Environmental Epidemiology 22, 248—257. Available from: https://doi.org/10.1038/jes.2012.15.

Podgorski, J., Berg, 2020. Global threat of arsenic in groundwater. Science 368 (6493), 845−850. Available from: https://doi.org/10.1126/science.aba1510.

Podgorski, J., Wu, R., Chakravorty, B., Polya, D.A., 2020. Groundwater arsenic distribution in India by machine learning geospatial modelling. International Journal of Environmental Research and Public Health 17 (19), 7119. Available from: https://doi.org/10.3390/ijerph17197119.

Polya, D.A., Watts, M.J., 2017. Chapter 5: Sampling and analysis for monitoring arsenic in drinking water. In: Bhattacharya, P., Polya, D.A., Jovanovic, D. (Eds.), Best Practice Guide on the Control of Arsenic in Drinking Water. IWA Publishing. Available from: https://doi.org/10.1016/B978-0-12-418688-0.00008-310.2166/9781780404929_049. (Accessed 05.03.2021).

Reimann, C., Matschullat, J., Birke, M., Salminen, R., 2009. Arsenic distribution in the environment: the effects of scale. Applied Geochemistry 24, 1147−1167. Available from: https://doi.org/10.1016/j.apgeochem.2009.03.013.

Rosenbaum, B.E., Klein, R., Hagan, P.G., Seady, M.-Y., Quarcoo, N.L., Hoffmann, R., et al., 2017. Dermatology in Ghana: a retrospective review of skin disease at the Korle Bu Teaching Hospital Dermatology Clini c. The Pan African Medical Journal 26, 125.

Rossiter, H.M.A., Owusu, P.A., Awuah, E., MacDonald, A.M., Schaefer, A.I., 2010. Chemical drinking water quality in Ghana: water costs and scope for advanced treatment. Science of the Total Environment 408 (11), 2378−2386. Available from: https://doi.org/10.1016/j.scitotenv.2010.01.053.

Roy, P., Saha, A., 2002. Metabolism and toxicity of arsenic: a human carcinogen. Current Science 82 (1), 38−45.

SA DWAF (South Africa Department of Water Affairs and Forestry), 1996. South Africa Water Quality Guidelines, Volume 1: Domestic Water Use, second ed. Available from: https://www.iwa-network.org/file-manager-uploads/WQ_Compendium/Database/Selected_guidelines/041.pdf. (Accessed 08 March 2021).

Sanyal, T., Bhattacharjee, P., Paul, S., Bhattacharjee, P., 2020. Recent advances in arsenic research: significance of differential susceptibility and sustainable strategies for mitigation. Frontiers in Public Health 8, 464. Available from: https://doi.org/10.3389/fpubh.2020.00464.

Sharpless, G.R., Metzger, M., 1941. Arsenic and goiter. The Journal of Nutrition 21 (4), 341−346. Available from: https://doi.org/10.1093/jn/21.4.341.

Shi, H., Shi, X., Liu, K.J., 2004. Oxidative mechanism of arsenic toxicity and carcinogenesis. Molecular and Cellular Biochemistry 255, 67−78.

Singh, A.P., Goel, R.K., Kaur, T., 2011. Mechanisms pertaining to arsenic toxicity. Toxicology International 18 (2), 87−93.

Skalny, A.V., Skalnaya, M.G., Nikonorov, A.A., Tinkov, A.A., 2016. Selenium antagonism with mercury and arsenic: from chemistry to population health and demography. In: Hatfield, D., Schweizer, U., Tsuji, P., Gladyshev, V. (Eds.), Selenium. Springer, Cham. Available from: https://doi.org/10.1007/978-3-319-41283-2_34.

Smedley, P.L., Kinniburgh, D.G., 2001. Chapter 1: Source and behaviour of arsenic in natural waters. United Nations Synthesis Report on Arsenic in Drinking-Water: Water and Sanitation - Protection of the Human Environment. Available from: https://www.ircwash.org/sites/default/files/UNACC_2001-United.pdf. (Accessed 15 March 2021).

Smedley, P.L., Kinniburgh, D.G., 2002. A review of the source, behaviour and distribution of arsenic in natural waters. Applied Geochemistry 17, 517−568. Available from: https://doi.org/10.1016/S0883-2927(02)00018-5.

Smedley, P.L., Kinniburgh, D.G., 2013. Arsenic in groundwater and the environment, In: Selinus, O., et al. (Ed.), Essentials of Medical Geology, Revised Edition. British Geological Survey, Oxfordshire, pp. 279−310. Available from: https://doi.org/10.1007/978-94-007-4375-5_12. (Accessed 03 March 2021).

Smedley, P.L., Knudsen, J., Maiga, D., 2007. Arsenic in groundwater from mineralised Proterozoic basement rocks of Burkina Faso. Applied Geochemistry 22 (5), 1074−1092. Available from: https://doi.org/10.1016/j.apgeochem.2007.01.001.

SON (Standards Organisation of Nigeria), 2007. Nigerian Standard for Drinking Water Quality. NIS 554: 2007/ ICS 13.060.20. SON Governing Council, Price group D. Available from: https://www.health.gov.ng/doc/ StandardWaterQuality.pdf. (Accessed 09 March 2021).

Stolz, J.F., Basu, P., Santini, J.M., Oremland, R.S., 2006. Arsenic and selenium in microbial metabolism. Annual Review of Microbiology 60 (1), 107−130.

Tchounwou, P.B., Yedjou, C.G., Udensi, U.K., Pacurari, M., Stevens, J.J., Patlolla, A.K., et al., 2019. State of the science review of the health effects of inorganic arsenic: perspectives for future research. Environmental Toxicology 34 (2), 188−202. Available from: https://doi.org/10.1002/tox.22673.

Tolins, M., Ruchirawat, M., Landrigan, P., 2014. The developmental neurotoxicity of arsenic: cognitive and behavioral consequences of early life exposure. Annals of Global Health 80 (4), 303−314.

Torres, E., Auleda, M., 2013. A sequential extraction procedure for sediments affected by acid mine drainage. Journal of Geochemical Exploration 128, 35−41.

Tsai, C.-W., Chang, N.-W., Tsai, R.-Y., Wang, R.F., Hsu, C.M., Lin, S.S., et al., 2010. Synergistic cytotoxic effects of arsenic trioxide plus dithiothreitol on mice oral cancer cells. Anticancer Research 30 (9), 3655−3660.

Tseng, C.H., 2005. Blackfoot disease and arsenic: a never-ending story. Journal of Environmental Science and Health, Part C: Environmental Carcinogenesis and Ecotoxicology Reviews 23, 55−74. Available from: https://doi.org/10.1081/GNC-200051860.

US ATSDR (Agency for Toxic Substances and Disease Registry), 2007. Toxicological Profile for Arsenic (Update). U.S. Public Health Service, U.S. Department of Health and Human Services, Atlanta, GA. Available from: https://www.atsdr.cdc.gov/toxprofiles/tp2.pdf (Accessed 05 March 2021).

US NRC (US National Research Council), 1999. Arsenic in Drinking Water. The National Acdemies Press, Washington, DC. Available from: https://doi.org/10.17226/644 (Accessed 05 March 2021).

US NRS (US National Research Council, Committee on Medical and Biological Effects of Environmental Pollutants), 1977. Chapter 3: Arsenic: Medical and Biologic Effects of Environmental Pollutants. In: NAS (Ed.), Distribution of Arsenic in the Environment. US National Academies Press, Washington (DC). Available from: https://doi.org/10.17226/9003, http://www.ncbi.nlm.nih.gov/books/NBK231018/. (Accessed March 2021).

Uthus, E.O., 1992. Evidence for arsenic essentiality. Environmental Geochemistry and Health 14 (2), 55−58. Available from: https://doi.org/10.1007/BF01783629.

Uthus, E.O., 2003. Arsenic essentiality: a role affecting methionine metabolism. Journal of Trace Elements in Experimental Medicine 16, 345−355. Available from: https://doi.org/10.1002/jtra.10044.

UW (University of Washington), 2019. Warmer temperatures will increase arsenic levels in rice. ScienceDaily. Available from: http://www.sciencedaily.com/releases/2019/12/191204152827.htm. (Accessed 09 March 2021).

Vellinger, C., Parant, M., Rousselle, P., Usseglio-Polate, P., 2012. Antagonistic toxicity of arsenate and cadmium in a freshwater amphipod (*Gammarus pulex*). Ecotoxicology 21, 1817−1827. Available from: https://doi.org/10.1007/s10646-012-0916-1.

Wang, H.Y., Chen, S., Xue, R.Y., Lin, X.Y., Yang, J.L., Zhang, Y.S., Li, S.W., 2023. Arsenic ingested early in life is more readily absorbed: Mechanistic insights from gut microbiota, gut metabolites and intestinal morphology and functions. Environmental Science and Technology 57 (2), 1017−1027. Available from: https://doi.org/10.1021/acs.est.2c04584.

Wang, T.S., Shu, Y.F., Liu, Y.C., Jan, K.Y., Huang, H., 1997. Glutathione peroxidase and catalase modulate the genotoxicity of arsenite. Toxicology 121 (3), 229−237. Available from: https://doi.org/10.1016/s0300-483x (97)00071-1.

WHO/World Health Organisation, 1993. Guidelines for Drinking-Water Quality, second ed., Water, Sanitation, Hygiene. Vol. 1 - Recommendations. Available from: https://www.who.int/water_sanitation_health/dwq/ 2edvol1i.pdf. (Accessed 15 March 2021).

WHO (World Health Organisation), 2000. Arsenic. In: Air quality guidelines for Europe, second ed. WHO Regional Publications, European Series, No. 91. Copenhagen, World Health Organisation Regional Office for Europe. Available from: http://www.euro.who.int/_data/assets/pdf_file/0005/74732/E71922.pdf. (Accessed 28 February 2021).

WHO (World Health Organization), 2001. Arsenic and arsenic compounds (environmental health criteria 224), second ed. Geneva: World Health Organization, International Programme on Chemical Safety. Available from: https://www.who.int/ipcs/publications/ehc/ehc_224/en/.

WHO (World Health Organisation), 2018. Arsenic. WHO Fact Sheet. Geneva, World Health Organization. Available from: https://www.who.int/en/news-room/fact-sheets/detail/arsenic. (Accessed 28 February 2021).

WHO (World Health Organisation), 2019. Exposure to arsenic: a major public health concern. In: Preventing Disease Through Healthy Environments, Document: WHO/CED/PHE/EPE/19.4.1. Available from: https://apps.who.int/iris/bitstream/handle/10665/329482/WHO-CED-PHE-EPE-19.4.1-eng.pdf?ua = 1. (Accessed 28 February 2021).

WHO (World Health Organization), 2003. Arsenicosis case-detection, management and surveillance: report of a regional consultation, New Delhi, India, 5−9 November 2002. In: Arsenicosis Case-Detection, Management and Surveillance. WHO Regional Office for South-East Asia, New Delhi. NAWASIS, Jakarta, Indonesia. Available from: http://nawasis.org/portal/digilib/read/arsenicosis-case-detection-management-and-surveillance-report-of-a-regional-consultation-new-delhi-india-5-9-november-2002/1402. (Accessed 15 March 21).

WHO, 2017. Guidelines for drinking-water quality, fourth ed. incorporating the first addendum. World Health Organization, Geneva, pp. 315−318. Available from: https://apps.who.int/iris/bitstream/handle/10665/254637/9789241549950-eng.pdf. (Accessed 28 February 2021).

Wichelns, D., 2016. Managing water and soils to achieve adaptation and reduce methane emissions and arsenic contamination in Asian rice production. Water 8 (4), 141. Available from: https://doi.org/10.3390/w8040141.

Wysocki, R., Rodrigues, J.I., Litwin, I., Tamás, M.J., 2023. Mechanisms of genotoxicity and proteotoxicity induced by the metalloids arsenic and antimony. Cellular and Molecular Life Sciences: CMLS 80 (11), 342. Available from: https://doi.org/10.1007/s00018-023-04992-5.

Yotsu, R.R., Kouadio, K., Vagamon, B., N'guessan, K., Akpa, A.J., Yao, A., et al., 2018. Skin disease prevalence study in schoolchildren in rural Côte d'Ivoire: implications for integration of neglected skin diseases (skin NTDs). PLoS Neglected Tropical Diseases 12 (5), e0006489. Available from: https://doi.org/10.1371/journal.pntd.0006489.

Yunus, F.M., Khan, S., Chowdhury, P., Milton, A.H., Hussain, S., and Rahman, M., 2016. A review of groundwater arsenic contamination in Bangladesh: the millennium development goal era and beyond. International Journal of Environmental Research and Public Health 13, 215. Available from: https://doi.org/10.3390/ijerph13020215.

Zawierucha, I., Nowik-Zajac, A., Malina, G., 2020. Selective removal of As(V) ions from acid mine drainage using polymer inclusion membranes. Minerals 10 (10), 909. Available from: https://doi.org/10.3390/min10100909.

Zhu, J., Zhang, P., Yuan, S., Tong, M., 2020. Arsenic oxidation and immobilization in acid mine drainage in karst areas. Science of the Total Environment 727, 138629. Available from: https://doi.org/10.1016/j.scitotenv.2020.138629.

Further reading

Alves, E.W., de Meis, L., 1987. Effects of arsenate on the Ca^{2+} ATPase of sarcoplasmic reticulum. European Journal of Biochemistry 166 (3), 647−651. Available from: https://doi.org/10.1111/j.1432-1033.1987.tb13562.x.

Bhattacharya, P., Frisbie, S., 2002. Chapter 6: Arsenic in the environment: a global perspective In: Sarkar, B. (Ed.), Heavy Metals in the Environment, pp. 147−216.

Biswas, A., Hendry, M.J., Essilfie-Dughan, J., 2016. Geochemistry of arsenic in low sulfide-high carbonate coal waste rock, Elk Valley, British Columbia, Canada. Science of the Total Environment 579, 396–408. Available from: https://doi.org/10.1016/j.scitotenv.2016.11.084.

Duker, A.A., Carranza, E.J.M., Hale, M., 2005. Arsenic geochemistry and health. Environment International 31 (5), 631–641.

Gyasi, S.F., Awuah, E., Larbi, J.A., Koffuor, G.A., 2012. Arsenic in water and soil: a possible contributory factor in *Mycobacterium ulcerans* infection in Buruli Ulcer Endemic areas. Asian Journal of Biological Sciences 5, 66–75. Available from: https://doi.org/10.3923/ajbs.2012.66.75.

Herath, I., Vithanage, M., Bundschuh, J., Maity, J.P., Bhattacharya, P., 2016. Natural arsenic in global groundwaters: distribution and geochemical triggers for mobilization. Current Pollution Reports 2, 68–89. Available from: https://doi.org/10.1007/s40726-016-0028-2.

Health Canada, 2017. Guidelines for Canadian drinking water quality - Summary Table. Water and Air Quality Bureau, Healthy Environments and Consumer Safety Branch, Health Canada, Ottawa, Ontario. Available from: https://www.canada.ca/en/health-canada/services/environmental-workplace-health/reports-publications/water-quality/guidelines-canadian-drinkingwater-quality-summary-table.html. (Accessed 01 March 2021).

Hindmarsh, J.T., Abernathy, C.O., Peters, G.R., McCurdy, R.F., 2002. Chapter 7: Environmental aspects of arsenic toxicity. In: Sarkar, B. (Ed.), Heavy Metals in the Environment. CRC Press, Boca Raton, pp. 217–230.

IPCS (International Programme on Chemical Safety), 2001. Arsenic and arsenic compounds, second ed. World Health Organization, International Programme on Chemical Safety. Environmental Health Criteria 224. Available from: http://whqlibdoc.who.int/ehc/WHO_EHC_224.pdf. (Accessed 28 February 2021).

Nordstrom, D.K., Archer, D.G., 2003. Arsenic thermodynamic data and environmental geochemistry. In: Welch, A.H., Stollenwerk, K.G. (Eds.), Arsenic in Groundwater: Geochemistry and Occurrence. Kluwer Academic Publishers, Boston, pp. 1–25.

Nordstrom, D.K., Alpers, C.N., Ptacek, C.J., Blowes, D.W., 2000. Negative pH and extremely acidic mine waters from Iron Mountain California. Environmental Science and Technology 34, 254–258.

Ofori-Sarpong, G., Amankwah, R.K., 2019. Acid drainage potential of rocks in south-western Ghana. International Journal of Environmental Protection and Policy 7 (1), 9–16. Available from: https://doi.org/10.11648/j.ijepp.20190701.12.

Plant, J., Smith, D., Smith, B., Williams, L., 2000. Environmental geochemistry at the global scale. Journal of the Geological Society, Special Issue 157 (4), 837–849. Available from: https://doi.org/10.1144/jgs.157.4.837.

Polya, D.A., Middleton, D.R., 2017. Chapter 1: Arsenic in drinking water: sources and human exposure. In: Bhattacharya, P., Polya, D.A., Jovanovic, D. (Eds.), Best Practice Guide on the Control of Arsenic in Drinking Water, first ed. International Water Association Publishing, London, UK. Available from: https://doi.org/10.2166/9781780404929_049.

US EPA (US Environmental Protection Agency), 2018. (Updated 2021) Groundwater and Drinking Water, National Primary Drinking water regulations. Available from: https://www.epa.gov/ground-water-and-drinking water/national-primary-drinking-water-regulations#Inorganic. (Accessed 01 March 2021).

WHO (World Health Organization), 2006. Guideline for drinking-water quality [electronic resource]: incorporating first addendum. Vol. 1, Recommendations, third ed. WHO, Geneva, Switzerland, p. 515.

WHO (World Health Organization), 2008. Guidelines for drinking-water quality, third ed. Incorporating the First and Second Addenda, Volume 1: Recommendations. Available from: https://www.who.int/water_sanitation_health/dwq/fulltext.pdf. (Accessed 15 March 2021).

Wikipedia, 2020. Inorganic phosphate transporter family. Available from: https://en.wikipedia.org/wiki/Inorganic_phosphate_transporter_family. (Accessed 16 March 2021).

Zeneli, L., Sekovanić, A., Ajvazi, M., Kurti, L., Daci, N., 2016. Alterations in antioxidant defense system of workers chronically exposed to arsenic, cadmium and mercury from coal flying ash. Environmental Geochemistry and Health 38, 65–72. Available from: https://doi.org/10.1007/s10653-015-9683-2.

9 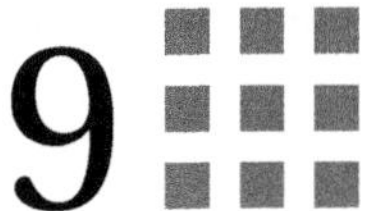

Medical geology of fluorine

"The East African Rift Valley presents an ideal setting for researching fluorine toxicity."

Davies, 2021

Key chapter features

1. Brief overviews of the most relevant and important characteristics of F in Medical Geology.
2. Metabolic interactions involving the F^- ion, and how these engender nutritional benefits (proper development of teeth and bone) or how they induce maladies (fluorosis).
3. Latest advances in research on disorders involving F ($>90\%$ of the references post-date the year 2000), and how they apply to the African situation.
4. Suggested areas of future F research in Africa.

Sources

Fluoride (F) is present in all soils, water, plants, and animals. The main sources of F intake are drinking water, other dietary constituents, and the unintentional swallowing of fluoridated dental products (*dentifrices*) (US NIH, 2023).

Much of the F in our environment is derived from geogenic sources, namely, normal weathering processes resulting in F release from rocks and minerals, volcanic activity, and marine aerosol emission (D'Alessandro, 2006; Lubojanski et al., 2023), together with the natural component of biomass burning. However, the total anthropogenic contribution to the environmental F budget is also substantial, and in some areas represents a threat to the biosphere (Fuge, 2019). Anthropogenic sources can be broadly categorized into those deriving from industrial processes, which include coal-fired power generation, brick making and ceramic manufacture, and Al production; and those deriving from agricultural practices, such as phosphate fertilizer production, sewage sludge application, and the use of herbicides and pesticides containing F^- (Fuge, 2019).

Igneous rocks

Felsic and volcanics: 864 ppm (average)—Gao et al. (1998). Felsic igneous rocks: 735 ppm (average)—Levinson (1980).

Medical Geology of Africa. DOI: https://doi.org/10.1016/B978-0-12-818748-7.00007-1

Granites: 520–850 ppm (range)—Kabata-Pendias and Pendias (2001); 533–692 ppm (range), 613 ppm (average)—Gao et al. (1998); 800 ppm (average)—Reimann and Caritat (1998); 735 ppm (average)—Levinson (1980).

Alkalic igneous rocks: >1000 ppm—Frencken (1992).

Intermediate igneous rocks: 500–1200 ppm (range)—Kabata-Pendias and Pendias (2001); 500 ppm (average)—Levinson (1980).

Diorite: 631 ppm (average)—Gao et al. (1998).

Mafic igneous rocks: 400 ppm (average)—Levinson (1980).

Gabbro: 13 ppm (average)—Ganapathy et al. (1970).

Ultramafic rocks: 50–100 ppm (range)—Kabata-Pendias and Pendias (2001); 100 ppm (average)—Frencken (1992); 20 ppm (average)—Reimann and Caritat (1998); 130 ppm (average)—Boyle (1976).

Sedimentary rocks

Carbonates: 50–350 ppm (average)—Kabata-Pendias and Pendias (2001); 454–465 ppm (range); 459 ppm (average)—Gao et al. (1998), Horn and Adams (1966), Turekian and Wedepohl (1961); Limestones: 330 ppm (average)—Levinson (1980).

Sandstones: 50–270 ppm (range)—Kabata-Pendias and Pendias (2001); 482–489 ppm (range), 485.5 ppm (average)—Gao et al. (1998); 200 ppm (average)—Reimann and Caritat (1998); 270 ppm (average)—Levinson (1980).

Shales: 500–800 ppm (range)—Kabata-Pendias and Pendias (2001); 700vppm (average)—Reimann and Caritat (1998); 740 ppm (average)—Levinson (1980); Phosphatic shale: 2.99–32,600 ppm (range); 16,301.5 ppm (average)—Gulbrandsen (1966).

Metamorphic rocks

In 2017 Hayes et al. noted the paucity of data on the geochemistry of F in metamorphic rocks. Earlier studies (2015) by García and Borgnino had shown the tendency of F to be concentrated in biotite and muscovite, among primary minerals, during metamorphism, as water is lost during prograde reactions.

Minerals

Being the most electronegative of all known chemical elements ($\chi = 4.0$; Pauling, 1960), F is also the most reactive. In nature, F hardly ever occurs as a free element but combines readily to form F^-, which is the form generally encountered. The only relatively common rock-forming minerals that have F as an essential constituent are fluorite (CaF_2) and topaz [$Al_2SiO_4(F, OH)_2$]. The other minerals in which F is an essential component are rare accessory minerals: fluorapatite [$Ca_5(PO_4)_3F$], fluormica (phlogopite) [$KMg_3AlSi_3O_{10}(F,OH)_2$], cryolite ($Na_3AlF_6$ or AlF_6Na_3), and villiaumite (NaF).

Fluorine and Cl can substitute for the diatomic hydroxide anion (OH^-) in apatite [(Ca_5 $PO_4)_3$ (OH, F, C1)] and (F) may even occupy practically all of the OH^- positions in certain instances to form the fluorapatite-end member (See Marincea et al., 2022).

Villiaumite is found within *trona* ($Na_2CO_3 \cdot NaHCO_30.2H_2O$) in closed-basin alkaline lakes, such as in the East African Rift Valley. Trona is the raw material from which sodium carbonate (i.e., washing soda or soda ash) is produced. Should concentrations of villiaumite in the trona ore reach a certain level, it could become a feasible byproduct from which F may be recovered (Hayes et al., 2017).

Soils

The F^- content of most soils is derived primarily from the weathering of common or relatively common rock forming minerals such as biotite, hornblende, fluorite or cryolite (Whelton, 2009; Jha et al., 2011). More than 90% of the F^- in soil is bound to soil particles, and is undissolved. As a soil constituent, F^- is located largely in the colloidal or clay fraction of the soil, as adsorbed ions and compounds; with the movement of F^- being strongly influenced by the sorption capacity of the soil, the soil pH, the type of soil sorbent present, and the salinity (See, e.g., Pickering, 1985; Prabhu et al., 2023; Wehr et al., 2023). A further source of F^- addition to soils is from phosphate fertilizer application (Cronin et al., 2000; Moirana et al., 2021). Other factors influencing the soil F content through soil-forming processes include the soil texture and climatic conditions (Kabata-Pendias and Mukherjee, 2007; Fuge, 2019).

The F^- concentration in most soils ranges between <10 and $1000\,ppm$ (Kabata-Pendias and Mukherjee, 2007; Fuge, 2019), with an average figure of 200 ppm as given earlier (1980) by Levinson. However, due to the downward leaching of soluble F^- in the soil during heavy precipitation or floods, an increase of inorganic F^- with depth is realized, especially when the soil is slightly acidic (pH of 5.5$-$6.5) (Barrow and Ellis, 1986; WHO, 2002; Metcalfe-Smith et al., 2003; Davison and Weinstein, 2006), or rich in clay minerals (Liu et al., 2014).

In soils of the East African Rift Valley, F^- content can be particularly high (See Gaciri and Davies, 1993; Johansen, 2013), as was also observed long ago with phosphate-derived soils in North Africa (Manley et al., 1975). The lowest F contents are generally observed in sandy soils under humid climate conditions (Kabata-Pendias and Mukherjee, 2007).

High content of F in soil is thought to be harmful to essential soil organisms, such as those that "fix" atmospheric N (e.g., *Rhizobium*) by transforming the nutrients into a usable form for the host plant (See, e.g., Geretharan et al., 2020).

Plants

We now know that the levels of F in plants is generally quite low. In 1966 Brewer gave a range as low as 2$-$20 ppm for the F content of plant materials on a dry weight basis. A few species, though, such as tea, are known to build up high levels of F, with dried tea leaves having up to 400 ppm F (See Whelton, 2009; Fordyce, 2011; Waugh et al., 2016).

Several overviews have appeared on the effect of high F concentrations in the substrate or the atmosphere on plant life (e.g., MacIntire et al., 1942; Jacobson et al., 1966; Cronin et al., 2000; Hayes et al., 2017; Fuge, 2019; Pant et al., 2008; Gheorghe and Ion, 2011; Bhat et al., 2015; Banerjee and Roychoudhury, 2019; Choudhary et al., 2019). Earlier studies by Underwood (1981) in the parts of North Africa have shown that chronic fluorosis is caused by the contamination of plant foliage and water by fluoride-rich phosphatic dust blown from phosphate rock deposits. More recent researches (e.g., Chatterjee et al., 2020; Geretharan et al., 2020; Jarosz and Pitura, 2021; Chahine et al., 2023) also show that high F concentrations in soil adversely affect plant life, destroy soil microbial activity, disrupt soil ecology, and cause soil and water pollution.

Atmospheric F absorbed by the plant leaves, eventually enters plant tissues where it becomes involved in a variety of plant physiological processes (Choudhary et al., 2019; Chahine et al., 2023). The amount of F^- uptake by plants largely depends upon the plant species, the soil type, the soil F^- content, and the kind of F compounds present in the soil (See, e.g., Wang et al., 2023).

Food crops and other major dietary sources

While almost all foodstuffs are known to contain at least traces of F^-, water and nondairy beverages have been listed by Lennon et al. (2004) as being the main sources of ingested F^-. Using another format for this listing, Guo (2009) itemized: fluoridated water, tea, coffee, soybeans, and marine fish with bones, such as canned salmon and mackerel, as being some of the most important F^- sources. In many cases, these fluoride-rich foods can contribute to the overall body load of F (See, e.g., Ozsvath, 2009, quoted in Fuge, 2019). Fordyce (2011) noted the likelihood of low-protein diets being an exacerbator of the problem of fluorosis. Guo (2009) gives the recommended dietary allowance for adults of 1.5−4 mg/day; for children up to 6 months, 0.1−0.5 mg/day; ages 6−11 months, 0.2−1 mg/day, and ages 1−3 years, 0.5−1.5 mg/day (See also Waugh et al., 2016).

Natural waters

Most natural waters like seawater, groundwater, lakes, rivers, and so on, contain some level of F. In the vicinity of certain industrial areas, even rainwater can contain significant amounts of F. Edmunds and Smedley (2013) assert that fluorite (CaF_2) solubility, by and large, controls the F content of waters, and hence the Ca content of the water. These authors give a typical range of 0.1−10 ppm for F^- in natural waters, generally.

The F content of seawater is usually around 1 ppm, while rivers and lakes contain less than 0.5 ppm (Lubojanski et al., 2023). In seawater, F abundance falls in the range of 0.03−1.4 ppm, giving an average value of approximately 1.3 ppm (See Rao, 2003; Whelton, 2009 García and Borgnino 2015; Schulz et al., 2017). Marine aquatic organisms are known to accumulate F commonly at the concentration levels an order of magnitude higher than in freshwater fish and land organisms, due to the generally higher F content in seawater compared to freshwater (Ermakov, 2004).

The average F^- content of surface waters was given as 0.1 ppm by Levinson way back in 1980; but later measurements reported by De Vos et al. in 2006 put F^- values for unpolluted surface waters in the range of $<0.05-1.6$ ppm. Schulz et al. (2017) gave a range of between 0.05 and 2.7 ppm for the normal F content of river waters. Evaporation concentrates F in surface waters, locally elevating the F concentration to several hundreds of ppm in closed basin lakes.

Strong F^- enrichment of groundwaters is commonly observed, too, with some values for global groundwaters quoted by Ali et al. (2016) ranging up to over 20 ppm. Edmunds and Smedley (2013) and Ali et al. (2016) have generally categorized F^--rich groundwaters into three groups whose global occurrence they have thoroughly reviewed.

The F content of both groundwaters and surface waters impacted by fluorine-rich hydrothermal waters can become very high. In the East African Rift Valley, for instance, F^- in groundwater, lakes and rivers derived mainly from mixing with fluids from hot springs and volcanic gases, can contain concentrations of several tens to hundreds of ppm. The highest natural F^- concentration ever found in water, 2800 ppm (Bakshi, 1974), was in the Lake Nakuru in the Kenya Rift Valley.

Drinking water

The most important pathway for F (as F^-) entry into the body is through drinking water (Fuge, 2019). Earlier observations by Dissanayake (1991), and later by several other authors (e.g., Edmunds and Smedley, 2013; Ali et al., 2016; Çakır and Şahin, 2023) have shown that drinking water F^- concentrations of $0.5-1.5$ ppm will promote healthy teeth development; concentrations over and above 1.5 ppm causing dental fluorosis, and concentrations exceeding 4 ppm resulting in skeletal fluorosis. In Africa, these conclusions are buttressed by the observation that major problems with regard to human fluorosis are manifested in areas where the domestic water supply derives from fluoride-rich groundwaters (Malago et al., 2017; Maghanga et al., 2022; See also the Section "Current status of geomedical research on fluoride in Africa," in this chapter).

In 1993 the World Health Organization (WHO) recommended the drinking water quality "guideline value" of 1.5 ppm for F^-; but cautioned that in setting national standards (*cf.,* Africa and other developing regions) it will be particularly important to consider climatic conditions, volumes of water intake, and intake of F^- from other sources such as food and air. The WHO, 1993 also noted that in areas with high natural F^- levels (*cf.,* the East African Rift Valley) the "guideline value" might be difficult to achieve in some circumstances, due to the limited technology.

In another review in 2006, the WHO, 1993 recommended "maximum permissible level (MPL)" of F^- in drinking water of 1.5 ppm, and an optimum range of $0.7-1.2$ ppm, clearly restating that this figure would vary with geographical locations and body weight; with the lower concentration applying to warmer climates where the water consumption is higher, and the higher concentration to colder climates.

The WHO's most recent review (2017, as at the time of writing), of their recommended "guideline value" for F^- in drinking water, again put the figure at 1.5 ppm, restating the caveat that: "However, it is apparent that there are modifying influences, such as the amount of water consumed on a daily basis, which generally reflects climate." In tropical climatic conditions such as exist in Ghana, West Africa, for instance, it has been shown (e.g., Craig et al., 2015) that this figure (1.5 ppm of F^-) is in agreement with that in areas of the country where dental fluorosis is manifested, namely, areas where the domestic water supply derives from fluoride-rich groundwaters. In the majority of cases, the source of F in these waters is natural (Edmunds and Smedley, 2013).

Air

Fluoride enters the atmosphere from both natural and anthropogenic sources, with the latter sources being predominant. Natural or geogenic sources of F compounds include volcanic eruptions, fumeroles, rock dust, forest fires, and marine aerosols. The main anthropogenic sources are Al smelters, fertilizer factories, coal-burning power plants and other industrial activities, such as brick, tile, pottery and cement works, ceramics, glass manufacture and oil refineries (Cape et al., 2003; Fuge, 2019). Because of its extensive industrial use, hydrogen fluoride (HF) is probably the single most serious atmospheric F^- contaminant (WHO, 1984). The air around Al smelters can contain F^- particles with diameters varying from 0.1 μm to around 10 μm (Less et al., 1975), a size range that makes possible their penetration into the lung *alveoli*, with resultant health effects.

The contribution of volcanic activity to the content of F^- in the Earth's atmosphere is given by US EPA (1980) as $1-7 \times 10^6$ tons per year. Fluoride emission into the air occurs both as (gaseous HF and SiF_4), and as solid particulate form (WHO, 2000).

Gaseous HF is taken up by the vegetation (DeMille et al., 2023). According to Weinstein (1977), F^- is the most phytotoxic air pollutant, the dominant pathway of entry of atmospheric F into vegetation being through direct aerial uptake via the leaves (Davison and Weinstein, 2006). If this vegetation includes forage crops that are fed to cattle, sheep, horses, or pigs, serious health consequences may ensue. Aerial F^- particles can also fall and remain at the surface of soils and become bioavailable as water-soluble F^- to plants.

Association (additive/synergistic/antagonistic)

The relationship between Ca and F^- is intriguing. An association exists between the varying levels of F^- present in drinking water and serum levels of Ca and the related hormones: vitamin D and parathyroid hormone in pregnant women and newborn infants (Patel et al., 2017). Stimulation of osteoblastic activity by F^- in conjunction with Ca is also known (Mousny et al., 2008). Fluoride gets integrated into the bone matrix as fluorapatite, which in turn increases the hardness of bones. In contrast, an antagonisitic relationship is inherent in the alteration of Ca^{2+} signaling and mitochondrial function in enamel cells through exposure to F^- (See Aulestia et al., 2020).

The possible synergistic or antagonistic effect of F^- with a host of other species has been documented by several researchers. Examples include: the inhibitory effect of F^- on the nitrification process (See Beg and Atiqullah, 1983); the combined toxicity of As and F^- (e.g., Zeng et al., 2014; Mondal and Chattopadhyay, 2020); wide range of health effects induced by Al-F complexes (Barbier et al., 2010); F^- in synergy with Al acting as a false signal in *G protein* cascades of hormonal and neuronal regulations (Strunecka and Strunecký, 2020); the significant relationship of serum F^- concentration with thyroid hormone activity (Singh et al., 2014); the inhibition of the *sodium iodide symporter (NIS)* expression and functionality by F^- (Waugh, 2019); the relationship between water fluoridation and elevated blood lead (PbB) concentrations in children (Macek et al., 2006).

Chemical form or speciation in soil and aqueous phases

Only one chemical valence is known for F. With an ionic radius of 1.36 angstroms (Å), F can substitute readily for the hydroxyl ion (OH^-) (ionic radius, 1.40 Å) in mineral structures (Fleischer and Robinson, 1963). Substitution of F for Cl^- and O^{-2} is less common.

As already indicated, F usually occurs as the fluoride ion (F^-) in aqueous solution. In marine waters, MgF^+ ions are also found. Fluoride and MgF^+ ions account for about 51% and 47%, respectively, of the total F concentration in seawater (García and Borgnino, 2015; El-Said et al., 2023).

"Fluoride forms its most stable bonds with Fe, Al, and Ca, and labile F is held by soil components that contain these elements, including clay minerals, calcium and magnesium compounds, and iron and aluminum compounds" (several authors quoted by Cronin et al., 2000). The F^- minerals are sparingly soluble in water, fluorite (CaF_2) being known to be the least soluble and exerting the main control on aqueous concentrations of F in the natural environment (Edmunds and Smedley, 2013).

Environmental circulation and impact of climate change

Most of the F circulating in the Earth's surface environment stems from natural weathering processes, whereby F-rich parent rocks can produce soils containing over 1 wt.% F (Fuge, 2019). Volcanic activities, wind-blown dust, a small component from the marine realm and (natural) biomass burning are also significant sources. Several human activities, notably in urban/industrialized areas contribute substantial amounts of F to the environmental budget. The largest of these sources is considered to be the application of phosphate fertilizers, which probably adds a global average of over 2.3 Mt of F annually to soils (Fuge, 2019). Much of this contribution is strongly retained in soils, but some of it is transferred to groundwater (Fig. 9—1). Rivers deliver approximately 3.6 Tg/yr of dissolved F to the oceans and it (F) is removed from the oceans by the deposition of terrigenous and authigenic sediments (Schlesinger et al., 2020).

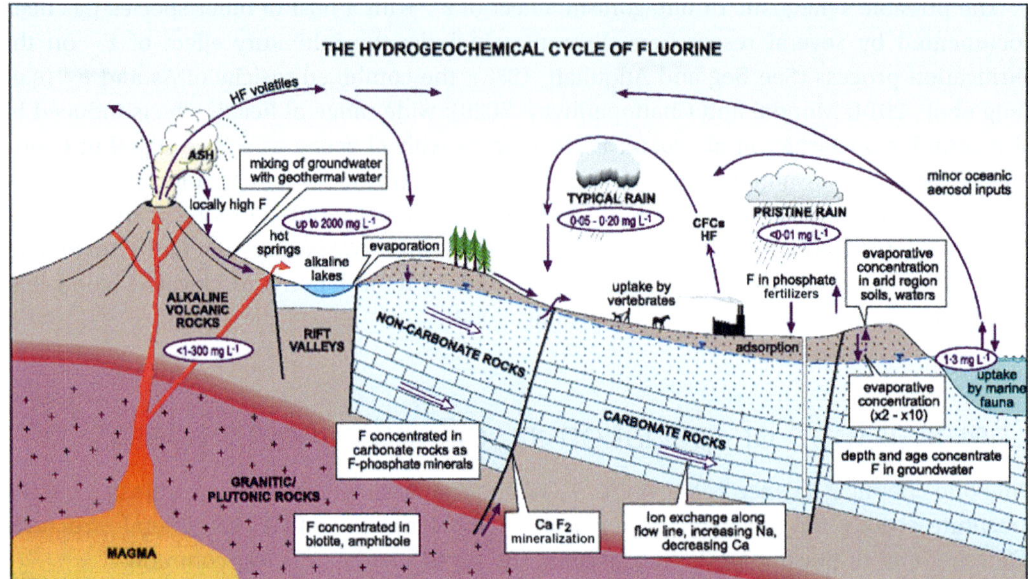

FIGURE 9–1 Schematic diagram showing the fluorine hydrogeochemical cycle. *Reproduced from Edmunds, W.M., Smedley, P.L., 2013. Fluoride in natural waters. In: Selinus, O. (Ed.), Essentials of Medical Geology. Springer, Dordrecht. https://doi.org/10.1007/978-94-007-4375-5_13.*

The rapid pace of industrialization witnessed during the last five to six decades has engendered the infusion of substantial quantities of F into the environment through the everyday use of *fluorocarbon compounds,* many of whose degradation, together with pyrolysis of *fluoropolymers* and household refuse combustion has resulted in the deposition of *organofluorine compounds* (e.g., *trifluoroacetic acid*) in the environment (See Fuge, 2019).

Fluorocarbons

Fluorocarbons (CFs) are chain polymer-containing F and C. They have applications in the manufacture of a wide range of household and commercial products, such as waterproofing agents, leather conditioners, lubricants, sealants, solvents, and anesthetics. Use of these products has been associated with occasional outbreaks of a number of respiratory conditions, such as *acute respiratory distress syndrome* and *pneumonitis* (Walters et al., 2017; Hays et al., 2020). The mechanism of fluoropolymer-related respiratory illness in humans is not as yet well understood, mainly because of the many challenges posed to researchers involved in this study (See Hays et al., 2020; and "Suggested areas for further fluoride research in Africa," in this Chapter).

Fluorinated gases

Fluorinated greenhouse gases, hydrofluorocarbons (HFCs), perfluorocarbons, and sulfur hexafluoride are a family of gases containing F. They are entirely from anthropogenic releases, unlike many other greenhouse gases. Their emissions stem from the use as substitutes for ozone-depleting substances such as refrigerants, and also from certain industrial processes, such as aluminum and semiconductor manufacturing. Compared to other greenhouse gases, the global warming potentials of fluorinated gases are extremely high. Their atmospheric lifetimes can also be quite long, in some cases, lasting thousands of years (US EPA, 2020, 2021).

The good news is that fluorinated gases are known to be ozone-friendly, enable energy efficiency; and, owing to their low levels of toxicity and flammability, are considered to be relatively safe for use by the public (US EPA, 2021).

Another very welcome news in the 2023 Inter-Governmental Panel on Climate Change (IPCC) Data Description Paper is "…there is evidence that increases in greenhouse gas emissions have slowed, and depending on societal choices, a continued series of these annual updates over the critical 2020s decade could track a change of direction for human influence on climate." [*Sic.*] (Forster at al., 2023).

Mobility, routes of exposure/mode of entry into food chains, and bioavailability

Fluoride compounds are generally soluble; some, only sparingly so [e.g., CaF_2 (0.016 g/L), MgF_2 (0.13 g/L), and Na_3AlF_6 (0.42 g/L)]; others, quite highly soluble [e.g., HF, SiF_4 (hydrates), and NaF (40 g/L) (Cronin et al., 2000)].

Fluoride is generally not highly mobile, its mobility in aqueous media being dependent on the interplay of several biogeochemical factors, chief among which are pH, temperature, the dissolution, and precipitation of F^- bearing minerals, the adsorption/desorption of/from metal (hydr-)oxides and clay minerals and the concentrations of Al, Ca, Mg, P, and hydroxides (García and Borgnino, 2015; Hayes et al., 2017).

Under acidic conditions, mobility is limited by retention/absorption on clayey substrates (Pickering, 1985). Under alkaline conditions, desorption from clays occurs (Saxena and Ahmed, 2001; Saxena and Ahmed, 2003), but availability is again limited when carbonates are present (Kabata-Pendias and Mukherjee, 2007). In carbonate-free substrates, however, alkaline evaporitic conditions produce the greatest known solubility of F^- at ambient temperature (Hayes et al., 2017).

According to Weinstein and Davison (2004), the bioavailabilty of soil F is generally low, especially in soils having a pH of 5.5−6.5; thus little soil F is taken up by the plants. Incorporation of soil-derived F^- into the aerial parts of plants is, in any case, obviated through exclusion by the roots (Davison and Weinstein, 2006).

The most important primary exposure route for F compounds is oral ingestion of drinking water, food, and fluoride-containing dental products. Inhalation and skin exposure to HF is an acute occupational exposure that can occur (US ASTDR, 2003).

Fluoride absorption and distribution

Certain dietary factors account for either a decrease or an increase in F^- absorption and its utilization (Cerklewski, 1997; Rocha et al., 2015). For instance, in the absence of high dietary concentrations of Ca and some other associated cations with which F may form insoluble and poorly absorbed compounds, as much as 80% or more is typically absorbed. According to the USIM, 1997, some 50% or so of dietary F^- taken in orally (from food and water) is absorbed from the gastrointestinal tract after approximately 30 minutes. Once absorbed into the blood, F^- readily distributes throughout the body, with the greatest amount retained in calcium-rich areas, such as bone and teeth (dentine and enamel).

The passage of F^- through the gastrointestinal tract and its subsequent absorption are the processes that determine the quantity of F^- is finally available for systemic circulation and is retained in the target tissues; and hence, the beneficial or undesirable outcome associated with exposure to this element (Rocha et al., 2015).

Almost all of the F^- that enters the body (about 99%) gets into calcified tissues to which it is strongly, but not irreversibly bound (USIM US Institute of Medicine, 1997; Zohoori and Duckworth, 2016). About half of this amount is quickly taken-up by developing bone and teeth, and the remainder is excreted in the urine, an amount that increases as bone growth slows (Cerklewski, 1997).

Metabolic function (essentiality/clinical toxicity)

The burden of F^- on the body is strongly controlled and has a direct impact at both cellular and systemic levels. Fluoride metabolism can be influenced by a whole array of systemic, geochemical, metabolic, and genetic interactions (Nagendra et al., 2021; Zohoori and Buzalaf, 2021). Also, the sheer diversity of fluoride-containing compounds renders complex, the process of deciphering their various interactions within the body, which depends on their reactivity and structure, solubility, acid-base disturbance, ability to release F^- ions, and a number of other factors (Martínez-Mier, 2012; Zohoori and Duckworth, 2016). Metabolism of the F^- ion is illustrated in Fig. 9–2.

After ingestion, absorption of F^- takes place in the stomach, followed by its distribution through soft and hard tissues, and eventual urine excretion through various reactions that are all largely pH dependent (Buzalaf et al., 2015); though several other factors influence the rate of absorption, including the solubility of the ingested F^- compound.

It is well-documented that there are two forms of F^- in plasma (Martínez-Mier, 2012). One fraction, termed *ionic fluoride* and the second, termed *nonionic* or *bound F^-*, composed of lipid-soluble organic fluoro-compounds. *The biological significance of the nonionic fraction is not well understood as noted by Ekstrand and Ehrnebo back in 1983.* In 2012 Martínez-Mier noted that when fluoride reaches the plasma, rapid deposition takes place in the skeleton or excretion occurs via the kidneys.

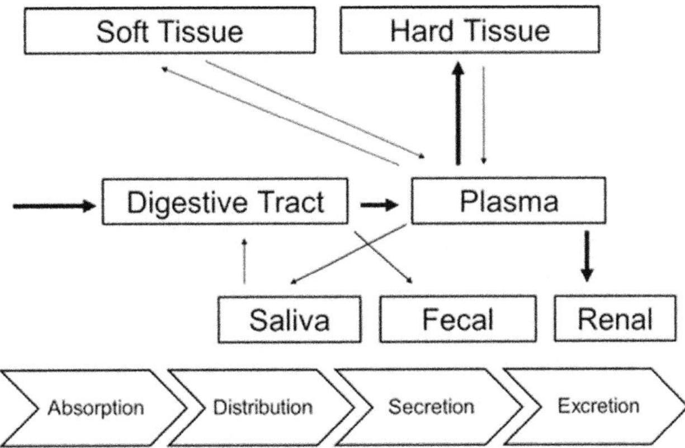

FIGURE 9–2 Flowchart of fluoride metabolism. *From Martinez-Mier, E., 2012. Fluoride its metabolism, toxicity, and role in dental health. Journal of Evidence-Based Complementary and Alternative Medicine 17, 28–32.*

Essentiality

Fluorine is not considered an essential element *per se*, although its beneficial effects on human oral health and bone development (when ingested at levels of *ca.* 1.5 ppm), are now well established and documented (See Section "Drinking water," in this Chapter).

Fluorine is *anticariogenic* (effectiveness in reducing the risk and prevalence of dental caries or tooth decay), through mechanisms, such as by reduction of tooth demineralization or enhancement of the remineralization. It can also stimulate bone cell proliferation, and hence is of benefit in optimizing bone mineral density. This (feature) is important in the maintenance of bone health throughout the life; and is one that underscores its recommended use as a therapeutic agent in the treatment of *osteoporosis*. Fluorine compounds are also added to many pharmaceuticals to increase their efficacy.

Fluoride deficiency (<0.5 ppm) in Africa is a continent-wide problem. Dental caries, dental cavities, weak bones and teeth are widespread all around the African Rift Valley as well as in other regions of both active and dormant volcanoes (See Section "Current status of geomedical research on fluoride in Africa," in this chapter).

Toxicity

The mechanism of action by which F toxicity is produced, has, to this day, remained poorly understood; which underscores the spate of new research we are currently witnessing on the topic (See, e.g., Ullah et al., 2017; Grandjean, 2019; WHO, 2019; Aulestia et al., 2020; Guth et al., 2020; Strunecka and Strunecký, 2020; Chlubek et al., 2021; Das et al., 2021; Nasiri et al., 2021; Tausta et al., 2021; Thippeswamy et al., 2021; Chahine et al., 2023).

The most well-known health problems arising from ingestion of excess F^- are dental and skeletal fluorosis. Other undesirable health conditions have also been postulated. These outcomes include cancer of various organs, low IQs in children, renal problems, and interference with endocrine systems (Ozsvath, 2009; Fordyce, 2011). However, there is little substantive evidence to support the association of these health conditions with F^- intake. Detailed accounts on the clinical toxicology of F^- is given in the several recent reviews given at the start of the current Section ["Metabolic function (essentiality/clinical toxicity)"], Subsection "Toxicology."

(Box 9—1)

Finally, one other area of F toxicity that should be considered important, is the exposure of miners to dust or to radiation from the mining and processing of some types of ores, including fluorite (CaF_2). Inhalation of high amounts of fluorite dust during processing of such ores can lead to health issues that include silicosis, gastric, intestinal, circulatory system, and nervous system problems, and in the long term, anemia, weight loss, bone and teeth defects, pulmonary lesions, and bronchitis (Osinsky and Stellman, 1998) as well as cancer (US NASEM, 2006; Barbier et al., 2010). For example, an excess of mesotheliomas found among people in a rural area near the Mt. Etna volcano, Sicily, was attributed to inhalation of a white dust, found to be rich in fluoro-edenite [$NaCa_2Mg_5(Si_7Al)O_{22}F_2$], a relatively new asbestiform mineral (Burragato et al., 2005; Fubini and Fenoglio, 2007; Bruno et al., 2015). These people had never had an occupational exposure to asbestos.

BOX 9—1   The stages of tooth decay (The Open University, 2020)

The main structural chemical in the enamel of teeth hydroxyapatite [$Ca_{10}(PO_4)_6(OH)_2$] is the same as that in bone.

Acid dissolves the hydroxyapatite through the process of *demineralization*. Once the acid has been neutralized by the saliva, the minerals can be restored to the tooth surface through the process of *remineralisation*. However, too many sugary foods mean that there is insufficient time for this remineralisation to occur completely and the tooth begins to decay. It is thought that fluoride helps to prevent this decay in several different ways:

- As the enamel is developing in children's teeth, if fluoride is present it replaces the OH (hydroxy-) part of hydroxyapatite, forming fluoroapatite, which is harder and more resistant to decay.
- When the remineralisation process is occurring in the presence of fluoride, again the newly formed enamel is stronger.
- Fluoride becomes concentrated inside the plaque bacteria, which reduces their ability to produce acid, so less demineralisation of the teeth occurs.
- There is some evidence that children who grow up in areas where fluoride is present in the water have shallower grooves in the biting surfaces of their teeth, thus reducing the places where bacteria can lodge to form plaque.

It seems likely that the remineralisation effect (second bullet point) is the most important and so the control of sugars in the diet and the regular use of fluoride toothpaste, to supplement fluoride in the water, are the best preventative measures.

Paraphrased from: The Open University, 2020. Nutrition: Vitamins and minerals, Section 2.7- Fluorine (F). https://www.open.edu/openlearn/science-maths-technology/biology/nutrition-vitamins-and-minerals/content-section-2.7. (Accessed 21.03.20).

Supplementation

Supplementation of F^- can generally be achieved by controlled direct additions to drinking water or to selected food items, as well as by the use of dentifrices. The ADA (2021) considers the use of F^- and community water fluoridation as a safe and effective way of preventing tooth decay in both children and adults. However, Sauerheber's 2013 study indicates: "... that industrial fluoride added to drinking water forms intact corrosive hydrofluoric acid under acidic conditions that prevail in the stomach of man (pH 1.5−3) and animals. Ingested fluoride from water enters the bloodstream as an artificial component, not a normal constituent, and disrupts intermolecular hydrogen bonding, forming interatomic hydrogen bonding." [*Sic.*].

Guo (2009) and US NIH (2020) list common food additives as sodium fluoride, sodium fluorophosphate, and hexafluorosilicic acid. Sodium fluoride is taken mostly in the form of multivitamin/multimineral supplements, multivitamins plus F^-, or supplements containing trace minerals only (US NIH, 2020). Yet other supplements are in the form of dental products (dentifrices) (e.g., toothpaste, mouth rinses, and gels) containing one or a combination of the compounds sodium fluoride (NaF), sodium monofluorophosphate (Na_2PO_3F) and stannous fluoride (SnF_2) (Pader, 2012); or medications, such as voriconazole (VFEND or Vfend), an oral antifungal medication used to treat several infectious conditions (US FDA, 2015).

The status of fluoride supplementation in Africa is not clear; different countries apparently having their own set of guidelines. Mulder (2018) has put forward updated guidelines for systemic fluoride supplementation for practitioners in South Africa.

According to Shike (2006), adding F^- to agricultural soils for food crop production is thought to be ineffective because F^- in the soil is inactivated and does not usually find its way into the food chain. Early studies on the use of supplements containing phosphate rock materials in ruminant feedstuffs have been shown to sometimes lead to the incidence of chronic and acute fluorosis (Suttie, 1969; Jubb et al., 1993; Schultheiss and Godley, 1995); so does the extraction of deep fluoride-rich groundwater to maintain production during droughts and summer periods (Botha et al., 1993).

Reduction of body burden

There is very little information in the literature regarding reduction of the body burden of F^-. Back in 1994, Whitford showed that a high Ca diet results in a negative F^- balance in rats due to a significant increase in fecal excretion of F^-. Although alterations in plasma F^- levels were found to be insignificant, the high Ca diet produced a significant decrease in F^- levels in the *femur epiphysis*. More recently, Srivastava et al. (2017) found an increase in the serum Ca and significant decrease in urinary F^- 6 months after administering Ca supplementation to a study population of 5000 in Uttar Pradesh in India.

Mediating the body burden of F^- through lowering the intake level generally involves changing the source of water supply to one having a satisfactory F^- concentration. However, for many communities, especially those in Africa and other developing regions, this may be impossible or impracticable. Thankfully, there is now a large number of relatively simple and readily available techniques for the removal of F^- from drinking water (See Section "Defluoridation techniques," in this chapter).

Analytical determination

The following quotation by the TUM, 2019 regarding the structure of F underlines the complexities of accurately determining F in biological and environmental samples, namely: "Investigations with neutrons settle scientific dispute about the structure of solid fluorine" (See also Ivlev et al., 2019).

Determination of F^- in biological materials and other environmental samples is normally done by two standard methods, (1) by potentiometric [ion selective electrode (ISE)] and (2) by gas chromatographic (GC) methods. Some researchers have employed colorimetric methods (e.g., Zhu et al., 2005; Gao et al., 2007; Amereih et al., 2013), but colorimetry generally suffers from limitations such as protracted operational procedures and low sensitivity.

Perhaps the most commonly used technique for routine F^- analyses and surveillance is that of ISEs, which are electrochemical ion sensors used to determine the activity of ions in aqueous solution by measuring the electrical potential. The method has many advantages compared to other F^- measurement techniques; but, like many other analytical methods, certain limitations are inherent; among which are its relatively low sensitivity (e.g., Han et al., 2019) and interference from other ions in solution (Dimeski et al., 2009). Thankfully, modern analytical chemistry research now makes possible the introduction of exquisite refinements that aim to obviate earlier limitations of the technique (See, e.g., Antes et al., 2012; Bahar et al., 2019; Radić et al., 2020; Mendes et al., 2020; Yamada et al., 2023).

GC is a laboratory technique used for separating and analyzing compounds that can be vaporized without decomposition. Thus the method can detect bound F, which is an advantage over ISE. The GC method also has a high sensitivity, being able to detect nanogram quantities of F^- in a milliliter of urine or plasma (Kage et al., 2008; Kwon and Shin, 2014). This method is also useful for estimating the F^- released from F-containing drugs in biological fluids.

Like most other analytical methods, however, there are a few disadvantages with the use of GC, such as occurrence of interference with Al ion under the operating conditions of the GC (See, e.g., Ikenishi and Kitagawa, 1988), as is also the case with the ISE method. Modern analytical chemical research now makes possible, the elimination of the disadvantages of the method (e.g., Pagliano et al., 2013; Du et al., 2021).

Defluoridation techniques

Several well-developed techniques are now available for the removal of F^- from contaminated water supplies (See, e.g., Renuka and Pushpandali, 2013; Ingle et al., 2014; Darchen et al., 2016; Swarnakar et al., 2016; Mobeen and Kumar, 2017; Maghanga et al., 2022; Dar and Kurella, 2023). Of these techniques, the one commonly used in African countries can be broadly grouped into the following categories: (1) precipitation and coagulation, and (2) adsorption and ion exchange. Each of these technologies has its own advantages as well as drawbacks, that must be considered when deciding on the most suitable one to use. In making such decisions, environmental health authorities have to take into account, the suitability of the selected method in terms of the ambient conditions, cost involved, and whether or not it is environmentally benign or ecofriendly. Excellent reviews on these techniques have appeared in several recent publications (e.g., Fan et al., 2003; Akafu et al., 2019; Kumar et al., 2019; Nyangi et al., 2020; Pillai et al., 2021; Lacson et al., 2021; Geleta et al., 2021; Maghanga et al., 2022; Mohandas et al., 2023), stemming from a flurry of ongoing researches, testament to the abundance, and severity of F^- maladies worldwide.

Of these techniques the "so-called" *Nalgonda technique* (Fig. 9–3) is the one commonly used in many African countries, such as Kenya, Senegal, and Tanzania (Pillai et al., 2021; Renuka and Pushpandali, 2013; Yami et al., 2018; Dahi et al., 1996). This technique, developed by the National Environmental Engineering Research Institute (NEERI) in Nagpur, India in the mid-1970s is an economical and simple method of defluoridation used at both the domestic and community levels (Nawlakhe et al., 1975). The removal process was originally thought to be by coprecipitation (Dahi et al., 1996), via a mechanistic sequence that comprises of the addition of alum [aluminum sulfate $(KAl(SO_4)_2 \cdot 12H_2O)$], lime [CaO/Ca

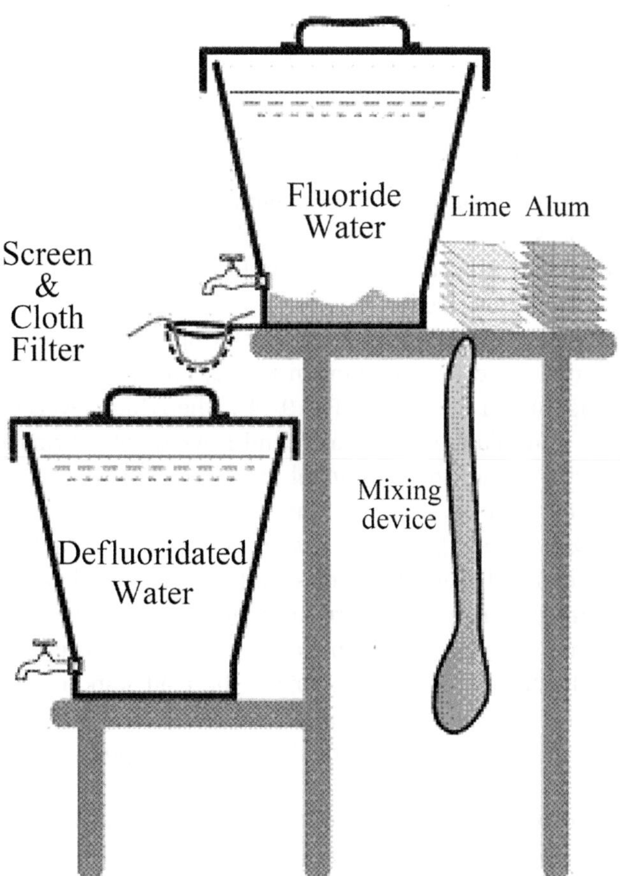

FIGURE 9–3 The Nalgonda process as optimized for use in the Rift Valley. Source: *Akvopedia, 2017. De-fluoridation. https://akvopedia.org/wiki/De-fluoridation. (Accessed 23.03.2024).*

(OH$_2$)], and bleaching powder [Ca(OCl)$_2$] to the raw water followed by flocculation, sedimentation, filtration, disinfection, and distribution (Hess et al., 2012) (Fig. 9–3).

Just like other defluoridation techniques, the Nalgonda technique is not without some demerits. The advantages and disadvantages of the technique are outlined in Renuka and Pushpandali, 2013. Certain modifications have been made to the technique since its inception, that have resulted in improvement in efficiency (e.g., Dahi, 2016; Suneetha et al., 2008; Suneetha et al., 2015; Zewge, 2017; Yami et al., 2018). However, *more research is still needed in developing low cost, simple, household defluoridation units based on the Nalgonda technique for African villages* (Woldeyes et al., 2004; Teshome et al., 2018) (See also: "Suggested areas for further fluoride research in Africa," in this chapter).

Current status of geomedical research on fluoride in Africa

In Africa, high F$^-$ intake is likely by populations living in areas of high F$^-$ bedrock and groundwaters. These are areas situated along or within the Great Rift Valley from north to south and associated volcanic centers (in Egypt, Ethiopia, Somalia, Uganda, Kenya, Tanzania, Malawi, Mozambique, and South Africa) (Van Alstine and Schruben, 1980; Johansen, 2013; Kut et al., 2016; Malago et al., 2017; Fig. 9–4). Volcanic areas also occur in other parts of Africa, such as along the Cameroon Volcanic Line in West Africa and some countries of North Africa. Certain areas of crystalline bedrock outside the Rift System, such as in Ghana, and some sedimentary basins with high-fluoride groundwaters (e.g., Senegal and parts of North Africa) also constitute high F$^-$ sources of exposure to populations (Smedley et al., 2002; Edmunds, 2008).

Skeletal deformations and the crippling bone deformities of *genu valgum* (that is, *knock knee*) have been reported to occur in regions in Kenya, Ethiopia, Tanzania, and other countries in and around the African rift system (Fig. 9–4), where high levels of F$^-$ (sometimes up to 8 ppm) are present in drinking water (Gaciri and Davies, 1993; Ermakov, 2004; Kut et al., 2016; Malago et al., 2017); and, in exceptional cases, F$^-$ levels can reach up to 305 ppm, such as in surface water, groundwater, and mine water from Kerio Valley, Kenya, in Africa's Eastern Rift Valley (Davies, 1994). In such instances (of exceptionally high F$^-$ concentrations), tooth decay, dental and skeletal fluorosis, and osteochondral conditions are common.

Fluorite deposits, ranging in age from Precambrian through tertiary, are also found along or near the rift systems, such as the Kerio Valley Fluorite deposits in Kenya. Deposits in South Africa include the Nokeng Fluorspar Project (the Plattekop deposits and the Outwash Fan deposits) located on a property that borders the Vergenoeg Mine in Northern Gauteng Province. Also found in South Africa is the Doornhoek deposit in North West Province (Hayes et al., 2017).

In Africa, environmental considerations of F$^-$ mining focus especially on drinking water, where high F concentrations can lead to tooth decay, dental and skeletal fluorosis, and bone and cartilage conditions including *genu valgum* (knock-knee). Fluoride toxicity (greater than

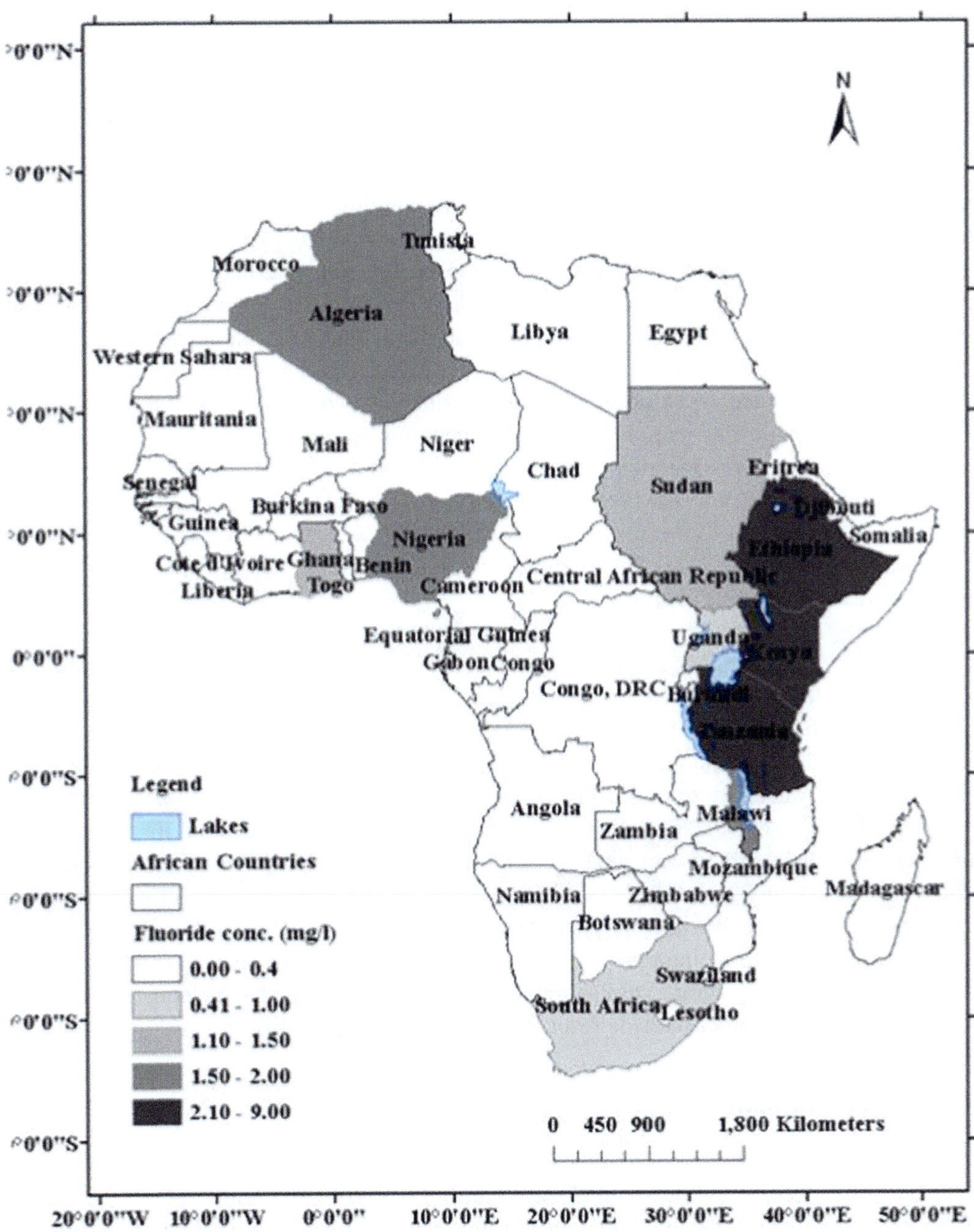

FIGURE 9—4 African countries with high-fluoride levels in ground and surface waters. *Adapted from Malago, J., Makoba, E., Muzuka, A.N.N., 2017. Fluoride levels in surface and groundwater in Africa: a review, American Journal of Water Science and Engineering 3 (1), 1—17. https://doi.org/10.11648/j.ajwse.20170301.11.*

1.5 ppm) causes mottling of teeth and dental fluorosis. At higher concentrations, skeletal fluorosis may occur (Davies, 2008; Edmunds, 2008). Trace amounts of other elements in fluorite ores are a concern at some deposits.

Although most of the reported cases of fluorosis have been among residents of high F^- areas, as far back as 1986, Manji et al. pointed out the existence of a significant number of individuals around the Rift Valley zones and other volcanic centers with fluorosis from exposure to relatively low levels of F^-. In an attempt to explain this apparent paradox, Davies (2013) invoked the interplay of metabolic and nutritional factors hinged on the biophysiochemistry of the F^- ion, which largely determines its bioavailability in mammals. As for now, this question, and that of whether fluorosis is the cause or the result of metabolic dysfunction, remains open-ended and forms an important addition to the multitude of diagnostic problems that could potentially be unraveled from a complete geochemical baseline map for Africa (See: Chapter 1, Part I).

The widespread occurrence of undesirable physiological consequences of F^- exposure in so many African countries, coupled with improvements in analytical techniques for F^- in biological and other environmental samples, has triggered a flurry of research on the subject in the last one to two decades (See Grobler et al., 2001; Ncube and Schutte, 2005; Vasak et al., 2010; Thole, 2013; Kut et al., 2016; Addison et al., 2020; Demelash et al., 2019; Onipe et al., 2020; Onipe et al., 2021) and pertinent references under: "List of selected references for further reading," in this chapter.

Back in 1997, in a study on F^- content of selected food items from East Africa, Malde et al. found F^- levels higher than safe limits recommended by the US National Research Council (US NRC, 1989) in certain foods (e.g., marine fish) and beverages (e.g., tea). However, this study did not establish the relationship between F^- level of the soil or aquatic environment and the actual amount of F^- levels that get into biological tissues, underlining the need for bioavailability studies on F^- as well as other micronutrients or toxic agents.

The relatively recent (2020) field experiments conducted by Rizzu et al. in fluoride-contaminated areas of northern Tanzania revealed the potential F^- exposure from diet and the related health risk for the local population by assessing the content of F^- in soil and plant (food crop) tissues. Results from these studies led the authors to suggest that food crops can substantially contribute to fluorosis in African countries, especially during childhood, reaffirming the necessity for more studies on F^- contents in African dietary constituents (See: "Suggested areas for further fluoride research in Africa," in this chapter).

Regulatory implications

In view of the difficulty posed in some circumstances as a result of the limited technology available, most African countries operate on the WHO drinking water quality "guideline value" for F^- of 1.5 ppm (WHO, 1993), taking into account the caveat by WHO, 1993 on the use of this value (See Section "Drinking water," in this chapter).

By the year 2000, no maximum tolerable level for F in soils existed (Cronin et al., 2000), partly because there were very few data on the content of various forms of F in natural soils globally available; and the very large disparities that existed in reported levels of F in the limited data sets documented. To date (2021), no guideline value for tolerable F⁻ levels in soils have been found for African countries. Similarly, no data could be found on recommendations of maximum tolerable F⁻ content in dietary constituents. Early studies by Malde et al. (1997) and by Kjellevoid et al. also in 1997, and later studies by Kebede et al. (2016) have all advocated for the setting of national standards and recommendations for dietary constituents.

Wastewaters from industry can contain high amounts of F, which they carry to nearby rivers, streams, or other adjoining waterways, posing deleterious effects on aquatic organisms (Davies, 1994; US ASTDR, 2003); but water quality criteria and stream-sediment guidelines for F in waterways for ensuring the protection of aquatic life, are generally lacking. In 1996 Suter proposed the use of an estimate of the highest concentration in surface water to which an aquatic community can be exposed briefly (i.e., *acute exposure*) and indefinitely (i.e., *chronic exposure*) without resulting in an undesirable effect. These secondary acute and chronic aquatic ecosystem guidelines, established with fewer data than regulatory guidelines are 19.2 and 1.18 ppm F, respectively (Suter, 1996), which as noted by Hayes et al. (2017), were higher than those found in the vast majority of natural surface waters in the United States. As far as is known, no guideline values for F⁻ in aquatic ecosystems in Africa currently exist.

Suggested areas for further fluoride research in Africa

- In 1997 Malde et al. called for further research on the relationship between the F⁻ level of the growth media (soil, aquatic milieu) and the F⁻ content of agricultural products in East Africa, as well as bioavailability studies on F⁻ intake from food and beverages derived from these growth media.
- By 2018, there had hardly been any studies on the reaction or association of HF and atmospheric particulate matter, a knowledge gap that Cheng (2018) surmised could be because the low concentration of HF commonly found in the atmosphere renders this pathway of little significance. Cheng (2018) further called for more research on the mechanism of deposition HF onto various surfaces, as this knowledge could provide useful clues in developing effective ways for the removal of HF in the atmosphere.
- Antagonistic, synergistic and independent effects of F⁻ have all been reported, but many of the results of these studies remain controversial (See, e.g., Mittal and Flora, 2006). Well-designed experimental studies are needed to clarify these relationships.
- In studying the dynamics of migration and health effects of F on grazed-pasture systems in New Zealand, Cronin et al. (2000) noted that "Further research also needs to ascertain rates of plant uptake of F under differing pH and consequent F speciation conditions, and potential impacts of elevated F levels on microbiological processes, such as nitrogen fixation and nitrification" and "Factors controlling F uptake in acidic soils require further clarification, particularly in areas subject to soil acidification."
- Since it is generally believed that there are many areas worldwide having yet undiscovered F deposits (See, e.g., Hayes et al., 2017), diagnosis of health problems

analogous to known F maladies should not be discounted on the basis of unavailability of data on F^- occurrences in any particular area.

- According to Barbier et al. (2010): "... Fluoride can interact with a wide range of cellular processes, such as gene expression, cell cycle, proliferation, and migration, respiration, metabolism, ion transport, secretion, endocytosis, apoptosis/necrosis, and oxidative stress ... many of the targets and the exact mechanisms/pathways taking part in these events are still unknown."
- Systematic analyses of the F^- content of foods, beverages, and water for human consumption using standard methods are still limited; these are measurements that would enable more realistic assessments to be made of total F^- intake and of F^- intake from different sources. The situation remains extant (as at 2020), even for EU countries, and not least for Africa (See, e.g., Guth et al., 2020).
- Although tea is a well-known F^- hyperaccumulator, the risks of F^- intake via tea consumption are still not well-understood (Pattaravisitsate et al., 2021; Zhang et al., 2023a,b). More research effort is therefore needed in elucidating the health risks posed by F^- intake from drinking tea, to make possible, the formulation of appropriate measures for mitigating these risks and minimizing the total body burden of F^- in humans.
- Because of the lack of data and documentation regarding the likely involvement of F^- nutrition/metabolism in the onset of conditions, such as *prolapsed intervertebral disk (PID)*, spinal stenosis, and Paget's disease of bone (See Davies, 2013), research to determine reliable biomarkers for the burden of F^- in the body of those afflicted with these diseases is highly warranted.
- According to Hays et al. (2020), it has been challenging to conduct experiments aimed at clarifying the mechanism of fluoropolymer-related respiratory illness in humans, for several reasons, such as their proprietary nature, the intermittently reformulated chemical formulations used, and the nonreporting of the presence of fluorocarbons.
- According to Whelton (2009), amalgamation of F^- into sucrose, which is the most commonly recognized cariogenic food, would appear to have a lot of potential for targeted caries prevention. Research in this area has not, however, been advanced in recent years.
- Several well-developed techniques are now available for the removal of F^- from contaminated water supplies (See references under: "Defluoridation techniques," in this chapter). However, parallel methods for removal of F^- from foodstuffs are yet to be developed (See, e.g., Mondal and Chattopadhyay, 2020).
- The abundance and low cost of clays and related minerals, whether modified or in their natural state, enhance their potential as alternatives for adsorption of F^-. However, as noted by Woldeyes et al. in 2004: "... further research is required to improve the low permeability and relatively low adsorption capacity of clays that prevails in their natural state." [*Sic.*].
- The above-listed knowledge gaps are well-encapsulated in the conclusive statements of Barbier et al. (2010): "Finally, in drinking water, fluoride is frequently used with other elements (metals and/or metalloids), which does not necessarily lead to more

pronounced toxicity; in some particular cases, antagonistic effects have been reported. Therefore in the absence of clear proof to counter the known toxic effects of fluoride in combination with metalloids and metals, extensive studies are needed to conclusively determine the effects of such combinations on relevant cell types." [*Sic.*]; and, of Buzalaf (2018): "An "optimal range" of fluoride intake is, however, desirable at the population level to guide programs of community fluoridation, but further research is necessary to provide additional support for future decisions on guidance in this area. This list includes the effect of factors affecting fluoride metabolism, clinical trials on the effectiveness of low-fluoride dentifrices to prevent caries in the primary dentition, and validation of biomarkers of exposure to fluoride." [*Sic.*].

The following additional listings, excerpted [*Sic.*] from the USIM's (1997) research recommendations regarding "dietary intakes of fluoride," are still largely extant today:

- Epidemiological studies (especially analytical studies) of the relationships among fluoride exposures from all major sources and the prevalence of dental caries and enamel fluorosis at specific-life stages should continue for the purposes of detecting trends and determining the contribution of each source to the effects demonstrated.
- Epidemiological and basic laboratory studies should further refine our understanding of the effects of fluoride on the quality and biomechanical properties of bone and on the calcification of soft tissue.
- Studies are needed to define the effects of metabolic and environmental variables on the absorption, excretion, retention, and biological effects of fluoride. Such variables would include the composition of the diet (e.g., calcium content), acid-base balance, and the altitude of residence." [*Sic.*].
- A much improved understanding through research, of the geomedical parameters of F reviewed in this chapter (namely, sources, migration pathways, uptake levels, bioavailability, distribution and metabolism, etc.) would aid in the elucidation of the mechanism of the element's participation in teeth and skeletal development or induction of fluorosis; knowledge considered vital in the improvement of current diagnostic, and therapeutic measures for disorders involving F.

Glossary of terms

- *Acute exposure* is a short (up to 14 days) single contact with a chemical/toxic substance that results in severe biological harm or death.
- *Alveoli* (Singular: *Alveolus*): Any of the many tiny air sacs of the lungs, which allow for rapid gaseous exchange.
- *Anticariogenic*: Effective in suppressing caries production.
- *Chronic exposure* is continuous or repeated contact with a chemical/toxic substance over a long period of time (months or years), leading to adverse health effects.

- The *epiphysis* is the expanded end of a long bone at the upper end of the *femur* (the only bone within the human thigh). It is the growth center that eventually becomes the femoral head.
- *G proteins*, also known as *guanine nucleotide-binding proteins*, are a family of proteins that act as molecular switches inside the cells, and are involved in transmitting signals from a variety of stimuli outside a cell to its interior.
- *HFC* refers to any of several organic compounds composed of hydrogen, fluorine, and carbon. HFCs are produced synthetically and are used primarily as refrigerants. While HFCs have an ozone depletion potential of zero, they are potent greenhouse gases, and thus their manufacture and use became increasingly regulated in the 21st century (Kara, 2019).
- *Osteoporosis* is a condition in which bones become weak and brittle.
- *Prolapsed intervertebral disk*: refers to sudden, severe lower back pain that occurs when the soft center of a spinal disk pushes through a crack in the tougher exterior casing.
- "The *sodium iodide symporter (NIS)* is the plasma membrane glycoprotein that mediates active iodide transport in the thyroid and other tissues, such as the salivary, gastric mucosa, rectal mucosa, bronchial mucosa, placenta, and mammary glands." Waugh (2019).

References

ADA (American Dental Association), 2021. Oral Health Topics: Fluoride — Topical and Systematic Supplements. https://www.ada.org/en/public-programs/advocating-for-the-public/fluoride-and-fluori-dation. (Accessed 28 March 2021).

Addison, M.J., Rivett, M.O., Robinson, H., Fraser, A., Miller, A.M., Phiri, P., et al., 2020. Fluoride occurrence in the lower East African Rift System, southern Malawi. Science of the Total Environment 712, 136260. Available from: https://doi.org/10.1016/j.scitotenv.2019.136260.

Akafu, T., Chimdi, A., Gomoro, K., 2019. Removal of fluoride from drinking water by sorption using diatomite modified with aluminum hydroxide. Journal of Analytical Methods in Chemistry 2019, 4831926. Available from: https://doi.org/10.1155/2019/4831926.

Ali, S., Thakur, S.K., Sarkar, A., Shekhar, S., 2016. Worldwide contamination of water by fluoride. Environmental Chemistry Letters 14, 291−315. Available from: https://doi.org/10.1007/s10311-016-0563-5.

Amereih, S., Barghouthi, Z., Majjiad, L., 2013. Colorimetric determination of fluoride in drinking groundwater using a polymeric zirconium complex of 5-(2-carboxyphenylazo)-8-hydroxyquinoline. Journal of Advances in Chemistry 12, 460−465. Available from: https://doi.org/10.24297/jac.v12i7.6703.

Antes, F.G., Pereira, J.S.F., Spadoa, L.C., Muller, E.I., Flores, Erico, M.M., et al., 2012. Fluoride determination in carbon nanotubes by ion selective electrode. Journal of the Brazilian Chemical Society 23 (6), 1193−1198. Available from: https://doi.org/10.1590/S0103-50532012000600026.

Aulestia, F.J., Groeling, J., Bomfim, G.H.S., Costiniti, V., Manikandan, V., Chaloemtoem, A., et al., 2020. Fluoride exposure alters Ca2 signaling and mitochondrial function in enamel cells. Science Signaling 13 (619), eaay0086. Available from: https://doi.org/10.1126/scisignal.aay0086.

Bahar, D.U., Topku, C., Ozimen, D., Isildak, I., 2019. A novel borate ion selective electrode based on carbon nanotube-silver borate. International Journal of Electrochemical Science 15, 899−914. Available from: https://doi.org/10.20964/2020.01.40.

Bakshi, A.K., 1974. Dental conditions and dental health. In: Vogel, L.C., et al., (Eds.), Health and Diseases in Kenya. East African Literature Bureau, Nairobi, pp. 519−522.

Banerjee, A., Roychoudhury, A., 2019. Fluorine: a biohazardous agent for plants and phytoremediation strategies for its removal from the environment. Biologia Plantarum 63, 104–112. Available from: https://doi.org/10.32615/bp.2019.013.

Barbier, O., Arreola-Mendoza, L., Del Razo, L.M., 2010. Molecular mechanisms of fluoride toxicity, Chemico-Biological Interactions, 188. pp. 319–333. Available from: https://doi.org/10.1016/j.cbi.2010.07.011.

Barrow, N.J., Ellis, A.S., 1986. Testing a mechanistic model–III. The effects of pH on fluoride retention by a soil. Journal of Soil Science 37, 287–293.

Beg, S.A., Atiqullah, M., 1983. Synergism and antagonism of arsenic, chromium and fluoride on nitrification process. Journal of Environmental Science and Health, Part A: Environmental Science and Engineering 18 (5), 633–650. Available from: https://doi.org/10.1080/10934528309375130.

Bhat, N., Jain, S., Asawa, K., Tak, M., Shinde, K., Singh, A., 2015. Assessment of fluoride concentration of soil and vegetables in vicinity of zinc smelter, Debari, Udaipur, Rajasthan. Journal of Clinical Diagnostic Research 9 (10), 63–66.

Botha, C.J., Naude, T.W., Minnaar, P.P., Van Amstel, S.R., Janse Van Rensburg, S.D., 1993. Two outbreaks of fluorosis in cattle and sheep. Journal of the South African Veterinary Association 64, 165–168.

Boyle, D.R., 1976. The Geochemistry of Fluorine and Its Applications in Mineral Exploration (Ph.D. thesis). University of London, Royal School of Mines, Imperial College of Science and Technology, London. file:///D:/Boyle-DR-1976-PhD-Thesis.pdf. (Accessed 20 March 2021).

Bruno, C., Bruni, B., Scondotto, S., Comba, P., 2015. Prevention of disease caused by fluoro-edenite fibrous amphibole: the way forward. Annali dell'Istituto Superiore di Sanita 51 (2), 90–92. Available from: https://doi.org/10.4415/ANN_15_02_02.

Burragato, F., Comba, P., Baiocchi, V., Palladino, D.M., Simei, S., Gianfagna, A., et al., 2005. Geo-volcanological, mineralogical and environmental aspects of quarry materials related to pleural neoplasm in the area of Biancavilla, Mount Etna (Eastern Sicily, Italy). Environmental Geology 47, 855–868.

Buzalaf, C.P., Leite, A.de L., Buzalaf, M.A.R., 2015. Chapter 4: Fluoride metabolism. In: Preedy, V.R. (Ed.), Fluorine: Chemistry, Analysis, Function and Effects. Published by Royal Society of Chemistry, pp. 54–74. Available from: https://doi.org/10.1039/9781782628507-00054.

Buzalaf, M.A.R., 2018. Review of fluoride intake and appropriateness of current guidelines. Advances in Dental Research 29 (2), 157–166. Available from: https://doi.org/10.1177/0022034517750850.

Çakır, A., Şahin, T.N., 2023. Evaluation of the impact of fluoride in drinking water and tea on the enamel of deciduous and permanent teeth. BMC Oral Health 23, 565. Available from: https://doi.org/10.1186/s12903-023-03267-6.

Cape, J.N., Fowler, D., Davison, A., 2003. Ecological effects sulphur dioxide, fluorides, and minor air pollutants: recent trends and research needs. Environment International 29, 201–211. Available from: https://doi.org/10.1016/s0160-4120(02)00180-0.

Cerklewski, F.L., 1997. Fluoride bioavailability. Nutrition Research 17 (5), 907–929.

Chahine, S., Melito, S., Giannini, V., Seddaiu, G., Roggero, P.P., 2023. Fluoride stress affects seed germination and seedling growth by altering the morpho-physiology of an African local bean variety. Plant Growth Regulation . Available from: https://doi.org/10.1007/s10725-023-01064-3.

Chatterjee, N., Sahu, G., Bag, A.G., Plal, B., Hazra, G.C., 2020. Role of fluoride on soil, plant and human health: a review on its sources, toxicity and mitigation strategies. International Journal of Environment and Climate Change 10 (8), 77–90. Available from: https://doi.org/10.9734/ijecc/2020/v10i830220.

Cheng, M.D., 2018. Atmospheric chemistry of hydrogen fluoride. Journal of Atmospheric Chemistry 75, 1–16. Available from: https://doi.org/10.1007/s10874-017-9359-7.

Chlubek, D., Goschorska, M., Sikora, M. (Guest Eds.), 2021. Toxic Effects of Fluoride. Special Issue, MDPI Applied Sciences. ISSN 2076-3417. https://www.mdpi.com/journal/applsci/special_issues/Toxic_Effects_Fluoride. (Accessed 27 March 2021).

Choudhary, S., Rani, S., Devika, O.S., Patra, A., Singh, R.K., Prasad, S.K., 2019. Impact of fluoride on agriculture: a review on its sources, toxicity in plants and mitigation strategies. International Journal of Chemical Studies 7 (2), 1675–1680.

Craig, L., Lutz, A., Berry, K.A., Yang, W., 2015. Recommendations for fluoride limits in drinking water based on estimated daily fluoride intake in the Upper East Region, Ghana. Science of the Total Environment 532, 127–137. Available from: https://doi.org/10.1016/j.scitotenv.2015.05.126.

Cronin, S.J., Manoharan, V., Hedley, M.J., Loganathan, P., 2000. Fluoride: a review of its fate, bioavailability, and risks of fluorosis in grazed-pasture systems in New Zealand. New Zealand Journal of Agricultural Research 43 (3), 295–321. Available from: https://doi.org/10.1080/00288233.2000.9513430.

D'Alessandro, W., 2006. Human fluorosis related to volcanic activity: a review. In: Kungolos, A.G., Brebbia, C. A., Samaras, C.P., Popov, V. (Eds.), Environmental Toxicology, 21. WIT Press, Southampton, pp. 21–30. Available from: https://doi.org/10.2495/ETOX060031.

Dahi, E., 2016. Africa's U-Turn in defluoridation policy: from the Nalgonda technique to bone char. Research Report. Fluoride 49 (4 Part 1), 401–416.

Dahi, E., Mtalo, F., Njau, B., Bregnhj, H., 1996. Reaching the unreached: challenges for the 21st century. Defluoridation using the Nalgonda Technique in Tanzania. In: Proceedings of 22nd WEDC Conference, New Delhi, India, 1996. http://www.lboro.ac.uk/departments/cv/wedc/papers/22/groupf/dahi.pdf. (Accessed 01.04.21).

Dar, F.A., Kurella, S., 2023. Fluoride in drinking water: an in-depth analysis of its prevalence, health effects, advances in detection and treatment. Materials Today Proceedings . Available from: https://doi.org/ 10.1016/j.matpr.2023.05.645.

Darchen, A., Sivasankar, V., Chaabane, T., Prabhakaran, M., 2016. Methods of defluoridation: adsorption and regeneration of adsorbents. In: Sivasankar, V. (Ed.), Surface Modified Carbons as Scavengers for Fluoride from Water. Springer, Cham. Available from: https://doi.org/10.1007/978-3-319-40686-2_4. (Accessed 03 March 2021).

Das, M., Maity, D., Acharya, T.K., Sau, S., Giri, C., Goswami, C., et al., 2021. Lowest aqueous picomolar fluoride ions and in vivo aluminum toxicity detection by an aluminum (III) binding chemosensor. Dalton Transactions (Cambridge, England: 2003) 50 (8), 3027–3036. Available from: https://doi.org/10.1039/ d0dt03901b.

Davies, T.C., 1994. Water quality characteristics associated with fluorite mining in the Kerio Valley area of western Kenya. International Journal of Environmental Health Research 4 (3), 165–175. Available from: https://doi.org/10.1080/09603129409356814.

Davies, T.C., 2008. Environmental health impacts of East African Rift volcanism. Environmental Geochemistry and Health 30 (4), 325–338.

Davies, T.C., 2013. Geochemical variables as plausible aetiological cofactors in the incidence of some common environmental diseases in Africa. Journal of African Earth Sciences 79, 24–49. Available from: https://doi.org/10.1016/j.

Davies, T.C., 2021. "The East African Rift Valley represents an ideal field laboratory for researching fluorine toxicity." - Chapter author's own quotation (unpublished).

Davison, L.H., Weinstein, L.H., 2006. Chapter 8: Some problems relating to fluorides in the environment: effects on plants and animals. In: Tressaud, A. (Ed.), Fluorine and the Environment, Atmospheric Chemistry, Emissions and Lithosphere, Vol. 1. Elsevier, Amsterdam, pp. 251–298.

Demelash, H., Beyene, A., Abebe, Z., Melese, A., 2019. Fluoride concentration in ground water and prevalence of dental fluorosis in Ethiopian Rift Valley: systematic review and meta-analysis. BMC Public Health 19, 1298. Available from: https://doi.org/10.1186/s12889-019-7646-8.

DeMille, K.F., Emsbo-Mattingly, S.D., Krieger, G., Howard, M., Webster, K.B., DaCosta, M., 2023. Novel gas exposure system for the controlled exposure of plants to gaseous hydrogen fluoride. Environmental Monitoring and Assessment 195 (6), 752. Available from: https://doi.org/10.1007/s10661-023-11382-8.

Dimeski, G., Badrick, T., John, A.S., 2009. Ion selective electrodes (ISEs) and interferences — a review. Clinica Chimica Acta 411 (5—6), 309—317. Available from: https://doi.org/10.1016/j.cca.2009.12.005.

Dissanayake, C.B., 1991. The fluoride problem in the ground water of Sri Lanka — environmental management and health. International Journal of Environmental Studies 38 (2—3), 137—155. Available from: https://doi.org/10.1080/00207239108710658.

Du, J., Sheng, C., Wang, Y., Zhang, H., Jiang, K., 2021. Determination of trace fluoride in water samples by silylation and gas chromatography/mass spectrometry analysis. Rapid Communications in Mass Spectrometry 35, e9089. Available from: https://doi.org/10.1002/rcm.9089.

Edmunds, W.M., 2008. Groundwater in Africa: palaeowater, climate and modern recharge. In: Adelana, S., MacDonald, A., Alemayehu, T., Tindimugaya, C. (Eds.), Applied Groundwater Studies in Africa. Taylor and Francis, London, pp. 305—322.

Edmunds, W.M., Smedley, P.L., 2013. Fluoride in natural waters. In: Selinus, O. (Ed.), Essentials of Medical Geology. Springer, Dordrecht. Available from: https://doi.org/10.1007/978-94-007-4375-5_13.

El-Said, G.F., Sinoussy, K.S., Abdel Kawy, S.M.H., Khedawy, M., 2023. Abnormal fluoride distribution, human health risk assessment, predicted no-effect concentration (PNEC) and environmental hazards in an Egyptian lake connected to the Mediterranean Sea. Marine Environmental Research 188, 106029. Available from: https://doi.org/10.1016/j.marenvres.2023.106029.

Ermakov, V.V., 2004. Chapter 9.1: Fluorine. In: Merian, E., Anke, M., Ihnat, M., Stoeppler, M. (Eds.), Elements and Their Compounds in the Environment — Occurrence, Analysis and Biological Relevance, second ed. Wiley-VCH, Weinheim, Germany, pp. 1415—1421. Available from: http://doi.org/10.1002/9783527619634.ch60a.

Fan, X., Parker, D.J., Smith, M., 2003. Adsorption kinetics of fluoride on low cost materials. Water Research 37 (20), 4929—4937. Available from: https://doi.org/10.1016/j.watres.2003.08.014.

Fleischer, M., Robinson, W.O., 1963. Some problems of the geochemistry of fluorine. In: Shaw, D.M. (Ed.), Studies in Analytical Geochemistry, University of Toronto Press in Cooperation with the Royal Society of Canada Special Publication No. 6, Toronto, Ontario, Canada, pp. 58—75.

Fordyce, F.M., 2011. Fluorine: human health risks. In: Nriagu, J.O. (Ed.), Encyclopedia of Environmental Health, vol. 2, pp. 776—785. https://doi.org/10.1016/B978-0-444-52272-6.00697-8.

Forster, P.M., Smith, C.J., Walsh, T., Lamb, W.F., Lamboll, R., Hauser, M., et al., 2023. Indicators of global climate change 2022: annual update of large-scale indicators of the state of the climate system and human influence. Earth Systems Science Data 15 (6), 2295—2327. Available from: https://doi.org/10.5194/essd-15-2295-2023.

Frencken, J.E. (Ed.), 1992. Endemic Fluorosis in Developing Countries: Causes, Effects and Possible Solutions. Publication Number 91.082, NIPG-TNO, Leiden, The Netherlands, 1992.

Fubini, B., Fenoglio, I., 2007. Toxic potential of mineral dusts. Elements 3, 407—414. Available from: https://www.uvm.edu/~gdrusche/Classes/HCOL%20195/Elements%20article%20on%20toxicity%20of%20mineral%20dusts%202007.pdf..

Fuge, R., 2019. Fluorine in the environment, a review of its sources and geochemistry. Applied Geochemistry 100, 393—406. Available from: https://doi.org/10.1016/j.apgeochem.2018.12.016.

Gaciri, S.J., Davies, T.C., 1993. The occurrence and geochemistry of fluoride in some natural waters of Kenya. Journal of Hydrology 143, 395—412.

Ganapathy,R., Keays, R.R., Laul, J.C., Anders, E., 1970. Trace elements in Apollo 11 lunar rocks: implications for meteorite influx and origin of moon. In: Apollo 11 Lunar Science Conference, Houston, Texas. Proceedings Volume 2, Geochimica et Cosmochimica Acta, Supp. 1, pp. 1117—1142.

Gao, S., Luo, T.-C., Zhang, B.-R., Zhang, H.F., Han, Y.-W., et al., 1998. Chemical composition of the continental crust as revealed by studies in East Asia. Geochimica et Cosmochimica Acta 62, 1959—1975. Available from: https://doi.org/10.1016/S0016-7037(98)00121-5.

Gao, X., Zheng, H., Shang, G.-q, Xu, J.-G., 2007. Colorimetric detection of fluoride in an aqueous solution using Zr (IV)-EDTA complex and a novel hemicyanine dye. Talanta 73 (4), 770−775. Available from: https://doi.org/10.1016/j.talanta.2007.04.063.

García, M.G. and Borgnino, L., 2015. Chapter 1: Fluoride in the context of the environment. In: Fluorine: Chemistry, Analysis, Function and Effects. Food and Nutritional Components in Focus, pp. 3−21. doi: 10.1039/9781782628507-00003.

Geleta, W.S., Alemayehu, E., Lennartz, B., 2021. Volcanic rock materials for defluoridation of water in fixed-bed column systems. Molecules (Basel, Switzerland) 26, 977. Available from: https://doi.org/10.3390/molecules26040977.

Geretharan, T., Jeyakumar, P., Bretherton, M., Anderson, C.W.N., 2020. Fluorine and white clover: assessing fluorine's impact on Rhizobium leguminosarum. Journal of Environmental Quality 49, 987−999. Available from: https://doi.org/10.1002/jeq2.20089.

Gheorghe, I.F., Ion, B., 2011. The effects of air pollutants on vegetation and the role of vegetation in reducing atmospheric pollution. IntechOpen. Available from: 10.5772/17660.

Grandjean, P., 2019. Developmental fluoride neurotoxicity: an updated review. Environmental Health 18, 110. Available from: https://doi.org/10.1186/s12940-019-0551-x.

Grobler, S.R., Dreyer, A.G., Blignaut, R.J., 2001. Drinking water in South Africa: implications for fluoride supplementation. SADJ: Journal of the South African Dental Association = Tydskrif van die Suid-Afrikaanse Tandheelkundige Vereniging 56 (11), 557−559.

Gulbrandsen, R.A., 1966. Chemical composition of phosphorites of the phosphoria formation. Geochimica et Cosmochimica Acta 30 (8), 769−778. Available from: https://doi.org/10.1016/0016-7037(66)90131-1.

Guo, M., 2009. Chapter 6: Vitamins and minerals as functional ingredients. In: Guo, M. (Ed.), Functional Foods: Principles and Technology − A Volume in Woodhead Publishing Series in Food Science, Technology and Nutrition. Woodhead Publishing, pp. 197−236. . Available from: https://www.sciencedirect.com/book/9781845695927/functional-foods#book-info. (Accessed 24 March 2021).

Guth, S., Hüser, S., Roth, A., Degen, G., Diel, P., Edlund, K., et al., 2020. Toxicity of fluoride: critical evaluation of evidence for human developmental neurotoxicity in epidemiological studies, animal experiments and in vitro analyses. Archives of Toxicology 94 (5), 1375−1415. Available from: https://doi.org/10.1007/s00204-020-02725-2.

Han, T., Mattinen, U., Bobabka, J., 2019. Improving the sensitivity of solid-contact ion-selected electrodes by using coulometric signal transduction. ACS Sensors 4 (4), 900−906. Available from: https://doi.org/10.1021/acssensors.8b01649.

Hayes, T.S., Miller, M.M., Orris, G.J., Piatak, N.M., 2017. Chapter G: Fluorine. In: Schulz, K.J., DeYoung, J.H., Jr., Seal, R.R., II, Bradley, D.C. (Eds.), Critical Mineral Resources of the United States − Economic and Environmental Geology and Prospects for Future Supply: U.S. Geological Survey Professional Paper 1802, pp. G1−G80. https://doi.org/10.3133/pp1802G.

Hays, H.L., Mathew, D., Chapman, J., 2020. Fluorides and fluorocarbons toxicity [Updated 2020 Jul 20]. StatPearls [Internet]. StatPearls Publishing, Treasure Island (FL). Available from: https://www.ncbi.nlm.nih.gov/books/NBK430799/.

Hess, J., Kadambi, H., Zhou, X., Venort, T., 2012. Precipitation methods and the Nalgonda technique. PPT − Precipitation Methods and the Nalgonda Technique PowerPoint Presentation - ID:2168496 (slideserve.com). (Accessed 26 March 2021).

Horn, M.K., Adams, J.A.S., 1966. Computer-derived geochemical balances and element abundances. Geochimica et Cosmochimica Acta 30 (3), 279−297. Available from: https://doi.org/10.1016/0016-7037(66)90003-2.

Ikenishi, R., Kitagawa, T., 1988. Gas chromatographic method for the determination of fluoride ion in biological samples. II. Stability of fluorine-containing drugs and compounds in human plasma. Chemical and Pharmaceutical Bulletin (Tokyo) 36 (2), 810−814. Available from: https://doi.org/10.1248/cpb.36.810.

Ingle, N., Dubey, H., Kaur, N., Sharma, I., 2014. Defluoridation techniques: which one to choose. Journal of Health Research and Reviews 1 (1), 1−4.

Ivlev, S.I., Karttunen, A.J., Hoelzel, M., Conrad, M., Kraus, F., 2019. The crystal structures of α- and β-F_2 revisited. Chemistry 25, 3310. Available from: https://doi.org/10.1002/chem.201805298.

Jacobson, J.S., Weinstein, L.H., Mccune, D.C., Hitchcock, A.E., 1966. The accumulation of fluorine by plants. Journal of the Air Pollution Control Association 16 (8), 412−417. Available from: https://doi.org/10.1080/00022470.1966.10468494.

Jarosz, Z., Pitura, K., 2021. Fluoride toxicity limit − can the element exert a positive effect on plants? Sustainability 13 (21), 12065. Available from: https://doi.org/10.3390/su132112065.

Jha, S.K., Mishra, V.K., Sharma, D.K., Damodaran, T., 2011. Fluoride in the environment and its metabolism in humans. In: Whitacre, D. (Ed.), Reviews of Environmental Contamination and Toxicology Volume 211. Reviews of Environmental Contamination and Toxicology (Continuation of Residue Reviews), vol. 211. Springer, New York, NY. Available from: https://doi.org/10.1007/978-1-4419-8011-3_4.

Johansen, E.S., 2013. The Effects of Fluoride on Human Health in Eastern Rift Valley, Northern Tanzania (Thesis). (*Prosjektoppgave ved Det medisinske fakultet*). Department of Medicine, University of Oslo, Norway. https://core.ac.uk/download/pdf/30848945.pdf. (Accessed 03 March 2021).

Jubb, T.F., Annand, T.E., Main, D.C., Murphy, G.M., 1993. Phosphorus supplements and fluorosis in cattle − a northern Australian experience. Australian Veterinary Journal 70, 379−383.

Kabata-Pendias, A., Pendias, A., 2001. Trace Elements in Soils and Plants, third ed. CRC Press.

Kabata-Pendias, A., Mukherjee, A.B., 2007. Trace Elements From Soil to Human. Springer Verlag, Berlin, Germany.

Kage, S., Kudo, K., Nishida, N., Ikeda, H., Yoshioka, N., Ikeda, N., 2008. Determination of fluoride in human whole blood and urine by gas chromatography-mass spectrometry. Forensic Toxicology 26 (1), 23−26.

Kara, R., 2019. Hydrofluorocarbon. Encyclopedia Britannica. https://www.britannica.com/science/hydrofluorocarbon. (Accessed 26 March 2021).

Kebede, A., Retta, N., Abuye, C., Whiting, S.J., Kassaw, M., Zeru, T., et al., 2016. Dietary fluoride intake and associated skeletal and dental fluorosis in school age children in rural Ethiopian Rift Valley. International Journal of Environmental Research and Public Health 13 (8), 756. Available from: https://doi.org/10.3390/ijerph13080756.

Kumar, P.S., Suganya, S., Srinivas, S., Priyadharshini, S., Karthika, M., et al., 2019. Treatment of fluoride contaminated water. A review. Environmental Chemistry Letters 17 (4), 1707−1726.

Kut, K., Sarswat, A., Srivastava, A., Pittman, C.U., Mohan, D., 2016. A review of fluoride in African groundwater and local remediation methods. Groundwater for Sustainable Development 2-3 (12), 190−212. Available from: https://doi.org/10.1016/j.gsd.2016.09.001.

Kwon, S.M., Shin, H.S., 2014. Sensitive determination of fluoride in biological samples by gas chromatography-mass spectrometry after derivatization with 2-(bromomethyl)naphthalene. Analitica Chimica Acta 852, 162−167. Available from: https://doi.org/10.1016/j.aca.2014.09.035.

Lacson, C., Francis, Z., Lu, M., Huang, Y., 2021. Fluoride-containing water: a global perspective and a pursuit to sustainable water defluoridation management: an overview. Journal of Cleaner Production 280, 124236. Available from: https://doi.org/10.1016/j.jclepro.2020.124236.

Lennon, M.A., Whelton, H., O'Mullane, D., Ekstrand, J., 2004. Rolling Revision of the WHO Guidelines for Drinking-Water Quality: Fluoride. WHO, Geneva. Available from: https://www.who.int/water_sanitation_health/dwq/nutfluoride.pdf. (Accessed 23 March 2021).

Less, L.N., Mcgregor, A., Jones, L.H.P., Cowling, D.W., Leafe, E.L., 1975. Fluorine uptake by grass from aluminum smelter fume. International Journal of Environmental Studies 7 (3), 153−160. Available from: https://doi.org/10.1080/00207237508709687.

Levinson, A.A., 1980. Introduction to Exploration Geochemistry. Applied Publishing, Wilmette, Illinois.

Liu, X., Wang, B., Zheng, B., 2014. Geochemical process of fluorine in soil. Chinese Journal of Geochemistry 33, 277−279. Available from: https://doi.org/10.1007/s11631-014-0688-9.

Lubojanski, A., Piesiak-Panczyszyn, D., Zakrzewski, W., Dobrzynski, W., Szymonowicz, M., Rybak, Z., et al., 2023. The safety of fluoride compounds and their effect on the human body-a narrative review. Materials (Basel, Switzerland) 16 (3), 1242. Available from: https://doi.org/10.3390/ma16031242.

Macek, M.D., Matte, T.D., Sinks, T., Malvitz, D.M., 2006. Blood lead concentrations in children and method of water fluoridation in the United States, 1988−1994. Environmental Health Perspectives 114 (1), 130−134. Available from: https://doi.org/10.1289/ehp.8319.

MacIntire, W.H., Winterberg, S.H., Thompson, J.G., Hatcher, B.W., 1942. Fluorine content of plants. Industrial and Engineering Chemistry Research 34 (12), 1469−1479. Available from: https://doi.org/10.1021/ie50396a011.

Maghanga, J., Okello, V., Michira, J., Ojwang, L., Mati, B., Segor, F., 2022. Fluoride in water, health implications and plant-based remediation strategies. Physical Sciences Reviews 17, 100751. Available from: https://doi.org/10.1515/psr-2022-0123.

Malago, J., Makoba, E., Muzuka, A.N.N., 2017. Fluoride levels in surface and groundwater in Africa: a review. American Journal of Water Science and Engineering 3 (1), 1−17. Available from: https://doi.org/10.11648/j.ajwse.20170301.11.

Malde, M.K., Maage, A., Macha, E., Julshamn, K., Bjorvatn, K., 1997. Fluoride content in selected food items from five areas in East Africa. Journal of Food Composition and Analysis 10 (3), 233−245. Available from: https://doi.org/10.1006/jfca.1997.0537.

Manley, T.R., Stewart, D.J., White, D.A., Harrison, D.L., 1975. Natural fluorine levels in the Bluff area, New Zealand. New Zealand Journal of Science 18, 433−440.

Marincea, Ş., Dumitraş, D.-G., Sava Ghineţ, C., Dal Bo, F., 2022. Carbonate-bearing, F-overcompensated fluorapatite in magnesian exoskarns from Valea Rea, Budureasa, Romania. Minerals 12 (9), 1083. Available from: https://doi.org/10.3390/min12091083.

Martínez-Mier, E., 2012. Fluoride its metabolism, toxicity, and role in dental health. Journal of Evidence-Based Complementary and Alternative Medicine 17, 28−32.

Mendes, A.L.G., Nascimento, M.S., Picoloto, R.S., Flores, E.M.M., Mello, P.A., 2020. A sample preparation method for fluoride detection by potentiometry with ion-selective electrode in medicinal plants. Journal of Fluorine Chemistry 231, 109459. Available from: https://doi.org/10.1016/j.jfluchem.2020.109459.

Metcalfe-Smith, J.L., Holtze, K.E., Sirota, G.R., Reid, J.J., de Solla, S.R., 2003. Toxicity of aqueous and sediment-associated fluoride to freshwater organisms. Environmental Toxicology and Chemistry 22 (1), 161−166.

Mittal, M., Flora, S.J., 2006. Effects of individual and combined exposure to sodium arsenite and sodium fluoride on tissue oxidative stress, arsenic and fluoride levels in male mice. Chemical and Biological Interactions 162, 128−139.

Mobeen, N., Kumar, P., 2017. Defluoridation techniques − a critical review. Asian Journal of Pharmaceutical and Clinical Research 10 (6), 64−71. Available from: https://doi.org/10.22159/ajpcr.2017.v10i6.13942.

Mohandas, S.A., Janardhanan, S., Rasheed, P.A., Gangadharan, P., 2023. Improved defluoridation and energy production using dimethyl sulfoxide modified carbon cloth as bioanode in microbial desalination cell. Heliyon 9 (6), e16614. Available from: https://doi.org/10.1016/j.heliyon.2023.e16614.

Moirana, R., Mkunda, J., Paradelo, M., Machunda, R., Mtei, K., 2021. The influence of fertilizers on the behavior of fluoride fractions in the alkaline soil. Journal of Fluorine Chemistry 250, 109883. Available from: https://doi.org/10.1016/j.jfluchem.2021.109883.

Mondal, P., Chattopadhyay, A., 2020. Environmental exposure of arsenic and fluoride and their combined toxicity: a recent update. Journal Applied Toxicology 40, 552−566. Available from: https://doi.org/10.1002/jat.3931.

Mousny, M., Omelon, S., Wise, L., Everett, E.T., Dumitriu, M., Holmyard, D.P., et al., 2008. Fluoride effects on bone formation and mineralization are influenced by genetics. Bone 43 (6), 1067–1074. Available from: https://doi.org/10.1016/j.bone.2008.07.248.

Mulder, R., 2018. Systemic fluoride supplementation in South Africa — updated guidelines for practitioners. International Dentistry — African Edition 8, 6. Available from: https://www.moderndentistrymedia.com/dec_jan2019/mulder.pdf.

Nagendra, A.H., Bose, B., Shenoy, P.S., 2021. Recent advances in cellular effects of fluoride: an update on its signalling pathway and targeted therapeutic approaches. Molecular Biology Reports 48, 5661–5673. Available from: https://doi.org/10.1007/s11033-021-06523-6.

Nasiri, P., Malekzadeh, S.A., Elyassi, G.N., Arab-Nozari, M., Nahvi, A., 2021. Efficacy and safety of fluoride in children: a narrative review. Journal of Pediatrics Review 9 (1), 37–46. Available from: http://jpr.mazums.ac.ir/article-1-296-en.html.

Nawlakhe, W.G., Kulkarni, D.N., Pathak, B.N., Bulusu, K.R., 1975. Defluoridation of water by Nalgonda technique. Indian Journal of Environmental Health 17, 26–65.

Ncube, E.J., Schutte, Cf, 2005. The occurrence of fluoride in South African groundwater: a water quality and health problem. Water SA 31 (1), 35–40. Available from: https://doi.org/10.4314/wsa.v31i1.5118.

Nyangi, M.J., Chebude, Y., Kilulya, K.F., 2020. Fluoride removal efficiencies of Al-EC and Fe-EC reactors: process optimization using Box—Behnken design of the surface response methodology. Applied Water Sciences 10, 214. Available from: https://doi.org/10.1007/s13201-020-01297-x.

Onipe, T., Edokpayi, J.N., Odiyo, J.O., 2020. A review on the potential sources and health implications of fluoride in groundwater of Sub-Saharan Africa. Journal of Environmental Science and Health, Part A 55 (9), 1078–1093. Available from: https://doi.org/10.1080/10934529.2020.1770516.

Onipe, T., Edokpayi, J.N., Odiyo, J.O., 2021. Geochemical characterization and assessment of fluoride sources in groundwater of Siloam area, Limpopo Province, South Africa. Scientific Reports 11, 14000. Available from: https://doi.org/10.1038/s41598-021-93385-4.

Osinsky, D., Stellman, J.M., 1998. Chapter 62: Minerals and agricultural chemicals. In: Stellman, J.M. (Ed.), Encyclopaedia of Occupational Health and Safety, fourth ed., vol. 3, Part 9, pp. 62.1–62.41. http://www.ilocis.org/documents/chpt62e.htm. (Accessed 10 April 2021).

Ozsvath, D.L., 2009. Fluoride and environmental health: a review. Reviews in Environmental Science and Biotechnology 8, 59–79. Available from: https://doi.org/10.1007/s11157-008-9136-9.

Pader, M., 2012. Dentifrices. In: Kirk-Othmer Encyclopaedia of Chemical Technology. https://doi.org/10.1002/0471238961.0405142016010405.a001.pub2. (Accessed 10 April 2021).

Pagliano, E., Meija, J., Ding, J., Sturgeon, R.E., D'Ulivo, A., Mester, Z., 2013. Novel ethyl-derivatization approach for the determination of fluoride by headspace gas chromatography/mass spectrometry. Analytical Chemistry 85 (2), 877–881. Available from: https://doi.org/10.1021/ac302303r.

Pant, S., Pant, P., Bhiravamurthy, P.V., 2008. Effects of fluoride on early root and shoot growth of typical crop plants of India. Fluoride 40, 57–60.

Patel, P.P., Patel, P.A., Zulf, M.M., Yagnik, B., Kajale, N., Mandlik, R., et al., 2017. Association of dental and skeletal fluorosis with calcium intake and serum vitamin D concentration in adolescents from a region endemic for fluorosis. Indian Journal of Endocrinology and Metabolism 21 (1), 190–195. Available from: https://doi.org/10.4103/2230-8210.196013.

Pattaravisitsate, N., Phetrak, A., Denpetkul, T., Kittipongvises, S., Kuroda, K., 2021. Effects of brewing conditions on infusible fluoride levels in tea and herbal products and probabilistic health risk assessment. Scientific Reports 11, 14115. Available from: https://doi.org/10.1038/s41598-021-93548-3.

Pauling, L., 1960. The Nature of the Chemical Bond, third ed. Cornell University Press, Ithaca, New York.

Pickering, W.F., 1985. The mobility of soluble fluoride in soils. Environmental Pollution (Series B) 9, 281–308.

Pillai, P., Dharaskar, S., Pandian, S., Panchal, H.D., 2021. Overview of fluoride removal from water using separation techniques. Environmental Technology and Innovation 21, 101246. Available from: https://doi.org/10.1016/j.eti.2020.101246.

Prabhu, S.M., Yusuf, M., Ahn, Y., Park, H.B., Choi, J., Amin, M.A., et al., 2023. Fluoride occurrence in environment, regulations, and remediation methods for soil: a comprehensive review. Chemosphere 324, 138334. Available from: https://doi.org/10.1016/j.chemosphere.2023.138334.

Radić, J., Bralić, M., Kolar, M., Genorio, B., Prkić, A., Mitar, I., 2020. Development of the new fluoride ion-selective electrode modified with Fe_xO_y nanoparticles. Molecules (Basel, Switzerland) 25 (21), 5213. Available from: https://doi.org/10.3390/molecules25215213.

Rao, N.C.R., 2003. Fluoride and environment: a review. In: Martin, J., Bunch, V., Suresh, M., Kumaran, T.V. (Eds.), Proceedings of the Third International Conference on Environment and Health, Chennai, India, 15−17 December 2003. Chennai: Department of Geography, University of Madras and Faculty of Environmental Studies, York University, pp. 386−399.

Reimann, C., Caritat, P.de, 1998. Chemical elements in the environment. Factsheets for the Geochemist and Environmental Scientist. Springer.

Rocha, R., Dinoraz, V., Vicenta, 2015. Chapter 9: Bioavailability of fluoride: factors and mechanisms involved. In: Preedy, V.R. (Ed.), Fluorine: Chemistry, Analysis, Function and Effects, pp. 155−172. doi: 10.1039/9781782628507-00155.

Renuka, P., Pushpandali, K., 2013. Review on defluoridation techniques of water. The International Journal of Engineering and Science 2 (3), 86−94.

Sauerheber, R., 2013. Physiologic conditions affect toxicity of ingested industrial fluoride. Journal of Environmental and Public Health 2013, 439490. Available from: https://doi.org/10.1155/2013/439490.

Saxena, V., Ahmed, S., 2001. Dissolution of fluoride in groundwater: a water-rock interaction study. Environmental Geology 40, 1084−1087.

Saxena, V., Ahmed, S., 2003. Inferring the chemical parameters for the dissolution of fluoride in groundwater. Environmental Geology 43, 731−736.

Schlesinger, W.H., Klein, E.M., Vengosh, A., 2020. Global biogeochemical cycle of fluorine. Global Biogeochemical Cycles 34, . Available from: https://doi.org/10.1029/2020GB006722. e2020GB006722.

Schultheiss, W.A., Godley, G.A., 1995. Chronic fluorosis in cattle due to the ingestion of a commercial lick. Journal of the South African Veterinary Association 66, 83−84.

Schulz, K.J., DeYoung, J.H., Jr., Seal, R.R., II, Bradley, D.C., (Eds.), 2017. Critical Mineral Resources of the United States − Economic and Environmental Geology and Prospects for Future Supply. U.S. Geological Survey Professional Paper 1802. Available from: https://doi.org/10.3133/pp1802.

Shike, M., 2006. Modern Nutrition in Health and Disease. Lippincott Williams and Wilkins, Baltimore, MD.

Singh, N., Verma, K.G., Verma, P., Sidhu, G.K., Sachdeva, S., 2014. A comparative study of fluoride ingestion levels, serum thyroid hormone and TSH level derangements, dental fluorosis status among school children from endemic and non-endemic fluorosis areas. Springer Plus 3, 7. Available from: https://doi.org/10.1186/2193-1801-3-7.

Smedley, P.L., Nkotagu, H., Pelig-Ba, K., MacDonald, A.M., Tyler-Whittle, R., Whitehead, E.J., et al., 2002. Fluoride in Groundwater from High-fluoride Areas of Ghana and Tanzania Commissioned Report, R8033. CR/02/316, Phase I (Final Report): 'Minimising Fluoride in Drinking Water in Problem Aquifers'. https://resources.bgs.ac.uk/sadcreports/tanzania2002smedleyfluoridecr02316n.pdf. (Accessed 31 March 2021).

Srivastava, A.P., Singh, A., Tripathi, P., Mathur, A., 2017. Prevalence of dental fluorosis and the role of calcium supplementation in its prevention. Journal of Medical Sciences 17, 156−161. Available from: https://doi.org/10.3923/jms.2017.156.161, https://scialert.net/abstract/?doi = jms.2017.156.161.

Strunecka, A., Strunecký, O., 2020. Mechanisms of fluoride toxicity: from enzymes to underlying integrative networks. Applied Sciences 10, 7100. Available from: https://doi.org/10.3390/app10207100.

Suneetha, N., Rupa, K., Sabitha, V., Kumar, K., Mohanty, S., Kanagasabapathy, et al., 2008. Defluoridation of water by a one-step modification of the Nalgonda technique. Annals of Tropical Medicine and Public Health 1 (2), 56−58.

Suneetha, M., Syama Sundar, B., Ravindhranath, K., 2015. Studies on defluoridation techniques: a critical review. International Journal of ChemTech Research 8 (8), 295−309.

Suter II, G.W., 1996. Toxicological benchmarks for screening contaminants of potential concern for effects on freshwater biota. Environmental Toxicology and Chemistry 15 (7), 1232−1241. Available from: https://doi.org/10.1002/etc.5620150731.

Suttie, J.W., 1969. Fluoride content of commercial dairy concentrates and alfalfa forage. Journal of Agricultural and Food Chemistry 17, 1350−1352.

Swarnakar, A.K., Choubey, S., Sar, S.K., 2016. Defluoridation of water by various technique. International Journal of Innovative Research in Science, Engineering and Technology: A Review 5 (7), 13174−13178. Available from: https://doi.org/10.15680/IJIRSET.2016.0507192.

Tausta, S.L., Berbasova, T., Peverelli, M., Strobel, S.A., 2021. The fluoride transporter FLUORIDE EXPORTER (FEX) is the major mechanism of tolerance to fluoride toxicity in plants. Plant Physiology 131. Available from: https://doi.org/10.1093/plphys/kiab131.

Teshome, L.Y., Jim, F.C., Feleke, Z.B., David, A.S., 2018. Performance enhancement of Nalgonda technique and pilot testing electrolytic defluoridation system for removing fluoride from drinking water in East Africa. African Journal of Environmental Science and Technology 12, 357−369. Available from: https://doi.org/10.5897/ajest2018.2545.

The Open University, 2020. Nutrition: Vitamins and minerals, Section 2.7- Fluorine (F). https://www.open.edu/openlearn/science-maths-technology/biology/nutrition-vitamins-and-minerals/content-section-2.7. (Accessed 21 March 2020).

Thippeswamy, H.M., Devananda, D., Nanditha Kumar, M., Wormald, M.M., Prashanth, S.N., 2021. The association of fluoride in drinking water with serum calcium, vitamin D and parathyroid hormone in pregnant women and newborn infants. European Journal of Clinical Nutrition 75, 151−159. Available from: https://doi.org/10.1038/s41430-020-00707-2.

Thole, B., 2013. Ground water contamination with fluoride and potential fluoride removal technologies for East and Southern Africa. In: Ahmad, I., Dar, M.A. (Eds.), Perspectives in Water Pollution. IntechOpen. Available from: https://doi.org/10.5772/54985, https://www.intechopen.com/books/perspectives-in-water-pollution/ground-water-contamination-with-fluoride-and-potential-fluoride-removal-technologies-for-east-and-so.

TUM (Technical University of Munich), 2019. Fluorine: Toxic and aggressive, but widely used: Investigations with neutrons settle scientific dispute about the structure of solid fluorine. ScienceDaily. http://www.sciencedaily.com/releases/2019/03/190327112637.htm. (Accessed 31 March 2021).

Turekian, K.K., Wedepohl, K.H., 1961. Distribution of the elements in some major units of the Earth's crust. Geological Society of America Bulletin 72, 175−192. Available from: https://doi.org/10.1130/0016-7606(1961)72[175:DOTEIS]2.0.CO;2.

Ullah, R., Zafar, M.S., Shahani, N., 2017. Potential fluoride toxicity from oral medicaments: a review. Iranian Journal of Basic Medical Sciences 20 (8), 841−848. Available from: https://doi.org/10.22038/IJBMS.2017.9104.

Underwood, E.J., 1981. The Mineral Nutrition of Livestock, second ed. Commonwealth Agricultural Bureaux, Slough, England, p. 180.

US ASTDR (US Agency for Toxic Substances and Disease Registry), 2003. Toxicological Profile for Fluorides, Hydrogen Fluoride, and Fluorine. U.S. Department of Health and Human Services, Public Health Service Agency for Toxic Substances and Disease Registry. https://www.atsdr.cdc.gov/toxprofiles/tp11.pdf. (Accessed 25 March 2021).

US EPA (US Environmental Protection Agency), 1980. Reviews of the Environmental Effects of Pollutants: Section IX, Fluoride. Document EPA-600/1-78-050. Cincinnati, OH.

US EPA (US Environmental Protection Agency), 2020. Greenhouse Gas Emissions: Overview of Greenhouse Gases. https://www.epa.gov/ghgemissions/overview-greenhouse-gases#f-gas. (Accessed 07 April 2020).

US EPA (US Environmental Protection Agency), 2021. Greenhouse Gas Reporting Program (GHGRP): Fluorinated Greenhouse Gas Emissions and Supplies Reported to the GHGRP. https://www.epa.gov/ghgreporting/fluorinated-greenhouse-gas-emissions-and-supplies-reported-ghgrp. (Accessed 07 April 2020).

US FDA (U.S. Food and Drug Administration), 2015. VFEND: Highlights of Prescribing Information. https://www.accessdata.fda.gov/drugsatfda_docs/label/2015/021266s038,021267s047,021630s028lbl. pdf. (Accessed 11 April 2021).

US NASEM, 2006. Chapter 10: Genotoxicity and carcinogenicity. Fluoride in Drinking Water: A Scientific Review of EPA's Standards. The Unites States National Academies of Sciences, Engineering and Medicine, pp. 304−339. The National Academies Press. https://www.nap.edu/read/11571/chapter/12. (Accessed 18 April 2021).

US NIH (National Institutes of Health), 2023. Fluorine. Fact Sheet for Health Professionals. https://ods.od.nih.gov/factsheets/Fluoride-HealthProfessional/. (Accessed 31 October 2023).

US NIH (US National Institutes of Health), 2020. US Supplement Database (DSLD). https://ods.od.nih.gov/Research/Dietary_Supplement_Label_Database.aspx#:∼:text = The%20Dietary%20Supplement%20Label%20Database,available%20in%20the%20U.S.%20marketplace. (Accessed 11 April 2021).

US NRC (US National Research Council), 1989. Subcommittee on the Tenth Edition of the Recommended Dietary Allowances: Recommended Dietary Allowances, tenth ed. National Academies Press (US), Washington DC. https://pubmed.ncbi.nlm.nih.gov/25144070/. (Accessed 22 March 2021).

USIM (US Institute of Medicine), 1997. Chapter 8: Fluoride. Standing Committee on the Scientific Evaluation of Dietary Reference Intakes: Dietary Reference Intakes for Calcium, Phosphorus, Magnesium, Vitamin D, and Fluoride. National Academies Press (US), Washington (DC). https://www.ncbi.nlm.nih.gov/books/NBK109832/. (Accessed 24 March 2021).

Van Alstine, R.E., Schruben, P.G., 1980. Fluorspar Resources of Africa. United States Geological Survey (USGS) Bulletin 1487. https://doi.org/10.3133/b1487. (Accessed 31 March 2021).

Vasak, S., Griffioen, J., Feenstra, L., 2010. Fluoride in African groundwater: occurrence and mitigation. In: Sustainable Groundwater Resources in Africa, first ed. CRC Press, UNESCO, IHP, pp. 153−168. eBook ISBN: 9780203859452.

Walters, G.I., Trotter, S., Sinha, B., Richmond, Z., Burge, P.S., 2017. Biopsy-proven hypersensitivity pneumonitis caused by a fluorocarbon waterproofing spray. Occupational Medicine (Lond) 67 (4), 308−310. Available from: https://doi.org/10.1093/occmed/kqx039.

Wang, M., Wang, H., Lei, G., Yang, B., Hu, T., Ye, Y., et al., 2023. Current progress on fluoride occurrence in the soil environment: sources, transformation, regulations and remediation. Chemosphere 341, 139901. Available from: https://doi.org/10.1016/j.chemosphere.2023.139901.

Waugh, D.T., 2019. Fluoride exposure induces inhibition of sodium/Iodide symporter (NIS) contributing to impaired iodine absorption and iodine deficiency: molecular mechanisms of inhibition and implications for public health. International Journal of Environmental Research and Public Health 16 (6), 1086. Available from: https://doi.org/10.3390/ijerph16061086.

Waugh, D.T., Potter, W., Limeback, H., Godfrey, M., 2016. Risk assessment of fluoride intake from tea in the Republic of Ireland and its implications for public health and water fluoridation. International Journal of Environmental Research and Public Health 13 (3), 259. Available from: https://doi.org/10.3390/ijerph13030259.

Wehr, J.B., Dalzell, S.A., Menzies, N.W., 2023. Predicting and modelling availability of fluoride in soil from sorption properties. Soil Use and Management 39, 521−534. Available from: https://doi.org/10.1111/sum.12854.

Weinstein, L.H., 1977. Fluoride and plant life. Journal of Occupational Medicine 19, 49−78.

Weinstein, L.H., Davison, L.H., 2004. Fluorides in the Environment: Effects on Plants and Animals. CABI, Wallingford, UK.

Whelton, H., 2009. Chapter 24: Functional foods and oral health. In: Wilson, M. (Ed.), Food Constituents and Oral Health: Current Status and Future Prospects. A Volume in Woodhead Publishing Series in Food Science, Technology and Nutrition. Woodhead Publishing, pp. 488–528. https://doi.org/10.1533/9781845696290.3.488. (Accessed 24 March 2020).

WHO (World Health Organization), 2000. Chapter 6.5: Fluorides. Air quality guidelines, second ed. WHO Regional Office for Europe, World Health Organization, Copenhagen, Denmark.

WHO (World Health Organization), 2002. Fluorides. Environmental Health Criteria 227. Geneva. https://www.who.int/ipcs/publications/ehc/en/ehc227.pdf. (Accessed 04 March 2021).

WHO (World Health Organization), 2019. Inadequate or excess fluoride: a major public health concern: preventing disease through healthy environments. https://www.who.int/publications/i/item/WHO-CED-PHE-EPE-19.4.5. (Accessed 27 March 2021).

WHO (World Health Organization), 1993. Guidelines for Drinking-Water Quality, second ed. WHO, Geneva. https://www.who.int/water_sanitation_health/publications/gdwq2v1/en/. (Accessed 11 April 2021).

WHO (World Health Organization) 1984. Fluorine and fluorides. Environmental Health Criteria, No. 36. Geneva. https://iris.who.int/bitstream/handle/10665/37288/9241540966-eng.pdf. (Accessed 03 February 2024).

Woldeyes, B., Demma, L., Tefera, N., 2004. Potentially low-cost drinking water defluoridation methods: a review. Journal of EEA 21, 29–38. Available from: file:///C:/Users/theo%20davies/Downloads/123926-Article%20Text-338744-1-10-20151015%20(3).pdf.

Yamada, T., Kanda, K., Yanagida, Y., Mayanagi, G., Washio, J., Takahashi, N., 2023. All-solid-state fluoride ion-selective electrode using LaF_3 single crystal with poly(3,4-ethylenedioxythiophene) as solid contact layer. Electroanalysis 35, e202200103. Available from: https://doi.org/10.1002/elan.202200103.

Yami, T.L., Chamberlain, J.F., Beshah, F.Z., Sabatini, D., 2018. Performance enhancement of Nalgonda technique and pilot testing electrolytic defluoridation system for removing fluoride from drinking water in East Africa. African Journal of Environmental Science and Technology 12, 357–369.

Zeng, Q., Xu, Y., Yu, X., Yang, J., Hong, F., Zhang, A., 2014. The combined effects of fluorine and arsenic on renal function in a Chinese population. Toxicology Research 3 (5), 359–366. Available from: https://doi.org/10.1039/c4tx00038b.

Zewge, F., 2017. Combined aluminium sulfate/hydroxide process for fluoride removal from drinking water. African Journals Online 30 (3). Available from: https://doi.org/10.4314/bcse.v30i3.7.

Zhang, D., Xu, X., Wu, X., Lin, Y., Li, B., Chen, Y., et al., 2023a. Monitoring fluorine levels in tea leaves from major producing areas in China and the relative health risk. Journal of Food Composition and Analysis 118, 105205. Available from: https://doi.org/10.1016/j.jfca.2023.105205.

Zhang, K., Lu, Z., Guo, X., 2023b. Advances in epidemiological status and pathogenesis of dental fluorosis. Frontiers in Cell and Developmental Biology 11, 1168215. Available from: https://doi.org/10.3389/fcell.2023.1168215.

Zhu, C.-Q., Chen, J.-L., Hong, Z., Wu, Y.-Q., Xu, J.-G., 2005. A colorimetric method for fluoride determination in aqueous samples based on the hydroxyl deprotection reaction of a cyanine dye. Analytica Chimica Acta 539 (1 and 2), 311–316. Available from: https://doi.org/10.1016/j.aca.2005.03.002.

Zohoori, F., Duckworth, R.M., 2016. Fluoride: intake and metabolism, therapeutic and toxicological consequences. In: Collins, J. (Ed.), Molecular, Genetic, and Nutritional Aspects of Major and Trace Minerals. Elsevier, pp. 539–550. Available from: https://doi.org/10.1016/B978-0-12-802168-2.00044-0.

Zohoori, F.V., Buzalaf, B., 2021. Fluoride Metabolism and the Effects of Fluoride on Metabolic Pathways and Diseases. Metabolites, Special Issue. https://www.mdpi.com/journal/metabolites/special_issues/Fluoride_Metabolism. (Accessed 01 November 2023).

Further readings

Bombik, E., Bombik, A., Rymuza, K., 2020. The influence of environmental pollution with fluorine compounds on the level of fluoride in soil, feed and eggs of laying hens in Central Pomerania, Poland. Environmental Monitoring and Assessment 192, 178. Available from: https://doi.org/10.1007/s10661-020-8143-3.

Brewer, R.F., 1966. Fluorine. In: Chapman, H.D. (Ed.), Diagnostic Criteria for Plants and Soils. University of California, Los Angeles, pp. 180−196.

Carpenter, R., 1969. Factors controlling the marine geochemistry of fluorine. Geochimica et Cosmochimica Acta 33 (10), 1153−1167.

De Vos, W., Tarvainen, T., Salminen, R., Reeder, S., De Vivo, B., Demetriades, A., et al., 2006. Geochemical Atlas of Europe. Part 2 − Interpretation of Geochemical Maps, Additional Tables, Figures, Maps, and Related Publications. Geological Survey of Finland, Otamedia Oy, Espoo Europe. ISBN-13: 978-9516909564. Geochemical Atlas of Europe: Part 2: Interpretation of Geochemical Maps, Additional Tables, Figures, Maps, and Related Publications − CROSBI (irb.hr). (Accessed 10 April 2021).

Duffin, S., Duffin, M., Grootveld, M., 2022. Revisiting fluoride in the twenty-first century: safety and efficacy considerations. Frontiers in Oral Health 3, 873157. Available from: https://doi.org/10.3389/froh.2022.873157.

Ekstrand, J., Ehrnebo, M., 1983. The relationship between plasma fluoride, urinary excretion rate and urine fluoride concentration in man. Journal of Occupational Medicine 25, 745−748.

Frant, M.S., 1997. Where did ion selective electrodes come from? The story of their development and commercialization. Chemical Education 2, 159. Available from: https://doi.org/10.1021/ed074p159.

Fuge, R., Andrews, M.J., 1988. Fluorine in the UK environment. Environmental Geochemistry and Health 10, 96−104. Available from: https://doi.org/10.1007/BF01758677.

Kjellevold, M., Maage, A., Macha, E., Julshamn, K., Bjorvatn, K., 1997. Fluoride content in selected food items from five areas in East Africa. Journal of Food Composition and Analysis 10, 233−245. Available from: https://doi.org/10.1006/jfca.1997.0537.

Koga, K., Rose-Koga, E., 2018. Fluorine in the Earth and the solar system: where does it come from and can it be found? Comptes Rendus Chimie. Elsevier Masson.

Koritnig, S., 1972. Chapter 9: Fluorine. In: Wedepohl, K.H. (Ed.), Handbook of Geochemistry. Springer, Berlin.

Liberti, A., Mascini, M., 1969. Anion determination with ion selective electrodes using Gran's plots. Application to fluoride. Analytical Chemistry 41 (4), 676−679. Available from: https://doi.org/10.1021/ac60273a031.

Manji, F., Baelum, V., Fejerskov, O., 1986. Dental fluorosis in an area of Kenya with 2 ppm fluoride in drinking water. Journal of Dental Research 65, 659−662. Available from: https://doi.org/10.1177/00220345860650050501.

Plant, J., Smith, D., Smith, B., Williams, L., 2000. Environmental geochemistry at the global scale. Journal of the Geological Society 157 (4), 837−849. Available from: https://doi.org/10.1144/jgs.157.4.837.

Rizzu, M., Tanda, A., Canu, L., Masawe, K., Mtei, K., Deroma, M.A., et al., 2020. Fluoride uptake and translocation in food crops grown in fluoride-rich soils. Journal of the Science of Food Agriculture 100, 5498−5509. Available from: https://doi.org/10.1002/jsfa.10601.

Sawangjang, B., Takizawa, S., 2023. Re-evaluating fluoride intake from food and drinking water: effect of boiling and fluoride adsorption on food. Journal of Hazardous Materials 443 (Part A), 130162. Available from: https://doi.org/10.1016/j.jhazmat.2022.130162.

Schultheiss, W.A., Van Niekerk, J.C., 1994. Suspected chronic fluorosis in a sheep flock. Journal of the South African Veterinary Association 65, 84−85.

Symonds, R.B., Rose, W.J., Reed, M.H., 1988. Contribution of Cl^- and F^- bearing gases to the atmosphere by volcanoes. Nature 334, 415−418.

Wedepohl, K.H., 1978. Fluorine: abundance in common igneous rocks. In Wedepohl, K.H. (Ed.), Handbook of Geochemistry Elements Scet. 9E, 1−9.

Wedepohl, K.H., 1995. The composition of the continental crust. Geochimica et Cosmochimica Acta 59, 1217−1232.

Wei, W., Kastner, M., Deyhle, A., Spivack, A.J., 2006. Chapter 6: Geochemical cycling of fluorine, chlorine, bromine and boron, and implications for fluid-rock reactions in Mariana Forearc, South Chamorro Seamount, ODP Leg 1951. In: Shinohara, M., Salisbury, M.H., Richter, C. (Eds.), Proceedings of the Ocean Drilling Program, Scientific Results, vol. 195. http://www-odp.tamu.edu/PUBLICATIONS/195_SR/VOLUME/CHAPTERS/106.PDF. (Accessed 26. March 2021).

Weinstein, P., Horwell, C.J., Cook, A., 2013. Volcanic emissions and health. In: Selinus, O., Alloway, B., Centeno, J.A., Finkelman, R.B., Fuge, R., Lindh, U., Smedley, P. (Eds.), Essentials of Medical Geology, Revised, (Eds.) . Springer, pp. 217−238.

Whitford, G.M., 1994. Effects of plasma fluoride and dietary calcium concentrations on GI absorption and secretion of fluoride in the rat. Calcified Tissue International 54, 421−425.

WHO (World Health Organization), 2006. Guidelines for Drinking-Water Quality, Third Ed. WHO, Geneva. https://reliefweb.int/report/world/guidelines-drinking-water-quality-third-edition-incorporating-first-and-second-addenda (Accessed 04.02.2024).

WHO (World Health Organization), 2017. Guidelines for Drinking-Water Quality, Fourth Ed. Incorporating the 1st Addendum. Geneva. https://www.who.int/publications/i/item/9789241549950. (Accessed 11 April 2021).

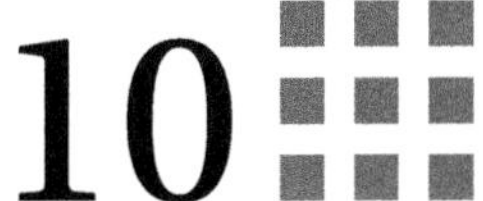

10

Medical geology of iodine

"Although the number of iodine-deficient countries has been reduced by almost 50% over the last decade, it remains a frequently misunderstood health problem"

Speeckaert et al. (2011)

Key chapter features

1. Presents a look at parts of the I geochemical cycle that are most pertinent in studying environmental and metabolic controls on I deficiency disorders (IDD) and their sequelae.
2. Highlights latest advances in I research in Medical Geology ($>70\%$ of the references post-date the year 2000), and how they apply to the African situation.
3. Suggests areas of future I research in Africa to serve as a starting point of investigation for graduate Medical Geology researchers and others from allied disciplines.

Introduction

Iodine is a lustrous, near-black, nonmetallic element [(symbol I; atomic number 53; atomic weight 126.904; electronegativity 2.66 (Pauling, 1960))]. It belongs to the *halogen* group (Group VIIA) of the Periodic Table. The heaviest of the stable halogens, I exists as a solid at standard conditions but melts to form a deep violet liquid at 114°C, and boils to a violet gas at 184°C.

Iodine can be found naturally in air, rocks, soils, waters, plants, animal tissues, and some foodstuffs. Iodine-bearing minerals are rare, the principal sources being seawater, old salt brines, and salt wells, as well as nitrate deposits of Chile. It occurs in large quantities in seaweeds, such as *kelp*.

Iodine has long been known (since the middle of the 19th century) to be an important element environmentally (Fuge and Johnson, 2015), and its role in endemic goiter was the first recognized association between a trace element in the environment and human health (Johnson, 2003). In the thyroid gland, the primary function of I is as a constituent of the thyroid hormones, *thyroxine (T4) (molecular formula, $C_{15}H_{12}I_3NO_4$)* and *triiodothyronine (T3) (molecular formula, $C_{15}H_{11}I_4NO_4$).*

Iodine deficiency disorders (IDD) remain a global health threat to individuals and societies worldwide. Over the last 25 years or so, tremendous progress has been made to control IDDs, largely through the "Universal salt iodization (USI)" program. However, several gaps still persist in our knowledge of I dynamics in the geochemical cycle as well as its role in

Medical Geology of Africa. DOI: https://doi.org/10.1016/B978-0-12-818748-7.00008-3

481

thyroid hormone metabolism (See Section "Suggested areas for further iodine research," in this Chapter). This holds true also for Europe, where there exist fragmentation of available data and diversity of approaches that are not yet fully harmonized (See, e.g., Völzke et al., 2018). In other developed regions of the world, too, the reemergence of IDD is still a cause for concern, particularly with regard to subclinical mental impairment (See: Li and Eastman, 2012; Zimmermann, 2009). In these regions, further research on these fronts are continuing (See, e.g., Krela-Kaźmierczak et al., 2021; Moreno-Reyes, 2021; Ittermann et al., 2020).

The verity underlining these gaps in knowledge is that the USI program, albeit successful in reducing the risks from IDD in many parts of the world, have gone on without a sound understanding of its principal cause a deficiency of I in the environment and related geo-chemobiological interactions (Johnson, 2003), the dynamics of which determine the form and quantity of I that enters the body and the outcome of its involvement in thyroid hormone synthesis.

A substantial amount of information on the geochemobiology of I, up to the time of writing of this Chapter (2021), has appeared in the recent IDD literature of Africa (See Section "Current status of geomedical research on iodine in Africa," in this Chapter). For example, aside from goiter and cretinism, pediatricians working in Africa now know a lot more about the effect of I deficiency in pregnancy and infancy, such as poor neurodevelopment in the child and reduction in cognitive and motor function (See, e.g., Businge et al., 2020; Toloza et al., 2020; Businge et al., 2019).

The purpose of this Chapter, therefore, is to highlight the pertinent geomedical characteristics of I and briefly review the most current information (at the time of writing, 2021) on the status of IDD in Africa, to serve as a starting point for graduate Medical Geology research on IDD in Africa and as a reference point for other researchers on IDD from allied disciplines. Although there has been a virtual elimination of these conditions in many parts of the Continent, as is the case in other regions of the world, knowledge gaps are exposed, and the reasons given why a complete eradication of IDD is still possibly a distant dream, and how a reemergence of IDD can be prevented in areas of the Continent that now experience an almost complete elimination of these disorders.

Sources

Iodine is present in the Earth's crust to the tune of approximately 0.5 ppm, and in the human body about 0.4 ppm (Prothero, 2015). It ranges between 45 and 60 ppb in the oceans; and $10-20$ ng/m^3 in the atmosphere, depending on how close the area is to the sea, and the kind of soil (ATSDR, 2004). From these figures, one can glean that the primary natural sources of I are the oceans, with accumulation of the element occurring in the tissues of fish, shellfish, and seaweed (Moreda-Piñeiro et al., 2011). Muramatsu et al. (2004) have suggested that the major reservoirs of terrestrial I are oceanic sediment and continental sedimentary rocks, of which the weathered equivalents have higher I concentrations than their pristine equivalents, due, presumably, to groundwater interaction (BGS, 2000). The highest concentrations in rocks are

typically found in muds and shales. The most important associations of I are volcanic emanations and sea salt; but significant source concentrations are also held in formation waters and
fluid inclusions (Davies, 2008).

Igneous rocks

The concentration of I in igneous rocks is fairly uniform, with an average content of
0.24 ppm (Fuge and Johnson, 1986). The following values for I content of common
(unweathered and unaltered) igneous rocks are from Johnson (1980), unless otherwise
stipulated.

Intrusives
Granite: <0.02−0.72 ppm (range); 0.25 ppm (average) (Fuge and Ander, 1998).
   Dolerite: 0.04 ppm (average).
   Gabbro: 0.32 ppm (average).
   Other intrusives: 0.22 ppm (average) (Fuge and Ander, 1998).

Extrusives
Rhyolite: 0.16−0.24 ppm (range); 0.20 ppm (average).
   Andesite: 0.12−0.40 ppm (range); 0.25 ppm (average).
   Basalt: 0.12 ppm−0.50 ppm (range); 0.27 ppm (average).
   Other volcanics: 0.24 ppm (average) (Fuge and Ander, 1998).

Sedimentary rocks

The concentration of I in sedimentary rocks tends to be higher than in igneous and meta-
morphic rocks. Organic-rich sediments are particularly enriched in I (Fuge and Johnson,
1986). Of the sedimentary rocks, muds, and shales typically have the highest concentrations.
   The following values are from Fuge and Johnson (1986), unless otherwise stipulated.
   Recent sediments: 5−200 ppm (range).
   Carbonates (Limestones): 2.7 ppm (average).
   Shales: 2.3 ppm (average).
   Organic-rich shales: 16.7 ppm (average) (Fuge and Ander, 1998).
   Sandstones: 0.8 ppm.

Minerals

Owing to its large size [ionic radius of the I^- ion is 220 pm (picometers)], I is incompatible
with most rock-forming (silicate) minerals. Hence, it is of low concentration in most rocks,
and I-bearing minerals, quite uncommon. Examples of such minerals are *iodargyrite* or
iodyrite and lautarite. Iodargyrite or iodyrite (AgI) is a natural form of silver iodide. Lautarite
$[Ca(IO_3)_2]$ consists of calcium iodate and occurs as prismatic crystals. Other known I-bearing
minerals are *perroudite* $[Hg_5Ag_4S_5(I, Br)_2Cl_2]$ and *schwartzembergite* $[Pb_5H_2(IO_2)O_4Cl_3]$. Sulfide

minerals, organic matter, and iron oxides may contain appreciable amounts of I, and it is also relatively enriched in mineral veins (rich in sulfide minerals) as well as in hydrothermal solutions.

Soils

Over the last two decades or so, several studies have been conducted on the geochemistry of I in soils, following recognition of the vital link between I in the food chain and environmental I (e.g., Fuge and Johnson, 2015; Hu et al., 2005; Muramatsu et al., 2004; Bostock et al., 2003; Johnson, 2003; Johanson, 2000). Johnson's (2003) review shows that the I content of soils varies widely, quoting a range of range of <0.1−150 ppm, and mean of all values of 5.1 ppm. More recently, Smyth and Johnson (2011) recorded values of up to 600 ppm in soils from Northern Ireland.

Back in 1986, Fuge and Johnson observed that most I in soils is derived from the atmosphere where, in turn, it has been derived from the oceans. No surprise, then, that soils in coastal areas are generally found to be more I-rich than those from inland areas. It should be noted, however, that unraveling the correlation between the I content of soil and its distance from the sea is no simple matter (See Johanson, 2000). This is because of the complex nature of the geochemobiological interactions engendered in the transfer of I from the ocean to inland areas.

Fuge and Johnson also stated in 1986, that soils are generally found to be much richer in I than the parent rocks, with the actual I content dependent on soil type, including soil texture (one with the ability to fix the I, i.e., *high I fixation potential*) and locality as well as a host of other physical- (e.g., climate, topography, parent material, and water-logging) and biogeochemical parameters (e.g., microbial activity and Eh and pH). In a later publication, Fuge and Johnson (2015) noted that transfer factors of I from soil to plants are generally small, resulting in only limited uptake of I through the plant system. Organic matter plays an important role in I retention. As soil I is hardly water-soluble, much I is thought to be bound with (sorbed onto) organic matter, clays, and Al and Fe oxides (See: Watts et al., 2015; Yamaguchi et al., 2010). This phenomenon can thus potentially affect the mobility of I in the soil solution and hence its availability for plant uptake (Hu et al., 2005; Dai et al., 2009). It is the mobile or bioavailable I present in the environment, and that gets into the metabolic centers that should be of most interest.

Plants

The oceans being the dominant reservoir of mobile I, it is perhaps unsurprising that marine plants are frequently enriched in I while terrestrial plants, on the other hand, are found to have generally low contents of the element (Fuge and Johnson, 1986).

Kohlmeier (2003) gave a value of around 1 μg I/g dry weight for plants from I-sufficient soil, and only a fraction of that amount for plants from areas with I-depleted soils (old mountain ranges and coastal areas).

Researching on whether I is important in the growth of plants is a process fraught with some difficulty, as it (I) is always present in variable amounts in the soil, water, and atmosphere (See Kiferle et al., 2021). Indeed, as recently as 2021, little knowledge had accrued regarding the role of I in plant physiology (See Kiferle et al., 2021), despite the existence at that time (2021) of evidence of its (I) involvement in plant nutrition. Iodine is a structural component of several different proteins in both the roots and shoots of *phylogenetically* distant species (See Section: "Glossary of Terms," in this Chapter).

As early as 1958, Lehr et al., and later, during 1970 and 1971, Umaly and Poel, respectively, (Umaly and Poel, 1970; Umaly and Poel, 1971) determined that I may be either beneficial or harmful to plant growth, with the outcome depending on the concentration of I and the form of I (as iodide or iodate) that gets into plant tissues. To these variables, Borst Pauwels (1961) earlier included "the stage of development of the plant" as a determining factor of plant benefit or harm. Indeed, in sand-culture experiments in pots, Borst Pauwels (1961) had observed that various crops reacted to minute applications of iodide or iodate; and that, in general, iodate had a more favorable effect on growth than iodide, particularly in the initial stages of development.

According to Medrano-Macías et al. (2016a), for land plants, I is not considered an essential nutrient; however, in some aquatic plants, I plays an important role in antioxidant metabolism (See also Leija-Martínez et al., 2016b).

As soil I is known to be mainly strongly bound, with only small transfer factors from soil to plants (Fuge and Johnson, 2015), only limited amounts of I would be taken up through the root system of plants. As Kiferle et al. (2021) noted, "… plants can also assimilate I from the air or absorb it through the leaves if dissolved in salt solutions or in rain." However, in areas where *volatilization* is not an important process, little I would be taken up by the plants (See Bostock et al., 2003).

As Fuge and Johnson (1986) and later, Ashworth (2009) noted, all of these processes occur naturally, underscoring the observation that plant cannot grow in the complete absence of I.

Food crops and other major dietary sources

Because the body does not produce I, it needs to be supplied in the diet. The wide variation in I content of the Earth's soils closely correlates with the I content of crops, such that a high variability of unsupplemented I content is seen both between food categories as well as within each food category (Kohlmeier, 2003). Unfortunately, it is now well-established that I-deficient soils are common in many regions of the world, increasing the ever-present risk of IDDs among people who consume foods primarily from those areas (US NIH, 2021).

The richest dietary sources of I as given in the recent (post-2000) literature are marine products, such as seaweed (e.g., *kelp, nori, kombu,* and *wakame*), shellfish, and molluscs (US NIH (United States National Institutes of Health), 2021; Zimmermann, 2009). Other food sources with high I content include dairy products, such as eggs, milk, yogurt, and cheese; vegetables and other food crops (Bouga et al., 2018). Other authors, e.g., IOM (2001) and

Ershow et al. (2018) report, however, that most fruits and vegetables are poor sources of I, although agreeing that the amounts of I these foods contain are affected by the I content of the soil, fertilizer use, and irrigation practices. Groundwater, in areas where this is a major source of drinking water, may also supply appreciable quantities of I.

Natural waters

Water solubility of I depends on pH and temperature, and is reported to be 0.03 ppm at 20°C, 0.78 ppm at 50°C, and 4.45 ppm at 100°C (West, 1984).

Iodine in seawater

Wong (1991) gave a figure of 60 ppb for the mean I content of seawater. Fuge and Johnson earlier (1986) noted that lower and very variable concentration levels of I are usually found in nonsaline surface waters. Subsurface brines and mineral waters are also observed to be highly enriched in I.

The importance of I in the ocean is underscored by the observation that small emissions of I species to the atmosphere have a significant impact on ozone and air quality (See Section "Environmental circulation and impact of climate change," in this Chapter).

In a recent review of the distribution of I in marine surface waters, Chance et al. (2014) state that I is generally highest in low latitudes with pronounced increases between 50 degrees and 20 degrees (See also, the more recent paper by Chance et al., 2019). In researching the global cycling of I, reduction of the stable iodate ion to the metastable iodide ion in the oceans is an important process that should be considered (Pound et al., 2023).

Surface waters

Earlier literature compilations of data for I in surface waters suggest that they generally contain <20 ppb of I, with many values falling within the range of 0.5−5 ppb (Fuge and Johnson, 1986). More recent data, however, reveal a much broader range of values, which is consistent with the observation that the I content of surface waters is influenced by a myriad of factors. One such factor, *viz.*, proximity to the marine environment has, for instance, been shown to influence the I content of stream waters, with those occurring in near coastal localities being somewhat richer in I than those from further inland (See Negri et al., 2013).

The I content of surface waters can also reflect human activities, such as agricultural practices (Fuge, 1989), biomass burning (Negri et al., 2013), and irrigation (Moran et al., 2002). In addition, thermal waters, many of which are I-rich, can have an impact on surface waters as at El Tatio, Chile, where geyser waters containing up to 783 ppm I influence surface runoff (Cortecci et al., 2005).

Groundwaters

The main sources of I in groundwater are aquifers, soils, and the atmosphere. According to BGS (2000) the typical concentration range of I in potable groundwater is from <1 to 70 ppb, with extremes of up to 400 ppb. Saline waters, such as are found in coastal and arid or semiarid areas generally show higher values in this range.

According to Xu et al. (2013) and Duan et al. (2016), I is released in groundwater through the processes of reductive dissolution and microbial decomposition of organic matter, followed by concentration through the process of evaporation in areas where sediments are found to have a low hydraulic conductivity. The actual I content of groundwaters is therefore dependent on geographical location, local geology, type of soil, as well as a number of other biophysicochemical factors.

Geographical location

If we consider the effect of variations in topography alone on I content of groundwater, mountainous areas are the more likely to have a higher prevalence of IDD than coastal and other low-lying areas (e.g., Fuge and Johnson, 2015). It should be pointed out, though, that I deficiency endemias have, in the past, been recorded in nonmountainous regions of Africa, e.g., the belt extending from the Cameroon grasslands across northern Zaire and the Central African Republic to the borders of Uganda and Rwanda (Eastman and Zimmermann, 2018).

Geology

In a review of earlier studies (pre-2000), BGS (2000) considered that if geology alone is taken into account, groundwater in basement and limestone aquifers are the most likely to have low I concentrations. As a probable result of interaction with groundwater, weathered rocks are often seen to have higher I concentrations than their fresh rock counterparts (See Fuge and Johnson, 2015). Smedley et al. (1995) found I concentrations in groundwaters from certain goiter prevalent areas overlying basement rocks of the Upper East Region of Ghana, in the range $<1-10$ ppb (median; 3.1 ppb, most of the areas having I contents of <6 ppb).

Biophysicochemical factors

Soil pH is an example of a biophysicochemical variable that can influence I release to groundwater; binding being much stronger at low pH (less than around 5; Whitehead, 1973), such that I release is less in acid conditions. As noted later by Xu et al. (2013), evaporation of shallow groundwater and the oxidation by microorganisms of elevated organic matter content in deep groundwater also contribute to the enrichment of I.

Taken together, these factors combined with nutritional factors greatly complicate the expected picture of regional distributions of IDDs in relation to the groundwater regime. Having said this, although accurate diagnosis of IDD risk is not guaranteed, analyzing I content in groundwater from a particular area will no doubt help delineate potential areas with IDD prevalence, as its (I) content in groundwater will broadly reflect its contents in local soils, rocks and food crops and therefore, the overall dietary intake.

Iodine in rainwater

The I content of rainwater has been shown to vary during a rainfall event, rising to peak values during the early stages and then declining through the event, probably as a result of progressive washout (See, e.g., Truesdale and Jones, 1996). According to WHO (2019), the I content of rainwater is mostly within the range 0.5–5.0 ppb.

The principal natural source of I being seawater [average value 58 ppm (Fuge and Johnson, 1986)], "… Iodine in rainfall in coastal areas is therefore generally higher than over continental areas, and hence recharge of water to soils and aquifers will be higher in coastal areas." (BGS, 2000). This is consistent with the observation that goiter and other IDDs are often concentrated in inland areas far removed from the sea. However, as noted by Dissanayake and Chandradith in 1993, IDDs have also been reported in some coastal areas with relatively high water-, soil, and food I concentrations (See: Section "Determinants of worldwide goiter distribution," in this Chapter, for possible explanations).

Drinking water

The content of I in drinking water is known to be highly variable (WHO, 2018; Andersen et al., 2008). Figures recorded for I-deficient areas are generally below 2 ppb (See, e.g., Kohlmeier, 2003). However, it is stressed that prevalence of IDD relates not only to drinking water intake but to additional factors, such as quality of nutrition (e.g., presence of *goitrogens* in the diet) and other factors militating against bioavailability of the element to the centers where it is needed for pertinent metabolic processes.

The WHO "Guidelines for Drinking-water Quality" published in 2017 did not formally establish a guideline value for I. In a previous review by WHO (WHO, 1993) it was suggested, on the basis of available data on the effects of I, that derivation of a guideline value was inappropriate.

Human activities are known to cause the content of I in drinking waters and soils to increase (BGS, 2000). Such activities include its use in herbicides, fungicides, dairy sterilants, detergents, table salt, pharmaceuticals, and bread-making. Whitehead (1979) had earlier suggested that surface waters draining industrialized and urban areas contain much more I than those draining rural areas. The content of I is also enhanced in the atmosphere as a result of fossil-fuel combustion, vehicular emissions, and sewage disposal.

Iodine in air

Iodine is *atmophile* in nature, a characteristic considered to be of immense significance in the geochemical cycle. It volatilizes from the marine environment to the atmosphere, from where it is absorbed onto plant leaves.

Association (additive/synergistic/antagonistic)

In 2012 Schoendorfer and Davies wrote, "Many nutrients function in harmony to complement digestive function and assimilation. Some may hinder these processes and compete for uptake, while others may also be required in tandem to assist in metabolism, which may ultimately affect a number of biochemical cycles."

Associations of iodine

Here, examples are given of chemical elements/substances that have been reported as having possible synergistic or antagonistic associations with I in producing IDDs and related conditions:

Selenium

Early reports of Se deficiency acting in concert with I deficiency in producing IDDs and related conditions are given in the works of: Thilly et al. (1992) and Arthur and Beckett (1999), Wang et al. (2023), among others. Thomson et al. (2011) reported a minimal impact of excess iodate intake on thyroid hormones and Se status in older New Zealanders; and Köhrle (2015) presented some evidence to show that the interaction of Se with I and Fe together contribute to adequate thyroid hormone synthesis.

Derumeaux et al. (2003) observed that severe Se deficiency in the endemic goiter belt of northern Democratic Republic of Congo (DRC) occurs *in tandem* with I deficiency, giving rise to *myxoedematous cretinism* (See Section: "Glossary of Terms," this Chapter). These researchers also found an inverse association between Se status and thyroid volume, thyroid tissue damage, and goiter in French women.

Other trace elements

In 1984 Aihara et al. suggested that the metabolism of Zn, Cu, Mn, and Se is abnormal in thyroid diseases. Later (in 1997), Taye and Argaw found statistically significant associations between goiter and Fe deficiency anemia, duration of consumption of kale and millet, and maternal literacy status (All $\rho < 0.05$). These authors also found significant association between goiter grade and duration of stay in their area. Arthur and Beckett (1999) similarly reported that suboptimal or supraoptimal dietary intakes of Fe, Zn, and Cu can adversely affect thyroid hormone metabolism.

The work of Zimmermann et al. (2000) represents another example of an antagonistic reaction of I with Fe, whereby the therapeutic response to oral I was impaired in goitrous children with Fe deficiency anemia, suggesting that the presence of Fe deficiency anemia in children limits the effectiveness of I intervention programs.

The impact of Fe and Se deficiencies on I and thyroid metabolism and the biochemistry and relevance to public health were considered by Zimmermann and Köhrle (2002). Zimmermann (2006) noted that several minerals and trace elements are essential for normal thyroid hormone metabolism, naming I, Fe, Se, and Zn as examples; and reiterated that coexisting deficiencies of these elements can impair thyroid function. In that same year, Zimmermann (2006) showed that Fe supplementation improves the Fe status in areas of overlapping deficiency (anemia and endemic goiter) as well as increases the efficacy of I prophylaxis. Eftekhari et al. (2006), while investigating whether Fe supplementation can improve the thyroid hormone function in Fe-deficient adolescent girls, found evidence indicating that improvement of Fe status was accompanied by an improvement in some indices of thyroid hormones.

Ertek et al. (2010) found significant correlation of serum Zn levels with thyroid volume in nodular goiter patients, with thyroid *autoantibodies* in autoimmune thyroid disease (AITD) and with free T3 in patients with normal thyroid.

Betsy et al. (2013) commented that hypothyroidism is a common and well-recognized cause of diffuse hair loss; and that Zn and other trace elements, such as Cu and Se are required for the synthesis of thyroid hormones, with their deficiencies resulting in hypothyroidism. Çelik et al. (2014) investigating I deficiency in urine and Se, Zn, Cu, or Mo deficiency, which may accompany this in children aged between 6 and 12 years in two schools in the province of Hatay (endemic goiter region), concluded that Se and Zn deficiency may accompany I deficiency; and that Se and Zn deficiency should be considered in individuals who are found to have I deficiency especially in endemic goiter regions.

Llop et al. (2015) studying the relationship between Hg and thyroid hormones among pregnant women found that prenatal Hg exposure was inversely associated with total triiodothyronine levels among women who took I supplements during pregnancy. Further conclusion by these authors was that the results could be of public health concern, although they noted that "… *further research is needed.*" In that same year, Sinha et al. (2015) showed that significant positive correlation exists between *TSH* (also known as *thyrotropin-releasing hormone*) and Zn levels.

Fluoride-iodine antagonism

The fluoride-iodine antagonism has a long history. As far back as 1854, Maumené suggested that high F- in water might be the cause of goiter. Several investigations relating to this issue have gone on ever since; and have culminated in a lengthy list of published articles (See, e.g., Wang et al., 2021; Yadykina et al., 2021) in support of this *antagonism*. However, in spite of this apparently overwhelming early support (during the 1850s) of the "antagonism thesis," later reviews of data published up to 1984 failed to support the view that F^- in doses recommended for caries prevention, adversely affects the thyroid (See, e.g., Bürgi et al., 1984).

Recent works (e.g., Waugh, 2019 and Lin et al., 2018) provide evidence strongly suggesting that F^- ingestion contributes to pathological states by impeding the absorption of I and lessening the ability of I to concentrate. This finding has policy implications in that the fluoridation of drinking water may inadvertently be contributing to the pathogenesis of neurodevelopmental disorders, impaired immune responses, inflammatory diseases, and cancer. At the same time, the latest researches (e.g., Mascarenhas, 2021; Turska-Szybka et al., 2021; Clark and Slayton, 2020; Šket et al., 2017), continue to affirm a proven efficiency of F^- in caries reduction and its ubiquitous recommendation as the standard for caries prevention.

Iodine is readily inactivated by starch to convert it to iodide (Olson, 2004). The content of starch in cassava root, which is widely consumed in Africa, could engender a goitrogenic action, which could provide one of the reasons why a complete eradication of IDD in some parts of Africa would be hard to realize (See Bourdoux et al., 1978).

Goitrogens

In 2009 Pearce and Braverman reported that more than 100 naturally occurring and synthetic substances affect thyroid function or thyroid hormone metabolism; but by that time (2009), *most of these substances had not yet been fully characterized in human studies.* Some of these substances are dietary constituents, known collectively as *"goitrogens."* They appear to exert their deleterious effects mainly by competitively inhibiting the active transport of iodide both in the thyroid and extrathyroidal I-concentrating tissues, thus producing a deficiency. Several mechanisms for this have been proposed (See, e.g., Eastman and Zimmermann, 2018; Gaitan, 2004).

Examples of goitrogens are vegetables, such as crops belonging to the family *Brassicaceae* [e.g., kale or *sukumawiki* (in East Africa), cabbage, radish, and broccoli]. Other goitrogens that may directly or indirectly account for some of the goiter cases in Africa include the univalent complex ions, thiocyanate (SCN^-), and tetrafluoroborate (BF_4^-), whose ionic sizes are similar to that of iodide; these may competitively inhibit the active transport of iodide both in the thyroid and extrathyroidal I-concentrating tissues. Thiocyanate overload originating from consumption of poorly detoxified cassava results in such deficiency (Tamer et al., 2019; Gaitan, 2004; Vanderpas, 2003: Thilly et al., 1992). Thus as already indicated, despite the laudable effort by various agencies (e.g., WHO, UNICEF, ICCIDD) in combating IDD in Africa, through the USI program, a foolproof solution is thought to be unlikely, because of the presence of these goitrogens in some African diets (See also, Adebayo et al., 2018; Wenhold and Faber, 2008) as well as other geochemobiological factors, such as crop cultivation on a predominantly limestone lithology or on organic-rich soils, impeding I uptake, and transport to metabolic centers.

Chemical form or speciation of iodine in soil and aqueous phases

It is important to determine the factors that affect I speciation to understand the element's global biogeochemical cycle (Gonzales et al., 2017). It also makes possible, predictions and modeling [such as use of the biosphere model (*BIOTRAC*)] of the processes of distribution, exposure, and consequences. In soils, in particular, knowledge of the chemical form of I is crucial, since its mobility, bioavailability and revolatilization all depend on its chemical form. Knowledge of the chemical form also enables us make credible dose estimations (mole fraction) of I that enters the body and the mechanism that transports it to the centers (of the body) where it is needed for vital reactions involving the thyroid hormones.

In addition to all this, the form and speciation of I determines its dynamics of transport in nuclear reactor experiments, both inside containment, as well as on release to the atmosphere outside (Takeda et al., 2015).

There are several oxidation states of I, of which $+7$, $+5$, $+1$, and -1 are the known stable ones. The most common chemical species include iodate (IO_3^-), iodide (I^-), and organo-iodine (Bagwell et al., 2019). The iodides of sodium and potassium (NaI and KI, respectively) and the iodates ($NaIO_3$ and KIO_3, respectively) are the most common compounds of I. It had earlier been shown that soil I is mainly in an organic form with variable amounts of I^- and IO_3^- (Hu et al., 2005, 2012; Moreda-Piñeiro et al., 2011; Patrick, 2008; Shimamoto et al., 2011).

As has already been established in this Chapter, the oceans are the largest reservoir in the global I cycle. In this setting, I occurs mainly as the anions, I^- and IO_3^-, with variable amounts of organically bound forms (Wong, 1991). The suggestion that I occurs concurrently with humic substances of marine origin is emphasized in the work of Andersen et al. (2002). More recent work by Li et al. (2017) account for the occurrence of high I groundwater with IO_3^- and I^- as the main species in shallow (oxidizing conditions) and deep (reducing conditions) groundwater.

Iodine is reactive, taking part in both oxidation and reduction reactions, as well as in iodinating organic matter. In soil systems the speciation and behavior of I is primarily governed by the prevailing redox potential and microbial activity (Koch-Steindl and Pröhl, 2001; Muramatsu et al., 2004; Hu et al., 2005; Yamaguchi et al., 2006; Xu et al., 2011). Söderlund et al. (2017), in a study of humus and mineral soil samples on Olkiluoto (in Finland) mineral soils, found that reduction of IO_3^- to I^- occurred to a significant extent especially in anaerobic soil conditions and in low pH. Foods contain I mainly as I^-, IO^{3-}, or thyroxine (Kohlmeier, 2003).

Radioisotopes of iodine

It is generally recognized that radioiodine is potentially one of the most hazardous fission products that could be released following *loss of coolant accident (LOCA)*. The reasons for the extremely hazardous nature of radioiodine is captured in this statement by McFarlane et al. in 2002, "It has a large inventory in irradiated nuclear fuel; its isotope of greatest radiological significance, ^{131}I, has a half-life of 8.04 days; it has a complex chemistry that may lead to the formation of volatile species; and it can cause hazardous biological effects." [*Sic*]. These reasons underscore the necessity for being able to accurately predict the dynamics of I release from containment to the external atmosphere, and hence public dose in case of accidents.

Thirty-seven isotopes of I are recognized, from ^{108}I to ^{144}I. All undergo radioactive decay except ^{127}I, which is stable. The most important ones being ^{123}I (synthetic)—half-life, 13 hours; ^{124}I (synthetic)—half-life, 4.176 day; ^{125}I (synthetic)—59.40 days; ^{127}I (100% nature); ^{129}I (trace)—half-life, 1.57×10^7 years; ^{131}I (synthetic)—half-life, 8.0(2) days. Of these, only one stable isotope, ^{127}I is found in nature.

^{129}I is considered a highly important radionuclide with regard to its possible future dose for man (Hjerpe et al., 2010) based on its long physical half-life (1.57×10^7 years) and its presence as an anionic species.

It has recently been noted that I deficient environments will have a significant impact on the migration of radioactive isotopes through the local food chain following nuclear accidents (See, e.g., Korobova et al., 2010). For instance, ^{131}I, a short-lived [half-life, 8.0(7) days] artificial isotope of I, and existing in the atmosphere in several forms (Yoshida, 1999) is a fission product of U and Pu, which can be released to the environment during a severe nuclear power plant accident, or from a radionuclide production facility.

Because of their bioavailability, especially given their propensity to concentrate in the thyroid gland, isotopes of I are important for health and in radiological studies. Iodine-131, for instance, has been used in treating the thyroid maladies, despite its hazardous biological potential (See McFarlane et al. in 2002).

Environmental circulation and impact of climate change

The concentration of most elements in soils and natural waters is determined largely by the amounts released from weathering processes of crustal rocks and surficial sediments, which are then taken up by plants (Ahmad et al., 2021; Jensen et al., 2019; Fuge and Johnson, 2015). In the case of I, however, although the weathering of rocks releases I in the surface environment, only very small quantities of the element are derived from weathering of most rocks. Under these circumstances, the dynamics of I distribution in the environment can be considered unique, its contents in crustal rocks being generally low [range, 0.004−0.028 ppm (Yoshida, 1999). Even organic-rich shales average only about 0.02 ppm I]. Soils and waters therefore derive only very small quantities of I from weathering of most rocks. It follows then, that, only a small fraction of the soil I available to plants is derived directly from rock weathering (Fuge and Johnson, 2015; Jensen et al., 2019; Ahmad et al., 2021). Instead, as we shall see later, the most important feature of the biogeochemical cycling of I is its volatilization from the ocean (hydrosphere) into the atmosphere (Fig. 10−1), which results in an uneven distribution of I in the terrestrial environment. In contrast, seawater is relatively rich in I, containing on average about 60 ppb (Wong, 1991) and, as such, represents a vital store of I on the Earth and the major source in its environmental cycle.

Taking a look at the geochemical cycle of I, one can see that its transfer from the oceans to the atmosphere through volatilization is probably the most important process in its geochemistry (Fuge and Johnson, 2015; Fig. 10−2). This process is one that involves organo-I compounds and elemental I from biological and nonbiological sources in the oceans (Carpenter, 2003; Fuge and Johnson, 2015; Stemmler et al., 2013; Yokouchi et al., 2012). According to Ahad and Ganie (2010), this process is thought to be aided by ultraviolet radiation. However, *the extent to which different mechanisms and I species contribute to this transfer has not been firmly established* (See: Carpenter et al., 2013; Prados-Roman et al., 2015; Sellegri et al., 2016; Wadley et al., 2020).

Migration of I inland is then followed by deposition in wet or dry precipitation, which explains the observation that soils in coastal areas are generally more I-rich than those from areas removed from the sea. There is also increasing evidence of terrestrial volatilization from soils, micro-organisms, and plants (Weng et al., 2013; Redeker et al., 2004; Amachi, 2008).

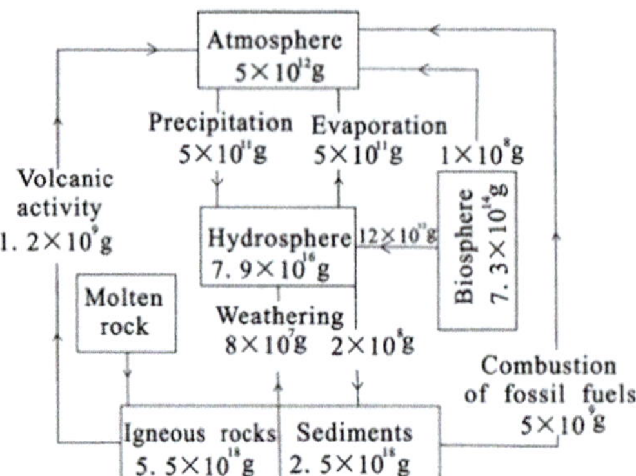

FIGURE 10–1 Annual cycle of I in various spheres of the Earth. *From Johnson, C.C., 2003. The Geochemistry of Iodine and its Application to Environmental Strategies for Reducing the Risks of Iodine deficiency Disorders. British Geological Survey Commissioned Report Number CR/03/057N, DfID KAR Project R7411. file:///C:/Users/theo% 20davies/Downloads/The_Geochemistry_of_Iodine_and_its_Application_to_%20(1).pdf. (Accessed 21 May 2021).*

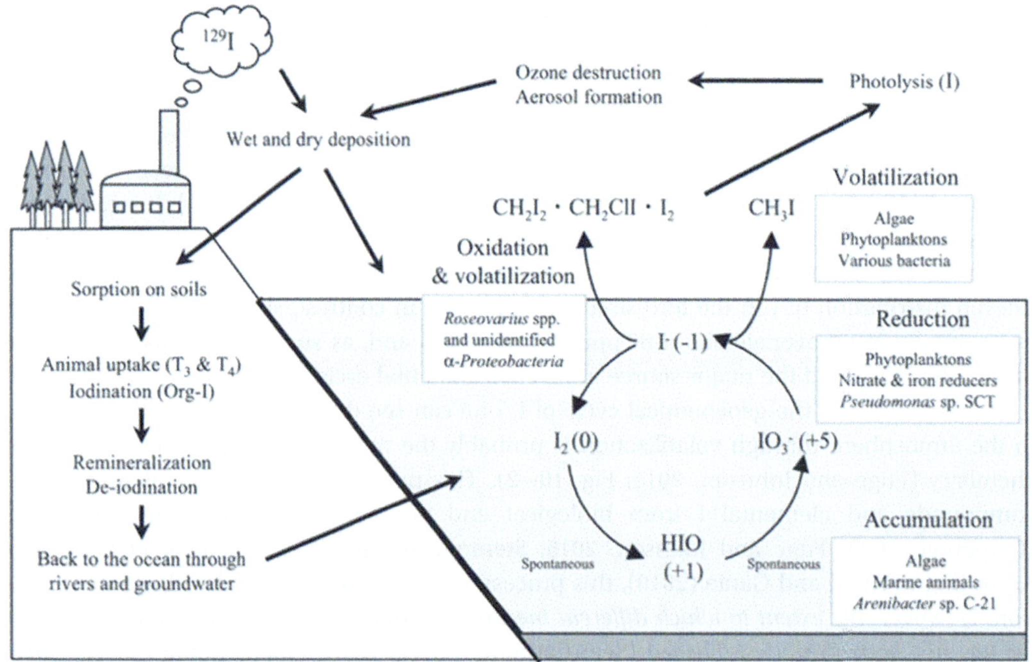

FIGURE 10–2 The volatilization process in the framework of biogeochemical cycling of I. "… Oxidation states of iodine are shown in parentheses. Since ^{129}I has a very long half-life (1.6 × 10^7 years), it will enter the cycle over long periods. Methyl iodide (CH$_3$I) is volatilized not only from the ocean but from terrestrial soils by the actions of fungi, plants and bacteria. Redox changes in inorganic iodine species may also occur in the terrestrial environment, *but the microbial effect is still unclear.* T3 and T4 represent triiodothyronine and thyroxine, respectively." [*Sic*] *From Amachi, S., 2008. Microbial contribution to global iodine cycling: volatilization, accumulation, reduction, oxidation, and sorption of iodine. Microbes and Environments 23 (4), 269–276. https://doi.org/10.1264/jsme2.me08548.*

Iodine and global change

Another important aspect of the biogeochemistry of I is the way it affects global climate. According to Carpenter et al. (2000), the marine environment is likely the major source of emission of both molecular I and iodinated organic compounds into the atmosphere. The complex interactions between these compounds can lead to a destruction of tropospheric ozone, modification of nitrogen oxide and hydrogen oxide cycles, and increase in *cloud condensation* (nuclei), a group of aerosol particles able to form cloud droplets (O'Dowd et al., 2002).

It has also been proposed that in Arctic regions, the reactive species I oxide (IO) and atomic I take part in reactions that lead to depletion in atmospheric Hg whereby gaseous elemental Hg is converted to reactive Hg (HgII) which then precipitates (Auzmendi-Murua et al., 2014; Calvert and Lindberg, 2005; Calvert and Lindberg, 2004; Raofie et al., 2008). The notion that organic halogenated compounds produced in the oceans constitute an important link between ocean biology, atmospheric composition, and climate has been reviewed by Gonzales et al. (2017).

Finally, as noted earlier (2002) by Kolb, if the conversion of I-containing emissions from marine algae into aerosol particles by sunlight occurs on a large scale, we could well-realize a significant effect on climate. A model produced by O'Dowd et al. that same year (2002), suggests that IOs produced from volatile organic iodide compounds, such as CH_2I_2 must be added to the list of precursors for secondary aerosol formation.

Mode of entry into food chains/absorption, bioavailability and distribution

The intake of I, its mode of absorption, mechanisms of transport in thyroid cells, as well as disorders linked to abnormal I transport, have been exquisitely described by Pesce and Kopp in 2014. Work carried out in northern China a little earlier by Li et al. (2013) had indicated that the mobilization processes of I comprise of the reductive dissolution of Fe oxyhydroxides and interactions among iodide, iodate, and organic I, reactions whose courses are determined by microbial activities operating under alkaline and reducing conditions. Zhang et al. (2013) have provided critical information for predicting the long-term fate and transport of ^{129}I, a highly important radionuclide with regard to its possible future dose for man (Hjerpe et al., 2010).

Subsequent to deposition, I can be taken up by plants and enter the food chain, largely through food consumption, though some populations also obtain reasonable amounts of I from drinking water (Fuge and Johnson, 2015). Dietary I derived from plants is generally insufficient for requisite thyroid hormone interactions, as is noted from the low dietary I supplied by strict vegetarian diets. As already noted (See Section "Food crops and other major dietary sources," this Chapter), seafoods provide most of the dietary I input, though other inputs, such as from iodized salt, dairy produce and the use of I-containing sterilants can supplement I.

Early on, in 1997, Phillips noted that the use of I and *iodophors* for sanitizing purposes can release significant amounts of I into the food chain. As the use of I for long-term disinfection is currently not recommended, lifetime exposure to I concentrations, such as might occur from drinking-water disinfection is considered unlikely (See WHO, 2003).

The Provisional Maximum Tolerable Daily Intake for I of 1 mg/day [17 μg/kg body weight (bw) per day] from all sources, set by the Joint FAO/WHO Expert Committee on Food Additives in 1988 (JECFA, 1989) was reaffirmed by WHO (World Health Organization) in 1994. This value was arrived at, based on the tolerance of high doses of I in healthy I-sufficient adults, but excluded neonates and young infants. Values published later (2001) by the United States National Academies Press (US IM, 2001), updated in US NIH, (2021) were specifically with respect to the United States population. See also US FNBIM (2006).

Revised values of Recommended Daily Intake (RDI) of I for the different age groups are given from time to time by WHO and its collaborators, e.g., for pregnant and lactating women, and children less than 2-years-old, see Andersson et al. (2007); for recommended dosages of daily and annual I supplementation for pregnant women, lactating women, women of reproductive age (15−49 years) and children <2 years old, see: WHO/UNICEF/ICCIDD (2008).

Bioavailability

In 2018 WHO noted the paucity of data (both from human and animal studies) regarding the bioavailability of I from foods and water; but it (I bioavailability) is thought to be generally high, with inorganic I (usually in the form of iodide) being readily absorbed (to the tune of 92%) from the small intestine.

As already indicated, the bioavailability of I for metabolic processes depends not only on the environmental abundance of I, but also on its existence in a chemical form suitable for its uptake in sufficient quantities as well as the efficiency of the mechanism that transports it to the centers where it is required in the body for pertinent metabolic processes. This mechanism can be modulated by a number of biophysicochemical factors, including the activity of *goitrogens* (See under: "Goitrogens", this Chapter) such that cases of IDDs can occur even in regions known to be I sufficient.

Iodine absorption and distribution

Aspects of I absorption, metabolism, and homeostasis were reviewed by Rohner et al. in 2014. Ingestion and inhalation are the principal modes of absorption (of I), with absorption through the skin known to be very low (<1% of applied dose) (WHO, 2018; Gad, 2014).

Iodine is rapidly inactivated by food in the digestive system where it is afterward converted to the relatively harmless iodide. It is absorbed throughout the *gastrointestinal (GI) tract* after its conversion to iodide (Fisher and Delange, 1998). Earlier studies by Ogborn et al. (1960) and Morrison et al. (1963), and later work by Zbigniew (2017) have shown that from the GI tract, I is rapidly distributed, including across the placenta and mammary glands in pregnant and breast-feeding women; and then stored in the thyroid gland for the synthesis of thyroid hormones (T4 and T3). The I distribution process is considered to be similar in adults, adolescents, children, and older infants, but with uptake in neonates being of the order of 2−20% lower (See, e.g., Zbigniew, 2017). The principal mode of excretion of excess I is through urine; but very small quantities can also be expelled in sweat, feces, and in expiration, as well as into human breast milk (WHO, 2018; Gad, 2014).

Metabolic function (essentiality/clinical toxicity)

In 2010 Ahad and Ganie wrote "Recent advances in physiology and molecular science have revolutionized our understanding of iodine metabolism at the cellular and subcellular level. This in turn has improved our knowledge of IDD, their prevention, management, and control." [*Sic*]. Today (2021) this understanding has gone even further.

The regulation of thyroid function in the human body is a complex process. It takes place in a stepwise fashion involving the *hypothalamus, pituitary*, and thyroid glands as well as the blood (Ahad and Ganie, 2010) (Fig. 10−3).

The first step involves the trapping of I in a process that begins with the uptake of iodide from the capillary into the *follicular cell* of the gland by an active transport system. This occurs against chemical and electrical gradients by the *sodium iodide symporter* (*NIS*) protein found in the *basolateral membrane* of the follicular cell (Ahad and Ganie, 2010) (See Section: "Glossary of Terms," of this Chapter for explanation of unfamiliar terminology). Competition with NIS transport of I occurs from exposure to numerous anions including perchlorate, chlorate, nitrate, and thiocyanate.

A host of other important biochemical reactions are regulated by thyroid hormones, including protein synthesis and enzymatic activity. One such primary regulatory function is performed by TSH, which is secreted by the pituitary gland. TSH is the master regulator of thyroid gland growth and function, protecting the body from *hypothyroidism* and *hyperthyroidism* (US NIH, 2021; US NRC, 2005). If sufficient I is not produced, TSH levels remain high, leading to goiter, reflecting the body's attempt to trap more I from the circulation and produce thyroid hormones. These interactions under optimum conditions (of I supply) are all critical determinants of metabolic activity (US NRC, 2005). (US IM, 2001), as well as requirements for proper skeletal and central nervous system development in fetuses and infants (US NRC, 2005).

A healthy adult's body contains up to 20 mg of I of which 70%−80% resides in the thyroid (Fisher and Oddie, 1969). In chronic I deficiency, the I content of the thyroid may fall to <20 μg (Zimmermann, 2011). On the other hand, a temporary decrease in the production of thyroid hormones in the thyroid gland can occur due to an acute excess of iodide (over and above previous dietary intake levels) (ATSDR, 2004). This is known as the *Wolff-Chaikoff effect* (See Section: "Glossary of Terms," this Chapter). The presumed mechanisms by which I excess can affect thyroid function and the effects produced have been highlighted by Rohner et al. in 2014. More

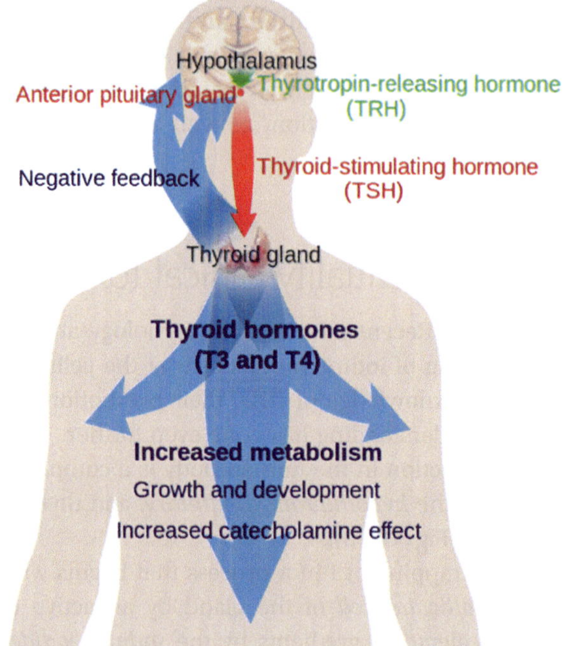

FIGURE 10–3 Thyroid homeostasis. *Used by Mikael Häggström. https://en.wikipedia.org/wiki/File:Thyroid_system. png#/media/File:Thyroid_system.svg. (Accessed 19 July 2021). See: Section L: "Glossary of Terms," this Chapter, for explanation of new terminology.*

recently, US NIH (2021) has commented on records of other physiological functions of I in the human body; e.g., its role in immune response —*mammary dysplasia* and *fibrocystic breast disease* (See Section: "Glossary of Terms," this Chapter), being two conditions for which I might have a beneficial effect (US NIH, 2021).

Essentiality

Iodine is a critically important micronutrient. It is required at all stages of life, with adult mammals requiring trace amounts of the order of 0.0004 wt.% for proper functioning of the thyroid gland (Eskin, 1977). The most critical phases of this requirement are fetal life and early childhood. Among the functions of thyroid hormones for the developing nervous system are normal neuronal migration and *myelination* of the brain (Rohner et al., 2014).

Iodine deficiency disorders

The absence of adequate amounts of I in the body leads to a spectrum of disorders collectively referred to as *IDDs*, which include goiter (Fig. 10–4), cretinism, low psychomotor performance, reduced cognitive function, still births, and many other sequelae Table 10−1 illustrates the stages of life where IDDs can be most graphically expressed.

Determinants of worldwide goiter distribution

A survey of goiter cases and sequelae in a population and its limitations are vexing problems. Surveys of the global distribution of goiter in 1960 by Kelly and Snedden and more recently by Samuel and Devi (2015) revealed marked differences in goiter rates in different endemic regions and even in districts adjacent to each other. In attempting to account for the variability in the expression of endemic goiter from one locality to the next, the availability of I should be investigated before searching for some other subtle dietary or genetic factors. Also, in reaching conclusions on the dynamics of I bioavailability with regard to the distribution of goiter, one must remember that an observed goiter rate may not necessarily reflect the existing situation, but rather the footprint of a preexisting I deficiency that has not yet been entirely obviated.

Toxicity

Iodine has widely been used as an antiseptic and disinfectant product. Accidental or suicidal poisonings are reported ways in which toxic effects can occur in humans. Toxic effects can also occur through ingestion of seaweed, *mucolytic* expectorants or X-ray contrast. However, according to US NLM (n.d.), the replacement of I by *iodophors* has resulted in a much-reduced prevalence of I toxicity.

Besides the development of goiter, chronic excessive I intake can result in early onset of subclinical thyroid disorders—hyperthyroidism and hypothyroidism—an increased incidence

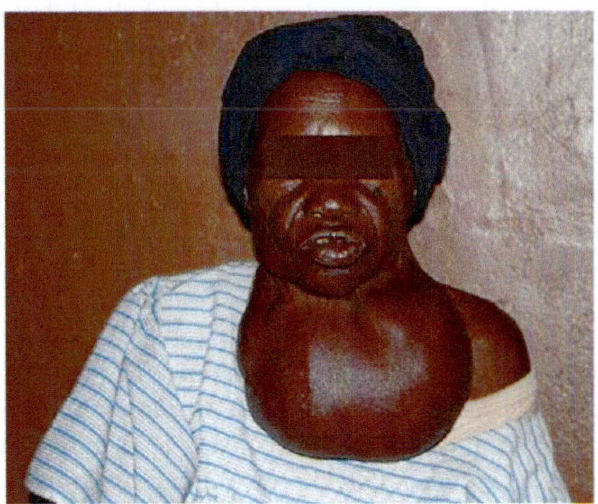

FIGURE 10–4 Patient with large bilateral goiter, Burkina Faso (Rumstadt et al., 2008). *From Rumstadt, B., Klein, B., Kirr, H., Kaltenbach, N., Homenu, W., Schilling, D., 2008. Thyroid surgery in Burkina Faso, West Africa: experience from a surgical help programme. World Journal of Surgery 32, 2627–2630. https://www.readcube.com/articles/ 10.1007%2Fs00268-008-9775-6.*

Table 10–1 The spectrum of iodine deficiency disorders (FAO/WHO, 2001).

Stage in life	Effects
Fetus	*Abortions
	*Stillbirths
	*Congenital anomalies
	*Increased perinatal mortality
	*Increased infant mortality
	*Neurological cretinism: mental deficiency, deaf mutism, spastic diplegia, and squint
	*Myxedematous cretinism: mental deficiency and dwarfism
	*Psychomotor defects
Neonate	*Neonatal goiter
	*Neonatal hypothyroidism
Child and adolescent	*Goiter
	*Juvenile hypothyroidism
	*Impaired mental function
	*Retarded physical development
Adult	*Goiter with its complications
	*Hypothyroidism
	*Impaired mental function

Source: From FAO/WHO (Food and Agricultural Organisation/World Health Organisation), 2001. Human vitamin and mineral requirements. Chapter 12: Iodine. Expert Consultation on Human Vitamin and Mineral Requirements. Food and Nutrition Division, Rome. http://www.fao.org/3/Y2809E/y2809e0i.htm. (Accessed 21 May 2021).

of autoimmune *thyroiditis* and increased risk of thyroid cancer (TC) (Chaker et al., 2022; Liu et al., 2017; Resende de Paiva et al., 2017; Teng et al., 2006). Chronic exposure to excess I can lead to reproductive impairments secondary to thyroid gland dysfunction, as well as the potential to produce neurological effects (Boyages, 2000a). Acute oral toxicity is mainly due to irritation of the GI tract, marked loss of fluid, and shock occurring in severe cases (WHO, 2019).

Protection against acute I toxicity can be achieved by several biological mechanisms, of which reduction in I uptake and preferential production of the more heavily iodinated thyroid hormone appear to be the most common mechanisms [ATSDR, 2004; EVM, 2003].

Immunotoxicity

Our earlier discussions have established the importance of I in thyroid hormone synthesis and in reproduction. *However, its direct effects on the immune system are still largely unknown (See, e.g., Bilal et al., 2017). The existence of a huge data gap regarding the immunotoxic effects of I in humans following repeated exposure was reiterated by WHO in 2018.*

Studies by Chen et al. in 2007 have suggested that disorders of T lymphocyte immune function can be induced in mice following exposure to excessive I as exogenous chemical materials. According to Zhao et al. (2006) I might exert an influence on the level of *CD4/CD8*

(See Section: "Glossary of Terms," this Chapter), and thus the production of thyroid antibodies might directly or indirectly participate in the process of thyroid autoimmunity.

Genotoxicity/carcinogenicity

As recently as 2018, WHO reported that no data could be located regarding genotoxic effects in humans following repeated exposure to I. Short-term tests by Ilin and Nersesyan (2013) have not confirmed carcinogenic potency for I based on the absence of genotoxicity data of I and I-containing compounds. Also, no significant associations between I and TC have been found in studies of populations in which I intakes are known to be sufficient (Horn-Ross et al., 2001; Kolonel et al., 1990).

Radiation-induced thyroid cancer

Radioactive I uptake into the thyroid is higher in people with I deficiency than in people with I sufficiency. This puts I-deficient individuals in a particularly high-risk group in terms of developing radiation-induced TC when exposed to radioactive I.

Nauman and Wolff (1993) and later, WHO (1999) have noted that radioactive I released into the environment by nuclear accidents increases the risk of TC in exposed individuals, especially children. The results of Pfinder et al. (2015) suggest that administration of potassium iodide (KI) after a nuclear accident could be an effective means of reducing the risk of TC, more particularly so, in children.

Animal studies

A potential reproductive and developmental toxic effect of I has been indicated in a number of animal studies, e.g., Yang et al. (2006), Boyages (2000a), and Boyages (2000b). Toxicity studies of laboratory animals such as poultry, pigs and cattle have shown a high tolerance to large I intakes. However, animal data are of limited value to humans because of species differences in *basal metabolic rate (BMR)* and in I metabolism (See IOM, 2001).

Iodine supplementation

According to Krela-Kaźmierczak et al. (2021), although the USI program went a long way in the fight to eradicate endemic goiter, it seems that the program did not resolve all the issues, and the issue of I deficiency is still of relevance today (See also: WHO/UNICEF/ICCIDD, 2008; Temech et al., 2023; US IM, 2001). New, preventive solutions are therefore necessary. It also appears that there is no diet that would fully cover the I requirements. In addition, the consumption of processed foods, which do not necessarily include iodized salt, also affects the I intake and the I status. Under these considerations, food supplementation is inevitable.

The administration of dietary supplements has so far proved to be the best way of reducing the prevalence rate of IDD in endemic populations. The most common I supplements to date include the use of iodized salt (*cf.*, USI programme) or the use of slow-release oral or intramuscular I supplements, such as iodized oil (See: Temple et al., 2020; Eastman and

Zimmermann, 2018; Völzke et al., 2018). By 2007, the USI program had been implemented in approximately 120 countries around the world (WHO, 2017).

In 2011 Speeckaert et al. reviewed the multiple beneficial effects of supplementation, as well as presented a critical look at the possible side effects, such as I-induced *hyperthyroidism* (See also: Krohn et al., 2005; Johnson, 2003; Todd et al., 1995; Galofre et al., 1994). Much more recently, Drewnowski et al. (2021) and Duborská et al. (2020), respectively, looked at the current status of I supplementation and the most recent alternative methods of application of I in agriculture, including its effect on the quality of a variety of plant species (See also: Smyth et al., 2016; Voogt et al., 2014; Kiferle et al., 2013; Landini et al., 2011). Weng et al. (2013) have shown how exposure to a seaweed-rich environment can result in intake of I by respiration of gaseous I derived from seaweed.

The interaction or interference of I with medicines is well-known (e.g., Santos-Oliveira, 2008). An example is the interaction of I supplements with antithyroid medications, such as methimazole (Tapazole) used to treat hyperthyroidism (See, e.g., Singh and Correa, 2021).

Such an interaction is known to cause too little thyroid hormone production by the body (See also, Chung, 2014).

KI taken in combination with medicines for high blood pressure known as *ACE inhibitors* (See Section: "Glossary of Terms," this Chapter) could raise the amount of K in the blood to unacceptable levels (See Torti and Correa, 2021). ACE inhibitors include benazepril (Lotensin), lisinopril (Prinivil and Zestril), and fosinopril (Monopril). The amount of blood K can also get extremely high if KI is taken with potassium-sparing *diuretics* (See Section: "Glossary of Terms," in this Chapter), such as spironolactone (Aldactone) and amiloride (Midamor) (See Torti and Correa, 2021).

Excess supplementation in pregnancy

In 2016 Aronson wrote that I can readily cross the placenta, with the fetal concentration usually exceeding the maternal concentration, when I supplementation is administered in excess. Since the placenta does not seem to have a regulating transfer mechanism (Aronson, 2016), this raises the possibility of I intoxication of the fetus due to excessive maternal intake of I, with consequent hypothyroidism and development of goiter.

Reduction of body burden/elimination

The thyroid gland is store of approximately 90% of the I in the human body (ATSDR, 2004). According to earlier investigations by Larsen et al. (1998) and later ones by Hays (2001), about 97% of this I is excreted in the urine as iodide, with fecal elimination being between 1 and 2%. Breast milk, saliva, sweat, tears, and exhaled air are other routes by which absorbed I can be excreted (See also Section "Iodine absorption and distribution," in this Chapter; and Cavalieri, 1997).

The thyroid is also the key target organ of radioiodine (ATSDR, 2004). Early methods for decreasing the absorption and retention of radioiodine I (^{131}I) in rats were described by

Kostial et al. (1980) and Kostial et al. (1981). More recently, Ravichandran et al. (2014) and Ravichandran et al. (2010) have described methods for estimating radioactive body burden of I (^{131}I). The elimination half-life of absorbed I varies widely between individuals, with Hays (2001), and Van Dilla and Fulwyler (1963), respectively, giving an estimate of 31 days for healthy adult males.

KI and potassium perchlorate are compounds that can be used to protect the thyroid from exposure to ^{131}I after accidental release from nuclear reactors to prevent hypothyroidism and TC (Reiners and Schneider, 2013).

Methods for reducing toxic effects

Specific information regarding treatment following exposures to I has been provided by Braverman in 2020. Earlier descriptions of research and clinical practice for reducing the toxic effects of exposure to I were given by ATSDR in 2004. Treatment of toxicity from exposure to excess I is aimed at reducing exposure and, if clinical hypothyroidism or hyperthyroidism continues, correcting the thyroid dysfunction. The treatment of toxicity from exposure to radioiodine is also directed at reducing thyroid gland uptakes of absorbed I, e.g., by administration of KI.

Measurement of iodine concentration

Assessment of iodine status

Hussain et al. (2019), Wainwright and Cook (2019) and Pearce and Caldwell (2016) have, respectively, presented overviews of the available biomarkers of I status and their relative utility or suitability at the level of both populations and individuals for the investigation of I deficiency and I excess.

Besides the determination of thyroid size, urinary I concentration, newborn TSH, and serum thyroglobulin have been shown to be useful biomarkers to assess I status (See: Zimmermann, 2020; Speeckaert et al., 2011; Ristic-Medic et al., 2009). The assessment of I status in pregnancy, however, has often proved to be difficult; and it remains unclear whether I intakes are sufficient in this group, which uncertainty has led to calls for I supplementation during pregnancy in several industrialized countries (See Zimmermann, 2009).

Determination of iodine in biomarkers of iodine status

Recent references [at time of writing (2021)] giving excellent accounts on determination of I in biomarkers of I status include the works of: Nikulin et al. (2020), Mika et al. (2019), Costa et al. (2018), Judprasong et al. (2016), Jooste and Strydom (2010). Earlier studies by Fisher and Delange in 1998 had already established that urine was a good measure of I intake, since urinary I excretion can reflect the average daily I requirement in a normal population showing no evidence of clinical I deficiency either in the form of endemic goiter or any of its sequelae. On this basis, therefore, Fisher and Delange (1998) considered serum T4 and TSH levels (indicating normal thyroid status) and urinary I excretion as the important indices for the determination of the I requirements.

The biomarkers of nutrition for development (BOND) project

The "Biomarkers of Nutrition for Development (BOND) Project" (Rohner et al., 2014) was specifically aimed at providing up-to-date information and service on the selection, use, and interpretation of biomarkers of nutrient exposure, status, function, and effect. The Project sought to develop agreement on accurate methods of assessment that researchers (laboratory/clinical/surveillance), clinicians, programmers, and policy makers (data consumers) could apply. The BOND project was also designed to develop well focused research schedules that could support the discovery and development of biomarkers through a better understanding of nutrient biology within the framework of pertinent biological systems (Rohner et al., 2014).

Quantitative determination of iodine in environmental media and in biomarkers of iodine status

The sheer importance of I as a micronutrient element has brought about considerable interest in its environmental geochemistry, its biochemical cycling, and not least, its accurate quantitative determination the various environmental media as well as in the biomarkers of iodine status. However, analytical capacity for these determinations has up to recently been quite limited. Much of the early data (on I measurements) were not up to the level of accuracy needed for making tangible conclusions regarding its (I) migration and uptake pathways, nor of the mechanisms that produce IDDs and related maladies.

In 2011 Moreda-Piñeiro et al. reviewed the different sample pretreatments to extract total I as well as specific iodinated forms. According to these authors, "... the most used analytical techniques for assessing total iodine cathodic stripping voltammetry, differential pulse voltammetry, neutron activation analysis (NAA), catalytic spectrophotometry, and atomic spectrometric techniques, mainly inductively coupled plasma-mass spectrometry (ICP-Ms) and iodine speciation (nonchromatographic methods, and mainly high-performance liquid chromatography, gas chromatography, and capillary electrophoresis coupled to electrochemical, UV spectrophotometric, and mass spectrometric detection)." In that same year (2011), Shelor and Dasgupta provided a similar review, this time directed particularly at quantification of I in complex matrices. These authors noted then (2011) that NAA and X-ray fluorescence were the commonly used methods for in vivo determination of stable I in the thyroid.

Up to the start of the last decade (2010−20), the most frequently used method for I in urine was colorimetry of the *Sandell-Koltoff reaction* (See Section: "Glossary of Terms," this Chapter), while this method was gradually being overtaken by the newer ICP-Ms method, which was much more accurate and simple, compared to the other available methods (See Hou and Ding, 2009), and has remained so, since (See, e.g., Haap et al., 2017; Todorov and Gray, 2016; Tagami et al., 2006). Outstanding among the recent advancements in analytical techniques for I, is the development and use of the ICP-Ms method (See: Druzian et al., 2021; Jerše et al., 2021; Quarles et al., 2020), which has engendered the generation of more reliable data resulting in a massive volume of published works on the various aspects of the geochemistry of I and its sources in human and animal diets.

One of the age-old restrictions encountered (apparently now resolved) in the accurate quantitative determination of I pertained to its volatility. The relative volatility of many I species means they can be readily oxidized or reduced; hence, care must be taken to prevent volatility losses during analysis. Knowledge of the chemical form and oxidation state of I under various conditions (See Section "Chemical form or speciation of iodine in soil and aqueous phases," in this Chapter) is therefore important to avoid losses during measurements (See: Baker and Yodle, 2021; Kim et al., 2018; Reid et al., 2008).

Determination of I in soils and natural waters

By 2018, advancements in measurement of total I in soils as well as in other environmental media had indicated that the ICP-Ms was probably the best technique, a conclusion drawn largely on the basis of improvements in extraction procedures and sensitivity, with extremely good detection limits of typically $<\mu g/L$ now readily attainable (See, e.g., Humphrey et al., 2018).

Insights into methods that have been used for quantitative determination of I in natural waters from the year 2000 onward are provided in the works of Mika et al. (2019), Takeda et al. (2016), Gong and Zhang (2013), Jones et al. (2023), Marczenko and Balcerzak (2000), among others; and in drinking water in particular: Sayess et al. (2021), Zeng and Zhu (2021).

Quantitative determination of radioisotopes of iodine

Determination of radioisotopes of I is usually carried out by radiometric methods employing the γ-*detector technique*. Only NAA and accelerator mass spectrometry are sensitive enough for determination of environmental levels of ^{129}I (Hou and Ding, 2009). ^{131}I decays with a half-life of about 8.02 days by β^- decay to ^{131}Xe, accompanied by the emission of γ rays with energy of 364.5 keV (81.7%), 637 keV (7.2%), and 284.3 keV (6.1%) (L'Annunziata, 2020). Therefore ^{131}I can also be determined relatively easily by γ spectrometry, which is the method used to monitor it around nuclear facilities. A higher concentration of ^{131}I was registered by many laboratories in Japan using this technique after the accidents in Chernobyl and Fukushima (See, e.g., Martin et al., 2020).

Field testing for iodine

By 2000, the USI program, which had already realized considerable success, had provided a wide variety of field test kits; but these were generally designed for measurement of samples with higher concentrations (greater than around 0.5 ppm of I) than those found typically in natural groundwaters, such as in swimming pools (BGS, 2000). Iodine determinations were therefore best carried out in the laboratory by the methods listed in Section

"Quantitative determination of iodine in environmental media and in biomarkers of I status," this Chapter.

By 2015, further considerable improvements had been made to these earlier field measurement devices, such that they showed acceptable laboratory performance in terms of consistency and ease of use (Rohner et al., 2015). These user- and field-friendly devices were readily available during the 2000s and the most appropriate selection could be made depending on the number of samples and the budget available (Rohner et al., 2015). Most recently, UNICEF (2019) has described even better designed Salt Test Kits for detection of I in salt that has been fortified either with potassium iodate (KIO_3) or KI, supplied by UNICEF.

Current status of geomedical research on iodine in Africa

Africa is characterized by huge differences in elevation, the Continent having undergone a complex tectonic episode. The mountainous areas thus constitute a barrier to the transport of I in sea spray to central continental areas. Another important source of I, *viz.*, the leaching of acid mineralized soils originally supplied by volcanic activity, is a determinant of the element's geographical distribution. Thus the most notable goiter areas in Africa are the lee sides of high-altitude areas such as the Kerio district in the Kenya Rift Valley, which records the highest goiter prevalence in the region, with 72% of children having goiter as at 1974 (See Hanegraaf and McGill, 1974). And where heavy I-poor rain has depleted the soil of I through leaching, lateritization, podsolization, and gleying processes typical of the humid tropical terrain in Africa.

Conversely, traverses from the Moroccan coast show that strong soil-I enrichment has occurred, albeit limited to about 100−200 km from the coast (Johnson, 2003); though some would say that the link between I in soils and distance from the coast is not that strong (See, e.g., Fuge, 2005).

Since the early 1990s, there has, as already stated, been a gradual decline in goiter cases in Africa. However, the devastating effect of IDDs in Africa up to this time (the early 1990s) and knowledge that a complete elimination of IDDs and sequelae may be unattainable, led to a vast array of research activities on the subject, which continues to the present day. These activities range from studies of the relationship of IDD to African dietary types (nutritional status), reported in several publications, some of the more recent of which include the works of: Venance et al. (2020), Smuts and Baumgartner (2019), El-Bashir et al. (2016), Watts et al. (2015), Seal et al. (2006), Jooste et al. (2005); the inhibition of I transport mechanism by: goitrogens [e.g., thiocyanate (SCN^-) in cassava], reported in several publications, including the works of: Gaitan (2004), Vanderpas (2003), Delange and Ahiuwalia (1983), Bourdoux et al. (1978); vegetables of the genus *Brassica*, reported in several publications, including the works of: Eastman and Zimmermann (2018), Miller (2017); the masking effect (competitiveness) of the tetrafluoroborate ion (HBF_4^-) over iodide, reported in several publications, including the works of: Jiang and DeGrado (2018), O'Doherty et al. (2017), Jauregui-Osoro et al. (2010), Anbar and Inbar (1964); and IDD in pregnancy, reported in several publications, some of the more recent of which include the works of: Kabthymer et al. (2021), Businge et al. (2020), Toloza et al. (2020), Businge et al. (2019), Mabasa et al. (2019); policy considerations on iodized salt administration, reported in several publications, some of the more recent of which include the works of: Drewnowski et al. (2021), Houston et al. (2021), Appiah et al. (2020); I supplementation, reported in several publications, some of the more recent of which include the works of: Elias et al. (2021), Menyanu et al. (2021), Yeshaw et al. (2021), Ajema et al. (2020), Dida et al. (2020), Charlton et al. (2018a), USICST (2017); Jooste and Zimmermann (2008), Okosieme (2006); measurement of I concentrations and assessment of I status, reported in several publications, some of the more recent of which include the works of: Gyamfi et al. (2020), Charlton et al. (2018b), Duressa et al. (2014), Shelor and Dasgupta (2011) and WHO/UNICEF/ICCIDD (2008); estimating the health and economic benefits of the USI program for combatting IDDs (e.g., Gorstein et al., 2020; Aburto et al., 2014); regional monitoring and surveillance, reported in several publications, some of the more recent of which include the works of: Kassim et al. (2014), Traoré and Torheim (2009), Seal et al. (2006), Bimenya et al. (2002).

Findings from regional monitoring and surveillance in Africa have established that the best strategy to control I deficiency in populations is carefully administered iodization of salt (See, e.g., Charlton et al., 2018a; Okosieme, 2006). However, a few remarks garnered from the above-listed works are perhaps instructive in understanding the trends in effort to eliminate the IDD problem in Africa through the USI program alone (See also: Terefe et al. (2023)).

In 2020, Businge et al. assessed the burden of I deficiency in pregnancy in Africa using estimated *pregnancy median urinary I concentration*, and found that there is the likelihood of a high prevalence of insufficient I intake in pregnancy (Fig. 10−5), including in some countries classified as having adequate I intake in the general population. These authors noted that the I status of school age children (SAC) having an mUIC ≤ 175 ppb portends insufficient I intake among pregnant women in these settings.

Various studies in Africa have indeed revealed that areas of apparent I deficiency (which had experienced high rates of IDD) are not the areas that are necessarily I deficient, but where I supply to metabolic centers has been restricted by the activity of any number of bio-physicochemical factors, such as the activity of goitrogens, adsorption of I by organic or other particulate matter, and so on (See, e.g., Crewdson et al., 2019; Davies, 2008; Thilly et al., 1992); (See also, Section "Bioavailability," in this Chapter and Section "Determinants of worldwide goiter distribution," in this Chapter). As a corollary to this observation, there is the existence of areas of the Continent where I sufficiency in the substrate has been established, and where over-consumption of I can occur, with relatively high rates of visible goiter. Examples of such areas can be found in Somalia (Kassim et al., 2014), Kericho in Kenya (References in Davies, 2008), and Luwero in Uganda (Bimenya et al., 2002). Such cases may be an expression of I-induced hyperthyroidism (Bimenya et al., 2002) or the result of goitrogenic substances in the diet (Adebayo et al., 2018; Wenhold and Faber, 2008). *This observation warrants further investigations.*

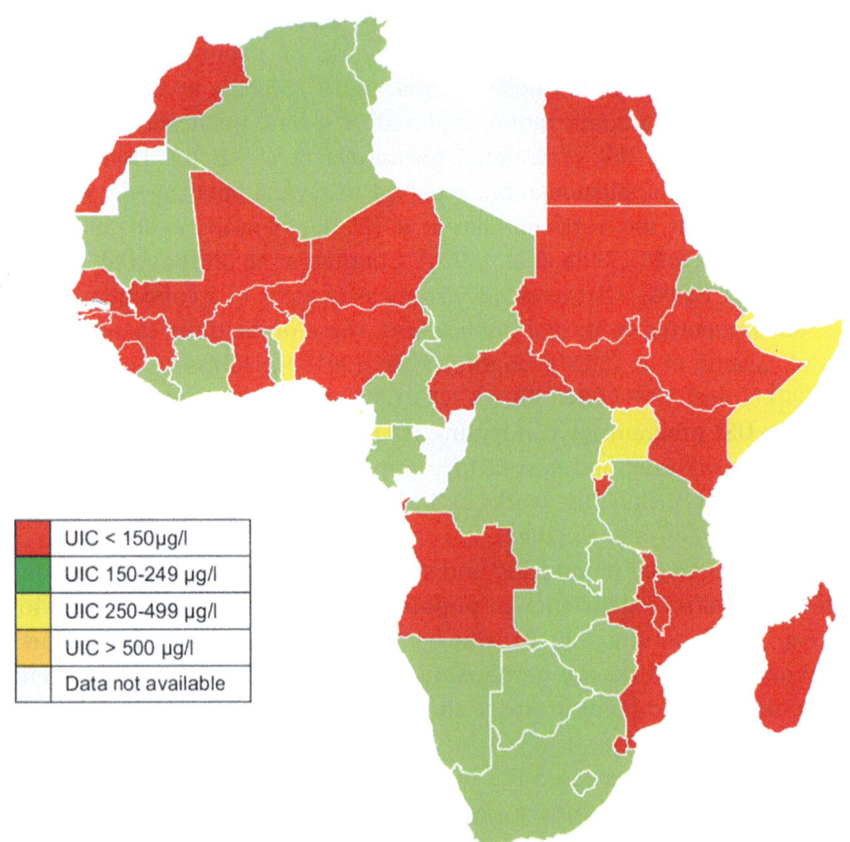

FIGURE 10–5 Map of Africa showing country-specific I nutrition status during pregnancy (pregnancy median *urinary iodine concentration (UIC)* as estimated from SAC median UIC). *SAC*, School age children. *From Businge, C.B., Longo-Mbenza, B., Kengne, A.P., 2020. Iodine nutrition status in Africa: potentially high prevalence of iodine deficiency in pregnancy even in countries classified as iodine sufficient. Public Health Nutrition. 24 (12), 3581–3586. https://doi.org/10.1017/S1368980020002384.*

Similarly, in 2006, Seal et al. found excessive I consumption as well as excessive I excretion occurring in several surveyed populations living in refugee camps in East, North, and Southern Africa. These authors commented that the full causes of these excesses remained unknown (as at 2006) prompting calls for *further investigation to urgently identify the cause (s), assess any health impact(s), and proffer remedial measures, that would almost certainly include revision of the level of salt iodization to meet current WHO recommendations.*

Suggested areas for further iodine research

Why do we need to do further research on IDDs in Africa? The salt iodization programs, albeit successful in reducing the risks from IDD in many parts of the world, including some African countries, have gone on without a sound understanding of their principal cause, a deficiency of I in the environment and related biogeochemical interactions (See Johnson, 2003). This observation should underpin further IDD research if we are to attain a foolproof solution and prevent a reemergence of the problem.

- In 2003 Johnson emphasized that there was a perceived need for a better understanding of I geochemistry, so that we can ensure that the small amounts of I that are available in the environment are used in the most efficient way. Again too, I added through environmental supplementation techniques, (such as the addition of I to irrigation waters) needs to be managed in an effective way to ensure optimum use is made of the added I through an improved knowledge of its geochemical behavior.
- According to Carpenter et al. (2021), "While substantial progress has been made in the last decades on developing atmospheric models of iodine cycling, *global ocean iodine modeling* is in its infancy (Wadley et al., 2020). There are still large gaps in our basic knowledge that significantly limit how iodine biogeochemistry can be represented." [*Sic*]. For example, *our knowledge of environmental transformation reactions is incomplete* (See Bagwell et al., 2019), *greatly limiting our ability to accurately inventory and predictably model the fate of radioiodine.*
- For the effects of KI after release of radioiodine in nuclear accidents on health outcomes (hypothyroidism and benign thyroid nodules), Pfinder et al. (2015) have pointed out the necessity for high-quality studies to provide an evidence-based appraisal, targeting sensitive subgroups, such as pregnant women and children. According to these authors, *future research should consider the timing and dosage when investigating effects of KI after release of radioiodine on health outcomes.*
- Organic I compounds have been shown to play an important role in the global atmospheric I cycle and the atmospheric I sink (See, e.g., Gilfedder et al., 2007). *Future tropospheric I models must therefore consider organic-I reactions in a much more intuitive manner.*
- In 1996 Sheppard noted: "Nonetheless, detailed understanding of the behavior of I in the environment is essential to the credibility of models, such as BIOTRAC. *There is substantial room for improved knowledge of the speciation of I*, especially in freshwater and soil environments." [*Sic*].
- The literature data on potential anthropogenic influences on I in rainwater is scanty (See: Gilfedder et al., 2007; Fuge and Johnson, 2015). Since rainwater harvesting is a common practice in many water-scarce African countries, *research on the behavior of I in rainwater regimes in African countries and possible links with IDDs and associated maladies is highly warranted.*
- Kiferle et al. (2021) have called for further research on the importance of organification of I in proteins on their catalytic and/or regulatory function to help clarify the functional role of I as a plant nutrient.
- The construction of maps showing the geochemical distribution of I in soil, natural waters, and the African atmosphere should be considered with some urgency, within the

framework of the "Africa Geochemical Database Program" (See: GEO, 2016 and Davies, 2008).

- The relationship between Se and I deficiency, which has for long been reported in several areas of Africa has not yet (as at 2020) been well-clarified (See, e.g., Ligowe et al., 2020).
- Long after the call of Fisher and Delange (1998), the role of T_4 in brain development at the molecular level has not yet been fully elaborated (See, e.g., Talhada et al., 2019).
- The possible interference of infections and other systemic illnesses with I or thyroid hormone has not yet been well-elucidated on a population basis (See, e.g., Franco et al., 2013).
- In 2012 Lauberg et al. noted that: "The interaction between weight control and therapy of thyroid disease is important to many patients and it should be studied in more detail."
- More research is needed in understanding I uptake from I-enriched fertilizers in various crops under different cultivation methods and growing conditions, in order for marketing of I-biofortified foods to be done in a much more cost-effective manner (See WIA, n.d.).
- In 2017 Resende de Paiva et al. wrote "The incidence of TC is increasing although explanatory causes are lacking. A link between cancer and inflammation is well-documented *but unclear for AITDs and TC.*"
- By 2019, greater access to iodized salt and awareness of its importance have been realized in Africa. However, according to Saha et al. (2019), the emerging problem of excess I intakes of the different population groups in Africa, that results in hypothyroidism (*cf.,* *Graves' disease, atrophic auto-immune hypothyroidism,* and TC) *needs further research* (See also Sidibé, 2007a). Of especial importance in this regard is the I status of pregnant women, which remains a major public health problem in many African countries, such as Ethiopia and South Africa (See Kabthymer et al., 2021; Siro et al., 2022). The results of such studies ought to be taken into consideration by policy makers and program implementers.
- Indeed, about a decade ago, Negro et al. (2011) wrote: "Progress has been made in our knowledge of isolated *hypothyroxinemia* during pregnancy, *but serious gaps remain.* The optimal clinical test, which is accurate, inexpensive, and widely available, does not exist yet. Similarly, trimester-specific reference ranges in iodine-sufficient and thyroperoxidase antibody-negative populations are yet to be determined." [*Sic*].
- Writing on the mode of development of *ontogenesis* of the *hypothalamo-pituitary-thyroid axis* of the fetus of the mother in cases of hypothyroidism, Sidibe wrote in (2007b) "An important role is played by type III deiodinase, which is especially active in the placenta during pregnancy, probably involving the T3 activity on nuclear and also mitochondrial receptors." *Up till today (2021), the mode of maturation of these receptors has not been fully elucidated.*
- Further investigation is required to identify the full cause of excessive I consumption and excretion, such as was found in some African refugee camps by Seal et al. in 2006. A thorough assessment of the IDD status of such populations as well as associated health impact is warranted and remedial actions proffered (See also, Section "Current status of geomedical research on iodine in Africa," this Chapter).
- Waugh (2019) has provided evidence that strongly suggests that F ingestion contributes to pathological states by impairing I absorption and diminishing I concentrating ability. Kassim et al. had earlier (2014) observed that in Africa, although I intake from drinking

water sources may be important in preventing I deficiency, it may also contribute to excessive intakes and adverse effects, e.g., I deficiency resulting from co-occurrence of I with F in water.

The extent to which F may be inhibiting the activity of I in areas that are apparently I sufficient (e.g., in Uganda, Somalia, and Kenya), has great implication on the policy of fluoridation of drinking water. *Further research on these associations, as well as the molecular mechanisms involved, remains unclear.*

- Researching other options, such as the delivery of processed foods that is already taking place in some countries such as Ethiopia, Ghana, and Senegal (Kabthymer et al., 2021; Businge et al., 2019) might boost the effect of iodized salt particularly among pregnant women. However, for the general population, Krela-Kaźmierczak et al. (2021) consider that the search for alternative diets, other than iodized salt, must continue, in order to prevent IDDs.
- Working on I biofortification protocols of the tomato plant, Kiferle et al. (2013) have called for a more extensive evaluation of the production aspects, such as exist in open field conditions. This means that a tangible improvement should be demonstrated in the I status of the consumer of biofortified tomatoes, such that the I accumulated must be sufficiently bioavailable.
- Further research on elucidating the molecular events governing iodide uptake is warranted, because of the implications of these events in targeting the NIS, e.g., *SLC5A5*, for redifferentiation therapies and the importance of their use in nonthyroidal tissues (See Pesce and Kopp, 2014).
- In 2014, Rohner et al. remarked: "Even though we have learned much about the mechanisms and impact of I deficiency it should be noted that the presentation and severity of IDD (e.g., goiter, thyroid dysfunction) are extremely variable between and within the populations even in endemic areas. Our ability to understand why this might be the case is limited by *an incomplete appreciation of the dietary, environmental, and/or genetic factors that might contribute to this variability.*"
- As already indicated, Africa, has, since the early 1990s, witnessed a gradual decline in goiter cases. However, there still exist areas of I sufficiency in the substrate (where overconsumption of I occurs), with relatively high rates of visible goiter, e.g., Kericho in Kenya (Davies, 2008), Somalia (Kassim et al., 2014) and Luwero in Uganda (Bimenya et al., 2002); and may be an expression of I-induced hyperthyroidism (Bimenya et al., 2002) or the result of goitrogenic substances in the diet (Adebayo et al., 2018; Wenhold and Faber, 2008), or indeed, as already noted, the inhibitory (masking) effect of F over I uptake (See, e.g., Waugh, 2019). These observations warrant further in-depth investigations.
- As a sequel to the success of the USI program, proposals on exploring the feasibility of adding Fe to iodized salt to produce double fortified salt (DFS) has been generating a great deal of interest. The objective of DFS is to provide supplementary Fe in cases where Fe deficiency is widespread and contributes to an increased risk of anemia. Drewnowski

et al. (2021) have reviewed the technology and program implementation of the DFS proposal and have stressed the potential benefits to be accrued from *further exploration of this evolving technology.*

- Although the WHO/UNICEF/ICCIDD increased the *RDI* for I from 200 to 250 μg/day for pregnant women and lactating mothers (See Utoro et al., 2007), the need for more data on the level of I intake that ensures maternal and newborn *euthyroidism* (See Section "Glossary of terms," in this Chapter) was also duly emphasized (WHO/UNICEF/ICCIDD, 2008). *More research on this aspect is therefore desired*, particularly directed at the most vulnerable groups; and also an assessment made of possible adverse effects of high intake of I in these groups (See also Zimmermann and Boelaert, 2015). In administering prophylaxis, care should be taken to ensure adequate I intake during pregnancy and infancy without causing excess intakes in the remainder of the population (See Lisco et al., 2023; Zimmermann and Boelaert, 2015).
- According to Zimmermann and Boelaert (2015), "Future research priorities in iodine nutrition should include correlation of community iodine intake with long-term risk of thyroid disorders, and more precise definition of the limits of deficient and excessive iodine intakes that increase risk of thyroid disorders in populations. *The modulation of this relation by genetic and environmental factors (pollutants and other micronutrient deficiencies) needs to be clarified.*" [Sic].
- Surely, the personalization of I requirements based on age, gender, ethnicity, environment, and health status is an aspect worthy of consideration in any further revision of the RDI.
- Because the relationship between thyroid hormone metabolism and trace element levels has not always been in conformity, judging from previous reports that compared trace element levels in patients with hypothyroidism and healthy individuals, Talebi et al. (2020) have *recommended that high-quality studies with large sample sizes be performed to clarify the link between trace element status and hypothyroidism.*
- As Ristic-Medic et al. concluded in their 2009 paper: "Despite the high risk of bias in many of the included studies, the results suggested that urinary iodine, *thyroglobin*, serum thyroxine, and thyroid-stimulating hormone are useful biomarkers of iodine status, at least in some groups. *High-quality controlled studies measuring relevant long-term outcomes are needed to address, which biomarker is the most appropriate for assessing iodine intake in some population groups and settings.*" [Sic]. Development *of new biomarkers of exposure and function at the individual level is a subject for future research, likely through the new "-omics" technologies* (See Rohner et al. in 2014).
- The accuracy and reliability of quantitative measurements of the I content of salt or of any of the recommended biomarkers of I status, require suitably equipped public or private laboratories. Many African countries now have central laboratories for this purpose. Others depend on regional analytical facilities. As noted by Rohner et al. in 2014, "... it is critical that the laboratory or the network of laboratories provides reliable data from continuous quality assessments of the results obtained with their salt iodine measurements."

- All said and done, the long-standing recommendation of Fisher and Delange in 1998 regarding the establishment of quality assurance procedures at iodized salt production sites, as well as the tracking of progress of IDD elimination through the implementation of cyclic monitoring *has not yet been fully implemented in many African countries.*
- Finally, the statements of Rohner et al. in 2014, still remain true today, *viz*: "Despite much success in the global efforts to control and eliminate iodine deficiency, including the elimination of severe deficiency and endemic goiter in most countries, mild to moderate iodine deficiency continues to affect 32 countries, more than half of them in the industrialized world. Where efforts have succeeded, ensuring the sustainability of those efforts becomes a high priority, requiring regular surveillance and vigilance to ensure the adequacy of salt iodine content. Conversely, as salt iodization efforts become more commonplace and effective, attention must be paid to the potential for excess iodine intakes and consequent health effects." [*Sic*].

Glossary of terms

- *ACE inhibitors* (*Angiotensin-converting-enzyme inhibitors*) "…are a class of medication used primarily for the treatment of high blood pressure and heart failure. They work by causing relaxation of blood vessels as well as a decrease in blood volume, which leads to lower blood pressure and decreased oxygen demand from the heart." (Wikipedia, 2021a).
- *Atmophile* (literally "gas-loving") elements are extremely volatile chemical elements with an affinity for the atmosphere.
- *Basolateral*: situated below and toward the side: located in or on the base and one or more sides.
- The *BIOTRAC model*: BIOsphere Transport and Assessment Code (BIOTRAC) is a model developed by the Atomic Energy of Canada Limited (AECL) for use in postclosure assessments of Canada's nuclear fuel waste industry and writing of "environmental impact statements" (Zach, 1997).
- *BMR* refers to the *"basal metabolic rate"* which is the amount of energy that is expended at rest in a neutral environment after the digestive system has been inactive for about 12 hours.
- *Catecholamines* are hormones produced by the brain, nerve tissues, and adrenal glands. They act both as neurotransmitters and hormones vital to the maintenance of homeostasis through the nervous system.
- *CD4 cells* are white blood cells that play an important role in the immune system.
- *CD8 + T cells* (often called *cytotoxic T lymphocytes*, or *CTLs*) are very important cells for immune defense against intracellular pathogens, including viruses and bacteria; as well as for tumor surveillance.
- A *diuretic* is any substance that promotes *diuresis*, the increased production of urine.

- *Euthyroidism* is the state of having normal thyroid gland function, with normal serum levels of TSH and T4, as opposed to *hyperthyroid* (overactive thyroid) or *hypothyroid* (underactive thyroid).
- *Fibrocystic breast disease*, commonly called *fibrocystic breasts* or *fibrocystic change*, is a benign (noncancerous) condition which gives the breasts a lumpy or ropelike texture.
- *Follicular cell*: An epithelial cell lining a follicle, such as that of the thyroid or ovary.
- *Gamma spectroscopy* is the technique of identifying and/or quantifying of radionuclides by analysis of the gamma-ray energy spectrum produced in a gamma-ray spectrometer. A gamma *ray*, also known as *gamma radiation*, is a penetrating form of electromagnetic radiation arising from the radioactive decay of atomic nuclei.
- The *GI tract*, also called *digestive tract* or *alimentary canal* is the tract from the mouth to the anus, through which food enters the body and solid wastes are expelled. The GI tract includes all the organs of the digestive system in humans and other animals.
- *Geochemobiological pathways* refers to the rock-soil-plant-animal/human pathway that describes the various ways through which humans and other animals can come in contact with elements originally derived from rocks (AAPG, 2019).
- *Graves' disease* is an immune system disorder that results in the overproduction of thyroid hormones (hyperthyroidism).
- *Homeostasis* is any self-regulating process by which living systems tend to maintain state of steady internal, physical, and chemical conditions that are best for survival.
- "The *hypothalamic-pituitary-adrenal axis* (*HPA axis* or *HTPA axis*) is a complex set of direct influences and feedback interactions among three components: the hypothalamus, the pituitary gland (a pea-shaped structure located below the thalamus), and the adrenal (also called "suprarenal") glands (small, conical organs on top of the kidneys)." [Wikipedia (2021b)].
- *Hypothyroxinemia* refers to a subnormal thyroxine concentration in the blood.
- *International Council for the Control of Iodine Deficiency Disorders (ICCIDD)*: This is "a nonprofit, nongovernment organization for the sustainable elimination of iodine deficiency and the promotion of optimal iodine nutrition worldwide" (ICCIDD Global Network, 2021).
- *Iodophors* are substances that have elemental iodine attached to a high-molecular-weight moiety.
- A *LOCA* is a mode of failure for a nuclear reactor; if not managed effectively, the results of a LOCA could result in reactor core damage.
- *Mammary dysplasia*: A group of conditions marked by changes in breast tissue that are benign (noncancerous).
- A *mucolytic* is any agent which dissolves thick mucus, used to help relieve breathing difficulties.
- *Myelination* is process that takes place when a substance called *myelin*, which is made up of fatty lipids and proteins, accumulates around nerve cells, or neurons, in the brain.
- *Myxedema* or *myxedema* generally refers to severe hypothyroidism caused by decreased activity of the thyroid gland in adults and characterized by dry skin, swellings around the lips and nose, mental retardation and a subnormal BMR.

- *Ontogenesis* refers to the course of development of an individual organism or anatomical or behavioral feature from the earliest stage to maturity. In the case of an organism, this usually engenders the time from fertilization of the egg to adult.
- *Omics technology*: The branches of science known informally as *omics* are various disciplines in biology whose names end in the suffix *-omics*, such as genomics, proteomics, metabolomics, metagenomics and transcriptomics.
- *Phylogenetically*: In a way that relates to the evolutionary development and diversification of a species or group of organisms.
- The *Sandell-Koltoff reaction*: "In 1937 Sandell and Kolthoff described a sensitive procedure for the determination of quantities of iodide of the order of $0.05-3 \, \gamma \, (\mu g)$ in 1 mL solution or in a suitable amount of solid sample (sodium chloride), known in the literature as the *Sandell-Kolthoff-reaction* or method (Sandell and Kolthoff, 1937)." (Haap et al., 2017).
- "*NIS* is the plasma membrane glycoprotein that mediates active iodide transport in the thyroid and other tissues, such as the salivary, gastric mucosa, rectal mucosa, bronchial mucosa, placenta and mammary glands." (Waugh, 2019). *SLC5A5* is a gene that provides instructions for making the protein, sodium (Na)-iodide symporter or NIS.
- The *transcriptome* is the complete set of *transcripts* in a specific type of cell or tissue.
- *Thyroglobulin* is a protein made by the follicular cells of the thyroid gland. It is used by the thyroid gland to produce T3 and T4.
- *Thyroiditis* is swelling (inflammation) of the thyroid gland, causing either unusually high or low levels of thyroid hormones in the blood; and often presenting as a result of thyroid gland autoimmunity.
- *Thyrotropin* is a substance produced by cells called thyrotrophs in the anterior pituitary gland; it binds to specific receptors on the surface of cells.
- *Transcripts* are multiple different RNAs produced from a copy of a gene's DNA sequence.
- *TSH* refers to thyroid stimulating hormone. A TSH test is a blood test that measures this hormone.
- The *Wolff-Chaikoff effect* is an effective means of rejecting the large quantities of iodide and therefore preventing the thyroid from synthesizing large quantities of thyroid hormones (Markou et al., 2001).

References

AAPG (American Association of Petroleum Geologists), 2019. Medical geology. https://wiki.aapg.org/Medical_geology. (Accessed 14.07.2021).

Aburto, N.J., Abudou, M., Candeias, V., Wu, T., 2014. Effect and safety of salt iodisation to prevent iodine deficiency disorders: a systematic review with meta-analyses. World Health Organisation. https://apps.who.int/iris/handle/10665/148175. (Accessed 08.07.2021).

Adebayo, O.A., Abiola, O.M., Samuel, O.A., Folake, J.K., Augustine, K.J., 2018. Evaluation of goitrogeniccontent of common vegetables in southwest Nigeria. Asian Food Science Journal 4 (1), 1−6. Available from: https://doi.org/10.9734/AFSJ/2018/42750.

Ahad, F., Ganie, S.A., 2010. Iodine, iodine metabolism and iodine deficiency disorders revisited. Indian Journal of Endocrinology and Metabolism 14 (1), 13–17.

Ahmad, S., Bailey, E.H., Arshad, M., Ahmed, S., Watts, M., Young, S., 2021. Multiple geochemical factors may cause iodine and selenium deficiency in Gilgit-Baltistan, Pakistan. Environmental Geochemistry and Health 43, 4493–4513. Available from: https://doi.org/10.1007/s10653-021-00936-9.

Aihara, K., Nishi, Y., Hatano, S., Kihara, M., Yoshimitsu, K., Takeichi, N., et al., 1984. Zinc, copper, manganese, and selenium metabolism in thyroid disease. American Journal of Clinical Nutrition 40 (1), 26–35. Available from: https://doi.org/10.1093/ajcn/40.1.26.

Ajema, D., Bekele, M., Yihune, M., Tadesse, H., Gebremichael, G., Mengesha, M.M., 2020. Socio-demographic correlates of availability of adequate iodine in household salt: a community-based cross-sectional study. BMC Research Notes 13, 125. Available from: https://doi.org/10.1186/s13104-020-04983-w.

Amachi, S., 2008. Microbial contribution to global iodine cycling: volatilization, accumulation, reduction, oxidation, and sorption of iodine. Microbes and Environments 23 (4), 269–276. Available from: https://doi.org/10.1264/jsme2.me08548.

Anbar, M., Inbar, M., 1964. Studies on the accumulation of fluoroborate ions in the thyroid of mice. Acta Endocrinologica, European Journal of Endocrinology 46 (4), 639–642. Available from: https://doi.org/10.1530/acta.0.0460639.

Andersen, S., Petersen, S.B., Laurberg, P., 2002. Iodine in drinking water in Denmark is bound in humic substances. European Journal of Endocrinology 147, 663–670.

Andersen, S., Pedersen, K.M., Iversen, F., et al., 2008. Naturally occurring iodine in humic substances in drinking water in Denmark is bioavailable and determines population iodine intake. British Journal of Nutrition 99, 319–325.

Andersson, M., de Benoist, B., Delange, F., Zupan, J.WHO Secretariat, 2007. Prevention and control of iodine deficiency in pregnant and lactating women and in children less than 2-years-old: conclusions and recommendations of the Technical Consultation. Public Health Nutrition 10 (12A), 1606–1611. Available from: https://doi.org/10.1017/S1368980007361004.

Appiah, P.K., Yanbom, C.T., Ayanore, M.A., Bapula, A., 2020. Iodine content of salt use after years of universal iodization policy and Knowledge on Iodized Salt among Households in the Sissala East Municipality in Upper West Region of Ghana. Journal of Food Quality 2020, 9437365. Available from: https://doi.org/10.1155/2020/9437365.

Aronson, J.K. 2016. Iodine-containing medicaments. In: Aronson, J.K. (Editor), Meyler's Side Effects of Drugs, sixteenth ed., The International Encyclopaedia of Adverse Drug Reactions and Interactions, pp. 298 - 304. Elsevier. Available from: https://doi.org/10.1016/B978-0-444-53717-1.00911-2. (Accessed 30.04.2021).

Arthur, J.R., Beckett, G.J., 1999. Thyroid function. British Medical Bulletin 55 (3), 658–668. Available from: https://doi.org/10.1258/0007142991902538.

Ashworth, D.J., 2009. Transfers of Iodine in the Soil-Plant-Air System: Solid—Liquid Partitioning, Migration, Plant Uptake and Volatilization. In: Preedy, V.R., Burrow, G.N (Eds.).

ATSDR (Agency for Toxic Substances and Disease Registry), 2004. Chapter 6: Potential for human exposure: toxicological profile for iodine. CAS# 7553-56-2. http://www.atsdr.cdc.gov/toxprofiles/tp.asp?id = 479&tid = 85. (Accessed 27.05.2021).

Auzmendi-Murua, I., Castillo, Á., Bozzelli, J.W., 2014. Mercury oxidation via chlorine, bromine, and iodine under atmospheric conditions: thermochemistry and kinetics. The Journal of Physical chemistry, Part A 118 (16), 2959–2975. Available from: https://doi.org/10.1021/jp412654s.

Bagwell, C.E., Zhong, L., Wells, J.R., Mitroshkov, A.V., Qafoku, N.P., 2019. Microbial methylation of iodide in unconfined aquifer sediments at the Hanford Site, USA. Frontiers in Microbiology 10, 2460. Available from: https://doi.org/10.3389/fmicb.2019.02460.

Baker, A., Yodle, C., 2021. Measurement report: indirect evidence for the controlling influence of acidity on the speciation of iodine in Atlantic aerosols. Atmospheric Chemistry and Physics 21 (7), 13067–13076.

Betsy, A., Binitha, M., Sarita, S., 2013. Zinc deficiency associated with hypothyroidism: an overlooked cause of severe alopecia. International Journal of Trichology 5 (1), 40−42. Available from: https://doi.org/10.4103/0974-7753.114714.

BGS (British Geological Survey), 2000. Water quality fact sheet: iodine. https://www.ign.org/cm_data/2000_Iodine_in_water_Br_geological_survey.pdf. (Accessed 21.05.2021).

Bilal, M.Y., Dambaeva, S., Kwak-Kim, J., Gilman-Sachs, A., Beaman, K.D., 2017. A role for iodide and thyroglobulin in modulating the function of human immune cells. Frontiers in Immunology 8, 1573. Available from: https://doi.org/10.3389/fimmu.2017.01573.

Bimenya, G.S., Olico-Okui, Kaviri, D., Mbona, N., Byarugaba, W., 2002. Monitoring the severity of iodine deficiency disorders in Uganda. African Health Sciences 2 (2), 63−68.

Borst Pauwels, G.W.F.H., 1961. Iodine as a micronutrient for plants. Plant and Soil 14, 377−392. Available from: https://doi.org/10.1007/BF01666295.

Bostock, A., Shaw, G., Bell, J.N., 2003. The volatilisation and sorption of (129)I in coniferous forest, grassland and frozen soils. Journal of Environmental Radioactivity 70 (1 - 2), 29−42.

Bouga, M., Lean, M.E.J., Combet, E., 2018. Contemporary challenges to iodine status and nutrition: the role of foods, dietary recommendations, fortification and supplementation. Proceedings of the Nutrition Society 77 (3), 302−313. Available from: https://doi.org/10.1017/S0029665118000137.

Bourdoux, P., Delange, F., Gerard, M., Mafuta, M., Hanson, A., Ermans, A.M., 1978. Evidence that cassava ingestion increases thiocyanate formation: a possible etiologic factor in endemic goiter. The Journal of Clinical Endocrinology and Metabolism 46 (4), 613−621. Available from: https://doi.org/10.1210/jcem-46-4-613.

Boyages, S.C., 2000a. The neuromuscular system and brain in hypothyroidism. In: Braverman, L.E., Utiger, R.D. (Eds.), Werner and Ingbar's the Thyroid: A Fundamental and Clinical Text, eighth ed. Lippincott-Raven, Philadelphia, PA, pp. 803−810.

Boyages, S.C., 2000b. The neuromuscular system and brain in thyrotoxicosis. In: Braverman, L.E., Utiger, R.D. (Eds.), Werner and Ingbar's the Thyroid: A Fundamental and Clinical Text, eighth ed Lippincott-Raven, Philadelphia, PA, pp. 631−633.

Braverman, L.E., 2020. Werner and Ingbar's The Thyroid: A Fundamental and Clinical Text, eleventh ed. Lippincott Williams and Wilkins (LWW), p. 912.

Bürgi, H., Siebenhüner, L., Miloni, E., 1984. Fluorine and thyroid gland function: a review of the literature. Wiener Klinische Wochenschrift 62 (12), 564−569. Available from: https://doi.org/10.1007/BF01728174.

Businge, C.B., Longo-Mbenza, B., Kengne, A.P., 2019. The prevalence of insufficient iodine intake in pregnancy in Africa: protocol for a systematic review and meta-analysis. Systematic Reviews 8, 209. Available from: https://doi.org/10.1186/s13643-019-1092-7.

Businge, C.B., Longo-Mbenza, B., Kengne, A.P., 2020. Iodine nutrition status in Africa: Potentially high prevalence of iodine deficiency in pregnancy even in countries classified as iodine sufficient. 24 (12) 3581−3586. Public Health Nutrition. Available from: https://doi.org/10.1017/S1368980020002384.

Calvert, J.G., Lindberg, S.E., 2004. The potential influence of iodine-containing compounds on the chemistry of the troposphere in the polar spring, II. Mercury depletion. Atmospheric Environment 38 (30), 5105−5116.

Calvert, J.G., Lindberg, S.E., 2005. Mechanisms of mercury removal by O_3 and OH in the atmosphere. Atmospheric Environment 39 (18), 3355−3367. Available from: https://doi.org/10.1016/j.atmosenv.2005.01.055.

Carpenter, L.J., 2003. Iodine in the marine boundary layer. Chemical Reviews 103 (12), 4953−4962. Available from: https://doi.org/10.1021/cr0206465.

Carpenter, L.J., Malin, G., Liss, P.S., Küpper, F.C., 2000. Novel biogenic iodine-containing trihalomethanes and other short-lived halocarbons in the coastal east Atlantic. Global Biogeochemical Cycles 14 (4), 1191−1204.

Carpenter, L.J., MacDonald, S.M., Shaw, M.D., Kumar, R., Saunders, R.W., Parthipan, R., et al., 2013. Atmospheric iodine levels influenced by sea surface emissions of inorganic iodine. Nature Geoscience 6, 108−111. Available from: https://doi.org/10.1038/ngeo1687.

Carpenter, L., Chance, R., Sherwen, T., Adams, T., Ball, S., Evans, M., et al., 2021. Marine iodine emissions in a changing world. Proceedings of the Royal Society A 477, 20200824. Available from: https://doi.org/10.1098/rspa.2020.0824.

Cavalieri, R.R., 1997. Iodine metabolism and thyroid physiology: current concepts. Thyroid 7 (2), 177−181.

Çelik, T., Savaş, N., Kurtoğlu, S., Sangün, Ö., Aydın, Z., Mustafa, D., et al., 2014. Iodine, copper, zinc, selenium and molybdenum levels in children aged between 6 and 12 years in the rural area with iodine deficiency and in the city center without iodine deficiency in Hatay. Turkish Archives of Paediatrics 49 (2), 111−116. Available from: https://doi.org/10.5152/tpa.2014.1209.

Chaker, L., Razvi, S., Bensenor, I.M., Fereidom, A., Pearce, E.N., Peeters, R.P., 2022. Hypothyroidism. Nature Reviews Disease Primers 8 (30). Available from: https://doi.org/10.1038/s41572-022-00357-7 (accessed 05.02.2024).

Chance, R., Baker, A.R., Carpenter, L., Jickells, T.D., 2014. The distribution of iodide at the sea surface. Environmental science: Processes and Impacts 16 (8), 1841−1859. Available from: https://doi.org/10.1039/c4em00139g.

Chance, R.J., Tinel, L., Sherwen, T., Baker, A.R., Bell, T., Brindle, J., et al., 2019. Global sea-surface iodide observations, 1967 - 2018. Scientific Data (Nature) 6, 286. Available from: https://doi.org/10.1038/s41597-019-0288-y.

Charlton, K., Ware, L.J., Baumgartner, J., Cockeran, M., Schutte, A.E., Naidoo, N., et al., 2018a. How will South Africa's mandatory salt reduction policy affect its salt iodisation programme? A cross-sectional analysis from the WHO-SAGE Wave 2 Salt and Tobacco study. BMJ Open 8 (3), e020404. Available from: https://doi.org/10.1136/bmjopen-2017-020404.

Charlton, K.E., Ware, L.J., Baumgartner, J., Cockeran, M., Schutte, A.E., Naidoo, N., et al., 2018b. Iodine status assessment in South African adults according to spot urinary iodine concentrations, prediction equations, and measured 24-h iodine excretion. Nutrients 10 (6), 736. Available from: https://doi.org/10.3390/nu10060736.

Chen, X., Liu, L., Yao, P., Yu, D., Hao, L., Sun, X., 2007. Effect of excessive iodine on immune function of lymphocytes and intervention with selenium. Journal of Huazhong University of Science and Technology (Medical Sciences) 27 (4), 422−425. Available from: https://doi.org/10.1007/s11596-007-0418-1.

Chung, H.R., 2014. Iodine and thyroid function. Annals of Pediatric Endocrinology and Metabolism 19 (1), 8−12. Available from: https://doi.org/10.6065/apem.2014.19.1.8.

Clark, M.B., Slayton, R.L., 2020. Fluoride use in caries prevention in the primary care setting. AAP Section on Oral Health: Pediatrics 146 (6), e2020034637.

Cortecci, G., Boschetti, T., Mussi, M., Herrera, L.C., Mucchino, C., Maurizio, B., 2005. New chemical and original isotopic data on waters from El Tatio geothermal field, northern Chile. Geochemical Journal (Geochemical Society of Japan) 39 (6), 547−571.

Costa, de O.G., Feiteira, F.N., de, M., Schuenck, H., Pacheco, W.F., 2018. Iodine determination in table salts by digital images analysis. Analytical Methods 10, 4463. Available from: https://doi.org/10.1039/C8AY01248B.

Crewdson, E., Smedley, P.L., MacDonald, A.M., 2019 Iodine in drinking water from East African groundwater sources. [Poster] In: Early Career Hydrogeologists' Conference 2019, Leeds, UK, October 25, 2019. British Geological Survey (Unpublished).

Dai, J.L., Zhang, M., Hu, Q.H., Huang, Y.Z., Wang, R.Q., Zhu, Y.G., 2009. Adsorption and desorption of iodine by various Chinese soils: II. Iodide and iodate. Geoderma 153 (1-2), 130−135. Available from: https://doi.org/10.1016/j.geoderma.2009.07.020.

Davies, T.C., 2008. Environmental health impacts of East African rift volcanism. Environmental Geochemistry and Health 30 (4), 325−338. Available from: https://doi.org/10.1007/s10653-008-9168-7.

Delange, F., Ahiuwalia, R., 1983. Cassava toxicity and thyroid: research and public health issues. In: Proceedings of a Workshop Held in Ottawa, Canada, May 31−June 02, 1982, Ottawa, Ontario, IDRC, CA

IDRC-207e 1983. 148 p. https://idl-bnc-idrc.dspacedirect.org/bitstream/handle/10625/20837/IDL-20837. pdf?sequence = 1. (Accessed 10.07.2021).

Derumeaux, H., Valeix, P., Castetbon, K., Bensimon, M., Boutron-Ruault, M.C., Arnaud, J., et al., 2003. Association of selenium with thyroid volume and echostructure in 35- to 60-year-old French adults. European Journal of Endocrinology 148 (3), 309−315. Available from: https://doi.org/10.1530/eje.0.1480309.

Dida, N., Legese, A., Aman, A., Muhamed, B., Damise, T., Birhanu, T., et al., 2020. Availability of adequately iodised salt at household level and its associated factors in Robe town, Bale Zone, South East Ethiopia: community-based cross-sectional study. South African Journal of Clinical Nutrition 33 (3), 58−63. Available from: https://doi.org/10.1080/16070658.2018.1551767.

Dissanayake, C.B., Chandrajith, R.L.R., 1993. Geochemistry of endemic goitre, Sri Lanka. Applied Geochemistry 8 (Suppl. 2), 211−213.

Drewnowski, A., Garrett, G.S., Kansagra, R., Khan, N., Kupka, R., Kurpad, A.V., et al., 2021. Double fortified salt consultation (DFS) Steering Group. Key considerations for policymakers - iodized salt as a vehicle for iron fortification: current evidence, challenges, and knowledge gaps. Journal of Nutrition 151 (Supplement_1), 64S−73S. Available from: https://doi.org/10.1093/jn/nxaa377.

Druzian, G.T., Nascimento, M.S., Cerqueira, U.M.F.M., Novaes, C.G., Bezerra, M.A., Duarte, F.A., et al., 2021. Determination of Cl, Br and I in granola: development of an accurate analytical method using ICP-MS. Food Chemistry 344, 128677. Available from: https://doi.org/10.1016/j.foodchem.2020.128677.

Duan, L., Wang, W., Sun, Y.,, Sun, I., Zhang, C., 2016. Iodine in groundwater of the Guanzhong Basin, China: sources and hydrogeochemical controls on its distribution. Environmental Earth Sciences 75, 970. Available from: https://doi.org/10.1007/s12665-016-5781-4.

Duborská, E., Urík, M., Šeda, M., 2020. Iodine biofortification of vegetables could improve iodine supplementation status. Agronomy 10 (10), 1574. Available from: https://doi.org/10.3390/agronomy10101574.

Duressa, T.F., Mohammed, A.Y., Feyissa, G.R., Tufa, L.T., Siraj, K., 2014. Comparative analysis of iodine concentration in water, soil, cereals and table salt of Horaboka, Mio and Besaso Towns of Bale Robe, South East Ethiopia. Journal of Environment Pollution and Human Health 2 (1), 27−33. Available from: https://doi.org/10.12691/jephh-2-1-6.

Eastman, C.J. and Zimmermann, M.B., (2018). The iodine deficiency disorders. In: Feingold, K.R., Anawalt, B., Boyce, A., Chrousos, G., de Herder, W.W., Dhatariya, K., Dungan, K., et al. (Eds.), Endotext (Internet): Table 5, Dietary Goitrogens. South Dartmouth (MA). https://www.ncbi.nlm.nih.gov/books/NBK285556/. (Accessed 31.05.2021).

Eftekhari, M.H., Simondon, K.B., Jalali, M., Keshavarz, S.A., Elguero, E., Eshraghian, M.R., et al., 2006. Effects of administration of iron, iodine and simultaneous iron-plus-iodine on the thyroid hormone profile in iron-deficient adolescent Iranian girls. European Journal of Clinical Nutrition 60, 545−552. Available from: https://doi.org/10.1038/sj.ejcn.1602349.

El-Bashir, J.M., Abbiyesuku, F.M., Aliyu, I.S., Randawa, A.J., Adamu, R., Akuyam, S.A., et al., 2016. Nutritional iodine status of pregnant women in Zaria, North-Western Nigeria. Sub-Saharan African Journal of Medicine 3 (1), 41−44.

Elias, E., Tsegaye, W., Stoecker, B.J., Gebreegziabhe, T., 2021. Excessive intake of iodine and low prevalence of goiter in school age children five years after implementation of national salt iodization in Shebedino woreda, southern Ethiopia. BMC Public Health 21, 165. Available from: https://doi.org/10.1186/s12889-021-10215-y.

Ershow, A.G., Skeaff, S.A., Merkel, J.M., Pehrsson, P.R., 2018. Development of databases on iodine in foods and dietary supplements. Nutrients 10 (100), 1−20.

Ertek, S., Cicero, A.F., Caglar, O., Erdogan, G., 2010. Relationship between serum zinc levels, thyroid hormones and thyroid volume following successful iodine supplementation. Hormones (Athens, Greece) 9 (3), 263−268. Available from: https://doi.org/10.14310/horm.2002.1276.

Eskin, B.A., 1977. Iodine and mammary cancer. Advances in Experimental and Tropical Medicine 91, 293−304.

EVM (Expert Group on Vitamins and Minerals), 2003. Risk Assessments: Iodine. Expert Group on Vitamins and Minerals. http://cot.food.gov.uk/sites/default/files/vitmin2003.pdf. (Accessed 28.05.2021).

FAO/WHO (Food and Agricultural Organisation/World Health Organisation), 2001. Human vitamin and mineral requirements. Chapter 12: Iodine. Expert Consultation on Human Vitamin and Mineral Requirements. Food and Nutrition Division, Rome. Available from: http://www.fao.org/3/Y2809E/y2809e0i.htm (Accessed 21.05.2021).

Fisher, D.A., Oddie, T.H., 1969. Thyroid iodine content and turnover in euthyroid subjects - validity of estimation of thyroid iodine accumulation from shortterm clearance studies. Journal of Clinical Endocrinology and Metabolism 29, 721−727.

Fisher, D.A., Delange, F.M., 1998. Thyroid hormone and iodine requirements in man during brain development. In: Stanbury, J.B., Delange, F.M., Dunn, J.T., Pandav, C.S. (Eds.), Iodine in Pregnancy. Oxford University Publication, New Delhi.

Franco, J.S., Amaya-Amaya, J., Anaya, J.M., 2013. Chapter 30: Thyroid disease and autoimmune diseases. In: Anaya, J.M., Shoenfeld, Y., Rojas-Villarraga, A., et al.,Autoimmunity: From Bench to Bedside [Internet]. El Rosario University Press, Bogota (Colombia). Available from: https://www.ncbi.nlm.nih.gov/books/NBK459466/ (Accessed 11.07.2021).

Fuge, R., 1989. Iodine in waters: possible links with endemic goitre. Applied Geochemistry 4 (2), 203−208. Available from: https://doi.org/10.1016/0883-2927(89)90051-6.

Fuge, R., 2005. Soils and iodine deficiency. In: Selinus, O., Alloway, B., Centeno, J.A., Finkelman, R.B., Fuge, R., Lindh, U., et al.,Essentials of Medical Geology. Elsevier Academic Press, New York, NY, pp. 417−433.

Fuge, R., and Ander, E., 1998. Geochemical barriers and the distribution of iodine in the secondary environment: Implications for radio-iodine. In: Nicholson, K. (Editor), Energy and the Environment—Geochemistry of Fossil, Nuclear and Renewable Resources. MacGregor Science, pp. 163−170.

Fuge, R., Johnson, C.C., 1986. The geochemistry of iodine—a review. Environmental Geochemistry and Health 8 (2), 31−54. Available from: https://doi.org/10.1007/BF02311063.

Fuge, R., Johnson, C.C., 2015. Iodine and human health, the role of environmental geochemistry and diet: a review. Applied Geochemistry 63, 282−302.

Gad, S.C., 2014. Iodine. In: Wexler, P. (Ed.), Encyclopaedia of Toxicology, third ed. Elsevier.

Gaitan, E., 2004. Goitrogens, environmental. In: Martini, L., 1n: Martini, L. (Eds.), The Encyclopedia of Endocrine Diseases. Academic Press, pp. 286−294.

Galofre, J.C., Fernandez-Calvet, L., Rios, M., García-Mayor, R.V., 1994. Increased incidence of thyrotoxicosis after iodine supplementation in an iodine sufficient area. Journal of Endocrinological Investigation 17, 23−27.

GEO (Group on Earth Observations), 2016. 2017−2019 Work Programme: African geochemical baselines. GEO-XIII-5.3, pp. 12−13. https://www.earthobservations.org/documents/geo_xiii/GEO-XIII-5-3_2017-19_Work%20Programme.pdf. (Accessed 07.01.2021).

Gilfedder, B.S., Lai, S.C., Petri, M., Biester, H., Hoffmann, T., 2007. Iodine speciation in rain, snow and aerosols. Atmospheric Chemistry and Physics 8, 6069−6084.

Gong, T., Zhang, X., 2013. Determination of iodide, iodate and organo-iodine in waters with a new total organic iodine measurement approach. Water Research 47 (17), 6660−6669. Available from: https://doi.org/10.1016/j.watres.2013.08.039.

Gonzales, J., Tymon, T., Küpper, F.C., Edwards, M.S., Carrano, C.J., 2017. The potential role of kelp forests on iodine speciation in coastal seawater. PLoS One 12 (8), e0180755. Available from: https://doi.org/10.1371/journal.pone.0180755.

Gorstein, J., Bagriansky, J., Pearce, E., Kupka, R., Zimmermann, M., 2020. Estimating the health and economic benefits of universal salt iodisation programmes to correct iodine deficiency disorders. Thyroid: Official Journal of the American Thyroid Association 30, 1802−1809.

Gyamfi, D., Wiafe, Y.A., Yaw, A., Ofori Awuah, E., Adu, E.A., Boadi, E.K., 2020. Goitre prevalence and urinary iodine concentration in school-aged children in the Ashanti Region of Ghana. International Journal of Endocrinology 2020, 3759786. Available from: https://doi.org/10.1155/2020/3759786.

Haap, M., Roth, H.J., Huber, T., Dittmann, H., Wahl, R., 2017. Urinary iodine: comparison of a simple method for its determination in microplates with measurement by inductively-coupled plasma mass spectrometry. Scientific Reports 7, 39835. Available from: https://doi.org/10.1038/srep39835.

Hanegraaf, T.A.C., McGill, P.E., 1974. Thyroid diseases: population based studies of endemic goitre. In: Vogel, L.C., Muller, A.S., Odingo, R.S., Onyango, Z., de Geus, A. (Eds.), Health and Disease in Kenya. East African Literature Bureau, Nairobi, Kenya, pp. 395−403.

Hays, M.T., 2001. Estimation of total body iodine content in normal young men. Thyroid (The Official Journal of The American Thyroid Association) 11 (7), 671−675. Available from: https://doi.org/10.1089/105072501750362745.

Hjerpe, T., Ikonen, A.T.K., Broed, R., 2010. Biosphere assessment report 2009. Posiva Report 2010−03.

Horn-Ross, P.L., Morris, J.S., Lee, M., West, D.W., Whittemore, A.S., McDougall, I.R., et al., 2001. Iodine and thyroid cancer risk among women in a multiethnic population: The Bay Area Thyroid Cancer Study. Cancer Epidemiology Biomarkers and Prevention 10 (9), 979−985.

Hou, X., Ding, W., 2009. Chapter 46: Isotopes of iodine in thyroid and urine: source, application, level and determination. In: Preedy, V., Burrow, G., Watson, R. (Eds.), Comprehensive Handbook of Iodine, first ed., pp. 437−448. https://doi.org/10.1016/B978-0-12-374135-6.00046-7.

Houston, R., Tsang, B.L., Gorstein, J., 2021. The double-fortified salt (iodized salt with iron) consultation: a process for developing evidence-based considerations for countries. The Journal of Nutrition 151 (Supplement_1), 1S−2S. Available from: https://doi.org/10.1093/jn/nxaa157.

Hu, Q., Zhao, P., Moran, J.E., Seaman, J.C., 2005. Sorption and transport of iodine species in sediments from the Savannah River and Hanford Sites. Journal of Contaminant Hydrology 78 (3), 185−205.

Hu, Q., Moran, J., Gan, J., 2012. Sorption, degradation and transport of methyl iodide and other iodine species in geologic media. Applied Geochemistry 27 (3), 774−781. Available from: https://doi.org/10.1016/j.apgeochem.2011.12.022.

Humphrey, O.S., Young, S.D., Bailey, E.H., Crout, N.M.J., Ander, E.L., Watts, M.J., 2018. Iodine soil dynamics and methods of measurement: a review. Environmental Science: Processes and Impacts 20 (2), 288−310.

Hussain, H., Selamat, R., Kuay, L.K., Zain, F., Jalaludin, M.Y., 2019. Urinary iodine: biomarker for population iodine nutrition. Biochemical Testing—Clinical correlation and Diagnosis [Working Title]. Available from: http://doi.org/10.5772/intechopen.84969, https://www.intechopen.com/books/biochemical-testing-clinical-correlation-and-diagnosis/urinary-iodine-biomarker-for-population-iodine-nutrition.

ICCIDD Global Network, 2021. International Council for the Control of Iodine Deficiency Disorders. https://iccidd.org/. (Accessed 18.07.2021).

Ilin, A., Nersesyan, A., 2013. Toxicology of iodine: a mini review. Archive of Oncology 21, 65−71.

IOM (Institute of Medicine), 2001. Dietary Reference Intakes for Vitamin A, Vitamin K, Arsenic, Boron, Chromium, Copper, Iodine, Iron, Manganese, Molybdenum, Nickel, Silicon, Vanadium, and zinc. Washington (DC): National Academy Press, Institute of Medicine, Food and Nutrition Board. http://www.nap.edu/download.php?record_id = 10026#. (Accessed 28.05.2021).

Ittermann, T., Albrecht, D., Arohonka, P., Bilek, R., de Castro, J.J., Dahl, L., et al., 2020. Standardized map of iodine status in Europe. Thyroid (The Official Journal of: American Thyroid Association) 30 (9), 1346−1354.

Jauregui-Osoro, M., Sunassee, K., Weeks, A.J., Berry, D.J., Rowena, L.P., et al., 2010. Synthesis and biological evaluation of [18F] tetrafluoroborate: a PET imaging agent for thyroid disease and reporter gene imaging of the sodium/iodide symporter. European Journal of Nuclear Medicine and Molecular Imaging 37, 2108−2116. Available from: https://doi.org/10.1007/s00259-010-1523-0.

JECFA (Joint FAO/WHO Expert Committee on Food Additives), 1989. Toxicological Evaluation of Certain Food Additives and Contaminants. WHO Food Additives Series, No. 24. World Health Organisation, Geneva, pp. 267−294.

Jensen, H., Orth, B., Reiser, R., Bürge, D., Lehto, N.J., Almond, P., et al., 2019. Environmental parameters affecting the concentration of iodine in New Zealand pasture. Journal of Environmental Quality 48, 1517−1523. Available from: https://doi.org/10.2134/jeq2019.03.0128.

Jerše, A., Amlund, H., Rasmussen, R.R., Sloth, J.J., 2021. Iodine determination in animal feed by inductively coupled plasma mass spectrometry − results of a collaborative study. Food Additives and Contaminants: Part A 38 (2), 261−267. Available from: https://doi.org/10.1080/19440049.2020.1856942.

Jiang, H., DeGrado, T.R., 2018. [18F]Tetrafluoroborate ([18F]TFB) and its analogs for PET imaging of the sodium/iodide symporter. Theranostics 8 (14), 3918−3931. Available from: https://doi.org/10.7150/thno.24997.

Johnson, C.C., 1980. The geochemistry of iodine and a preliminary investigation into its potential use as a pathfinder element in geochemical exploration (Ph.D. thesis). University College of Wales, Aberystwyth.

Johanson, K.J., 2000. Iodine in soil. Technical Report TR-00−21. Uppsala. https://www.osti.gov/etdeweb/servlets/purl/20153356. (Accessed 04.05.2021).

Johnson, C.C., 2003. The Geochemistry of Iodine and its Application to Environmental Strategies for Reducing the Risks of Iodine deficiency Disorders. British Geological Survey Commissioned Report Number CR/03/057N, DfID KAR Project R7411. file:///C:/Users/theo%20davies/Downloads/The_Geochemistry_of_Iodine_and_its_Application_to_%20(1).pdf. (Accessed 21.05.2021).

Jones, M.R., Chance, R., Dadic, R., Hannula, H.R., May, R., Ward, M., Carpenter, L.J., 2023. Environmental iodine speciation quantification in seawater and snow using ion exchange chromatography and UV spectrophotometric detection. Analytica Chimica Acta 1239, 340700. Available from: https://doi.org/10.1016/j.aca.2022.340700. (Accessed 07.02.2021).

Jooste, P.L., Strydom, E., 2010. Methods for determination of iodine in urine and salt. Best Practice and Research Clinical Endocrinology and Metabolism 24 (1), 77−88. Available from: https://doi.org/10.1016/j.beem.2009.08.006.

Jooste, P.L., Zimmermann, M.B., 2008. Progress towards eliminating iodine deficiency in South Africa. South African Journal of Clinical Nutrition 21 (1), 8−14. Available from: https://doi.org/10.1080/16070658.2008.11734145.

Jooste, P.L., Upson, N., Charlton, K.E., 2005. Knowledge of iodine nutrition in the South African adult population. Public Health Nutrition 8 (4), 382−386. Available from: https://doi.org/10.1079/phn2004696.

Judprasong, K., Jongjaithet, N., Chavasit, V., 2016. Comparison of methods for iodine analysis in foods. Food Chemistry 193, 12−17. Available from: https://doi.org/10.1016/j.foodchem.2015.04.058.

Kabthymer, R.H., Shaka, M.F., Ayele, G.M., Malako, B.G., 2021. Systematic review and meta-analysis of iodine deficiency and its associated factors among pregnant women in Ethiopia. BMC Pregnancy and Childbirth 21 (1), 106. Available from: https://doi.org/10.1186/s12884-021-03584-0.

Kassim, I.A., Moloney, G., Busili, A., Nur, A.Y., Paron, P., Jooste, P., et al., 2014. Iodine intake in Somalia is excessive and associated with the source of household drinking water. Journal of Nutrition 144 (3), 375−381. Available from: https://doi.org/10.3945/jn.113.176693.

Kelly, F.C., Snedden, W.W., 1960. Prevalence and geographical distribution of endemic goitre. Endemic Goitre. World Health Organization Publication, Geneva, pp. 27−234.

Kiferle, C., Gonzali, S., Holwerda, H.T., Ibaceta, R.R., Perata, P., 2013. Tomato fruits: a good target for iodine biofortification. Frontiers in Plant Science 4, 205. Available from: https://doi.org/10.3389/fpls.2013.00205.

Kiferle, C., Martinelli, M., Salzano, A.M., Gonzali, S., Beltrami, S., Salvadori, P.A., et al., 2021. Evidences for a nutritional role of iodine in plants. Frontiers in Plant Science 12, 616868. Available from: https://doi.org/10.3389/fpls.2021.616868.

Kim, T., Kim, M., Kim, D., Jung, S.-H., Yeon, J.-W., 2018. Concentration determination of volatile molecular iodine and methyl iodide. Bulletin of the Korean Chemical Society 39, 824−828. Available from: https://doi.org/10.1002/bkcs.11484.

Koch-Steindl, H., Pröhl, G., 2001. Considerations on the behaviour of long-lived radionuclides in the soil. Radiation and Environmental Biophysics 40 (2), 93−104. Available from: https://doi.org/10.1007/s004110100098.

Kohlmeier, M., 2003. Iodine. In: Kohlmeier, M. (Ed.), Food Science and Technology: Nutrient Metabolism. Academic Press, London, pp. 712−718. Available from: https://doi.org/10.1016/B978-012417762-8.50101-0.

Köhrle, J., 2015. Selenium and the thyroid. Current Opinion in Endocrinology and Diabetes and Obesity 22 (5), 392−401. Available from: https://doi.org/10.1097/MED.0000000000000190.

Kolb, C., 2002. Iodine's air of importance. Nature 417, 597−598. Available from: https://doi.org/10.1038/417597a.

Kolonel, L.N., Hankin, J.H., Wilkens, L.R., Fukunaga, F.H., Winds, M.W., 1990. An epidemiologic study of thyroid cancer in Hawaii. Cancer Causes and Control 1, 223−234. Available from: https://doi.org/10.1007/BF00117474.

Korobova, E., Anoshko, Y., Kesminiene, A., Kouvyline, A., Romanov, S., Tenet, V., et al., 2010. Evaluation of stable iodine status of the areas affected by the Chernobyl accident in an epidemiological study in Belarus and the Russian Federation. Journal of Geochemical Exploration 107 (2), 124−135. Available from: https://doi.org/10.1016/j.gexplo.2010.08.005.

Kostial, K., Vnučec, M., Tominac, Č., Šimonović, I., 1980. A method for a simultaneous decrease of strontium, caesium and iodine retention after oral exposure in rats. International Journal of Radiation Biology and Related Studies in Physics, Chemistry, and Medicine 37 (3), 347−350. Available from: https://doi.org/10.1080/09553008014550411.

Kostial, K., Kargacin, B., Rabar, I., Blanusa, M., Maljković, T., Matković, V., et al., 1981. Simultaneous reduction of radioactive strontium, caesium and iodine retention by single treatment in rats. Science of the Total Environment (1), 1−10. Available from: https://doi.org/10.1016/0048-9697(81)90076-0.

Krela-Kaźmierczak, I., Czarnywojtek, A., Skoracka, K., Rychter, A.M., Ratajczak, A.E., Szymczak-Tomczak, A., et al., 2021. Is there an ideal diet to protect against iodine deficiency? Nutrients 13 (2), 513. Available from: https://doi.org/10.3390/nu13020513.

Krohn, K., Führer, D., Bayer, Y., Eszlinger, M., Brauer, V., Neumann, S., et al., 2005. Molecular patho-genesis of euthyroid and toxic multinodular goiter. Endocrine Reviews 26, 504−524. Available from: https://doi.org/10.1210/er.2004-0005.

L'Annunziata, M.F. (Ed.), 2020. Handbook of Radioactivity Analysis, Volume 2: Radioanalytical Applications. fourth ed. Elsevier Inc. Available from: https://doi.org/10.1016/C2016-0-04811-8.

Landini, M., Gonzali, S., Perata, P., 2011. Iodine biofortification in tomato. Journal of Plant Nutrition and Soil Science 174, 480−486. Available from: https://doi.org/10.1002/jpln.201000395.

Larsen, P.R., Davies, T.F., Hay, I.D., 1998. The thyroid gland. In: Wilson, J.D., Foster, D.W., Kronenberg, H.M., et al., (Eds.), Williams Textbook of Endocrinology. W.B. Saunders Company, Philadelphia (PA), pp. 390−515.

Lehr, J.J., Wybenga, J.M., Rosanow, M., 1958. Iodine as a micronutrient for tomatoes. Plant Physiology 33 (6), 421−427. Available from: https://doi.org/10.1104/pp.33.6.421.

Li, M., Eastman, C.J., 2012. The changing epidemiology of iodine deficiency. Nature Reviews Endocrinology 8 (7), 434−440. Available from: https://doi.org/10.1038/nrendo.2012.43.

Li, J., Wang, Y., Xie, X., Zang, L., Guo, W., 2013. Hydrogeochemistry of high iodine groundwater: a case study at the Datong Basin, northern China. Environmental Science: Processes and Impacts 15 (4), 848−859.

Li, J., Zhou, H., Wang, Y., Xie, X., Qian, K., 2017. Sorption and speciation of iodine in groundwater system: the roles of organic matter and organic-mineral complexes. Journal of Contaminant Hydrology 201, 39−47. Available from: https://doi.org/10.1016/j.jconhyd.2017.04.008.

Ligowe, I.S., Phiri, F.P., Ander, E.L., Bailey, E.H., Chilimba, A., Gashu, D., et al., 2020. Selenium deficiency risks in sub-Saharan African food systems and their geospatial linkages. Proceedings of the Nutrition Society 79 (4), 457−467. Available from: https://doi.org/10.1017/S0029665120006904.

Lin, Y.S., Rothen, M.L., Milgrom, P., 2018. Pharmacokinetics of iodine and fluoride following application of an anticaries varnish in adults. JDR Clinical and Translational Research 3 (3), 238−245. Available from: https://doi.org/10.1177/2380084418771930.

Lisco, G., De Tullio, A., Triggiani,, D., Zupo, R., Giagulli, V.A., De Pergola, G., et al., 2023. Iodine deficiency and iodine prophylaxis: an overview and update. Nutrients 15 (1), 1004. Available from: https://doi.org/10.3390/nu15041004. (Accessed 05.02.2021).

Liu, Y., Su, L., Xiao, H., 2017. Review of factors related to the thyroid cancer epidemic. International Journal of Endocrinology 2017, 5308635. Available from: https://doi.org/10.1155/2017/5308635.

Llop, S., Lopez-Espinosa, M., Murcia, M., Alvarez-Pedrerol, M., Vioque, J., Aguinagalde, X., et al., 2015. Synergism between exposure to mercury and use of iodine supplements on thyroid hormones in pregnant women. Environmental Research 138, 298−305. Available from: https://doi.org/10.1016/J.ENVRES.2015.02.026.

Mabasa, E., Mabapa, N.S., Jooste, P.L., Mbhenyane, X.G., 2019. Iodine status of pregnant women and children age 6 to 12 years feeding from the same food basket in Mopani district, Limpopo province, South Africa. South African Journal of Clinical Nutrition 32 (3), 76−82. Available from: https://doi.org/10.1080/16070658.2018.1449370.

Marczenko, Z., Balcerzak, M., 2000. Chapter 25: Iodine. Analytical Spectroscopy Library 10, 222−225. Available from: https://doi.org/10.1016/S0926-4345(00)80089-9.

Markou, K., Georgopoulos, N., Kyriazopoulou, V., Vagenakis, A.G., 2001. Iodine-induced hypothyroidism. Thyroid: Official Journal of the American Thyroid Association 11 (5), 501−510. Available from: https://doi.org/10.1089/105072501300176462.

Martin, P.G., Jones, C.P., Cipiccia, S., Batey, D.J., Hallam, K.R., Satou, Y., 2020. Compositional and structural analysis of Fukushima-derived particulates using high-resolution x-ray imaging and synchrotron characterisation techniques. Scientific Reports 10, 1636.

Mascarenhas, A.K., 2021. Is fluoride varnish safe? Validating the safety of fluoride varnish. Journal of the American Dental Association 152 (5), 364−368. Available from: https://doi.org/10.1016/j.adaj.2021.01.013.

Maumené, E., 1854. Experience pour determiner l'action des fluores sur l'economie animale. Comptes rendus de l'Académie des Sciences 39, 538.

McFarlane, J., Wren, J.C., R. Lemire, R., 2002. Chemical speciation of iodine source term to containment. Nuclear Technology 138, 162−178.

Medrano-Macías, J., Leija-Martinez, S., González-Morales, Maldonado, A.J., 2016a. Use of iodine to biofortify and promote growth and stress tolerance in crops. Frontiers in Plant Science 7, 1146.

Leija-Martínez, P., JuárezMaldonado, A., Rocha-Estrada, A., Benavides-Mendoza, A., 2016b. Effect of iodine application on antioxidants in tomato seedlings. Revista Chapingo Serie Horticultura 22 (2), 133−143. Available from: https://doi.org/10.5154/r.rchsh.2015.12.025.

Menyanu, E., Corso, B., Minicuci, N., Rocco, I., Zandberg, L., Baumgartner, J., et al., 2021. Salt-reduction strategies may compromise salt iodization programs: learnings from South Africa and Ghana. Nutrition (Burbank, Los Angeles County, Calif.) 84, 111065. Available from: https://doi.org/10.1016/j.nut.2020.111065.

Mika, A., Wątor, K., Kmiecik, E., Sekuła, K., 2019. Determination of iodine in geothermal water samples - preliminary ICP-MS method validation results. Water Supply 19 (4), 1264−1270. Available from: https://doi.org/10.2166/ws.2018.186.

Miller, M.A., 2017. Chapter 12: Goitrogens. In: Zachary, J.F. (Ed.), Pathologic Basis of Veterinary Disease, sixth ed., Endocrine System 1. pp. 682−723. . Available from: https://doi.org/10.1016/B978-0-323-35775-3.00012-6. (Accessed 10.07.2021).

Moran, J.E., Oktay, S.D., Santschi, P.H., 2002. Sources of iodine and iodine 129 in rivers. Water Resources Research 38, 1149. Available from: https://doi.org/10.1029/2001WR000622.

Moreda-Piñeiro, A., Romaris Hortas, V., Bermejo-Barrera, P., 2011. A review on iodine speciation for environmental, biological and nutrition fields. Journal of Analytical Atomic Spectrometry 26, 2107−2152. Available from: https://doi.org/10.1039/C0JA00272K.

Moreno-Reyes, R., 2021. The Little-known (but large) Problem of Iodine Deficiency in Europe. The Iodine Global Network. Available from: https://www.ign.org/the-iodine-blog-june-2021.htm. (Accessed 08.06.2021).

Morrison, R.T., Birkbeck, J.A., Evans, T.C., et al., 1963. Radioiodine uptake studies in newborn infants. Journal of Nuclear Medicine 4, 162−166.

Muramatsu, Y., Yoshida, S., Fehn, U., Amachi, S., Ohmomo, Y., 2004. Studies with natural and anthropogenic iodine isotopes: iodine distribution and cycling in the global environment. Journal of Environmental Radioactivity 74 (1-3), 221−232. Available from: https://doi.org/10.1016/j.jenvrad.2004.01.011.

Nauman, J., Wolff, J., 1993. Iodide prophylaxis in Poland after the Chernobyl reactor accident: benefits and risks. American Journal of Medicine 94, 524−532.

Negri, A.E., Niello, J.O.F., Wallner, A., Arazi, A., Fified, L.K., Tims, S.G., 2013. [129]I dispersion in Argentina: concentrations in fresh and marine water and deposition fluences in Patagonia. Environmental Science and Technology 47 (17), 9693−9698. Available from: https://doi.org/10.1021/es400610h.

Negro, R., Soldin, O.P., Obregon, M.J., Stagnaro-Green, A., 2011. Hypothyroxinemia and pregnancy. Endocrine Practice: Official Journal of the American College of Endocrinology and the American Association of Clinical Endocrinologists 17 (3), 422−429. Available from: https://doi.org/10.4158/EP10309.RA.

Nikulin, A., Potanina, O., Okuneva, M., Abramovich, R., Bokov, D., Smyslova, O., 2020. Development and validation of the quantitative determination procedure of iodine in the iodides form in the kelp thallus by the ionometry method. Journal of Pharmacy and Bioallied Sciences 12 (3), 277−283. Available from: https://doi.org/10.4103/jpbs.JPBS_198_20.

O'Doherty, J., Jauregui-Osoro, M., Brothwood, T., Szyszko, T., Marsden, P.K., O'Doherty, M.J., et al., 2017. 18F-Tetrafluoroborate, a PET probe for imaging sodium/iodide symporter expression: whole-body biodistribution, safety, and radiation dosimetry in thyroid cancer patients. Journal of Nuclear Medicine 58 (10), 1666−1671. Available from: https://doi.org/10.2967/jnumed.117.192252.

O'Dowd, C.D., Jimenez, J.L., Bahreini, R., Flagan, R.C., Seinfeld, J.H., Hämeri, K., et al., 2002. Marine aerosol formation from biogenic iodine emissions. Nature 417, 632−636. Available from: https://doi.org/10.1038/nature00775.

Ogborn, R.E., Waggener, R.E., VanHove, E., 1960. Radioactive-iodine concentration in thyroid glands of newborn infants. Paediatrics 26, 771−776.

Okosieme, O.E., 2006. Impact of iodination on thyroid pathology in Africa. Journal of the Royal Society of Medicine 99 (8), 396−401. Available from: https://doi.org/10.1258/jrsm.99.8.396.

Olson, K.R. (Ed.), 2004. Poisoning and Drug Overdose. fourth ed. Lange Medical Books/McGraw-Hill, New York, NY, p. 226.

Patrick, L., 2008. Iodine: deficiency and therapeutic considerations. Alternative Medicine Review 13, 116−127.

Pauling, L., 1960. The Nature of the Chemical Bond, third ed. Cornell University Press. Available from: file:///C:/Users/davies.theophilus/Downloads/Pauling_L_The_nature_of_the_chemical_bon.pdf. (Accessed 14.07.2021.

Pearce, E.N., Braverman, L.E., 2009. Environmental pollutants and the thyroid. Best Practice Research in Clinical Endocrinology and Metabolism 23, 801−813.

Pearce, E.N., Caldwell, K.L., 2016. Urinary iodine, thyroid function, and thyroglobulin as biomarkers of iodine status. The American Journal of Clinical Nutrition 104 (Supplement 3), 898S–901S. Available from: https://doi.org/10.3945/ajcn.115.110395.

Pesce, L., Kopp, P., 2014. Iodide transport: implications for health and disease. International Journal of Paediatric Endocrinology 2014 2014, 8.

Pfinder, M., Dreger, S., Christianson, L., Lhachimi, S.K., Zeeb, H., 2015. The effects of iodine blocking on thyroid cancer, hypothyroidism and benign thyroid nodules following nuclear accidents: A systematic review. Journal of Radiological Protection 36 (4), 112–130. Available from: https://doi.org/10.1088/0952-4746/36/4/R112.

Phillips, D.I., 1997. Iodine, milk, and the elimination of endemic goitre in Britain: the story of an accidental public health triumph. Journal of Epidemiology and Community Health 51, 391–393.

Pound, R.J., Brown, L.V., Evans, M.J., Carpenter, L.J., 2023. An improved estimate of inorganic iodine emissions from the ocean using a coupled surface microlayer box model. EGUsphere [preprint]. Available from: https://doi.org/10.5194/egusphere-2023-2447.

Prados-Roman, C., Cuevas, C.A., Fernandez, R.P., Kinnison, D.E., Lamarque, J.-F., Saiz-Lopez, A., 2015. A negative feedback between anthropogenic ozone pollution and enhanced ocean emissions of iodine. Atmospheric Chemistry and Physics 15, 2215–2224. Available from: https://doi.org/10.5194/acp-15-2215-2015.

Prothero, J.W., 2015. The Design of Mammals: A Scaling Approach. Cambridge University Press. Available from: https://doi.org/10.1017/CBO9781316275108.

Quarles (Jr.), C.D., Toms, A.D., Smith, R.D., Sullivan, P., Bass, D., Leone, J., 2020. Automated ICP-MS method to measure bromine, chlorine, and iodine species and total metals content in drinking water. Talanta Open 1, 100002. Available from: https://doi.org/10.1016/j.talo.2020.100002.

Raofie, F., Snider, G., Ariya, P.A., 2008. Reaction of gaseous mercury with molecular iodine, atomic iodine, and iodine oxide radicals—kinetics, product studies, and atmospheric implications. Canadian Journal of Chemistry 86, 811–820.

Ravichandran, R., Binukumar, J., Saadi, A.A., 2010. Estimation of effective half-life of clearance of radioactive Iodine (I) in patients treated for hyperthyroidism and carcinoma thyroid. Indian Journal of Nuclear Medicine (IJNM): The Official Journal of the Society of Nuclear Medicine, India 25 (2), 49–52. Available from: https://doi.org/10.4103/0972-3919.72686.

Ravichandran, R., Al Saadi, A., Al Balushi, N., 2014. Radioactive body burden measurements in (131)iodine therapy for differentiated thyroid cancer: effect of recombinant thyroid stimulating hormone in whole body (131)iodine clearance. World Journal of Nuclear Medicine 13 (1), 56–61. Available from: https://doi.org/10.4103/1450-1147.138576.

Redeker, K.R., Manley, S.L., Walser, M., Cicerone, R.J., 2004. Physiological and biochemical controls over methyl halide emissions from rice plants. Global Biogeochemical Cycles 18, GB1007. Available from: https://doi.org/10.1029/2003GB002042.

Reid, H., Bashammakh, A.A., Goodall, P., Landon, M.R., O'Connor, C., Sharp, B., 2008. Determination of iodine and molybdenum in milk by quadrupole ICP-MS. Talanta 75 (1), 189–197.

Reiners, C., Schneider, R., 2013. Potassium iodide (KI) to block the thyroid from exposure to I-131: current questions and answers to be discussed. Radiation and Environmental Biophysics 52 (2), 189–193.

Resende de Paiva, C., Grønhøj, C., Feldt-Rasmussen, U., von Buchwald, C., 2017. Association between Hashimoto's thyroiditis and thyroid cancer in 64,628 patients. Frontiers in Oncology 7, 53. Available from: https://doi.org/10.3389/fonc.2017.00053.

Ristic-Medic, D., Piskackova, Z., Hooper, L., Ruprich, J., Casgrain, A., Ashton, K., et al., 2009. Methods of assessment of iodine status in humans: a systematic review. American Journal of Clinical Nutrition 89 (6), 2052S–2069S. Available from: https://doi.org/10.3945/ajcn.2009.27230H.

Rohner, F., Zimmermann, M., Jooste, P., Pandav, C., Caldwell, K., Raghavan, R., et al., 2014. Biomarkers of nutrition for development—iodine review. The Journal of Nutrition 144 (8), 1322S–1342S. Available from: https://doi.org/10.3945/jn.113.181974.

Rohner, F., Kangambèga, M.O., Khan, N., Kargougou, R., Garnier, D., Sanou, I., et al., 2015. Comparative validation of five quantitative rapid test kits for the analysis of salt iodine content: laboratory performance, user- and field-friendliness. PLoS One 10 (9), e0138530. Available from: https://doi.org/10.1371/journal.pone.0138530.

Rumstadt, B., Klein, B., Kirr, H., Kaltenbach, N., Homenu, W., Schilling, D., 2008. Thyroid surgery in Burkina Faso, West Africa: experience from a surgical help programme. World Journal of Surgery 32, 2627–2630.

Saha, S., Abu, B.A.Z., Jamshidi-Naeini, Y., Mukherjee, U., Miller, M., Peng, L.-L., et al., 2019. Is iodine deficiency still a problem in sub-Saharan Africa?: A review. Proceedings of the Nutrition Society 78 (4), 554–566. Available from: https://www.cambridge.org/core/journals/proceedings-of-the-nutrition-society/article/is-iodine-deficiency-still-a-problem-in-subsaharan-africa-a-review/6C87E944AF05DEE3B7821D986D2F1B77. (Accessed 07.02.2024).

Samuel, A.R., Devi, G., 2015. Geographical distribution and occurrence of endemic goitre. Research Journal of Pharmacy and Technology 8, 973–978.

Sandell, E.B., Kolthoff, I.M., 1937. Micro determination of iodine by a catalytic method. Microchimica Acta 1, 9–25.

Santos-Oliveira, R., 2008. Radiopharmaceutical drug interactions. Revista de Salud Pública (Bogota) 10 (3), 477–487. Available from: https://doi.org/10.1590/s0124-00642008000300013.

Sayess, R., Eyring, A.M., Reckhow, D.A., 2021. Source and drinking water organic and total iodine and correlation with water quality parameters. Water Research 190, 116686. Available from: https://doi.org/10.1016/j.watres.2020.116686.

Schoendorfer, N.C., Davies, P.S.W., 2012. Chapter VII: Micronutrient interrelationships: synergism and antagonism. In: Betancourt, A.I., Gaitan, H.F. (Eds.), Micronutrients: Sources, Properties and Health Effects, Series on Nutrition and Diet Research Progress. Nova Science Publishers, New York, NY, United States, pp. 159–177.

Seal, A.J., Creeke, P.I., Gnat, D., Abdalla, F., Mirghani, Z., 2006. Excess dietary iodine intake in long-term African refugees. Public Health Nutrition 9 (1), 35–39. Available from: https://doi.org/10.1079/phn2005830.

Sellegri, K., Pey, J., Rose, C., Culot, A., Dewitt, H.L., Mas, S., et al., 2016. Evidence of atmospheric nanoparticle formation from emissions of marine microorganisms. Geophysical Research Letters 43, 6596–6603. Available from: https://doi.org/10.1002/2016GL069389.

Shelor, C.P., Dasgupta, P.K., 2011. Review of analytical methods for the quantification of iodine in complex matrices. Analytica Chimica Acta 702 (1), 16–36. Available from: https://doi.org/10.1016/j.aca.2011.05.039.

Sheppard, S.C., 1996. Importance of Chemical Speciation of Iodine in Relation to Dose Estimates from 129i. Report AECL-11669. Whiteshell Laboratories, Pinawa. https://www.osti.gov/etdeweb/servlets/purl/612934. (Accessed 01 May 2021).

Shimamoto, Y.S., Takahashi, Y., Terada, Y., 2011. Formation of organic iodine supplied as iodide in a soil-water system in Chiba, Japan. Environmental Science and Technology 45 (6), 2086–2092. Available from: https://doi.org/10.1021/es1032162. (Accessed 04.02.2024).

Sidibé, El. H., 2007a. Thyréopathies en Afrique subsaharienne [Thyroid diseases in sub-Saharan Africa]. Sante (Montrouge, France) 17 (1), 33–39.

Sidibé, El. H., 2007b. Réflexions sur le retard mental et le crétinisme de l'hypothyroïdie congénitale et de la carence des minéraux à l'état de traces [Reflections on mental retardation and congenital hypothyroidism: effects of trace mineral deficiencies]. Sante (Montrouge, France) 1, 41–50.

Singh, G., Correa, R., 2021. Methimazole. StatPearls [Internet]. StatPearls Publishing, Treasure Island (FL). Available from: https://www.ncbi.nlm.nih.gov/books/NBK545223/. (Accessed 02.07.2021).

Sinha, S., Kar, K., Dasgupta, A., Basu, S., Sen, S., 2015. Correlation of serum zinc with TSH in hyperthyroidism. Asian Journal of Medical Sciences 7 (1), 66−69. Available from: https://doi.org/10.3126/ajms.v7i1.12895.

Siro, S.S., Zandberg, L., Ngounda, J., Wise, A., Symington, E.A., Malan, L., et al., 2022. Iodine status of pregnant women living in urban Johannesburg, South Africa. Maternal and Child Nutrition 18 (1), e13236. Available from: https://doi.org/10.1111/mcn.13236. (Accessed 05.02.2024).

Šket, T., Kukec, A., Kosem, R., Artnik, B., 2017. The history of public health use of fluorides in caries prevention. Zdravstveno Varstvo 56 (2), 140−146. Available from: https://doi.org/10.1515/sjph-2017-0018.

Smedley, P.L., Edmunds, W.M., West, J.M. Gardner, S.J., Pelig-Ba, K.B., 1995. Vulnerability of shallow groundwater quality due to natural geochemical environment: 2: Health problems related to groundwater in the Obuasi and Bolgatanga areas, Ghana. British Geological Survey Technical Report, WC95/43, 122 pp. http://nora.nerc.ac.uk/id/eprint/16470/1/WC9543.pdf. (Accessed 11.06.2021).

Smuts, C.M., Baumgartner, J., 2019. Are we neglecting iodine nutrition in South Africa? South African Journal of Clinical Nutrition 32 (3), 3−4. Available from: https://journals.co.za/doi/pdf/10.10520/EJC-1811e6467d.

Smyth, D., Johnson, C.C., 2011. Distribution of iodine in soils of Northern Ireland. Geochemistry Exploration Environment Analysis 11 (1), 25−39. Available from: https://doi.org/10.1144/1467-7873/09-015.

Smyth, P., Burns, R., Casey, M., Mullan, K., O'Herlihy, C., O'Dowd, C., 2016. Iodine status over two decades: influence of seaweed exposure. Ireland Medical Journal 109 (6), 421.

Söderlund, M., Virkanen, J., Aromaa, H., Gracheva, N., Lehto, J., 2017. Sorption and specisation of iodine in boreal forest soil. Journal of Radioanalytical and Nuclear Chemistry 311, 549−564. Available from: https://doi.org/10.1007/s10967-016-5022-z.

Speeckaert, M.M., Speeckaert, R., Wierckx, K., Delanghe, J.R., Kaufman, J.-M., 2011. Value and pitfalls in iodine fortification and supplementation in the 21st century. British Journal of Nutrition 106 (7), 964−973. Available from: https://doi.org/10.1017/S000711451100273X.

Stemmler, I, Rothe, M., Hense, I., Hepach, H., 2013. Numerical modelling of methyl iodide in the eastern tropical Atlantic. Biogeosciences 10, 4211−4225. Available from: https://doi.org/10.5194/bg-10-4211-2013 (accessed 04.02.2024).

Tagami, K., Uchida, S., Hirai, I., Tsukada, H., Takeda, H., 2006. Determination of chlorine, bromine and iodine in plant samples by inductively coupled plasma mass spectrometry after leaching with tetramethyl ammonium hydroxide under a mild temperature condition. Analytica Chimica Acta 570, 88−92.

Takeda, A., Tsukada, H., Takahashi, M., Takaku, Y., Hisamatsu, S., 2015. Changes in the chemical form of exogenous iodine in forest soils and their extracts. Radiation Protection Dosimetry 167 (1-3), 181−186. Available from: https://doi.org/10.1093/rpd/ncv240.

Takeda, A., Tsukada, H., Takaku, Y., Satta, N., Baba, M., Shibata, T., et al., 2016. Determination of iodide, iodate and total iodine in natural water samples by HPLC with amperometric and spectrophotometric detection, and off-line UV irradiation. Analytical Sciences 32 (8), 839−845. Available from: https://doi.org/10.2116/analsci.32.839.

Talebi, S., Ghaedi, E., Sadeghi, E., et al., 2020. Trace element status and hypothyroidism: a systematic review and meta-analysis. Biology of Trace Element Research 197, 1−14. Available from: https://doi.org/10.1007/s12011-019-01963-5.

Talhada, D., Santos, C.R.A., Gonçalves, I., Ruscher, K., 2019. Thyroid hormones in the brain and their impact in recovery mechanisms after stroke. Frontiers in Neurology 10, 1103. Available from: https://www.frontiersin.org/article/10.3389/fneur.2019.01103.

Tamer, C., Suna, S. and Özcan-Sinir, G., 2019. Chapter 14: Toxicological aspects of ingredients used in nonalcoholic beverages. In: Grumezescu, A.M. and Holban, A.M., Non-Alcoholic Beverages—Volume 6: The Science of Beverages, pp. 441 - 481. Woodhead Publishing/Elsevier. Available from: https://doi.org/10.1016/B978-0-12-815270-6.00014-1.

Taye, A., Argaw, H., 1997. Prevalence and prominent factors for iodine deficiency disorders in Shebe area, Seka Chekorsa district, South-Western Ethiopia. Ethiopian Journal of Health Sciences 7, 63−76.

Temple, V.J., Zimmermann, M.B., Karoke, W., Dogimab, E., 2020. Iodized oil to combat severe IDD and cretinism in PNG. Universal Salt Iodisation Project. https://www.researchgate.net/publication/344224770_Iodized_oil_to_combat_severe_IDD_and_cretinism_in_PNG. (Accessed 30.06.2021).

Temech, E.C., Said, O., Endalik, G., Demilew, Y.M., Belay, M.A., Kassie, T.D., Dessie, A.M., 2023. Adequacy of iodized salt and its associated factors among households in the Bahir Dar Zuria district, Northwest Ethiopia, 2022. Frontiers in Nutrition 10, 1215613. Available from: https://doi.org/10.3389/fnut.2023.1215613. (Accessed 05.02.2024).

Teng, W., Shan, Z., Teng, X., Guan, H., Li, Y., Teng, D., et al., 2006. Effect of iodine intake on thyroid diseases in China. The New England Journal of Medicine 354 (26), 2783−2793.

Terefe, B., Jembere, M.M., Assimamaw, N.T., 2023. Iodized household salt utilization and associated factors among households in East Africa: a multilevel modelling analysis using recent national health surveys. BMC Public Health 23 (1), 2387. Available from: https://doi.org/10.1186/s12889-023-17296. (Accessed 04.02.2024).

Thilly, C.H., Vanderpas, J.B., Bebe, N., Ntambue, K., Contempre, B., Swennen, B., et al., 1992. Iodine deficiency, other trace elements, and goitrogenic factors in the etiopathogeny of iodine deficiency disorders (IDD). Biological Trace Element Research 32, 229. Available from: https://doi.org/10.1007/BF02784606.

Thomson, C.D., Campbell, J.M., Miller, J., Skeaff, S.A., 2011. Minimal impact of excess iodate intake on thyroid hormones and selenium status in older New Zealanders. European Journal of Endocrinology 165 (5), 745−752. Available from: https://doi.org/10.1530/EJE-11-0575.

Todd, C.H., Allain, T., Gomo, Z.A., Hasler, J.A., Ndiweni, M., Oken, E., 1995. Increased thyrotoxicosis associated with iodine supplements in Zimbabwe. Lancet 346 (8989), 1563−1564. Available from: https://doi.org/10.1016/s0140-6736(95)92095-1.

Todorov, T.I., Gray, P.J., 2016. Analysis of iodine in food samples by inductively coupled plasma-mass spectrometry. Food Additives and Contaminants. Part A, Chemistry, Analysis, Control, Exposure and Risk Assessment 33 (2), 282−290. Available from: https://doi.org/10.1080/19440049.2015.1131337.

Toloza, F.J.K., Motahari, H., Maraka, S., 2020. Consequences of severe iodine deficiency in pregnancy: evidence in humans. Frontiers in Endocrinology 11, 409. Available from: https://doi.org/10.3389/fendo.2020.00409.

Torti, J.F., Correa, R., 2021. Potassium Iodide. [Updated 2021 May 15]. StatPearls [Internet]. StatPearls Publishing, Treasure Island (FL). Available from: https://www.ncbi.nlm.nih.gov/books/NBK542320/. (Accessed 02.07.2021).

Traoré, A.K. and Torheim, L.E., 2009. Chapter 13: Iodine deficiency disorders in Mali. In: Preedy, V.R., Burrow, G.N. and Watson, R. (Eds.), Comprehensive Handbook of Iodine, first ed., pp. 1265−1270, Elsevier Inc. Academic Press. Available from: https://doi.org/10.1016/B978-0-12-374135-6.00131-X.

Truesdale, V.W., Jones, S.D., 1996. The variation of iodate and total iodine in some UK rainwater samples during 1980−1981. Journal of Hydrology 179, 67−86.

Turska-Szybka, A., Gozdowski, D., Twetman, S., Olczak-Kowalczyk, D., 2021. Clinical effect of two fluoride varnishes in caries-active preschool children: a randomized controlled trial. Caries Research 55, 137−143. Available from: https://doi.org/10.1159/000514168.

Umaly, R.C., Poel, L.W., 1970. Effects of various concentrations of iodine as potassium iodide on the growth of barley, tomato and pea in nutrient solution culture. Annals of Botany 34 (4), 919−926. Available from: https://doi.org/10.1093/oxfordjournals.aob.a084423.

Umaly, R.C., Poel, L.W., 1971. Effects of iodine in various formulations on the growth of barley and pea plants in nutrient solution culture. Annals of Botany 35 (1), 127−131. Available from: https://doi.org/10.1093/oxfordjournals.aob.a084451.

UNICEF (United Nations Childrens Fund), 2019. Product Update for Salt Test Kits. Technical Bulletin No. 26 (July 2019), UNICEF. https://www.unicef.org/supply/media/1451/file/Update-for-salt-test-kits-technical-bulletin.pdf. (Accessed 07.07.2021).

US FNBIM (United States Food and Nutrition Board, Institute of Medicine), 2006. Dietary Reference Intakes: The Essential Guide to Nutrient Requirements. National Academy Press, Washington, DC. Available from: http://doi.org/10.17226/11537. (Accessed 27.07.2021).

US IM (United States Institute of Medicine, Food and Nutrition Board), 2001. Dietary Reference Intakes for Vitamin A, Vitamin K, Arsenic, Boron, Chromium, Copper, Iodine, Iron, Manganese, Molybdenum, Nickel, Silicon, Vanadium, and Zinc. National Academy Press, Washington, DC. Available from: https://doi.org/10.17226/10026. (Accessed 27.07.2021).

US NIH (United States National Institutes of Health), 2021. Iodine: Fact Sheet for Health Professionals. https://ods.od.nih.gov/factsheets/Iodine-HealthProfessional/. (Accessed 30.04.2021).

US NLM (US National Library of Medicine), (n.d.). Iodine, Radioactive. https://webwiser.nlm.nih.gov/substance?substanceId = 414&identifier = Iodine,%20Radioactive&identifierType = name&menuItemId = 62&catId = 83. (Accessed 28.05.2021).

US NRC (United States National Research Council), 2005. Committee to Assess the Health Implications of Perchlorate Ingestion. Health Implications of Perchlorate Ingestion. Washington, DC: The National Academies Press.

USICST (Universal Salt Iodization Coverage Survey Team), 2017. Household coverage with adequately iodized salt varies greatly between countries and by residence type and socioeconomic status within countries: results from 10 national coverage surveys. The Journal of Nutrition 147 (5), 1004S−1014S. Available from: https://doi.org/10.3945/jn.116.242586.

Utoro, J., Mangasaryan, N., de Benoist, B., Ian Darnton-Hill, I., 2007. Reaching optimal iodine nutrition in pregnant and lactating women and young children. Public Health Nutrition 10 (12A), 1527−1529. Available from: https://doi.org/10.1017/S1368980007705360.

Van Dilla, M.A., Fulwyler, M.J., 1963. Thyroid metabolism in children and adults using very small (nanocurie) doses of iodine125 and iodine131. Health Physics 9, 1325−1331.

Vanderpas, J., 2003. Goitrogens and antithyroid compounds. In: Caballero, B. (Ed.), Encyclopedia of Food Sciences and Nutrition, second ed. Academic Press, pp. 2949−2957. Available from: https://doi.org/10.1016/B0-12-227055-X/00566-6. (Accessed 10.07.2021).

Venance, M.S., Martin, H.D., Kimiywe, J., 2020. Iodine status and discretionary choices consumption among primary school children, Kinondoni, Tanzania. Paediatric Health Medicine and Therapeutics 11, 359−368. Available from: https://doi.org/10.2147/PHMT.S265117.

Völzke, H., Erlund, I., Hubalewska-Dydejczyk, A., Ittermann, T., Peeters, R.P., Rayman, M., et al., 2018. How do we improve the impact of iodine deficiency disorders prevention in Europe and beyond? European Thyroid Journal 7 (4), 193−200. Available from: https://doi.org/10.1159/000490347.

Voogt, W., Steenhuizen, J., Eveleens, B., 2014. Uptake and distribution of iodine in cucumber, sweet pepper, round, and cherry tomato. Report GTB-1329, Project number: 3242176100. Wageningen UR Glastuinbouw. https://edepot.wur.nl/328925. (Accessed 20.05.2021).

Wadley, M.R., Stevens, D.P., Jikells, T.D., Hughes, C., Chance, R., Hepach, H., et al., 2020. A global model for iodine speciation in the upper ocean. Global Biogeochemical Cycles 34. Available from: https://doi.org/10.1029/2019GB006467.

Wainwright, P., Cook, P., 2019. The assessment of iodine status—populations, individuals and limitations. Annals of Clinical Biochemistry 56 (1), 7−14. Available from: https://doi.org/10.1177/0004563218774816.

Wang, Y., Cui, Y., Zhang, D., Chen, C., Hou, C., Cao, L., 2021. Moderating role of TSHR and PTPN22 gene polymorphisms in effects of excessive fluoride on thyroid: a school-based cross-sectional study. Biology of Trace Elements Research 200 (3), 1104−1116. Available from: https://doi.org/10.1007/s12011-021-02753-8.

Wang, F., Li, C., Li, S., Cui, L., Zhao, J., Liao, L., 2023. Selenium and thyroid diseases. Frontiers in Endocrinology 14, 1133000. Available from: https://doi.org/10.3389/fendo.2023.1133000. (Accessed 05.02.2024).

Watts, M.J., Joy, E.J.M., Young, S.D., Broadley, M.R., Chilimba, A.D.C., Gibson, R.S., et al., 2015. Iodine source apportionment in the Malawian diet. Scientific Report 5, 15251. Available from: https://doi.org/10.1038/srep15251.

Waugh, D.T., 2019. Fluoride exposure induces inhibition of sodium/iodide symporter (NIS) contributing to Impaired iodine absorption and iodine deficiency: molecular mechanisms of inhibition and implications for public health. International Journal of Environmental Research and Public Health 16 (6), 1086. Available from: https://doi.org/10.3390/ijerph16061086.

Weng, H.X., Hong, C.L., Xia, T.H., Bao, L.T., Liu, H.P., Li, De. W., 2013. Iodine biofortification of vegetable plants—an innovative method for iodine supplementation. Chinese Science Bulletin 58, 2066–2072. Available from: https://doi.org/10.1007/s11434-013-5709-2.

Wenhold, F. and Faber, M., 2008. Nutritional Status of South Africans: Links to Agriculture and Water. Water Research Commission of South African. WRC.pdf (uct.ac.za). (Accessed 14.06.2021).

West, R.C., 1984. Handbook of Chemistry and Physics. CRC Press, Boca Raton (FL).

Whitehead, D.C., 1973. The sorption of iodide by soils as influenced by equilibrium conditions and soil properties. Journal of the Science Food and Agriculture 24, 547–556. Available from: https://doi.org/10.1002/jsfa.2740240508.

Whitehead, D., 1979. Iodine in the U.K. environment with particular reference to agriculture. Journal of Applied Ecology 16 (1), 269–279. Available from: https://doi.org/10.2307/2402746.

WHO (World Health Organisation), 1999. General Guidelines for Iodine Prophylaxis following Nuclear Accidents. Report: WHO/SDE/PHE/99.6. https://www.who.int/ionizing_radiation/pub_meet/Iodine_Prophylaxis_guide.pdf. (Accessed 31.05.2021).

WHO (World Health Organisation), 2003. Iodine in Drinking-Water: Background Document For Development of Guidelines for Drinking-water Quality. World Health Organization, Geneva. Available from: http://www.who.int/water_sanitation_health/dwq/chemicals/iodine.pdf. (Accessed 27.05.2021).

WHO (World Health Organisation), 2017. Guidelines for Drinking-Water Quality, fourth ed. World Health Organization, Geneva. Available from: http://www.who.int/water_sanitation_health/publications/drinkingwater-quality-guidelines-4-including-1st-addendum/en/. (Accessed 28.05.2021).

WHO (World Health Organisation), 2018. Iodine As a Drinking Water Disinfectant. License: CC BY-NC-SA 3.0 IGO. Geneva. https://www.who.int/water_sanitation_health/publications/iodine-02032018.pdf?ua = 1#:~: text = Iodine%2Dbased%20disinfection%20of%20water,1922%3B%20Vergnoux%2C%201915). (Accessed 27.05.2021).

WHO (World Health Organisation), 2019. Iodine in Drinking-Water—Draft Background Document for Development of WHO Guidelines for Drinking-Water Quality 26. Document: WHO/SDE/WSH/0x.xx/xx. https://www.who.int/water_sanitation_health/water-quality/guidelines/chemicals/draft-iodine-gdwq-190924.pdf. (Accessed 11.06.2021).

WHO (World Health Organization), 1993. Guidelines for Drinking-Water Quality, second ed., Volume 1 – Recommendations. Geneva. https://www.who.int/water_sanitation_health/dwq/2edvol1i.pdf. (Accessed 25.07.2021).

WHO (World Health Organization), 1994. Indicators for Assessing Iodine Deficiency Disorders and Their Control Through Salt Iodisation. WHO/NUT/94.6. World Health Organization/International Council for the Control of Iodine Deficiency Disorders (ICCIDD), Geneva.

WHO/UNICEF/ICCIDD (World Health Organisation/United Nations Children's Fund/International Council for the Control of the Iodine Deficiency Disorders (WHO/UNICEF/ICCIDD), 2008. Assessment of the Iodine Deficiency Disorders and Monitoring Their Elimination, third ed. WHO, Geneva. Available from: http://apps.who.int/iris/bitstream/handle/10665/43781/9789241595827_eng.pdf;jsessionid = 2EA1E53A1E6D21D0DC800E137B93E1BD?sequence = 1. (Accessed 30.04.2021).

WIA (World Iodine Association), n.d. Crops nutrition. https://www.worldiodineassociation.com/crops_n/. (Accessed 11.07.2021).

Wikipedia, 2021a. ACE inhibitor. https://en.wikipedia.org/wiki/ACE_inhibitor. (Accessed 02.07.2021).

Wikipedia, 2021b. Hypothalamic-pituitary-adrenal axis. https://en.wikipedia.org/wiki/Hypothalamic%E2%80%93pituitary%E2%80%93adrenal_axis. (Accessed 14.07.2021).

Wong, G.T.F., 1991. The marine geochemistry of iodine. Reviews in Aquatic Sciences 4 (1), 45−73.

Xu, C., Zhang, S., Ho, Y.-F., Miller, E.J., Roberts, K.A., Li, H.-P., et al., 2011. Is soil natural organic matter a sink or source for mobile radioiodine (129I) at the Savannah River site? Geochimica et Cosmochimica Acta 75 (19), 5716−5735. Available from: https://doi.org/10.1016/j.gca.2011.07.011.

Xu, F., Ma, T., Shi, L., Zhang, J.W., Wang, Y.Y., Dong, Y.H., 2013. The hydrogeochemical characteristics of high iodine and fluoride groundwater in the Hetao Plain, Inner Mongolia. Procedia Earth and Planetary Science 7, 908−911. Available from: https://doi.org/10.1016/j.proeps.2013.03.183.

Yadykina, T.K., Bugaeva, M.S., Kochergina, T.V., Mikhailova, N.N., 2021. Clinical and experimental studies of the effect of chronic fluoride intoxication on the hormonal status of the body and morphological changes in the thyroid gland. Russian Journal of Occupational Health and Industrial Ecology 61 (3), 173−180. Available from: https://doi.org/10.31089/1026-9428-2021-61-3-173-180.

Yamaguchi, N., Nakano, M., Tanida, H., Fujiwara, H., Kihou, N., 2006. Redox reaction of iodine in paddy soil investigated by field observation and the I K-Edge XANES fingerprinting method. Journal of Environmental Radioactivity 86 (2), 212−226. Available from: https://doi.org/10.1016/j.jenvrad.2005.09.001.

Yamaguchi, N., Nakano, M., Takamatsu, R., Tanida, H., 2010. Inorganic iodine incorporation into soil organic matter: evidence from iodine k-edge X-ray absorption near-edge structure. Journal of Environmental Radioactivity 101, 451−457. Available from: https://doi.org/10.1016/j.jenvrad.2008.06.003.

Yang, X.F., Xu, J., Hou, X.H., Guo, H.L., Hao, L.P., Yao, P., et al., 2006. Developmental toxic effects of chronic exposure to high doses of iodine in the mouse. Reproductive Toxicology 22 (4), 725−730. Available from: https://doi.org/10.1016/j.reprotox.2006.05.010.

Yeshaw, Y., Liyew, A.M., Teshale, A.B., Alamneh, T.S., Worku, M.G., Tessema, Z.T., et al., 2021. Individual and community level factors associated with use of iodized salt in sub-Saharan Africa: a multilevel analysis of demographic health surveys. PLoS One 16 (5), e0251854. Available from: https://doi.org/10.1371/journal.pone.0251854.

Yokouchi, Y., Nojiri, Y., Toom-Sauntry, D., Fraser, P., Inuzuka, Y., Tanimoto, H., et al., 2012. Long-term variation of atmospheric methyl iodide and its link to global environmental change. Geophysical Research Letters 39 (23), L23805. Available from: https://doi.org/10.1029/2012GL053695.

Yoshida, V.M.S., 1999. Effects of microorganisms on the fate of iodine in the soil environment. Geomicrobiology Journal 16, 85−93.

Zach, R., 1997. Nuclear Fuel Waste Management—Biosphere Program Highlights—1978 to 1996. Document AECL-11811. Environmental Science Branch, Pinawa, Canada. https://www.osti.gov/etdeweb/servlets/purl/681704. (Accessed 18.07.2021).

Zbigniew, S., 2017. Role of iodine in metabolism. Recent Patents on Endocrine, Metabolic and Immune Drug Discovery 10 (2), 123−126. Available from: https://doi.org/10.2174/1872214811666170119110618.

Zeng, Y., Zhu, X., 2021. Rapid determination of iodine content in drinking water by isopropyl sensitization and inductively coupled plasma mass spectrometry (ICP-MS). E3S Web of Conferences 245, 03004. Available from: https://doi.org/10.1051/e3sconf/202124503004.

Zhang, S., Xu, C., Creeley, D., Ho, Yi-F., Li, H.-P., Grandbois, R., et al., 2013. Iodine-129 and iodine-127 speciation in groundwater at the Hanford site, US: iodate incorporation into calcite. Environmental Science and Technology 47 (17), 9635−9642.

Zhao, S.J., Sun, F.J., Tian, E.J., Chen, Z.P., 2006. Experimental study on effects of iodine deficiency and excess on thyroid autoimmunity. Zhonghua Yu Fang Yi Xue Za Zhi [Chinese Journal of Preventive Medicine] 40 (1), 18−20. Available from: https://doi.org/10.3760/j:issn:0253-9624.2006.01.004 (in Chinese).

Zimmermann, M.B., 2006. The influence of iron status on iodine utilization and thyroid function. Annual Review of Nutrition 26 (1), 367−389.

Zimmermann, M.B., 2009. Iodine deficiency. Endocrine Reviews 30 (4), 376−408. Available from: https://doi.org/10.1210/er.2009-0011.

Zimmermann, M.B., 2011. The role of iodine in human growth and development. Seminars in Cell and Developmental Biology (6), 645−652. Available from: https://doi.org/10.1016/j.semcdb.2011.07.009.

Zimmermann, M.B., 2020. Chapter 25: Iodine and the iodine deficiency disorders. In: Marriott, B.P., Birt, D. F., Stallings, V.A., Yates, A.A. (Eds.), Present Knowledge in Nutrition, eleventh ed. Volume 1: Basic Nutrition and Metabolism. Academic Press, pp. 429−441. Available from: https://doi.org/10.1016/B978-0-323-66162-1.00025-1 (Accessed 30.05.2021).

Zimmermann, M.B., Köhrle, J., 2002. The impact of iron and selenium deficiencies on iodine and thyroid metabolism: Biochemistry and relevance to public health. Thyroid 12 (10), 867−878. Available from: https://doi.org/10.1089/105072502761016494.

Zimmermann, M.B., Boelaert, K., 2015. Iodine deficiency and thyroid disorders. Lancet Diabetes and Endocrinology 3 (4), 286−295. Available from: https://doi.org/10.1016/S2213-8587(14)70225-6.

Zimmermann, M., Adou, P., Torresani, T., Zeder, C., Hurrell, R., 2000. Persistence of goitre despite oral iodine supplementation in goitrous children with iron deficiency anaemia in Côte d'Ivoire. American Journal of Clinical Nutrition 71 (1), 88−93. Available from: https://doi.org/10.1093/ajcn/71.1.88.

Further reading

Alexander, W., Harden, R., Harrison, M., Shimmins, J., 1967. Some aspects of the absorption and concentration of iodide by the alimentary tract in man. Proceedings of the Nutrition Society 26, 62−66.

Amachi, S., Fujii, T., Shinoyama, H., Muramatsu, Y., 2005. Microbial influences on the mobility and transformation of radioactive iodine in the environment. Journal of Nuclear and Radiochemical Sciences 6 (1), 21−24. Available from: https://doi.org/10.14494/jnrs2000.6.21.

Andersson, M., Takkouche, B., Egli, I., Allen, H.E., de Benoist, B., 2005. Current global iodine status and progress over the last decade towards the elimination of iodine deficiency. Bulletin of the World Health Organization (WHO) 83, 518−525.

Arthur, J.R., Beckett, G.J., Mitchell, J.H., 1999. The interactions between selenium and iodine deficiencies in man and animals. Nutrition Research Reviews 12 (1), 55−73. Available from: https://doi.org/10.1079/095442299108728910.

Asfaw, A., Behailu, M., Oumer, A., Gebremariam, T., Asefa, K., 2023. Factors associated with recent iodine intake level among household food handlers in Southwest Ethiopia: a cross-sectional study. BMC Women's Health 1, 354. Available from: https://doi.org/10.1186/s12905-023-02516-8 (accessed 07.02.2024).

Beahm, E.C., Weber, C.F., Kress, T.S. Parker, G.W., 1992. Iodine Chemical Forms in LWR Severe Accidents. Report NUREG/CR-5732 ORNL/TM-11861. U.S. Nuclear Regulatory Commission Washington, DC. https://www.nrc.gov/docs/ML0037/ML003726825.pdf. (Accessed 03 May 2021).

Berbel, P., Obregón, M.J., Bernal, J., Escobar, del Rey F., Morreale de Escobar, G., 2007. Iodine supplementation during pregnancy: a public health challenge. Trends in Endocrinology and Metabolism 18 (9), 338−343. Available from: https://doi.org/10.1016/j.tem.2007.08.009.

Black, A., Hounam, R.F., 1968. Penetration of iodine vapour through the nose and mouth and the clearance and metabolism of the deposited iodine. Annals of Occupational Hygiene 11, 209−225.

Bowley, H.E., 2013. Iodine Dynamics in the Terrestrial Environment (Ph.D. thesis). University of Nottingham, Great Britain. http://eprints.nottingham.ac.uk/13241/1/Iodine_dynamics_in_the_terrestrial_environment_-_Hannah_Bowley.pdf. (Accessed 11 June 2021).

Brenchley, W., 2008. The effect of iodine on soils and plantsFirst published 1924Annals of Applied Biology 11 (1), 86−111. Available from: https://doi.org/10.1111/j.1744-7348.1924.tb05695.x.

CRN (Council for Responsible Nutrition), 2013. Vitamin and Mineral Safety, third ed. Council for Responsible Nutrition. http://www.crnusa.org/safety/CRN-SafetyBook-3rdEdition-2014-fullbook.pdf. (Accessed 27 May 2021).

Iodine status worldwide: WHO Global Database on Iodine Deficiency. NLM classification: WD 105In: de Benoist, B., Andersson, M., Egli, I., Takkouche, B., Allen, H. (Eds.), Department of Nutrition for Health and Development, World Health Organisation, Geneva. Available from: https://apps.who.int/iris/bitstream/handle/10665/43010/9241592001.pdf?sequence = 1. (Accessed 01 June 2021).

Dobosy, P., Vetési, V., Sandil, S., Endrédi, A., Kröpfl, K., Óvári, M., et al., 2020. Effect of irrigation water containing iodine on plant physiological processes and elemental concentrations of cabbage (*Brassica oleracea* L. var. capitata L.) and tomato (*Solanum lycopersicum* L.) cultivated in different soils. Agronomy 10 (5), 720. Available from: https://doi.org/10.3390/agronomy10050720.

EFSA (European Food Safety Authority), 2014. Scientific opinion on dietary reference values for iodine. EFSA Journal 12 (5), 3660. Available from: http://www.efsa.europa.eu/en/efsajournal/pub/3660.htm.

Fuge, R., 1996. Geochemistry of Iodine in Relation to Iodine Deficiency Diseases. Geological Society, London (Special Publications), vol. 113, pp. 201-211. https://doi.org/10.1144/GSL.SP.1996.113.01.16. (Accessed 03 April 2021).

Fuge, R., 2013. Chapter 17: Soils and iodine deficiency. In: Selinus, O., Alloway, B., Centeno, J.A., Finkelman, R.B., Fuge, R., et al.,Essentials of Medical Geology. Springer, Dordrecht, pp. 417−432. . Available from: https://doi.org/10.1007/978-94-007-4375-5_17. (Accessed 05 June 2021).

Gardas, A., 1991. The influence of iodine on the immunological properties of thyroglobulin and its immunological complexes. Autoimmunity 9 (4), 331−336. Available from: https://doi.org/10.3109/08916939108997135.

Gonzali, S., Kiferlyy, C., Perat, P., 2017. Iodine biofortification of crops: agronomic biofortification, metabolic engineering and iodine bioavailability. Current Opinion in Biotechnology 44, 16−26.

Gosselin, R.E., Smith, R.P., Hodge, H.C., 1984. Clinical Toxicology of Commercial Products, fifth ed. Williams and Wilkinsp. III-213, Baltimore.

Gunnarsdottir, I., Dahl, L., 2012. Iodine intake in human nutrition: a systematic literature review. Food and Nutrition Research 56, 19731. Available from: https://doi.org/10.3402/fnr.v56i0.19731.

Harjantini, U., Hanim, D., Lanti, Y., Dewi, R., Nurwati, I., 2021. Correlation of dietary zinc intake and serum zinc with thyroid stimulating hormone (TSH) and free thyroxine (FT4) levels in adult hyperthyroid patients. International Journal of Human and Health Sciences 5 (3), 341−346.

Kim, M.J., Kim, S.C., Chung, S., Kim, S., Yoon, J.W., Park, Y.J., 2020. Exploring the role of copper and selenium in the maintenance of normal thyroid function among healthy Koreans. Journal Trace Elements in Medicine and Biology 61, 126558. Available from: https://doi.org/10.1016/j.jtemb.2020.126558.

Korobova, E., 2010. Soil and landscape geochemical factors which contribute to iodine spatial distribution in the main environmental components and food chain in the central Russian plain. Journal of Geochemical Exploration 107 (2), 180−192.

Laurberg, P., Cerqueira, C., Ovesen, L., Rasmussen, L.B., Perrild, H., Andersen, S., et al., 2010. Iodine intake as a determinant of thyroid disorders in populations. Best Practice and Research Clinical Endocrinology and Metabolism 24 (1), 13−27. Available from: https://doi.org/10.1016/j.beem.2009.08.013.

Laurberg, P., Knudsen, N., Andersen, S., Carlé, A., Pedersen, I.B., Karmisholt, J., 2012. Thyroid function and obesity. European Thyroid Journal 1 (3), 159−167. Available from: https://doi.org/10.1159/000342994.

Mackowiak, C.L., Grossl, P.R., Cook, K.L., 2005. Iodine toxicity in a plant-solution system with and without humic acid. Plant and Soil 269, 141−150. Available from: https://doi.org/10.1007/s11104-004-0401-6.

Muramatsu, S., Yoshida, Y., 1999. Effects of microorganisms on the fate of iodine in the soil environment. Geomicrobiology Journal 16 (1), 85−93. Available from: https://doi.org/10.1080/014904599270776.

OSU (Oregon State University), 2021. Iodine. Micronutrient Information Centre, Linus Pauling Institute, OSU. Available from: https://lpi.oregonstate.edu/mic/minerals/iodine. (Accessed 30 May 2021).

RCA/WHO (Regional Committee for Africa/World Health Organisation), 58, 2008. Iodine Deficiency Disorders in the WHO African Region: Situation Analysis and Way Forward. Document Number: AFR/RC58/7. https://apps.who.int/iris/handle/10665/19986. (Accessed 03 June 2020).

Saller, B., Fink, H., Mann, K., 1998. Kinetics of acute and chronic iodine excess. Experimental and Clinical Endocrinology and Diabetes 106 (Supplement 3), S34−S38.

Suess, E., Aemisegger, F., Sonke, J.E., Sprenger, M., Wernli, H., Winkel, L.H.E., 2019. Marine versus continental sources of iodine and selenium in rainfall at two European high-altitude locations. Environmental Science and Technology 53 (4), 1905−1917. Available from: https://doi.org/10.1021/acs.est.8b05533.

USEPA (United States Environmental Protection Agency), 2006. Reregistration eligibility decision for iodine and iodophor complexes. http://www.epa.gov/opp00001/reregistration/REDs/iodine-red.pdf. (Accessed 27 May 2021).

Watson, R. (Ed.), 2009. Comprehensive Handbook of Iodine. Academic Press, Oxford, pp. 107−118.

WHO/FAO (World Health Organisation/Food and Agricultural Organisation), 2004. Vitamin and Mineral Requirements in Human Nutrition, second ed. World Health Organization/Food and Agriculture Organization, Geneva. Available from: http://apps.who.int/iris/bitstream/10665/42716/1/9241546123.pdf. (Accessed 27 May 2021).

Yeager, C.M., Amachi, S., Grandbois, R., Kaplan, D.I., Xu, C., Schwehr, K.A., et al., 2017. Microbial transformation of iodine: from radioisotopes to iodine deficiency. Advances in Applied Microbiology 101, 83−136. Available from: https://doi.org/10.1016/bs.aambs.2017.07.002.

Zicker, S., Schoenherr, B., 2012. The role of Iodine in Nutrition and Metabolism, 34. Compendium, Yardley, PA, pp. E1−E4.

Zimmermann, M.B., 2012. Chapter 36: Iodine and iodine deficiency disorders. In: Erdman, J.W., Macdonald, I.A., Zeisel, S.H. (Eds.), Present Knowledge in Nutrition. https://doi.org/10.1002/9781119946045.ch36. (Accessed 31 May 2021).

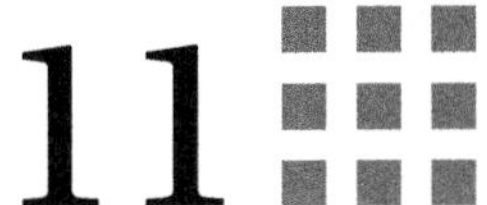

11

Medical geology of lead

"The World Health Organization's 2021 update of "The Public Health Impact of Chemicals: Knowns and Unknowns" estimate that nearly half of the 2 million lives lost to known chemicals exposure in 2019 were due to lead exposure ..."

WHO, 2022

Key chapter features

1. Brief overviews of the most relevant and important characteristics of Pb, that are of relevance to the study of Medical Geology.
2. Attributes that make Pb one of the most severe poisons in nature.
3. Summary of the latest advances in Pb research in Africa as reflected in key papers in reputable journals or book chapters, about 90% of which postdate the year 2000).
4. Suggested areas of future Pb research in Medical Geology and how these apply to the African situation.

Sources

Naturally occurring Pb ores comprise 0.002% (20 ppm) of the Earth's crust (ATSDR, 2019).

Igneous rocks

Average value given by Turekian and Wedepohl (1961) for igneous rocks is 13 ppm.

Average values given by Levinson (1980) are for Granite, 20 ppm; Basalt, 5 ppm.

Average value given for alkali rocks by Wedepohl (1956) and Turekian and Wedepohl (1961) is 12.0 ppm.

Average value given for rhyolite and obsidian given by Wedepohl (1956) is 21.0 ppm.

Average value given for diorite by Wedepohl (1956) is 10.0 ppm.

Average value given for ultramafic rocks by Wedepohl (1956) is 3.0 ppm.

Average value given for intermediate rocks (diorite and andesite) by Vinogradov (1956; 1962) is 15.0 ppm.

The Pb^{2+} ion [119 picometers (pm)], being intermediate in size between K^+ (138 pm) and Ca^{2+} (100 pm), replaces these ions in K-feldspar, mica, and, to a lesser extent, plagioclase, and apatite. So, felsic igneous rocks are enriched in Pb relative to mafic rocks, and, according to MacDonald et al. (1973), Pb is mobile in late-stage magmatic processes. More recent values for Pb in igneous rocks as quoted by Mielke (1979) (*cf.*: Vinogradov, 1956,

Medical Geology of Africa. DOI: https://doi.org/10.1016/B978-0-12-818748-7.00002-2

1962; Wedepohl, 1956; Turekian and Wedepohl, 1961) confirm the enrichment of Pb in felsic igneous rocks: ultramafic, 1 ppm; basaltic, 6 ppm; granitic, 15−19 ppm; syenite, 12 ppm; and a crustal abundance of 13 ppm.

Sedimentary rocks

Median (M) value and normal range (NR), respectively, given by Lovering (1976), are:
   Terrestrial sediments—M, 15 ppm; NR, 5−40 ppm.
   Marine sediments—M, 60 ppm; NR, 30−150 ppm.
   Coal—M, 10 ppm; NR, 3−30 ppm.
   Siltstone and shale—M, 15 ppm; NR, 30−50 ppm;
   Limestone—M, 5 ppm; NR, 3−15 ppm.
   Average value given earlier by Turekian and Wedepohl (1961) for shale is 20 ppm.

Metamorphic rocks

Median (M) value and NR, respectively, given by Lovering (1976) are:
   Amphibole—M, 10 ppm; NR, <1−25 ppm.
   Gneiss—12 ppm; NR, <1−40 ppm.
   Schist—15 ppm; NR, <1−50 ppm.

Soils

According to recent estimates, e.g., UMA (2019), Pb content in soils varies from 15 to 40 ppm. Lindsay (1979) gave an average value for soils of 10 ppm. Kabata-Pendias and Pendias (2001) estimated the baseline Pb content for surface soil on the global scale to be 25 ppm, with levels above this suggesting an anthropogenic influence.

The natural Pb content in soil is, as expected, related to the composition of the parent rock. Although the species of Pb vary considerably with soil type, it is mainly found in association with clay minerals, Mn oxides, Fe and Al hydroxides, and organic matter. In some soil types, high concentrations of Pb may be found in Ca carbonate particles or in phosphate concentrations (Kabata-Pendias and Pendias, 2001).

Speciation of Pb in soils greatly affects its bioavailability and thus, its toxicity on plants and microbes (Kushwaha et al., 2018). In the area of agricultural production the accumulation of Pb in soils poses great concern due to the toxic effects on soil microflora, crop growth, and food safety, despite the development by plants and bacteria of detoxification mechanisms to obviate the toxic effect of Pb. Several factors influence the speciation of Pb; these include the pH of the soil, organic matter content, presence of various amendments, clay minerals, and presence of organic colloids and iron oxides. However, as of relatively recently (2018), unlike other metals, our knowledge about the speciation and mobility of Pb in soil was still at its infancy (Kushwaha et al., 2018).

Plants

Kumar et al.'s (2020) computation from literature values of average Pb contents of different crop plants including vegetable crops, spices, fruit crops, cereals, and legumes ranged from [(0.003−0.05 ppm) (Banana, *Musa* sp.)] to 149.5 ppm (Sugar beet, *Beta vulgaris* L).

Atmospheric deposition largely accounts for Pb content in plants. This is because particulate matter is held strongly on plant surfaces and is difficult to remove through washing (US EPA, 1977). After Pb is absorbed into the soil directly from the external atmosphere, it enters the plant system (Colin et al., 2022). Uptake of Pb into plant tissue appears to involve a combination of factors involving characteristics of the leaf surface and root system, Pb speciation, and soil characteristics (See Angelova et al., 2010). After uptake, Pb mainly accumulates in root cells, because of the blockage by *Casparian strips* (See "Glossary of Terms" in this Article) within the endodermis. Lead is also trapped by the negative charges that exist on roots' cell walls (Pourrut et al., 2011).

Food crops and other major dietary sources

According to the EU (n.d.), cereal products and grains, vegetables (especially potatoes and leafy vegetables), and tap water are the most important contributors to lead dietary exposure in the general European population. Few such data for Pb in Africa exist, and what data exist are at best patchy.

Foods made from the organs of plants, such as vegetables and fruits, are contaminated with Pb because plants easily take up toxic metals from the soil and air.

Natural waters

Lead seldom occurs naturally in water supplies, such as rivers and lakes. In surface water, the Pb content is highly variable and depends upon sources of pollution, Pb content of sediments, and characteristics of the system (pH, temperature, etc.).

The content of Pb in unpolluted rainwater and snow according to Hem (1992) is about 1 ppb, which is also typical for most unpolluted surface and ground waters. Pb contents in surface water are generally higher in urban areas than in rural areas (Levin et al., 2021), and Pb measured in natural or "pristine" surface waters may be due to anthropogenic input.

Ocean water

Modern reviews on Pb concentrations in ocean water are still based on the values reported pre-1990. For example, Botté et al.'s (2022) review was based on Pb concentration values of 0.033 ppb for northeastern Atlantic Ocean (Copin-Montegut et al., 1986a, b); 0.0148 ppb for the Arctic Ocean (Mart et al., 1984); 0.014 ppb for northeastern Pacific Ocean (Schaule and Patterson, 1981); and 0.030 ppb for the Indian Ocean (Danielsson, 1980).

Concentrations of Pb in the oceans are dependent on wet deposition and the concentration of Pb present in atmosphere (Boyle et al., 2014). Ecosystems in the marine realm are sinks of terrestrial contaminations. Wet deposition takes away Pb from the atmosphere to the ocean surface. Precipitation leads to solubilization of aerosols and washout of particulates. The main sink for Pb is burial in marine sediments (Cullen and McAlister, 2017).

Drinking water

Source water contamination can account for Pb in drinking water; however, the more common source of Pb in drinking water is from internal corrosion or wearing away of materials containing Pb in the water distribution system piping and plumbing.

A provisional drinking water quality guideline value for 0.01 ppm (10 ppb) was recommended by WHO (2017).

Air

Emission sources of Pb vary from one area to another (USEPA, 2023). Major sources of Pb in the air at the national level include ore and metal processing and piston-engine aircraft operating on leaded aviation fuel. Other sources are lead-acid battery manufacturers, waste incinerators, and utilities. The highest air concentrations of Pb are usually recorded near Pb smelters.

Back in 1988, Nriagu and Pacyna estimated that some 0.33×109 kg of Pb per year are directly emitted into the atmosphere. Patterson (1965) had earlier estimated preindustrial levels of Pb in air from natural origins to be about 0.6 ng/m^3.

Minerals

Lead is usually recovered from its ores, chiefly, galena (PbS); but also, from anglesite (PbSO$_4$), cerussite (PbCO$_3$), and minium (Pb$_3$O$_4$), the naturally occurring form of lead tetroxide, also known as red lead. Lead is also extensively dispersed in trace amounts in a range of other minerals, including K-feldspar, plagioclase, mica, zircon, and magnetite.

Associations (synergism/antagonism)

Earlier evidence for an antagonism between Pb and nutritional levels of Se is inconclusive (Levander, 1979).

Recent examples of antagonism involving Pb are discussed in the works of Liu et al. (2017) on "Antagonistic effect of selenium on lead-induced inflammatory injury through inhibiting the nuclear factor-κB signaling pathway and stimulating selenoproteins in chicken hearts" and Miao et al. (2022) on "The antagonistic effect of selenium on lead-induced apoptosis and necroptosis via P38/JNK/ERK pathway in chicken kidney".

The synergistic effect of Pb and Hg can lead to extreme neurotoxicity and has been reported to be much worse than the single action of any of the metals (Wildemann et al., 2015).

Back in 1991, Hercovits and Pérez-Coll reported that the antagonism or synergism between the metals Pb and Zn is dose-dependent.

Jarosite [$KFe_3(SO_4)_2(OH)_6$] minerals are effective scavengers of potentially toxic elements (PTEs) and are abundant, for example, in acid rock/mine drainage scenarios. The retention process is highly relevant for environmental attenuation of toxic metals and metalloids since these are usually highly soluble and thus, mobile under acidic conditions.

Aguilar-Carrillo et al. (2018) looked at simultaneous structural incorporation behavior and mechanism of the synergistic arsenic(V) and lead(II) retention on synthetic jarosite; such retention processes being highly relevant for environmental attenuation of toxic metals and metalloids since these are usually highly soluble and thus mobile under acidic conditions.

Chemical form or speciation in soil and aqueous phases

The chemical form or speciation in which each element occurs determines the level of health risk that it poses. Chemical forms of the element affect its absorption, retention, and utilization from diets. Interactions within the transporting medium can alter the "form" of the element from low toxicity form to high toxicity form and vice versa.

Lead has two oxidation states ($+2$ and $+4$) and four relatively stable naturally occurring isotopes ^{204}Pb, ^{205}Pb, ^{207}Pb, and ^{208}Pb, the last three forms of which represent ends of decay chains.

Lead can occur in any one of the three states, *viz.*, metallic, organic, or inorganic, and is mainly found in the environment in the form of salts [$PbCO_3$, Pb (NO_3), $PbSO_4$], hydroxylated [$Pb(OH)_2$] or ionized (Pb^{2+}). Lead speciation in the form Pb^{2+} is very important in determining the bioavailability and behavior of this metal.

Potential exposure to organic forms of Pb is extremely dangerous, as organic Pb is absorbed through the skin and is highly toxic to the brain and central nervous system, much more so than inorganic lead (Raj and Das, 2023; ATSDR, 2019; See Section on "Toxicity," in this Chapter).

Environmental circulation and impact of climate change

Lead persists in the environment and can be deposited from sources of Pb air pollution onto soils, sediments, and vegetation. Other sources of Pb to ecosystems include direct discharge of waste streams to water bodies and mining. Elevated Pb levels in the environment can result in decreased growth and reproduction in plants and animals, and neurological effects in vertebrates (Fig. 11−1).

Climate change and lead

Very few recent studies are found in the global literature on the role of climate change and forest fires in the spread of Pb pollution in the environment. It is, however, evident that in many places, warmer temperatures will lead to increased dryness and drought, with drier soils making dirt dustier. If Pb is present in the soil, it becomes more mobile in that dust.

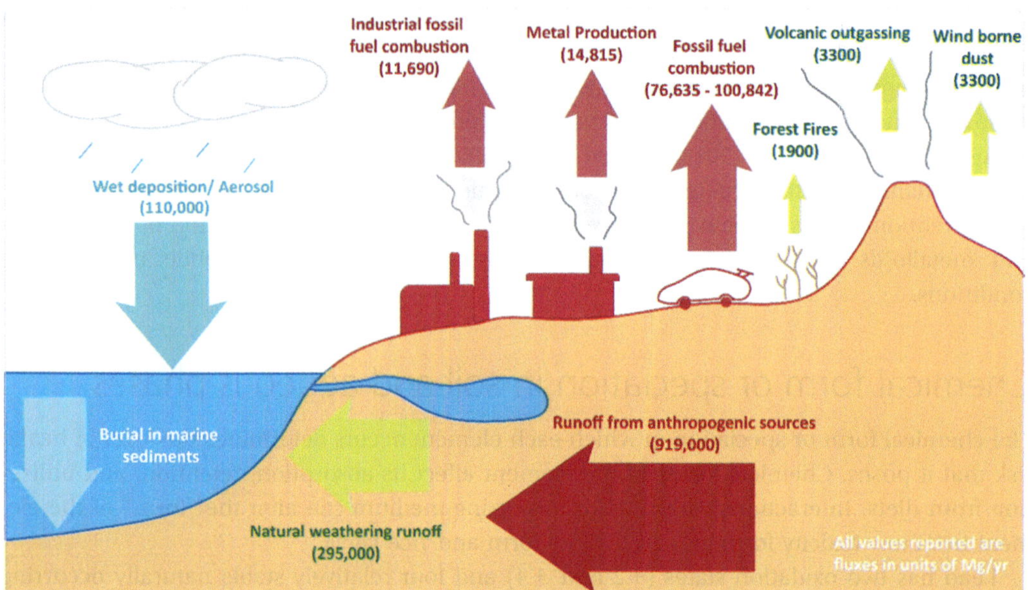

FIGURE 11–1 Simplified schematic of the lead cycle. All values indicated are fluxes with unit of Mg/year; the values have been obtained from Cullen and McAlister (2017). The size of the arrows are approximately proportional to their flux. The major reservoir for lead is the crust and mantle with a concentration of 11–14.8 ppm. The natural sources (green arrows) of lead in the atmosphere are volcanic eruptions, plant exudates, forest fires, extraterrestrial particles, radioactive decay, and physical and chemical weathering of rocks (Pacyna & Pacyna, 2011) The major anthropogenic sources (red arrows) are mining and smelting of ores, nonferrous metal production, stationary fossil fuel combustion platforms, and mobile fossil fuel combustion platforms (Cullen & McAlister, 2017). The sinks (blue arrows) of lead are wet deposition of aerosols on to the ocean water surface (Boyle et al., 2014) and the subsequent burial in deep sediments. Since lead is toxic to life, there are no predominant metabolic pathways (Cullen & McAlister, 2017).

Thus tiny Pb dust particles become pervasive in the environment. As the climate shifts, communities already contaminated with Pb will become even more exposed.

Forest fires

Available research data shows that forest fires remobilize Pb that was already in the environment, which can make Pb contamination more widespread (See Odigie and Flegal, 2011).

Mobility, mode of entry into food chains, and bioavailability

Mobility of lead

Unlike, many other metals, there is still room for more knowledge on the mobility of Pb in soil (Kushwaha et al., 2018). In sulfide-mineralized areas, Pb becomes mobile as a result of acidity that stems from the weathering of sulfide minerals, chiefly galena. Lead mobility is restricted by sorption on clay, organic matter, secondary iron and manganese oxides, and the formation of secondary minerals with low solubilities.

Entry into the food chain

Knowledge of the characteristics of Pb and its isotopic transfer in different compartments is scant, especially for the mobility of Pb isotopes in the geochemical cycle (Wang et al., 2013).

Lead can enter our food supply through a number of different ways, e.g.,: "Lead in the environment can settle on or be absorbed by fruits or vegetables or cereal grains used as food as well as in ingredients added to food, including dietary supplements. …. Lead can enter, inadvertently, through manufacturing processes. For example, plumbing that contains lead can contaminate water used in food production. …" (US FDA, 2022).

Bioavailability

Lead bioavailability is influenced by soil type, soil properties, and metal content. There are variations in methods for assessing Pb bioavailability. This, together with the differences in Pb source, limits our ability to conduct statistical analyses on influences of soil factors on Pb bioavailability (Yan et al., 2017).

As long ago as 1989, Chaney et al. suggested that bioavailability of soil Pb should be considered in setting limits for Pb levels in soils for areas where children live and play. This is because of children's innate tendency and their age-appropriate hand-to-mouth behavior of consuming nonfood items, such as contaminated soil or dust and flakes from decaying lead-containing paint.

Absorption and distribution

The main routes of exposure and absorption of Pb are inhalation, ingestion, and dermal contact (Ettinger et al., 2007). On entry into the body, Pb is distributed to organs such as the brain, kidneys, liver, and bones. The body stores Pb in the teeth and bones, in which organs it accumulates over time. Lead stored in bone may get into the blood during pregnancy, such that the fetus gets exposed. Young children are particularly at risk of Pb poisoning since they are capable of absorbing as much as four to five times of ingested Pb as adults from a given source (WHO, 2022).

Metabolic function (essentiality/clinical toxicity)

There are two main ways in which Pb interacts with human physiology. With its generally strong affinity for sulfhydryl groups and electron donor groups, Pb ends up getting bound to- and affecting, a wide range of proteins (Halmo and Nappe, 2022). Because Pb is so similar to other divalent cations like Ca and Zn, it interferes with the vast array of cellular mechanisms that are regulated by, and mediated by, these cations (Mitra et al., 2017).

Essentiality

Lead has no known biological role in plants or animals and is highly toxic to mammals and aquatic life; once it enters the body, it is known to cause severe health effects that might be irreversible.

Toxicity

The United Nations Environment Program (UNEP) has designated Pb as one of the most dangerous environmental pollutants (Morel, 2008). What makes Pb a particularly dangerous chemical, is that it can accumulate in individual organisms and in entire food chains. Thus lead pollution is a worldwide issue, not only in Africa. Lead is especially harmful to vulnerable populations, which include infants, young children, pregnant women, and their fetuses, as well as people with chronic health conditions (Chaurasia et al., 2023).

The main target organ for Pb toxicity is the central nervous system, the developing brain being more vulnerable to the neurotoxicity of Pb than the mature brain (UN FDA, 2022; Fig. 11−2). Beside developmental neurotoxicity in young children, Pb toxicity can also result in cardiovascular and nephrotoxicity disorders in adults. The effects of Pb poisoning are cumulative, can involve multiple body systems (WHO, 2022). As lead exposure increases, the range, and severity of symptoms and effects also increase (WHO, 2022).

There is no level of exposure to Pb that is known to be without harmful effects. In children, blood Pb concentrations as low as $5\,\mu g/dL$ have been shown to be associated with decreased intelligence, behavioral difficulties and learning problems (WHO, 2021). However, Pb exposure is known to be preventable (WHO, 2022.

Supplementation

Lead exposure is particularly harmful to fetal neurodevelopment and growth. Since more than 95% of Pb in the body during pregnancy is stored in bone (Ettinger et al., 2007), mobilization of cumulative maternal Pb stores into the circulation represents an endogenous source of exposure, which may pose a significant hazard for the fetus and infant. Maternal dietary Ca supplementation has been shown to reduce Pb levels in both animal and human experiments when administered during pregnancy and lactation. It is therefore considered that supplementation of the maternal diet with Ca may constitute a viable secondary prevention strategy in reducing circulating levels of Pb in the mother, and also reducing Pb exposure to the developing fetus and nursing infant (Ettinger et al., 2007).

In a randomized, placebo-controlled trial, Rosando et al. (2006) found that Fe supplementation of Pb-exposed children significantly improved Fe status but did not reduce blood Pb concentrations (PbBs) and that Zn supplementation did not reduce PbBs independently of Zn nutritional status. On this basis, Rosando et al. (2006) concluded that "... neither Fe nor Zn can be recommended as the sole treatment for lead-exposed school children." [Sic.]

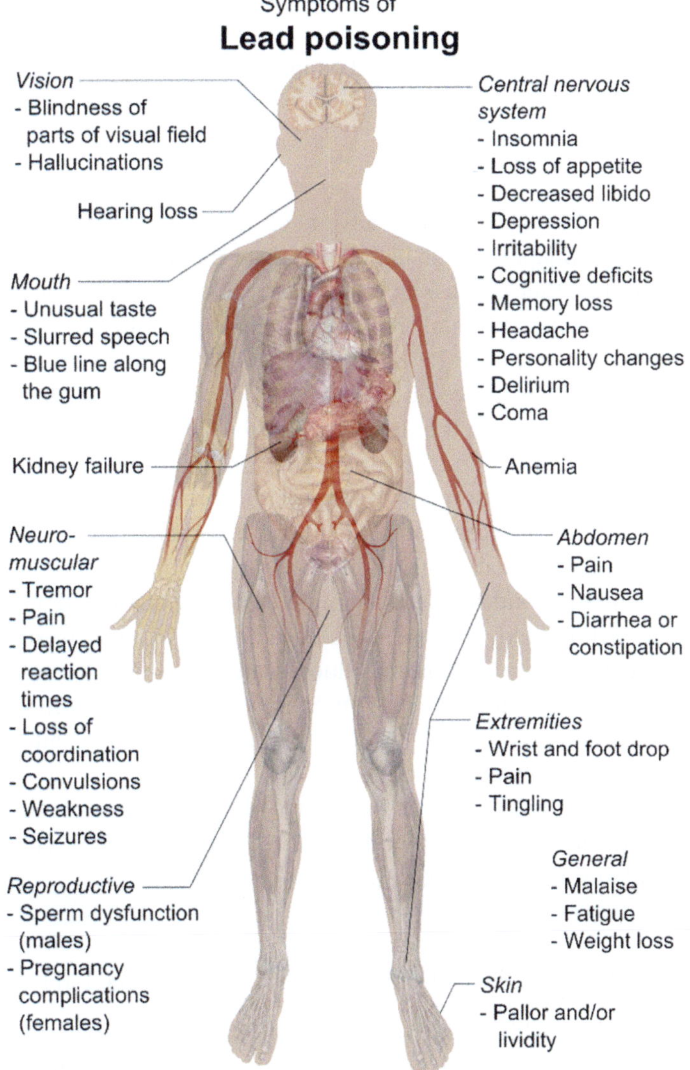

FIGURE 11–2 Symptoms of lead poisoning. Credit: Mikael Häggström.

Reduction of body burden

The main excretion pathway for most of the Pb absorbed into the body is via kidney clearance; excretion through other pathways, such as via sweat, saliva, hair, and fingernails, are, by comparison, negligible. *However, there are still many unknowns. Mechanisms by which inorganic Pb is excreted in urine, for instance, have not yet been fully studied (See*

Rădulescu and Lundgren, 2019). Such researches have been confronted by setbacks associated with measuring ultrafilterable Pb in plasma and thereby in measuring the rate of *glomerular filtration* of *Pb* (Larsen and Sánchez-Triana, 2023; Chamberlain et al., 1978).

Analytical determination

A number of deleterious health conditions have been linked to chronic Pb intoxication (See under "Toxicity," in this Chapter). Just like many toxic metals, Pb tends to accumulate in bone, but also in various other tissues, albeit in lesser amounts. This makes the determination of the Pb content of environmental and biological samples of great importance. The following specific reasons for accurate Pb determinations to be made on environmental and biological samples have been advanced by WHO (2020) "Identification of, and removal from, the source of exposure, or chelation therapy; to determine the effectiveness of risk mitigation measures; as part of a health screening or surveillance program to identify lead-exposed children; for exposure and risk assessment, for example, a prevalence study of lead exposure related to lead paint or other sources, and for occupational monitoring."

A wide range of methods exist (both destructive and nondestructive) for the measurement of Pb in environmental and biological samples. These include XRF and PIXE (particle-induced X-ray emission) methods and methods based on atomic spectroscopy, *viz.*, absorption and emission spectroscopy and atomic fluorescence methods (Sabol, 2020). Hauser's (2017) survey of the methods for the determination of Pb in environmental samples, is mainly based on sensitive atomic spectrometry, with discussions of alternative methods, such as electrochemical and separation methods.

Other noteworthy recent references on Pb determination in environmental and biological samples include the works of Beardsley et al. (2021), Forero-Mendieta et al. (2022), Gende and Schmeling (2022) and Karasiński et al. (2023).

Current status of geomedical research on lead in Africa

Lead exposure is a public health problem of a global dimension. It has been widely investigated in the developed countries but much less so in the developing countries, more particularly in Africa, where abundant evidence points to widespread and excessive childhood Pb exposure.

Most of the relatively few existing African studies have been conducted in South Africa; however, in 2010 and 2015, respectively, major episodes of Pb poisoning occurred in northern Nigeria that elicited considerable national and international attention, and triggered a number of research initiatives (e.g., Dooyema et al., 2012; Greig et al., 2014; Kaufman et al., 2016; Plumlee et al., 2013; Thurtle et al., 2014; Aguilar-Carrillo et al., 2018) to unravel the exact causes and to formulate credible mitigation and prevention measures.

The more recent South African studies on Pb exposure include that of Mathee et al. (2004), who presented an overview of Pb exposure in South African children over recent decades, with particular emphasis on sources of Pb exposure in the home environment. Again, in 2014, Mathee collated available information "… to illustrate that despite some progress, a wide range of sources of lead exist in South Africa, and that certain settings and groups continue to be at high risk of lead exposure." More recently, Barnes (2018) called for more rigorous epidemiological studies for strengthening the evidence "… to determine the magnitude of the association between child lead poisoning and psychological outcomes in South Africa." In 2023, Mathee et al. further their studies on environmental Pb in the South African environment by looking at the concentrations of Pb in ceramic tableware (Mathee et al., 2023).

Finally, in Ahmadi et al.'s (2020) study in Benin, West Africa, their prospective cohort confirmed the constantly high prevalence of elevated blood lead levels in children living in a rural sector south of the country, as well as the presence of multiple and continuous sources of Pb. These results underscore the need for prevention programs to reduce and eliminate Pb exposure in children.

Encouragingly, the successful phasing out of leaded gasoline in most countries, including some in Africa, together with other Pb control measures, has resulted in a significant decline in population-level blood lead concentrations (WHO, 2022). However, more needs to be done to phase out Pb paint; so far, only 41% of countries have introduced legally binding controls on lead paint (GHO, 2021).

Suggested areas for further research in Africa

There is a dearth of research on the various facets of Pb poisoning in Africa, especially among young children, the group that is most vulnerable—a statement that re-echoes the sentiment of Chukwuma (1997). For example, there are obvious gaps in knowledge of Pb poisoning prevention, especially for rural residents; and research and educational efforts focusing on decreasing these gaps must be considered urgent!

The past decade alone has seen the tragic deaths of hundreds of African children from Pb poisoning including in Nigeria as a result of artisanal Au mining, and in Senegal from battery recycling (WHO, n.d.). By 2006, however, most African countries had succeeded in phasing out of leaded petrol—a praiseworthy accomplishment! More work, though, is needed to determine basic measures required to effectively protect people against Pb from a variety of other sources—mines and smelters, automotive repair facilities, battery manufacturing and recycling plants, lead paint, hobbies and recreation involving Pb (such as fishing, jewelry making Pb-related cottage industries, and the use of Pb ammunition), geophagic practices, the consumption of adulterated alcohol "home brews," and the use of certain traditional medicines.

The WHO (n.d.) advocated the urgent need for a coordinated regional and country level program that would engender "… research and surveillance, regulation of the Pb content of paint, intensive public education and awareness campaigns, implementation of worker protection programs that also reduce para-occupational exposure, environmental assessments, the identification of high-risk settings, processes and groups, screening of vulnerable or high risk groups, and rehabilitation programs where necessary."

By 1996, the serious gaps existed in our knowledge of the chronic, subclinical toxicity of Pb, and dose-response relationships were still largely apparent. Little information was available on whether there existed threshold levels below which no toxic effects of Pb can be detected (Todd et al., 1996). Today, many of these knowledge gaps still exist in Africa where coordinated attempts at studying Pb toxicity, especially its epidemiology, and prevention are still lacking.

Additional specific target areas of Pb research in Africa include the following:

- As of relatively recently (2018), unlike other metals, our knowledge about the speciation and mobility of Pb in soil was still in its infancy (Kushwaha et al., 2018). Further studies on the several factors influencing the speciation of Pb, including pH of the soil, organic matter content, presence of various amendments, clay minerals, and presence of organic colloids and iron oxides are needed. This aspect of Pb research is of utmost importance, because the chemical form or speciation in which Pb occurs can determine the level of health risk that it poses by influencing its absorption, retention, and utilization from diets. Interactions within the transporting medium can alter the "form" of the element from low-toxicity form to high-toxicity form and vice versa.
- Lead intake by dust ingestion and by *geophagic practices* by children living in the neighborhood of present and past mining areas deserves further research.
- The provision of sufficient data and evidence through epidemiological studies to determine the magnitude of the association between child Pb poisoning and psychological outcomes in South Africa (See Barnes, 2018) and in Africa in general, is needed.
- Embracement of opportunities for psychological research in intervention studies and studies that are framed by an environmental justice agenda in South Africa (See Barnes, 2018) and in Africa in general is highly desirable.
- What is the possible contribution of Pb dose (both acute and chronic) to neuropsychological dysfunction in adults, to neurological degenerative diseases, such as dementias and movement disorders; renal dysfunction; hypertension; male reproductive impairment with reduction of sperm count? The question of whether or not lipid metabolism is involved in Pb-induced neuroinflammation still awaits a complete answer (See, e.g., Hu et al., 2023).
- Todd et al.'s 1996 review of "gaps in knowledge" in Pb research also raised the following questions regarding Pb poisoning during pregnancy "Is lead mobilized from bone during pregnancy a threat to mother or baby? Can it cause fetal neurotoxicity in utero? Should certain high-risk women be evaluated for body lead burden prior to becoming pregnant? If their levels are high, should they be chelated? Is chelation effective in reducing body lead burdens in women with chronic overexposure? Is such lead a legacy of environmental injustice?" [Sic.].
- Do occupational and environmental standards exist to protect against Pb toxicity in African countries? If they do, are they sufficiently stringent?

- In their 2006 examination of knowledge, attitudes, and practices related to Pb exposure in South Western Nigeria, Adebamowo concluded that: "... the findings of this study should be used, in conjunction with others, to develop appropriate health education intervention for lead exposure in the domestic environment."
- Kassy et al. (2021) concluded that for roadside and organized panel beaters in Enugu metropolis, Nigeria, "... there were general poor knowledge of hazards of lead exposure and high prevalence of elevated lead levels, which were more among organized panel beaters. Advocacy on standard organizational structures that support improved occupational health practices is needed and routine outreach by research institutions for health education and safety training." [Sic.]
- Zhang et al. (2013) suggested that, "... the control of lead poisoning should focus on primary prevention of lead exposure in children and development of special education programs for students with lead poisoning."

Glossary of Terms

- *Casparian strips* are a cellular feature found in the root endodermis of all higher plants. They are ring-like, hydrophobic cell wall impregnations.
- *Geophagy* (or *geophagia*), the habit of eating Earth materials (soil, clay, sediments, soft stone, wall scraping, sand, termite mound, anthill, dried-up stream sediment, etc.) is a practice that is common worldwide, not least in Africa, among members of the animal kingdom, including people.
- *Jarosite* [$KFe_3(SO_4)_2(OH)_6$] minerals are effective scavengers of PTEs and are abundant, for example, in acid rock/mine drainage scenarios.

References

Aguilar-Carrillo, J., Villalobos, M., Pi-Puig, T., Quiroz, I., Romero, F., 2018. Synergistic arsenic (V) and lead (II) retention on synthetic jarosite. I. Simultaneous structural incorporation behaviour and mechanism. Environmental Science: Processes and Impacts 20 (2), 354–369. Available from: https://pubs.rsc.org/en/content/articlelanding/2018/em/c7em00426e.

Ahmadi, S., Le Bot, B., Zoumenou, R., Durand, S., Fiévet, N., Ayotte, P., et al., 2020. Follow-up of elevated blood lead levels and sources in a cohort of children in Benin. International Journal of Environmental Research and Public Health 17 (22), 8689. Available from: https://doi.org/10.3390/ijerph17228689.

Angelova, V.R., Ivanova, R.V., Todorov, J.M., Ivanov, K., 2010. Lead, cadmium, zinc, and copper bioavailability in the soil-plant-animal system in a polluted area. Scientific World Journal 10, 273–285. Available from: https://doi.org/10.1100/tsw.2010.33.

ATSDR (Agency for Toxic Substances and Disease Registry), 2019. What is lead? https://www.atsdr.cdc.gov/csem/leadtoxicity/what_lead.html. (Accessed 31 August 2021).

Barnes, B.R., 2018. Child lead poisoning in South Africa: implications for psychological research. South African Journal of Psychology 48 (4), 410–419. Available from: https://doi.org/10.1177/0081246317714106.

Beardsley, C.A., Fuller, K.Z., Reilly 3rd, T.H., Henry, C.S., 2021. Method for analysis of environmental lead contamination in soils. The Analyst 146 (24), 7520−7527. Available from: https://doi.org/10.1039/d1an01744f.

Botté, A., Seguin, C., Nahrgang, J., Zaidi, M., Guery, J., Leignel, V., 2022. Lead in the marine environment: concentrations and effects on invertebrates. Ecotoxicology (London, England) 31 (2), 194−207. Available from: https://doi.org/10.1007/s10646-021-02504-4.

Boyle, E., Lee, J.-M., Echegoyen, Y., Noble, A., Moos, S., Carrasco, G., et al., 2014. Anthropogenic lead emissions in the ocean: the evolving global experiment. Oceanography 27 (1), 69−75. Available from: https://doi.org/10.5670/oceanog.2014.10.

Chamberlain, A.C., Heard, M.J., Little, P., Newton, D., Wells, A.C., Wiffen, R.D., 1978. Investigations into lead from motor vehicles. Technical report, Series AERE-R9198; Atomic Energy Research Establishment, Environmental and Medical Sciences Division, 151 p. (Accessed 25 September 2022).

Chaurasia, P., McClean, S.I., Mahdi, A.A., Yogarajah, P., Ansari, J.A., Kunwar, S., Ahmad, M.K., 2023. Automated lead toxicity prediction using computational modelling framework. Health Information Science and Systems 1156. Available from: https://doi.org/10.1007/s13755-023-00257-4.

Chukwuma, C., Sr, 1997. Environmental lead exposure in Africa. Ambio 26 (6), 399−403.

Collin, S., Baskar, A., Geevarghese, D.M., Ali, M.N.V.S., Bahubali, P., Choudhary, R., et al., 2022. Bioaccumulation of lead (Pb) and its effects in plants: a review. Journal of Hazardous Materials Letters 3, 100064. Available from: https://doi.org/10.1016/j.hazl.2022.100064.

Copin-Montegut, G., Courau, P.G., Laumond, F., 1986a. Occurrence of mercury in the atmosphere and waters of the Mediterranean. Papers presented at the FAO/UNEP/WHO/IOC/IAEA meeting on the Biogeochemical Cycle of Mercury in the Mediterranean, Siena, Italy, 27−31 August, 1984, FAO fisheries report. https://doi.org/10.1007/978-88-470-2105-1_14. (Accessed 21 September 2022)

Copin-Montegut, G., Courau, P.G., Nicolas, E., 1986b. Distribution and transfer of trace elements in the western Mediterranean. Marine Chemistry 18 (2−4), 189−195. Available from: https://doi.org/10.1016/0304-4203(86)90007-1.

Cullen, J., McAlister, J., 2017. 2. Biogeochemistry of lead. Its release to the environment and chemical speciation. In: Sigel, A., Sigel, H., Sigel, R.K.O. (Eds.), Lead: Its Effects on Environment and Health. De Gruyter, Berlin, Boston, pp. 21−48. Available from: https://doi.org/10.1515/9783110434330-002. (Accessed 31 August 2022).

Danielsson, L.G., 1980. Cadmium, cobalt, copper, iron, lead, nickel and zinc in Indian Ocean water. Marine Chemistry 8 (3), 199−215. Available from: https://doi.org/10.1016/0304-4203(80)90010-9.

Dooyema, C.A., Neri, A., Lo, Y.C., Durant, J., Dargan, P.I., Swarthout, T., et al., 2012. Outbreak of fatal childhood lead poisoning related to artisanal gold mining in northwestern Nigeria, 2010. Environmental Health Perspectives 120 (4), 601−607. Available from: https://doi.org/10.1289/ehp.1103965.

Ettinger, A.S., Hu, H., Hernandez-Avila, M., 2007. Dietary calcium supplementation to lower blood lead levels in pregnancy and lactation. The Journal of Nutritional Biochemistry 18 (3), 172−178. Available from: https://doi.org/10.1016/j.jnutbio.2006.12.007.

EU (European Union), n.d. Food safety: lead in food. https://food.ec.europa.eu/safety/chemical-safety/contaminants/catalogue/lead_en. (Accessed 21 September 2022).

Forero-Mendieta, J.R., Varón-Calderón, J.D., Varela-Martínez, D.A., Riaño-Herrera, D.A., Acosta-Velásquez, R.D., Benavides-Piracón, J.A., 2022. Validation of an analytical method for the determination of manganese and lead in human hair and nails using graphite furnace atomic absorption spectrometry. Separations 9, 158. Available from: https://doi.org/10.3390/separations9070158.

Gende, M., Schmeling, M., 2022. Development of an analytical method for determination of lead and cadmium in biological materials by GFAAS using Escherichia coli as model substance. PLoS One 17 (5), e0267775. Available from: https://doi.org/10.1371/journal.pone.0267775.

GHO (Global Health Observatory), 2021. Regulations and Controls on Lead Paint. World Health Organization, Geneva. Available from: https://www.who.int/data/gho/data/themes/topics/indicator-groups/legally-binding-controls-on-lead-paint. (Accessed 31 August 2021).

Greig, J., Thurtle, N., Cooney, L., Ariti, C., Ahmed, A.O., Ashagre, T., et al., 2014. Association of blood lead level with neurological features in 972 children affected by an acute severe lead poisoning outbreak in Zamfara State, Northern Nigeria. PLoS One 9 (4), e93716. Available from: https://doi.org/10.1371/journal.pone.0093716.

Halmo, L., Nappe, T.M., 2022. Lead toxicity. StatPearls [Internet]. Treasure Island (FL): StatPearls Publishing. National Library of Medicine. Available from: https://www.ncbi.nlm.nih.gov/books/NBK541097/.

Hauser, P.C., 2017. Chapter 3: Analytical methods for the determination of lead in the environment. In: Sigel, A., Sigel, H., Sigel, R.K.O. (Eds.), Lead: Its Effect on the Environment and Health. De Gruyter, Boston, pp. 49−60. Available from: https://doi.org/10.1515/9783110434330-003.

Hem, J.D., 1992. Study and Interpretation of the Chemical Characteristics of Natural Water, third ed. The United States Geological Survey U.S.G.S. Water-Supply Paper 2254. https://pubs.usgs.gov/wsp/wsp2254/pdf/intro.pdf. (Accessed 03 October 2022).

Hu, M., Zang, J., Wu, J., Su, P., 2023. Lead exposure induced lipid metabolism disorders by regulating the lipophagy process in microglia. Environmental Science and Pollution Research 30 (60), 125991−126008. Available from: https://doi.org/10.1007/s11356-023-31086-3.

Kabata-Pendias, A., Pendias, H., 2001. Trace Elements in Soils and Plants, third ed. CRC Press, Boca Raton, London, New York.

Karasiński, J., Bulska, E., Halicz, L., Tupys, A., Wagner, B., 2023. Precise determination of lead isotope ratios by MC-ICP-MS without matrix separation exemplified by unique samples of diverse origin and history. Journal of Analytical Atomic Spectrometry 38, 2468−2476. Available from: https://pubs.rsc.org/en/content/articlelanding/2023/ja/d3ja00259d/unauthn.

Kassy, C.W., Ochie, N.C., Ogugua, I.J., Aniemenam, C.R., Aniwada, C.E., Aguwa, E.N., 2021. Comparison of knowledge of occupational hazards of lead exposure and blood lead estimation among roadside and organized panel beaters in Enugu metropolis, Nigeria. The Pan African Medical Journal 40, 47. Available from: https://doi.org/10.11604/pamj.2021.40.47.28281.

Kaufman, J.A., Brown, M.J., Umar-Tsafe, N.T., Adbullahi, M.B., Getso, K.I., Kaita, I.M., et al., 2016. Prevalence and risk factors of elevated blood lead in children in gold ore processing communities, Zamfara, Nigeria, 2012. Journal of Health and Pollution 6 (11), 2−8. Available from: https://doi.org/10.5696/2156-9614-6-11.2.

Kumar, A., Kumar, A., Cabral-Pinto, M.M.S., Chaturvedi, A.K., Shabnam, A.A., Subrahmanyam, G., et al., 2020. Lead toxity: health hazards, influence on food chain, and sustainable remediation approaches. International Journal of Environmental Research and Public Health 17 (7), 2179. Available from: https://doi.org/10.3390/ijerph17072179.

Kushwaha, A., Hans, N., Kumar, S., Rani, R.A., 2018. Critical review on speciation, mobilization and toxicity of lead in soil-microbe-plant system and bioremediation strategies. Ecotoxicology and Environmental Safety 147, 1035−1045. Available from: https://doi.org/10.1016/j.ecoenv.2017.09.049.

Larsen, B., Sánchez-Triana, E., 2023. Global health burden and cost of lead exposure in children and adults: a health impact and economic modelling analysis. The Lancet. Planetary Health . Available from: https://api.semanticscholar.org/CorpusID:261763645. (Accessed 26 January 2024).

Levander, O.A., 1979. Lead toxicity and nutritional deficiencies. Environmental Health Perspectives 29, 115−125. Available from: https://doi.org/10.1289/ehp.7929115.

Levinson, A.A., 1980. Introduction to Exploration Geochemistry, second ed. Applied Publishing Ltd., Wilmette, IL, p. 924.

Levin, R., Zilli Vieira, C.L., Rosenbaum, M.H., Bischoff, K., Mordarski, D.C., Brown, M.J., 2021. The urban lead (Pb) burden in humans, animals and the natural environment. Environmental Research 193, 110377. Available from: https://doi.org/10.1016/j.envres.2020.110377.

Lindsay, W.L., 1979. Chemical Equilibria in Soils. John Wiley and Sons, New York, p. 449.

Liu, Y., Jiao, X., Teng, X., Gu, X., Teng, X., 2017. Antagonistic effect of selenium on lead-induced inflammatory injury through inhibiting the nuclear factor-κB signaling pathway and stimulating selenoproteins in chicken hearts. RSC Advances 7, 24878−24884. Available from: https://doi.org/10.1039/C7RA00034K.

Lovering, T.G., 1976 (Ed.). Lead in the Environment. US Geological Survey, Professional Paper, Issue 957. https://pubs.usgs.gov/pp/0957/report.pdf. (Accessed 20 September 2022).

MacDonald, R., Upton, B.G.T., Thomas, J.E., 1973. Potassium and fluorine-rich hydrous phase co-existing with peralkaline granite in south Greenland. Planetary Science Letters 18, 217−222.

Mart, L., Nurnberg, H.W., Dryssen, D., 1984. Trace metals levels in the eastern Artic Ocean. Science of the Total Environment 39 (1−2), 1−14. Available from: https://doi.org/10.1016/0048-9697(84)90020-2.

Mathee, A., Renton, L., Strreet, R., 2023. Concentrations of lead in ceramic tableware in South Africa. South African Journal of Science 119 (9/10)13853. Available from: https://doi.org/10.17159/sajs.2023/13853.

Mathee, A., von Schirnding, Y., Montgomery, M., Röllin, H., 2004. Lead poisoning in South African children: the hazard is at home. Reviews on Environmental Health 19 (3-4), 347−361.

Miao, Z., Miao, Z., Shi, X., Wu, H., Yao, Y., Xu, S., 2022. The antagonistic effect of selenium on lead-induced apoptosis and necroptosis via P38/JNK/ERK pathway in chicken kidney. Ecotoxicology and Environmental Safety 231, 113176. Available from: https://doi.org/10.1016/j.ecoenv.2022.113176. (Accessed 26 January 2024).

Mielke, J.E., 1979. Composition of the Earth's crust and distribution of the elements. In: Siegel, F.R. (Ed.), Review of Research on Modern Problems in Geochemistry, Earth Science Series, 16. International Association for Geochemistry and Cosmochemistry, Paris, pp. 13−37.

Mitra, P., Sharma, S., Purohit, P., Sharma, P., 2017. Clinical and molecular aspects of lead toxicity: An update, 2017. Critical Reviews in Clinical Laboratory Sciences 54 (7 - 8), 506−528. Available from: https://doi.org/10.1080/10408363.2017.1408.

Morel, F.M., 2008. The co-evolution of phytoplankton and trace element cycles in the oceans. Geobiology 6 (3), 318−324. Available from: https://doi.org/10.1111/j.1472-4669.2008.00144.x.

Odigie, K.O., Flegal, A.R., 2011. Pyrogenic remobilisation of historical industrial lead depositions. Environmental Science and Technology 45 (15), 6290−6295. Available from: https://doi.org/10.1021/es200944w.

Pacyna, J.M., Pacyna, E.G., 2011. An assessment of global and regional emissions of trace metals to the atmosphere from anthropogenic sources worldwide. Environmental Reviews 9 (4), 269−298. Available from: https://doi.org/10.1139/a01-012.

Patterson, C., 1965. Contaminated and natural lead environments of man. Archives of Environmental Health 11, 344−360.

Plumlee, G.S., Durant, J.T., Morman, S.A., Neri, A., Wolf, R.E., Dooyema, C.A., et al., 2013. Linking geological and health sciences to assess childhood lead poisoning from artisanal gold mining in Nigeria. Environmental Health Perspectives 121 (6), 744−750. Available from: https://doi.org/10.1289/ehp.1206051.

Pourrut, B., Shahid, M., Dumat, C., Winterton, P., Pinelli, E., 2011. Lead uptake, toxicity, and detoxification in plants. Reviews in Environmental Contamination and Toxicology (213), 113−136. Available from: https://doi.org/10.1007/978-1-4419-9860-6_4.

Raj, K., Das, A., 2023. Lead pollution: impact on environment and human health and approach for a sustainable solution. Environmental Chemistry and Ecotoxicology 5 (17), 79−85. Available from: https://doi.org/10.1016/j.enceco.2023.02.001.

Rosando, J.L., López, P., Kordas, K., García-Vargas, G., Ronquillo, D., Alatorre, J., et al., 2006. Iron and/or zinc supplementation did not reduce blood lead concentrations in children in a randomized, placebo-controlled trial. The Journal of Nutrition 136 (9), 2378−2383. Available from: https://doi.org/10.1093/jn/136.9.2378.

Rădulescu, A., Lundgren, S.A., 2019. A pharmacokinetic model of lead absorption and calcium competitive dynamics. Scientific Reports 9, 14225. Available from: https://doi.org/10.1038/s41598-019-50654-7.

Sabol, J., 2020. Major analytical methods for determining lead in environmental and biological samples. In: Gupta, D., Chatterjee, S., Walther, C. (Eds.), Lead in Plants and the Environment. Radionuclides and Heavy Metals in the Environment. Springer, Cham. Available from: https://doi.org/10.1007/978-3-030-21638-2_1. (Accessed 21 September 2022).

Schaule, B.K., Patterson, C.C., 1981. Lead concentrations in the northeast Pacific: evidence for global anthropogenic perturbations. Earth and Planetary Science Letters 54 (1), 97−116. Available from: https://doi.org/10.1016/0012-821X(81)90072-8.

Thurtle, N., Greig, J., Cooney, L., Amitai, Y., Ariti, C., Brown, M.J., et al., 2014. Description of 3,180 courses of chelation with dimercaptosuccinic acid in children ≤ 5 y with severe lead poisoning in Zamfara, northern Nigeria: a retrospective analysis of programme data. PLoS Medicine 11, e1001739. Available from: https://doi.org/10.1371/journal.pmed.1001739.

Todd, A.C., Wetmur, J.G., Moline, J.M., Godbold, J.H., Levin, S.M., Landrigan, P.J., 1996. Unraveling the chronic toxicity of lead: an essential priority for environmental health. Environmental Health Perspectives 104 (Supplement 1), 141−146. Available from: https://doi.org/10.1289/ehp.96104s1141.

Turekian, K., Wedepohl, K.H., 1961. Distribution of the elements in some major units of the Earth's crust. Geological Society of America Bulletin 72, 175−192.

UMA (University of Massachusetts Amherst), 2019. Soil lead fact sheet. https://ag.umass.edu/soil-plant-nutrient-testing-laboratory/fact-sheets/soil-lead-fact-sheet. (Accessed 31 August 2022).

US EPA. 1977. Standards of performance for secondary lead smelters. U.S. Environmental Protection Agency. Code of Federal Regulations. 40 CFR 60; Subpart L.

US FDA (The US Food and Drug Administration), 2022. Lead in Food, Foodwares, and Dietary Supplements. https://www.fda.gov/food/metals-and-your-food/lead-food-foodwares-and-dietary-supplements. (Accessed 21 September 2021).

USEPA (United States Environmental Protection Agency), 2023. Basic Information about Lead Air Pollution. https://www.epa.gov/lead-air-pollution/basic-information-about-lead-air-pollution. (Accessed 26 January 2023).

Vinogradov, A.P., 1956. Regularity of distribution of chemical elements in the Earth's crust. Geokhimiya I 6−52 (in Russian) Geochemistry 1, 1−43.

Vinogradov, A.P., 1962. Average contents of the chemical elements in the principal types of igneous rocks of the Earth's crust Geokhimiya 7, 555−571 (in Russian with English summary) [translated in Geochemistry 7, 641−664].

Wang, C., Wang, J., Yang, Z., Mao, C., Ji, J., 2013. Characteristics of lead geochemistry and the mobility of Pb isotopes in the system of pedogenic rock−pedosphere−irrigated riverwater−cereal−atmosphere from the Yangtze River delta region, China. Chemosphere 93, 1927−1935. Available from: https://doi.org/10.1016/j.chemosphere.2013.06.073.

Wedepohl, K.H., 1956. Untersuchungen zur Geochemie des Bleis [with English summary]. Geochimica et Cosmochimica Acta 10 (1−2), 69−148.

WHO (World Health Organisation), 2017. Guidelines for Drinking-water Quality, fourth ed. Incorporating the First Addendum, Geneva. 9789241549950-eng.pdf;jsessionid = 19FDD6C86618A662A93E152B51AB037A (who.int). (Accessed 21 September 2022).

WHO (World Health Organisation), 2020. Brief Guide to Analytical Methods for Measuring Lead in Blood, second ed. Geneva. https://apps.who.int/iris/bitstream/handle/10665/333914/9789240009776-eng.pdf?sequence = 1&isAllowed = y. (Accessed 01 October 2022).

WHO (The World Health Organization), 2021. The public health impact of chemicals: knowns and unknowns - data addendum for 2019. The Public health impact of chemicals: knowns and unknowns - data addendum for 2019 (who.int). (Accessed 27 September 2022).

WHO (World Health Organisation), 2022. Lead Poisoning. World Health Organisation, Geneva. Available from: https://www.who.int/news-room/fact-sheets/detail/lead-poisoning-and-health. (Accessed 02 October 2022).

Wildemann, T.M., Weber, L.P., Siciliano, S.D., 2015. Combined exposure to lead, inorganic mercury and methylmercury shows deviation from additivity for cardiovascular toxicity in rats. Applied Toxicology 35, 918−926. Available from: https://doi.org/10.1002/jat.3092.

Yan, K., Dong, Z., Wijayawardena, M., Liu, Y., Naidu, R., Semple, K., 2017. Measurement of soil lead bioavailability and influence of soil types and properties: a review. Chemosphere 184, 27−42. Available from: https://doi.org/10.1016/j.chemosphere.2017.05.143.

Zhang, N., Baker, H.W., Tufts, M., Raymond, R.E., Salihu, H., Elliott, M.R., 2013. Early childhood lead exposure and academic achievement: evidence from Detroit public schools, 2008-2010. American Journal of Public Health 103 (3), e72−e77.

Further reading

Adebamowo, E.O., Agbede, O.A., Sridhar, M.K., Adebamowo, C.A., 2006. An examination of knowledge, attitudes and practices related to lead exposure in South Western Nigeria. BMC Public Health 6, 82. Available from: https://doi.org/10.1186/1471-2458-6-82.

ATSDR (Agency for Toxic Substances and Disease Registry), 2007. A Toxicological Profile for Lead. U.S. Department of Health and Human Services. https://semspub.epa.gov/work/05/930045.pdf. (Accessed 03 October 2022).

Chaney, R.L., Mielke, H.W., Sterrett, S.B., 1989. Speciation, mobility, and bioavailability of soil lead. In: Davies, B. E., Wixson, B.G. (Eds.), Proc. Intern. Conf. Lead in Soils: Issues and Guidelines. Environmental Geochemistry and Health 11 (Supplement), 105−129. https://www.researchgate.net/profile/Howard-Mielke/publication/284048911_Speciation_Mobility_and_Bioavailability_of_Soil_Lead/links/5c51c55f458515a4c74afbd5/Speciation-Mobility-and-Bioavailability-of-Soil-Lead.pdf. (Accessed 31 August 2022).

Herkovits, J., Pérez-Coll, C.S., 1991. Antagonism and synergism between lead and zinc in amphibian larvae. Environmental Pollution (Barking, Essex: 1987) 69 (2-3), 217−221. Available from: https://doi.org/10.1016/0269-7491(91)90145-m.

Mathee, A., 2014. Towards the prevention of lead exposure in South Africa: contemporary and emerging challenges. Neurotoxicology 45, 220−223. Available from: https://doi.org/10.1016/j.neuro.2014.07.007.

Nriagu, J.O., Pacyna, J.M., 1988. Quantitative assessment of worldwide contamination of air, water and soils by trace metals. Nature 333, 134−139. Available from: https://doi.org/10.1038/333134a0.

Tirima, S., Bartrem, C., von Lindern, I., von Braun, M., Lind, D., Anka, S.M., et al., 2018. Food contamination as a pathway for lead exposure in children during the 2010 - 2013 lead poisoning epidemic in Zamfara, Nigeria. Journal of Environmental Sciences (China) 67, 260−272. Available from: https://doi.org/10.1016/j.jes.2017.09.007.

12

Medical geology of manganese

"Despite decades of interrogation, we still have much to learn about the players, mechanisms, and consequences of the Mn cycle and new and exciting discoveries are being made at a rapid rate."

Hansel (2017)

Key chapter features

1. Presents a look at important geomedical aspects influencing Mn metabolism in humans with beneficial endpoints or resulting toxicity;
2. Highlights latest advances in Mn research in Medical Geology (> 70% of the references postdating the year 2000);
3. Suggests areas of future Mn research to serve as a starting point of investigation for graduate Medical Geology researchers and others from allied disciplines.

Introduction

Manganese (Mn), a silvery white, hard, but very brittle metallic element, belongs to Group 7 (VIIb) of the Periodic Table. Its atomic number is 25, and atomic mass, 54.9380 g/mol. Manganese occupies 12th position in terms of elemental abundance in the Earth's crust and is found mostly in the range of 0.1%−0.2% in common crustal rocks (Cannon et al., 2017). Manganese often occurs in solid form, but under acidic conditions, can become soluble, which enhances its ability to migrate to various compartments of the geosphere, where it can be amenable to interactions with virtually all life forms including humans. For this reason, its release from mining and ore processing, which can produce very high concentrations, must be monitored throughout the life cycle of Mn production and use (Cannon et al., 2017).

Manganese can also be released from other anthropogenic sources, such as agriculture, traffic, and engineering; but its release from geogenic sources lead to greater exposures than from anthropogenic sources (Yang and Sañudo-Wilhelmy, 1998).

Manganese is an important nutrient for most living organisms, including humans, being required in small amounts for the proper functioning of numerous organ systems and the manufacture of enzymes that are necessary for the metabolism of proteins and fats. According to Aschner and Erikson (2017), "The beneficial effects of manganese are due to

Medical Geology of Africa. DOI: https://doi.org/10.1016/B978-0-12-818748-7.00001-0

the incorporation of the metal into metalloproteins." Horning et al. (2015) gave the range of Mn in mammalian tissue in the order of 0.3−2.9 µg Mn/g wet tissue weight, which makes Mn one of the most common metals in tissues.

Though Mn toxicity (*manganism*) seldom occurs, cases of overexposure have been reported, mainly in occupational settings such as mining, smelting, and welding, and *iatrogenic* conditions or accidental situations. The main effects of Mn toxicity are issues concerning the central nervous system, which can present as tremors, muscle spasms, *tinnitus*, loss of hearing, and a semblance of unsteadiness on one's feet (Nielsen, 2012; Buchman, 2014). Other marked effects include hepatotoxicity, cardiotoxicity and increased mortality in infants (O'Neal and Zheng, 2015).

Sources

Igneous rocks

The content of Mn in igneous rocks as given by Mielke (1979) are, for:

Granitic rocks, 390−540 ppm.
Syenite, 850 ppm.
Basaltic rocks, 1500 ppm.
Ultramafic igneous rocks: 1600 ppm.
In most igneous rocks, Mn correlates closely with ferrous Fe, giving Mn:Fe ratios that generally lie within the range of 0.015−0.02 (Everett et al., 2019).

Sedimentary rocks

Limestone or dolomite; sandstone; shale; coal.

The Mn content of sedimentary rocks is controlled by the geochemistry of the source rock and the redox conditions of the depositional environment (Wedepohl, 1978).

Shale and greywacke generally have higher levels of Mn (c. 700 ppm) relative to coarser quartzitic sandstone and grits (c. 170 ppm) (Everett et al., 2019).

Carbonate rocks, particularly dolomite, may also contain high concentrations of Mn, on average c. 550 ppm (Wedepohl, 1978).

Loess has an average concentration of 560 ppm Mn (McLennan and Murray, 1999).

Minerals

Trace amounts of Mn are found in many minerals. In the ore minerals, Mn occurs as oxides (e.g., *pyrolusite* (MnO_2) and *cryptomelane* [$K(Mn^{4+}Mn^{2+})_8O_{16}$]), carbonates (e.g., *rhodochrosite* ($MnCO_3$) and *kutnahorite* [$CaMn^{2+}(CO3)_2$]), hydroxides, e.g., manganite [$MnO(OH)$], and silicates (e.g., *braunite*, [$Mn^{2+}Mn^{3+}6(SiO_4)O_8$]). The carbonate mineral rhodochrosite is the most widely recognized Mn mineral in known deposits (Cannon et al., 2017).

Soils

Manganese contents in soils range from 7 to 9200 ppm (Kabata-Pendias and Mukherjee, 2007). Higher Mn contents commonly occur in soils developed from mafic rocks, soils rich in Fe and/or organic matter, and soils developed in arid or semiarid regions.

Plants

Mn is an important plant micronutrient, which sustains various metabolic roles, such as in the process of photosynthesis, respiration, and nitrogen assimilation (Alejandro et al., 2020). It is required in small amounts for healthy plant growth, but can also be toxic to plant tissue if taken up in excess. In many ways, Mn activity in plants is similar to that of Fe (See: Section B "Associations (additive/synergistic/antagonistic)," in this Chapter).

Despite the importance of Mn for photosynthesis and other processes, Alejandro et al. (2020) consider that the physiological importance of Mn uptake and compartmentation in plants has been underestimated. Mn deficiency is particularly a serious and widespread plant nutritional disorder in dry, well-aerated, and calcareous soils; but also in soils with high organic matter content, where bioavailability of Mn can decrease to an extent far less than that required for normal plant growth. As a result, plants have evolved an efficient mechanism to regulate the uptake of Mn as well as its trafficking, and storage. Alejandro et al. (2020) provides a comprehensive overview on knowledge up to the year 2020, of the multiple functions of transporters involved in Mn *homeostasis*, as well as their regulatory mechanisms in the plant's response to different conditions of Mn availability.

Food crops and other major dietary sources

A wide range of foodstuffs contain Mn. These include whole grains, soybeans, and other legumes, rice, leafy vegetables, clams, oysters, mussels, nuts, coffee, tea, and various spices, such as black pepper (Buchman, 2014; Nielsen, 2012; Aschner and Aschner, 2005; Ansari et al., 2004; US DAAR (US Department of Agriculture, Agricultural Research Service), 2019). Optimal nutritional sources of Mn are mainly from the plant-derived nutritional sources, such as whole grains, soybeans, but supplementation through vitamins or health products constitutes another notable contribution. Drinking water also provides significant amounts of Mn at concentrations of $1-100$ µg/L (Aschner and Aschner, 2005) with specific cut-offs measured to prevent toxic exposures. Breast milk and formula milk are other beneficial sources for infants, which can ward off deficiency during critical stages of their development.

Natural waters

Seawater

Manganese is widely distributed throughout the oceans of the world. Dust, sediments, and, more locally, rivers constitute the most important sources, giving a mainly homogeneous background concentration of dissolved Mn of about $0.10 \times 109 - 0.15 \times 109$ mol/L throughout most of the deep ocean (van Hulten et al., 2016). In the oceans, Mn occurs in the form of Mn^{2+}, Mn^{3+}, and Mn oxides. In the oxide form, Mn sequesters, along with Fe, very high levels of trace metals and nutrients compared to the surrounding seawater (Hansel, 2017), which are precipitated as Mn and ferromanganese (FeMn) mineral deposits that occur extensively in the deep ocean (Mizell and Hein, 2021).

Because fish are thought to be sensitive to low levels of dissolved metals, they are considered useful indicators of contamination in aquatic systems (Cannon et al., 2017).

Surface waters

Migration of Mn in the surface water environment takes place through its attachment to particulate matter that is either transported downstream or deposited as sediment.

Gaillardet et al. (2003) and references therein give a range of 0.4−113 ppb for dissolved Mn contents in large rivers worldwide, whereas Viers et al. (2009) give a figure of 1679 ppm for the average concentration of Mn in the suspended sediment of world rivers. US ATSDR (2000) gave a range of 0.4−10 ppb for ambient Mn concentrations in seawater, with an average of about 2 ppb given by Barceloux in 1999.

According to US ATSDR (2000), Mn occurs in both dissolved and suspended forms in surface waters, but this depends on factors, such as pH, anions present, and oxidation-reduction potential.

Groundwater

Concentrations of Fe and Mn *in groundwater* are often higher than those recorded in surface waters, a feature attributable to the reducing conditions found in groundwater. This may also be the case in some lakes and reservoirs. The US ATSDR (2000) reports concentrations of up to 1300 ppb in neutral groundwater and 9600 ppb in acidic groundwater. High concentrations of Mn can occur in municipal water as well as water from private wells. Factors determining the extent to which Fe and Mn dissolve in groundwater are the amount of oxygen in the water and, to a lesser extent, the degree of acidity, i.e., its pH.

Rainwater

Manganese is released into the atmosphere by both natural and anthropogenic sources. The main mechanism for the removal of Mn from the atmosphere is thought to be via rainwater (Inscore, 2009). The main atmospheric forms are divalent Mn (Mn^{2+}) and tetravalent Mn (Mn^{4+}). According to Inscore (2009), Mn^{2+} is a soluble oxidation state whereas Mn^{4+} is found only as a particulate. Divalent Mn is therefore found in higher concentrations in rainwater.

Drinking water

The normal range of Mn in drinking water is from 1 to 100 ppb, but many sources are found to have contents of <10 ppb (Keen and Zidenberg-Cherr, 1994). In 2013 US EPA (US Environmental Protection Agency) provided a (US) nonenforceable national secondary drinking water regulation for Mn—a guideline value set at 50 ppb. Up until 2011, the WHO included Mn at a maximum allowable concentration in drinking water at 400 ppb in its "Guidelines for Drinking Water Quality." The inclusion of Mn was subsequently discontinued in consideration that the recommended value was well above normal concentrations of Mn in drinking water (WHO (World Health Organisation), 2011a). Frisbie et al. (2012) suggest that "The toxic effects and geographic distribution of Mn in drinking-water supplies justify a re-evaluation by the WHO of its decision to discontinue its drinking-water guideline for Mn."

The bioavailability of Mn in drinking water may be greater than it is in food. However, very few studies have measured dietary Mn, *so total Mn intake in many cases remains largely unknown.* In the United States, the EPA recommends 0.05 ppm as the maximum allowable Mn concentration in drinking water (US EPA (United States Environmental Protection Agency), 2021), but the WHO, did not, by 2018, have a health-based limit for Mn in drinking water (Kullar et al., 2019; Lucchini et al., 2018).

Air

The concentration levels of Mn in the air vary widely depending on the proximity of point sources. Air erosion of dust or soil has been identified as the major source of Mn in the atmosphere (US ATSDR (US Agency for Toxic Substances and Disease Registry), 2012; Nriagu, 1989). This is primarily a natural source contribution, estimated to be 317 million kilograms per year (Mkg per year) by Nriagu in 1989. Another major source contribution to atmospheric Mn are emissions coming from anthropogenic activities such as mining and ore processing, as well as from iron and steel production facilities, coke ovens, and power plants.

In the United States, the mean concentration of Mn in ambient air was given as $0.02\,\mu g/m^3$ by the US ATSDR in 2012. However, in the vicinity of industrial sources, ambient levels of Mn can range from 0.22 to $0.3\,\mu g/m^3$ (US ATSDR (US Agency for Toxic Substances and Disease Registry), 2012), whereas Mn levels in urban and rural areas without point sources can range from 10 to $70\,ng/m^3$ (Barceloux, 1999).

In remote regions over the South Pole, air Mn concentrations range from 0.004 to $0.02\,ng/m^3$ (nanograms per cubic meter) and from 2.8 to $4.5\,ng/m^3$ over Greenland (Kabata-Pendias and Pendias, 2001).

Associations (additive/synergistic/antagonistic)

Manganese and Fe behave in a similar fashion, both in terms of their chemistry and biochemistry. Indeed, the mechanisms of Mn distribution both in health and disease are thought to be determined by Fe *homeostasis* (Ye et al., 2017; Aschner, 2000). This similarity engenders an inverse correlation between Fe intake and Mn absorption, since both metals compete for the same divalent metal transporter, *DMT1* (Tuschl et al., 2013). Thus a deficiency in Fe brings about a markedly increased intake of Mn (See also, Kim et al., 2012). In plants, too, uptake of Mn is in competition with uptake of Fe and, to a lesser extent, with Zn, Cu, Mg, and Ca (Alejandro et al., 2020).

Some studies indicate that Ca levels regulate Mn uptake in a synergistic manner (e.g., Ji et al., 2006); but earlier studies have reported little effect of supplemental Ca on Mn metabolism (e.g., Johnson and Lykken, 1991).

Another association of Mn is that seen with Se, with recent reports such as that of Zhang et al. (2021), indicating a link between higher levels of Se or Mn in the mother's blood, with lower blood pressure readings in children.

Chemical form or speciation in soil and aqueous phases

Both inorganic and organic forms of Mn exist with the most common oxidation states being: $+2$, $+3$, $+4$, $+6$, and $+7$. With reference to health effects, compounds containing inorganic Mn in the $+2$, $+3$, or $+4$ oxidation states are the most relevant, since these are the forms that are most commonly found in the environment and in the workplace (Williams et al., 2012). The Mn^{3+} and Mn^{4+} phases are thought to be some of the strongest oxidants in the environment, where the reactivity of Mn is dominated by its tendency to participate in redox reactions (Tebo et al., 2004).

Environmental circulation and impact of climate change

Hansel (2017) has noted the connection between the Mn cycle (Fig. 12−1) and nearly all other elemental cycles, and its intricate involvement in the health, metabolism, and function of the ocean's *microbiome*. Similarities between the Mn cycle and the Fe cycle, especially the geochemical and microbiological interactions within it are discussed by Geszvain et al. (2012) and Canfield et al. (2005).

According to Cannon et al. (2017), most Mn on the continents occurs in trace amounts in reduced ($+2$) form in silicate and carbonate minerals. Continental weathering accounts for release of Mn, some of which is dissolved in surface water and subsequently transported by rivers to the oceans of the world. Due to the generally oxidizing nature of surface water, the solubility of Mn is extremely low.

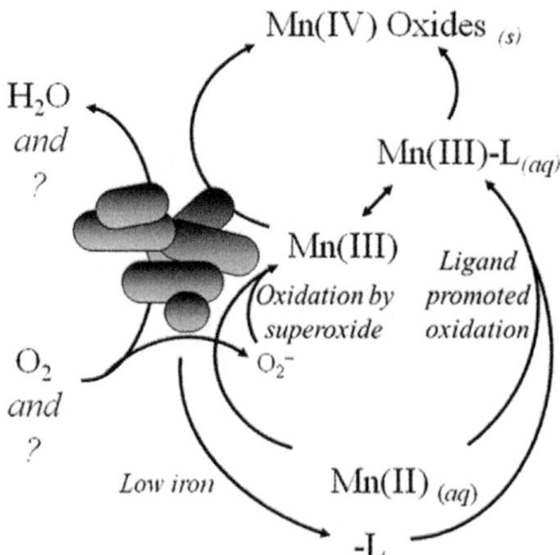

FIGURE 12−1 The updated manganese cycle. *After Geszvain, K., Butterfield, C., Davis, R.E., Madison, A.S., Lee, S.W., Parker, D.L., et al., 2012. The molecular biogeochemistry of manganese (II) oxidation. Biochemical Society Transactions 40 (6), 1244−1248. https://doi.org/10.1042/BST20120229.*

Manganese and climate change

One relationship that Mn has to climate change stems from the release of CO_2 from soil into the atmosphere as leaf litter decomposes, e.g., in boreal forest systems. Manganese is required by the fungi for them to be able to break down leaf litter. The rate of litter decomposition bears some relationship to the amount of carbon emitted by soils, such that if the rate of decomposition increases, the amount of CO_2 given out by soils into the atmosphere also increases (See: Li et al., 2021).

Recent researches on "manganese" having implications for climate change include the works of Ye et al. (2019) (Mn enrichments during interglacial-type conditions with high sea levels); Stendahl et al. (2017) (negative correlation between Mn availability and C accumulation in northern coniferous forest humus layers); Fernando and Lynch (2015) (the role of heightened risk of Mn toxicity in evaluating the potential impacts of global climate change on vegetation) and Keiluweit et al. (2015) (litter decomposition and Mn cycling).

Mobility, mode of entry into food chains and bioavailability

As already indicated, Mn is a potentially mobile element. Factors controlling the mobility as well as the solubility of Mn in various solutions include the acidity (pH) and oxidation potential (Eh) of Mn-bearing solutions (surface water, ocean water, and high-temperature fluids). The bioavailability of ingested Mn in humans has long ago been estimated to be to be in the region of 3%−5% (Finley et al., 1994). Both the bioavailability and absorption of Mn are known to be affected by an individual's nutrition status and other dietary factors (See LeMoon, 2019).

Absorption and distribution

Through the gastrointestinal tract in human adults, Mn is readily absorbed by either active transport or possibly facilitated diffusion when intakes are high (Buchman, 2014; Nielsen, 2012; Bai et al., 2008). The amount of Mn absorbed depends on the amount of dietary intake; but its typical average stands at about 3%−8% (EFSA (European Food Safety Authority), 2009). Subsequent to absorption, some Mn remains free, but most is bound to *transferrin*, *albumin*, and *plasma alpha-2-macroglobulin* (Fig. 12−2). Manganese is then taken up by the liver and other tissues, *but the mechanism of this process is not yet well understood* (Nielsen, 2012). Accumulation of Mn largely occurs in the *basal ganglia*, leading to manganism, with symptoms much akin to those of *Parkinson's disease* (e.g., Racette et al., 2021).

Metabolic function (essentiality/clinical toxicity)

A wealth of information exists on the effects of Mn on human health because of being both a hugely essential micronutrient and a potential neurotoxin at high intake levels

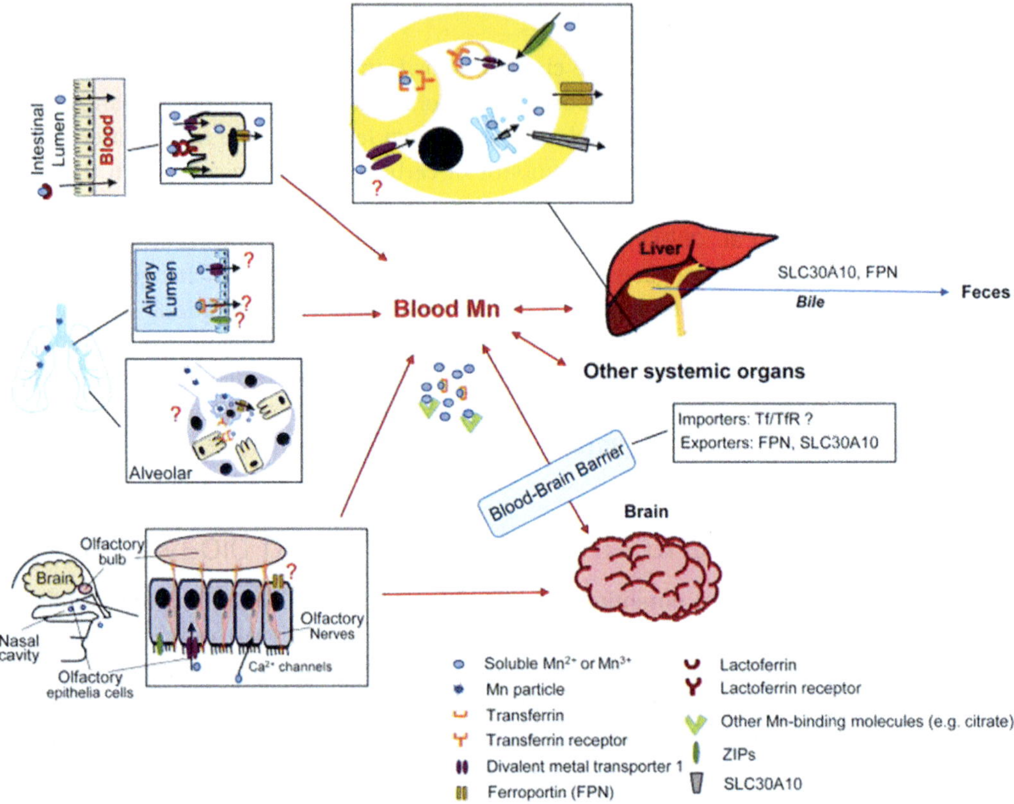

FIGURE 12–2 Absorption, distribution, and disposal of Mn. *From: Ye, Q., Park, J.E., Gugnani, K., Betharia, S., Pino-Figueroa, A., Kim, J., 2017. Influence of iron metabolism on manganese transport and toxicity. Metallomics 9 (8), 1028–1046. https://pubs.rsc.org/hy/content/articlelanding/2017/mt/c7mt00079k.*

(Cannon et al., 2017; Neal and Guilarte, 2013). For these reasons, tight regulation of Mn homeostasis by the body is crucial; but we are just at the beginning of our understanding of the exact mechanisms that govern Mn *homeostasis* (See, e.g., Aschner and Aschner, 2005; Avila et al., 2022).

Essentiality

The importance of Mn as a micronutrient in humans, is underscored by its involvement in the manufacture of enzymes necessary for the metabolism of proteins and fats, maintenance of normal immune function, and a host of other metabolic activities. Recent evidence presented by Rondanelli et al. (2021) based on literature study, appears to buttress the notion that the Mn plays an essential role in the synthesis of cartilage and bone collagen, as well as in bone mineralization.

Manganese is also an essential plant nutrient, playing a major role in a number of physiological processes, more especially, photosynthesis.

Deficiency

There is very limited evidence on the effect of Mn deficiency in humans, but it is thought to be rare (Buchman, 2014; Nielsen, 2012). Studies by Nielsen in 2012 and the earlier work by US IM in 2001 have indicated that bone demineralization and poor growth in children are two conditions whose causes, inter alia, may be attributed to Mn deficiency. Skin rashes, hair depigmentation, decreased serum cholesterol, and increased alkaline phosphatase activity in men are other disorders that can possibly be linked to Mn deficiency. In women, altered mood and increased premenstrual pain have been attributed to Mn deficiency (Nielsen, 2012; US IM (US Institute of Medicine, Food and Nutrition Board), 2001). More recently, Li and Yang (2018) have noted that Mn deficiency might also alter lipid and carbohydrate metabolism and cause abnormal glucose tolerance.

In plants, deficiency symptoms of Mn often resemble those of Fe, and present as interveinal chlorosis (yellow leaves with green veins) on young foliage.

Toxicity

Modes of exposure of the public to Mn are through consumption of food and water, inhalation of air, skin contact with air, water and soil, and handling and use of consumer products that contain Mn (See US ATSDR (US Agency for Toxic Substances and Disease Registry), 2012). By 2003, there was little or no evidence to suggest that the consumption of a Mn-rich plant-based diet will result in Mn toxicity (See, e.g., Finley et al., 2003). More recent work by LeMoon (2019), however, has revealed that Mn levels in plant-based diets may exceed U.S. dietary recommendations. An "Adequate Intake" set by US IM in 2001 for adult men and women is 2.3 and 1.8 mg/day, respectively. Severe neurological conditions can occur if there is oral exposure over and above this level. Workers at Mn mining and processing facilities are most likely to inhale Mn-rich dust. Inhaled Mn can lead to a permanent neurological disorder: *manganism* (*locura manganica*) (Bjørklund et al., 2017). The effects of Mn toxicity are evident mainly in the central nervous system, causing tremors, *tinnitus*, facial muscle spasms, hearing loss, and the feeling of being unsteady on one's feet (Buchman, 2014; Nielsen, 2012). *Hypermanganesemia* with dystonia type 1 is a disorder caused by *biallelic* mutations *in SLC30A10*, which encodes for a Mn exporter (Leyva-Illades et al., 2014). It mainly expresses in the small intestine, liver and brain (Bosomworth et al., 2012).

Exposure of children to high levels of Mn from environmental sources (airborne, drinking water, dietary) may lead to variety of adverse neurodevelopmental conditions, which may include changes in behavior, memory, and learning ability (Bjørklund et al., 2017; Kippler and Oskarsson, 2024). The ability of children to excrete Mn is low, and their brain is selectively permeable to the metal ion making them especially vulnerable to Mn excess.

Supplementation

Data on Mn intakes, including from dietary supplements, are very hard to come by but the (data) that presently exist indicate that most people obtain adequate amounts of Mn

(US NIH (US National Institutes of Health), 2021; US NIH (National Institutes of Health), 2018). Manganese is present in a variety of forms in dietary supplements including amino acid chelates (e.g., Mn bisglycinate chelate, Mn glycinate chelate, and manganese aspartate) (US NIH (National Institutes of Health), 2018), but by 2018, the relative bioavailability of the different forms of supplemental Mn remained unknown.

Research conducted on animals by Li and Yang in 2018 suggests that Mn supplementation can improve glucose tolerance, reduce oxidative stress, and improve endothelial dysfunction in diabetes; however, there were no clinical trials conducted in humans. *More research is therefore needed to determine whether Mn plays any role in the development of diabetes (US NIH (National Institutes of Health), 2018).*

No clinically relevant interactions with medications have, by 2018, been reported for Mn (US NIH (National Institutes of Health), 2018).

Reduction of body burden

About 10–20 mg Mn is contained in the human body, with 25%–40% residing in bone (Ferreira and Gahl, 2017; Buchman, 2014; Nielsen, 2012). Early estimates of the daily turnover of Mn by Brandt and Schramm (1986) put the figure at 20 μg, with dietary intake ranging from 2 to 22 mg/day. As already indicated, absorption to the tune of 2%–10% occurs throughout the intestinal tract (See, e.g., Hutchens et al., 2023). Recent work by Evans and Masullo (2021) reveals that excretion can be modulated by metabolic reactions in the *biliary system* and pancreas; and discharges (>90%) via *bile* into feces. Only a minuscule amount is reabsorbed (See: Buchman, 2014; Nielsen, 2012; Aschner and Aschner, 2005; US IM (US Institute of Medicine, Food and Nutrition Board), 2001), and an even smaller amount discharged in urine (US NIH (US National Institutes of Health), 2021).

Analytical determination

By 1999, we already had sensitive methods [e.g., atomic absorption spectrometry (AAS)], replacing earlier titrimetric methods (e.g., Lingane and Karplus, 1946) for determining total Mn in biological and environmental samples; although there were still challenges in distinguishing between different Mn oxidation states (Crapnell and Banks, 2022; IPCS, 1999). In 2000 US ATSDR reported detection limit for Mn as low as 1 ppb for urine, 200 ppb for hair, and 0.01 ppb for water samples using the AAS technique (US ATSDR, 2000). The use of inductively coupled argon-plasma optical emission spectrometry during this time also gave very low detection limits for Mn in biological and environmental samples (around 1–2 ppb for liquids and 5 $\mu g/m^3$ for air) (US ATSDR, 2000). Colorimetric methods that were used in water analysis also gave good pretty good detection limits (about 10 ppb) (ISO (International Organisation for Standardisation), 1986).

There has since been further improvements in the various methods for determination of Mn, including spectroscopic methods (e.g., Bernardini et al., 2021; Schuh et al., 2021;

Soylak et al., 2021; Kostova, 2010), voltammetry (e.g., Elias et al., 2020); and the introduction of novel techniques, such as the use of submersible microfluidic lab-on-chip analyzer (Geißler et al., 2021).

Treatment methods and performance

Common treatment methods, such as oxidation and filtration are available for lowering Mn concentrations in drinking water (WHO (World Health Organisation), 2011b). These methods are usually adequate to achieve a Mn concentration of 0.05 mg/L in drinking water.

Current status of geomedical research on manganese in Africa

In Africa, most of the research on environmental health impacts of Mn has been done with respect to exposures of workers and nearby residents of Mn mining centers in South Africa. This is unsurprising given that South Africa has one of the world's most abundant Mn ore deposits (Bam et al., 2016). Recent studies on Mn exposure in South Africa include the works of Dlamini et al. (2020), Röllin et al. (2014), Nelson et al. (2012), Röllin et al. (2005), Myers et al. (2003), and Hermanus (2000).

More recently, Cusick et al. (2018) assessed blood levels of a number of toxic metals including Mn in healthy children in Uganda and found that the prevalence of high blood levels could be correlated with the elevated levels of Mn.

Suggested areas for further research on manganese

Despite the appearance of voluminous literature reflecting the existence of a wealth of information on the effects of Mn on human health (as both an essential nutrient and a potential toxin in cases of overexposure), there are still several knowledge gaps to be filled. The complete picture of Mn metabolism and homeostasis needs to be definitively mapped out (Chen et al., 2018; US NIH (US National Institutes of Health), 2021). Starting with a better understanding of the reactions involved in the mode of entry of Mn into biological systems, the mechanism of Mn metabolic interactions that produce beneficial or toxic effects, are areas where the contributions of medical geologists would be greatly desired. Here are a few details regarding these knowledge gaps.

- Characterization of airborne and dietary Mn exposures, the deposition and fate of Mn, and its impact on central nervous system function are areas that need more research.
- Several microorganisms (especially bacteria and some fungi) could form Mn(III, IV) minerals by the oxidation of dissolved Mn(II) (See, e.g., Tebo et al., 2004). Though Mn (II)-oxidizing bacteria are so widely distributed in nature, we still do not fully understand the biochemical mechanism of Mn(II) oxidation (Nakama et al., 2014; Butterfield et al., 2013).

- The effect of long-term exposure to low levels of methylcyclopentadienyl manganese tricarbonyl, a manganese-containing antiknock agent used as an additive in gasoline combustion products, is not yet fully elucidated and warrants further research (Aschner, 2000).
- Research so far has not unequivocally established Mn to be a cause of epilepsy in humans; So, the relationship between Mn metabolism and epilepsy deserves further research (See, e.g., Gonzalez-Reyes et al., 2007; Keen and Zidenberg-Cherr, 1996; Carl et al., 1993).
- Little is known about the mechanisms that control Mn absorption efficiency in humans with respect to intake levels (See, e.g., Buchman, 2014). We now know that Mn is taken up by the liver and other tissues (e.g., Nielsen, 2012); however, the mechanism of this process is not yet fully established (Buchman, 2014).
- Lin et al. wrote in (2020) "Manganese exposure produces Parkinson's-like neurologic symptoms, suggesting a selective dysregulation of *dopamine* transmission. It is unknown, however, how manganese accumulates in dopaminergic brain regions or how it regulates the activity of dopamine neurons." Almost a decade ago now, Donaldson (2001) pointed out the lack of understanding of the reason for the loss of dopamine producing cells in the *corpus striatum* region of brain. Unfortunately, in 2020, the reason for this phenomenon still remains elusive.
- As of 2021, the role of Mn in bone construction had still not been fully elucidated (See, e.g., Chon et al., 2017; Grados et al., 2003 for possible role of Mn in bone construction). Research is needed to decipher whether associations exist between circulating Mn levels, bone mineral density, and osteoporosis in humans, since present evidence is very limited and inconsistent (See, e.g., US NIH (US National Institutes of Health), 2021). More studies are needed to establish whether Mn supplementation affects bone health in humans (US NIH (US National Institutes of Health), 2021).
- Whether Mn plays any role in the development of diabetes is still unclear (See US NIH (US National Institutes of Health), 2021). Although some research indicates that Mn supplements may help a person with diabetes produce more insulin naturally (See, e.g., Gong et al., 2020; Du et al., 2018), more research in humans is necessary to confirm these effects.
- The relationship between Mn and the development of cancer cells is still unclear. In 2012 US ATSDR reported that there was no evidence that Mn causes cancer in humans. However, more recent research has shown both positive (e.g., Zhang et al., 2014) and negative (e.g., Henroth et al., 2018) relationship, respectively, between Mn activity and cancer cell production and proliferation.
- By the year 2000, the studies on Mn transport across the blood-brain barrier as well as the mechanisms of its neurotoxicity were still inconclusive (See Aschner, 2000; Pajarillo et al., 2021).
- As far as drinking water sources are concerned, Ghosh et al.'s recent (2020) plea, regarding the need further research aimed at acquiring a better understanding of the individual health risks associated with Fe and Mn in drinking water sources, is worthy of consideration.

- For further areas that require more research, readers are referred to research objectives of the Manganese Health Research Program of Aschner et al. (2006), many of which are still being researched today.
- Finally, the concluding remarks in Donaldson's (2001) article is worth reiterating "Whatever the eventual outcome of manganese role in the neurobiology of aggression, there is little doubt that unraveling its role in the geochemical environment and its relation to health will prove a stimulating challenge for the emerging field of medical geology."

Glossary of terms

- *Biallelic*: Of, relating to, or affecting both alleles (variant forms) of a gene.
- The *biliary system*, also called the *biliary tract* or *biliary tree*, consists of the organs and ducts (bile ducts, gallbladder, and related structures) that are involved in the production and transportation of *bile* (a fluid produced and released by the liver and stored in the gallbladder).
- *Corpus Striatum*, also called *striatum*, is an important nucleus present in the forebrain.
- *DMT1* refers to the divalent metal transporter 1, which is a multimetal transporter with an important function in the Fe transport mechanism (Cheli et al., 2018).
- *Dopamine* refers to a type of neurotransmitter that plays several important roles in the brain and body.
- *Iatrogenic* refers to the causation of a disease or symptoms inadvertently induced in a patient by the medical treatment or other medical activity, including diagnosis.
- *Homeostasis* is any self-regulating process by which living systems tend to maintain state of steady internal, physical, and chemical conditions that are best for survival.
- *SLC30A10* is a cell surface-localized Mn efflux transporter.
- *Tinnitus*, commonly described as "ringing in the ears" is intermittent or continuous internal sound, in one or both ears, that is not coming from the outside world. This means that you can hear it, but other people cannot.

References

Alejandro, S., Höller, S., Meier, B., Peiter, E., 2020. Manganese in plants: from acquisition to subcellular allocation. Frontiers in Plant Science 11, 300. Available from: https://doi.org/10.3389/fpls.2020.00300.

Ansari, T.M., Ikram, N., Najam-Ul-Haq, M., Fayyaz, I., Fayyaz, Q., Ghafoor, I., et al., 2004. Essential trace metal (zinc, manganese, copper and iron) levels in plants of medicinal importance. Journal of Biological Sciences 4 (2), 95−99. Available from: https://doi.org/10.3923/jbs.2004.95.99.

Aschner, M., 2000. Manganese: brain transport and emerging research needs. Environmental Health Perspectives 108 (supplement 3), 429−432. Available from: https://doi.org/10.1289/ehp.00108s3429.

Aschner, J.L., Aschner, M., 2005. Nutritional aspects of manganese homeostasis. Molecular Aspects of Medicine 26, 353−362.

Aschner, M., Erikson, K., 2017. Manganese. Advances in Nutrition (Bethesda, MD) 8 (3), 520–521. Available from: https://doi.org/10.3945/an.117.015305.

Aschner, M., Lukey, B., Tremblay, A., 2006. The Manganese Health Research Programme (MHRP): status report and future research needs and directions. Neurotoxicology 27 (5), 733–736. Available from: https://doi.org/10.1016/j.neuro.2005.10.005.

Avila, D.S., Rocha, J.B.T., Tizabi, Y., dos Santos, A.P.M., Santamaria, A., Bowman, A.B., Aschner, M., 2022. Manganese Neurotoxicity. In: Kostrzewa, R.M. (Ed.), Handbook of Neurotoxicity. Springer, Cham, pp. 2305–2329.

Bai, S.P., Lu, L., Luo, X.G., Liu, B., 2008. Kinetics of manganese absorption in ligated small intestinal segments of broilers. Poultry Science 87 (12), 2596–2604. Available from: https://doi.org/10.3382/ps.2008-00117.

Bam, W., Zyl, H.V., Steenkamp, J., 2016. Identifying barriers faced by key role players in the South African manganese industry. In: SAIIE27 Proceedings, October 25–27, 2016. Stonehenge, South Africa. <https://www.mintek.co.za/Pyromet/Files/2016VanZyl.pdf> (Accessed 13 September 2021).

Barceloux, D.G., 1999. Manganese. Journal of Toxicology: Clinical Toxicology 37, 293–307. Available from: https://doi.org/10.1081/CLT-100102427.

Bernardini, S., Bellatreccia, F., Della Ventura, G., Sodo, A., 2021. A reliable method for determining the oxidation state of manganese at the microscale in Mn oxides *via* Raman spectroscopy. Geostandards and Geoanalytical Research 45, 223–244. Available from: https://doi.org/10.1111/ggr.12361.

Bjørklund, G., Chartrand, M.S., Aaseth, J., 2017. Manganese exposure and neurotoxic effects in children. Environmental Research 155, 380–384. Available from: https://doi.org/10.1016/j.envres.2017.03.003.

Bosomworth, H.J., Thornton, J.K., Coneyworth, L.J., Ford, D., Valentine, R.A., 2012. Efflux function, tissue-specific expression and intracellular trafficking of the Zn transporter ZnT10 indicate roles in adult Zn homeostasis. Metallomics 4 (8), 771–779.

Brandt, M., Schramm, V.L., 1986. Chapter 1: Mammalian manganese metabolism and manganese uptake and distribution in rat hepatocytes. In: Wedler, V.L.S.C. (Ed.), Manganese in Metabolism and Enzyme Function. Academic Press, pp. 3–16. < http://www.sciencedirect.com/science/article/pii/B9780126290509500053 > (Accessed 11 September 2021).

Buchman, A.R., 2014. Manganese. In: Catharine Ross, A., Cousins, R.J., Tucker, K.L., Ziegler, T.R. (Eds.), Modern Nutrition in Health and Disease, eleventh ed. Lippincott Williams and Wilkins, Baltimore, MD, pp. 238–244.

Butterfield, C.N., Soldatova, A.V., Lee, S.-W., Spiro, T.G., Tebo, B.M., 2013. Mn (II, III) oxidation and MnO_2 mineralization by an expressed bacterial multicopper oxidase. Proceedings of the National Academy of Sciences of the USA 110 (29), 11731–11735. Available from: https://doi.org/10.1073/pnas.1303677110.

Canfield, D.E., Kristensen, E., Thamdrup, B., 2005. The iron and manganese cycles. Advances in Marine Biology 48, 269–312. Available from: https://doi.org/10.1016/S0065-2881(05)48008-6.

Cannon, W.F., Kimball, B.E., Corathers, L.A., 2017, Chapter L: Manganese. In: Schulz, K.J., DeYoung, J.H., Jr., Seal, R.R., II, Bradley, D.C. (Eds.), Critical Mineral Resources of the United States—Economic and Environmental Geology and Prospects for Future Supply: U.S. Geological Survey Professional Paper 1802, pp. L1–L28, <https://doi.org/10.3133/pp1802L> (Accessed 10 September 2021).

Carl, G.F., Blackwell, L.K., Barnett, F.C., Thompson, L.A., Rissinger, C.J., Olin, K.L., et al., 1993. Manganese and epilepsy: brain glutamine synthetase and liver arginase activities in genetically epilepsy prone and chronically seizured rats. Epilepsia 34 (3), 441–446. Available from: https://doi.org/10.1111/j.1528-1157.1993.tb02584.x.

Cheli, V.T., Santiago González, D.A., Marziali, L.N., Zamora, N.N., Guitart, M.E., Spreuer, V., et al., 2018. The divalent metal transporter 1 (DMT1) is required for iron uptake and normal development of oligodendro-cyte progenitor cells. The Journal of Neuroscience: The Official Journal of the Society for Neuroscience 38 (43), 9142–9159. Available from: https://doi.org/10.1523/JNEUROSCI.1447-18.2018.

Chen, P., Bornhorst, J., Aschner, M., 2018. Manganese metabolism in humans. Frontiers in Bioscience-Landmark 23 (9), 1655−1679. Available from: https://doi.org/10.2741/4665, https://fbscience.com/Landmark/articles/.

Chon, S.J., Koh, Y.K., Heo, J.Y., Lee, J., Kim, M.K., Yun, B.H., et al., 2017. Effects of vitamin D deficiency and daily calcium intake on bone mineral density and osteoporosis in Korean postmenopausal woman. Obstetrics and Gynecology Science 60 (1), 53−62. Available from: https://doi.org/10.5468/ogs.2017.60.1.53.

Crapnell, R.D., Banks, C.E., 2022. Electroanalytical overview: the determination of manganese. Sensors and Actuators Reports 100110 (ISSN 2666-0539). Available from: https://e-space.mmu.ac.uk/630837/ (accessed 05.02.24).

Cusick, S.E., Jaramillo, E.G., Moody, E.C., Ssemata, A.S., Bitwayi, D., Lund, T.C., et al., 2018. Assessment of blood levels of heavy metals including lead and manganese in healthy children living in the Katanga settlement of Kampala, Uganda. BMC Public Health 18, 717. Available from: https://doi.org/10.1186/s12889-018-5589-0.

Dlamini, W.W., Nelson, G., Nielsen, S.S., Racette, B.A., 2020. Manganese exposure, parkinsonian signs, and quality of life in South African mine workers. American Journal of Industrial Medicine 63, 36−43. Available from: https://doi.org/10.1002/ajim.23060.

Donaldson, J., 2001. Manganese madness clues to the etiology of human brain disease emerges from a geological anomaly. Medical Geology Newsletter 4, 8−10.

Du, S., Wu, X., Han, T., Duan, W., Liu, L., Qi, J., et al., 2018. Dietary manganese and type 2 diabetes mellitus: two prospective cohort studies in China. Diabetologia 61, 1985−1995. Available from: https://doi.org/10.1007/s00125-018-4674-3.

EFSA (European Food Safety Authority), 2009. Scientific Opinion of the Panel on Food Additives and Nutrient Sources added to Food on manganese ascorbate, manganese aspartate, manganese bisglycinate and manganese pidolate as sources of manganese added for nutritional purposes to food supplements following a request from the European Commission. The EFSA Journal 1114, 1−23.

Elias, A.C., Castro, S.V.F., Muñoz, R.A.A., Silva, S.G., 2020. Voltammetric determination of free and total manganese in tea infusions. Journal of the Brazilian Chemical Society 31 (7), 1485−1491. Available from: https://doi.org/10.21577/0103-5053.20200035.

Evans, G.R., Masullo, L.N., 2021. Manganese toxicity. StatPearls. StatPearls Publishing, < https://www.ncbi.nlm.nih.gov/books/NBK560903/ > (Accessed 16 September 2021).

Everett, P., Lister, T., Fordyce, F., Ferreira, A., Donald, A., Gowing, C., et al., 2019. Stream Sediment Geochemical Atlas of the United Kingdom. Geoanalytics and Modelling Programme. Open Report OR/18/048. ISSN: 978-0-85272-787-4. <file:///D:/STREAM.pdf> (Accessed 24 September 2021).

Fernando, D.R., Lynch, J.P., 2015. Manganese phytotoxicity: new light on an old problem. Annals of Botany 116 (3), 313−319. Available from: https://doi.org/10.1093/aob/mcv111.

Ferreira, C.R., Gahl, W.A., 2017. Disorders of metal metabolism. Translational Science of Rare Diseases 2 (3-4), 101−139. Available from: https://doi.org/10.3233/TRD-170015.

Finley, J.W., Johnson, P.E., Johnson, L.K., 1994. Sex affects manganese absorption and retention by humans from a diet adequate in manganese. The American Journal of Clinical Nutrition 60, 949−955. Available from: https://doi.org/10.1093/ajcn/60.6.949.

Finley, J.W., Penland, J.G., Pettit, R.E., Davis, C.D., 2003. Dietary manganese intake and type of lipid do not affect clinical or neuropsychological measures in healthy young women. Journal of Nutrition 133 (9), 2849−2856. Available from: https://doi.org/10.1093/jn/133.9.2849.

Frisbie, S.H., Mitchell, E.J., Dustin, H., Maynard, D.M., Sarkar, B., 2012. World Health Organisation discontinues its drinking-water guideline for manganese. Environmental Health Perspectives 120, 775−778. Available from: https://doi.org/10.1289/ehp.1104693.

Gaillardet, J., Viers, J., Dupré, B., 2003. Trace elements in river waters. In: Drever, J.I. (Ed.), Surface and Ground Water, Weathering, and Soils, vol. 5 of: Holland, H.D., Turekian, K.K. (Executive Eds.), Treatise on Geochemistry, 5. Elsevier-Pergamon, pp. 225—272. Available from: https://doi.org/10.1016/B0-08-043751-6/05165-3.

Geißler, F., Achterberg, E.P., Beaton, A.D., Hopwood, M.J., Esposito, M., Mowlem, M.C., et al., 2021. Lab-on-chip analyser for the *in situ* determination of dissolved manganese in seawater. Scientific Reports 11, 2382. Available from: https://doi.org/10.1038/s41598-021-81779-3.

Geszvain, K., Butterfield, C., Davis, R.E., Madison, A.S., Lee, S.W., Parker, D.L., et al., 2012. The molecular bio-geochemistry of manganese (II) oxidation. Biochemical Society Transactions 40 (6), 1244—1248. Available from: https://doi.org/10.1042/BST20120229.

Ghosh, G.C., Khan, M.J.H., Chakraborty, T.K., Zaman, S., Enamul Kabir, A.H.M., 2020. Human health risk assessment of elevated and variable iron and manganese intake with arsenic-safe groundwater in Jashore, Bangladesh. Scientific Reports 10, 5206. Available from: https://doi.org/10.1038/s41598-020-62187-5.

Gong, J.H., Lo, K., Liu, Q., Li, J., Lai, S., Shadyab, A.H., et al., 2020. Dietary manganese, plasma markers of inflammation, and the development of type 2 diabetes in postmenopausal women: findings from the Women's Health Initiative. Diabetes Care 43 (6), 1344—1351. Available from: https://doi.org/10.2337/dc20-0243.

Gonzalez-Reyes, R.E., Gutierrez-Alvarez, A.M., Moreno, C.B., 2007. Manganese and epilepsy: a systematic review of the literature. Brain Research Reviews 53 (2), 332—336. Available from: https://doi.org/10.1016/j.brainresrev.2006.10.002.

Grados, F., Brazier, M., Kamel, S., Duver, S., Heurtebize, N., Maamer, M., et al., 2003. Effects on bone mineral density of calcium and vitamin D supplementation in elderly women with vitamin D deficiency. Joint, Bone, Spine: Revue du Rhumatisme 70 (3), 203—208. Available from: https://doi.org/10.1016/S1297-319X(03)00046-0.

Hansel, C.M., 2017. Manganese in marine microbiology. Advances in Microbial Physiology 70, 37—83. Available from: https://doi.org/10.1016/bs.ampbs.2017.01.005.

Henroth, B., Holm, I., Gondikas, A., Tassidis, H., 2018. Manganese inhibits viability of prostate cancer cells. Anticancer Research 38 (1), 137—145. Available from: https://doi.org/10.21873/anticanres.12201.

Hermanus, M.A., 2000. Manganese—a public health concern: Its relevance for occupational health and safety policy and regulation in South Africa. International Journal of Occupational and Environmental Health 6 (2), 151—160. Available from: https://doi.org/10.1179/oeh.2000.6.2.151.

Horning, K.J., Caito, S.W., Tipps, K.G., Bowman, A.B., Aschner, M., 2015. Manganese is essential for neuronal health. Annual Review of Nutrition 35, 71—108. Available from: https://doi.org/10.1146/annurev-nutr-071714-034419.

Hutchens, S., Jursa, T.P., Melkote, A., Grant, S.M., Smith, D.R., Mukhopadhyay, S., 2023. Hepatic and intesti-nal manganese excretion are both required to regulate brain manganese during elevated manganese exposure. American Journal of Physiology. Gastrointestinal and Liver Physiology 325 (3), G251—G264. Available from: https://doi.org/10.1152/ajpgi.00047.2023 (accessed 05.02.24).

Inscore, M.T., 2009 Manganese Concentration and Speciation in Coastal Rainwater, Southeastern North Carolina. A Thesis Submitted to the University of North Carolina Wilmington, in Partial Fulfillment of the Requirements for the Degree of Master of Science. <https://libres.uncg.edu/ir/uncw/listing.aspx?id = 1656> (Accessed 11 September 2021).

IPCS (The International Programme on Chemical Safety), 1999. Manganese and Its Compounds. Geneva, World Health Organization, International Programme on Chemical Safety (Concise International Chemical Assessment Document) 12. Available from: http://www.who.int/entity/ipcs/publications/cicad/en/cicad12.pdf (accessed 01.10.21).

ISO (International Organisation for Standardisation), 1986. Water quality—determination of manganese: for-maldoxime spectrometric method. Geneva, International Organisation for Standardisation (ISO 6333:1986). <https://www.iso.org/standard/12631.html> (Accessed 01 October 2021).

Ji, F., Luo, X., Lu, L., Liu, B., Yu, S., 2006. Effects of manganese source and calcium on manganese uptake by in vitro everted gut sacs of broiler's intestinal segments. Poultry Science 85, 1217−1225. Available from: https://doi.org/10.1093/ps/85.7.1217.

Johnson, P., Lykken, G., 1991. Manganese and calcium absorption and balance in young women fed diets with varying amounts of manganese and calcium. Journal of Trace Elements in Experimental Medicine 4, 19−35. Available from: https://agris.fao.org/agris-search/search.do?recordID = US201301956127.

Kabata-Pendias, A., Pendias, H., 2001. Trace Elements in Soils and Plants, third ed. CRC Press, Boca Raton, FL, p. 413.

Kabata-Pendias, A., Mukherjee, A.B., 2007. Trace Elements From Soil to Human. Springer Verlag, Berlin, Germany, p. 550.

Keen, C.L., Zidenberg-Cherr, S., 1994. Manganese toxicity in humans and experimental animals. In: Klimis-Tavantzis, D.J. (Ed.), Manganese in Health and Disease. CRC Press, Inc., Boca Raton, pp. 193−205.

Keen, C.L., Zidenberg-Cherr, S., 1996. Manganese. In: Ziegler, E.E., Filer, L.J. (Eds.), Present Knowledge in Nutrition, seventh ed. ILSI Press, Washington DC, pp. 334−343. Available from: Textbooks.com.

Keiluweit, M., Nico, P., Harmon, M.E., Mao, J., Pett-Ridge, J., Kleber, M., 2015. Litter decomposition and manganese cycling. Proceedings of the National Academy of Sciences 112 (38), E5253−E5260. < https://www.pnas.org/content/112/38/E5253 > .

Kim, J., Li, Y., Buckett, P.D., Böhlke, M., Thompson, K.J., Takahashi, M., et al., 2012. Iron-responsive olfactory uptake of manganese improves motor function deficits associated with iron deficiency. PLoS One 7 (3), e33533. Available from: https://doi.org/10.1371/journal.pone.0033533.

Kippler, M., Oskarsson, A., 2024. Manganese − a scoping review for nordic nutrition Recommendations 2023. Food and Nutrition Research 68. Available from: https://doi.org/10.29219/fnr.v68.10367 (accessed 05.02.24).

Kostova, D., 2010. Determination of manganese by a new spectrophotometry method using toluidine blue. Journal of Analytical Chemistry 65, 159−163. Available from: https://doi.org/10.1134/S1061934810020103.

Kullar, S.S., Shao, K., Surette, C., Foucher, D., Mergler, D., Cormier, P., et al., 2019. A benchmark concentration analysis for manganese in drinking water and IQ deficits in children. Environment International 130, 104889. Available from: https://doi.org/10.1016/j.envint.2019.05.083.

LeMoon, T., 2019. Manganese levels in plant-based diets may exceed U.S. dietary recommendations—an examination of the Ornish and DASH diets (P24-011-19). Current Developments in Nutrition 3 (Suppl. 1). Available from: https://doi.org/10.1093/cdn/nzz044.P24-011-19.

Leyva-Illades, D., Chen, P., Zogzas, C.E., Hutchens, S., Mercado, J.M., Swaim, C.D., et al., 2014. SLC30A10 is a cell surface-1329 localized manganese efflux transporter, and Parkinsonism-causing mutations block its intracellular trafficking1330 and efflux activity. Journal of Neuroscience 34 (42), 14079−14095.

Li, H., Santos, F., Butler, K., Herndon, E., 2021. A critical review on the multiple roles of manganese in stabilizing and destabilizing soil organic matter. Environmental Science and Technology 55 (18), 12136−12152. Available from: https://doi.org/10.1021/acs.est.1c00299 (accessed 05.02.24).

Li, L., Yang, X., 2018. The essential element manganese, oxidative stress, and metabolic diseases: links and interactions. Oxidative Medicine and Cell Longevity 2018, 7580707. Available from: https://doi.org/10.1155/2018/7580707.

Lin, M., Colon-Perez, L.M., Sambo, D.O., Miller, D.R., Lebowitz, J.J., Jimenez-Rondan, F., et al., 2020. Mechanism of manganese dysregulation of dopamine neuronal activity. The Journal of Neuroscience: The Official Journal of the Society for Neuroscience 40 (30), 5871−5891. Available from: https://doi.org/10.1523/JNEUROSCI.2830-19.2020.

Lingane, J.J., Karplus, R., 1946. New method for determination of manganese. Industrial and Engineering Chemistry, Analytical Edition 18 (3), 191−194. Available from: https://doi.org/10.1021/i560151a010.

Lucchini, R.G., Aschner, M., Landrigan, P.J., Cranmer, J.M., 2018. Neurotoxicity of manganese: indications for future research and public health intervention from the Manganese 2016 Conference. Neurotoxicology 64, 1−4. Available from: https://doi.org/10.1016/j.neuro.2018.01.002.

McLennan, S.M., Murray, R.W., 1999. Geochemistry of sediments. In: McDonough, W.F., Marshall, C.P., Fairbridge, R.F. (Eds.), Encyclopaedia of Geochemistry. Kluwer Academic Publishers, Dordrecht, pp. 282−292.

Mielke, J.E., 1979. Composition of the earth's crust and distribution of the elements. In: Siegel, F.R. (Ed.), Review of Research on Modern Problems in Geochemistry. Earth Science Series 16. New York, International Association of Chemistry and Cosmochemistry. UNESCO, pp. 13−37.

Mizell, K., Hein, J.R., 2021. Ocean floor manganese deposits. Reference Module in Earth Systems and Environmental Sciences, second ed. Elsevier, pp. 993−1001. Available from: https://doi.org/10.1016/B978-0-08-102908-4.00030-8. (Accessed 11 September 2021).

Myers, J.E., teWaterNaude, J., Fourie, M., Zogoe, H.B., Naik, I., Theodorou, P., et al., 2003. Nervous system effects of occupational manganese exposure on South African manganese mineworkers. Neurotoxicology 24 (4-5), 649−656. Available from: https://doi.org/10.1016/S0161-813X(03)00035-4.

Nakama, K., Medina, M., Lien, A., Ruggieri, J., Collins, K., Johnson, H.A., 2014. Heterologous expression and characterization of the manganese-oxidizing protein from *Erythrobacter* sp. strain SD21. Applied Environmental Microbiology 80 (21), 6837−6842. Available from: https://doi.org/10.1128/AEM.01873-14.

Neal, A.P., Guilarte, T.R., 2013. Mechanisms of lead and manganese neurotoxicity. Toxicology Research 2 (2), 99−114. Available from: https://doi.org/10.1039/C2TX20064C.

Nelson, G., Criswell, S.R., Zhang, J., Murray, J., Racette, B.A., 2012. Research capacity development in South African manganese mines to bridge exposure and neuropathologic outcomes. Neurotoxicology 33 (4), 683−686. Available from: https://doi.org/10.1016/j.neuro.2012.01.003.

Nielsen, F.H., 2012. Manganese, molybdenum, boron, chromium, and other trace elements. In: Erdman Jr., J.W., Steven, I.A.M., Zeisel, S.H. (Eds.), Present Knowledge in Nutrition, tenth ed. Wiley-Blackwell, pp. 586−607.

Nriagu, J.O., 1989. A global assessment of natural sources of atmospheric trace metals. Nature 338, 47−49. Available from: https://doi.org/10.1038/338047a0.

O'Neal, S.L., Zheng, W., 2015. Manganese toxicity upon overexposure: a decade in review. Current Environmental Health Reports 2 (3), 315−328. Available from: https://doi.org/10.1007/s40572-015-0056-x.

Pajarillo, E., Nyarko-Danquah, I., Adinew, G., Rizor, A., Aschner, M., Lee, E., 2021. Neurotoxicity mechanisms of manganese in the central nervous system. Advances in Neurotoxicology, 5, 215−238. Available from: https://doi.org/10.1016/bs.ant.2020.11.003.

Racette, B.A., Nelson, G., Dlamini, W.W., Prathibha, P., Turner, J.R., Ushe, M., et al., 2021. Severity of parkinsonism associated with environmental manganese exposure. Environmental Health 20, 27. Available from: https://doi.org/10.1186/s12940-021-00712-3.

Röllin, H., Mathee, A., Levin, J., Theodorou, P., Wewers, F., 2005. Blood manganese concentrations among first-grade schoolchildren in two South African cities. Environmental Research 97 (1), 93−99. Available from: https://doi.org/10.1016/j.envres.2004.05.003.

Röllin, H.B., Kootbodien, T., Theodorou, P., Odland, J.Ø., 2014. Prenatal exposure to manganese in South African coastal communities. Environmental Science: Processes and Impacts 16, 1903−1912. Available from: https://doi.org/10.1039/C4EM00131A.

Rondanelli, M., Faliva, M.A., Peroni, G., Infantino, V., Gasparri, C., Iannello, G., et al., 2021. Essentiality of manganese for bone health: an overview and update. Natural Product Communications 16 (5). Available from: https://doi.org/10.1177/1934578X211016649.

Schuh, C., Kirchner, M., Hebisch, R., Brock, T., Hartwig, A.MAK Commission, 2021. Manganese—determination of manganese and its inorganic compounds in workplace air using atomic absorption spectrometry (AAS). Air Monitoring Method 6 (2). Available from: https://doi.org/10.34865/am743996e6_2or (Translation of the German version from 2021).

Soylak, M., Agirbas, M., Yilmaz, E., 2021. A new strategy for the combination of supramolecular liquid phase microextraction and UV-Vis spectrophotometric determination for traces of maneb in food and water samples. Food Chemistry 338, 128068. Available from: https://doi.org/10.1016/j.foodchem.2020.128068.

Stendahl, J., Berg, B., Lindahl, B.D., 2017. Manganese availability is negatively associated with carbon storage in northern coniferous forest humus layers. Scientific Reports 7, 15487. Available from: https://doi.org/10.1038/s41598-017-15801-y.

Tebo, B.M., Bargar, J.R., Clement, B.G., Dick, G.J., Murray, K.J., Parker, D., et al., 2004. Biogenic manganese oxides: properties and mechanisms of formation. Annual Review of Earth and Planetary Sciences 32, 287–328. Available from: https://doi.org/10.1146/annurev.earth.32.101802.120213.

Tuschl, K., Mills, P.B., Clayton, P.T., 2013. Manganese and the brain. International Reviews on Neurobiology 110, 277–312.

US ATSDR, 2000. Toxicological Profile for Manganese. Atlanta, GA, United States Department of Health and Human Services, Public Health Service, Agency for Toxic Substances and Disease Registry.

US ATSDR (US Agency for Toxic Substances and Disease Registry), 2012, Toxicological Profile for Manganese: Atlanta, GA, U.S. Department of Health and Human Services, Public Health Services, September, 506 p. Plus 3 Appendixes. <http://www.atsdr.cdc.gov/ToxProfiles/tp151.pdf> (Accessed 10 September 2021).

US DAAR (US Department of Agriculture, Agricultural Research Service), 2019. FoodData Central. <https://fdc.nal.usda.gov/data-documentation.html> (Accessed 01 October 2021).

US EPA (US Environmental Protection Agency), 2013. National Primary Drinking Water Regulations: U.S. Environmental Protection Agency Web page: <http://water.epa.gov/drink/contaminants/index.cfm> (Accessed 10 September 2021).

US EPA (United States Environmental Protection Agency), 2021. Secondary Drinking Water Standards: Guidance for Nuisance Chemicals. <https://www.epa.gov/sdwa/secondary-drinking-water-standards-guidance-nuisance-chemicals> (Accessed 25 September 2021).

US IM (US Institute of Medicine, Food and Nutrition Board), 2001. Dietary Reference Intakes for Vitamin A, Vitamin K, Arsenic, Boron, Chromium, Copper, Iodine, Iron, Manganese, Molybdenum, Nickel, Silicon, Vanadium, and Zinc. National Academy Press, Washington, DC, <https://www.ncbi.nlm.nih.gov/books/NBK222310/>.

US NIH (National Institutes of Health), 2018. Dietary Supplement Label Database 2018. <https://dsld.od.nih.gov/> (Accessed 12 September 2021).

US NIH (US National Institutes of Health), 2021. Manganese. <https://ods.od.nih.gov/factsheets/Manganese-HealthProfessional/> (Accessed 13 September 2021).

van Hulten, M., Dutay, J., Middag, R., Baar, H.D., Roy-Barman, M., Gehlen, M., et al., 2016. Manganese in the world ocean: a first global model. Biogeosciences Discussions 1–38. Available from: https://doi.org/10.5194/bg-2016-282.

Viers, J., Dupré, B., Gaillardet, J., 2009. Chemical composition of suspended sediments in world rivers—new insights from a new database. Science of the Total Environment 407 (2), 853–868. Available from: https://doi.org/10.1016/j.scitotenv.2008.09.053.

Wedepohl, K.H., 1978. Handbook of Geochemistry. Springer-Verlag, Berlin.

WHO (World Health Organisation), 2011a. Guidelines for Drinking Water Quality, fourth ed. World Health Organisation, Geneva, Switzerland, p. 541, <http://www.who.int/water_sanitation_health/publications/2011/dwq_guidelines/en/>.

WHO (World Health Organisation), 2011b. Manganese in Drinking-water Background document for development of WHO Guidelines for Drinking-water Quality. WHO/SDE/WSH/03.04/104/Rev/1, Geneva, Switzerland. <https://www.who.int/water_sanitation_health/dwq/chemicals/manganese.pdf> (Accessed 28 September 2021).

Williams, M., Todd, G.D., Roney, N., Crawford, J., Coles, C., McClure, P.R., et al., 2012. Toxicological Profile for Manganese. Atlanta (GA): Agency for Toxic Substances and Disease Registry (US);

September 4, 2012, Chemical and Physical Information. <https://www.ncbi.nlm.nih.gov/books/NBK158876/> (Accessed 26 February 2021).

Yang, M., Sañudo-Wilhelmy, S.A., 1998. Cadmium and manganese distributions in the Hudson River estuary: interannual and seasonal variability. Earth and Planetary Science Letters 160 (3-4), 403−418. Available from: https://doi.org/10.1016/S0012-821X(98)00100-9.

Ye, Q., Park, J.E., Gugnani, K., Betharia, S., Pino-Figueroa, A., Kim, J., 2017. Influence of iron metabolism on manganese transport and toxicity. Metallomics 9 (8), 1028−1046. Available from: https://pubs.rsc.org/hy/content/articlelanding/2017/mt/c7mt00079k.

Ye, L., März, C., Polyak, L., Yu, X., Zhang, W., 2019. Dynamics of manganese and cerium enrichments in Arctic Ocean sediments: a case study from the Alpha Ridge. Frontiers in Earth Science 6, 236. < https://www.frontiersin.org/article/10.3389/feart.2018.00236 > .

Zhang, Q., Pan, E., Liu, L., Hu, W., He, Y., Xu, Q., et al., 2014. Study on the relationship between manganese concentrations in rural drinking water and incidence and mortality caused by cancer in Huai'an City. BioMed Research International 2014, 645056. Available from: https://doi.org/10.1155/2014/645056.

Zhang, M., Liu, T., Wang, G., Buckley, J.P., Guallar, E., Hong, X., et al., 2021. *In utero* exposure to heavy metals and trace elements and childhood blood pressure in a U.S. urban, low-Income, minority birth cohort. Environmental Health Perspectives 129 (6), 067005. Available from: https://doi.org/10.1289/EHP8325.

Further reading

Bosomworth, H.J., Thornton, J.K., Coneyworth, L.J., Ford, D., Valentine, R.A., 2012. Efflux function, tissue-specific expression and intracellular trafficking of the Zn transporter ZnT10 indicate roles in adult Zn homeostasis. Metallomics 4 (8), 771−779. Available from: https://doi.org/10.1039/c2mt20088k.

Kies, C., 1994. Bioavailability of manganese. In: Klimis-Tavantzis, D. (Ed.), Manganese in Health and Disease. CRC Press, Inc., Boca Raton, pp. 39−58.

Lin, W., Vann, D.R., Doulias, P.-T., Wang, T., Landesberg, G., Li, X., et al., 2017. Hepatic metal ion transporter ZIP8 regulates manganese homeostasis and manganese-dependent enzyme activity. Journal of Clinical Investigation 127 (6), 2407−2417. Available from: https://doi.org/10.1172/JCI90896.

US NRC (US National Research Council), 1989. Recommended Dietary Allowances, tenth ed. The National Academies Press, Washington, DC, p. 285, < http://www.nap.edu/catalog/1349/recommended-dietary-allowances-10th-edition > .

Wedler, F.C., 1994. Biochemical and nutritional role of manganese: an overview. In: Klimis-Tavantzis, D.J. (Ed.), Manganese in Health and Disease, 1994. CRC Press, Boca Raton, LA, pp. 1−36.

13 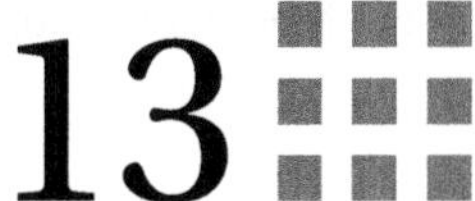

Medical geology of mercury

"Interpretation of mercury (Hg) geochemistry in environmental systems remains a challenge."

McLagan et al. (2022)

Sources

Mercury is emitted by natural sources, such as volcanoes, geothermal springs, geologic deposits, and the ocean, and by human activities. Human-related sources primarily include coal combustion, waste incineration, industrial uses, and mining. Mercury content of broad categories of rocks in the Earth's crust range from 10 to 20,000 ppb (USGS, 1970) (Box 13−1).

In the environment, Hg occurs mainly in the form of metallic Hg and inorganic Hg compounds, respectively. Metallic and inorganic Hg enter the air from the mining of deposits of ores that contain Hg, from the emissions of coal-fired power plants, from burning municipal and medical waste, from the production of cement, and from uncontrolled releases in factories that use Hg (ATSDR, 2015).

Igneous rocks

Igneous rocks are the basic sources of Hg (USGS, 1970). These (igneous) rocks generally contain less than 200 ppb of Hg and 100 ppb on average, except for alkalic igneous rocks and deep-seated eclogites and kimberlites that average several hundred ppb Hg.

Key chapter features

- Brief overviews of the most relevant and important characteristics of Hg, that are of relevance to the study of Medical Geology.
- Attributes that make Hg one of the most severe poisons in nature.
- Latest advances in Hg research in Africa as reflected in key papers in reputable journals or book chapters, about 90% of which postdate the year 2000.
- Suggested areas of future Hg research in Medical Geology and how these apply to the African situation.

Medical Geology of Africa. DOI: https://doi.org/10.1016/B978-0-12-818748-7.00004-6

The following world averages are from Jonasson and Boyle (1972): Granite, granodiorite, 62 ppb; Acid extrusives, 62 ppb; Alkali-rich rocks (syenite, nepheline syenite, and phonolite), 150 ppb; Diorite, 38 ppb; Andesite, 66 ppb; Basalt, 20 ppb; and Ultrabasic rocks (dunite and peridotite), 168 ppb.

Sedimentary rocks

Most sedimentary rocks have Hg contents less than 200 ppb Hg, except for shales, clays, and soils, for which the data show considerable variation with average contents of a few hundred ppb Hg.

Sedimentary rocks generally average less than 100 ppb of Hg and seldom exceed 200 ppb except for certain organic-rich shales, which may reach concentrations of 10,000 ppb or more (USGS, 1970).

The following world averages are from Jonasson and Boyle (1972), except the value for "coal," which is from Swaine (1990): Sandstone, arkose, 55 ppb; Shale, 67 ppb; Limestone/dolomite, 40 ppb; Coal (coal burnt in power stations of EU countries—domestic and imported), 300 ppb.

In addition to organic-rich shales, other rocks with abnormally high Hg contents are known to exist (USGS, 1970).

Metamorphic rocks

The Hg content of metamorphic rocks vary very widely. The following values for Hg content of metamorphic rocks are from Jonasson and Boyle (1972): Amphibolite, 50 ppb; Schist, 100 ppb; Gneiss, 50 ppb.

Minerals

Mercury occurs as a liquid at room temperature, and does not satisfy the normal criteria to be a valid mineral. There are quite a number of Hg-bearing minerals, but only a few occur abundantly in nature. The most common mineral containing Hg in ore deposits is *cinnabar*, or Hg sulfide (HgS), but naturally occurring elemental Hg, or *quicksilver* is also found in some Hg deposits, e.g., some cinnabar deposits. *Metacinnabar*, a Hg sulfide mineral and a polymorph of cinnabar (has the same chemical composition as cinnabar) is also sometimes found. In the minerals cinnabar and metacinnabar, Hg occurs mainly in its inorganic form, and as impurities in other minerals. Mercury readily combines with Cl, S, and other elements, and subsequently weather to form inorganic salts. Calomel, a Hg chloride mineral with formula Hg_2Cl_2 also occurs as a secondary mineral, which forms as an alteration product in some Hg deposits.

Soils

Mercury is naturally present in soils at concentrations ranging between 0.003 and 4.6 ppm (Steinnes, 1997)—in most cases is present below 0.5 ppm (Schluter, 1993)—whereas in contaminated sites, concentrations of up to 11,500 and 14,000 ppm have been reported (Gray et al., 2002; Neculita et al., 2005). An earlier (1970) average value of 100 ppb, and range,

30−500 ppb, were given for the Hg content of typical soils by USGS, who also noted that this value varies within relatively narrow limits (USGS, 1970). Soil concentrations of Hg are extremely high say 40,000 ppb or more in the vicinity of HgS deposits, but plants growing in them actually are likely to have Hg contents far below the level of their environment, say, in the range of 1000−3500 ppb.

Plants

According to USGS (1970), most plants are likely to contain less than 500 ppb of Hg. Terrestrial plants, just like aquatic organisms, absorb minor elements (including Hg) from the soils in which they grow at rates depending on the quality of the environment and the genetic characteristics of the plants. Unlike aquatic organisms, there seems to be little tendency for terrestrial plants to concentrate Hg above environmental levels. Even in circumstances where plants are growing in soils over (HgS$^-$) mineralized substrate, it is primarily the plants, which are rooted through the surface soil into the Hg ore, which have high Hg contents; shallower rooted plants are likely to show much lower levels. A few plants, *hyperaccumulators* (See: "Glossary of terms," in this Chapter), however, have an unusual capacity to concentrate Hg and even to separate it in metallic form (See, e.g., Ssenku et al., 2023).

Food crops and other major dietary sources

The Provisional Tolerable Weekly Intake (PTWI) of Hg for adults recommended by the European Environment and Health Information System (ENHIS) in 2007, is 0.0016 (ppm) body weight (not isolated dietary intake value).

The content of Hg in most foodstuffs is normally below the level of detection (usually 20 ng/g fresh weight) (WHO, 2000). Fish and marine mammals are the primary sources, Hg being present in the form of methyl Hg compounds (70%−90% of the total) (Chen and Evers, 2023; WHO, 2000). Mercury content of edible fish tissues for various species can vary from 50 to 1400 ppb fresh weight depending on factors, such as pH, the redox potential of the water and the species, age, and size of the fish (WHO, 2000).

Natural waters

Mercury content in rainwater generally falls within the range of 0.005−0.1 ppb, but mean levels as low as 0.001 ppb have been reported (IPCS, 1990). Rain washes Hg from the atmosphere just as it does certain other atmospheric components.

Naturally occurring levels of Hg in groundwater and surface waters are generally <0.5 ppb, although in areas where we have Hg mineralization (HgS) higher levels are usually encountered. A lower value of less than 0.1 ppb was given by USGS (1970) for surface waters, barring the influence of recent anthropogenic pollution, or extraneous geological conditions. This figure reflects the relatively low concentration of Hg in rainwater and the relatively tight bonding of Hg in organic and inorganic materials over which the water passes in its environmental cycling (USGS, 1970).

Higher concentrations of Hg are generally found in groundwaters because of the longer contact times with mineral grains, as well as the influence of other geoenvironmental factors.

In the study of bays by Soerensen et al. (2017), MeHg was found to represent from only 4%−4.5% of total Hg in freshwater, in general; and only 2.1% in offshore saltwater. In 2014 Lambourg et al. estimated the total amount of anthropogenic Hg in the global ocean to be 290 ± 80 million moles, about two-thirds of which resides in water shallower than a thousand meters.

Earlier on, in 1983, Nishimura et al. had determined the total Hg content of 342 seawater samples collected in the Bering Sea, North and South Pacific, Japan Sea, East and South China Seas, and Indian Ocean, and obtained concentrations of "total" Hg ranging from 0.003 to 0.006 ppb with an arithmetic mean of 0.0053 ppb, and a geometric mean of 0.005 ppb, for 70% of the total number of samples analyzed.

Drinking water

Mercury in drinking water is usually in the range 0.005−0.1 ppb (the same range as for rain water), the average value being about 0.25 ppb (IPCS, 1990). The forms of Hg occurring in drinking water are not known with certainty, but in unpolluted drinking water, most of the Hg is thought to be in the form of inorganic Hg (Hg^{2+}) (See, e.g., WHO, 2005), but this amount is unlikely to engender direct risk to human health, in terms of the MAC for drinking water (2 ppb, US EPA, 2021) set by recognized institutions, such as US EPA. However, certain individuals who drink water containing Hg^{2+} considerably above the MAC for many years could experience a variety of symptoms, which according to Teixeira et al. (2018) could include oxidative stress as well as *necrosis* and *apoptosis* processes, which kill cells. The cell death is considered the main factor responsible for fine motor control changes (Fig. 13−1).

Air

The primary transport pathway of Hg emissions is the atmosphere, whereas the redistribution of Hg in terrestrial, freshwater, and marine ecosystems and the production of MeHg are driven by land and ocean processes (Driscoll et al., 2013). The range of Hg concentrations in air has been put at 2−10 ng/m^3 by WHO (2003). Because of the tendency of Hg to vaporize, the atmosphere measured at ground level near Hg ore deposits (*cf.*, HgS) may contain as much as 20,000 ng/m^3 of Hg in air (USGS, 1970). This airborne mercury can fall to the ground in raindrops, in dust, or simply due to gravity (known as "air deposition").

The amount of Hg deposited in any particular area depends on the amount of Hg is released from local, regional, national, and international sources.

Few data are available on the speciation of Hg in the atmosphere, but practically all of the Hg emitted or re-emitted into the atmosphere is thought to be gas phase elemental Hg, with very small amounts of gas phase Hg^{2+} compounds and particulate bound Hg(II) (Friedli et al., 2003; ATSDR, 2022).

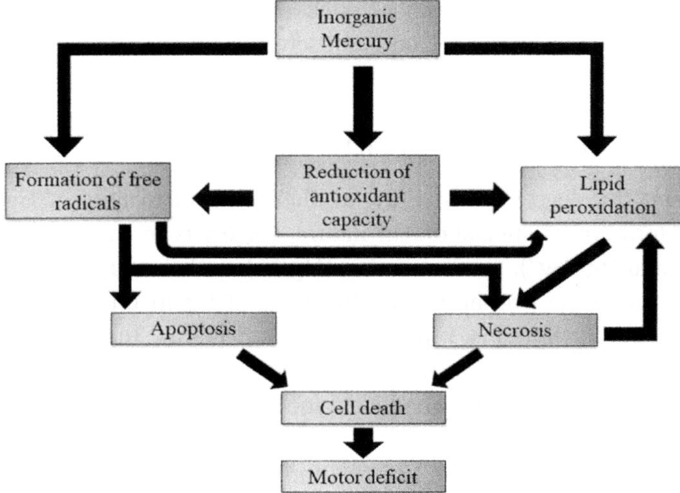

FIGURE 13–1 Long-term effects of excessive amounts of inorganic Hg. *From: Teixeira, F.B., de Oliveira, A., Leão, L., Fagundes, N., Fernandes, R.M., Fernandes, L., et al. (2018). Exposure to inorganic mercury causes oxidative stress, cell death, and functional deficits in the motor cortex. Frontiers in Molecular Neuroscience 11, 125. https://doi.org/10.3389/fnmol.2018.00125 (Permission being sought).*

Associations (synergism/antagonism)

Elemental Hg is a liquid at room temperature. Thus its gas phase is important geochemically, since Hg and some of its compounds have relatively high vapor pressures.

Mercury is a *chalcophile* element (having affinity for a sulfide phase) and extremely active biologically. A complex relationship exists between Hg and Se (also primarily chalcophilic in character) with reference to toxicity effects, with coexposure leading to reduced toxicity, enhanced toxicity, or no effect at all (See Wyatt et al., 2016). Different interactions may be related to chemical speciation, concentration, and method of administration (Dang and Wang, 2011); but these have not been thoroughly studied.

According to Timmerman and Omaye (2021), it is likely that the toxicity of Hg can result in Se deficiency, in that the affinity of Hg for Se is strong enough to cause deactivation of Se-dependent enzymes and proteins. Conversely, Se compounds can have prophylactic or antidotal effects, whereby the adverse toxicity action of Hg exposure can be prevented or reversed (Timmerman and Omaye, 2021).

Other metals/metalloids shown to have synergistic and/or antagonistic relationship with Hg include—Cr: e.g., synergistic action with Hg on fish (See: Selvan et al., 2012; Dwivedi et al., 2012); As: e.g., synergistic action in attenuating canonical and noncanonical *NLRP3* (See: "Glossary of terms," this Chapter) inflammasome activation (See "Glossary of terms," this article) (See Ahn et al., 2018); Pb: combined exposure (synergism) to Pb, inorganic Hg and MeHg showing deviation from additivity for cardiovascular toxicity in rats

(See Dwivedi et al., 2012) and Ni: synergistic effect of Ni and Hg on fatty acid composition in the muscle of fish *Lates calcarifer* (See Senthamilselvan et al., 2016).

Chemical form or speciation in soil and aqueous phases

Before working out the degree of mobility, availability, and toxicity of Hg, it is of critical importance to determine its speciation. In nature, Hg exists in three oxidation states, *viz.*, metallic or elemental Hg (Hg^0), mercurous ion (Hg_2^{2+}), and mercuric ion ($Hg^{2+)}$. These oxidation states determine the properties and behavior of Hg in the environment as well as in metabolic processes in humans, other animals, and plants. Inorganic Hg^{1+} species can also occur but are generally believed to play an inconsiderable role in natural systems; it is found mainly as shorter-lived intermediates in redox cycling between Hg^0 and Hg^{2+} species.

All three forms of Hg convey some degree of toxicity to life forms—including humans—but it is the Hg compounds containing the mercuric ion that are the most toxic, especially the organic-mercury compound, methyl mercury $[CH_3Hg^+]^+$, or simply, MeHg.

Mercury has seven stable isotopes, making it unique among the heavy elements—these isotopes are: Hg^{196}, Hg^{198}, Hg^{199}, Hg^{200}, Hg^{201}, Hg^{202} and Hg^{204}, and they span a relative mass difference of 4%. On this score, Bergquist and Blum (2009) recognized the great promise held by Hg isotope systematics—both mass-dependent and mass-independent fractionations—for tracking this toxin's geochemical transformations, and by extension, its adverse metabolic interactions (See Section: "Metabolic function (clinical toxicity)," in this Chapter).

Environmental circulation and impact of climate change

A lot has been written about the biogeochemical cycling of Hg, due it its intractability in terms of our understanding of several aspects of the cycle, such as oxidation processes in the atmosphere, land-atmosphere, and ocean-atmosphere cycling, and methylation processes in the ocean. Some of the more recent accounts include those of Gustin et al. (2020), Chételat et al. (2022), Wang et al. (2022), Sonke et al. (2023), Schneider et al. (2023).

Most of the Hg that enters the atmosphere (from mining deposits of ores that contain Hg, from the emissions of coal-fired power plants, from burning municipal and medical waste, from cement production, and from uncontrolled releases in factories that use Hg), is in the form of metallic (H^0) and inorganic compounds (ATSDR, 2015). The fate of Hg after it is emitted depends on several factors, such as the form of Hg emitted, location of the emission source, the height above the landscape from which the Hg is emitted (e.g., the height of a power-plant stack), the surrounding terrain, and the weather. Redeposition of Hg^0 takes place on terrestrial surfaces by dry deposition, including foliar uptake, sorption, and gas dissolution (Enrico et al., 2016). The ultimate deposition of Hg, probably as HgS ore, is believed to be in ocean sediments, where part of the inorganic Hg emitted becomes oxidized to Hg^{++} and then methylated or in other ways transformed into organomercurials (WHO, 2000).

Climate change

Some of the effects of climate change on the environmental perturbations of Hg include its impact on Hg deposition, *viz.*, a synchronous change in global Hg deposition, as reported by Li et al. (2020); changes in cycling of Hg, engendering an increase in MeHg and hence increased contamination (See: Fahnestock et al., 2019; Chételat et al., 2022); increase in Hg uptake in fish and marine mammals in regions experiencing glacial retreat (See, e.g., Wang et al., 2020; McKinney et al., 2022).

Mobility, mode of entry into food chains, and bioavailability

The mobility as well as availability of Hg in soils is quite low, and dependent primarily on the factors, such as soil texture, the amount of organic matter, and soil pH (See, e.g., Różański et al., 2016). A wide array of physical, chemical, biological, and isotopic exchange methods has been developed to estimate the (bio)-available pools of Hg in soil in an attempt to offer a plausible assessment of its risks. Unfortunately, many of these methods do not mirror the (bio)-available pools of soil Hg and suffer from technical drawbacks (Huang et al., 2020).

In aqueous systems and anoxic sediments, the geochemical forms of Hg (and subsequent bioavailability) are controlled primarily by reactions between Hg(II), inorganic sulfides, and natural organic matter (Hsu-Kim et al., 2013). Inorganic Hg compounds are, in general water soluble with a bioavailability of 7%–15% after ingestion (Park and Zheng, 2012).

Plants differ (from species to species) in their ability to take up Hg, and can develop a tolerance to high concentrations on contaminated sites, with corresponding elevated concentrations in edible parts compared with natural soils (Kabata-Pendias and Mukherjee, 2007).

Studying the process of methylation of inorganic forms of Hg constitutes the first step in understanding the process of bioaccumulation of MeHg in aquatic food webs (Fig. 13–2). Because the process is primarily mediated by anaerobic bacteria (Hsu-Kim et al., 2013) knowledge about the bioavailability of inorganic Hg to these microbes is important in understanding of MeHg bioaccumulation in fish and other high trophic-level species.

The dominant route of exposure of humans and other higher-order wildlife to MeHg is through the ingestion of fish and seafood products, with most fish—both freshwater and saltwater—containing varying levels of MeHg, which can enter and accumulate in the food chain

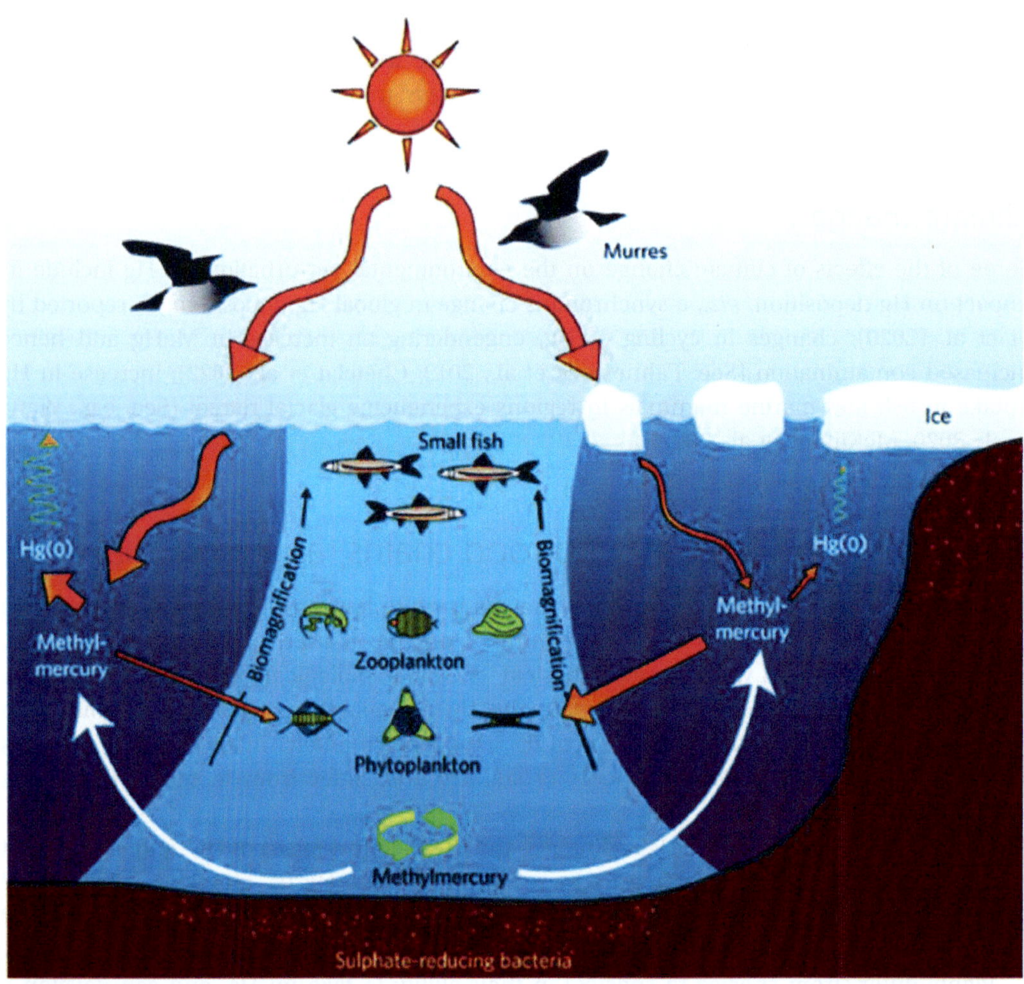

FIGURE 13–2 Marine Hg breakdown (Blum, 2011). *From: Blum, J.D., 2011. Marine mercury breakdown. Nature Geoscience 4, 139–140. https://doi.org/10.1038/ngeo1093.*

(Ullrich et al., 2001). Although Hg is not known to be an essential part of the food chain, it is assimilated by organisms living in environments that contain it.

Absorption and distribution

The dynamics of absorption, transport, and detoxification of Hg is dependent on the Hg species. Inhalation is the main entry route of elemental Hg into the body, while oral exposure is the primary route for inorganic Hg salts. Dermal penetration is usually not a significant route of exposure to inorganic Hg.

For elemental and inorganic Hg, inhalation of Hg vapors is also the most significant source of absorption, with up to 80% of the inhaled vapors directly entering the bloodstream (Martinez-Finley and Aschner, 2014). The main route of absorption is the gastrointestinal (GI) tract but some absorption of MeHg through the skin and the lungs also occurs. Absorption of elemental Hg by the GI tract; however, is weak ($<0.01\%$ in rats) (WHO, 2000). Upon absorption, elemental Hg enters all tissues and accumulates in the central nervous system and kidneys. Inorganic Hg has low lipid solubility and does not readily cross cell membranes. Once absorbed, the majority of Hg^{2+} accumulates in the kidney and liver (Martinez-Finley and Aschner, 2014).

Subsequent to absorption into the circulation, MeHg enters *erythrocytes* (See "Glossary of terms," in this Chapter) where a substantial amount binds to hemoglobin and a lesser amount to plasma proteins. A small amount of Hg (some 10% of the burden of MeHg) gets to the brain where demethylation to an inorganic Hg form proceeds slowly. Methylmercury readily crosses the placenta to the fetus, where deposition within the developing fetal brain can take place (See: ATSDR, 2015; Bjørklund et al., 2019) with serious neurological consequences for the newborn.

Metabolic function (clinical toxicity)

The toxicity of Hg is strongly influenced by microbially mediated biotransformation between its organic (MeHg) and inorganic [Hg (II) and Hg^0] forms.

Mechanism of toxicity

Several mechanisms have been invoked in the study of Hg toxicity. These include alteration or disruption in the regulation of intracellular Ca homeostasis, cytoskeleton, mitochondrial function, oxidative stress, neurotransmitter release, and DNA methylation (ATSDR, 2022), with Hg binding to thiolate anions thought to underlie many of these alterations or disruptions. Once absorbed, elemental and inorganic Hg enter an oxidation-reduction cycle. Absorbed elemental Hg is oxidized to the mercuric form (Hg^{++}) in the red blood cells and tissues, in a process that takes only minutes (Park and Zheng, 2012).

According to Rand Lab URMC (2022), evaluating the potential for MeHg toxicity relies on the accurate prediction of the body burden of Hg that results with eating fish (Fig. 13-3). The Hg body burden is directly determined by the slow elimination kinetics of MeHg in the human body ($t_{1/2} \sim 50$days) (Rand Lab URMC, 2022). Existing studies are limited in size and number, but sufficient evidence now exists to show that the biological half-life of MeHg in humans can vary widely ($t_{1/2} = 30$ to >150 days).

As at 2022, the Rand Lab studies continue to characterize MeHg metabolism and kinetics in people, focusing on the most vulnerable populations, viz., pregnant women and young children. In parallel, these researchers are investigating the gut microbiome in more depth for its role in the biotransformation process. Rand Lab data are being used to inform development of a physiologically based pharmacokinetic model in silico capable of predicting MeHg kinetics in humans (Rand Lab URMC, 2022).

Supplementation

Mercury is not classified as an essential element and no known benefits have been reported for life processes in animals and plants. However, a number of studies have revealed the presence of Hg in significant quantities in herbal dietary supplements and in several therapeutic preparations (See, e.g., Puścion-Jakubik et al., 2021; Brodziak-Dopierała et al., 2018), such as to warrant further research that would possibly lead to implementation of a stricter quality control of dietary supplements. The use of Hg in traditional medicine is also becoming widespread (See, e.g., Street et al., 2015).

Reported therapeutic uses appearing in the literature include various medications, ointments, dental fillings, contact lenses, and cosmetics. A number of instruments for medical applications, such as thermometers and *sphygmomanometers* also make use of Hg (See Iqbal and Asmat, 2012).

Reduction of body burden/elimination

Excretion of Hg takes place primarily through the urine and feces, with smaller amounts leaving the body through exhaled breath, sweat, and saliva.

Following short-term exposure to Hg vapor [elemental Hg (Hg^0)], about a third of the absorbed Hg will be eliminated in unchanged form through exhalation, while the rest will be eliminated mainly through the feces (Hursh et al., 1976). About equal amounts of Hg is excreted through the urine and feces after long-term exposure (several years) (Tejning and Ohman, 1966). Another form of excretion, albeit an undesirable one, is through lactation (ATSDR, 2015).

Some studies, e.g., the one by Li et al. (2012) have noted that organic Se supplementation could increase Hg excretion and decrease urinary *malondialdehyde* and *8-hydroxy-2-deoxyguanosine* levels in local residents. For synergistic/antagonistic relationships between Se and Hg, see Section "Associations (synergism/antagonism)," in this Chapter.

Analytical determination

Mercury concentrations in humans and other animals have been determined in blood, urine, body tissues, hair, breast milk, and umbilical cord blood. Most of the earlier methods used atomic absorption spectrometry (AAS), atomic fluorescence spectrometry (AFS), or neutron activation analysis, although mass spectrometry (Ms), spectrophotometry, and anodic stripping voltammetry have also been employed. The most commonly used method in the pre-2000s was cold vapor (CV) AAS, introduced in 1968 by Hatch and Ott. As this technique (CV AAS) went through further development, it became possible to reliably determine Hg

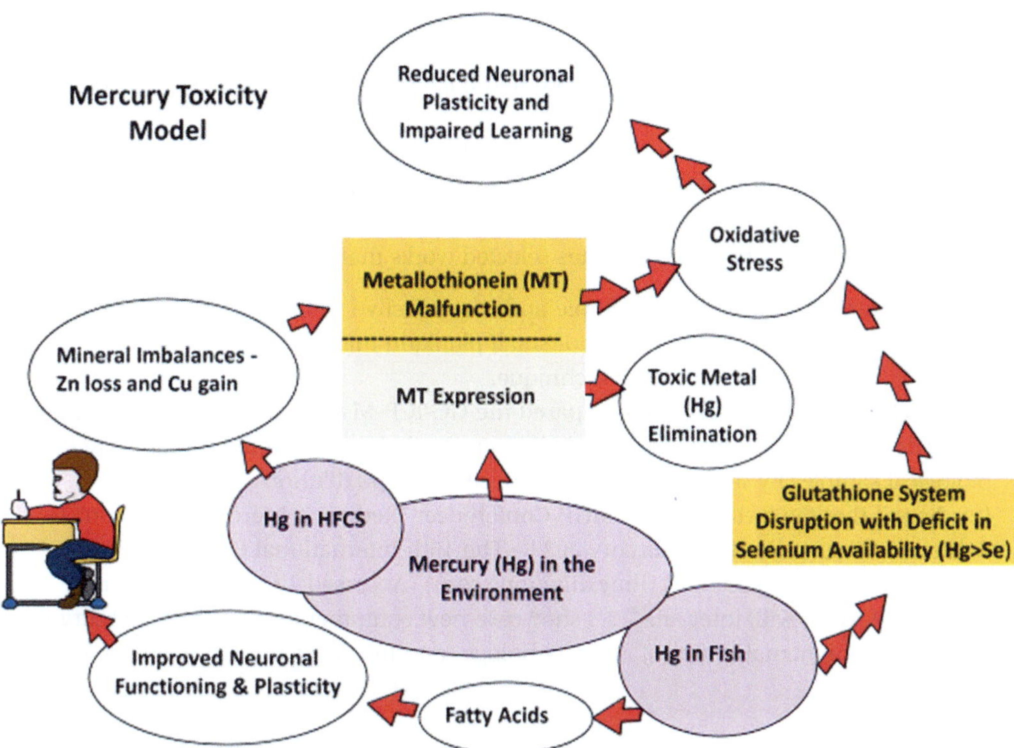

FIGURE 13–3 The original mercury toxicity model. *Reproduced from Dufault et al. (2009), Figure 1. Dufault, R., Schnoll, R., Lukiw, W.J., LeBlanc, B., Cornett, C., Patrick, L., Wallinga D, Gilbert S.G., Crider, R., 2009. Mercury exposure, nutritional deficiencies and metabolic disruptions may affect learning in children. Behavioral and Brain Functions 5, 44. https://doi.org/10.1186/1744-9081-5-44.*

concentrations below the ppb level ($>76\%$ recovery) through either direct reduction of the sample or reduction subsequent to predigestion (See Chirita et al., 2023; Risher, 2003).

Today, the concentration of Hg can be accurately determined in air, water, soil, and biological samples (blood, urine, tissue, hair, breast milk, and breath) by a variety of analytical methods. Most of these methods are for total Hg (inorganic as well as organic Hg compounds) and are based on wet oxidation followed by a reduction step. However, relatively modern methods also exist for the separate determination of inorganic Hg compounds and organic Hg compounds (See, e.g., Risher, 2003).

There continues to be a huge and continuous interest in improvements on the determination of Hg and its speciation, mainly due to the element's high toxicity and biomagnification. Developments in analytical procedures for Hg, its speciation analysis and its applications in environmental and biological studies since the year 2008 were extensively reviewed by Gao et al. (2012). Despite these recent advances in analytical techniques, interpretation of Hg geochemistry in environmental systems still remains a challenge (See, e.g., McLagan et al., 2022).

This is largely due to our inability to identify specific Hg transformation processes and species using established analytical methods in Hg geochemistry (total Hg and Hg speciation) [See, e.g., Izadi et al. (2024) for an improved ratiometric colorimetric determination of mercury(II) ion in water samples].

In the post-2010 years, we continue to see further improvements in analytical techniques for determining Hg in biological and environmental samples, particularly in the attainment of lower detection limits and ease of operation of analytical instrumentation, as exemplified in the following overview, which encapsulates selected works that represent these developments:

- In 2022 Santos et al. described a simple and cost-effective wet digestion method for the determination of total Hg in samples of small plankton material using a cold vapor atomic fluorescence spectroscopic technique.
- Suárez-Criado et al. (2022) have compared the GC-ICP-Ms, GC-EI-Ms and GC-EI-Ms/MS methods for the determination of methylmercury, ethylmercury and inorganic Hg in biological samples by triple spike species-specific isotope dilution Ms.
- It is hoped that the forthcoming CMGP Conference: "Reducing Mercury Emissions to Achieve a Greener World": [Mercury 2022—The 15th International Conference on Mercury as a Global Pollutant (ilmexhibitions.com) (Accessed 12 July 2022)] during July 24−29, 2022, will, inter alia," … showcase new equipment to measure mercury in various environmental samples, and technology to reduce mercury emissions and exposure." [Sic].

Current status of geomedical research on mercury in Africa

In this section, a brief mention is made of selected recent geomedical research on Hg in Africa (Davies, 2023), focusing on work done in the last two decades (post-2000) or so. Most of the research on Hg exposure in Africa has been in South Africa (See, e.g., George et al., 2023; Laker, 2023), despite the existence of multiple sources of Hg pollution in several other African countries (Brunnschweiler et al., 2024; Ondayo et al., 2023). Much of this research centers around the impact of Hg exposure during alluvial Au mining, which is a common activity in a number of countries across the Continent. A few selected examples follow:

Bonzongo et al. (2004) present a case study on health impacts due to Hg exposure in artisanal Au mining in Ghana, West Africa. Oosthuizen et al. (2010) studied Hg exposure in a low-income community in South Africa associated with Au mining activities to determine whether significant exposure to Hg occurs. Spiegel (2009) looked at socioeconomic dimensions of Hg pollution abatement in artisanal mining communities in Sub-Saharan Africa, and in 2013, Lusilao-Makiese et al. assessed the impact of the historical use of Hg for the extraction of Au in watersheds from an abandoned mine in Randfontein, a town located some 45 km west of Johannesburg (South Africa).

In 2007 Leaner reported on the development of the South African Mercury Assessment (SAMA) Program during the Stockholm Water Week that year. The research areas addressed in the SAMA Program included "(1) regulatory framework; (2) analytical methods; (3) source, speciation, fate, and transport; and (4) impacts (ecological and human health)." Papu-Zamxaka et al. (2010) looked at the consequences of Hg contamination in the vicinity of a Hg processing plant in the KwaZulu-Natal Province of South Africa.

Walters et al. (2011) reviewed the status of Hg as a pollutant in the aquatic ecosystems of South Africa, followed by the work of Street et al. in 2015, which looked at the perceptions and practices in the use of metallic Hg by South African traditional health practitioners, stressing the need for putting in place, public education messages, and regulatory measures.

Garnham and Langerman (2016) calculated the amount of Hg emitted from each of the South Africa Electricity Supply Commission (Eskom) coal-fired power stations, based on the amount of coal burnt, and the Hg content in the coal. These authors calculated Emission Reduction Factors from two sources, in working out the cobenefit derived from the emission control technologies at the stations.

Belelie et al. (2019) performed a pilot study to characterize ambient total gaseous Hg concentrations over the South African Highveld (a portion of the South African inland plateau). In the same year (2019), Panichev et al. carried out large-scale assessment of atmospheric air pollution by Hg using the lichen *Parmelia caperata* as biological indicator. In the study by Panichev et al. (2019), five provinces of South Africa were sampled between 2013 and 2017, and the lichens analyzed to provide time-integrated data, which correspond to the mean Hg concentration in air at a specific location over a long time period. Nevondo et al. (2019) suggested that Hg coming from contaminated groundwater from landfill leachate seepage could carry significant amounts of Hg, especially in landfills that are not lined with a geomembrane.

Bredenkamp's 2021 "Thesis" (North-West University, South Africa) pointed out the paucity of long-term measurements of atmospheric Hg at inland background sites in South Africa; and held that such measurements, in conjunction with data from the coastal *CP GAW* station (See: "Glossary of terms," in this Chapter), will be essential in providing credible estimates on background atmospheric Hg concentrations in South Africa; as well as contribute to formulation of cogent policies and standards on Hg emission limits.

Elumalai et al. (2022) performed a chemometric analysis of exposure and human health on tourist beaches in Durban, South Africa, coming up with observations that lend credence to the criticality of erecting long-term monitoring schemes for Hg to provide requisite data for formulating improved public health management strategies.

Case description

• Cato-Ridge is a small, industrial village in Kwazulu-Natal Province in eastern South Africa. Mercury waste shipments from other countries have been received at a Cato-Ridge plant owned by Thor Chemicals in South Africa, a subsidiary of Thor Chemicals, Inc. of Great Britain (See, e.g., Phalane and Steady, 2009). Investigation into the international shipment of hazardous wastes has focused attention on the consignment and its delivery to South Africa.

This Firm has been accused of poisoning workers and putting surrounding communities at risk from Hg exposure. As of December, 2000, the Hg wastes are still stockpiled at Thor Chemicals, Inc., in Cato-Ridge. *Research on the fate of residual Hg at Cato-Ridge is perhaps a worthwhile proposition.*

Other suggested areas for further research in Africa

Mercury pollution is a global problem, not least in the Continent of Africa. Despite the prolificity of research in the past two decades or so, there are still several uncertainties and variabilities in our knowledge of both the element's exposure dynamics and its health effects.

Understanding the intricacies of the element's emissions-to-impact path is rendered intractable by its varied environmental fate and the overarching influences of environmental, geological/geochemical, biological, and socioeconomic drivers (See Budnik and Casteleyn, 2019); as this knowledge could sway the course of diagnosis and treatment preferred by medics for Hg-induced maladies.

Here, a list comprised of key uncertainties, future research needs, and recommendations is presented to assist graduate students and researchers in allied fields in their Hg project selection exercise. A comprehensive list of the more recent references is given, to aid the search process of researchers who seek additional information on areas not covered in this Chapter.

- Attempts at controlling Hg emissions have been ongoing since the mid-1980s. However, we do not know whether reductions achieved by some countries have been offset by the increase in emissions from rapidly industrializing countries. In addition, a major challenge facing scientists now is to ascertain links between Hg in the atmosphere, deposition, and ecosystem contamination (See: Pirrone et al., 2010; Sundseth et al., 2017).
- Interpretation of Hg geochemistry in environmental systems continues to be problematic (See, e.g., McLagan et al., 2022). This is largely due to our inability to identify specific Hg transformation processes and species using modern analytical tools in Hg geochemistry (total Hg and Hg speciation) (See Section "Analytical Determination," in this Chapter). Research on these processes is vital to our understanding of Hg speciation, bioavailability, *bioaccessibility* and distribution of Hg that enters the body.
- Links between fully coupled biogeochemical models with biological interactions in food webs should be unveiled in order to provide the basis for designing future policies.
- There is an extreme paucity of measurements of reactive Hg worldwide (See AMAP/ UNEP, 2019); in many countries, including those in Africa these measurements are virtually nonexistent.
- There is an urgent need for research on the effectiveness of implementation of solutions to reduce the emissions and exposure to Hg as a global environmental pollutant. Mechanisms for testing the efficiency of implementation of the *Minamata Convention* in Africa and other parts of the world, are still needed. Such research data should be

translated swiftly into management tools for local policy makers and health professionals, with particular attention paid to the major differences in Hg contamination across Africa and other parts of the world (See Budnik and Casteleyn, 2019).

- The Minamata Convention includes provisions to reduce emissions of Hg to the atmosphere with the overarching goal of reducing Hg in the environment. A major setback for scientists and policy makers is the inability to gauze the effectuality of actions put in place by policy makers in response to the Convention, which relies on firm understanding of the element's sources, distribution, transformation, and fate.

- Many faunal groups are clearly vulnerable to Hg contamination, but the myriad of relationships existing between Hg and faunal biology remain poorly construed (See Seewagen, 2010). From a conservation standpoint, we need a proper understanding of the threats facing faunal populations. The taxonomic groups, geographic ranges, life history stages, habitat associations, and foraging behaviors of several fauna that are significantly threatened by Hg pollution need to be more precisely identified. Successful conservation strategies are dependent on a comprehensive understanding of the threats facing populations. Additional research into how Hg is impacting faunal groups will hugely benefit our efforts at conservation.

- As for biota, there is a need to further elucidate the physiological mechanisms that control the accumulation, distribution, and toxicity of Hg by coupling new techniques with Hg-stable isotopes (See, e.g., Mi-Ling et al., 2022), taking full cognizance of the fact that in biotic systems, processes do not operate in isolation. Many of the processes affecting the conversion of inorganic Hg to MeHg and subsequent uptake in biota are poorly understood, particularly in marine ecosystems (Hsu-Kim et al., 2013; Lamborg et al., 2014).

- The chemistry of Hg in oceans is still not fully understood, *cf.*, precise mechanisms behind the methylation process. Basic questions remain unanswered regarding the geochemical forms of Hg that persist in anoxic settings, the mode of assimilation by methylating bacteria in the ocean, and the biochemical pathway by which these microorganisms produce and break down MeHg (See Hsu-Kim et al., 2013). These are questions whose answers are needed to inform policy makers and develop long-term strategies for controlling Hg contamination.

- Mercury science and policy will likely benefit from furthering the development of Hg research at the food-environment nexus and integrating seafood safety and food security themes and approaches into the framework of the Minamata Convention (See Pacyna, 2020).

- Recent research has established the uniqueness of the chemical interactions between Hg and Se. The Hg capturing capacity of Se is several times higher than that of S compounds and results in inactive complexes that can be exploited in preventing or reversing the severe toxic action of Hg. Future work can target engineering technology that can exploit the Hg capturing ability of Se in reducing the toxicity of Hg in the environment (See, e.g., Castriotta et al., 2020; Timmerman and Omaye, 2021). The role of technological development in contributing to the reduction of Hg exposure and improvement of environmental responsibility need further study.

- The suggestion by Nevondo et al. (2019) that Hg coming from contaminated groundwater from landfill leachate seepage could carry significant amounts of Hg deserves further research, since several African rural communities living near landfill sites depend on groundwater for drinking purposes. Additional studies to identify pathways of release from contaminated sites to water and air, are needed.
- Since consumption of seafood, particularly fish, has been established as the most common form of human exposure to MeHg, the topic has ignited a surge in research, especially the relationship between maternal fish intake and the health effects on the developing fetus. Despite these research efforts, the Hg concentrations in fish in Africa and in many other parts of the world remain unknown (see, e.g., Pandey and Shrivastav, 2012). There is therefore the need for additional data on Hg content in fish to provide consumers and government regulators with accurate information upon which to base consumption guidelines. A major achievement in the effort to reduce Hg in the environment would be to provide policymakers with reliable data on the relationship between anthropogenic emissions and the resulting concentrations of Hg in seafood eaten by humans.
- The magnitude of uncertainty and variability of Hg and MeHg toxicokinetics in highly exposed population groups, such as pregnant women and children need further study.
- Abundant evidence now exists to support the thesis that Hg toxicity is complex; hence, Martinez-Finley and Aschner (2014) have proposed that genetic analyses and testing with low-level Hg species could likely unveil new insights into the mechanism of toxicity. Studies (both genetic and behavioral) suggest that the extent of disease may be modulated not only by levels of Hg uptake, but also by age (both of exposure and at testing) and genetic background (Martinez-Finley and Aschner, 2014).
- Further gaps in our knowledge of Hg cycling were pointed out by Cossa et al. (2022), in their work on the Mediterranean Sea, in particular. These gaps include "... temporal variations in air-sea exchange, hydrothermal, and cold seep inputs, point sources, submarine groundwater discharge, and exchanges between margins and the open sea." These authors also pointed out the need for long-term observations and dedicated high-resolution Earth System Models for future assessment of global change impacts under the Minamata Convention Hg policy.
- Innovative and holistic ways of addressing the above "bulleted" issues regarding Hg pollution and health effects will be required within the framework of the Minamata Convention. This is so, because the highly complex nature of Hg cycling and exposure dynamics require that multitiered risk assessments that focus on environmental health and human health separately will need to be brought together.
- As this Chapter was being written (2022), work was still ongoing on evaluating the completeness of our understanding of Hg as a global environmental pollutant; and on the best approaches for developing and implementing solutions to reduce the emissions and exposure to this pollutant. This was the principal mandate of the International

Conference on Mercury as a Global Pollutant (ICMGP), the 14th ICMGP, which took place in Krakow, Poland in September 2019. Major subject of discussions at this Conference centered around the improvement of our knowledge on Hg sources, fate, impacts, and emission control options under the broad topic: "Bridging knowledge on global mercury with environmental responsibility, human welfare and policy response." Sadly, though, at this meeting, it was consensual that global anthropogenic emissions of Hg to the environment were still rising.

Glossary of terms

Apoptosis, also called *programmed cell death*, is an orderly process in which the cell's contents break down (self-destruct) through a series of molecular steps when stimulated by the appropriate trigger.

Bioaccessibility is defined as the amount of an ingested nutrient that is available for absorption in the gut after digestion; whereas *bioavailability* refers to the fraction of an ingested nutrient that reaches the systemic circulation and the specific sites where it can exert its biological action.

The CP GAW Station refers to the Cape Point Global Atmosphere Watch (CP GAW) station at the southern tip of the Cape Peninsula in South Africa.

Deoxyguanosine, also known as *dG*, is an organic compound belonging to the purine $2'$-deoxyribonucleosides class. It is one of the four deoxyribonucleosides that make up DNA, and exists in all living species, from bacteria to plants to humans.

Erythrocyte: A red blood cell, which (in humans) is typically a biconcave disk without a nucleus. Erythrocytes contain the pigment hemoglobin, which imparts the red color to blood, and transports oxygen and carbon dioxide to and from the tissues.

Hyperaccumulators are unusual plants growing in soil or water, and capable of accumulating very high concentrations of particular metals or metalloids in their living tissues at levels that may be hundreds or thousands of times greater than is normal for most plants, and are toxic to closely related species not adapted to growing on the metalliferous soils (See Reeves et al., 2018).

8-Hydroxy-2'-deoxyguanosine (8-OhdG) is a product of oxidatively damaged DNA formed by hydroxyl radical, singlet oxygen, and direct photodynamic action.

In silico is a relatively new term used to describe an emerging area of study, denoting experimental work "performed on computer or via computer simulation."

Minamata disease (first reported in Japan) is a MeHg poisoning that occurred in humans, with neurological signs and symptoms (ataxia, numbness in the hands and feet, and so on). In severe cases, insanity, paralysis, coma, and death may follow. The disease is caused by the consumption of large amounts of fish and shellfish that were heavily contaminated by MeHg generated in a chemical factory (Chisso Co. Ltd.) and then discharged into the sea.

The Minamata Convention: "The Minamata Convention on Mercury" is an international treaty designed to protect human health and the environment from anthropogenic emissions and releases of Hg and Hg compounds. The aim of the global Minamata Convention is to sustain an overall reduction in Hg levels in the environment over time thus protecting human health and the environment from anthropogenic emissions and releases of Hg and Hg compounds. The Convention was propounded in 2017, on realization of the global nature of the Hg problem. It includes provisions to reduce emissions of Hg to the atmosphere with the overarching goal of reducing HG in the environment.

Malondialdehyde (MDA) is the organic compound with the nominal formula $CH_2(CHO)_2$.

Necrosis is the death of cells in living tissue caused by external factors, such as toxins, infection or trauma.

*The *NLRP3 inflammasome* is a unique innate immune sensor that can be activated in response to a wide array of endogenous metabolic "danger signals."

*A *sphygmomanometer* is a medical instrument that detects and measures blood pressure.

Bibliography

The bibliography comprises a list of cited references (Over 90% of the references post-date the year 2000), and a list of other pertinent references for further reading.

References

Ahn, H., Kim, J., Kang, S.G., Yoon, S., Ko, H.-J., Kim, P.H., et al., 2018. Mercury and arsenic attenuate canonical and non-canonical NLRP3 inflammasome activation. Scientific Reports 8, 13659. Available from: https://doi.org/10.1038/s41598-018-31717-7.

AMAP/UNEP (Arctic Monitoring and Assessment Programme/United Nations Environment Programme), 2019. Technical Background Report to the Global Mercury Assessment 2018. Oslo, Norway, pp. 1−430. gma_tech (2).pdf. (Accessed 13 July 2022).

ATSDR (Agency for Toxic Substances and Disease Registry), 2015. Mercury: Mercury Overviews, ToxZine. https://www.atsdr.cdc.gov/sites/toxzine/mercury_toxzine.html#: ∼ :text = Most%20of%20the%20metallic%20mercury,body%20in%20the%20exhaled%20breath. (Accessed 11 July 2022).

ATSDR (Agency for Toxic Substances and Disease Registry), 2022. Toxicological Profile for Mercury (Draft for Public Comment). https://www.atsdr.cdc.gov/toxprofiles/tp46.pdf. (Accessed 11 July 2022).

Belelie, M.D., Piketh, S.J., Burger, R.P., Venter, A.D., Naidoo, M., 2019. Characterisation of ambient total gaseous mercury concentrations over the South African Highveld. Atmospheric Pollution Research 10, 12−23. Available from: https://doi.org/10.1016/j.apr.2018.06.001.

Bergquist, B.A., Blum, J.D., 2009. The odds and evens of mercury isotopes: applications of mass-dependent and mass-independent isotope fractionation. Elements 5, 353−357. Available from: https://doi.org/10.2113/gselements.5.6.353.

Bjørklund, G., Chirumbolo, S., Dadar, M., Pivina, L., Lindh, Butnariu, M., et al., 2019. Mercury exposure and its effects on fertility and pregnancy outcome. Basic Clinical Pharmacology and Toxicology 125, 317−327. Available from: https://doi.org/10.1111/bcpt.13264.

Blum, J.D., 2011. Marine mercury breakdown. Nature Geoscience 4, 139−140. Available from: https://doi.org/10.1038/ngeo1093.

Bonzongo, J.C.J., Donkor, A.K., Nartey, V.K., Lacerda, L.D., 2004. Mercury pollution in Ghana: a case study of environmental impacts of artisanal gold mining in Sub-Saharan Africa. In: Drude de Lacerda, L., Santelli, R.E., Duursma, E.K., Abrão, J.J. (Eds.), Environmental Geochemistry in Tropical and Subtropical Environments. Environmental Science. Springer, Berlin, Heidelberg, https://doi.org/10.1007/978-3-662−07060-4_12.

Bredenkamp, L., 2021. Background ambient atmospheric mercury concentrations for the South African interior (M.S. dissertation). Environmental Sciences, North-West University. https://repository.nwu.ac.za/bitstream/handle/10394/37866/Bredenkamp%20L%2023488697.pdf?sequence = 1. (Accessed 13 July 2022).

Brodziak-Dopierała, B., Fischer, A., Szczelina, W., Stojko, J., 2018. The content of mercury in herbal dietary supplements. Biological Trace Element Research 185 (1), 236−243. Available from: https://doi.org/10.1007/s12011-018-1240-2.

Brunnschweiler, C.N., Karapetyan, D., Lujala, P., 2024. Opportunities and risks of small-scale and artisanal gold mining for local communities: survey evidence from Ghana. The Extractive Industries and Society 17101403. Available from: https://doi.org/10.1016/j.exis.2024.101403.

Budnik, L.T., Casteleyn, L., 2019. Mercury pollution in modern times and its socio-medical consequences. The Science of the Total Environment 654, 720−734. Available from: https://doi.org/10.1016/j.scitotenv.2018.10.408.

Castriotta, L., Rosolen, V., Biggeri, A., Ronfani, L., Catelan, D., Mariuz, M., et al., 2020. The role of mercury, selenium and the Se-Hg antagonism on cognitive neurodevelopment: a 40-month follow-up of the Italian mother-child PHIME cohort. International Journal of Hygiene and Environmental Health 230, 113604. Available from: https://doi.org/10.1016/j.ijheh.2020.113604.

Chen, C.Y., Evers, D.C., 2023. Global mercury impact synthesis: processes in the Southern Hemisphere. Ambio 52, 827−832. Available from: https://doi.org/10.1007/s13280-023-01842-3 (Accessed 07 February 2024).

Chételat, J., McKinney, M.A., Amyot, M., Dastoor, A., Douglas, T.A., Heimbürger-Boavida, L.E., et al., 2022. Climate change and mercury in the Arctic: abiotic interactions. The Science of the Total Environment 824, 153715. Available from: https://doi.org/10.1016/j.scitotenv.2022.153715.

Chirita, L., Covaci, E., Ponta, M., Frentiu, T., 2023. Mercury determination in various environmental, food and material complex matrices using unified operating conditions for a cold vapor generation high-resolution continuum source quartz tube atomic absorption spectrometry method. Analytical Methods/Royal Society of Chemistry 15, 6294−6301. Available from: https://pubs.rsc.org/en/content/articlepdf/2023/ay/d3ay01468a (Accessed 07 February 2024).

Cossa, D., Knoery, J., Bǎnaru, D., Harmelin-Vivien, M., Sonke, J.E., Hedgecock, I.M., et al., 2022. Mediterranean mercury assessment 2022: an updated budget, health consequences, and research perspectives. Environmental Science and Technology (American Chemical Society) 56 (7), 3840−3862. Available from: https://doi.org/10.1021/acs.est.1c03044.

Dang, F., Wang, W.-X., 2011. Antagonistic interaction of mercury and selenium in a marine fish is dependent on their chemical species. Environmental Science and Technology 45 (7), 3116−3122. Available from: https://doi.org/10.1021/es103705a.

Davies, T.C., 2023. An updated review of the salient geomedical aspects of mercury for enhancement of data quality in simulation modelling and other prognostic applications: Africa case descriptions. Frontiers in Analytical Science 3, 1069678. Available from: https://doi.org/10.3389/frans.2023.1069678.

Driscoll, C.T., Mason, R.P., Chan, H.M., Jacob, D.J., Pirrone, N., 2013. Mercury as a global pollutant: sources, pathways, and effects. Environmental Science & Technology 47 (10), 4967−4983. Available from: https://doi.org/10.1021/es305071v.

Dwivedi, S., Chezhian, A., Kabilan, N., Kumar, T.S., 2012. Synergistic effect of mercury and chromium on the histology and physiology of fish, *Tilapia Mossambica* (Peters, 1852) and *Lates calcarifer Calcarifer* (Bloch, 1790). Toxicology International 19 (3), 235−240. Available from: https://doi.org/10.4103/0971-6580.103655.

Elumalai, V., Sujitha, S.B., Jonathan, M.P., 2022. Mercury pollution on tourist beaches in Durban, South Africa: a chemometric analysis of exposure and human health. Marine Pollution Bulletin 180, 113742. Available from: https://doi.org/10.1016/j.marpolbul.2022.113742.

ENHIS (European Environment and Health Information System), 2007. Exposure of children to chemical hazards in food. Fact Sheet No. 4.4, May 2007. CODE: RPG4_Food_Ex1. https://www.euro.who.int/__data/assets/pdf_file/0003/97446/4.4.pdf. (Accessed 17 July 2022).

Enrico, M., Le Roux, G., Marusczak, N., Heimbürger, L.E., Claustres, A., Fu, X., et al., 2016. Atmospheric mercury transfer to peat bogs dominated by gaseous elemental mercury dry deposition. Environmental Sciience and Technology 50, 2405−2412. Available from: https://doi.org/10.1021/acs.est5b06058.

Fahnestock, M.F., Bryce, J.G., McCalley, C.K., Montesdeoca, M., Bai, S., Li, Y., et al., 2019. Mercury reallocation in thawing subarctic peatlands. Geochemical Perspectives Letters 33. Available from: https://doi.org/10.7185/geochemlet.1922.

Friedli, H.R., Radke, L.F., Prescott, R., Hobbs, P.V.,, Sinha, P., 2003. Mercury emissions from the Aug 2001 wildfires in Washington State and an agricultural waste fire in Oregon and atmospheric mercury budget estimates. Global Biogeochemical Cycles 17, 1039−1047.

Gao, Y., Shi, Z., Long, Z., Wu, P., Zheng, C., Hou, X., 2012. Determination and speciation of mercury in environmental and biological samples by analytical atomic spectrometry. Microchemical Journal 103, 1−14. Available from: https://doi.org/10.1016/j.microc.2012.02.001.

Garnham, B., Langerman, K., 2016. Mercury emissions from South Africa's coal-fired power stations. Clean Air Journal 26 (2), 14−20. Available from: https://doi.org/10.17159/2410-972X/2016/v26n2a8.

George, J., Sadiq, E., Moola, I., Maharaj, S., Mochan, A., 2023. Informal gold miners with mercury toxicity: novel asymmetrical neurological presentations. South African Medical Journal 113 (12), 1522−1525. Available from: https://doi.org/10.7196/SAMJ.2023.v113i11.1127.

Gray, J.E., Crock, J.G., Lasorsa, B.K., 2002. Mercury methylation at mercury mines in the Humboldt River Basin, Nevada, USA. Geochemistry: Exploration, Environment, Analysis 2 (2), 143−149. Available from: https://doi.org/10.1144/1467-787302-017.

Gustin, M., Bank, M., Bishop, K., Bowman, K., Branfireun, B., Chételat, J., et al., 2020. Mercury biogeochemical cycling: a synthesis of recent scientific advances. Science of the Total Environment 737, 139619. Available from: https://doi.org/10.1016/j.scitotenv.2020.139619.

Hatch, W.R., Ott, W.L., 1968. Determination of sub-cold vapour atomic absorption spectrophotometry. Analytical Chemistry 40 (14), 2085−2087. Available from: https://doi.org/10.1021/ac50158a025.

Hsu-Kim, H., Kucharzyk, K.H., Zhang, T., Deshusses, M.A., 2013. Mechanisms regulating mercury bioavailability for methylating microorganisms in the aquatic environment: a critical review. Environmental Science and Technology 47 (6), 2441−2456. Available from: https://doi.org/10.1021/es304370g.

Huang, J.H., Shetaya, W.H., Osterwalder, S., 2020. Determination of (bio)-available mercury in soils: a review. Environmental Pollution 263, 114323. Available from: https://doi.org/10.1016/j.envpol.2020.114323.

Hursh, J.B., Cherian, M.G., Clarkson, T.W., Vostal, J.J., Mallie, R.V., 1976. Clearance of mercury (HG-197, HG-203) vapor inhaled by human subjects. Archives of Environmental Health 31 (6), 302−309. Available from: https://doi.org/10.1080/00039896.1976.10667240.

IPCS (International Programme on Chemical Safety), 1990. Methylmercury; Environmental Health Criteria 101. International Programme on Chemical Safety World Health Organisation, Geneva. Methylmercury (EHC 101, 1990). inchem.org. (Accessed 17 July 2022).

Iqbal, K., Asmat, M., 2012. Uses and effects of mercury in medicine and dentistry. Journal of Ayub Medical College, Abbottabad (JAMC) 24 (3−4), 204−207.

Izadi, S., Tashkhourian, J., Alireza Hosseini Hafshejani, S., 2024. Ecofriendly ratiometric colorimetric determination of mercury(II) ion in environmental water samples using gallic acid-capped gold nanoparticles. Spectrochimica Acta 308 (Part A), 123778. Available from: https://doi.org/10.1016/j.saa.2023.123778 (Accessed 07 February 2024).

Jonasson, I.R., Boyle, R.W., 1972. Geochemistry of mercury and origins of natural contamination in the environment. CIM Bulletin 32−39.

Kabata-Pendias, A., Mukherjee, A.B., 2007. Plants. Trace Elements from Soil to Human. Springer, Berlin, Heidelberg, https://doi.org/10.1007/978-3-540-32714-1_6. (Accessed 10 July 2022).

Laker, M.C., 2023. Environmental impacts of gold mining—with special reference to South Africa. Mining 3 (2), 205−220. Available from: https://doi.org/10.3390/mining3020012 (Accessed 07 February 2024).

Lamborg, C., Bowman, K., Hammerschmidt, C., Gilmour, C., Munson, K., Selin, N., et al., 2014. Mercury in the anthropocene ocean. Oceanography 27 (1), 76−87. Available from: https://doi.org/10.5670/oceanog.2014.11.

Leaner, J. (2007). Focus on CSIR research in pollution waste: South African mercury assessment (SAMA) programme. CSIR Natural Resources and the Environment 2007. 2007 Stockholm World Water Week, August 13−17, 2007, p. 1. http://hdl.handle.net/10204/1123. (Accessed 12 July 2022).

Li, Y.F., Dong, Z., Chen, C., Li, B., Gao, Y., Qu, L., et al., 2012. Organic selenium supplementation increases mercury excretion and decreases oxidative damage in long-term mercury-exposed residents from Wanshan, China. Environmental Science and Technology 46 (20), 11313−11318. Available from: https://doi.org/10.1021/es302241v.

Li, F., Ma, C., Zhang, P., 2020. Mercury deposition, climate change and anthropogenic activities: A review. Frontiers in Earth Sciences. Sec. Quaternary Science, Geomorphology and Paleoenvironment. https://doi.org/10.3389/feart.2020.00316. (Accessed 18 July 2022).

Lusilao-Makiese, J., Cukrowska, E., Tessier, E., Amouroux, D., Weiersbye, I., 2013. The impact of post gold mining on mercury pollution in the West Rand region, Gauteng, South Africa. Journal of Geochemical Exploration 134, 111−119. Available from: https://doi.org/10.1016/j.gexplo.2013.08.010.

Martinez-Finley, E.J., Aschner, M., 2014. Recent advances in mercury research. Current Environmental Health Reports 1 (2), 163−171. Available from: https://doi.org/10.1007/s40572-014-0014-z.

McKinney, M.A., Chételat, J., Burke, S.M., Elliott, K.H., Fernie, K.J., Houde, M., et al., 2022. Climate change and mercury in the Arctic: biotic interactions. The Science of the Total Environment 834, 155221. Available from: https://doi.org/10.1016/j.scitotenv.2022.155221.

McLagan, D.S., Schwab, L., Wiederhold, J.G., Chen, L., Pietrucha, J., Kraemer, S.M.,, et al., 2022. Demystifying mercury geochemistry in contaminated soil-groundwater systems with complementary mercury stable isotope, concentration, and speciation analyses. Environmental Science: Processes & Impacts 24, 1406−1429. Available from: https://doi.org/10.1039/d1em00368b.

Mi-Ling, L., Kwon, S.Y., Poulin, B.A., Tsui, M.T., Motta, L.C., Cho, M., 2022. Internal dynamics and metabolism of mercury in biota: a review of insights from mercury stable isotopes. Environmental Science & Technology 56 (13), 9182−9195. Available from: https://pubs.acs.org/doi/10.1021/acs.est.1c08631.

Neculita, C.-M., Zagury, G.J., Deschenes, L., 2005. Mercury speciation in highly contaminated soils from chlor-alkali plants using chemical extractions. Journal of Environmental Quality 34 (1), 255−262. Available from: https://doi.org/10.2134/jeq2005.0255a.

Nevondo, V., Malehase, T., Daso, A.P., Okonkwo, O.J., 2019. Leachate seepage from landfill: a source of groundwater mercury contamination in South Africa. Water SA 45 (2), 225−231. Available from: https://doi.org/10.4314/wsa.v45i2.09.

Nishimura, M., Konishi, S., Matsunaga, K., Hata, K., Kosuga, T., 1983. Mercury concentration in the ocean. Journal of the Oceanographical Society of Japan 39, 295−300. Available from: https://doi.org/10.1007/BF02071824.

Ondayo, M.A., Watts, M.J., Mitchell, C.J., King, D., Osano, O., 2023. Review: artisanal gold mining in Africa - environmental pollution and human health implications. Exposure and Health. Available from: https://doi.org/10.1007/s12403-023-00611-7.

Oosthuizen, M.A., John, J., Somerset, V., 2010. Mercury exposure in a low-income community in South Africa. South African Medical Journal 100 (6), 366−371.

Pacyna, J.M., 2020. Recent advances in mercury research. The Science of the Total Environment 738, 139955. Available from: https://doi.org/10.1016/j.scitotenv.2020.139955.

Pandey, G., Shrivastav, A.B., 2012. Contamination of mercury in fish and its toxicity to both fish and humans: an overview. International Research Journal of Pharmacy 3, 44−47.

Panichev, N., Mokgalaka, N., Panicheva, S., 2019. Assessment of air pollution by mercury in South African provinces using lichens Parmelia caperata as bioindicators. Environmental Geochemistry and Health 41, 2239–2250. Available from: https://doi.org/10.1007/s10653-019-00283-w.

Papu-Zamxaka, V., Mathee, A., Harpham, T., Barnes, B., Röllin, H., Lyons, M., et al., 2010. Elevated mercury exposure in communities living alongside the Inanda Dam, South Africa. Journal of Environmental Monitoring 12 (2), 472–477. Available from: https://doi.org/10.1039/b917452d.

Park, J.D., Zheng, W., 2012. Human exposure and health effects of inorganic and elemental mercury. Journal of Preventive Medicine and Public Health. Korean Society for Preventive Medicine 45 (6), 344–352. Available from: https://doi.org/10.3961/jpmph.2012.45.6.344.

Phalane, M.F., Steady, F.C., 2009. Chapter 8: Nuclear energy, hazardous waste, health, and environmental justice in South Africa. In: Steady, F.C. (Ed.), Environmental Justice in the New Millennium: Global Perspectives on Race, Ethnicity and Human Rights. Palgrave Macmillan, New York, pp. 189–201.

Pirrone, N., Cinnirella, S., Feng, X., Finkelman, R.B., Friedli, H.R., Leaner, J., et al., 2010. Global mercury emissions to the atmosphere from anthropogenic and natural sources. Atmospheric Chemistry and Phyics 10 (13), 5951–5964.

Puścion-Jakubik, A., Mielech, A., Abramiuk, D., Iwaniuk, M., Grabia, M., Bielecka, J., et al., 2021. Mercury content in dietary supplements From Poland containing ingredients of plant origin: a safety assessment. Frontiers in Pharmacology 12, 738549. Available from: https://doi.org/10.3389/fphar.2021.738549.

Rand Lab (URMC) (University of Rochester Medical Centre), 2022. Methylmercury Metabolism and Elimination Status (MerMES) in Humans. https://www.urmc.rochester.edu/labs/rand/projects/methyl-mercury-metabolism-elimination.aspx. (Accessed 11 July 2022).

Reeves, R.D., Baker, A.J.M., Jaffré, T., Erskine, P.D., Echevarria, G., van der Ent, A., 2018. A global database for plants that hyperaccumulate metal and metalloid trace elements. New Phytologist 218, 407–411. Available from: https://doi.org/10.1111/nph.14907.

Risher, J.F., 2003. Elemental Mercury and Inorganic Mercury Compounds: Human Health Aspects. Concise International Chemical Assessment Document 50. World Health Organisation, Geneva. https://apps.who.int/iris/bitstream/handle/10665/42607/9241530502.pdf?sequence = 1&isAllowed = y. (Accessed 11 July 2022).

Różański, S., Castejón, J.M.P., Fernández, G.G., 2016. Bioavailability and mobility of mercury in selected soil profiles. Environmental Earth Science 75, 1065. Available from: https://doi.org/10.1007/s12665-016-5863-3 (accessed 18.07.2022).

Santos, J.P., Mehmeti, L., Slaveykova, V.I., 2022. Simple acid digestion procedure for the determination of total mercury in plankton by cold vapor atomic fluorescence spectroscopy. Methods and Protocols 5 (2), 29. Available from: https://doi.org/10.3390/mps5020029.

Schluter, K., 1993. The fate of mercury in soil: a review of current knowledge. Soil and Groundwater Research Report IV, Technical Report EUR 14666 EN, Commission of the European Communities, Luxembourg, UK.

Schneider, L., Fisher, J.A., Diéguez, M.C., Fostier, A-H., Guimaraes, J.R.D., Leaner, J.J., Mason, R., 2023. A synthesis of mercury research in the Southern Hemisphere, part 1: natural processes. Ambio 52, 897–917. Available from: https://doi.org/10.1007/s13280-023-01832-5 (Accessed 07 February 2024).

Seewagen, C.L., 2010. Threats of environmental mercury to birds: Knowledge gaps and priorities for future research. Bird Conservation International 20, 112–123. Available from: https://doi.org/10.1017/S095927090999030X.

Selvan, D.S., Chezhian, A., Kabilan, N., Kumar, T., 2012. Synergistic effect of mercury and chromium on the histology and physiology of fish, *Tilapia Mossambica* (Peters, 1852) and *Lates calcarifer Calcarifer* (Bloch, 1790). Toxicology International 19, 235–240. Available from: https://doi.org/10.4103/0971-6580.103655.

Senthamilselvan, D., Chezhian, A., Suresh, E., 2016. Synergistic effect of nickel and mercury on fatty acid composition in the muscle of fish Lates calcarifer. Journal of Fisheries and Aquatic Science 11, 77–84.

Soerensen, A.L., Schartup, A.T., Skrobonja, A., Björn, E., 2017. Organic matter drives high interannual variability in methylmercury concentrations in a subarctic coastal sea. Environmental Pollution 229, 531−538. Available from: https://doi.org/10.1016/j.envpol.2017.06.008.

Sonke, J.E., Angot, H., Zhang, Y., Poulain, A., Björn, E., Schartup, A., 2023. Global change effects on biogeochemical mercury cycling. Ambio 52 (5), 853−876. Available from: https://doi.org/10.1007/s13280-023-01855-y (Accessed 07 February 2024).

Spiegel, S.J., 2009. Socioeconomic dimensions of mercury pollution abatement: engaging artisanal mining communities in Sub-Saharan Africa. Ecological Economics 68, 3072−3083. Available from: https://doi.org/10.1016/j.ecolecon.2009.07.015.

Ssenku, J.E., Naziriwo, B., Kutesakwe, J., Mustafa, A.S., Kayeera, D., Tebandeke, E., 2023. Mercury accumulation in food crops and phytoremediation potential of wild plants thriving in artisanal and small-scale gold mining areas in Uganda. Pollutants 3, 181−196. Available from: https://doi.org/10.3390/pollutants3020014 (Accessed 07 February 2024).

Steinnes, E., 1997. Mercury. In: Alloway, B.J. (Ed.), Heavy Metals in Soils, 2nd edition Blackie Academics and Professional Press, London, UK, pp. 245−259.

Street, R.A., Kabera, G.M., Connolly, C., 2015. Metallic mercury use by South African traditional health practitioners: perceptions and practices. Environmental Health 14, 67. Available from: https://doi.org/10.1186/s12940-015-0053-4.

Suárez-Criado, L., Queipo-Abad, S., Cea, A.,R., Rodríguez-González, Alonso, J.I.C., 2022. Comparison of GC-ICP-MS, GC-EI-MS and GC-EI-MS/MS for the determination of methylmercury, ethylmercury and inorganic mercury in biological samples by triple spike species-specific isotope dilution mass spectrometry. Journal of Analytical Atomic Spectrometry 37, 1462−1470.

Sundseth, K., Pacyna, J.M., Pacyna, E.G., Pirrone, N., Thorne, R.J., 2017. Global sources and pathways of mercury in the context of human health. International Journal of Environmental Research and Public Health 14 (1), 105. Available from: https://doi.org/10.3390/ijerph14010105.

Swaine, D.J., 1990. Trace Elements in Coal. Butterworths, London, ISBN: 10 0408033096.

Teixeira, F.B., de Oliveira, A., Leão, L., Fagundes, N., Fernandes, R.M., Fernandes, L., et al., 2018. Exposure to inorganic mercury causes oxidative stress, cell death, and functional deficits in the motor cortex. Frontiers in Molecular Neuroscience 11, 125. Available from: https://doi.org/10.3389/fnmol.2018.00125.

Tejning, S., Ohman, H., 1966. Uptake, excretion and retention of metallic mercury in chloralkali workers. In: Proceedings of the 15th International Congress on Occupational Health. London, Permanent Commission and International Association on Occupational Health, p. 239.

Timmerman, R., Omaye, S., 2021. Selenium's utility in mercury toxicity: a mini-review. Food and Nutrition Sciences 12, 124−137. Available from: https://doi.org/10.4236/fns.2021.122011.

Ullrich, S.M., Tanton, T.W., Abdrashitova, S.A., 2001. Mercury in the aquatic environment: a review of factors affecting methylation. Critical Reviews in Environmental Science and Technology 31 (3), 241−293. Available from: https://doi.org/10.1080/20016491089226.

US EPA (United States Environmental Protection Agency), 2021. What EPA is doing to reduce mercury pollution, and exposures to mercury. https://www.epa.gov/mercury/what-epa-doing-reduce-mercury-pollution-and-exposures-mercury#understanding. (Accessed 23 July 2022).

USGS (United States Geological Survey), 1970. Mercury in the Environment. Geological Survey Professional Paper 713. https://pubs.usgs.gov/pp/0713/report.pdf. (Accessed 09 July 2022).

Walters, C.R., Somerset, V.S., Leaner, J.J., Nel, J.M., 2011. A review of mercury pollution in South Africa: current status. Journal of Environmental Science and Health. Part A, Toxic/hazardous Substances &Eenvironmental Engineering 46 (10), 1129−1137. Available from: https://doi.org/10.1080/10934529.2011.590.

Wang, X., Luo, J., Yuan, W., Lin, C.-J., Wang, F., Liu, C., et al., 2020. Global warming accelerates uptake of atmospheric mercury in regions experiencing glacier retreat. Proceedings of the National Academy of

Sciences of the United States of America 117 (4), 2049−2055. Available from: https://doi.org/10.1073/pnas.1906930117.

Wang, J., Ma, L.Q., Letcher, R., Bradford, S.A., Feng, F.,, Rinklebe, J., 2022. Biogeochemical cycle of mercury and controlling technologies: Publications in Critical Reviews in Environmental Science & Technology in the period of 2017−2021. Critical Reviews in Environmental Science and Technology 52, 4325−4330. Available from: https://doi.org/10.1080/10643389.2022.2071210.

WHO (World Health Organisation), 2000. Mercury. Chapter 6.9: Air Quality Guidelines, second ed. Regional Office for Europe, Copenhagen, Denmark, Microsoft Word - 6.9-Mercury.doc (who.int). (Accessed 10 July 2022).

WHO (World Health Organisation), 2003. Mercury in Drinking-water; Background Document for Development of WHO Guidelines for Drinking-water Quality. Document WHO/SDE/WSH/03.04/10. Geneva. https://cdn.who.int/media/docs/default-source/wash-documents/wash-chemicals/mercury-2003.pdf?sfvrsn = f2804f45_3. (Accessed 17 July 2022).

WHO (World Health Organisation), 2005. Mercury in Drinking-water. Background Document for Development of WHO Guidelines for Drinking-water Quality. WHO/SDE/WSH/05.08/10. Geneva, Switzerland. https://www.who.int/docs/default-source/wash-documents/wash-chemicals/mercury-background-document.pdf?sfvrsn = 9b117325_4#:~:text = If%20a%20level%20in%20drinking.the%20range%202%2E%80%9320%20%C2%B5g. (Accessed 23 July 2022).

Wyatt, L.H., Diringer, S.E., Rogers, L.A., Hsu-Kim, H., Pan, W.K., Meyer, J.N., 2016. Antagonistic growth effects of mercury and selenium in *Caenorhabditis elegans* are chemical-species-dependent and do not depend on internal Hg/Se ratios. Environmental Science & Technology 50 (6), 3256−3264. Available from: https://doi.org/10.1021/acs.est.5b06044.

14

Medical geology of selenium

Of all the elements, selenium has one of the narrowest ranges between dietary deficiency ($<40\ \mu g/day$) and toxic levels ($>400\ \mu g\,day^{-1}$)...

WHO (1996)

Key chapter features

1. Brief overviews of the most relevant and important characteristics of Se in Medical Geology.
2. Role of the Medical Geologist in teams investigating causality of disorders involving Se metabolism and selenoprotein function.
3. Latest advances in Se research in Medical Geology ($>90\%$ of the references postdate the year 2000), and how they apply to the African situation.
4. Suggested areas of future Se research in Africa.

Sources

Selenium is widely distributed in minute amounts in virtually all materials of the Earth's crust. The crustal abundance has been estimated in various ways [(e.g., Taylor (1964): 0.05 ppm; Brunfelt and Steinnes (1967): 0.04 ppm; Lakin (1972): 0.09 ppm; Taylor and McLennan (1985): 0.05−0.09 ppm; Maryland et al. (1989): <0.1 ppm; Greenwood and Earnshaw (1997): 0.05 ppm; Alexander (2015): 0.05−0.09 ppm)]. It can be seen from these figures why the range of 0.05−0.09 given by Alexander (2015) can justifiably be regarded as the accepted estimate of the crustal abundance of Se.

Selenium is brought to Earth's surface through magmatic intrusions and volcanic emissions (Stillings, 2017). Rocks, from which Se can be mobilized via weathering and leaching are the primary source of Se in the terrestrial realm. The content of Se in various rock types from the Earth's crust is as follows.

Igneous rocks: 0.004−0.110 ppm (Brunfelt and Steinnes, 1967); Granite: 0.005 ppm (Brunfelt and Steinnes, 1967); 0.01−0.05 ppm (Kabata-Pendias and Pendias, 1984); Diabase: 0.110 ppm (Brunfelt and Steinnes, 1967); and Ulmaf: 0.013 ppm (Brunfelt and Steinnes, 1967).

Sedimentary rocks: Limestone: 0.03−0.10 ppm (Kabata-Pendias and Pendias, 1984); Sandstone: 0.05−0.08 ppm (Kabata-Pendias and Pendias, 1984); Shale: 0.6 (average) (Adriano, 1986); and Coal (world avg.): 3 ppm (Minkin et al., 1984).

Medical Geology of Africa. DOI: https://doi.org/10.1016/B978-0-12-818748-7.00011-3

Higher Se concentrations averaging 120 ppm have reportedly been found in volcanic rocks (Alexander, 2015). In certain sedimentary rocks, such as shales and in sandstone uranium deposits, estimates of 1000 ppm are given; and in some carbonate rocks, 30 ppm (Alexander, 2015).

Minerals: Selenium is found most commonly in igneous rocks as *selenites* and *selenides*, alongside sulfide minerals bearing *chalcophile* (sulfur-loving) elements, such as Bi, Cu, Pb, Hg, and Ag (Shamberger, 1981; Shamberger, 1983; US ATSDR, 2003; Högberg and Alexander, 2007). Some selenides, such as berzelianite (Cu_2 Se), clausthalite (PbSe), guanajuatite (Bi_2Se_3) form from Se-rich hydrothermal fluids (Pirri, 2002). The composition of other Se-bearing sulfides, such as aguilarite (Ag_4SeS) is consistent with the observation that Se commonly substitutes for S due to similarities in crystallography.

Soils: The concentration range of Se in most soils is given as $<0.01-2$ ppm (Rosenfeld and Beath, 1964); $0.1-10$ ppm (Alexander, 2015). Higher soil Se concentrations of up to 1200 ppm have been reported in seleniferous areas of the world (Winkel et al., 2012; Garcia Moreno et al., 2013; Alexander, 2015).

The geology of the area controls to some extent, the concentrations of Se in the soils on which we grow our crops and animals that form the human food chain (Fordyce, 2013). For example, it is known that sandy soils have lower Se contents compared to organic and calcareous soils (Kabata-Pendias, 2011; El-Ramady et al., 2015a). However, the Se content of crops grown on (seleniferous) soils is determined not only by the Se content of the soil, but also by a number of biophysiochemical parameters, such as the pH, soil texture, mineralogy, and organic matter (OM) content (Fordyce, 2007; Xu et al., 2024a). Since diet is the most important source of Se in humans, understanding the biogeochemical controls on the distribution and mobility of environmental Se is of utmost importance in the assessment of Se-related health risks (Fordyce, 2007).

Plants: As an essential micronutrient, Se is incorporated into the plant structure (Uden, 2005) and enhances the growth and development of the plant when present in tracce amounts (Khan et al., 2023). The Se content in most plants that grow on normal soils is <3 ppm plant dry weight (Whanger, 2002). But plants can accumulate and tolerate Se to varying degrees. Most plants grown in seleniferous soils show Se levels in the range of $1-10$ ppm plant dry weight; and in the case of Se-hyperaccumulator plants, Se levels can be anywhere between 1000 and 15,000 ppm dry weight (Pilon-Smits, 2019).

Selenium concentrations in plants generally reflect the levels of Se in the environment. However, just as in the case of soils on which they grow, there are many biophysiochemical factors that determine the concentration of Se in plants. These include, but not limited to, the type of vegetation, chemical form of Se in the soil, pH of the soil, moisture content of soil, and the total concentration of Se in the soil.

Soils containing high concentrations of Se are commonly found in many parts of the world, but in Africa, Se deficiency is more common, especially in its southern region (See Section: "Selenium Research in Africa," in this chapter).

Food crops and other major dietary sources: The richest sources of Se (not necessarily in order of abundance of Se) are Brazil nuts, seafoods, organ meats, poultry, offal, cereals, and vegetables (Sunde, 2012; Ibrahim et al., 2019). The content in animal tissue (domestic stock) has been given as: 0.44−4.0 ppm (Frost, 1972); whole fish (freshwater): 0.05−2.87 ppm (wet weight) (Frost, 1972); and whole fish (marine): 0.1−20 ppm (Hall et al., 1978). Later reviews by Kohlmeier (2003) gave Se values of 0.40−1.5 ppm for seafood and liver, and 0.01−0.4 ppm for meat; and revealed that grains, vegetables, and fruits usually contain much more Se.

Among the group of elements classified as micronutrients, Se is known to have one of the narrowest ranges between dietary deficiency ($<40\,\mu g$/day) and toxic levels ($>400\,\mu g$/day) (WHO, 2006; reviewed by Kabata-Pendias and Mukherjee, 2007 and Fordyce, 2007), making it necessary to carefully control intakes by humans and other animals. The recommended daily allowance for Se is 55 μg/day for people aged 14 years old and over; but during breast feeding it can go up to 70 μg/day (Duyff, 2012). The knowledge that diet constitutes the main source of Se uptake by humans underscores the importance of understanding the biogeochemical controls on the distribution and mobility of environmental Se in assessing Se-related health risks (See Fordyce, 2007; Werkneh et al., 2023).

Selenium is mostly found in an organic form, *selenomethionine* (SeMet) in plant-based foods, and as *selenocysteine* (SeCys) ($C_3H_7NO_2Se$) in animal sources (Ferreira and Gahl, 2017). In dietary supplements, it is found in inorganic forms (selenite and selenite) (Kasaikina et al., 2012).

Natural waters: Earlier research (pre-1990s) has given the following Se concentration levels for: Sea water: 60−120 ppm (surface layers); 200 ppm (deeper layers) (Sugimura et al., 1976); Lake water (Lake Michigan, USA): 0.083 ppb (Robberecht et al., 1982); Rainwater (Amazon River): 0.00021 ppm (Kharkar et al., 1968).

Later studies have yielded Se concentration range of: 0.06 ppb to about 400 ppb for groundwater and surface water (See references in WHO, 2011), and an average value of approximately 2000 ppb for marine fishmeal (Uden, 2019).

The dynamics of Se transport in natural waters is complex, as in the case of S (See, e.g., Tostevin et al., 2016). The entire gamut of inorganic species of Se (-2, 0, $+4$, and $+6$) as well as its organic species (monomethylated and dimethylated) have been reported in aquatic systems (Pettine et al., 2015). Biogeochemical cycling of Se from inflowing waters in oceans, seas, and lakes leads to some deposition of the element in sediments, the dynamics of the process are governed by ambient biophysicochemical conditions (Wells and Stolz, 2020), whose operation we still do not fully understand.

Leaching of Se species from soil to groundwater has significant consequences, as groundwater is a major source of drinking water, e.g., for many African populations. *However, the mechanisms involved in such leaching processes are poorly understood, owing to intricacies in water-rock interactions and heterogeneity of soil and sediment interfaces (See Dhillon et al., 2008).*

Drinking water: Traces of Se ranging from 0.001 to 10 ppb are found in drinking water (Gad and Pham, 2014; Uden, 2005). The range and *maximum concentration levels* (MAC) in drinking water for different countries are usually based on recommended values provided by authoritative regulatory bodies, such as the World Health Organization (WHO) and the United States Environmental Protection Agency (US EPA). In 2011 WHO gave an *upper tolerable intake level* for drinking water of 400 μg/day and a *provisional* guideline value of 40 ppb, designated as provisional because of the uncertainties inherent in the scientific database. The US EPA (Reviewed, 2021) set the *maximum contaminant level* and the *maximum contaminant level goal* in drinking water for Se at 0.05 ppm, which is the same as the MAC for total Se established by Health Canada in 2014. Such minuscule amounts established for Se in drinking water makes it unlikely for this source to be of particular significance from either a nutritional or a toxicological standpoint.

Air: Contributions to the atmospheric Se budget comprise natural emissions, including crustal weathering, sea spray, volcanic exhalations, and biogenic activity on land and sea. Emissions from human activities are mainly due to fossil fuel combustion, nonferrous metal production and manufacturing (Johnson et al., 2009; Wen and Carignan, 2007).

Estimates by Feinberg et al. (2020) put the annual mass of Se cycled (based on calculations from previous studies) at 13−19 Gg, which includes: volcanic outgassing (approximately 5% of total emissions), marine biota (35%), and terrestrial biota (15%)—and anthropogenic sources (40%)—including coal combustion, metal smelting, and biomass burning.

Estimates of Se input into the atmosphere due to volatilization have been put at 6×106 kg, the primary carriers being dimethyl selenide and dimethyl diselenide (Nriagu and Pacyna, 1988) volatilized from soil, plants, fresh and sea water, and volcanic activity. However, as noted by Mayland in 1989, appreciable quantities of gaseous Se can also be absorbed from the atmosphere by plants.

Of the other natural sources of atmospheric Se, such as wind-blown dust and sea salt, little contribution to the total budget is made, as these have shorter atmospheric lifetimes than those of the volatile forms of Se (Feinberg et al., 2020).

Associations (synergism/antagonism)

Several versions of the Periodic Table now indicate elements that are geomedically relevant, either as nutritional elements or as potentially toxic elements (see, e.g., Table 7−1 in "Introduction to Part II," this Volume). Elements within these groupings that can act in concert, either synergistically or antagonistically, are indicated.

The similarity in geochemical behavior of the Group VI elements is well exemplified by the distinctiveness of the S/Se ratios in different rock types, which can be used to elucidate the origin of rocks, ores, and sediments (Malisa, 2001).

Being a chalcophile element, Se is present in higher than background amounts along with other chalcophiles (e.g., Ag, As, Au, Cu, Hg, Mo, Pb, Sb, and Zn) in metal sulfide-mineralized areas, such as the African greenstone belts (See Davies, 2013). This association also manifests itself in metabolic interactions involving Se. For instance, Se can modulate the activity of these elements as potential toxicants or as anticarcinogenic agents by forming inert complexes with them or by changing the antioxidant potential of animals (including humans) through composing selenoproteins. On the other hand, associated elements can modulate Se activity, Hg and Cu, for instance, being known to ameliorate Se toxicity. *However, the mechanisms underlying these interactions, e.g., detoxification of Hg by Se, commonly described as "antagonistic,"* remain unclear (See, e.g., Zhao et al., 2013). Conversely, *synergistic* effects have also been observed (See, e.g., Penglase et al., 2014). The combined effect of Si and Se as stress mitigator in agricultue is currently being investigated (see, e.g., (Luís Oliveira Cunha and de Mello Prado, 2023; Moulick et al., 2024)).

Chemical form or speciation in soil and aqueous phases

Knowledge of the chemical forms or speciation in which Se occurs is important (as is the case with most of the other elements in the Periodic Table) because it determines its behavior in absorption and retention processes, and thus, its utilization from diets, including the level of health risk posed (Fordyce, 2013).

Just like S (with which Se shares the same group in the Periodic Table), Se exists in several different chemical forms/oxidation states. There are six natural stable isotopes of Se: ^{74}Se (0.9%), ^{76}Se (9.4%), ^{77}Se (7.6%), ^{78}Se (23.8), ^{80}Se (49.6%), and ^{82}Se (8.7%) (Uden, 2005). Selenium usually occurs in the sulfide ores of metals in the chalcophile group (See Section: "Minerals," in this chapter), such as Bi, Cu, Pb, Hg, and Ag, but rarely as the free element.

The known oxidation states are selenide (Se^{2-}), elemental selenium (Se^{0}), selenite (Se^{4+}), and selenate (Se^{6+}). In natural soil conditions, selenite (Se^{4+}) and selenate (Se^{6+}) are the main inorganic forms (of Se). In alkaline soils, selenate is the dominant form, with its mobility being higher than that of selenite, which commonly occurs in neutral or acid soils, and is easily sorbed onto oxyhydroxides (Kabata-Pendias, 2011). The major chemical form of Se in plants is SeMet, while that in animal tissues is SeCys.

Environmental circulation, and impact of climate change

Most biological and nonbiological materials in the environment (aquatic and terrestrial organisms, water, air, and soil) contain varying quantities of Se (Fig. 14−1). At the global scale, Se cycling occurs through the atmosphere, marine, and terrestrial systems. Since the middle of the 1970s, anthropogenic activities (mainly coal combustion) have made new Se inputs into the ecosystem (Maryland et al., 1989).

Complex climate-soil interactions are predicted to lead to a substantial decrease in total soil Se, especially in areas under agricultural production in southern Africa (El-Ramady et al., 2015a; Jones et al., 2017), a region that is already known to be Se deficient (Hurst et al., 2013). Against this background, Jones et al. (2017) have projected that "... total soil Se concentration, and thereby crop Se concentration and dietary Se intakes, will likely decrease under moderate climate-change scenarios by the years 2080−99."

Mobility, mode of entry into food chains, and bioavailability

The primary control of Se fluxes in the soil is the underlying geology. However, the mobility and uptake into plants and animals, *viz.*, the *bioavailability*, are determined by several biophysiochemical parameters, which include pH and redox conditions; the chemical form or speciation of

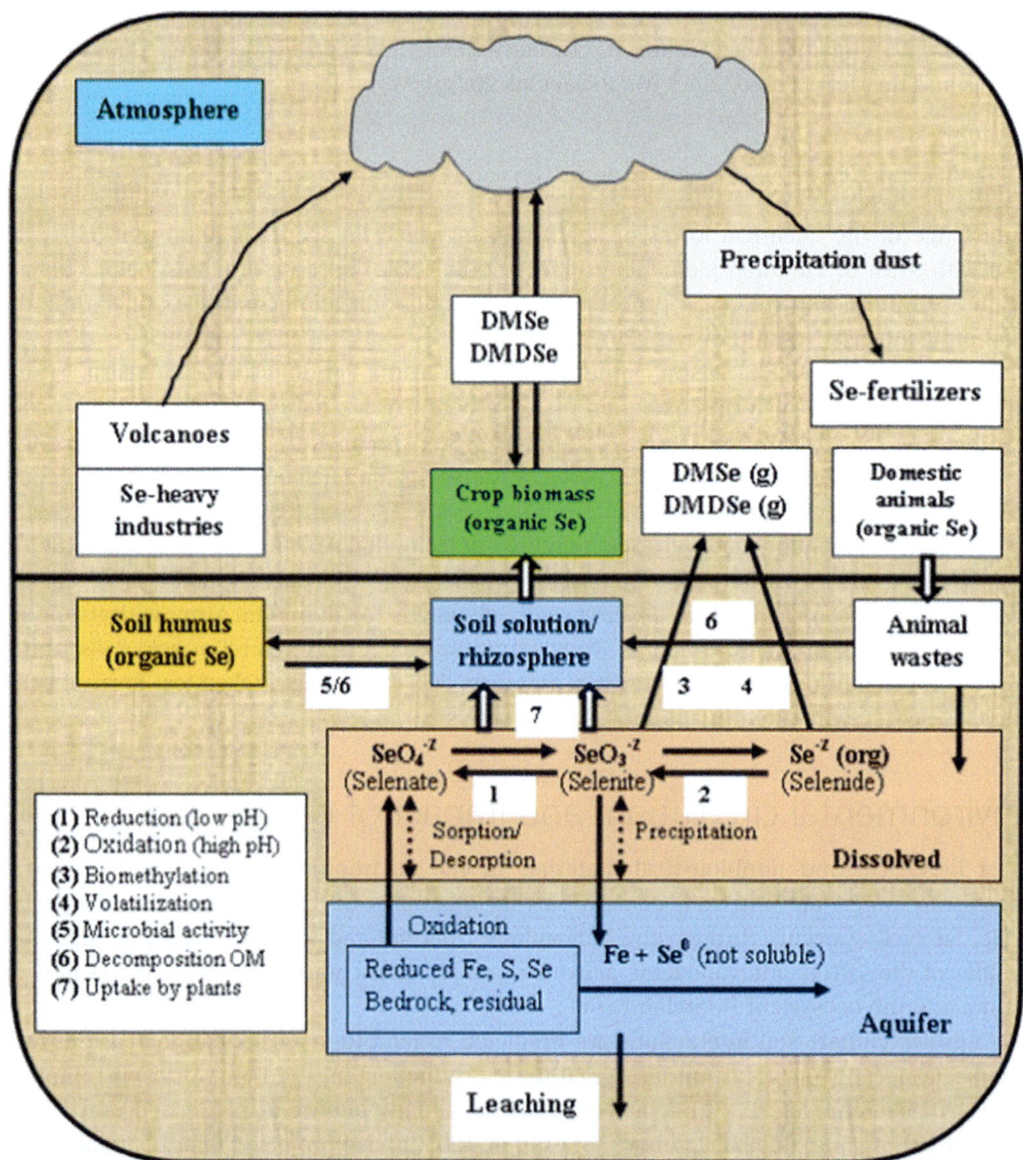

FIGURE 14–1 Biogeochemical cycle of Se under field conditions. *Adapted from El-Ramady, H., Abdalla, N., El-Henawy, A., Faizy, Salah, E.-D.A., Shams, M.S., 2015b. Selenium and its role in higher plants. In: Lichtfouse, E., Schwarzbauer, J., Robert, D. (Eds.), Pollutants in Buildings, Water and Living Organisms. Environmental Chemistry for a Sustainable World, vol. 7. Springer, Cham. <https://doi.org/10.1007/978-3-319-19276-5_6> (Accessed 24 February 2021).*

Se; soil texture and mineralogy; OM content; and the presence of competitive ions (Fordyce, 2007; Li et al., 2017). High Se mobility might be expected in soils of high pH and Eh, while soils with high contents of oxyhydroxide, clay, and OM lead to low Se mobility (Lopes et al., 2017).

Of the different oxidation states of Se (See Section: "Chemical forms or speciation in soil and aqueous phases," in this chapter), selenate (Se^{6+}), and $^+$selenite (Se^{4+}) are the predominant mobile forms, with Se^{2-} (selenide) and Se^0 being insoluble (Mayland et al., 1991). Selenate and selenite compete with other anions, such as phosphate, sulfate, oxalate, and molybdate for adsorption sites.

It is therefore important to acquire a thorough understanding of these controls to be able to reliably predict and remediate health risks from Se, as even soils that contain adequate total Se concentrations can result in Se deficiency if the element is not in readily bioavailable form (Fordyce, 2007).

Absorption and distribution

Selenium is an essential trace element at the level of approximately 0.1 ppm in diets. For the general population, oral ingestion is the primary exposure pathway/route of entry into the body (Gad and Pham, 2014).

Selenium in the form of SeMet and selenate is highly bioavailable to animals and humans. Optimum Se intake can be achieved through the consumption of foodstuffs that contain high proportions of these forms. The richest food sources of Se are Brazil nuts, seafoods, and organ meats, poultry, offal, milk products, cereals, grains, and vegetables (Sunde, 2012; Ibrahim et al., 2019) in addition to dietary supplements.

Dietary Se is known to be efficiently absorbed, with the retention of organic forms being higher than that of inorganic forms (Fairweather-Tait et al., 2010). Absorption of the inorganic forms falls within the 50%–90% of efficiency range (Bates, 2005).

The mechanisms of intestinal absorption of Se vary in a manner that depends on the chemical form of the element (See Section: "Chemical Forms or Speciation in Soil and Aqueous Phases," in this chapter). In animals, absorption takes place largely through the duodenum, but also in the cecum by active transport through a Na pump (Bates, 2005; Preda et al., 2015). Once inside the body, the distribution of Se compounds to major organs (of the body) takes place rapidly. Reduction and biotransformation of many Se compounds into excretable metabolites takes place in the liver (Alexander, 2015). According to Bates (2005), recent intake of Se tends to be reflected in urinary Se rather than tissue status, albeit a useful guide regarding the possibility of Se excess.

Bates (2005) summarizes the main pathways of interconversion of Se in mammalian tissues and comments that the element does not appear to be essential for plants.

Metabolic function (essentiality/clinical toxicity)

Essentiality

Selenium is an integral part of a number of *enzymes* [e.g., *glutathione peroxidase enzymes* (*GSHPx or simply GPx*); Fig. 14–2] that perform important functions in animals and humans. At least 25 essential selenoproteins have been identified (Burlingame et al., 1996; Rayman, 2005;

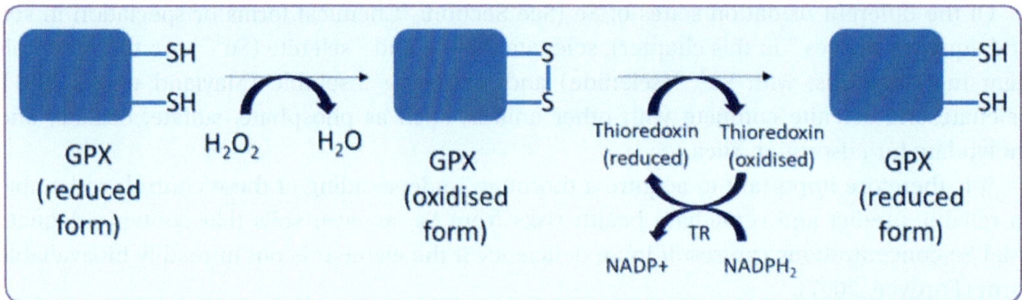

FIGURE 14–2 Reaction mechanism of glutathione peroxides (GPX). *Credit: Vishnu D. Rajput. Source: Rajput et al. (2021).*

Lin et al., 2015). The range of functions of selenoproteins is broad and includes a role in immune response, particularly cellular immunity (see, e.g., Minich, 2022). No doubt, determining the way Se intake differentially affects various types of immune responses and elucidating the mechanisms by which this (intake) occurs will lead to a better understanding of how Se-supplementation should be applied for human diseases involving Se interactions in the immune system (Hoffmann and Berry, 2008). There are numerous recent studies describing the role of Se in the function of immune cells, and the various immune deficiencies and diseases that result from inadequate/excessive dietary intake (e.g., Gill and Walker, 2008; Huang et al., 2012; Chisenga and Kelly, 2014).

Deficiency

It is well-documented that Se deficiency is associated with a spectrum of diseases and disorders, many of which occur in low-Se areas that correspond to areas where Se deficiency disorders occur. These diseases and disorders include:

- *Keshan disease*: An endemic *cardiomyopathy*, which was prevalent in a zone that stretches from northeast to southwest China, where Se contents in the soil and crops are extremely low (See Fordyce, 2013).
- Other *cardiovascular diseases*: Numerous epidemiological studies have been performed to examine whether Se deficiency in humans also enhances the risk of other heart diseases (See, e.g., Benstoem et al., 2015). Most studies have focused on the association between Se deficiency and coronary heart disease since the development of atherosclerosis is known to be influenced by Se deficiency in lipoprotein metabolism [(excess formation of reactive oxygen species (*ROS*) and oxidation of low-density *lipoproteins* (See "Glossary of terms," in this chapter)].
- *White muscle disease* is a disorder —chalky muscular appearance due to fibrosis and abnormal Ca deposition—observed in livestock raised on land with low Se levels. Extensive epidemiologic studies have shown that low Se levels in the soil and in local foodstuffs correlated with low Se levels in whole-blood and hair samples from people

living in areas where Keshan disease was endemic, as compared with other areas in China (Chen, 2012).

- *Kashin-Beck disease*: An endemic and chronic *osteochondropathy* is a disorder of the bones and joints of the hands and fingers, elbows, knees of children and adolescents who slowly develop stiff deformed joints, shortened limb length, and short stature due to *necrosis* (death) of the growth plates and articular cartilage. It is prevalent in areas of low Se intake. In a geographical area that stretches from southeastern Siberia to China in Asia (Yu et al., 2019; Wang et al., 2020).

- The general observation that Se deficiency is closely associated with I deficiency in certain KBD areas in China (e.g., Yu et al., 2019) can be explained as follows: In addition to GSHPx Se is an important component of another enzyme, *iodothyronine deiodinase*. In 1999 an editorial appeared in "The Archivist" postulating that combined Se deficiency and I deficiency could bring about tissue oxidative damage due to a lack of GSHPx; and this, together with tissue thyroid hormone deficiency, could produce joint damage. This apparent joint Se and I deficiency exists in the DRC (formerly Zaire) and the Central African Republic and was reviewed by Vanderpas et al. (1990). See Utiger (1998) on how the association of Se and I provides an expansion of the diagnostic spectrum of iodine deficiency disorders.

- It is well-documented that Se deficiency is associated with higher susceptibility to *RNA viral infections* and more severe disease outcomes (Hoffmann and Berry, 2008; Hiffler and Rakotoambinina, 2020). Evidence exists to show that Se deficiency is implicated in the promotion of mutations, replication, and virulence of RNA viruses; and that Se might be beneficial via restoration of host antioxidant capacity, reduction of *apoptosis*, and endothelial cell damages as well as *platelet* aggregation. It also appears that low Se status is a common finding in conditions considered at risk of severe COVID-19, especially in the elderly (Hiffler and Rakotoambinina, 2020).

- Selenium deficiency is known to cause *impaired motility* and structural disorder of sperm, leading to male infertility in animals (Qazi et al., 2019). Since a high concentration of Se is detected in the midpiece of sperm where mitochondria are abundant, Se-containing protein in the midpiece has been considered to be associated with the maintenance of structure and/or function of sperm.

- *HIV Infection*—Increasing amounts of evidence have indicated moderate Se deficiency in AIDS patients, with the involvement of Se both in the pathogenesis and progression of the deficiency syndrome (Kamwesiga et al., 2011; Stone et al., 2010). Although computer analyses of HIV genome have provided some intriguing explanation for the disturbance in Se metabolism in AIDS patients, *there is as yet no corroborative biological evidence to support this* (See, e.g., Stone et al., 2010).

As to the mechanism by which Se deficiency promotes the replication of HIV, both the suppression of immunological responses and the activation of *NFκB* (See "Glossary of terms," in this chapter) caused by enhanced oxidative stress may be involved (See: Huang et al., 2012).

- *Selenium and cancer:* Over the last 30 years or so, a number of epidemiological studies have been conducted to examine the relationship between dietary Se intake and the risk of cancer (e.g., Vinceti et al., 2018). To investigate the causal relationship between Se and cancer, prospective studies are considered to be a better approach (See, e.g., Vineis et al., 2017). In terms of evaluating preventive effects of Se on cancer occurrence, intervention studies seem to be the right way to go (See Meyskens et al., 2017).

Toxicity

The ecotoxicity of Se is complex and the *mechanism of its action "in vivo," remains poorly understood.* A number of reasons can be advanced for this, such as the lack of a specific target organ of Se toxicity and the existence of Se in many chemical forms (See Section: "Chemical forms or speciation in soil and aqueous phases," in this chapter), each one having a different bioavailability and ecotoxic potential. *And, as already indicated, an understanding of the mechanistic biochemistry of Se is predicated on the acquisition of a complete grasp of the processes of speciation, biotransformation, and accumulation in the food chain* (See, e.g., Uden, 2005).

Human studies

Selenium poisoning in humans expresses itself in diverse ways. Several cases of *acute*, sometimes fatal, Se poisoning in humans have been reported. *Chronic* exposures to high levels of environmental Se have occurred in South Dakota in the United States, Venezuela, Enshi County in Hubei Province of China, the Rivalta region of Italy, the Brazilian Amazon, Mexico, and India (See Vinceti et al., 2014 and references therein); but the literature hardly documents cases of Se poisoning in humans in Africa (Christophersen et al., 2013), though this does not imply that there are none. Selenium poisoning in humans expresses itself in diverse ways. According to Vinceti et al. (2014) acute Se intoxication have engendered "… suicide attempts, consumption of Se-containing dietary supplements, intake of food sources with very high Se content, like Brazil nuts, occupational exposures, and rarer etiologies."

Reported symptoms include morphological changes in fingernails and toenails, skin lesions, decay, and discoloration of teeth, loss of hair, garlic odor of the breath, gastrointestinal tract distress, and neurological symptoms, such as *paresthesia* in severe cases (Gad and Pham, 2014; Vinceti et al., 2014).

Reduction of body burden

Available clinical protocols for treatment of element toxicities are at the moment, generally poorly construed. Fordyce (2013) lists the metals Hg, Cu, and Cd as agents that may be effective in reducing the toxicity of Se, due to their reactions with Se in the intestinal tract to form insoluble Se compounds. In terms of human diets, dietary diversification may also help in reducing Se toxicity.

Selenium poisoning in animals

The most graphic manifestations (of Se poisoning) have been in farm animals in the United States, in the form of *blind staggers* and *alkali disease*, both of which have been observed since the first half of the 20th century and are thought to be caused by ingestion of Se-accumulating (seleniferous) plants (O'Toole and Raisbeck, 1995). Blind staggers is an acute syndrome in cattle and sheep characterized by wandering and stumbling of animals with impaired vision. However, as Jeffery Hall wrote in 2014, *"Blind staggers is no longer believed to be caused by selenium but by sulfate toxicity due to consumption of high-sulfate alkali water and/or high sulfur-containing forages. Excess sulfate (>1% of diet) leads to polioencephalomalacia and the classical signs of blind staggers."* [Sic.].

Alkali disease is a chronic form of selenosis of cattle, horses, mules, and hogs characterized by emaciation, lameness, loss of hair, and disorganized forms of hooves. A number of studies have reported both on Se deficiency and toxicity in animals in Africa. Pertinent references are given in the Section: "Selenium research in Africa," of this chapter.

Supplementation

The *nutritional supplementation* of Se for humans or animals whose Se intake is at low levels includes the addition of sodium selenite in table salts. Such supplementation is capable of achieving recovery of the physiological functions of Se (e.g., Himeno and Imura, 2002; Wang et al., 2017). Selenium-based drugs show some promise in several diseases, such as in orally active antihypertensive drugs, as well as in anticancer, antiviral, antimicrobial, and immunosuppressive drugs (Aronson, 2016). Organoselenium compounds have been shown to reduce oxidative tissue damage and *edema* (e.g., Toklu and Tümer, 2015).

A number of studies have shown that urinary Se reflects Se intake in a dose-dependent manner and is influenced by the chemical nature of dietary Se (Burk et al., 2006; Combs et al., 2011; Rayman, 2012; Kokarnig et al., 2015). Physiological responses are evaluated by the elevation of selenoenzymes, such as *c-GPx*.

Selenium-based drugs show some promise in a number of diseases [(e.g., Se supplementation with lower doses of IL-2 (a cytokine) to reduce the strength of the side effects)], and have recently been shown to have some promise in the treatment of AIDS patients (e.g., Baum et al., 2000).

Selenium has also been shown to have cancer-preventive potential, but this depends on the baseline Se status of the individual, the chemical form of Se that is utilized, gender, and genetic polymorphisms in selenoproteins or cancer-related processes (See Davis, 2012; Del Castillo Busto et al., 2024; Nie et al., 2023; Xu et al., 2024b).

The richest sources of Se, as already mentioned are Brazil nuts, seafoods, and organ meats, poultry, offal, cereals, and vegetables are the richest food sources of selenium (Sunde, 2012; Ibrahim et al., 2019).

In establishing supplementation regimes, strict monitoring of Se toxicity should be observed, as the *effective dose* (ED) and *toxic dose* of Se are very close.

Soil amendment

According to Fordyce (2007), one approach in enhancing Se uptake into agricultural crops is to alter the species of crops grown on deficient soil to plant types that take up more Se. Another approach is to apply Se-rich fertilizers to the soil to increase the amount of Se being taken up by plants, animals, and humans. However, the development of a safe and effective Se fertilizer must be done judiciously, taking into account the chemical form of Se, its solubility under the range of soil pH and redox potentials, sorption-desorption reactions with minerals and OM, and the rate and amount of Se uptake by the plant (See Lopes et al., 2017).

Use of fly ashes as soil amendments could affect the availability of Se and other elements, both in the ash and the soil; so, bioavailability constraints should be carefully evaluated (See, e.g., Kumar et al., 2017). Other factors to be taken into account in applying fly ashes as soil amendment include the pH of the ash, concentration of total soluble salts, and presence of elements toxic to plants. These are factors that affect growth and elemental concentration in the plant.

Removal of soil selenium

The most common method used today for the removal of excess soil selenium is perhaps, *phytoremediation*, the use of various types of plants to remove, transfer, stabilize, and/or destroy contaminants in the soil. Some researchers consider the presence of phosphate and sulfate in soils can inhibit the uptake of Se in plants, implying that application of these minerals as soil treatments could counter the effects of Se toxicity in agricultural crops (See, e.g., Fordyce, 2013).

Hyperaccumulating plants, such as *Brassica juncea* can also be used to remove Se from contaminated soils by phytoremediation.

Analytical determination

The many Se species present in biological and other environmental materials present a great analytical challenge. The levels of Se in these materials can be very low, varying from a few parts per billion to a few percent. Although the determination of *total element* is essential for determining the element mass balance, it provides insufficient information and must be accompanied by speciation of Se compounds, as related to biochemical cycles (Uden, 2019). Analysis of reduced species such as Se^0, polyselenides, and S-Se mixed species in the environmental samples also presents some limitations.

In accord with the low levels of Se in biological and other environmental samples, procedures for the quantitative determination of Se cover a wide range of concentrations. Of these, those that give good limits of detection in different matrices include atomic fluorescence spectrometry (AFS), neutron activation analysis (NAA), atomic absorption spectroscopy (AAS), inductively coupled plasma-atomic emission spectroscopy, inductively coupled plasma-mass spectrometry (ICP-MS) (see, e.g., Yang et al., 2023), gas chromatography, and X-ray fluorescence spectrometry. This list is not exhaustive; and is meant only for identification of the well-established methods that are used as the standard methods of analysis.

Of these methods, AFS is often preferred for natural materials, such as foods, plants, and soils. The NAA method is normally employed when separation of the different isotopic components of Se is required, such as in tracer studies, since it gives good detection limits, albeit a more specialized form of analysis. In 2003 Niedzielski and Siepak and later, Kumkrong et al. (2018) and Zam et al. (2019) discussed the limitations of these methods with regard to both detection limits and interfering sample matrix components *and pointed out the need for developing more sensitive and robust methods of analysis for regulatory compliance.*

Biological samples that are normally analyzed to indicate Se status include whole blood, plasma or serum, hair, toe-nail, and urine content. Of these, hair has been used extensively as it is easier to collect. The quantification of serum or plasma Se levels is also commonly used as biomarker of Se exposure. Whole blood Se or erythrocyte Se content is used much less frequently (See Vinceti et al., 2018 and references therein).

Selenium research in Africa

Selenium is a *chalcogen* (ore former). It is present in higher than background concentrations in metal sulfide-mineralized areas of the extensive African greenstone belts (Davies and Mundalamo, 2010). But, because of complexities involved in the operation of a nexus of bio-physicochemical processes governing rock weathering and element dispersion in the surficial environment, large areas of Se deficiency do exist in soils, especially in the southern parts of the Continent (Van Ryssen, 2001; Courtman et al., 2012; Hurst et al., 2013; Fig. 14−3).

Realization of the importance of Se in nutrition and its deficiency in soils of Sub-Saharan Africa (SSA), especially southern Africa, have triggered a surge in research effort in this domain. Recent publications on the environmental fluxes of Se in SSA and its nutritional and health implications include the works of Van Ryssen (2001), Webb et al. (2001), Kupka et al. (2004), Melse-Boonstra et al. (2007), Barnes et al. (2009), Courtman et al. (2012), Courtman (2013), Kamwesiga et al. (2011), Kishosha et al. (2011), Hurst et al. (2013), Mabeyo et al. (2015), Davis and Myburgh (2016), Gashu et al. (2016), Lyons et al. (2017), Adeniyi and Agoreyo (2018), Gashu et al. (2018), Swart et al. (2018), Joy et al. (2019), Gashu et al. (2019), Swart (2019), Belay et al. (2020), Gashu et al. (2020), Ligowe et al. (2020), Lukusa et al. (2021), Röllin et al. (2021), Takata et al. (2021), Zyambo et al. (2022), and Mutonhodza et al. (2023). These references are not exhaustive, but they are the examples of human (and other animal) health conditions including cancer, that have been related to Se deficiency. In the Tanzanian study, for instance, Kupka et al. (2004) observed a significant decrease in mortality with increase in plasma Se (> 85 ppb) over a 5-year follow-up period, out of a cohort of 949 HIV-positive pregnant women.

Going through these references, it could be noted that there is hardly any documentation of cases of Se poisoning in humans in Africa (Christophersen et al., 2013), though

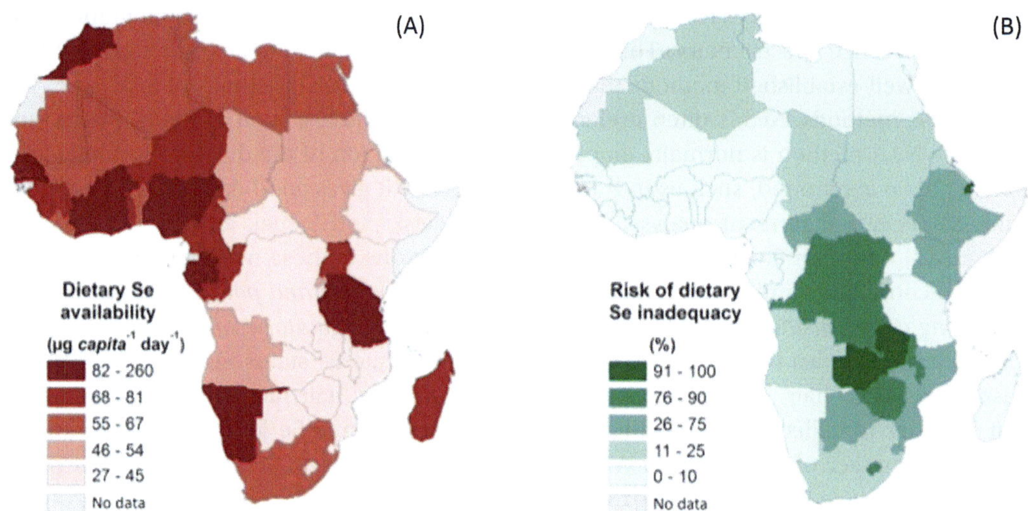

FIGURE 14–3 Mapping dietary Se availability in Africa. (A) Mean dietary Se availability and (B) Estimated risk of inadequate Se intake based on US Estimated Average Requirement (EAR) for different age groups. *Adapted from Hurst, R., Siyame, E.W., Young, S.D., Chilimba, A.D., Joy, E.J., Black, C.R., et al., 2013. Soil-type influences human selenium status and underlies widespread selenium deficiency risks in Malawi. Scientific Reports 3, 1425. https://doi. org/10.1038/srep01425.*

this does not imply that there are none. It may just be that overt clinical symptoms of Se toxicity and deficiency are not recognized as such and that these cases are rarely reported.

Also notable among these studies in terms of the number of African countries covered, is the work of Joy et al. (2014), which documented the mean and median national Se supplies of 50 and 55 µg/capita per day, respectively, for 46 countries across Africa, using FAO food supply and regional food composition tables and population-weighted data (See Joy et al., 2014). National Se supply in this study (Joy et al., 2014) ranged from 23 µg/capita per day (Liberia) to 93 µg/capita per day (Burkina Faso).

The Bill and Melinda Gates Foundation sponsored the GeoNutrition project (2017−21) (See Ligowe et al., 2020), whose specific aim was the mapping of soil-crop-human micronutrient linkages and their uncertainties including aspects of the environmental circulation of Se. The project also tested the effectiveness of increasing the Se concentration of cereal starch using Se fertilizers (agro-fortification) (Chilimba et al., 2012). The targeted study areas were in Malawi and Ethiopia, respectively, where previous studies have shown that a high prevalence of Se deficiency is likely.

Regulatory implications

Concern about the adverse effects resulting from exposure to excessive levels of Se has led to the establishment of optimum intake levels as well as upper limits of exposure for the element in diet and drinking water (See relevant sections, this chapter) by various national and international organizations. *Given Africa's unique geo-environmental setting, the precise biophysicochemical conditions of Se circulation in the environment, including those that determine its uptake and bioavailability apparently differ from those in other parts of the world, making it is necessary to establish regulatory standards of its own within the region.*

Other suggested areas of further research on geomedical characteristics of selenium

Selenium accumulates from the environment via plants/food crops and algae through the food chain to man and (other) animals. In most normal circumstances, therefore, food forms the major exposure route for Se. We still have much to learn about the uptake of Se in humans and (other) animals.

- Much research is clearly needed for setting reliable lower and upper safe Se levels, and current standards need to be quickly updated with reference to the upper levels, taking into account recent results of the epidemiologic studies, i.e., the high-quality randomized controlled trials and the environmental studies, and also considering the opportunity to set species-specific standards for this element (Vinceti et al., 2017). All said, there is the overarching need to appropriately revise the Se standards for dietary intake, drinking water, outdoor and indoor air levels, taking into account recently discovered adverse health effects of low-dose Se overexposure, and carefully assessing the significance of Se-induced *proteomic* changes.
- Precise knowledge regarding the biogeochemical controls that determine the distribution and mobility of environmental Se is essential for assessing health risks that are related to Se. Although we seldom have reports of overt clinical symptoms of Se toxicity and deficiency, there is little understanding about the possible subclinical effects, which should not be underestimated as medical science continues to uncover new essential functions for this biologically important element (See Fordyce, 2007).
- Information on bioavailability of Se is scanty. The need for robust data on this phenomenon is of immense importance, in view of the potential health benefits or toxicity associated with different intakes of Se, the need to derive dietary recommendations and to evaluate and improve the quality of food products (See, e.g., Fairweather-Tait et al., 2010).
- The possible impact that Se deficiency has on HIV patients is well-described, but must be a subject of further study (See, e.g., Melse-Boonstra et al., 2007; Barnes et al., 2009), as absolute certainty still does not yet exist on many aspects of this subject.
- Recent researches, yielding both successes and controversies, have focused on possible benefits of Se supplementation in preventing cancer and cardiovascular diseases, and in maintenance of an optimal immune system. Additional research is needed to determine whether Se supplements can help prevent or treat other immune-related conditions, such as thyroid disease and type 2 diabetes.

- Vinceti et al. (2014) have emphasized that "Additional research efforts are needed to better elucidate the neurotoxic effects exerted by selenium overexposure."
- We know that excess Se that does not undergo selenoprotein synthesis is transformed into methylated metabolites and excreted via urine or breath (See, e.g., Hu and Chan, 2020); and, as the uptake level and bioavailable fraction increase, the toxicological potential of Se does so too. Much more understanding is needed on the impact of this dynamic on selenoprotein pools in specific tissues/cells and nonspecific incorporation of Se, and the mechanistic bases in appropriate animal and cell models (See Cheng and Prabhu, 2019).
- Soil OM is known to influence the retention of Se in soils. However, as noted by Bruggeman et al. in (2007), "... the mechanisms of Se-OM interactions are poorly understood."
- We also need to research further the biological role of nonselenoprotein and selenium metabolites that are regulated by Se deficiency and/or supranutritional levels of selenium (Zhang et al., 2018).
- In 2002 Lemly stated that: "Few environmental contaminants have the potential to affect aquatic resources on such a broad scale, and still fewer exhibit the complex aquatic cycling pathways and range of toxic effects that are characteristic of selenium." These observations underscore the importance of the need for developing techniques for accurately identifying, diagnosing, and responding to problems associated with Se circulation in natural waters, before they become ubiquitous and intractable episodes.
- Atmospheric deposition of Se, especially through precipitation (rainfall), is an important source of Se to soils. However, according to Feinberg et al. (2018), the atmospheric Se cycle has not been extensively studied. Information regarding the distance that Se can travel in the atmosphere before being deposited, is scarce, and hence, our understanding of the spatial distribution of atmospheric Se is incomplete.
- Ligowe et al. remarked in 2020: "Evidence of Se or other micronutrient deficiency linkages across the agriculture−nutrition−health domains remains scarce in SSA.... There is clear scope for increasing the quality of geospatially defined information on the Se status of soil, crop, livestock, and human subjects in SSA."
- According to Ligowe et al. (2020), geospatial variation in Se status has not been studied widely in SSA; and outcomes of the GeoNutrition project (2017−21) (ICRAF (World Agroforestry), 2017) are expected to include findings on the most apposite use of geospatial information in support of policies to alleviate Se and other micronutrient deficiencies. Hence, according to Ligowe et al. (2020), "... new baseline maps and evidence (and uncertainties therein) will help to integrate evidence across the agriculture−nutrition−health domains to support policy decisions that are cost-effective and most expedient. For example, national micronutrient surveys, which currently focus solely on biomarkers and proxy outcomes of micronutrient status, could be integrated with geospatially resolved food composition/consumption surveys."
- The "Africa Geochemical Database" Program (See GEO, 2016 and Chapter 4), which has been ongoing during the past years is based on systematic collection of materials, such as

soil, sediment, water, rock and vegetation, which are then analyzed for a range of elements; and the data used to produce maps of element distributions in the environment. The use of correlation maps from a completed "Africa Geochemical Database" (See Chapter 4) would, among other applications, enable the depiction of areas where *disease clusters* overlie *anomalous* Se distribution (in water, soil, or air), and so permit an evidence-based statistical assessment of the magnitude of any geochemical component in the Se disease-causative web.

- Crafting of appropriate micronutrient surveillance programs for SSA where Se deficient soils are known to occur (See, e.g., Ligowe et al., 2020) is urgently needed.
- Despite increasing realization of the connections between Se and health under climate perturbations, there is still a paucity of research on this important linkage (See El-Ramady et al., 2015b).
- According to Zhang et al. (2020), "Recognition of the myriad mechanisms by which selenium might potentially benefit COVID-19 patients provides a rationale for randomized controlled trials of Se supplementation in SARS-CoV-2 infection." [Sic]. See also Kieliszek and Lipinski (2020).
- There is a need to develop or refine analytical methods that would obviate present complications in the assessment and quantification of Se species within foodstuffs. Currently available extraction techniques give only partial recovery, and the process can alter the form or forms present in the food. Current available methods need to be improved, standardized, and made more widely available for species quantification (See, e.g., Fairweather-Tait et al., 2010; Rievaj et al., 2021).
- Very few data exist regarding the environmental circulation of Se in the North African countries. As Ibrahim et al. wrote in 2019, "This warrants the need for rigorous and well-constructed studies to determine Se intake and status among the general population and specific populations, such as children and pregnant women, in those countries."
- Finally, the need for interdisciplinary collaboration in investigating Se in nutrition and health studies, can never be overemphasized. With reference to the Clark et al. study of 1996, for instance, Ip (1998), pointed out *the existence of a huge void in our understanding of the mechanism of action of Se in cancer prevention and cancer chemotherapy research,* and emphasized the need "…to develop a close interaction among chemists, biochemists, pharmacologists, oncologists, pathologists, toxicologists, cell biologists, and molecular biologists." Today, although there is a significant progress in this direction, there is still a lack of awareness by medical personnel, public health specialists, health policy formulators in government, and other stakeholders of the power of such collaboration, especially in light of the need for evaluating the status of the hundreds of geochemicals for anticancer activities (See, e.g., the conclusions in Johnson et al.'s paper of 2009).

As Ip observed in 1998, "Unless the community as a whole (including both commercial and public sectors) is willing to prioritize and commit the necessary resources for targeted research, the work on these hundreds of chemicals will proceed at the same agonizingly slow pace as we cross into the 21st century." All said, closer collaboration between medical and environmental scientists will be required to evaluate the real environmental health impact of this remarkable element in diseases, such as cancer, AIDS, heart disease, and COVID-19.

Glossary of terms

- *Acute toxicity* describes the adverse effects of a substance that result either from a single exposure or multiple exposures in a short period of time (usually less than 24 hours).
- *Apoptosis*, in biology, a mechanism that allows cells to self-destruct when stimulated by the appropriate trigger.
- *Cardiomyopathy* is a progressive disease of the heart muscle that makes it difficult for the heart to pump blood to the rest of the body and can lead to heart failure.
- c-GPx
- *Chronic toxicity* refers to the development of adverse effects as a result of long-term exposure to a contaminant or other stressor.
- *DNA* is an organic chemical that contains genetic information and instructions for protein synthesis.
- *Edema* is the medical term for fluid trapped in the body's tissues, usually occurring in the feet, legs, or ankles.
- *Effective dose*—In pharmacology, an ED or effective concentration is a dose or concentration of a drug that produces a biological response.
- *Enzymes* are proteins that act as biological catalysts (biocatalysts).
- *GSHPx* is the general name of an enzyme family with peroxidase activity whose main biological role is to protect the organism from oxidative damage. The human cellular glutathione peroxidase, *cGpx*, is the classical cytosolic version of the (Se-containing) enzyme Gpx1.
- *Lipoprotein*: Any member of a group of substances containing both lipid (fat) and protein.
- *NF-κB* (nuclear factor kappa light chain enhancer of activated B cells) is a transcription factor found in all nucleated cell types.
- *Paresthesia* is an abnormal sensation of the skin with no apparent physical cause.
- *Platelets* are tiny blood cells that help your body form clots to stop bleeding.
- Proteomics is the large-scale study of *proteins*, which are vital parts of living organisms, with many functions.
- *RNA* is *ribonucleic acid*, an important biological macromolecule that is present in all biological cells. It is principally involved in the synthesis of proteins, carrying the messenger instructions from DNA. mRNA is messenger RNA, a single-stranded RNA molecule that is complementary to one of the DNA strands of a gene.
- *Superoxide dismutases* are ubiquitous metalloenzymes that constitute the first line of defense against ROS, which are themselves highly reactive molecules and free radicals derived from molecular oxygen.

- *The iodothyronine deiodinases* constitute a family of selenoenzymes that selectively remove iodide from thyroxine and its derivatives, thus, activating or inactivating these hormones.

References

Adeniyi, M.J., Agoreyo, F.O., 2018. Nigeria and the selenium micronutrient: a review. Annals of Medical and Health Sciences Research 8, 5−11.

Adriano, D.C., 1986. Trace Elements in the Terrestrial Environment. Springer-Verlag, New York, NY, p. 533.

Alexander, J., 2015. Chapter 52: Selenium. In: Nordberg, G.F., Fowler, B.A., Nordberg, M. (Eds.), Handbook on the Toxicology of Metals, fourth ed., pp. 1175−1208. <https://doi.org/10.1016/B978-0-444-59453-2.00052-4> (Accessed 17 February 2021).

Aronson, J.K., 2016. Meyler's Side Effects of Drugs: The International Encyclopaedia of Adverse Drug Reactions and Interactions, sixteenth ed. Elsevier B.V, p. 7674.

Barnes, S.J., Savard, D., Bédard, L.P., Maier, W.D., 2009. Selenium and sulphur concentrations in the Bushveld Complex of South Africa and implications for formation of the platinum-group element deposits. Mineralium Deposita 44 (6), 647−663. Available from: https://doi.org/10.1007/s00126-009-0237-3.

Bates, C.J., 2005. Selenium. In: Caballero, B. (Ed.), Encyclopedia of Human Nutrition, second ed., pp. 118−125.

Baum, M.K., Miguez-Burbano, M.J., Campa, A., Shor-Posner, G., 2000. Selenium and interleukins in persons infected with human immunodeficiency virus type 1. Journal of Infectious Diseases 182 (Supplement 1), S69−S73. Available from: https://doi.org/10.1086/315911.

Belay, A., Joy, E.J.M., Chagumaira, C., Zerfu, D., Ander, E.L., Young, S.D., et al., 2020. Selenium deficiency is widespread and spatially dependent in Ethiopia. Nutrients 12 (6), 1565. Available from: https://doi.org/10.3390/nu12061565, https://doi.org/10.3390/nu12061565.

Benstoem, C., Goetzenich, A., Kraemer, S., Borosch, S., Manzanares, W., Hardy, G., et al., 2015. Selenium and its supplementation in cardiovascular disease - what do we know? Nutrients 7 (5), 3094−3118. Available from: https://doi.org/10.3390/nu7053094.

Brunfelt, A.O., Steinnes, E., 1967. Determination of selenium in standard rocks by neutron activation analysis. Geochimica et Cosmochimica Acta 31 (2), 283−285.

Burk, R.F., Norsworthy, B.K., Hill, K.E., Motley, A.K., Byrne, D.W., 2006. Effects of chemical form of selenium on plasma biomarkers in a high-dose human supplementation trial. Cancer Epidemiological Biomarkers Preview 15, 804−810.

Burlingame, A.L., Boyd, R.K., Gaskell, S.J., 1996. Mass spectrometry. Analytical Chemistry 68 (12), 599R−651R. Available from: https://doi.org/10.1021/a1960021u.

Chen, J., 2012. An original discovery: selenium deficiency and Keshan disease (an endemic heart disease). Asia Pacific Journal of Clinical Nutrition 21 (3), 320−326.

Cheng, W.H., Prabhu, K.S., 2019. Special issue of "Optimal selenium status and selenoproteins in health.". Biological Trace Element Research 192, 1−2. Available from: https://doi.org/10.1007/s12011-019-01898-x.

Chilimba, A.D.C., Young, S.D., Black, C.R., et al., 2012. Agronomic biofortification of maize with selenium (Se) in Malawi. Field Crops Research 125, 118−128.

Chisenga, C.C., Kelly, P., 2014. The role of selenium in human immunity. Medical Journal of Zambia 41 (4), 181−185.

Christophersen, O.A., Lyons, G., Haug, A., Steinnes, E., 2013. Selenium. In: Alloway, B. (Ed.), Heavy Metals in Soils. Environmental Pollution 22. Springer, Dordrecht. Available from: https://doi.org/10.1007/978-94-007-4470-7_16. (Accessed 17 February 2021).

Combs Jr, G.F., Watts, J.C., Jackson, M.I., Johnson, L.K., Zeng, H., Scheett, A.J., et al., 2011. Determinants of selenium status in healthy adults. Nutrition Journal 18 (10), 75. Available from: https://doi.org/10.1186/1475-2891-10-75.

Courtman, C.-C., 2013. Selenium Concentration of Maize Grain in South Africa and the Effect of Three Selenium Sources on the Selenium Concentration of Eggs and Egg Quality (M.Sc. (Agriculture) dissertation). University of Pretoria, South Africa. <http://hdl.handle.net/2263/33150> (Accessed 24 February 2021).

Courtman, C., Van Ryssen, J.B.J., Oelofse, A., 2012. Selenium concentration of maize grain in South Africa and possible factors influencing the concentration. South African Journal of Animal Science 42 (5), 454—458. Available from: https://doi.org/10.4314/sajas.v42i5.2.

Davies, T.C., 2013. Geochemical variables as plausible aetiological cofactors in the incidence of some common environmental diseases in Africa. Journal of African Earth Sciences 79, 24—49. Available from: https://doi.org/10.1016/j.

Davies, T.C., Mundalamo, H.R., 2010. Environmental health impacts of dispersed mineralisation in South Africa. Journal of African Earth Sciences 58 (4), 652—666. Available from: https://doi.org/10.1016/j.jafrearsci.2010.08.009.

Davis, C., 2012. Selenium supplementation and cancer prevention. Current Nutrition Reports 1 (1), 16—23. Available from: https://link.springer.com/article/10.1007/s13668-011-0003-x. (Accessed 15 March 2024).

Davis, A.J., Myburgh, J.G., 2016. Investigation of stillbirths, perinatal mortality and weakness in beef calves with low-selenium whole blood concentrations. Journal of the South African Veterinary Association 87 (1), a1336. Available from: https://jsava.co.za/index.php/jsava/rt/printerFriendly/1336/1753.

Del Castillo Busto, M.E., Ward-Deitrich, C., Evans, S.O., Rayman, M.P., Jameson, M.B., Goenaga-Infante, H., 2024. Selenium speciation studies in cancer patients to evaluate the responses of biomarkers of selenium status to different selenium compounds. Analytical and Bioanalytical Chemistry. Available from: https://doi.org/10.1007/s00216-024-05141-y.

Dhillon, S.K., Dhillon, K.S., Kohli, A., Khera, K.L., 2008. Evaluation of leaching and runoff losses of selenium from seleniferous soils through simulated rainfall. Zeitschrift Für Pflanzenernährung und Bodenkunde (Journal of Plant Nutrition and Soil Science) 171, 187—192.

Duyff, R.L., 2012. American Dietetic Association Complete Food and Nutrition Guide, fourth ed. John Wiley and Sons Inc, New Jersey, p. 720.

El-Ramady, H., Abdalla, N., Alshaal, T., Domokos-Szabolcsy, E., Elhawat, N., Prokisch, J., et al., 2015a. Selenium in soils under climate change, implication for human health. Environmental Chemistry Letters 13, 1—19. Available from: https://doi.org/10.1007/s10311-014-0480-4.

El-Ramady, H., Abdalla, N., El-Henawy, A., Faizy, Salah E.-D.A., Shams, M.S., 2015b. Selenium and its role in higher plants. In: Lichtfouse, E., Schwarzbauer, J., Robert, D. (Eds.), Pollutants in Buildings, Water and Living Organisms. Environmental Chemistry for a Sustainable World, vol. 7. Springer, Cham. Available from: https://doi.org/10.1007/978-3-319-19276-5_6. (Accessed 24 February 2021).

Fairweather-Tait, S.J., Collings, R., Hurst, R., 2010. Selenium bioavailability: current knowledge and future research requirements. American Journal of Clinical Nutrition 91 (5), 1484S—1491S. Available from: https://doi.org/10.3945/ajcn.2010.28674J. Erratum in: American Journal of Clinical Nutrition 92 (4), 1002.

Feinberg, A., Stenke, A., Peter, T., Winkel, L., 2018. The importance of marine selenium emissions for continental selenium deposition. In: 20th EGU General Assembly, EGU2018, Proceedings from the conference held April 4—13, 2018 in Vienna, Austria, p. 10240.

Feinberg, A., Stenke, A., Peter, T., Winkel, L.H.E., 2020. Constraining atmospheric selenium emissions using observations, global modeling, and Bayesian inference. Environmental Science and Technology 54 (12), 7146—7155. Available from: https://doi.org/10.1021/acs.est.0c01408.

Ferreira, C.R., Gahl, W.A., 2017. Disorders of metal metabolism. Translational Science of Rare Diseases 2 (3—4), 101—139. Available from: https://doi.org/10.3233/TRD-170015.

Fordyce, F.M., 2007. Selenium geochemistry and health. Ambio 36 (1), 94—97. Available from: https://doi.org/10.2307/4315793.

Fordyce, F.M., 2013. Selenium deficiency and toxicity in the environment. In: Selinus, O., et al., (Eds.), Essentials of Medical Geology. Springer, Dordrecht. Available from: https://doi.org/10.1007/978-94-007-4375-5_16 (Accessed 13 February 2021).

Frost, D.V., 1972. The two faces of selenium - can selenophobia be cured? CRC Critical Reviews in Toxicology 1 (4), 467−514.

Gad, S.C., Pham, T., 2014. Selenium, Encyclopedia of toxicology, third ed. 3. Elsevier, Amsterdam, pp. 232−235.

Garcia Moreno, R., Burdock, R., Alvarez, M.C.D., Crawford, J., 2013. Managing the selenium content in soils in semiarid environments through the recycling of organic matter. Applied and Environmental Soil Science, Special Issue: Soil Management for Sustainable Agriculture 2013, 283468. Available from: https://www.hindawi.com/journals/aess/2013/283468/.

Gashu, D., Stoecker, B.J., Adish, A., Haki, G.D., Bougma, K., Aboud, F.E., et al., 2016. Association of serum selenium with thyroxin in severely iodine-deficient young children from the Amhara region of Ethiopia. European Journal of Clinical Nutrition 70 (8), 929−934. Available from: https://doi.org/10.1038/ejcn.2016.27.

Gashu, D., Marquis, G.S., Bougma, K., Stoecker, B.J., 2018. Selenium inadequacy hampers thyroid response of young children after iodine repletion. Journal of Trace Elements in Medicine and Biology 50, 291−295. Available from: https://doi.org/10.1016/j.jtemb.2018.07.021.

Gashu, D., Marquis, G.S., Bougma, K., Stoecker, 2019. Spatial variation of human selenium in Ethiopia. Biology of Trace Element Research 189, 354−360. Available from: https://doi.org/10.1007/s12011-018-1489-5.

Gashu, D., Lark, R.M., Milne, A.E., Amede, T., Bailey, E.H., Chagumaira, C., et al., 2020. Spatial prediction of the concentration of selenium (Se) in grain across part of Amhara Region, Ethiopia. Science of the Total Environment 733, 139231. Available from: https://doi.org/10.1016/j.scitotenv.2020.139231.

GEO (Group on Earth Observations), 2016. "2017−2019 Work Programme: African Geochemical Baselines." GEO-XIII-5.3, pp 12−13. <https://www.earthobservations.org/documents/geo_xiii/GEO-XIII-5-3_2017-19_Work%20Programme.pdf> (Accessed 07 January 2021).

Gill, H., Walker, G., 2008. Selenium, immune function and resistance to viral infections. Nutrition and Dietetics - The Journal of Dietitians Australia 65, S41−S47. Available from: https://doi.org/10.1111/j.1747-0080.2008.00260.x.

Greenwood, N.N., Earnshaw, A., 1997. second ed. Chemistry of the Elements, 1. Butterworth Heinemann, Oxford, United Kingdom, p. 341.

Hall, R.A., Zook, E.G., Meaburn, G.M., 1978. National marine fisheries service survey of trace elements in the fishery resource. NOAA Technical Report MNFS SSRF-721 317. Available from: https://spo.nmfs.noaa.gov/sites/default/files/legacy-pdfs/SSRF721.pdf (Accessed 24 February 2021).

Hiffler, L., Rakotoambinina, B., 2020. Selenium and RNA virus interactions: potential implications for SARS-CoV-2 infection (COVID-19). Frontiers in Nutrition 7, 164.

Himeno, S., Imura, N., 2002. Chapter 17: Selenium in nutrition and toxicology. In: Sarkar, B. (Ed.), Heavy metals in the environment. CRC Press, Boca Raton, pp. 587−630.

Hoffmann, P.R., Berry, M.J., 2008. The influence of selenium on immune responses. Molecular Nutrition and Food Research 52 (11), 1273−1280. Available from: https://doi.org/10.1002/mnfr.200700330.

Högberg, J., Alexander, J., 2007. Chapter 38: Selenium. In: Nordberg, G.F., Fowler, B.A., Nordberg, M. (Eds.), Handbook on the toxicology of metals, third ed. Elsevier, Amsterdam, pp. 783−807.

Hu, X.F., Chan, H.M., 2020. Chapter 7: Selenium. In: Prasad, A.S., Brewer, G.J. (Eds.), Essential and toxic trace elements and vitamins in human health. Elsevier B.V., Amsterdam, pp. 113−125.

Huang, Z., Rose, A.H., Hoffmann, P.R., 2012. The role of selenium in inflammation and immunity: from molecular mechanisms to therapeutic opportunities. Antioxidants and Redox Signaling 16 (7), 705−743. Available from: https://doi.org/10.1089/ars.2011.4145.

Hurst, R., Siyame, E.W., Young, S.D., Chilimba, A.D., Joy, E.J., Black, C.R., et al., 2013. Soil-type influences human selenium status and underlies widespread selenium deficiency risks in Malawi. Scientific Reports 3, 1425. Available from: https://doi.org/10.1038/srep01425.

Ibrahim, S.A.Z., Kerkadi, A., Agouni, A., 2019. Selenium and health: an update on the situation in the Middle East and North Africa. Nutrients 11 (7), 1457. Available from: https://doi.org/10.3390/nu11071457.

ICRAF (World Agroforestry), 2017. GeoNutrition. <https://www.worldagroforestry.org/project/geonutrition> (accessed 15.02.21).

Ip, C., 1998. Lessons from basic research in selenium and cancer prevention. The Journal of Nutrition 128 (11), 1845−1854. Available from: https://doi.org/10.1093/jn/128.11.1845.

Johnson, C.C., Fordyce, F.M., Rayman, M.P., 2009. Symposium on geographical and geological influences on nutrition: Factors controlling the distribution of selenium in the environment and their impact on health and nutrition. The Proceedings of the Nutrition Society 69 (1), 119−132. Available from: https://doi.org/10.1017/S0029665109991807.

Jones, G.D., Droz, B., Greve, P., Gottschalk, P., Poffet, D., McGrath, S.P., et al., 2017. Selenium deficiency risk predicted to increase under future climate change. Proceedings of the National Academy of Sciences of the United States of America 114 (11), 2848−2853. Available from: https://doi.org/10.1073/pnas.1611576114.

Joy, E.J.M., Ander, E.L., Young, S.D., Black, C.R., Watts, M.J., Chilimba, A.D.C., et al., 2014. Dietary mineral supplies in Africa. Physiologia Plantarum 151, 208−229.

Joy, E., Kalimbira, A.A., Gashu, D., Ferguson, E.L., Sturgess, J., Dangour, A.D., et al., 2019. Can selenium deficiency in Malawi be alleviated through consumption of agro-biofortified maize flour? Study protocol for a randomised, double-blind, controlled trial. Trials 20 (1), 795. Available from: https://doi.org/10.1186/s13063-019-3894-2.

Kabata-Pendias, A., 2011. Trace Elements in Soils and Plants, fourth ed. CRC Press, Boca Raton, FL, p. 520.

Kabata-Pendias, A., Mukherjee, A.B., 2007. Trace Elements from Soil to Human. Springer, Berlin/Heidelberg/New York.

Kabata-Pendias, A., Pendias, H., 1984. Trace Elements in Soils and Plants. CRC Press, Boca Raton, Fla, p. 413.

Kamwesiga, J., Mutabazi, V., Kayumba, J., Tayari, J.-C.K., Smyth, R., Fay, H., et al., 2011. Effect of selenium supplementation on CD4 T-cell recovery, viral suppression, morbidity and quality of life of HIV-infected patients in Rwanda: Study protocol for a randomized controlled trial. Trials 12, 192. Available from: https://doi.org/10.1186/1745-6215-12-192.

Kasaikina, M.V., Hatfield, D.L., Gladyshev, V.N., 2012. Understanding selenoprotein function and regulation through the use of rodent models. Biochimica et Biophysica Acta 1823, 1633−1642. Available from: https://doi.org/10.1016/j.bbamcr.2012.02.018.

Khan, Z., Thounaojam, T.C., Chowdhury, D., Upadhyaya,, H., 2023. The role of selenium and nano selenium on physiological responses in plant: A review. Plant Growth Regulation 100 (2), 409−433. Available from: https://doi.org/10.1007/s10725-023-00988-0 (accessed 10.02.2024).

Kharkar, D.P., Turekian, K.K., Bertine, K.K., 1968. Stream supply of dissolved silver, molybdenum, antimony, selenium, chromium, rubidium and cesium to the oceans. Geochimica et Cosmochimica Acta 32, 285−298.

Kieliszek, M., Lipinski, B., 2020. Selenium supplementation in the prevention of coronavirus infections (COVID-19). Medical Hypotheses 143, 109878. Available from: https://doi.org/10.1016/j.mehy.2020.109878.

Kishosha, P.A., Galukande, M., Gakwaya, A.M., 2011. Selenium deficiency a factor in endemic goiter persistence in sub-Saharan Africa. World Journal of Surgery 35 (7), 1540−1545. Available from: https://doi.org/10.1007/s00268-011-1096-5.

Kohlmeier, M., 2003. Nutrient Metabolism. Elsevier Science, 840 p.

Kokarnig, S., Tsirigotaki, A., Wisenhofer, T., Lackner, V., Francesconi, K.A., Pergantis, S.A., et al., 2015. Concurrent quantitative HPLC−mass spectrometry profiling of small selenium species in human serum and urine after ingestion of selenium supplements. Journal of Trace Elements in Medicine and Biology 29, 83−90.

Kumar, K., Kumar, S., Gupta, M., Garg, H.C., 2017. Characteristics of fly ash in relation of soil amendment. Materials Today: Proceedings 4 (2), 527−532. Available from: https://doi.org/10.1016/j.matpr.2017.01.053.

Kumkrong, P., LeBlanc, K.L., Mercier, P.H.K., Mester, Z., 2018. Selenium analysis in waters. Part 1: regulations and standard methods. Science of The Total Environment 640-641, 1611−1634. Available from: https://doi.org/10.1016/j.scitotenv.2018.05.392.

Kupka, R., Msamanga, G.I., Spiegelman, D., 2004. Se status is associated with accelerated HIV disease progression among HIV-infected pregnant women in Tanzania. Journal of Nutrition 134, 2556−2560.

Lakin, H.W., 1972. Selenium accumulation in soils and its absorption by plants and animals. Geological Society of America Bulletin 83, 181.

Li, Z., Liang, D., Peng, Q., Cui, Z., Huang, J., Lin, Z., 2017. Interaction between selenium and soil organic matter and its impact on soil selenium bioavailability: A review. Geoderma 295, 69−79. Available from: https://doi.org/10.1016/j.geoderma.2017.02.019.

Ligowe, I.S., Phiri, F.P., Ander, E.L., Bailey, E.H., Chilimba, A.D.C., Gashu, D., et al., 2020. Selenium deficiency risks in sub-Saharan African food systems and their geospatial linkages. Proceedings of the Nutrition Society 79 (4), 457−467. Available from: https://doi.org/10.1017/S0029665120006904.

Lin, H.-C., Ho, S.-C., Chen, Y.-Y., Khoo, K.-H., Hsu, P.-H., Yen, H.-C.S., 2015. CRL2 aids elimination of truncated selenoproteins produced by failed UGA/Sec decoding. Science (New York, NY) 349 (6243), 91−95. Available from: https://doi.org/10.1126/science.aab0515.

Lopes, G., Ávila, F.W., Guilherme, L.R.G., 2017. Selenium behaviour in the soil environment and its implication for human health. Ciência e Agrotecnologia 41 (6), 605−615. Available from: https://doi.org/10.1590/1413-70542017416000517.

Luís Oliveira Cunha, M., de Mello Prado, R., 2023. Synergy of selenium and silicon to mitigate abiotic stresses: A review. Gesunde Pflanzen 75, 1461−1474. Available from: https://doi.org/10.1007/s10343-022-00826-9 (accessed 10.02.2024).

Lukusa, K., Abubeker, H., Lehloenya, K.C., 2021. Dietary selenium supplementation, clarified egg yolk extender and slow cooling improve cryopreserved sperm characteristics of Saanen buck. Asian Pacific Journal of Reproduction 10 (1), 43−48. Available from: https://doi.org/10.4103/2305-0500.306437.

Lyons, G., Gondwe, C., Banuelos, G., Mendoza, C., Haug, A., Christophersen, O., et al., 2017. Drumstick tree (Moringa oleifera) leaves as a source of dietary selenium, sulphur and pro-vitamin A. Acta Horticultura 1158, 287−292. Available from: https://doi.org/10.17660/ActaHortic.2017.1158.32.

Mabeyo, P.E., Manoko, M.L.K., Gruhonjic, A., Fitzpatrick, P., Landberg, G., Erdélyi, M., et al., 2015. Selenium accumulating leafy vegetables are a potential source of functional foods. International Journal of Food Science 2015, 549676. Available from: https://doi.org/10.1155/2015/549676.

Malisa, E.P., 2001. The behaviour of selenium in geological processes. Environmental Geochemistry and Health 23 (2), 137−158. Available from: https://doi.org/10.1023/A:1010908615486.

Maryland, H.F., James, L.F., Panter, K.E., Sonderegger, J.L., 1989. Selenium in seleniferous environments. In: Jacobs, L.W. (Ed.). Selenium in Agriculture and the Environment, Chapter 2: pp. 15−50. American Society of Agronomy, Madison, USA. https://doi.org/10.2136/sssaspecpub23.c2 (Accessed 10 February 2021).

Mayland, H.F., Gough, L.P., Stewart, K.C., 1991. Chapter E: Selenium mobility in soils and its absorption, translocation, and metabolism in plants. In: Severson, R.C., Fisher Jr, S.E., Gough, L.P. (Eds.), Proceedings of the 1990 Billings Land Reclamation Symposium. USA-MT-Billings, 1991/03/25-30, pp. 55−64.

Melse-Boonstra, A., Hogenkamp, P., Lungu, O.I., 2007. Mitigating HIV/AIDS in Sub-Saharan Africa Through Selenium in food. Golden Valley Agricultural Research Trust, Lusaka, Zambia. Available from: https://citeseerx.ist.psu.edu/viewdoc/download?doi = 10.1.1.455.5576&rep = rep1&type = pdf. (Accessed 14 February 2021).

Meyskens (Jr.), F.L., Mukhtar, H., Rock, C.L., Cuzick, J., Kensler, T.W., Yang, C.S., et al., 2017. Cancer prevention: obstacles, challenges, and the road ahead. Journal of the National Cancer Institute 108 (2), djv309. Available from: https://doi.org/10.1093/jnci/djv309.

Minich, W.B., 2022. Selenium metabolism and biosynthesis of selenoproteins in the human body. Biochemistry (Biokhimiia) 87 (1), S168−S177. Available from: https://doi.org/10.1134/S0006297922140139 (accessed 11.02.2024).

Minkin, J.A., Finkelman, R.B., Thompson, C.L., Chao, E.C.T., Ruppert, L.F., Blank, H., et al., 1984. Microcharacterization of arsenic- and selenium-bearing pyrite in Upper Freeport coal, Indiana County, Pennsylvania. Scanning Electron Microscopy 4, 1515−1524.

Moulick, D., Mukherjee, A., Das, A., Roy, A., Majumdar, A., Dhar, A., Pattanaik, B.K., et al., 2024. Selenium - An environmentally friendly micronutrient in agroecosystem in the modern era: An overview of 50-year findings. Ecotoxicology and Environmental safety 270, , 115832. Available from: https://doi.org/10.1016/j.ecoenv.2023.115832 (accessed 10.02.2024).

Mutonhodza, B., Chagumaira, C., Dembedza, M.P., Joy, E.J.M., Manzeke-Kangara, M.G., Njovo, H., Nyadzayo, T.K., et al., 2023. A pilot survey of selenium status and its geospatial variation among children and women in three rural districts of Zimbabwe. Frontiers in Nutrition 10, 1235113. Available from: https://doi.org/10.3389/fnut.2023.1235113.

Nie, S., He, X., Sun, Z., Zhang, Y., Liu, T., Chen, T., Zhao, J., 2023. Selenium speciation-dependent cancer radiosensitization by induction of G2/M cell cycle arrest and apoptosis. Frontiers in Bioengineering and Biotechnology 11, , 1168827. Available from: https://doi.org/10.3389/fbioe.2023.1168827 (accessed 10.02.2024).

Niedzielski, P., Siepak, M., 2003. Analytical methods for determining arsenic, antimony and selenium in environmental samples. Polish Journal of Environmental Studies 12 (6), 653−667.

Nriagu, J.O., Pacyna, J.M., 1988. Quantitative assessment of worldwide contamination of air, water, and soils by trace metals. Nature 333, 134−139.

O'Toole, D., Raisbeck, M.F., 1995. Pathology of experimentally induced chronic selenosis (alkali disease) in yearling cattle. Journal of Veterinary Diagnostic Investigation 7 (3), 364−373. Available from: https://doi.org/10.1177/104063879500700312.

Penglase, S., Hamre, K., Ellingsen, S., 2014. Selenium and mercury have a synergistic negative effect on fish reproduction. Aquatic Toxicology 149, 16−24.

Pettine, M., McDonald, T.J., Sohn, M., Anquandah, G.A.K., Zboril, R., Sharma, V.K., 2015. A critical review of selenium analysis in natural water samples. Trends in Environmental Analytical Chemistry 5, 1−7.

Pilon-Smits, E.A.H., 2019. On the ecology of selenium accumulation in plants. Plants 8 (7), 197. Available from: https://doi.org/10.3390/plants8070197.

Pirri, I.V., 2002. On the occurrence of selenium in sulfides of the ore deposit of Baccu Locci (Gerrei, SE Sardinia). Neues Jahrbuch für Mineralogie - Monatshefte 5, 207−224. Available from: https://doi.org/10.1127/0028-3649/2002/2002-0207.

Preda, C., Vasiliu, I., Ciobanu, D., Ungureanu, M.C., Leustean, L., et al., 2015. Selenium in the environment: essential or toxic to human health? Environmental Engineering and Management Journal [S.l.] 15 (4), 913−921. Available from: http://eemj.eu/index.php/EEMJ/article/view/2909.

Qazi, I.H., Angel, C., Yang, H., Zoidis, E., Pan, B., Wu, Z., et al., 2019. Role of selenium and selenoproteins in male reproductive function: a review of past and present evidence. Antioxidants 8, 268. Available from: https://doi.org/10.3390/antiox8080268.

Rajput, V.D., Harish, Singh, R.K., Verma, K.K., Sharma, L., Quiroz-Figueroa, F.R., Meena, M., et al., 2021. Recent developments in enzymatic antioxidant defence mechanism in plants with special reference to abiotic stress. Biology 10 (4), 267. Available from: https://doi.org/10.3390/biology10040267 (accessed 22.03.2024).

Rayman, M., 2005. Selenium in cancer prevention: a review of the evidence and mechanism of action. Proceedings of the Nutrition Society 64 (4), 527−542.

Rayman, M.P., 2012. Selenium and human health. The Lancet 379, 1256−1268. Available from: https://doi.org/10.1016/S0140-.

Rievaj, M., Culková, E., Šandorová, D., Lukáčová-Chomisteková, Z., Bellová, R., Durdiak, J., Tomčík, P., 2021. Electroanalytical techniques for the detection of selenium as a biologically and environmentally significant analyte: A short review. Molecules (Basel, Switzerland) 26 (6), , 1768. Available from: https://doi.org/10.3390/molecules26061768 (accessed 10.02.2024).

Robberecht, H., Berghe, D.V., Deelstra, H., Van Greiken, R., 1982. Selenium in Belgian soils and its uptake by rye-grass. Science of the Total Environment 25, 61−69. Available from: https://doi.org/10.1016/0048-9697 (82)90042-0.

Röllin, H.B., Channa, K., Olutola, B., Odland, J.Ø., 2021. Selenium status, its interaction with selected essential and toxic elements, and a possible sex-dependent response *in utero*, in a South African birth cohort.. International Journal of Environmental Research and Public Health 18 (16), , 8344. Available from: https://doi.org/10.3390/ijerph18168344 (accessed 10.02.2024).

Rosenfeld, I., Beath, O.A., 1964. Selenium in relation to public health. Selenium: Geobotany, Biochemistry, Toxicity, and Nutrition. Academic Press, New York, NY, pp. 279−289.

Shamberger, R.J., 1981. Selenium in the environment. Science of the Total Environment5t 17 (1), 59−74. Available from: https://www.sciencedirect.com/science/article/abs/pii/004896978190108X.

Shamberger, R.J., 1983. Environmental occurrence of selenium. In: Shamberger, R.J. (Ed.), Biochemistry of Selenium. Part of Book - Biochemistry of the Elements, vol. 2. Springer, Boston, MA, pp. 167−183. Available from: https://doi.org/10.1007/978-1-4684-4313-4_6. (Accessed 24 February 2021).

Stillings, L.L., 2017, Chapter Q: Selenium. In: Schulz, K.J., DeYoung, J.H., Jr., Seal, R.R., II, Bradley, D.C. (Eds.), Critical Mineral Resources of the United States - Economic and Environmental Geology and Prospects for Future Supply: U.S. Geological Survey Professional Paper 1802, pp. Q1- Q55. <https://doi.org/10.3133/pp1802Q> (Accessed 18 February 2021).

Stone, C.A., Kawai, K., Kupka, R., Fawzi, W.W., 2010. Role of selenium in HIV infection. Nutrition Reviews 68 (11), 671−681. Available from: https://doi.org/10.1111/j.1753-4887.2010.00337.x.

Sugimura, Y., Suzuki, Y., Miyake, Y., 1976. The content of selenium and its chemical form in sea water. Journal of the Oceanographical Society of Japan 32, 235−241. Available from: https://doi.org/10.1007/BF02107126.

Sunde, R.A., 2012. Selenium. In: Ross, A.C., Caballero, B., Cousins, R.J., Tucker, K.L., Ziegler, T.R. (Eds.), Modern Nutrition in Health and Disease, eleventh ed. Lippincott Williams and Wilkins, Philadelphia, PA, pp. 225−237.

Swart, R., 2019. Serum Selenium Levels, the Selenoprotein Glutathione Peroxidase and Cardiovascular Function in Black South Africans (Ph.D. thesis (Physiology)). North-West University, South Africa. 245 p.

Swart, R., Schutte, A.E., van Rooyen, J.M., Mels, C.M.C., 2018. Serum selenium levels, the selenoprotein glutathione peroxidase and vascular protection: the SABPA study. Food Research International 104, 69−76. Available from: https://doi.org/10.1016/j.foodres.2017.06.054.

Takata, N., Myburgh, J., Botha, A., Nomngongo, P.N., 2021. The importance and status of the micronutrient selenium in South Africa: A review. Environmental Geochemistry and Health 44 (11), 3703−3723. Available from: https://doi.org/10.1007/s10653-021-01126-3 (accessed 10.02.2024).

Taylor, S.R., 1964. Abundance of chemical elements in the continental crust: a new table. Geochimica et Cosmochimica Acta 28 (8), 1273−1285. Available from: https://doi.org/10.1016/0016-7037(64)90129-2.

Taylor, S.R., McLennan, S.M., 1985. The Continental Crust: Its Composition and Evolution. Blackwell, Oxford.

Toklu, H.Z., Tümer, N., 2015. Chapter 5: Oxidative stress, brain edema, blood−brain barrier permeability, and autonomic dysfunction from traumatic brain injury. In: Kobeissy, F.H. (Ed.), Brain Neurotrauma: Molecular, Neuropsychological, and Rehabilitation Aspects. CRC Press/Taylor and Francis, Boca Raton (FL).

Tostevin, R., Craw, D., Hale, R., Vaughan, M., 2016. Sources of environmental sulfur in the groundwater system, southern New Zealand. Applied Geochemistry 70, 1−16. Available from: https://doi.org/10.1016/j.apgeochem.2016.05.005.

Uden, P.C., 2005. Selenium - selenium in the environment. In: Worsfold, P., Townshend, A., Poole, C., Miró, M. (Eds.), Encyclopaedia of Analytical Science, second ed. Elsevier Science Publishing Co Inc, pp. 216−223.

Uden, P.C., 2019. *Environmental analysis/food analysis- selenium in the environment.* In: Worsfold, P., Townshend, A., Poole, C., Miró, M. (Eds.), Encyclopaedia of Analytical Science, third ed. Elsevier, p. 5109.

US ATSDR (US Agency for Toxic Substances and Disease Registry), 2003. Toxicological profile for selenium. https://www.atsdr.cdc.gov/ToxProfiles/tp92.pdf. (Accessed 17 February 2021).

Utiger, R.D., 1998. Kashin-Beck disease - expanding the spectrum of iodine-deficiency disorders. New England Journal of Medicine 339 (16), 1156−1158. Available from: https://doi.org/10.1056/NEJM199810153391611.

Van Ryssen, J.B.J., 2001. Geographical distribution of the selenium status of herbivores in South Africa. South African Journal Of Animal Science 31, 1−8. Available from: http://hdl.handle.net/2263/76018.

Vinceti, M., Mandrioli, J., Borella, P., Michalke, B., Tsatsakis, A., Finkelstein, Y., 2014. Selenium neurotoxicity in humans: bridging laboratory and epidemiologic studies. Toxicology Letters 230 (2), 295−303. Available from: https://doi.org/10.1016/j.toxlet.2013.11.016.

Vinceti, M., Filippini, T., Cilloni, S., Bargellini, A., Vergoni, A.V., Tsatsakis, A., et al., 2017. Health risk assessment of environmental selenium: emerging evidence and challenges (Review). Molecular Medicine Reports 15 (5), 3323−3335. Available from: https://doi.org/10.3892/mmr.2017.6377.

Vinceti, M., Filippini, T., Wise, L.A., 2018. Environmental selenium and human health: an update. Current Environmental Health Reports 5, 464−485. Available from: https://doi.org/10.1007/s40572-018-0213-0.

Vineis, P., Illari, P., Russo, F., 2017. Causality in cancer research: a journey through models in molecular epidemiology and their philosophical interpretation. Emerging Themes in Epidemiology 14, 7. Available from: https://doi.org/10.1186/s12982-017-0061-7.

Wang, N., Tan, H.Y., Li, S., Xu, Y., Guo, W., Feng, Y., 2017. Supplementation of micronutrient selenium in metabolic diseases: Its role as an antioxidant. Oxidative Medicine and Cell Longevity 2017, 7478523. Available from: https://doi.org/10.1155/2017/7478523.

Wang, J., Zhao, S., Yang, L., Gong, H., Li, H., Nima, C., 2020. Assessing the health loss from Kashin-Beck disease and its relationship with environmental selenium in Qamdo District of Tibet, China. International Journal of Environmental Research and Public Health 18 (1), 11. Available from: https://doi.org/10.3390/ijerph18010011.

Webb, E.C., van Ryssen, J.B., Erasmus, M.E., McCrindle, C.M., 2001. Copper, manganese, cobalt and selenium concentrations in liver samples from African buffalo (*Syncerus caffer*) in the Kruger National Park. Journal of Environmental Monitoring 3 (6), 583−585. Available from: https://doi.org/10.1039/b106307n.

Wells, M., Stolz, J.F., 2020. Microbial selenium metabolism: a brief history, biogeochemistry and ecophysiology. FEMS Microbiology Ecology 96 (12), fiaa209. Available from: https://doi.org/10.1093/femsec/fiaa209.

Wen, H., Carignan, J., 2007. Reviews on atmospheric selenium: emissions, speciation and fate. Atmospheric Environment 41, 7151−7165.

Werkneh, A.A., Gebretsadik, G.G., Gebru, S.B., 2023. Review on environmental selenium: Occurrence, public health implications and biological treatment strategies. Environmental Challenges 11, 100698. Available from: https://doi.org/10.1016/j.envc.2023.100698.

Whanger, P.D., 2002. Selenocompounds in plants and animals and their biological significance. Journal of the American College of Nutrition 21, 223−232. Available from: https://doi.org/10.1080/07315724.2002.10719214.

WHO (World Health Organization), 1996. Trace Elements in Human Nutrition and Health. World Health Organization, Geneva.

WHO (World Health Organisation), 2006. Guidelines for Drinking-Water Quality, third ed. WHO, Geneva, 460 p.

WHO (World Health Organisation), 2011. Selenium in Drinking-water - Background document (WHO/HSE/WSH/10.01/14) for development of WHO Guidelines for Drinking-water Quality. https://www.who.int/water_sanitation_health/dwq/chemicals/selenium.pdf. (Accessed 11 February 2021).

Winkel, L.H.E., Johnson, C.A., Lenz, M., Grundl, T., Leupin, O.X., Amini, M., et al., 2012. Environmental selenium research: from microscopic processes to global understanding. Environmental Science and Technology 46 (2), 571−579. Available from: https://doi.org/10.1021/es203434d.

Xu, Z.N., Lin, Z.Q., Zhao, G.S., Guo, Y.B., 2024a. Biogeochemical behavior of selenium in soil-air-water environment and its effects on human health.. International Journal of Environmental Science and Technology 21 (1), 1159−1180. Available from: https://doi.org/10.1007/s13762-023-05169-0 (Accessed 10 February 2024).

Xu, Y., Lai, H., Pan, S., Pan, L., Liu, T., Yang, Z., Chen, T., Zhu, X., 2024b. Selenium promotes immunogenic radiotherapy against cervical cancer metastasis through evoking P53 activation.. Biomaterials 305, , 122452. Available from: https://doi.org/10.1016/j.biomaterials.2023.122452 (Accessed 10 February 2024).

Yang, Q., Li, C., Hu, J., Hou, X, 2023. Ultrasensitive determination of selenium in water and food samples by ICP-MS: UiO-66-NH$_2$ for preconcentration and direct slurry hydride generation. Analytica Chimica Acta 1283, , 341901. Available from: https://doi.org/10.1016/j.aca.2023.341901 (accessed 10.02.2024).

Yu, F.-F., Qi, X., Shang, Y.-N., Ping, Z.-G., Guo, X., 2019. Prevention and control strategies for children with Kashin-Beck disease in China: a systematic review and meta-analysis. Medicine 98 (36), e16823. Available from: https://doi.org/10.1097/MD.0000000000016823.

Zam, W., Alshahneh, M., Hasan, A., 2019. Methods of spectroscopy for selenium determination: a review. Research Journal of Pharmacy and Technology 12 (12), 6149−6152. Available from: https://doi.org/10.5958/0974-360X.2019.01068.0.

Zhang, L., Zeng, H., Cheng, W.H., 2018. Beneficial and paradoxical roles of selenium at nutritional levels of intake in healthspan and longevity. Free Radical Biology and Medicine 127, 3−13.

Zhang, J., Saad, R., Taylor, E.W., Rayman, M.P., 2020. Selenium and selenoproteins in viral infection with potential relevance to COVID-19. Redox Biology 37, 101715. Available from: https://doi.org/10.1016/j.redox.2020.101715.

Zhao, J., Hu, Y., Gao, Y., Li, Y., Li, B., Dong, Y., et al., 2013. Mercury modulates selenium activity via altering its accumulation and speciation in garlic (Allium sativum). Metallomics 5 (7), 896−903. Available from: https://doi.org/10.1039/c3mt20273a.

Zyambo, K., Hodges, P., Chandwe, K., Chisenga, C.C., Mayimbo, S., Amadi, B., Kelly, P., Kayamba, V., 2022. Selenium status in adults and children in Lusaka, Zambia. Heliyon 8 (6), , e09782. Available from: https://doi.org/10.1016/j.heliyon.2022.e09782 (accessed 11.02.2024).

Further reading

Anderson, M.S., Lakin, H.W., Beeson, K.C., Smith, F.F., Thacker, E., 1961. Selenium in Agriculture. USDA Agriculture Handbook 200. U.S. Government Printing Office, Washington, DC.

Archivist, 1999. Iodine, selenium, and joints: Kashin-Beck disease. Archives of Disease in Childhood 80, 261.

Arnér, E.S.J., Holmgren, A., 2000. Physiological functions of thioredoxin and thioredoxin reductase. European Journal of Biochemistry 267, 6102−6109. Available from: https://doi.org/10.1046/j.1432-1327.2000.01701.x.

Bruggeman, C., Maes, A., Vancluysen, J., 2007. The interaction of dissolved boom clay and gorleben humic substances with selenium oxyanions (selenite and selenate). Applied Geochemistry 22, 1371−1379.

Cheng, W.H., Lei, X.G., 2017. Chapter 37: Selenium. In: Collins, J.F. (Ed.), Molecular, Genetic, and Nutritional Aspects of Major and Trace Minerals. Part X. Academic Press, Cambridge, Massachusetts, pp. 449−461. . Available from: https://doi.org/10.1016/B978-0-12-802168-2.00037-3 (accessed 10.02.2024).

Clark, L.C., Combs, G.F., Turnbull, B.W., Slate, E.H., Chalker, D.K., Chow, J., et al., 1996. Effects of selenium supplementation for cancer prevention in patients with carcinoma of the skin: a randomized controlled trial. JAMA 276 (24), 1957−1963. Available from: https://doi.org/10.1001/jama.1996.03540240035027. Erratum in: JAMA, 1997 May 21; 277 (19), 1520.

Cutter, G.A., Bruland, K.W., 1984. The marine biogeochemistry of selenium - a re-evaluation. Limnology and Oceanography 29 (6), 1179−1192. Available from: https://doi.org/10.4319/lo.1984.29.6.1179.

FAO/WHO (Food and Agricultural Organisation/World Health Organization), 1998. Preparation and use of food-based dietary guidelines. Report of a joint FAO/WHO consultation: Technical Report Series No. 880. WHO, Geneva.

Hall, J., 2014. Chronic selenium toxicosis. MSD Veterinary Manual (Alkali disease). <https://www.msdvetmanual.com/toxicology/selenium-toxicosis/chronic-selenium-toxicosis#:~:text> (Accessed 24 February 2021).

Hasanuzzaman, M., Bhuyan, M.H.M.B., Raza, A., Hawrylak-Nowak, B., Matraszek-Gawron, R., Nahar, K., et al., 2020. Selenium toxicity in plants and environment: biogeochemistry and remediation possibilities. Plants 9 (12), 1711. Available from: https://doi.org/10.3390/plants9121711.

Health Canada, 2014. Guidelines for Canadian Drinking Water Quality: Guideline Technical Document - Selenium. Water and Air Quality Bureau, Healthy Environments and Consumer Safety Branch, Health Canada, Ottawa, Ontario (Catalogue No H144-13/4-2013E-PDF).

Kabata-Pendias, A., Pendias, H., 1992. Trace Elements in Soils and Plants, second ed. CRC Press, Inc, Boca Raton, Florida, p. 365.

Kabata-Pendias, A., Pendias, H., 1993. Biogeochemistry of Trace Elements. PWN, Warszawa.

Kang, D., Lee, J., Wu, C., Guo, X., Lee, B.J., Chun, J.-S., et al., 2020. The role of selenium metabolism and selenoproteins in cartilage homeostasis and arthropathies. Experimental and Molecular Medicine 52, 1198−1208. Available from: https://doi.org/10.1038/s12276-020-0408-y.

Kraus, V.B., 2014. Chapter 185: Rare osteoarthritis: ochronosis and Kashin-Beck disease. In: Hochberg, M.C., Silman, A.J., Smolen, J.S., Weinblatt, M.E., Weisman, M.H. (Eds.), Rheumatology, sixth ed. Mosby Elsevier, Philadelphia, pp. 1536−1540.

Lemly, A.D., 2002. Selenium pollution around the world. Selenium Assessment in Aquatic Ecosystems. Springer Series on Environmental Management, Springer, New York, NY. Available from: https://doi.org/10.1007/978-1-4613-0073-1_1, Accessed 11 February 2021.

Levander, O.A., 1989. Progress in establishing human nutritional requirements and dietary recommendations for selenium. In: Wendel, A. (Ed.), Selenium in Biology and Medicine. Springer-Verlag, New York, NY, pp. 205−209.

Levander, O.A., Burk, R.F., 2006. Update of human dietary standards for selenium. In: Hatfield, D.L., Berry, M.J., Gladyshev, V.N. (Eds.), Selenium. Springer, Boston, MA. Available from: https://doi.org/10.1007/0-387-33827-6_35. (Accessed 24 February 2021).

MMWR, 1984. Morbidity and Mortality Weekly Report 33, 157−158.

NAS (The United States National Academy of Sciences), 2000. Dietary reference intakes for vitamin C, vitamin E, selenium, and carotenoids. A report of the Panel on Dietary Antioxidants and Related Compounds, Subcommittees on Upper Reference Levels of Nutrients and Interpretation and Uses of Dietary Reference Intakes, and the Standing Committee on the Scientific Evaluation of Dietary Reference Intakes. Washington, DC, National Academy of Sciences, Institute of Medicine, Food and Nutrition Board.

Niedzielski, P., Siepak, M., 2003. Analytical methods for determining arsenic, antimony and selenium in environmental samples. Polish Journal of Environmental Studies 12 (6), 653−667.

Reuter, D.J., 1975. Selenium in soils and plant: a review in relation to selenium deficiency in South Australia. Agricultural Record 2, 44−50.

Riso, R.D., Waeles, M., Garbarino, S., Le Corre, P., 2004. Measurement of total selenium and selenium (IV) in seawater by stripping chronopotentiometry. Analytical and Bioanalytical Chemistry 379 (7-8), 1113−1119. Available from: https://doi.org/10.1007/s00216-004-2677-z.

Schrauzer, G.N., 2000. Selenomethionine: a review of its nutritional significance, metabolism and toxicity. Journal of Nutrition 130, 1653−1656.

Taylor, E.W., 1997. Selenium and viral diseases: facts and hypotheses. The Journal of Orthomolecular Medicine 12 (4).

Taylor, E.W., Ruzicka, J.A., Premadasa, L., 2015. Theoretical and experimental evidence for RNA:RNA antisense tethering of thioredoxin reductase mRNAs by Ebola and HIV-1 for viral selenoprotein synthesis. <http://rgdoi.net/10.13140/RG.2.2.10237.51683> (Accessed 09 February 2020).

Taylor, E.W., Ruzicka, J.A., Premadasa, L., Zhao, L., 2016. Cellular selenoprotein mRNA tethering *via* antisense interactions with ebola and HIV-1 mRNAs may impact host selenium biochemistry. Current Topics in Medicinal Chemistry 16, 1530−1535. Available from: https://doi.org/10.2174/1568026615666150915121633.

Thomson, C.D., Burton, C.E., Robinson, M.F., 1978. On supplementing the selenium intake of New Zealanders. 1. Short experiments with large doses of selenite or selenomethionine. British Journal of Nutrition 39, 579−587.

US EPA (United States Environmental Protection Agency), (Updated 2021). Ground water and drinking water: national primary drinking water regulations. <https://www.epa.gov/ground-water-and-drinking-water/national-primary-drinking-water-regulations> (Accessed 11 February 2021).

US NRC (National Research Council (US) Subcommittee on Selenium), 1983. 3, Distribution. Selenium in Nutrition: Revised Edition. National Academies Press (US), Washington (DC). Available from: https://www.ncbi.nlm.nih.gov/books/NBK216733/. (Accessed 09 February 2021).

Vanderpas, J.B., Contempré, B., Duale, N.L., Goossens, W., Bebe, N., Thorpe, R., et al., 1990. Iodine and selenium deficiency associated with cretinism in northern Zaire. The American Journal of Clinical Nutrition 52 (6), 1087−1093. Available from: https://doi.org/10.1093/ajcn/52.6.1087.

Zoller, W.H., Reamer, D.C., 1976. Selenium in the atmosphere. In: Proceedings of the Symposium on Selenium-Tellurium in the Environment. Pittsburgh, PA, Industrial Health Foundation, pp. 54−66.

15

Medical geology of zinc

Almost half of the world's cereal crops are grown on zinc-deficient soils; as a result, zinc deficiency in humans is a widespread problem

—Alloway (2008).

> **Key chapter features**
>
> 1. Brief overviews of the most relevant and important characteristics of Zn, that are of relevance to the study of Medical Geology.
> 2. Attributes that make Zn the second most important essential micronutrient after Fe.
> 3. Latest advances in Zn research in Medical Geology of Africa (> 90% of the references post-date the year 2000), and how they apply to the African situation.
> 4. Suggested areas of future Zn research in Medical Geology of Africa.

Sources

Zinc is the 25th most abundant element in the Earth's crust (Simon-Hettich et al., 2001) and is present in all key compartments of the biophysical environment—flora, fauna, rocks, soils and water. In the Earth's crust, the element occurs at an average concentration of 120 ppm (Sicius, 2021).

Igneous rocks

(Average values from Rose et al., 1979)
Granitic rocks: 51 ppm; Ultramafic rocks: 58 ppm; Mafic rocks: 94 ppm.

Sedimentary rocks

Limestones: (Average values from Rose et al., 1979, except, where indicated otherwise)—21 ppm; Sandstones: 40 ppm; Shale (Average value from Quinby-Hunt and Wilde, 1988)—85 ppm; Black shale (Average value from Quinby-Hunt and Wilde, 1988)— < 300 ppm.

Minerals

The most common ore of Zn in the lithosphere is zinc blende, also known by the mineral name, sphalerite (ZnS). Other important zinc ores are wurtzite, smithsonite, and hemimorphite. The main Zn mining areas are Canada, Russia, Australia, USA, and Peru.

Medical Geology of Africa. DOI: https://doi.org/10.1016/B978-0-12-818748-7.00010-1

Several other low-temperature Zn minerals exist including sulfides, silicates, oxides, carbonates, phosphates, and arsenates. Mcphail et al. (2003a,b) list a number of supergene minerals that were found at the Skorpion deposit in Namibia, including sauconite (zinc smectite, which is dominant), hemimorphite ($Zn_4Si_2O_7(OH)_2 \cdot H_2O$), smithsonite ($ZnCO_3$), hydrozincite ($Zn(CO_3)_2(OH)_2$), tarbuttite ($Zn_2(PO_4)(OH)$), and scholzite ($CaZn_2(PO_4)_2 \cdot 2H_2O$). Yet, other secondary minerals containing Zn are described by Borg et al. (2003).

Soils

Zinc is present naturally in all soils. Normal background concentrations are in the order of 10−100 ppm (Mertens and Smolders, 2013). Earlier estimates such as by Lindsay (1972) give a broader range of 10−300 ppm. For more recent data on Zn content in different types of soils, the reader is referred to the work of Noulas et al. (2018). Zinc has no typically occurring phases in soils. It forms very diverse compounds, a feature which, according to Vodyanitskii (2010), makes the identification of Zn by chemical fraction quite a difficult procedure (See Section on "Analytical Determination," this Chapter).

Zinc deficiency in agricultural soils is considered to be the most geographically widespread micronutrient deficiency (De Groote et al., 2021). Sandy, calcareous, saline, and wetland soils, compacted and rich in organic matter with high nitrogen and phosphate levels are particularly vulnerable to Zn deficiency (Alloway, 2008). Zinc deficiency can limit crop production (yield losses can exceed 40%), whereas its excess in soil may be either of geological or anthropogenic origin (Alloway, 2008; Singh et al., 2005; Sadeghzadeh, 2013).

Plants

Zinc is an essential plant nutrient. Most plants contain Zn to the tune of 30−100 ppm dry matter, with contents above 300 ppm considered to be generally toxic. Growing in a natural habitat, Zn hyperaccumulator plants can accumulate as much as 10,000 ppm (dry weight) of Zn in their above-ground parts (Balafrej et al., 2020).

The lower critical content of whole plant is 15 ppm in rice (Srivastava and Gupta, 1996), 15−22 ppm in maize, and 22 ppm in groundnut. Beans, citrus, grapes, maize, rice, and sorghum are considered highly susceptible to Zn deficiency (Noulas et al., 2018). Zinc, in common with the other micronutrients, can affect growth when its content is either lower or above a critical level due to deficiency and/or toxicity problems, respectively.

The important roles of Zn in plant growth, first demonstrated by Mazé (1915), are legion. A constant and continuous supply of the element is therefore of extreme necessity for optimum growth and maximum yields (See, e.g., Ahmed et al., 2023; Jalal et al., 2023). Common symptoms of Zn deficiency include chlorosis, development of necrotic spots, bronzing of leaves, rosetting of leaves, stunting of plants, and malformed leaves (See: Alloway, 2008).

Broadley et al. (2007) has reviewed the main fluxes of Zn in the soil-root-shoot continuum as a basis for formulating credible genetic strategies for addressing the widespread problem of Zn-limited crop growth.

Food crops and other major dietary sources

In most crops, the typical leaf Zn concentration ($[Zn]_{leaf}$) required for adequate growth, according to Marschner (1995), is approximately 15–20 ppm dry weight. Over and above this threshold, however, Zn can express toxic effects on plant growth, photosynthetic activity, and oxidative stress levels (See, e.g., Meng et al., 2023).

Our body does not produce Zn naturally, so, this micronutrient has to be obtained from the food we eat or through Zn supplements. Protein rich, such as meat and marine organisms, contain high concentrations of zinc (10–50 ppm wet weight), whereas grains, vegetables, and fruit are low in Zn (usually 10–50 ppm wet weight) (WHO, 2003).

Rich sources of Zn include sea foods (oysters, crab, lobsters, mussels, and shrimp), red meat, poultry, vegetables such as mushrooms and kale (which are low-calorie sources), legumes such as chickpeas, nuts (such as cashews and almonds) and seeds, whole grains, fortified cereals, milk and dairy foods and dark chocolate. Beans citrus, grapes, maize, rice, and sorghum are considered highly susceptible to Zn deficiency (Noulas et al., 2018).

Natural waters

According to Elinder, back in 1986, the concentration of Zn in natural surface waters is usually below 10 ppb, and in groundwaters, 10–40 ppb.

In relatively undisturbed river systems dissolved Zn concentrations are typically of the order of 10^{-9}–10^{-8} mol/kg with some dependence of concentration on pH (Shiller and Boyle, 1985).

The content of Zn in estuaries and coastal waters is often much higher than in the ocean, with observed concentrations as high as 4 ppb and occasionally as high as 25 ppb (See: Neff, 2002). A large proportion of the Zn in sediments, both contaminated and uncontaminated, is thought to be residual, rendering it nonbioavailable.

Seawater

Values given for average concentrations of Zn in seawater vary widely, an observation that Neff (2002) considers could be due both to variability in concentrations actually present and to sample contamination. The average Zn content of seawater, according to various unpublished sources, is given as 0.6–5 ppb. Zinc entering the oceans of the world is largely derived from aerial deposition. Neff (2002) gives an estimated figure of between 11,000 and 60,000 metric tons for the amount of dissolved and particulate Zn deposited from the air into the ocean each year.

Some fish living in Zn-contaminated waterways can accumulate Zn in their bodies. When this occurs, bio-magnification up the food chain takes place.

Drinking water

Zinc intake through drinking water is very low. Normal concentrations are generally well below those of natural surface waters and groundwaters [upper threshold, 40 ppb (Elinder, 1986)]. However, high concentrations of Zn occur in tap water as a result of corrosion or leaching of pipings and fittings. Under such circumstances, concentrations of Zn in drinking water can be higher than levels in surface waters, and tap water can provide up to 10% of the daily intake (Gillies and Paulin, 1982).

Available data suggest that Zn concentrations in drinking water are far less than levels required to meet a daily intake level of 11 mg/day (assuming an adult water consumption of 2 L/day) (See, e.g., IOM [Institute of Medicine Food and Nutrition Board (NRC)], 2002).

Air

Zinc concentrations in air are relatively low and fairly constant except near sources such as smelters (ATSDR (Agency for Toxic Substances and Disease Registry), 2014). Air near industrial areas may have higher levels of Zn.

Average data for Zn contents in the atmosphere are available for many regions of North America, but are limited for many other areas of the world.

Average levels of Zn in the air throughout the United States are given as <1 microgram of Zn per cubic meter ($\mu g/m^3$) of air, and ranges from 0.1 to 1.7 $\mu g/m^3$ in areas near cities. The average Zn concentration for a 1-year period is given as 5 $\mu g/m^3$ in one area near an industrial source (ATSDR (Agency for Toxic Substances and Disease Registry), 2014).

Associations (synergism/antagonism)

Zinc is a *chalcophile* element (ore/copper/bronze-loving), commonly occurring with other chalcophile elements, such as Cu, Pb, Ag, and Au.

Zinc exerts synergistic effects with certain vitamins, such as vitamin A (e.g., Rahman et al., 2002) and C (e.g., Gromova et al., 2019), as well as a number of metals such as Cu (e.g., Bae et al., 2022), Ti (e.g., Tang et al., 2013) and Pb (Ong et al., 2013) either in aspects that enhance growth or in producing toxicity.

Antagonistic effects of Zn have been observed with certain metals and metalloids, including Pb (Ong et al., 2013), As (Guzman-Rangel et al., 2018), and Cd (Apiamu and Asagba, 2021).

Chemical form or speciation in soil and aqueous phases

The metal Zn occurs very rarely in nature, for it requires extreme reducing conditions to exist. The $+2$ valence state is the predominant species in natural waters at pH ≤ 8.5 (Stumm and Morgan, 1996). Zinc exhibits amphoteric properties (reacting with both acids and bases). At higher pH (pH ≥ 8) such as obtains in the open sea, the hydrolyzed species, $ZnOH^+$, and $Zn(OH)_2$ are the dominant species (Young and Harvey, 1991). However, according to ATSDR (Agency for Toxic Substances and Disease Registry) (2014), at the pH of most natural waters, the formation of anionic Zn species is unlikely. Zn in drinking water occurs in the form of salts or organic complexes.

Sorption is the dominant reaction mechanism governing the fate and transport of Zn in soil and sediments at or just below circumneutral pH (Barrow, 1993; Smolders et al., 2004; Grafe and Sparks, 2005). Zinc reacts with all soil constituents—clay minerals, organic matter, metal oxides and so on—, with both the ionic species present in solution and the charge on the solid reactant being the main factors determining the amount of Zn that is sorbed. Zinc deficiency for agricultural crops is found in about one third of soils around the world due to low total Zn concentrations and/or high pH.

Environmental circulation and impact of climate change

Zinc is commonly found in the Earth's crust as sulfide minerals, for example, sphalerite and wurtzite and as carbonates, for example, smithsonite [See "Minerals," this article]. These minerals are released to the terrestrial environment through both natural processes (e.g., weathering) and anthropogenic activities (Rauch and Pacyna, 2009); with releases from anthropogenic sources being greater than those from natural sources. The primary anthropogenic sources of Zn in the environment (air, water, soil) are related to mining and metallurgical operations involving Zn and use of commercial products containing Zn. Zinc is taken up by plants and animals, including humans, and then enters aquatic systems where it either settles into sediments or eventually enter the oceans (Fig. 15−1).

On a global scale, releases to soil are considered to constitute the largest source of Zn in the environment. An excellent overview of the anthropogenic cycle of Zn is presented by Meylan and Reck (2017), whereas the "puzzling" [strong correlation observed between Zn and silicate (SiO_4^{4-}), but not phosphate (PO_4^{3-})] internal cycling of Zn in the ocean is interrogated by Roshan et al. (2018).

Zinc and climate change

As more greenhouse gas CO_2 enters the atmosphere, leading to climate change, crops might carry lesser amounts of nutrients such as Zn. This is reiterated in the recent finding by Beach et al. (2019), that climate change and higher CO_2 could significantly reduce the availability of Zn and other critical nutrients, bringing another setback to global development and the bid to end undernutrition.

Mobility, mode of entry into food chains and bioavailability

The mobility of Zn is moderate, but is influenced by various other constituents in soil and in aqueous media, such as organic matter, ferrous Fe compounds (e.g., limonite) and phosphate (See, e.g., Pérez-Novo et al., 2011 for the effect of phosphate on Zn mobility). Availability of Zn to plant is hampered in soils in which Zn mobility is limited by say, the presence of organic matter (Kaur et al., 2024). Thus Zn deficiency is observed even though high amounts may be available in soil.

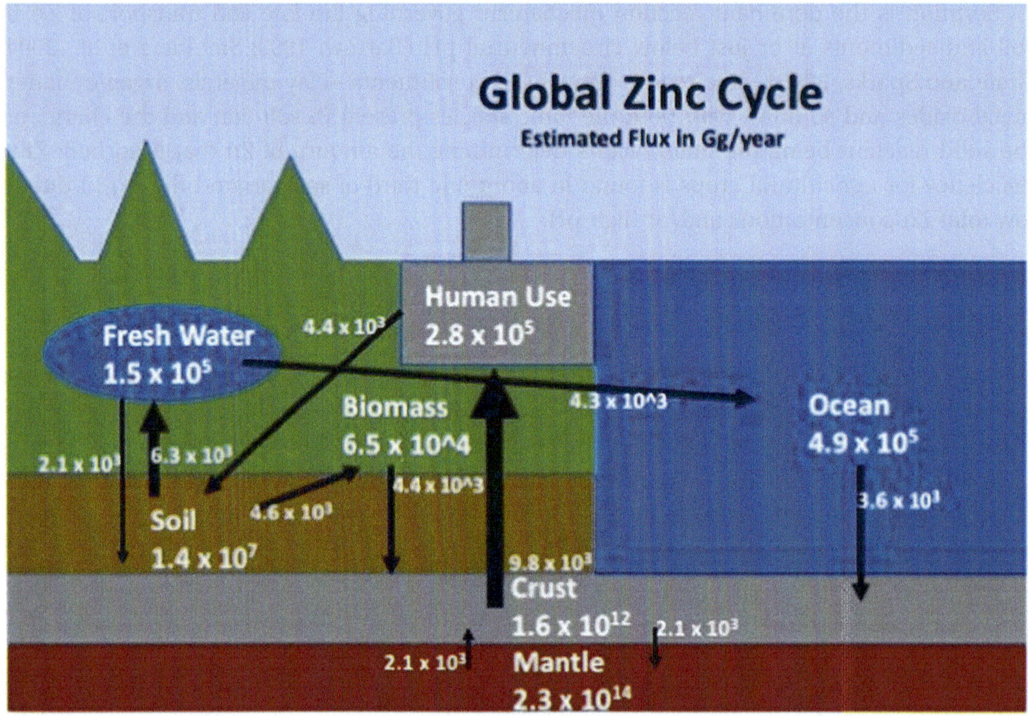

FIGURE 15–1 Zinc fluxes between the lithosphere and biosphere, through basins in soil, biomass, water systems, and industry. Estimated fluxes are shown as labeled arrows in Gg/year (Rauch and Pacyna, 2009). *From Rauch, J.N. and Pacyna, J.M., 2009. Earth's global Ag, Al, Cr, Cu, Fe, Ni, Pb, and Zn cycles. Global Biogeochemical Cycles 23(2), GB2001. https://doi.org/10.1029/2008GB003376.*

Absorption and distribution

Only 2−3 g of Zn is held in the human body for body weight above 60 kg (Hussain et al., 2022). About 60% of this amount is stored in skeletal muscle, 30% in bone, and some 5% in the skin and liver (To et al., 2020; Fig. 15−2). The rest of the Zn finds its way to other tissues, such as brain, kidneys, pancreas, and heart.

The systemic homeostasis of Zn is regulated largely through its absorption by the *duodenum* and *jejunum* (See "Glossary of terms," this Chapter for definitions) in the small intestine (Maares and Haase, 2020). The solubility and absorptive capacity of Zn is, however, strongly influenced by the environment within the gastrointestinal tract.

Zinc transporters, which are membrane transport proteins of the solute carrier family regulate the membrane transport of Zn by controlling Zn influx and efflux between extracellular and intracellular compartments; and so, modulate the Zn concentration and distribution (See, e.g., Chen et al., 2024; Fan et al., 2024). Zinc is excreted mainly through the gastrointestinal tract, with minor egestions through the urinary system.

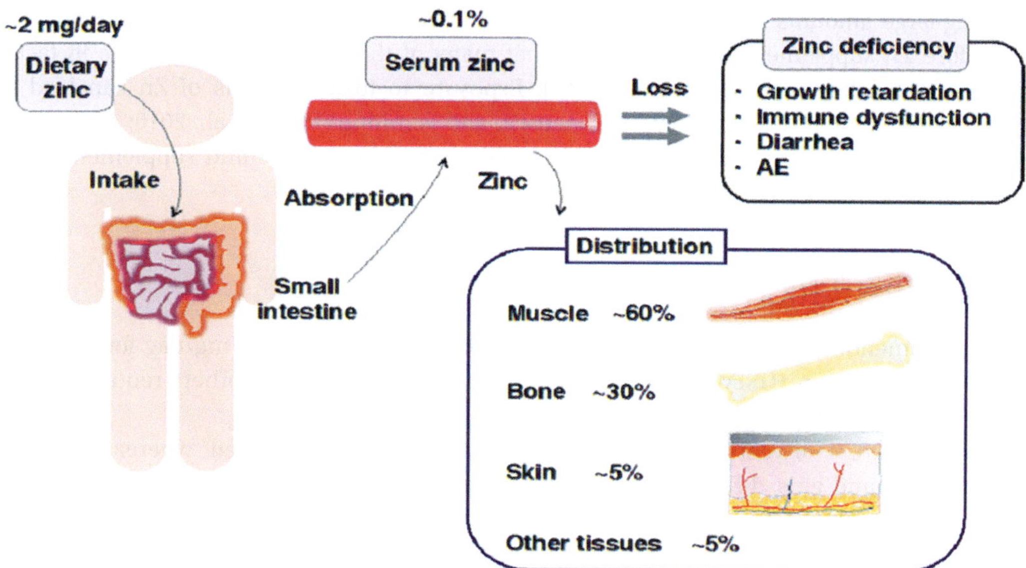

FIGURE 15–2 Illustration of Zn storage and distribution in the human body (Takagishi et al., 2017). *From Takagishi, T., Hara, T., Fukada, T., 2017. Recent advances in the role of SLC39A/ZIP zinc transporters in vivo. International Journal of Molecular Sciences 18(12), 2708. https://doi.org/10.3390/ijms18122708.*

Metabolic function (essentiality/clinical toxicity)

Zinc plays an essential role in multiple cellular metabolic processes, from enzyme catalysis, protein structural stabilization to immunity construction to gene expression (Hara et al., 2017). As Zn^{2+} is so abundant and mostly bound to proteins, several proteins in living organisms contain Zn. Zinc metallochaperone, named *ZNG1* (for Zn regulated GTPase metalloprotein activator 1) is the first identified protein that puts Zn into other proteins (Weiss et al., 2022), a strategy that may one day prove to be one of the most important regulatory strategies by which humans cope with severe Zn deficiency.

Essentiality

The World Health Organization ranks Zn deficiency as one of the main causes of disease and death (WHO, 2002). As a cofactor of major proteins, and regulator of several immunomodulatory functions, the essentiality of Zn for life can never be overstated. Zinc deficiency in humans is expressed in a range of disorders, including stunted growth, cell-mediated immune dysfunction, and cognitive impairment (Prasad, 2013).

Common symptoms of Zn deficiency in plants are given in "Plants" section, this chapter.

Toxicity

Converse to its essentiality, although relatively large concentrations of Zn can be handled by humans, consumption of too much Zn can disrupt the absorption of Cu and Fe, as well as creating large amounts of toxic free radicals. The interference with Cu uptake of long-term, high-dose Zn supplementation, can imply that many of the toxic effects of Zn are in fact a result of Cu deficiency (Plum et al., 2010). Exposure to toxic amounts of Zn can lead to symptoms that include abdominal pain, nausea, and vomiting (Plum et al., 2010; Agnew and Slesinger, 2022). As Zn is more readily absorbed from animal foods and supplements, it is easier to over-consume from these sources.

Supplementation

The recommended daily allowance of Zn for adults (19 + years) is 11 mg/day for men and 8 mg/day for women (USIM, 2001). Pregnant women and lactating mothers require slightly more at 11 and 12 mg/day, respectively.

For children, 4−8 years, a lower value of 4 mg/day is recommended, whereas for boys, 14−18 years, and girls, 14−18 years, 11 and 9 mg/day, respectively, are the recommended values (USIM, 2001). The work of Dembedza et al. (2023) looks at Zn nutriture of children aged 0−59 months in the Sub-Saharan Africa region.

Roohani et al. (2013) list four main intervention measures for obviating Zn deficiency, including in their list: supplementation, dietary modification/diversification, fortification, and bio-fortification.

Several different forms of Zn supplements are available, including zinc sulfate ($ZnSO_4$), zinc gluconate ($C_{12}H_{22}O_{14}Zn$), and zinc acetate ($ZnC_4H_6O_4$) (See, e.g., Stiles et al., 2024).

Wang et al. (2020) states that improving Zn content in wheat and its processed foods is an effective way to solve human Zn deficiency, although as Adams et al. (2023) noted for Burkina Faso, this will not fully close the dietary Zn gap.

Identifying Zn deficiency as a probably risk factor that can be added to the factors predisposing individuals to infection and detrimental progression of COVID-19, Wessels et al. (2020) consider that Zn supplementation may be beneficial for most of the at-risk population of COVID-19, particularly those with suboptimal Zn status.

Zinc supplements must be administered with great deal of care in view of the risk of excessive Zn intake and Zn toxicity.

Reduction of body burden

As already indicated in the Section on "Toxicity," this Chapter, reports of Zn intoxication are very rare; consequently, not much appears on reduction of Zn in the global literature.

One recent example of body burden reduction of Zn encountered in our literature search comes from Hungary (Bilics et al., 2020). A 65-year-old man ingested 125 g of Arvalin [containing 5 g ZnP (zinc protoporphyrin)] and presented to the Emergency Department, with respiratory deficit and limited consciousness. Hepatoprotective agents, alpha lipoic acid (Thiogamma Turbo-Set) with N-acetylcysteine, were administered for 6 consecutive days. After other subsidiary treatments, the patient was discharged 1 month later, fully recovered.

Analytical determination

Analytical techniques for determining micronutrient analysis in biological samples cover a wide range and include methods such as colorimetry, atomic absorption spectrophotometry (AAS), inductively coupled plasma optical emission spectroscopy, and advanced nondestructive estimation methods like X-ray fluorescence spectroscopy. Elango et al. (2021) provided an overview of these techniques, and their working principles.

Progress in unraveling the molecular functions of Zn, in particular, took place slowly as Zn is refractory to investigations with most spectroscopic techniques; and analysis in biological tissues proved difficult prior to the development of sufficiently sensitive analytical techniques such as flameless AAS (e.g., Aydinoglu, 2022). AAS has since remained one of the most widely used methods for the determination of Zn in biological samples, with an ever-improving detection limit (d) (e.g., attainment of 0.0032 ppm of Zn in determining total Zn in insulin by flame AAS) (Ata et al., 2015), and attainment of 0.1 ppb of Zn in determining Zn in water samples (Rezaee and Tajer-Mohammad-Ghazvini, 2022).

Modern methods for determining the speciation of zinc in water include physicochemical techniques (see, e.g., Odobasic et al., 2013; Kim et al., 2015) and theoretical techniques (see, e.g., Pearson et al., 2018).

Current status of geomedical research on zinc in Africa

Zinc deficiency has been reported in 17% of the world's population with the highest estimates given for Africa (24%) and Asia (19%), based on dietary supply estimates (Wessels and Brown, 2012; Kumssa et al., 2015; Fig. 15−3). The essentiality of Zn in nutrition (re: growth of the fungus *Aspergillus niger*) was reported by Raulin way back in 1869 (Raulin, 1869) and its essentiality in humans established by Prasad in 2013 (Prasad, 2013, 2014). However, it was not until this Century (2000s) that research on Zn deficiency in Africa [known to be a major public health problem on the Continent (Bailey et al., 2015; Endalifer et al., 2020; Gupta et al., 2020; Adams et al., 2023; Dembedza et al., 2023; Van Eynde et al., 2023)], gained traction. Most of the recent research (2000−present) on Zn essentiality in the various age categories and in pregnant women have been done in Ethiopia and South Africa.

In 2012 Lokuruka assessed dietary Zn intake among the African elderly, discussed the potential impact of current levels of Zn intake on their health, and recommended strategies for improving their dietary Zn intake. An earlier study by Oldewage-Theron et al. (2008) established the proneness of South African elderly to Zn deficiency because of nutritional and physiological susceptibilities associated with ageing.

Other examples in the most recent literature on Zn nutrition in Africa include the works of Harika et al. (2017), who examined the status and dietary intake of various micronutrients including Zn in women of reproductive age and pregnant women in Ethiopia, Kenya, Nigeria, and South Africa. Later works such as that of Berhe et al. (2019) looked at the prevalence and associated factors of Zn deficiency among pregnant women and children in Ethiopia, while the study by Endalifer et al. (2020) presented a systematic review protocol on

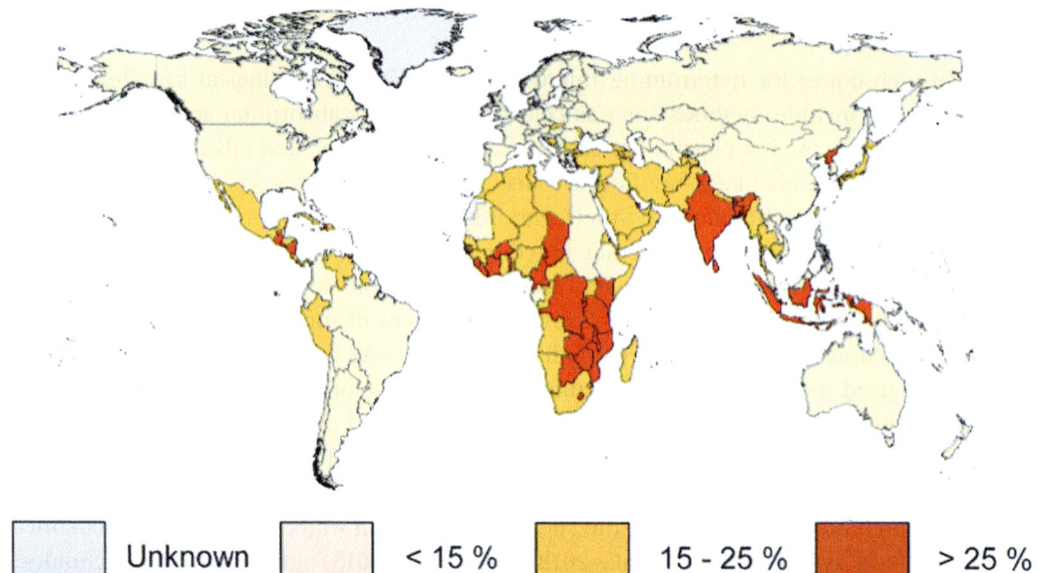

FIGURE 15–3 Estimated country-specific global prevalence of Zn deficiency, based on Zn availability in national food supplies and the prevalence of stunting (Wessels and Brown, 2012). See original publication for a full description of the database used for production of this image. *From Wessels, K.R. and Brown, K.H., 2012. Estimating the global prevalence of zinc deficiency: results based on zinc availability in national food supplies and the prevalence of stunting. PLoS One 7(11), e50568. https://doi.org/10.1371/journal.pone.0050568.*

the epidemiology of Zn deficiency and associated factors during pregnancy in the African Continent as a whole. Belay et al. (2021) explored demographic variation and spatial dependencies in the Zn status of the entire Ethiopian population. The work of Maciel et al. (2021) posits that higher energy and Zn intakes from complementary feeding are associated with decreased risk of undernutrition in African children, as well as children in other continents.

Suggested areas for further research

The critical role of Zn, though now firmly established, we still do not yet fully understand the mode of transport of the metal into the hundreds of proteins that use it or how our cells respond to Zn deficiency (Giedroc, 2022); (See, also: Weiss et al., 2022).

Based on the information presented in this chapter, a number of other research gaps in Zn nutrition still need to be filled (Hussain et al., 2022; Mohamad et al., 2023). This applies especially to Africa, one of the regions where Zn deficiency is most widespread (Fig. 15−3), but also to other Zn deficient regions of the world:

- There has been a massive void in our understanding of metal biology, whereby, up till 2022, we have, for example, been unable to identify and characterize *metallochaperones* (VUMC, 2022). The description and characterization of the first Zn metallochaperone (a protein that transports Zn into other "client" proteins) (Weiss et al., 2022) has thrown new light on the problem of Zn deficiency and opened up an entirely new area of metal biology for urgent research.

- Definition of Zn deficiency, recommended daily dietary intake level, doses for various age categories, and different health conditions must be precisely made and with international concordance as definition of these parameters by different regulatory bodies is not uniform.
- The effectiveness of Zn fortification programs in high-risk subpopulations, such as young children, adolescents, pregnant women and lactating mothers, and the elderly still needs to be established.
- In order to accurately chart the Zn status of African populations, it is necessary to obtain definitive data on the bioavailabilty of Zn from typical African diets.
- There is a need to explore the effectiveness of root Zn uptake mechanisms in contrasting soil environments in order to aid calculations on bioavailabilty of Zn derived from consumption of food crops.
- The prostate normally contains abundant intracellular Zn and Zn loss is a hallmark of the development of prostate cancer development. The underlying mechanism of this loss is not yet clearly understood (See: To et al., 2020).
- A great deal of research is still needed regarding the impact of maternal Zn supplementation during pregnancy on both maternal and infant health outcomes. The effect of Zn supplementation of undernourished women on their breast milk Zn concentrations needs further study.
- A massive data gap exists in Africa for biomarkers of micronutrient status (particularly for women), in addition to data gaps on fetal growth, including birth weight and gestational age assessments. The need for novel biomarkers of Zn status that are reliable, are of low-cost, and amenable to field implementation, can never be overstated (See: Solomons, 2013; Lowe et al., 2024).
- Genetic approaches to the attainment of effective supplementation of Zn through vegetables and fruits, and development of the effective Zn fertilizers to enhance Zn uptake efficiency in crop are areas in need of further research.
- More studies are needed on Zn fortification of staple foods of the sub-Saharan African region to avoid changing the dietary patterns of consumers in this region (See, e.g., Ohanenye et al., 2021).
- The effects of high Ca and Fe supplementation in humans on Zn absorption need further study so as to clarify the mechanisms, conditions and the reasons for depressed Zn and Cu absorption (See: Lokuruka, 2012).
- The precise role of Zn deficiency mechanisms in the pathogenesis of cardiovascular diseases is still not precisely known (See, e.g., Knez and Glibetic, 2021).
- Above all, African governments need to research ways of reorienting their food policies and systems to address gaps in nutrient availability and prioritize access and affordability. Awareness programs in areas suffering the most from Zn deficiency need to be developed.

Glossary of terms

- *Hypogonadism* occurs when the *gonads* (sex glands)—testes in men and ovaries in women—produce little, if any, sex hormones (*testosterone* in men and *estrogen* in women).
- *Jejunum:* The middle part of the small intestine, located between the *duodenum* (first part of the small intestine) and the *ileum* (last part of the small intestine).
- *Metallochaperones* are a distinct new family of soluble metal receptor proteins that facilitate intracellular trafficking of metal ions to metalloenzymes and metalloproteins in cells through protein—protein interactions.
- *Zinc transporter proteins*, or simply *Zn transporters*, are membrane transport proteins of the solute carrier family which control the membrane transport of Zn and regulate its intracellular and cytoplasmic concentrations.

References

Adams, K.P., Jarvis, M., Vosti, S.A., Manger, M.S., Tarini, A., Somé, J.W., Somda, H., McDonald, C.M., 2023. Estimating the cost and cost-effectiveness of adding zinc to, and improving the performance of, Burkina Faso's mandatory wheat flour fortification programme. Maternal and Child Nutrition 19 (3)e13515. Available from: https://doi.org/10.1111/mcn.13515.

Agnew, U.M., Slesinger, T.L., 2022. Zinc toxicity. StatPearls [Internet]. StatPearls Publishing, Treasure Island (FL). Available from: https://www.ncbi.nlm.nih.gov/books/NBK554548/.

Ahmed, R., Zia-Ur-Rehman, M., Sabir, M., Usman, M., Rizwan, M., Ahmad, Z., Alharby, H.F., Al-Zahrani, H. S., Alsamadany, H., Aldhebiani, A.Y., Alzahrani, Y.M., Bamagoos, A.A., 2023. Differential response of nano zinc sulphate with other conventional sources of Zn in mitigating salinity stress in rice grown on saline-sodic soil. Chemosphere 327, 138479. Available from: https://doi.org/10.1016/j.chemosphere.2023.138479.

Alloway, B.J., 2008. Zinc in Soils and Crop Nutrition, 2nd Edition IZA and IFA, Brussels Belgium, Paris. Available from: https://www.topsoils.co.nz/wp-content/uploads/2014/09/Zinc-in-Soils-and-Crop-Nutrition-Brian-J.-Alloway.pdf.

Apiamu, A., Asagba, S.O., 2021. Zinc-cadmium interactions instigated antagonistic alterations in lipid peroxidation, ascorbate peroxidase activity and chlorophyll synthesis in Phaseolus vulgaris leaves. Scientific African 11 (12), e00688. Available from: https://doi.org/10.1016/j.sciaf.2020.e00688.

Ata, S., Wattoo, F.H., Ahmed, M., Wattoo, M.H.S., Tirmizi, S.A., Wadood, A., 2015. A method optimization study for atomic absorption spectrophotometric determination of total zinc in insulin using direct aspiration technique. Alexandria Journal of Medicine 51 (1), 19—23. Available from: https://doi.org/10.1016/j.ajme.2014.03.004.

ATSDR (Agency for Toxic Substances and Disease Registry), Reviewed 2014. Toxicological Profile for Zinc, Chapter 6. Potential for Human Exposure. Report CAS#: 7440-66-6. https://wwwn.cdc.gov/TSP/ToxProfiles/ToxProfiles.aspx?id = 302&tid = 54 (Accessed 24 June 2022).

Aydinoglu, S., 2022. Iron and zinc determination in dietary supplements by flame atomic absorption spectrophotometry. Brazilian Journal of Pharmaceutical Sciences (Online) 58, e21094. Available from: https://doi.org/10.1590/s2175-97902022e21094.

Bae, B., Park, H., Kang, S., 2022. Quantitative estimation of synergistic toxicity of Cu and Zn on Growth of *Arabidopsis thaliana* by Isobolographic method. Toxics (MDPI) 10 (4), 195. Available from: https://doi.org/10.3390/toxics10040195.

Bailey, R.L., West Jr, K.P., Black, R.E., 2015. The epidemiology of global micronutrient deficiencies. Annals of Nutrition and Metabolism 66 (Suppl. 2), 22−33. Available from: https://doi.org/10.1159/000371618.

Balafrej, H., Bogusz, D., Triqui, Z.A., Guedira, A., Bendaou, N., Smouni, A., Fahr, M., 2020. Zinc hyperaccumulation in plants: a review. Plants (Basel, Switzerland) 9 (5), 562. Available from: https://doi.org/10.3390/plants9050562.

Barrow, N.J., 1993. Mechanisms of reaction of zinc with soil and soil components. In: Robson, A.D. (Ed.), Zinc in Soils and Plants. Developments in Plant and Soil Sciences, Vol 55. Springer, Dordrecht. Available from: https://doi.org/10.1007/978-94-011-0878-2_2.

Beach, R.H., Sulser, T.B., Crimmins, A., Cenacchi, N., Cole, J., et al., 2019. Combining the effects of increased atmospheric carbon dioxide on protein, iron, and zinc availability and projected climate change on global diets: a modelling study [published correction appears in Lancet Planet Health, 2020, 4(9): e385]Lancet Planet Health 3 (7), e307−e317. Available from: https://doi.org/10.1016/S2542-5196(19)30094-4.

Belay, A., Gashu, D., Joy, E.J.M., Lark, R.M., Chagumaira, C., Likoswe, B.H., Zerfu, D., Ander, E.L., Young, S. D., Bailey, E.H., Broadley, M.R., 2021. Zinc deficiency is highly prevalent and spatially dependent over short distances in Ethiopia. Scientific Reports 11, 6510. Available from: https://doi.org/10.1038/s41598-021-85977-x.

Berhe, K., Gebrearegay, F., Gebremariam, H., 2019. Prevalence and associated factors of zinc deficiency among pregnant women and children in Ethiopia: a systematic review and meta-analysis. BMC Public Health 19 (1), 1663. Available from: https://doi.org/10.1186/s12889-019-7979-3.

Bilics, G., Héger, J., Pozsgai, É., Bajzik, G., Nagy, C., Somoskövi, C., Varga, C., 2020. Successful management of zinc phosphide poisoning - a Hungarian case. International Journal of Emergency Medicine 13 (1), 48. Available from: https://doi.org/10.1186/s12245-020-00307-8.

Borg, G., Karner, K., Buxton, M., Armstrong, R., 2003. Geology of the Skorpion supergene zinc deposit, southern Namibia. Economic Geology 98, 749−771. Available from: https://doi.org/10.2113/98.4.749.

Broadley, M.R., Philip, J., White, J.P., Zeiko, H.I., Lux, A., 2007. Zinc in plants. New Phytologist 173 (4), 677−702. Available from: https://doi.org/10.1111/j.1469-8137.2007.01996.x.

Chen, B., Yu, P., Chan, W.N., Xie, F., Zhang, Y., Liang, L., Leung, K.T., Lo, K.W., Yu, J., Tse, G.M.K., Kang, W., To, K.F., 2024. Cellular zinc metabolism and zinc signaling: from biological functions to diseases and therapeutic targets. Signal Transduction and Targeted Therapy 9 (1), 6. Available from: https://doi.org/10.1038/s41392-023-01679-y.

De Groote, H., Tessema, M., Gameda, S., Gunaratna, N.S., 2021. Soil zinc, serum zinc, and the potential for agronomic biofortification to reduce human zinc deficiency in0 Ethiopia. Scientific Reports 11 (1), 8770. Available from: https://doi.org/10.1038/s41598-021-88304-6.

Dembedza, M.P., Chopera, P., Matsungo, T., 2023. Risk of zinc deficiency among children aged 0−59 months in sub-Saharan Africa: a narrative review. South African Journal of Clinical Nutrition . Available from: https://doi.org/10.1080/16070658.2023.2275932.

Elango, D., Kanatti, A., Wang, W., Devi, A.R., Ramachandran, M., Jabeen, A., 2021. Analytical methods for iron and zinc quantification in plant samples. Communications in Soil and Plant Analysis 52 (10), 1069−1075. Available from: https://doi.org/10.1080/00103624.2021.1872608.

Elinder, C.G., 1986. Zinc. In: Friberg, L., Nordberg, G.F., Vouk, V.B. (Eds.), Handbook on the Toxicology of Metals, 2nd Ed Elsevier Science Publishers, Amsterdam, pp. 664−679.

Endalifer, M.L., Azeze, G.G., Diress, G., Bizuneh, A.D., Demelash, H.A., 2020. A systematic review protocol on the epidemiology of zinc deficiency and associated factors during pregnancy in Africa. Journal of Global Health Reports 4, e2020021. Available from: https://doi.org/10.29392/001c.12501.

Fan, Y.G., Wu, T.Y., Zhao, L.X., Jia, R.J., Ren, H., Hou, W.J., Wang, Z.Y., 2024. From zinc homeostasis to disease progression: unveiling the neurodegenerative puzzle. Pharmacological Research 199, 107039. Available from: https://doi.org/10.1016/j.phrs.2023.107039.

Giedroc, D., 2022. Study identifies first cellular "chaperone" for zinc, sheds light on worldwide public health problem of zinc deficiency. Research Impact, Indiana University. https://research.impact.iu.edu/more-iu-research/stories/zinc.html. (Accessed 28 June 2022).

Gillies, M.E., Paulin, H.V., 1982. Estimations of daily mineral intakes from drinking water. Human Nutrition: Applied Nutrition 36A (4), 287−292. Available from: https://api.semanticscholar.org/CorpusID:28904650.

Grafe, M., Sparks, D.L., 2005. Kinetics of zinc and arsenate co-sorption at the goethite-water interface. Geochimica et Cosmochimica Acta 69, 4573−4595. Available from: https://studylib.net/doc/10765016/kinetics-of-zinc-and-arsenate-co-sorption-at-the-goethite.

Gromova, O.A., Torshin, I., Pronin, A.V., Kilchevsky, M.A., 2019. Synergistic application of zinc and vitamin C to support memory and attention and to decrease the risk of developing nervous system diseases. Neuroscience and Behavioral Physiology 49 (2). Available from: https://doi.org/10.1007/s11055-019-00740-0.

Gupta, S., Brazier, A.K.M., Lowe, N.M., 2020. Zinc deficiency in low- and middle-income countries: prevalence and approaches for mitigation. Journal of Human Nutrition and Dietetics 33, 624−643. Available from: https://doi.org/10.1111/jhn.12791.

Guzman-Rangel, G., Montalvo, D., Smolders, E., 2018. Pronounced antagonism of zinc and arsenate on toxicity to barley root elongation in soil. Environments 2018 (5), 83. Available from: https://doi.org/10.3390/environments5070083.

Hara, T., Takeda, T.A., Takagishi, T., Fukue, K., Kambe, T., Fukada, T., 2017. Physiological roles of zinc transporters: molecular and genetic importance in zinc homeostasis. The Journal of Physiological Sciences 67 (2), 283−301. Available from: https://doi.org/10.1007/s12576-017-0521-4.

Harika, R., Faber, M., Samuel, F., Kimiywe, J., Mulugeta, A., Eilander, A., 2017. Micronutrient status and dietary intake of iron, vitamin A, iodine, folate and zinc in women of reproductive age and pregnant women in Ethiopia, Kenya, Nigeria and South Africa: a systematic review of data from 2005 to 2015. Nutrients 9 (10), 1096. Available from: https://doi.org/10.3390/nu9101096.

Hussain, A., Jiang, W., Wang, X., Shahid, S., Saba, N., Ahmad, M., Dar, A., Masood, S.U., Imran, M., Mustafa, A., 2022. Mechanistic impact of zinc deficiency in human development. Frontiers in Nutrition 9, 717064. Available from: https://doi.org/10.3389/fnut.2022.717064.

IOM [Institute of Medicine Food and Nutrition Board (NRC)], 2002. Dietary reference intake of vitamin A, vitamin K, arsenic, chromium, copper, iodine, iron, manganese, molybdenum, nickel, silicon, vanadium, and zinc. Institute of Medicine, Food and Nutrition Board, NRC. National Academy Press, Washington DC, pp. 44−501.

Jalal, A., Oliveira, C.E.D.S., Bastos, A.C., Fernandes, G.C., de Lima, B.H., Furlani Junior, E., de Carvalho, P.H.G., Galindo, F.S., Gato, I.M.B., Teixeira Filho, M.C.M., 2023. Nanozinc and plant growth-promoting bacteria improve biochemical and metabolic attributes of maize in tropical Cerrado. Frontiers in Plant Science 13, 1046642. Available from: https://doi.org/10.3389/fpls.2022.1046642.

Kaur, H., Srivastava, S., Nandni, G., Walia, S., 2024. Behavior of zinc in soils and recent advances on strategies for ameliorating zinc phyto-toxicity. Environmental and Experimental Botany 220, 105676. Available from: https://doi.org/10.1016/j.envexpbot.2024.105676.

Kim, T., Obata, H., Gamo, T., 2015. Dissolved Zn and its speciation in the northeastern Indian Ocean and the Andaman Sea. Frontiers in Marine Science 2, 60. Available from: https://doi.org/10.3389/fmars.2015.00060.

Knez, M., Glibetic, M., 2021. Zinc as a biomarker of cardiovascular health. Frontiers in Nutrition 8, 686078. Available from: https://doi.org/10.3389/fnut.2021.686078.

Kumssa, D.B., Joy, E.J., Ander, E.L., Watts, M.J., Young, S.D., Walker, S., Broadley, M.R., 2015. Dietary calcium and zinc deficiency risks are decreasing but remain prevalent. Scientific Reports 5, 10974−10984. Available from: https://doi.org/10.1038/srep10974.

Lindsay, W.L., 1972. Zinc in soils and plant nutrition. In: Brady, N.C. (Ed.), Advances in Agronomy, 24. Academic Press, pp. 147–186. Available from: https://doi.org/10.1016/S0065-2113(08)60635-5.

Lokuruka, M.N.I., 2012. Role of zinc in human health with reference to African elderly: a review. African Journal of Food, Agriculture, Nutrition and Development 12, 6646–6664. 104314/AJFAND.V12I6.

Lowe, N.M., Hall, A.G., Broadley, M.R., Foley, J., Boy, E., Bhutta, Z.A., 2024. Preventing and controlling zinc deficiency across the life course: a call to action. Advances in Nutrition (Bethesda, MD) 15 (3), 100181. Available from: https://doi.org/10.1016/j.advnut.2024.100181.

Maares, M., Haase, H., 2020. A guide to human zinc absorption: general overview and recent advances of *in vitro* intestinal models. Nutrients 12 (3), 762. Available from: https://doi.org/10.3390/nu12030762.

Maciel, B., Costa, P.N., Filho, J.Q., Ribeiro, S.A., Rodrigues, F., Soares, A.M., Júnior, F.S., Ambikapathi, R., McQuade, E., Kosek, M., Ahmed, T., Bessong, P., Kang, G., Shresthra, S., Mduma, E., Bayo, E., Guerrant, R.L., Caulfield, L.E., Lima, A., 2021. Higher energy and zinc intakes from complementary feeding are associated with decreased risk of undernutrition in children from South America, Africa, and Asia. The Journal of Nutrition 151 (1), 170–178. Available from: https://doi.org/10.1093/jn/nxaa271.

Marschner, H., 1995. Mineral Nutrition of Higher Plants, 2nd Edition Academic Press, San Diego.

Mazé, P., 1915. Détermination des éléments minéraux rares nécessaires au développement du maïs. Comptes Rendus Hebdomadaires des Séances de L'académie des Sciences 160, 211–214.

Mcphail, D.C., Summerhayes, E., Welch, S., Brugger, J., 2003a. The geochemistry and mobility of zinc in the regolith. In: Roach, I.C. (Ed.), Advances in Regolith. CRC LEME, pp. 287–291. 2003. Available from: https://citeseerx.ist.psu.edu/document?repid = rep1&type = pdf&doi = 9b8dbb0c6fb71063b5 4a4eec1ca9eb3ece1bca5b.

Mcphail, D., Summerhayes, E., Welch, S., Brugger, J., 2003b. The geochemistry and mobility of zinc in the regolith. Advances in Regolith 287–291. Available from: https://www.academia.edu/29562826/ The_geochemistry_and_mobility_of_zinc_in_the_regolith.

Meng, Y., Xiang, C., Huo, J., Shen, S., Tang, Y., Wu, L., 2023. Toxicity effects of zinc supply on growth revealed by physiological and transcriptomic evidences in sweet potato (*Ipomoea batatas* (L.) Lam). Scientific Reports 13, 19203. Available from: https://doi.org/10.1038/s41598-023-46504-2.

Mertens, J., Smolders, E., 2013. Zinc. In: Alloway, B. (Ed.), Heavy Metals in Soils. Environmental Pollution, 22. Springer, Dordrecht, pp. 465–496. Available from: https://link.springer.com/chapter/10.1007/978-94-007-4470-7_17.

Meylan, G., Reck, B.K., 2017. The anthropogenic cycle of zinc: Status quo and perspectives. Resources Conservation and Recycling 123, 1–10. Available from: https://doi.org/10.1016/j.resconrec.2016.01.006.

Mohamad, N.S., Tan, L.L., Ali, N.I.M., Mazlan, N.F., Sage, E.E., Hassan, N.I., Goh, C.T., 2023. Zinc status in public health: exploring emerging research trends through bibliometric analysis of the historical context from 1978 to 2022. Environmental Science and Pollution Research International 30 (11), 28422–28445. Available from: https://doi.org/10.1007/s11356-023-25257-5.

Neff, J.M., 2002. Chapter 10: Zinc in the Ocean. In: Neff, J.M. (Ed.), Bioaccumulation in Marine Organisms: Effect of Contaminants from Oil Well Produced Water. Elsevier/Oxford, pp. 175–189. Available from: 10.1016/B978-008043716-3/50011-7.

Noulas, C., Tziouvalekas, M., Karyotis, T., 2018. Zinc in soils, water and food crops. Journal of Trace Elements in Medicine and Biology: Organ of the Society for Minerals and Trace Elements (GMS) 49, 252–260. Available from: https://doi.org/10.1016/j.jtemb.2018.02.009.

Odobasic, A., Šestan, I., Keran, H., Šestan, A., 2013. Chemical speciation of Zn in water of the Lake Modrac and assessment of toxicity of water. European Journal of Scientific Research 113, 616–624. Available from: https://www.researchgate.net/publication/259737651.

Ohanenye, I.C., Emenike, C.U., Mensi, A., Medina-Godoy, S., Jin, J., Ahmed, T., Sun, X., Udenigwe, C.C., 2021. Food fortification technologies: Influence on iron, zinc and vitamin A bioavailability and potential implications on micronutrient deficiency in sub-Saharan Africa. Scientific African 11, e00667. Available from: https://doi.org/10.1016/j.sciaf.2020.e0066.

Oldewage-Theron, W.H., Samuel, F.O., Venter, C.S., 2008. Zinc deficiency among the elderly attending a care centre in Sharpeville, South Africa. Journal of Human Nutrition and Dietetics: The Official Journal of the British Dietetic Association 21 (6), 566–574. Available from: https://doi.org/10.1111/j.1365-277X.2008.00914.x.

Ong, H., Yap, C., Maziah, M., Tan, S., 2013. Synergistic and antagonistic effects of zinc bioaccumulation with lead and antioxidant activities in Centella asiatica. Sains Malaysiana 42, 1549–1555. Available from: https://journalarticle.ukm.my/6617/1/02_G.H._Ong.pdf.

Pearson, H.B.C., Comber, S.D.W., Braungardt, C.B., Worsfold, P., Stockdale, A., Lofts, S., 2018. Determination and prediction of zinc speciation in estuaries. Environmental Science and Technology 52 (24), 14245–14255. Available from: https://doi.org/10.1021/acs.est.8b04372.

Plum, L.M., Rink, L., Haase, H., 2010. The essential toxin: impact of zinc on human health. International Journal of Environmental Research and Public Health 7 (4), 1342–1365. Available from: https://doi.org/10.3390/ijerph7041342.

Prasad, A.S., 2013. Discovery of human zinc deficiency: its impact on human health and disease. Advances in Nutrition 4 (2), 176–190. Available from: https://doi.org/10.3945/an.112.003210.

Prasad, A.S., 2014. Impact of the discovery of human zinc deficiency on health. Journal of Trace Elements in Medicine and Biology 28 (4), 357–363. Available from: https://doi.org/10.1016/j.jtemb.2014.09.002.

Pérez-Novo, C., Fernández-Calviño, D., Bermúdez-Couso, A., López-Periago, J.E., Arias-Estévez, M., 2011. Phosphorus effect on Zn adsorption-desorption kinetics in acid soils. Chemosphere 83 (7), 1028–1034. Available from: https://doi.org/10.1016/j.chemosphere.2011.01.064.

Quinby-Hunt, M.S., Wilde, P., 1988. Elemental geochemistry of black shales-statistical comparison of low-calcic shales with other shales. In: Grauch, R.I., Leventhal, J.S. (Eds.), Metalliferous Black Shales and Related Ore Deposits - Program and Abstracts. 1988 Meeting of the United States Working Group of International Geological Correlation Program Project 254. US Geological Survey Circular 1037. Available from: https://pubs.usgs.gov/circ/1989/1037/report.pdf.

Rahman, M.M., Wahed, M.A., Fuchs, G.J., Baqui, A.H., Alvarez, J.O., 2002. Synergistic effect of zinc and vitamin A on the biochemical indexes of vitamin A nutrition in children. The American Journal of Clinical Nutrition 75 (1), 92–98. Available from: https://doi.org/10.1093/ajcn/75.1.92.

Rauch, J.N., Pacyna, J.M., 2009. Earth's global Ag, Al, Cr, Cu, Fe, Ni, Pb, and Zn cycles. Global Biogeochemical Cycles 23 (2), GB2001. Available from: https://doi.org/10.1029/2008GB003376.

Raulin, J., 1869. Études cliniques sur la vegetation. Annales des Sciences Naturelles, Botanique et Biologie Vegetale 11 (Ser. 5, B), 93–99.

Rezaee, M., Tajer-Mohammad-Ghazvini, P., 2022. Rapid and efficient determination of zinc in water samples by graphite furnace atomic absorption spectrometry after homogeneous liquid-liquid microextraction via flotation assistance. Bulletin of the Chemical Society of Ethiopia 36 (1), 1–11. Available from: https://doi.org/10.4314/bcse.v36i1.1.

Roohani, N., Hurrell, R., Kelishadi, R., Schulin, R., 2013. Zinc and its importance for human health: an integrative review. Journal of Research in Medical Sciences: The Official Journal of Isfahan University of Medical Sciences 18 (2), 144–157. Available from: https://www.ncbi.nlm.nih.gov/pmc/articles/PMC3724376/.

Rose, A.W., Hawkes, H.E., Webb, J.S., 1979. Geochemistry in Mineral Exploration, 2nd Edition Academic Press, New York, p. 657.

Roshan, S., DeVries, T., Wu, J., Chen, G., 2018. The internal cycling of zinc in the ocean. Global Biogeochemical Cycles 32, 1833–1849. Available from: https://doi.org/10.1029/2018GB006045.

Sadeghzadeh, B., 2013. A review of zinc nutrition and plant breeding. Journal of Soil Science and Plant Nutrition 13 (4), 905–927. Available from: https://www.scielo.cl/scielo.php?script = sci_arttext&pid = S0718-95162013000400012.

Shiller, A., Boyle, E., 1985. Dissolved zinc in rivers. Nature 317, 49–52. Available from: https://doi.org/10.1038/317049a0.

Sicius, H., 2021. Zinkgruppe: Elemente der zweiten Nebengruppe. Handbuch der chemischen Elemente. Springer Spektrum, Berlin, Heidelberg. Available from: https://doi.org/10.1007/978-3-662-55939-0_17.

Simon-Hettich, B., Wibbertmann, A., Wagner, D., Tomaska, L., Malcolm, H., 2001. Environmental Health Criteria 221 Zinc. http://apps.who.int/iris/bitstream/handle/10665/42337/WHO_EHC_221.pdf?sequence = 1. (Accessed 02 February 2022).

Singh, B., Natesan, S.K.A., Singh, B.K., Ush, K., 2005. Improving zinc efficiency of cereals under zinc deficiency. Current Science 88, 36–44. Available from: http://eprints.icrisat.ac.in/46/1/Improving_zinc_efficiency_of_cereals_under_zinc.pdf.

Smolders, E., Buekers, J., Oliver, I., McLaughlin, M.J., 2004. Soil properties affecting toxicity of zinc to soil microbial properties in laboratory-spiked and field-contaminated soils. Environmental Toxicology and Chemistry 23, 2633–2640. Available from: https://doi.org/10.1897/04-27.

Solomons, N.W., 2013. Zinc. Editorial. Annals of Nutrition and Metabolism 62 (Suppl. 1), 5–6. Available from: https://doi.org/10.1159/000348577.

Srivastava, P.C., Gupta, U.C., 1996. Trace Elements in Crop Production. Science Publishers, Inc, p. 356.

Stiles, L.I., Ferrao, K., Mehta, K.J., 2024. Role of zinc in health and disease. Clinical and Experimental Medicine 24, 38. Available from: https://doi.org/10.1007/s10238-024-01302-6.

Stumm, W., Morgan, 1996. Aquatic Chemistry Chemical Equilibria and Rates in Natural Waters, 3rd. Edition John Wiley J.J. and Sons Inc., New York.

Takagishi, T., Hara, T., Fukada, T., 2017. Recent advances in the role of SLC39A/ZIP zinc transporters *in vivo*. International Journal of Molecular Sciences 18 (12), 2708. Available from: https://doi.org/10.3390/ijms18122708.

Tang, Y., Li, S., Qiao, J., Wang, H., Li, L., 2013. Synergistic effects of nano-sized titanium dioxide and zinc on the photosynthetic capacity and survival of Anabaena sp. International Journal of Molecular Sciences 14 (7), 14395–14407. Available from: https://doi.org/10.3390/ijms140714395.

To, P., Do, M., Cho, J.-H., Jung, C., 2020. Growth modulatory role of zinc in prostate cancer and application to cancer therapeutics. International Journal of Molecular Sciences 21 (8), 2991. Available from: https://doi.org/10.3390/ijms21082991.

USIM (United States Institute of Medicine), 2001. Food and Nutrition Board. Dietary Reference Intakes for Vitamin A, Vitamin K, Arsenic, Boron, Chromium, Copper, Iodine, Iron, Manganese, Molybdenum, Nickel, Silicon, Vanadium, and Zinc. Chapter 12: Zinc. The National Academies Press, Washington, DC. Available from: https://www.ncbi.nlm.nih.gov/books/NBK222310/.

Van Eynde, E., Breure, M.S., Chikowo, R., Njoroge, S., Comans, R.N.J., Hoffland, E., 2023. Soil zinc fertilisation does not increase maize yields in 17 out of 19 sites in Sub-Saharan Africa but improves nutritional maize quality in most sites. Plant Soil 490, 67–91. Available from: https://doi.org/10.1007/s11104-023-06050-2.

Vodyanitskii, Y.N., 2010. Zinc forms in soils (Review of publications). Eurasian Soil Science 43, 269–277. Available from: https://doi.org/10.1134/S106422931003004X.

VUMC (Vanderbilt University Medical Center), 2022. Study identifies cellular 'chaperone' for zinc: findings shed light on worldwide public health problem of zinc deficiency. ScienceDaily 17. Available from: http://www.sciencedaily.com/releases/2022/05/220517112206.htm.

Wang, M., Kong, F., Liu, R., Fan, Q., Zhang, X., 2020. Zinc in wheat grain, processing, and food. Frontiers in Nutrition 7, 124. Available from: https://doi.org/10.3389/fnut.2020.00124.

Weiss, A., Murdoch, C.C., Edmonds, K.A., Jordan, M.R., Monteith, A.J., Perera, Y.R., et al., 2022. Zn-regulated GTPase metalloprotein activator 1 modulates vertebrate zinc homeostasis. Cell 185 (12), 2148−2163.e27. Available from: https://doi.org/10.1016/j.cell.2022.04.011.

Wessels, K.R., Brown, K.H., 2012. Estimating the global prevalence of zinc deficiency: results based on zinc availability in national food supplies and the prevalence of stunting. PLoS One 7 (11), e50568. Available from: https://doi.org/10.1371/journal.pone.0050568.

Wessels, I., Rolles, B., Rink, L., 2020. The potential impact of zinc supplementation on COVID-19 pathogenesis. Frontiers in Immunology 11, 1712. Available from: https://www.frontiersin.org/article/10.3389/fimmu.2020.01712.

WHO (World Health Organization), 2002. World Health Report 2002 - Reducing Risks, Promoting Healthy Life. Geneva: World Health Organization. https://www.who.int/publications/i/item/9241562072.

WHO (World Health Organization), 2003. Zinc in Drinking-water. Background Document for Development of WHO Guidelines for Drinking-water Quality - WHO/SDE/WSH/03.04/17, Geneva. https://cdn.who.int/media/docs/default-source/wash-documents/wash-chemicals/zinc.pdf?sfvrsn = 9529d066_4. (Accessed 08 March 2024).

Young, L.B., Harvey, H.H., 1991. Metal concentrations in chironomids in relation to the geochemical characteristics of surficial sediments. Archives of Environmental Contamination and Toxicology 21, 202−211. Available from: https://doi.org/10.1007/BF01055338.

Further readings

World Health Organization, Guidelines for drinking-water quality, Originally published in 2nd. Ed. Vol. 2. Health Criteria and Other Supporting Information. World Health Organization, Geneva, 1996.

Kebede-Desta, M., Broadley, M., McGrath, S.P., Hernandez-Allica, J., Hassall, K.L., Gameda, S., et al., 2023. Linking soil adsorption-desorption characteristics with grain zinc concentrations and uptake by teff, wheat and maize in different landscape positions in Ethiopia. Frontiers in Agronomy 5, 1285880. Available from: https://doi.org/10.3389/fagro.2023.1285880.

Nriagu, J.O., 2011. Zinc toxicity in humans. In: Nriagu, J.O. (Ed.), Encyclopedia of Environmental Health. Elsevier, pp. 801−807. Available from: 10.1016/B978-0-444-52272-6.00675-9.

Vanga, N.R., Kesamsetty, V.R., Ummiti, K., Ratnakaram, V.N., Bapatu, H.R., 2024. Development and validation of a method for trace level zinc quantification in pharmaceutical zinc supplements using a carboxyl functional group packed column and refractive index detector. Journal of AOAC International 107 (1), 2−13. Available from: https://doi.org/10.1093/jaoacint/qsad101.

Zvěřina, O., Vychytilová, M., Riger, J., Goessler, W., 2023. Fast and simultaneous determination of zinc and iron using HR-CS GF-AAS in vegetables and plant material. Spectrochimica Acta, Part B Atomic Spectroscopy 201, 106616. Available from: https://doi.org/10.1016/j.sab.2023.106616.

Conclusion

16

General conclusions

A thorough understanding of how geoenvironmental variables, among other etiological factors, affect our health is essential in fulfilling the principal mandate of any public health intervention scheme, namely, to ascertain precisely the causes of disease so that we can effectively prevent and manage them (Paraphrased, after Davies, 2022).

Conclusions

Medical Geology studies the role of geoenvironmental factors on the geographical distribution of disease in man and (other) animals. This chapter summarizes the key conclusions from each chapter and briefly highlights the successes that have been attained, as well as the challenges that still confront the Medical Geologist in terms of many remaining gaps in our understanding of the position of the geoenvironmental causative factor of disease, and in the context of future work to be conducted.

Huge amounts of resources are being disbursed annually on apparently tractable questions in genetics, gene therapy, and immunology to decipher the origin of disease. But this Primer posits that, for diseases whose diagnoses are still uncertain, especially those that occur under geographically localized conditions, more attention should be given to the geoenvironmental hypotheses, and an examination made of their scientific strength.

The rapid development of Medical Geology followed the concretization and formalization of the study by the International Medical Geology Association (IMGA) meeting in Uppsala in the year 2000 (See Skinner and Berger, 2003) when it was agreed by international consensus that the former name of "Geomedicine" be replaced by "Medical Geology," a term that was deemed more likely to gain acceptance by the medical profession. A flurry of research efforts in this evolving scientific field has been witnessed since the Uppsala meeting.

A thorough understanding of how geoenvironmental variables among other etiological factors, affect our health is essential, in fulfilling the principal mandates of any public health intervention scheme, namely, to ascertain precisely the causes of diseases so that we can effectively prevent and manage them. These were the principles encapsulated in the technical recommendations promulgated at the first African Workshop on Medical Geology in 1999 (Davies, 1999), namely, (1) the desirability of increased collaborative efforts in research between scientists working in the fields of human health on the one hand, and animal health on the other, including physicians, nutritionists, geochemists, toxicologists, botanists and veterinarians; (2) the urgent need for raising awareness (even at grassroots level), build capacity and consensus, and immediately address the issue of the interrelationship between Medical Geology and the environment; (3) the great need for generating baseline data as well as effectively monitor

Medical Geology of Africa. DOI: https://doi.org/10.1016/B978-0-12-818748-7.00016-2

the flow of nutritional and toxic trace elements in the environment that lead to human exposure through the consumption of food and water, especially the elements As, Cd, Hg, Se, Rn, Pb, F, I, and Zn; and (4) the need for further research on speciation, bioavailability as well as bioaccessibility of trace elements in soils, plant materials, and in humans and (other) animals.

As is clearly evident from the main chapters of this Primer, much of these mandates have been attained since those early days, though admittedly, much more needs to be done. Over the last two decades or so, the gulf between geology and health has been significantly narrowed, lines of communication between allied disciplines working on disease origin, are opening up; and a mass of data is now available on the significance of various cross-cutting disease phenomena, such as defects in the transport medium during trace element entry into the body, and their involvement in metabolic processes (See, e.g., Davies, 2022).

Indeed, much of the recent research effort in Medical Geology in Africa over the last decade has focused on the determination of the circulation of both nutritional and potentially toxic elements in the soil-water- air-food crop nexus that enter the food chain. The principal rationale behind this approach has been the increasing realization of the significance of the entry (largely through the diet) of varying concentration levels of elements that may be bioavailable for negative interactions in metabolic processes that produce diseases, some of whose diagnoses, till now, are still ill-defined (diseases of unknown etiology) (Chapter 3, Part I). It is clear that most diseases of unknown etiology are multifactorial diseases, caused by a complex interplay between genetic factors (polygenic), immunological mediators (trace elements/metals/metalloids) and various environmental factors, none of which factors would cause the disease on its own.

The importance of the "'optimal range"' of intake for the essential elements, e.g., Se and F in metabolic processes is also emphasized in Part II of this Primer, and illustrations given of illnesses that can possibly occur on either side of this range; and how controlled intake of these elements may help combat a range of debilitating health conditions in Africans.

Although there has been some progress recently on study of the effects of ionizing radiation on the developing embryo, those often tasked with carrying out such studies are from a medical background, e.g., pediatric clinicians and radiologists. They go about their work with little cognizance taken of U's geogenic characteristics and other geodynamic intricacies that bear so heavily on the mode of radioactivity uptake, the chemical form of the radionuclide, the doses received, the dose rate, and dose distribution. Chapter 5, Part I considers the urgent need for the inclusion of Medical Geologists in teams investigating the effects of ionizing radiation exposure on pregnancy outcomes (neonatal and child radiation-related morbidity and mortality) in major U and Au mining African countries (such as Niger, Namibia and South Africa) is urgent, and can lead to much-improved prognosis, diagnosis, and therapy; as well as prevention and other forms of intervention, such as legal, for these conditions.

The "hot" topic of *geophagy* (or *geophagia*) (Chapter 6, Part I), the habit of eating Earth materials, is especially prominent among African women and children. To date, there is a great deal of misunderstanding of the phenomenon, but it is thought to have both beneficial and deleterious effects. As research on this phenomenon proliferated around the Continent, practical behavioral adjustments would emerge that would obviate the undesirable consequences of the practice as well as promote optimum derivation of its benefits.

Although the implementation of the iodized salt program by UNICEF, as well as other nutritional strategies (e.g., use of iodized oil), has resulted in a marked decline of iodine deficiency disorders (IDD) in Africa, goiter prevalence remains high in some countries. The thyroid hormone-inhibiting action of *goitrogens*, substances present in many African diets, means that IDD on the Continent can probably never be completely obliterated (See: Chapter 4, Part II). There is also the emerging problem of excess iodine intake (e.g., hyperthyroidism), which should be reckoned with by policymakers and program implementers.

The importance of collaborative studies between Medical Geologists and health professionals cannot be more graphically demonstrated than in deciphering the etiology of many cancers and other noncommunicable diseases in Africa, where a substantial percentage of global cancer deaths occur; and where, as already established, most of the population rely on local sources for their drinking water and food provisions.

More effective methods are needed to identify groups and individuals at greatest risk of cancer at a stage where intervention is possible. Much of the progress that has been made to date in cancer control has stemmed from epidemiological research that aims to understand environmental, genetic, and population risks for developing specific cancers. Since environmental factors have been judged likely to contribute to most human cancers, research should be strengthened to ascertain the efficacy potential of protective geochemical agents, such as Se and Zn supplementation in the long-term management of certain cancers, such as prostate cancer. The present uncertainties on the role of trace elements and cancer, mean that more studies appropriate to therapy are needed, that would especially take into consideration the Continent's unique geoenvironmental condition. Whether higher Mg intake, for instance might be protective against oncogens in humans as it is in some animal models is uncertain and deserves more study.

Although epidemiological research on cancer control has contributed to our understanding of environmental, genetic, and population risks of developing specific cancers, several challenges are inherent in attempts at interpreting epidemiological studies of human exposure and cancer, as they engender natural-, not experimental-, human exposure, and therefore, a multitude of factors that may affect cancer prevalence in addition to the exposure under study must be accounted for.

Studies on Kaposi's sarcoma a common tumor in Africa (Chapter 1, Part I) exemplifies these constraints. A multidisciplinary research effort from oncologists and Medical Geochemists should be organized on avoidable exposures to a wide variety of carcinogens contaminating the soil, water, and air environment of the African Continent.

Whatever uncertainties exist on the role of trace elements in cancer etiology, more studies appropriate to therapy would need to be designed that would especially take into consideration the continent's unique geoenvironmental condition as elucidated in Chapter 1, Part I.

Magnesium, the eighth most abundant element in the Earth's crust, plays many crucial roles in the body, being a cofactor in more than 300 enzymatic reactions needed for the structural function of proteins, nucleic acids, and mitochondria. Magnesium deficiency has been implicated in a broad range of cardiometabolic conditions, including diabetes, hypertension, and cardiovascular disease (CVD).

Resolution of uncertain questions regarding the implication of Mg deficiency in aspects of pathogenesis and treatment of neoplasms might lead to means to prevent conditions such as lympholeukemias or possibly of immuno-incompetence.

The role of Mg in *CVD* therapy as well as in psoriasis treatment is considered pertinent. Till today, uncertainties remain in our attempt at clarifying the relationship between water hardness (and Mg) and CVD. Clarifying this "correlation" in regions where CVD prevalence is high, is important. However, in making such investigations, the role of social, climatological, and environmental factors, other than the hardness of water, must be factored in. This makes further research along these lines highly desirable.

The health benefits of Mg and its role in disease prevention underlie the essentiality of this commonly overlooked micronutrient. As we continue to gain a better understanding of the numerous biochemical pathways and controls of Mg in the human body the importance of supplementation with this nutrient to maintaining our health will become ever more apparent.

Several types of dietary Mg supplements are available, such as Mg citrate, glycinate, and lactate. Other kinds have topical uses, such as in baths or on the skin (*cf.*, as remedy for psoriasis). All carry different reported effects and benefits. As the fields of Medical Geology, nutrition and medicine continue to reveal the benefits of Mg, it will become ever clearer that supplementation with this nutrient is quite crucial to maintenance of our health.

Several decades after the identification of *endomyocardial fibrosis* (EMF), its etiopathogenesis has not yet been conclusively established. Since none of the earlier hypotheses could convincingly explain the geographical prevalence of the disease in the tropics, the proposed geochemical postulate in accounting for cofactors, namely, increased dietary Ce and decreased Mg, seems highly plausible. Indeed, two decades after the postulation of the geochemical theory for an etiology cofactor, the prevalence of EMF has declined in both Africa and India, attributable to the better nutrition of poor children that corrects Mg deficiency. Despite this decline in prevalence, however, EMF remains endemic in Africa. Large, well-designed epidemiological studies are therefore needed in ascertaining not only the modifiable risk factors for cardiomyopathy, which would help in the formulation of preventive programs, but also the genetic and molecular epidemiology of cardiomyopathy on the Continent.

Podoconiosis (or *nonfilarial elephantiasis*) remains a neglected and under-researched condition in Africa. The involvement of geochemical cofactors in the etiology of podoconiosis is now well-established, namely, exposure to irritant red clay soil of volcanic origin; although other possible cofactors including genetic susceptibility remain poorly understood. Several trace elements in these soils have been implicated as being significantly related to podoconiosis including Fe, Be, Al, Ti, and Cr; however; the lack of quantification of these elements and the small sample sizes, preclude any quantitative support to this theory.

Podoconiosis can easily be prevented simply by initially providing shoes to peasant farmers. In the early and progressive stages of the disease, foot hygiene with soap, water, and antiseptics with the daily use of robust shoes and socks can result in significant improvement. Occupational rehabilitation is vital to provide patients with a social role and a means to earn a living that does not involve contact with irritant soil.

As at the time of the final stage of writing (2023) no precise therapy for *psoriasis* had been recorded; but there are several treatment options available to control the symptoms, though significant side effects are experienced with many of them. New management approaches should be explored to replace previous fortuitous strategies. Further studies on serum levels of trace elements, particularly metals, in psoriasis patients are required; and this may help in elucidating etiopathogenesis and potential precision treatment of this common and enigmatic disease.

In Africa, there are relatively few studies that have addressed the prevalence of *osteoporosis*, particularly among populations in the Sub-Saharan Africa region. Similarly, the paucity of high-quality relevant datasets in Africa, including both dietary Ca intake and bone health parameters, precludes the assumption of an ecological association between average Ca intake and the prevalence of osteoporosis or fracture; albeit evident that extremely poor populations that have very limited nourishment choices will be more susceptible to deficient Ca intake. There is also very little information regarding the incidence of osteoporotic vertebral fractures in African countries, but like hip fractures, they are thought to be uncommon.

The need to apply Ca, as well as the form in which it should best be applied, should be determined by considering factors such as the soil pH as well as the Ca status of the soil. In very acidic soils, Ca will need to be applied in a form that will raise the soil pH (e.g., lime), and at higher rates than are normally required to correct Ca deficiency. Formulation of strategies for preventing osteoporosis and fractures should aim at obviation of the risk factors such as poor Ca nutrition and adequate vitamin D levels [levels of 50 nmol/L (20 ng/mL) or above are adequate for most people for bone and overall health]; and high quality epidemiological data on prevalence and incidence of these conditions would be essential to be able to design appropriate strategies and monitor their effectiveness.

Low back pain (LBP) in Africa is clearly rising and is becoming a problem of great concern because its etiology is still largely indefinable. Further research into the most effective strategies to prevent and manage LBP in Africa is warranted. A circumspective approach to making a diagnosis for associated conditions (*Prolapsed Intervertebral Disk, spinal stenosis* and *Paget's disease of bone*) should involve history taking, physical examination, baseline, and specialized investigations. It is considered that further research that closely investigates the role of F^- balance (and possibly also Ca and Mg) in the nutrition of patients suffering from these maladies would go some way toward finding the most effective strategies to prevent and manage LBP and associated conditions in Africa.

Monitoring the levels and compositions of dust emissions in African mining and quarrying operations is an important exercise, but it is an endeavor that is, till now, fraught with many challenges, such as the lack of the sufficient number of monitoring stations, the poor distribution of available stations and the ability to discriminate between particles from different sources within mixtures, emissions, or deposited dust. This area of research calls for urgent attention, given the high prevalence of various respiratory diseases in the vicinity of mining centers, particularly in southern Africa. The question of whether or not there are links between the prevalence of Coal Workers' Pneumoconiosis CWP among various African coal miners and some physicochemical parameters remains open-ended, as is also the question of the possible existence of links between dust inhalation and the incidence of meningitis in Africa.

In spite of the recent increase in the use of satellites and remote sensing techniques to monitor the Saharan dust clouds over North Africa to North America and Europe, very little is known about the ability of these dust plumes to carry pathogenic microbes, kill coral and cause asthma; this is so, because the various experts working on these problems—medical professionals, geologists, climatologists and microbiologists—tend to do so in isolation. The development of strong collaborative ties between these professionals would assist in the characterization of the properties of geogenic particulates, their dispersal and the toxicological pathways of infectious agents they may transport.

During artisanal and small-scale gold mining activities in Africa, the use of Hg and $[SCN]^-$ (thiocyanate) in combination, to amalgamate the gold is common practice. Dissolution of gold and mercury occurs to form complexes that are easily mobilized by rain and often, due to poor containment practices, quickly reach stream waters.

The growth in Medical Geology research that is being experienced over last two decades is set to continue as "new" correlations between certain environmental health conditions with factors related to our interactions with geological materials and processes are unearthed. However, most of the more recent studies have so far tended (justifiably) to emphasize only the harmful health effects of geoenvironmental factors. This approach is apparently well-founded, given that a proper understanding of the deleterious health impacts has led to improved diagnosis and therapy in many cases. However, there are also health benefits accrued from the judiciously exploitation of geological materials and processes. This Primer therefore highlights recent developments in this emerging area of Medical Geology in Africa, exemplified by referencing several important medical applications of African clays, the therapeutic gains associated with hot springs, and balneology of peat deposits; and by drawing attention to other correlates that have not yet been fully covered in the previous literature. A comprehensive assemblage of the most current references is put together to serve as source material for those scientists wanting to deeply explore this fast-growing subject area (beneficial effects of geological materials and processes) in Africa.

Clay minerals have a wide range of applications, more so in the pharmaceutical and bio-medical fields. Their healing potential in *pelotherapy* is now well-documented. Despite the abundance of literature on healing clays, more research is needed on aspects, such as the degradation rate and the risk of accumulation in the body, since a few materials are relatively insoluble and could last longer, if implanted.

There are several geothermal systems in Africa that could be developed for balneological use to boost the health of citizens and bolster tourism. Despite the marked similarities that exist between the Dead Sea and some rift valley lakes in East Africa, such as Lake Magadi in Kenya, there is little evidence that these lakes are used to their full balneological potential. Experience gained from Dead Sea experiments and clinical trials on subjects having certain dermatological conditions can be applied in optimizing the balneological potential of these lakes.

It is quite evident that drawing up correlations between environmental variables and disease cannot be a simplistic affair, because of the involvement of several imponderables, such as methodological problems, lack of adequate control, confounding variables, the one-to-one comparison, health indicator definition, health indicator recall, the seasonality of impact variables, failure to analyze by age, and failure to record usage.

However, as a result of improved record-keeping over the last two decades or so, several extremely valuable records have become available that would allow progress in understanding etiological patterns in disease dynamics in Africa. But to ensure the viability of these correlations, such studies would need to be conducted within multidisciplinary and interdisciplinary settings involving cooperation between a battery of professionals—geoscientists, medical practitioners, social scientists, and so on. Elucidating the true environmental determinants of disease is of prime concern and professionals in the fields of medical geochemistry, physiochemistry, social, and other sciences, all have their roles. Social science research conducted at the macrosocietal level through behavioral and psychological levels will produce comprehensive research data needed for a fuller understanding of health and illness. This is so, because social science researchers working on the origin of diseases take cognizance of the fact that health may be affected by processes operating in different mechanistic pathways among and within existing social structures, underscoring the need for integrating this kind of research into interdisciplinary, multilevel studies of health.

The application of remotely sensing and Geographic Information Systems (GIS) technology in mapping environmental diseases over the last four decades has advanced the field of spatial epidemiology by a considerable distance. This potential of remote sensing to register health-related data is invaluable; particularly so because certain information cannot otherwise be captured, since they lie beyond the sensitivity of the human eye and conventional "panchromatic" cameras. A whole host of geoenvironmental variables related to soil, water and the atmosphere can be conveniently retrieved by remote sensing, which also allows the study of large areas at low cost. Images obtained by remote sensing, more specifically, satellite images, and aerial photographs, can be analyzed to produce important data that complement ground measurements of geoenvironmental variables.

Earth-observation satellites are now being exploited in modeling of transmission cycles of pathogens of relevance to epidemiology and public health. Such studies employ sensors with spatial, spectral, and geometric controls to address large-area questions regarding the epidemiology of certain diseases. Modern methods in epidemiology now make it possible to model large datasets with environmental variables remotely sensed at various resolution scales.

In our attempt to understand fully the role of geoenvironmental factors; however, it is clear that singling out a given influential factor from an array of complex environmental variables is not a simple task. The analysis and synthesis of data collected at different scales and in different spatial dimensions will always confront the environmental epidemiologist with difficult problems. Debates on uncompromised links between exposure and disease, for instance, are made in the midst of potential sources of bias and confounding, inherent in many epidemiological investigations. Confounding factors may, for example, prevent study groups from being comparable.

A discussion on the influence of confounding factors in drawing up of such correlations is given in Chapter 3 about deciphering the origins of diseases of diseases of unknown etiology. However, several examples exist where quantitative demonstrations can show convincingly, the connections between certain environmental diseases and geoenvironmental variability in Africa.

Public health informatics (PHI) is the systematic application of computers, clinical guidelines, communication, and information systems, to public health practice and research. Part of this emerging discipline, digital information technologies, are applied to aid in the detection and management of diseases and syndromes in individuals and populations, leading to faster, better, and more robust understanding and decision-making capabilities in the public health arena. As part of that effort, a GIS or more generally a Spatial Decision Support System (SDSS)—offers improved geographic visualization techniques, leading to faster, better, and more robust understanding and decision-making capabilities in the public health arena.

Two of the more well-known international publishers of research on PHI are: (1) The "International Journal of Health Geographics," which "… is fully dedicated to publishing novel, high quality and internationally significant manuscripts on all aspects of geospatial information systems and science applications in health and healthcare," and can be accessed at: https://ij-healthgeographics.biomedcentral.com/about (accessed 02 October 2023); (2) The Journal, "Geospacial Health" publishes research on: "… all aspects of the application of geographical information systems, remote sensing, global positioning systems, spatial statistics and other geospatial tools in human and veterinary health." This Journal can be accessed at: https://geospatialhealth.net/index.php/gh (accessed 02 October 2023).

It is considered that a complete "Africa Geochemical Database (AGD)" would constitute a fundamental basis for the construction of maps detecting abnormal spatial patterns of health outcomes and environmental risk factor associations, and therefore, spatial analysis techniques and methods will grow in importance.

In the above-stated context, it is thought that Africa will benefit immensely from a complete geochemical map of the entire region for all nongaseous elements and other chemical parameters in pertinent environmental media, based on agreed guidelines (See Chapter 3, Part 1 and Chapter 2, Part III).

It is suggested that correlation maps from a completed AGD analogous to what exists for China, England, and Wales, and a few other countries would, among other applications, enable the depiction of areas where disease clusters overlie anomalous element distribution (in soil, water, or air), and so permit an evidence-based statistical assessment of the magnitude of any geochemical component in the disease causative web.

This Primer is intended to serve as a provisional route map for postgraduate researchers of Medical Geology as well as research scientists from the medical profession and allied fields who want to explore the position and *modus operandi* of geoenvironmental variables as causal cofactors of disease. Despite the enormous strides made in medical research over the last three decades or so, and as the field of medical science continues to mature and expand, there remain several gaps in our knowledge engendering critical issues to address particularly about the role of metals/metalloids and isotopes as cofactors in diseases of unknown etiology, a debate that has, till relatively recently, been diminished by territorial disputes between disciplines, notably the gulf between medical practitioners and geoscientists.

It is hoped that the message articulated in this Primer would go some way towards narrowing the gulf between medical and public health professionals on the one hand, and geoscientists on the other, as they intensify multidisciplinary research efforts to decipher the true position of geoenvironmental factors in unknown etiologies.

References

Davies, T.C., 1999. The East and Southern Africa Regional Workshop on geomedicine. Episodes 22 (4), 303–304. Available from: file:///C:/Users/davies.theophilus/Downloads/IUGS022-04-09%20(4).pdf.

Davies, T.C., 2022. The position of geochemical variables as causal co-factors of diseases of unknown aetiology. S.N. Applied Sciences 4 (8), 236. Available from: https://doi.org/10.1007/s42452-022-05113-w.

Skinner, H.C.W., Berger, A.R. (Eds.), 2003. Geology and Health: Closing the Gap. New York. Oxford University Press, Online Edition, Oxford Academic. Available from: https://doi.org/10.1093/oso/9780195162042.001.0001.

17

Future perspectives and prospects

Just over two decades after the formalisation and concretisation of Medical Geology as a scientific discipline, the study is now well established around the world. In Africa, the future of Medical Geology is bright.

In Africa, activities in Medical Geology are reaching an unprecedented level. Research and study continue to evolve and thrive as new information is acquired, plausible "new" correlations between the geoenvironmental and disease emerge, and reinterpretation is given to old ones.

The key driver of this growth is attributable to a combination of factors that make the Continent a superb experimental station for studies in this field. Thus the likelihood of the discovery of further remarkable correlations between Earth materials and processes and human health will continue into the next decades, as research activities continue to proliferate.

The greater percentage of Medical Geology research over the last few years has been on investigations of the flow of nutritional and toxic trace elements in the environment that lead to human exposure through the consumption of food and water especially the elements As, Cd, Hg, Se, Rn, Pb, F, I, and Zn (See: Part II). Research on speciation, bioavailability, as well as bioaccessibility of the trace elements in soils, water, air, plant materials, and humans and animals, is progressing in various African laboratories (See Chapter 4, Part I).

This is in consonance with the growing awareness of the important role of nutritional elements in shaping the African diet, the criticality of their "'optimal range'" of intake in metabolic processes, and hence, in warding off disease. In the same vein, an understanding of the fluxes of potentially toxic elements (PTEs) in the soil, water, and air environments is essential for designing nutrient management strategies that would limit their (PTEs) entry into the food chain. This trend in research is set to continue.

In Africa, intense tropical weathering, erosion, leaching, lateritization, secondary mineralization, eluviation, podsolization, and gleying combined with urban and industrial activities have left an indelible imprint on the soils of the region. Given such a unique geochemical landscape, coupled with the reality that rural communities are still largely dependent on water and food sources that are locally derived (bringing them into direct contact with their local geomedical environment), a superb setting is created for studying the relationship between nutritional and PTE distribution (deficiency/excess) and disease.

However, drawing up tangible correlations between geoenvironmental variables and disease Medical Geologists is reliant on the availability of very high-quality analytical data for element concentrations and their distribution in environmental and biological samples. The content of certain critical elements (sometimes down to mg/L level) needs to be accurately

Medical Geology of Africa. DOI: https://doi.org/10.1016/B978-0-12-818748-7.00019-8

known. But the need for quality assurance has up till recently, bedeviled analytical work in many African geochemical laboratories. There have, till recently, been relatively few technicians with sufficient expertise for proper installation, maintenance, troubleshooting, and repair of today's analytical instrumentation.

Thankfully, this situation is rapidly changing, as many African laboratories are now acquiring smaller, more compact, accurate, easy-to-use analytical instrumentation, such as the portable X-ray spectrometer, which requires little technical expertise for its operation.

The development of new analytical techniques will continue to enhance current understanding of the behavior of trace elements in the geochemical cycle, chemical speciation, bioavailability, and bioaccumulation, thus obviating knowledge gaps on the detection of relationships between certain environmental diseases and trace element deficiency/excess in humans. Modern geoanalytical techniques, highly customized to the African situation, will generate accurate geochemical data whose judicious application offers promise for the development of innovative solutions to prevent or minimize exposure to potentially deleterious natural materials and geological processes.

It is considered that further success of Medical Geology in Africa would, to a large extent, be accelerated through a commitment by national Geological Surveys and relevant geoscience institutions to the completion of the Africa Geochemical Database Project (AGDP) (formerly referred to as the International Geological Correlation Program), Project 259 (IGCP 259) based on approved guidelines (See Chapter 4, Part I). As of today, a complete geochemical database does not exist for most of the countries of Africa. Geochemical coverage exists for a few countries, although there are considerable variations between datasets, precluding their use as a basis for studying processes operating in the region's surficial environment. This constraint also hinders, or does not justify, the assembly of most existing data into broader regional compilations. However, as analytical capacity continues to improve and collaborative links are strengthened with high-performance regional laboratories, these discrepancies will soon begin to disappear.

One other important consequence of the rather poor geochemical database is the fact that many environmental diseases that result from excess or deficiency of certain key elements are poorly diagnosed. The production of a high-precision geochemical map of Africa would be a cost-effective method of indirectly investigating the chemical composition of crops, and rural communities in Africa will constitute ideal subjects for researching the relationship between geochemistry and health. Such baseline data would be invaluable in understanding the hydrological, chemical and biological processes that determine the behavior of chemical elements in this part of the Earth's surface, to how they may affect the life of man and animals.

A high-quality AGD will enhance the credibility of correlations drawn between certain environmental diseases and the biochemical, biophysical, or geochemical interactions within the bedrock–groundwater–soil–food crop–human/animal tissue continuum, an area of research that has spiraled tremendously around the Continent during the last decade.

Significant correlations could then be conveyed to the medical profession, to heighten awareness of the huge potential significance that factors of the geoenvironment can have on disease causation and thereby serve to broaden the diagnostic spectrum and appropriate therapeutic interventions.

Realizing the huge potential of the AGD for characterizing the link between geochemical variables and disease, advocacy for the revival of the Program is getting stronger as national governments and relevant institutions continue to be made aware of the gains to be accrued from a better knowledge of the Continent's surface geochemistry. This initiative will increase the use of multielemental analysis for Medical Geology.

Arguments presented in this Book illustrate that correlation maps from a completed AGD would, among other applications, enable the depiction of areas where *disease clusters* overlie *anomalous* element distribution (in soil, water, or air), and so permit an evidence-based statistical assessment of the magnitude of any geochemical component in the disease causative web. It is thus possible to proffer that trace element dynamics/imbalances in metabolic processes may one day be found to play a far greater role in unraveling the etiology of diseases of unknown etiology than has hitherto been realized. Presently, however, an important question still lingers, *viz.*, on how best we can map geochemical information and correlate this with environmental epidemiology and disease registry data.

Till relatively recently, medical registry records were still inadequate in most African countries, besetting efforts to provide the evidence required for substantiating suspected correlations. But as efforts in medical record-keeping improve across the Continent, research in Medical Geology would in turn continue to gather pace, ushering the emergence of further exciting revelations in this field.

Considerable gains have been made regarding the resolutions taken at the first two African Workshops on Medical Geology (Davies and Schluter, 1999; Ceruti et al., 2001), but much more remains to be done. Remaining concerns for future Medical Geology research to address, include exceptionally high concentration levels of As in drinking water (Chapter 2, Part II); Hg emission from artisanal and small-scale gold mining, as well as from coal combustion, and its bioaccumulation in the environment (Chapter 7, Part II); the environmental health impacts of geological processes such as Sahara dust transport, volcanic emissions, earthquakes, tsunamis and climate change (Chapter 4, Part I). The input of Medical Geology researchers in teams working to avoid, prevent, mitigate or minimize health issues arising from these disasters will become increasingly realized in the future.

Large-scale research programs in Medical Geology will become increasingly common in medical and geoscience institutions seeking to understand the causes and interrelatedness of diseases. Aided by the power and promise of "Big Data," (from new data sources) Medical Geology research holds the potential to contribute toward changing the way medical data is collected, stored, manipulated, and analyzed to drive discovery.

Increased collaborative research is being developed between scientists working in the fields of human health on the one hand, and animal health on the other, including physicians, nutritionists, geochemists, botanists, veterinarians, and so on; and findings conveyed in a customized way to political decision makers and the public. The need to raise awareness (even at grassroots level), build capacity, and consensus is being successfully addressed and the importance of the interrelationship between Medical Geology and the environment is now being clearly appreciated by relevant professionals.

Several other Medical Geology activities such as the Biennial Conferences (MedGeo) of the Global Medical Geology Body, the International Medical Geology Association (IMGA), should maintain enthusiasm and momentum well into the future. The most recent Biennial Medical Geology Conference took place in the City of Monterrey in México from 6 to 9 August, 2023 [http://medgeomx.com/ (Accessed 12 October 2023)]. The six thematic sessions at this Meeting reflected the directions that global Medical Geology research is taking in terms of its relevance in addressing emerging geoenvironmental health issues of society. A number of plenary and keynote lectures, as well as short courses were held during this Meeting, underscoring the level of maturity the subject has attained.

However, the extent of involvement of professionals from the healthcare sector in Medical Geology is still relatively low, as is their attendance at major Medical Geology meetings. More participation of physicians in major Medical Geology meetings is desirable. Medical Geologists should actively seek collaboration in research projects by offering lectures and seminars at medical institutions and conferences, and getting more engaged with public health agencies.

Following a strong advocacy for the inclusion of Medical Geology into geoscience curricula in universities and institutes of higher learning (See Davies, 2019), several African universities now have dedicated Medical Geology modules in their graduate geoscience programs.

Several books on Medical Geology have appeared, since the original work: "Essentials of Medical Geology," which received a huge positive response from the medical profession and allied disciplines. The many colleges and universities in Africa that have incorporated Medical Geology into their curriculum have adopted this volume as their key reference in their course delivery; and customized course curricula drawn (Davies, 2019). The inclusion of coursework in Medical Geology for healthcare professionals in their academic training should now be given priority.

Several short courses on Medical Geology have been mounted around the world, not least in Africa, since those formative days. These courses are projected to continue well into the future, attracting enthusiastic Medical Geologists and future students.

Global funding partnerships have been increased and social media and online platforms are now in use to boost awareness and community engagement.

The main challenges in Medical Geology research development in the future would be tapping the potential of multidimensional and real-world evidence generation by retrieving, assimilating, and analyzing large datasets and the information generated by advanced scientific techniques. The involvement of artificial intelligence in future Medical Geology research is evident, as is the application of machine learning, and deep neural networks to improve recognition of correlation of disease with geoenvironmental factors.

In studies involving the correlation of disease distribution and geoenvironmental variables, "confounding" should always be addressed. Causality is difficult to prove conclusively and collocation does not unequivocally establish causal relationships. In studies concerned with causality, confounding can result in a biased estimate of the effect of exposure on disease. It is in such cases that traditional machine learning and statistical models would increasingly be used, to minimize the impact of confounders in matching disease distribution with geoenvironmental correlates (See Chapter 3, Part I). The field of Medical Geology will troubleshoot some of the confounding factors in diagnosis of diseases of unknown etiology.

Despite the substantial advances that have been realized in book and journal publication, curriculum development, teaching, research and advocacy, much more needs to be done in these domains. The number of existing Medical Geology degrees and certificate programs are nowhere near the numbers in other evolving scientific disciplines. Multidisciplinary publication outlets specifically devoted to the publication of Medical Geology papers are needed. Key publication outlets for a good proportion of Medical Geology research, are, at present: (1) The Springer Journal: "Environmental Geochemistry and Health," which is the Official Journal of the Society of Environmental Geochemistry and Health, https://www.springer.com/journal/10653/ (Accessed 07 October 2023), and (2) "GeoHealth," a transdisciplinary, Open Access Journal of the American Geophysical Union, published by John Wiley and Sons Inc., https://agupubs.onlinelibrary.wiley.com/hub/journal/24711403/open-access.html (Accessed 07 October 2023). As the science of Medical Geology continues to mature and expand, journal outlets specifically dedicated to this science will doubtlessly emerge.

The establishment of more seminars, conferences and workshops in Medical Geology will encourage people to meet and communicate. It is important for Medical Geologists to bring awareness of the value of their input to public health disaster managers and other environmental health personnel.

In the area of graduate employment, employers need to be aware of the huge potential contribution of Medical Geologists. For example, Medical Geologists are capable of effectively researching important societal issues such as water pollution, water purification, storage of nuclear waste, the spread of geogenic particulates, sources, and composition of dust that cause health problems in mines, health effects of climate change, and analysis of groundwater and soil for determining nutrient cycling. Medical Geologists are equipped with knowledge and practical skills to ensure an interesting and rewarding career in the specialist areas of contaminated land consultancy, regulation, and remediation.

Realizing the huge potential contribution of the Medical Geologist toward tracing the origin of emerging enigmatic diseases, donor agencies will become more inclined to provide adequate funding sources for researchers and internships and predoctoral and postdoctoral training opportunities:

Tracking of disease is often accompanied by maps. Such maps provide a visual representation of the geographical distribution of a disease within a population, and this could represent a useful guide in determining the origin of an outbreak or the spread of disease; and may take the form of Geographic Information Systems (GIS). GISs are modeling techniques, which, aside from many of the other tools available, are used by epidemiologists for defining and evaluating the place factor in a disease outbreak. GIS, no doubt, play an important role in the management of data and the process of visualization; however, spatial statistics are needed to determine whether the pattern on these maps is somehow unusual. GIS techniques can now be used to show a lack of correlation between causes and effects or between different effects. GIS displays will help inform proper understanding and drive better decisions.

Remote sensing, when combined with other data, such as meteorological, will offer the means to uncover the origin of several diseases of unknown etiology; and some of the ecological/geoenvironmental factors that determine their occurrence and distribution. This will prove valuable for community health workers in the majority of African countries where there is often a lack of detailed empirical survey data. However, the potential of such an approach will remain undefined until further studies are undertaken which consider issues such as spatial scale (See Chapter 3, Part I). Many ecological variables are scale-dependent and geoenvironmental variables are no exception, since they also vary as the scale changes.

The huge communication gap between the health science profession, toxicologists, environmental scientists, and physical scientists, evident during the pioneering days of Medical Geology in the late 1980s, is not yet completely bridged; yet, communication between these groups is crucial for the advancement of this emerging field (Medical Geology). In many instances, in our attempts to further advance the field, huge and expensive long-term studies would be needed just to narrow down the possibilities and in closing the gap.

Today, the number of diseases whose etiologies are ill-defined (diseases of unknown etiology) still remains extremely high. Most data on diseases of unknown etiology are locked in data silos, and multidisciplinary collaborative efforts are required to increase efficiency of unlocking likely geoenvironmental determinants and generating the best quality of evidence. Thankfully, these silos are now gradually being broken; digitalization, decentralization, and personalized, pragmatic, preventive, and precision medicine are now being applied to improve public health and address diversity, equity, and worldwide accessibility to therapeutics. Through multidisciplinary research, Medical Geology will continue to contribute to the discovery and application of solutions to diseases of unknown etiology ranging from Alzheimer's disease to nodding disease (See: Chapter 3, Part I). Such multifaceted research would enable us to eventually clarify certain complex metabolic reactions, such as the role of metals in "necroptosis" a form of cellular suicide that leads to Alzheimer's disease (See Balusu et al., 2023).

As multidisciplinary research intensifies and the boundaries between our disciplines grow ever thinner, the magnitude of the geoenvironmental factor in disease causation will become clearer, and the challenges in deciphering unknown etiologies will eventually disappear.

It is hoped that this Book would help generate ideas for the formulation of experimental studies that take into account the role of the geochemical environment, in an attempt to establish precisely the obscure etiology of some of the diseases of hitherto unknown etiology treated and uncover new pathways in their pathogenesis.

If the links between specific trace elements and carcinogenesis, cardiovascular diseases, psoriasis, spinal stenosis, Paget's disease of bone, endomyocardial fibrosis, Mseleni Joint Disease, and many other diseases yet to be explicated, could be independently verified by experimental studies, it would affirm the significance of the role of trace elements as cofactors in these health conditions; and thus, pave the way toward vastly improved diagnostic and therapeutic intervention schemes.

If crucial gaps pointed out in this Primer could be bridged by conscientious research by the next generation of Medical Geologists, we certainly would be in for clearing many of the unidentified causes of diseases that involve a geoenvironmental cofactor(s).

In the present perspective, it is envisioned that the future of Medical Geology research would engender some of the key areas emphasized at the end of each of the main chapters of this Primer, following an approach to designing, conducting, and generating evidence on which improved diagnoses and therapeutic interventions would be predicated. For instance, there still exists an exciting potential for finding novel health benefits of geologic materials and processes associated with hot springs, and balneology of peat deposits.

Transformation of medical science is taking place at an unprecedented pace that will dramatically advance the way we diagnose and treat different diseases. Breakthroughs in digital medicines, microbiometrics/metabolomics, genomics, nanomedicine, and others are occurring at a tremendous pace. The sheer rate of data capture and analysis will continue to shape future of the health ecosystem. The research of Medical Geologists has a role in this transformation.

Medical Geology is experiencing a renaissance at the moment, and a bright future beckons. This future lies in the hands of young scientists and enthusiastic students. They should be assisted and encouraged to stay the course.

References

Balusu, S., Horré, K., Thrupp, N., Craessaerts, K., Snellinx, A., Serneels, L., et al., 2023. MEG3 activates necroptosis in human neuron xenografts modelling Alzheimer's disease. Science (New York, NY) 381 (6663), 1176−1182. Available from: https://doi.org/10.1126/science.abp9556.

Ceruti, P.O., Davies, T.C., Selinus, O., 2001. GEOMED 2001. Medical geology: the African perspective. Episodes 24 (4), 268−270. ISSN: 0705-3797.

Davies, T.C., 2019. A Medical Geology curriculum for African Geoscience Institutions. Scientific African 6, e00131. Available from: https://doi.org/10.1016/j.sciaf.2019.e00131.

Davies, T.C., Schluter, T., 1999. The Eas77t and Southern Africa Regional Workshop on geomedicine Nairobi, Kenya, 23−27 June, 1999. Conference Report. Episodes 22 (4), 303−304. Available from: https://images. app.goo.gl/ZPj1TPAtKdcXC3eg8.

Further reading

Skinner, H.C.W., Berger, A.R. (Eds.), 2003. Geology and Health - Closing the Gap. Oxford University Press Inc., Oxford, New York, p. 192. , ISBN: 0-19-516204-8.

Author index

Note: Page numbers followed by *"b," "f,"* and *"t"* refer to boxes, figures, and tables, respectively.

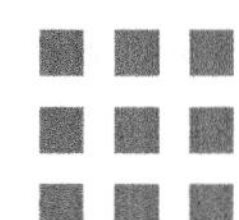

Subject index

Note: Page numbers followed by *"f"* and *"t"* refer to figures and tables, respectively.

Printed and bound by CPI Group (UK) Ltd, Croydon, CR0 4YY

06/08/2024

01023982-0009